Animal Species and Evolution

Animal Species and Evolution

Ernst Mayr

THE BELKNAP PRESS OF
HARVARD UNIVERSITY PRESS
Cambridge, Massachusetts
London, England

© *Copyright 1963 by the President and Fellows of Harvard College*
All rights reserved
Sixth printing, 1979

Library of Congress Catalog Card Number 63-9552
ISBN 0-674-03750-2

Printed in the United States of America

Preface

The steadily rising flood of scientific publications makes it increasingly difficult for a scientist to keep up with developments outside his own narrow area of specialization. The need for surveys of selected areas of science has never been greater than today. The present work is an attempt to summarize and review critically what we know about the biology and genetics of animal species and their role in evolution.

Friends have suggested that I go a step further and incorporate the relevant information on plants, expanding this account into a "species and evolution." I have resisted the temptation to yield to this suggestion. Having worked with animal species for some 35 years, in the field and in the laboratory, I believe that I have acquired some understanding of their species structure and evolutionary behavior. Lacking a similar familiarity with plants, I might come up with absurd generalizations if I tried to apply my findings to plants. Each of the kingdoms has its own evolutionary peculiarities and these must be worked out separately before a balanced synthesis can be attempted. Accordingly, I have also refrained from referring to numerous phenomena recently discovered in microorganisms that do not seem to have equivalents among higher organisms. When I speak of species, chromosomes, and gene pools, I refer to those of animals, particularly of the higher animals, even where this is not stated specifically. On the other hand, the findings derived from the higher animals concerning the population structure of their species and the mechanisms controlling their genetic variation are directly applicable to man. A study of the species of higher animals is, therefore, of the utmost importance, particularly in view of the impossibility of experimenting with man. An understanding of the biology of the species *Homo sapiens* is an indispensable requisite for the safeguarding of its future.

Evolutionary biology has been exceedingly active in recent years. I have endeavored in the present volume to concentrate on topics that have not been thoroughly discussed in recent works. Only a summary treatment is given here of cytology, of the more formal aspects of taxonomy, of paleontology, and of some areas of population genetics, because recent and comprehensive treatments are already available. This is a volume on the species and its role in evolution rather than on the evolutionary theory itself.

I have attempted to present a continuous story, with each chapter based on the preceding chapters. My aim has been to integrate and interpret rather than to present raw data. Interpretation is necessarily subjective; it requires the setting up of models and the testing of them with additional data. Where the issue is controversial I have not hesitated to choose the interpretation that seems most consistent with the picture of the evolutionary process as it now emerges. To take an unequivocal stand, it seems to me, is of greater heuristic value and far more likely to stimulate constructive criticism than to evade the issue. I have called attention whenever possible to unsolved problems. Where it helped the interpretation I have related the evolutionary subject matter to relevant material from other fields, such as physiology and biochemistry. Integration has been my major goal throughout.

Comparing two such different fields as, let us say, the evolutionary biology of species and enzyme chemistry brings home the enormous contrasts within science. In chemistry we deal with repeatable unit phenomena and with actions that, once correctly described, are known forever. In evolutionary biology we deal with unique phenomena, with intricate interactions and with balances of selection pressures—in short, with phenomena of such complexity that an exhaustive description is beyond our power. We can approach the truth only by a trial-and-error process of increasing accuracy. As in the humanities, and in contrast to many of the physical sciences, a thorough knowledge of the classical literature of the field is a prerequisite in evolutionary biology for a full understanding of the total conceptual framework.

I share the curiosity of those who are interested in the origin of the ideas in which we currently believe. It is for this reason that I have made an attempt in this work to trace whenever possible the history of concepts. Most references to publications antedating 1940 are included for historical reasons. It is well to remember that the main concepts of the biological role of species and of the process of species formation were established empirically by naturalists long before the turn of the century, but that a precise causal analysis became possible only after the rise of population genetics. The extraordinary vitality of the area of research covered by this book may seem surprising, considering that it has been active for more than a century. Yet anyone comparing the current interpretations with those prevailing even as recently as 1930, for example, will be struck by the clarification of ideas and change in emphasis.

The first draft of this work served as the text for a course given in 1949 at the University of Minnesota, and the ensuing years have been devoted to the seemingly never-ending task of improvement. The final version was completed in 1961, and it has not been feasible to include more than a fraction of the literature published since then.

I am deeply indebted to numerous friends and colleagues for encouragement, suggestions, and assistance of every kind. Having discussed almost every aspect of the subject matter with them, I find it quite impossible to separate their intellectual contribution from my own. Draft versions of certain chapters were read and criticized by Carleton Coon (20), J. F. Crow (part of 9), Th. Dobzhansky (1, 7, 8, 9, 10, 17), J. J. Hickey (4), L. B. Keith (4), R. H. MacArthur (4), F. A. Pitelka (4), G. G. Simpson (1), Bruce Wallace (7, 8, 9, 10, 17), M. J. D. White (15), and E. O. Wilson (5). All have made numerous suggestions, most of which though not all I have incorporated. Any remaining errors are strictly my own responsibility. I owe special thanks to Richard Lewontin for a penetrating analysis of chapters 1, 7, 8, 9, 10, and 17. I am indebted to Dr. Arthur Steinberg for making some unpublished data on the genetics of the Hutterites available to me. I am obliged to various authors and publishers for permission to republish illustrations, as acknowledged in the captions of these figures. Various assistants have helped in the preparation of the numerous drafts of the manuscript, particularly Sophie Prywata, Carmela Berrito Rosen, and, more recently, Lorna Levi and Emily Witte. Without their devoted services this volume would never have been completed.

E. M.

Contents

1. Evolutionary Biology   1
2. Species Concepts and Their Application   12
3. Morphological Species Characters and Sibling Species   31
4. Biological Properties of Species   59
5. Isolating Mechanisms   89
6. The Breakdown of Isolating Mechanisms (Hybridization)   110
7. The Population, Its Variation and Genetics   136
8. Factors Reducing the Genetic Variation of Populations   182
9. Storage and Protection of Genetic Variation   215
10. The Unity of the Genotype   263
11. Geographic Variation   297
12. The Polytypic Species of the Taxonomist   334
13. The Population Structure of Species   360
14. Kinds of Species   400
15. Multiplication of Species   424
16. Geographic Speciation   481
17. The Genetics of Speciation   516
18. The Ecology of Speciation   556
19. Species and Transpecific Evolution   586
20. Man as a Biological Species   622
    Glossary   663
    Bibliography   675
    Index   783

Tables

1–1. Theories of evolutionary change   2

2–1. Characteristics of four eastern North American species of *Catharus*   18

3–1. Biological differences among members of *Anopheles maculipennis* group of mosquitoes   36

3–2. Fertility of interspecific crosses in the *Anopheles maculipennis* group   37

3–3. Morphological differences of the sound file and characteristics of the song in the *Nemobius fasciatus* group of crickets   45

3–4. Some differences between two sibling species of sponges, *Halisarca*   49

3–5. Characteristics of three sibling species of the *Polistes fuscatus* group   51

3–6. Behavior differences between *Ammophila campestris* and *pubescens*   52

4–1. The relation between geographic distribution, breeding habits, and adaptive embryological characteristics in five species of North American frogs (*Rana*)   63

4–2. Average survival times of ants (*Formica*)   64

4–3. Species comparison in army ants (*Eciton*)   65

4–4. Biological differences of three species of *Peromyscus*   65

4–5. Two species of *Tribolium* in competition   77

4–6. Character divergence in *Geospiza* on the Galapagos   84

5–1. Classification of isolating mechanisms   92

5–2. Average bout length in *Drosophila melanogaster*   102

5–3. Number of offspring after a single insemination in *Drosophila*   104

7–1. Noninherited variation   140

7–2. Pleiotropic effects of genes a^+, a^k, and a in *Ephestia kuehniella*   160

7–3. Factors influencing the amount of genetic variation in a population   166

8–1. Differential mortality of *Biston betularia* released in different woodlands   192

8–2. Survival of Swiss Starlings (*Sturnus vulgaris*) after leaving nest 195
8–3. Temperature and polymorphism of *Cepaea nemoralis* in France 209
9–1. Size and production of experimental populations of *Drosophila melanogaster* 226
9–2. Seasonal change in frequency of yellow morphs among snails killed by thrushes 240
9–3. Inverse relation between frequency of mimic and precision of mimicry in *Pseudacraea eurytus* on islands in Lake Victoria 249
9–4. Protection of genetic variation against elimination by selection 255
11–1. Conspicuous geographic variation on an archipelago (*Zosterops rendovae*) 305
13–1. Species structure in birds of continental and island regions 385
13–2. Prevalence of polytypic species in a number of groups of animals 395
13–3. Subspeciation among 95 common lowland species of New Guinea songbirds 397
14–1. Classifying criteria for kinds of species 403
15–1. Potential modes of origin of species 428
15–2. Phenomena listed as biological races 454
15–3. Speciation in Eurasian fish 467
15–4. Nonrandom mating in Snow Geese 469
15–5. Speciation through two complementary factors 479
16–1. Geographic speciation in the beetle genus *Tristanodes*, Tristan da Cunha Islands 507
16–2. Frequency of various components of species structure in four families of birds 514
17–1. Selection for sexual isolation 553
18–1. Speciation in birds of the central Solomon Islands 559
19–1. Mixture of reptilian and avian characters in *Archaeopteryx* 597
20–1. Some of the differences between man and the anthropoids 625

Figures

2–1. Gradual speciation in time of the echinoid genus *Micraster*   25

3–1. Strict geographical replacement of five cytoplasmic races of *Culex pipiens* in Europe   43

3–2. The annual cycle of four sibling species of crickets (*Gryllus*) in North Carolina   46

3–3. Pattern of light flashes in North American fireflies (*Photuris*)   53

4–1. Carbon dioxide output of two species of buntings at various environmental temperatures   62

4–2. Geographic variation of bill length in two partially sympatric species of rock nuthatches, *Sitta neumayer* and *S. tephronota*   85

5–1. Quantitative differences in the major courtship components of two sibling species of *Drosophila*   97

6–1. Largely sympatric distribution of the House Sparrow, *Passer domesticus,* and the Willow Sparrow, *P. hispaniolensis*   120

6–2. Distribution of the Red-eyed Towhees (*Pipilo erythropthalmus* group) in Mexico   122

7–1. Growth rates of two *Coregonus* species in different Swedish lakes   143

7–2. Numbers of vertebrae in offspring of four samples of sea trout   145

7–3. Change of gene sequence by single inversion of a chromosome segment   155

7–4. Eight tail-pattern morphs in the fish *Xiphophorus maculatus*   156

7–5. Reciprocal, complementary, and diagonal pairs of chiasmata   180

8–1. Skull length in adult male and female moles (*Talpa europaea*) before and after a catastrophic winter kill   188

8–2. Begging responses in *Larus* chicks to natural and superoptimal models of head of feeding adult   196

8–3. ABO blood-group frequencies actually recorded in various populations throughout the world in relation to the complete possible range   211

9–1. Heritability of variation in nine phenotypic characters in a flock of Leghorn fowl   217

9–2. Fecundity of geographic populations of *Drosophila pseudoobscura* and of hybrids between them · 231

9–3. Frequency of the *succinea* morph of *Harmonia axyridis* on the Japanese islands and the adjacent Asiatic mainland · 243

9–4. Composition of *Cepaea nemoralis* populations in neighboring habitats at Wiltshire, England · 246

10–1. The product of a gene may affect many characters; a character may be affected by the products of many genes · 265

10–2. Bristle number in four lines of *Drosophila melanogaster,* derived from a single parental stock · 286

10–3. Divergent behavior of four experimental populations of *Drosophila pseudoobscura* · 292

11–1. Nesting area of Kirtland's Warbler (*Dendroica kirtlandi*) · 303

11–2. Geographic variation in three taxonomically important characters of the *Lalage aurea* group · 306

11–3. Geographic variation of color in two species of carpenter bees on Celebes and adjacent islands · 307

11–4. Mean wing length of four populations of the plover *Charadrius hiaticula* in relation to winter quarters · 322

11–5. Geographic variation of polymorphism in the squirrel *Sciurus vulgaris* in Finland · 330

12–1. Distribution pattern of the polytypic species *Passerella melodia* · 336

12–2. A polytypic species of terrestrial isopod (*Phymatoniscus*) from the south of France · 340

12–3. Distribution of two species of garter snakes (*Thamnophis*) in northern California and southern Oregon · 344

13–1. Character gradient concerning pigmentation of the upper side of the wing in females of *Pieris napi* from Fennoscandia · 363

13–2. Independent variations of two characters in *Paradisaea apoda* in New Guinea · 365

13–3. Course of the hybrid zone between the Carrion Crow (*Corvus c. corone*) and the Hooded Crow (*Corvus c. cornix*) in western Europe · 370

13–4. Tree runners (*Neositta*) from Australia · 373

13–5. Actual distribution of the Mountain Gorilla in East Africa · 383

13–6. Polytopic subspecies in the drongo *Dicrurus leucophaeus* · 388

13–7. Identical populations on different islands off Venezuela · 389

13–8. Number of gene arrangements in 24 populations of *Drosophila robusta* · 391

13–9. The pattern of distribution of populations of the halophilous land snail *Cerion* on the Banes Peninsula in eastern Cuba · 399

15–1. Change of a strongly isolated species and break-up of a species by geographic speciation and cross colonization (diagram) · 425

15–2. Distribution of some gene arrangements in *Drosophila americana, D. texana,* and *D. novamexicana* · 431

15–3. The chromosomes of the marine snail *Thais* (*Purpura*) *lapillus* · 445

15–4. Ecological races of the Song Sparrow in the San Francisco Bay region   457

15–5. Primary and subsidiary hosts of an essentially host-specific species   463

16–1. Distribution of the species and subspecies comprising the platyfish superspecies *Xiphophorus maculatus*   490

16–2. Branches of the polytypic species *Dicrurus hottentottus*   497

16–3. Isolation and speciation in savannah-dwelling tree creepers (*Climacteris picumnus*) of Australia   498

16–4. A superspecies of paradise magpies (*Astrapia*) in the mountains of New Guinea   500

16–5. Distribution of the kingfishers of the *Tanysiptera hydrocharis-galatea* group in the New Guinea region   503

16–6. Successive stages in the speciation of the Australian mallee thickheads (*Pachycephala*)   505

16–7. Circular overlap in gulls of the *Larus argentatus* group   509

16–8. Incomplete speciation in *Parus major*   511

16–9. Circular overlap in the bee *Hoplitis* (*Alcidamea*) *producta*   512

17–1. The frequency of *PP* chromosomes in 20 replicate experimental populations of mixed geographic origin   529

17–2. Diagrammatic representation of the changing adaptive value of genes on different genetic backgrounds   532

17–3. Loss and gradual recovery of genetic variation in a founder population   539

17–4. Niche utilization by two different species   547

18–1. Speciation in the white-eye *Zosterops rendovae* in the central Solomon Islands   558

18–2. Cosmopolitan distribution of the tardigrade *Macrobiotus hufelandii*   567

19–1. Geographic variation of bill function in the Hawaiian honey-creeper *Hemignathus lucidus*   591

19–2. Repeated and independent acquisition of mammalian characters by various lines of mammal-like reptiles (therapsids)   597

19–3. Typological diagram of the evolution of the mastodons   599

19–4. Graphs showing rate of loss of characters of ancestral type during the evolution of the Dipnoi   618

Animal Species and Evolution

1 ~ Evolutionary Biology

The theory of evolution is quite rightly called the greatest unifying theory in biology. The diversity of organisms, similarities and differences between kinds of organisms, patterns of distribution and behavior, adaptation and interaction, all this was merely a bewildering chaos of facts until given meaning by the evolutionary theory. There is no area in biology in which that theory has not served as an ordering principle. Yet this very universality of application has created difficulties. Evolution shows so many facets that it looks alike to no two persons. The more different the backgrounds of two biologists, the more different have been their attempts at causal explanation. At least, so it was through the history of evolutionary biology (Heuts 1952; Simpson 1949, 1960b; Eiseley 1958), until the many dissenting theories were almost suddenly fused, in the 1930's, into a broad unified theory, the "synthetic theory."

Many of the earlier evolutionary theories were characterized by heavy emphasis, if not exclusive reliance, on a single factor (Table 1-1). The synthetic theory has selected the best aspects from the earlier hypotheses and has combined them in a new and original manner. It attempts to evaluate the respective roles of the numerous interacting factors responsible for evolutionary change. In essence it is a two-factor theory, considering the diversity and harmonious adaptation of the organic world as the result of a steady production of variation and of the selective effects of the environment. It is thus basically a synthesis of mutationism and environmentalism.

Attempting to explain evolution by a single-factor theory was the fatal weakness of the pre-Darwinian and most 19th-century evolutionary theories. Lamarckism with its internal self-improvement principle, Geoffroyism with its induction of genetic change by the environment, Cuvier's catastrophism, Wagner's evolution by isolation, De Vries' mutationism, all were deficient through focusing on only one aspect of a complex set of interacting factors. These hypotheses tried to explain evolution by a single principle to the exclusion of all others. Even Darwin occasionally fell into this error, as when he wrote, late in his life, that natural selection

Table 1–1. Theories of evolutionary change (in part after Heuts 1952).

A. Monistic (single-factor explanations)
    1. Ectogenetic: changes directly induced by the environment
        (a) Random response (for example, radiation effects)
        (b) Adaptive response (Geoffroyism)
    2. Endogenetic: changes resulting from intrinsic forces
        (a) Finalistic (orthogenesis)
        (b) Volitional (genuine Lamarckism)
        (c) Mutational limitations
        (d) Epigenetic limitations
    3. Random events ("accidents")
        (a) Spontaneous mutations
    4. Natural selection
B. Synthetic (multiple-factor explanations)
    1(b) + 2(a) + 2(b) = most "Lamarckian-type" theories
    1(b) + 2(b) + 2(c) + 4 = some recent "Lamarckian" theories
    1(b) + 3 + 4 = late Darwin, Plate, most nonmutationists during first three decades of 20th century
    3 + 4 = early "Modern Synthesis"
    1(a) + 2(c) + 2(d) + 3 + 4 = recent "Modern Synthesis"

rather than isolation was responsible for the origin of species, as if the two forces were mutually exclusive (Mayr 1959c). Yet on the whole Darwin was the first to make a serious effort to present evolutionary events as due to a balance of conflicting forces. Indeed, he often went too far in compromising. It has been claimed, not without justification, that one can find support in Darwin's writings for almost any theory of evolution: speciation with geographic isolation or without it, direct effect of the environment or merely selection by the environment, evolutionary importance of large genetic changes or of small ones, and so on. This explains the paradox that the term "Darwinism" means such different things to an American, a Russian, or a French biologist. To be sure, the current theory of evolution—the "modern synthesis," as Huxley (1942) has called it— owes more to Darwin than to any other evolutionist and is built around Darwin's essential concepts. Yet it incorporates much that is distinctly post-Darwinian. The concepts of mutation, variation, population, inheritance, isolation, and species were still rather nebulous in Darwin's day. To avoid confusion, it has been suggested, particularly by Simpson

(1949, 1960b), that the term "neo-Darwinism," originally introduced into biology for Weismann's concepts of evolution, should be dropped.

The development of the modern theory was a slow process. Evolutionary biology was at first in the same situation as sociology, psychology, and other vast fields still are today: the available data were too voluminous and diversified to be organized at once into a single comprehensive theory. Individual, specialized theories had to spotlight selected aspects and assist in a preliminary sorting of the data before a complete synthetic interpretation of the field as a whole was feasible. Looking back over the history of the many false starts gives a valuable insight into the process of theory formation. One important lesson is that progress is stepwise and that some sets of data may not have significance until certain concepts are clarified or principles established. For instance, the true role of the environment in evolution could not be understood until the nature of small mutations and of selection were fully comprehended. Polygenes could not be analyzed and understood until the laws of inheritance had been clarified with the help of conspicuous mutations. The process of speciation could not be understood until after the nature of species and of geographic variation had been clarified. Discussions of variation among early evolutionists were utterly confused because they failed to make a clear distinction between geographical "variety" (geographical race) and individual variety. The replacement of the morphological by the biological species concept led to a reevaluation of the "biological race" and to a rather drastic shift in the study of speciation. The analysis of quantitative characters was futile until the principles of particulate inheritance had been fully understood. Genetics, morphology, biogeography, systematics, paleontology, embryology, physiology, ecology, and other branches of biology, all have illuminated some special aspect of evolution and have contributed to the total explanation where other special fields failed. In many branches of biology one can become a leader even though one's knowledge is essentially confined to an exceedingly limited area. This is unthinkable in evolutionary biology. A specialist can make valuable contributions to special aspects of the evolutionary theory, but only he who is well versed in most of the above-listed branches of biology can present a balanced picture of evolution as a whole. Whenever a narrow specialist has tried to develop a new theory of evolution, he has failed.

The importance of eliminating erroneous concepts is rarely given sufficient weight in discussions of theory formation. Only in some cases is it true that the new, better theory vanquishes the old, "bad" one. In many other instances it is the refutation of an erroneous theory that vacates the field for new ideas. An excellent illustration of this is Louis Agassiz's neglect of what seem to us most convincing evolutionary facts because they were inconsistent with his well-organized, harmonious world view (Mayr 1959d). Darwin, who had started the voyage of the *Beagle* with views similar to those of Agassiz, began to think seriously about evolution only after he had found overwhelming evidence that was completely irreconcilable with the idea of an origin of the world fauna and flora by creation. Or, to cite another example, as long as spontaneous generation and the instantaneous conversion of one species into another were universally believed in, even for higher animals and plants (Zirkle 1959), there was no room for a theory of evolution. By insisting on the fixity of species, Linnaeus did more to bring about the eclipse of the concept of spontaneous generation than did Redi and Spallanzani, who disproved it experimentally. Indirectly, Linnaeus did as much to prepare the ground for a theory of evolution as if he had proposed such a theory himself. Weismann, through his theoretical analysis of the relation between germ cells and soma cells, eliminated many of the misconceptions and errors that until then had prevented the recognition of the work of Mendel. These are merely a few illustrations of the importance of eliminating erroneous theories. The refutation of an erroneous idea thus is not a purely negative activity, and in this volume I often give considerable space to the analysis of that alternative of two opposing theories that I consider to be the less well-founded one.

More important for the development of the synthetic theory than the rejection of ill-founded special theories of evolution was the rejection of two basic philosophical concepts that were formerly widespread if not universally held: preformism and typological thinking. Preformism is the theory of development that postulates a preformed adult individual in miniature "boxed" into the egg or spermatozoon, ready to "unfold itself" during development. The term evolution is derived from this concept of unfolding, and this connotation continued well into the post-Darwinian period. It was perhaps the reason Darwin did not use the term "evolution" in his *Origin of Species*. Transferred from ontogeny to phylogeny, evolution meant the unfolding of a built-in plan. Evolution, according to this view, does not produce genuine change, but consists merely in the maturation of immanent potentialities. This, for instance, was Louis Agassiz's theory of evolution (Mayr 1959d). Some of the orthogenetic and finalistic theories of evolution are the last remnants of this type of thinking. The underlying erroneous assumption that the development of the "type" is essentially the same phenomenon as the development of the individual has also been the reason for much of the search for "phylogenetic laws." Mutationism was the extreme in the reaction to these orthogenetic concepts. The current theory compromises by admitting that genotype and phenotype of a given evolutionary line set severe limits to its evolutionary potential (Table 1–1, A2c,d), without, however, prescribing the pathway of future evolutionary change.

Typological thinking is the other major misconception that had to be eliminated before a sound theory of evolution could be proposed. Plato's concept of the *eidos* is the formal philosophical codification of this form of thinking. According to this concept the vast observed variability of the world has no more reality than the shadows of an object on a cave wall, as Plato puts it in his allegory. Fixed, unchangeable "ideas" underlying the observed variability are the only things that are permanent and real. Most of the great philosophers of the 17th, 18th, and 19th centuries were influenced by the idealistic philosophy of Plato and the modifications of it by Aristotle. The thinking of these schools dominated the natural sciences until well into the 19th century. The concepts of unchanging essences and of complete discontinuities between every *eidos* (type) and all others make genuine evolutionary thinking well-nigh impossible. I agree with those (such as Reiser 1958) who claim that the typological philosophies of Plato and Aristotle are incompatible with evolutionary thinking.

The assumptions of population thinking are diametrically opposed to those of the typologist. The populationist stresses the uniqueness of everything in the organic world. What is true for the human species, that no two individuals are alike, is equally true for all other species of animals and plants . . . All organisms and organic phenomena are composed of unique features and can be described collectively only in statistical terms. Individuals, or any kind of organic entities, form populations of which we can determine the arithmetic mean and the statistics of variation. Averages are merely statistical abstractions; only the individuals of which the populations are composed have reality. The ultimate conclusions of the population thinker and of the typologist are precisely the opposite. For the typologist, the type (*eidos*) is real and the variation an illusion, while for the populationist the type (average) is an abstraction and only the variation is real. No two ways of looking at nature could be more different (Mayr 1959c).

The replacement of typological thinking by population thinking is perhaps the greatest conceptual revolution that has taken place in biology. Many of the basic concepts of the synthetic theory, such as that of natural selection and that of the population, are meaningless for the typologist. Virtually every major controversy in the field of evolution has been between a typologist and a populationist. Even Darwin, who was more responsible than anyone else for the introduction of population thinking into biology, often slipped back into typological thinking, for instance in his discussions on varieties and species.

A new conceptual danger to the evolutionary theory at the present time comes not from metaphysics but from physics. Some physicists who believe in reductionism have attempted in recent years to express the laws of evolution in terms of the laws of physics. Statements to the effect that "evolution is one more expression of the general principle of irreversibility embodied in the second law of thermodynamics" are based on a facile analogy that has no operational value. Since every individual is unique, strict evolutionary reversibility is a logical impossibility. Yet acquired specializations may be lost again at later stages of evolution and a type may evolve that has in its essential structure reverted to a prior condition although it is obviously not the same as the ancestral type. The processes in physics and evolution labeled by the same term, "irreversibility," are fundamentally different. To drag the second law of thermodynamics into the discussion of evolutionary irreversibility confuses two distant levels of integration, the atomic level and the level of the phenotype. Those who try to explain the pathway of evolution in terms of the laws of physics do not seem to realize how dangerously close they sail to preformism.

CLARIFICATION OF EVOLUTIONARY CONCEPTS

A comparison of current evolutionary publications with those of only 20 or 25 years ago shows what great conceptual progress has been made in this short period. Since much of this volume is devoted to reporting on this progress, I will barely mention some of these advances in this introductory discussion. Our ideas on the relation between gene and character have been thoroughly revised and the phenotype is more and more considered not as a mosaic of individual gene-controlled characters but as the joint product of a complex interacting system, the total epigenotype (Waddington 1957). Interactions and balances among opposing forces are stressed to an increasing extent. Virtually every component of the phenotype is recognized as a compromise between opposing selection pressures.

The realization that the DNA of the chromosomes carries a code of information has led to great clarification. The phenomena of ontogeny and physiology are now interpreted as manifestations of the decoding of the information embodied in the genotype. Phylogeny, on the other hand, and all the phenomena involving evolutionary change are considered as the production of ever-new codes of information. Nothing could make the difference between ontogeny and phylogeny clearer than stating it in terms of codes of information. We shall come back to this later.

Let me cite some other advances in our understanding. Natural selection is no longer regarded as an all-or-none process but rather as a purely statistical concept. Isolation has been revealed as a dual phenomenon, either the separation of populations by environmental barriers or the maintenance of the genetic integrity of species by isolating mechanisms. The environment is restored to the place of one of the most important evolutionary factors but in a drastically different role than it held in the various "Lamarckian" theories. The new role of the environment is to serve as principal agent of natural selection.

OPEN PROBLEMS

The development of the evolutionary theory is a graphic illustration of the importance of the *Zeitgeist*. A particular constellation of available facts and prevailing concepts dominates the thinking of a given period to such an extent that it is very difficult for a heterodox viewpoint to get a fair hearing. Recalling this history should make us cautious about the validity of our current beliefs. The fact that the synthetic theory is now so universally accepted is not in itself proof of its correctness. It will serve as a warning to read with what scorn the mutationists in the first decade of this century attacked the contemporary naturalists for their belief in gradual changes and in the immense importance of the environment. It never occurred to the saltationists that their own typological and antiselectionist interpretation of evolution could be much further from the truth than the late Darwinian viewpoint of their adversaries. Unfortunately, the synthetic theory is still by some considered a form of mutationism. In my experience, every recent attack on the synthetic theory has really been an attack on crude mutationism rather than a reasoned argument against the actual tenets of the synthetic theory. Everything that is stated in the synthetic theory about mutations is in conflict with the claims of mutationism. We now believe that mutations do not guide evolution; the effect of a mutation is very often far too small to be visible.

Recombination produces far more selectively important phenotypes than does mutation, and the kinds of mutations and recombinations that can occur in a given organism are severely restricted. These statements are entirely consistent with the synthetic theory but they may be quite startling to those who are unaware of the modern developments and who are still fighting the fight of the last generation.

When we reread the volumes published in 1909, on the occasion of the 50th anniversary of the *Origin of Species*, we realize how little agreement there was at that time among the evolutionists. The change since then has been startling. Symposia and conferences were held all over the world in 1959 in honor of the Darwin centennial, and were attended by all the leading students of evolution. If we read the volumes resulting from these meetings at Cold Spring Harbor, Chicago, Philadelphia, London, Göttingen, Singapore, and Melbourne, we are almost startled at the complete unanimity in the interpretation of evolution presented by the participants. Nothing could show more clearly how internally consistent and firmly established the synthetic theory is. The few dissenters, the few who still operate with Lamarckian and finalistic concepts, display such colossal ignorance of the principles of genetics and of the entire modern literature that it would be a waste of time to refute them. The essentials of the modern theory are to such an extent consistent with the facts of genetics, systematics, and paleontology that one can hardly question their correctness. The basic framework of the theory is that evolution is a two-stage phenomenon: the production of variation and the sorting of the variants by natural selection. Yet agreement on this basic thesis does not mean that the work of the evolutionist is completed. The basic theory is in many instances hardly more than a postulate and its application raises numerous questions in almost every concrete case. The discussions throughout this volume are telling testimony to the truth of this statement.

Modern research is directed primarily toward three areas: evolutionary phenomena that do not yet appear to be adequately explained by the synthetic theory, such as stagnant or explosive evolution; the search for various subsidiary factors that, although inconspicuous at casual inspection, exercise unexpected selection pressures; and, perhaps most important, the interplay among genes and between genotype and environment resulting in the phenotype, the real "object" of natural selection.

Most contemporary arguments concern the relative importance of the various interacting factors. One will get highly diverse answers if one asks a number of contemporary evolutionists the following questions:

How important are random events in evolution?

How important is hybridization in evolution?

How important is interpopulation gene flow?

What proportion of new mutations is beneficial?

What proportion of genetic variability is due to balanced polymorphism?

Other areas in which there is still wide divergence of opinion are the importance of phenotypic plasticity, the pathway to adaptation, evolutionary mechanisms in higher and lower organisms, the origin of sexuality, and the origin of life. It must be stressed, for the benefit of nonevolutionists, that none of these arguments touches upon the basic principles of the synthetic theory. It is the application of the theory that is involved, not the theory itself. And with respect to application we still have a long way to go. There are vast areas of modern biology, for instance biochemistry and the study of behavior, in which the application of evolutionary principles is still in the most elementary stage.

THE MAJOR AREAS OF EVOLUTIONARY RESEARCH

Important contributions to our understanding of the evolutionary process have been made by virtually every branch of biology. During the past 100 years most of the research has been concerned with a number of discrete areas, progress within which has been unequal:

The fact of evolution,

The establishment of phylogenies,

The origin of discontinuities (speciation),

The material of evolution,

Rates of evolution,

Causes of evolution, and

The evolution of adaptation.

The amount of attention given to each of these areas has changed with time. To establish unequivocally the fact of evolution was after 1859 the first concern of the young science of evolutionary biology. The study of phylogeny soon became predominant, at least in zoology. Indeed, even today there still are some zoologists to whom the term "evolution" signifies little more than the determination of homologies, common ancestors, and phylogenetic trees. By far the majority of evolutionary biologists, however, have shifted their interest to a study of the causes and mechanisms of evolutionary change and to an attempt to determine the role and relative importance of various factors. A study of the difference in response to these factors displayed by different types of organisms is also receiving increasing attention. Evolutionary biology is beginning to become truly comparative.

The attack on evolutionary problems can be compared to an attack on a many-walled city by a number of separate armies. A breach in one of these walls made by the army of the geneticists improves the strategic situation of the paleontologists and systematists, and vice versa. Each factual or conceptual advance in one field will have an impact on the thinking in the others. Each branch of evolutionary biology occupies, so to speak, a special niche and is uniquely qualified to illuminate some special problem.

The geneticist is mainly concerned with the individual, the stability or mutability of loci, the modification of the phenotype, the interaction of parental genes in the production of the phenotype and the effect of this interaction on fitness, in short, all the problems concerning the gene and its interaction with other genes and with the environment. The development of population genetics led to an expansion from the gene to the gene pool of the population.

The contribution of genetics to the understanding of the process of evolution has not yet been evaluated objectively (Mayr 1959c). The assumption made by some geneticists, that it was quite impossible to have sensible ideas on evolution until the laws of inheritance had been worked out, is contradicted by the facts. Everyone admits that Darwin's evolutionary theories were essentially correct and yet his genetic theories were about as wrong as they could be. Conversely, the early Mendelians, the first biologists (except for Mendel himself) who truly understood genetics, misinterpreted just about every evolutionary phenomenon. Some of their contemporaries among the naturalists, on the other hand, though they did not understand genetics and even believed in some environmental induction (Geoffroyism), presented a remarkably correct picture of speciation, adaptation, and the role of natural selection. It would be going too far to claim that it is immaterial whether one believes the source of genetic variation to be De Vriesian or Lamarckian, yet it is true that it is less important for the understanding of evolution to know how genetic variation is manufactured than to know how natural selection deals with it (Waddington 1957). Replacing the erroneous belief in blending inheritance by the theory of particulate inheritance is the greatest single contribution of genetics. It has been the basis of all subsequent developments. The genetic material presented in Chapters 7–10 and 17 shows to what extent the modern genetic theory can explain many phenomena that the naturalist has long known and correctly described, but has been unable to interpret.

The study of long-term evolutionary phenomena is the domain of the paleontologist. He investigates rates and trends of evolution in time and is interested in the origin of new classes, phyla, and other higher categories. Evolution means change and yet it is only the paleontologist among all biologists who can properly study the time dimension. If the fossil record were not available, many evolutionary problems could not be solved; indeed, many of them would not even be apparent.

The taxonomist, who deals primarily with local populations, subspecies, species, and genera, is concerned with the region that lies between the areas of interest and competence of the geneticist and of the paleontologist, overlapping with both but approaching problems in the area of overlap from a somewhat different viewpoint. The species, the center of his interest, is one of the important levels of integration in the organic world. Neglect of this level in much of our biological curriculum is rather puzzling. We do not even have a special term for the study of the species, corresponding to cytology, the study of cells; histology, the study of tissues; and anatomy, the study of organs. Yet the species is not only the basic unit of classification, but also one of the most important units of interaction in ecology and ethology. The origin of new species, signifying the origin of essentially irreversible discontinuities with entirely new potentialities, is the most important single event in evolution. Darwin, who had devoted so much of his life to the systematics of species, fully appreciated the significance of this level, as he made clear in the choice of title for his classic *On the Origin of Species*.

The emphasis in the present volume is deliberately on those aspects of evolution that involve the species. Other aspects, of greater interest to the geneticist or paleontologist, and adequately treated by Dobzhansky (1951), Simpson (1953a), Rensch (1960a), and other modern writers, will be neglected or treated only incidentally. Evolutionary biology has become far too vast a field to be covered adequately in a single volume.

The basic structure of this volume is as follows. The *characteristics of species* will be dealt with in Chapters 1–6; the *structure and genetics of populations* in Chapters 7–10; the (population) *structure and variation of species* in Chapters 11–14; and the *multiplication of species* in Chapters 15–18. Chapter 19 will be devoted to a discussion of the role of *species in transpecific evolution* and Chapter 20 to a review of the possible consequences of our findings for *man*.

2 ~ Species Concepts and Their Application

Darwin's choice of title for his great evolutionary classic, *On the Origin of Species,* was no accident. The origin of new "varieties" within species had been known since the time of the Greeks. Likewise the occurrence of gradations, of "scales of perfection" among "higher" and "lower" organisms, was a familiar concept, though usually interpreted in a strictly static manner. The species remained the great fortress of stability and this stability was the crux of the antievolutionist argument. "Descent by modification," true biological evolution, could be proved only by demonstrating that one species could originate from another. It is a familiar and often-told story how Darwin succeeded in convincing the world of the occurrence of evolution and how—in natural selection—he found the mechanism that is responsible for evolutionary change and adaptation. It is not nearly so widely recognized that Darwin failed to solve the problem indicated by the title of his work. Although he demonstrated the modification of species in the time dimension, he never seriously attempted a rigorous analysis of the problem of the multiplication of species, of the splitting of one species into two. I have examined the reasons for this failure (Mayr 1959a) and found that among them Darwin's lack of understanding of the nature of species was foremost. The same can be said of those modern authors, for instance Goldschmidt (1940), who attempted to solve the problem of speciation by saltation or other heterodox hypotheses. They all failed to find solutions that are workable in the light of the modern appreciation of the population structure of species.

A thorough understanding of the biological properties of species is necessary not only for the evolutionist, but for every biologist. Every biologist, even the biochemist, constantly works with species or with compounds and systems that are species specific. Unawareness of this fact by some physiologists and ecologists has been the cause of much confusion in the literature. The failure properly to define and distinguish species has had far-reaching practical effects in the fields of economic entomology, parasitology, and medical entomology.

In view of this apparent importance of the species it is a curious paradox that since time immemorial it has been contended that species have no existence in nature. Such claims have been made by philosophers who think in terms of philosophical categories and consider species as something man-made (see Mayr 1957a for historical notes). They have come from taxonomists (including Darwin) who had unsuitable species criteria and became bewildered by the variability of their material. And finally, such claims have come from paleontologists, who work in the time dimension and who are therefore indeed unable in certain cases to delimit species against each other. From the discussions of the negators of species it is very apparent that they deal with only a few aspects of a very complex problem.

Among naturalists the attitude toward the species has greatly fluctuated in the course of history. For the ancients and for some naturalists as recently as the nineteenth century (Agassiz) or even the present, species are merely man-made "categories of thought." Folklore took it for granted that one species can change readily into another (Zirkle 1959), and a belief in the spontaneous generation even of higher animals and plants was almost universal far into the eighteenth century. In spite of Redi's and Spallanzani's experiments, spontaneous generation was used by the philosopher Schopenhauer as late as 1851 to explain the origin of species and higher categories. It is the great merit of Linnaeus (Mayr 1957a) to have challenged and decisively defeated the folklore belief of the easy mutability of species. The whole concept of evolution would be meaningless if species were the arbitrary and ephemeral units of the pre-Linnaean period. It was Linnaeus' insistence on the constancy and objectivity of species that posed the problem of the origin of species, a problem previously nonexistent in that form (see also Poulton 1903).

The supremacy of the Linnaean species extended from the 1750's to 1859. Two characteristics of species were stressed during this period: their constancy and their sharp delimitation against each other (their "objectivity"). The general assumption of the period was that these two characteristics "are strictly correlated and that one must make a choice of either believing in evolution (the 'inconstancy' of species) and then having to deny the existence of species except as purely subjective, arbitrary figments of the imagination, or, as most early naturalists have done, believing in the sharp delimitation of species but thinking that this necessitated denying evolution" (Mayr 1957a). Darwin, the evolutionist, simply denied the existence of nonarbitrary species: "I look at the term species as one arbitrarily given for the sake of convenience to a set of individuals closely resembling each other . . . it does not essentially differ from the term variety which is given to less distinct and more fluctuating forms" (Darwin 1859). By eliminating the species as a concrete natural unit, Darwin also neatly eliminated the need for a solution to the problem of how species multiply (Mayr 1959a). In retrospect it is apparent that Darwin's failure, as well as that of the antievolutionists, resulted to a large extent from a misunderstanding of the true nature of species. There is a great need, then, for an impartial investigation of species criteria, species characteristics, species concepts, and species definitions. For recent literature on the species problem see Mayr (1957a), Sylvester-Bradley (1956), Beaudry (1960), and Simpson (1961).

SPECIES CRITERIA

How to allocate the variable individuals in nature to species has been a problem for the naturalist since earliest times. The first author to express this in relatively modern terms was Ray (1686):

In order that an inventory of plants may be begun and a classification of them correctly established, we must try to discover criteria of some sort for distinguishing what are called "species." After a long and considerable investigation, no surer criterion for determining species has occurred to me than the distinguishing features that perpetuate themselves in propagation from seed. Thus, no matter what variations occur in the individual or the species, if they spring from the seed of one and the same plant, they are accidental variations and not such as to distinguish a species . . . Animals likewise that differ specifically preserve their distinct species permanently; one species never springs from the seed of another nor vice versa. (From Beddall 1957.)

The three themes of Ray's discussion dominated the thinking of the next 50 years: first, the implied rejection of spontaneous generation; second, the morphological definition of the species; and third, the endeavor to reconcile the observed variation with the typological species concept. Production by the same parents is the "common descent" of which such antievolutionists as Cuvier and von Baer spoke in their species definitions (Mayr 1957a).

Degree of morphological difference remained the dominant species criterion in taxonomy through the entire Linnaean period, and even in Darwin's writings. This concept was carried to the logical extreme by the proposal of mathematical formulas that would permit an unambiguous answer to the question whether or not a certain population is a species: "A given population is to be considered a species with respect to another closely related population when the degree of intergradation (overlap of the observed samples) is not more than 10 percent" (Ginsburg 1938:260). A revival of this sort of thinking can be found in the writings of some of the contemporary "numerical taxonomists." The weaknesses of this purely morphological approach will be discussed in more detail later (Chapter 3). It may be remarked here that nearly all of the older species definitions, including those of Buffon, Lamarck, and Cuvier, refer to the morphological similarity of the individuals of which species are composed.

Yet Ray's species characterization contained also the germ of the modern species concept, which considers reproductive relationship to be the principal species criterion. Instead of concentrating on the degree of difference in the offspring of a pair of parents, one can equally well concentrate on the degree of difference between potential mates and their ability to interbreed. As early as 1760 Koelreuter had stated that all those individuals belong to a species that are able to produce fertile offspring. With increasing frequency since then, interbreeding has been considered a decisive criterion in species definitions. Unfortunately, this criterion has often been narrowed down to a single aspect of successful interbreeding, that of fertility. Cross-fertility was accepted as the decisive species criterion in much of the genetic and botanical literature until naturalists pointed out that fully cross-fertile species of animals may live side by side without interbreeding because their reproductive isolation is maintained by isolating mechanisms other than the sterility barrier. Authors like myself, who have consistently advocated that the noninterbreeding of natural populations rather than the sterility of individuals be taken as the decisive species criterion, have nevertheless been accused by hasty readers of having a species concept "based on cross-sterility."

The history of the many attempts to find universally satisfactory species criteria has been given elsewhere (Mayr 1957a). The difficulties encountered are a strong indication not only of the great diversity of population phenomena and of types of species found in nature but, alas, also of much muddled thinking. It is evident that the word "species" has meant and still means different things to different people. There is no hope for ultimate unanimity unless one understands the various basic concepts to which the term has been applied.

Several different species concepts coexist in the biological literature. There are at least two categories of such concepts. Terms like "practical," "sterility," or "genetic," when used to describe species concepts, refer to criteria one may apply to concrete situations. They deal with the evidence that, as Simpson (1961) would say, one uses in order to determine whether a given taxon belongs to the category species or not. Yet these secondary, "applied" concepts are based on underlying primary or theoretical concepts. A study of all the species definitions published in recent years indicates that they are based on three theoretical concepts, neither more nor less. An understanding of these three concepts is a prerequisite for the investigation of the problem of speciation.

(1) *The Typological Species Concept.* This is the simplest and most widely held species concept. Species here means "a different thing," something that "looks different" (from the Latin *specere*, to look at, to regard), "a different kind." This is the concept the mineralogist has in mind when he speaks of "species of minerals" or the physicist who speaks of "nuclear species."

This simple concept of everyday life was made the basis of the *eidos* in Plato's philosophy (Chapter 1). Different authors have stressed different aspects of Plato's *eidos,* some its independence of perception, others its transcendent reality, and still others its eternity and immutability. All these concepts take for granted that there is an unchanging essence, an *eidos,* which alone has objective reality. Objects, on the other hand, are for Plato and his adherents merely varying manifestations ("shadows") of the *eidos.* The individuals of a natural species, being merely shadows of the same "type," do not stand in any special relation to each other. Variation, under this concept, is due to the imperfections in the visible manifestations of the "idea" implicit in each species.

There are, however, limits to the amount of variation that can be ascribed to the varying manifestations of a single *eidos.* Where it transgresses these limits, more than one *eidos* must be involved. Degree of morphological difference, thus, determines species status. The two aspects of the typological species concept, subjectivity and definition by degree of difference, depend on each other and are logical correlates. The typological species concept, translated into practical taxonomy, is the morphologically defined species.

In recent years most systematists have found this typological-morphological concept inadequate and have rejected it. They have pointed out that this concept treats the individuals of a species like an aggregation of inanimate objects, a singularly inappropriate treatment for a reproductive community. They have also called attention to the fact, discussed in Chapters 3 and 7, that the morphological species criterion is highly misleading in cases of polymorphic diversity within species or of morphologically extremely similar species. Where the taxonomist applies morphological criteria, he uses them as secondary indications of reproductive isolation.

(2) *The Nondimensional Species Concept.* This concept is based on the relation of two coexisting natural populations in a nondimensional system, that is, at a single locality and at the same time (sympatric and synchronous). This is the species concept of the local naturalist (Mayr 1946b). If one studies the birds, the mammals, the butterflies, or the snails near his home town, he finds each species clearly defined and sharply separated from all other species. This is sometimes better appreciated by primitive natives than by modern civilized man. Some 30 years ago I spent several months with a tribe of superb woodsmen and hunters in the Arfak Mountains of New Guinea. They had 136 different vernacular names for the 137 species of birds that occurred in the area, confusing only two species. It is not, of course, pure coincidence that these primitive woodsmen arrive at the same conclusion as the museum taxonomists, but an indication that both groups of observers deal with the same, nonarbitrary discontinuities of nature.

This striking discontinuity between sympatric populations is the basis of the species concept in biology. The two taxonomists who, more than anyone else, were responsible for the acceptance of species in biology were local naturalists, John Ray in England and Carolus Linnaeus in southern Sweden. But anyone can test the reality of these discontinuities for himself, even where the morphological differences are slight. In eastern North America, for instance, there are four rather similar species of the genus *Catharus* (Table 2–1), the Veery (*C. fuscescens*), the Hermit Thrush (*C. guttatus*), the Olive-backed or Swainson's Thrush (*C. ustulatus*), and the Gray-cheeked Thrush (*C. minimus*). These four species are sufficiently similar visually that they confuse not only the human observer, but also silent males of the other species. The species-specific songs and call notes, however, permit easy species discrimination, as experimentally substantiated by Dilger (1956a). Rarely more than two species breed in the same area and the overlapping species $f + g$, $g + u$,

Table 2–1. Characteristics of four eastern
North American species of *Catharus* (from Dilger 1956a).

Characteristic compared	C. fuscescens	C. guttatus	C. ustulatus	C. minimus
Breeding range	Southernmost	More northerly	Boreal	Arctic
Wintering area	No. South America	So. United States	C. America to Argentina	No. South America
Breeding habitat	Bottomland woods with lush undergrowth	Coniferous woods mixed with deciduous	Mixed or pure tall coniferous forests	Stunted northern fir and spruce forests
Foraging	Ground and arboreal (forest interior)	Ground (inner forest edges)	Largely arboreal (forest interior)	Ground (forest interior)
Nest	Ground	Ground	Trees	Trees
Spotting on eggs	Rare	Rare	Always	Always
Relative wing length	Medium	Short	Very long	Medium
Hostile call	*veer pheu*	*chuck seeeep*	*peep chuck-burr*	*beer*
Song	Very distinct	Very distinct	Very distinct	Very distinct
Flight song	Absent	Absent	Absent	Present

and *u + m* usually differ considerably in their foraging habits and niche
preference, so that competition is minimized with each other and with
two other thrushes, the Robin (*Turdus migratorius*) and the Wood
Thrush (*Hylocichla mustelina*), with which they share their geographic
range. In connection with their different foraging and migratory habits
the four species differ from each other (and from other thrushes) in the
relative length of wing and leg elements and in the shape of the bill. The
rather extraordinary number of small differences between these at first
sight very similar species has been worked out in detail by Dilger
(1956a,b). Most importantly, no hybrids or intermediates among these
four species have ever been found. Each is a separate genetic, behavioral.
and ecological system, separated from the others by a complete biological
discontinuity, a gap.

Indeed the most characteristic attribute of a species in such a nondimensional system is that it is separated by a gap from other units in
this system. The gap that surrounds a species is the core of the species
concept. The term "species" signifies a very definite mutual relation between sympatric populations, between units in a nondimensional system,
namely that of reproductive isolation. The great advantage of the cri-

terion of interbreeding between two populations in a nondimensional system is that its presence or absence can be determined unequivocally. Reproductive isolation thus supplies an objective yardstick, a completely nonarbitrary criterion, for the determination of species status of a population. The word "species" indicates a relationship, like the word "brother." Being a brother is not an inherent property of an individual, as hardness is the property of a stone. An individual is a brother only with respect to someone else. A population is a species only with respect to other populations. To be a different species is not a matter of difference but of distinctness.

(3) *The Interbreeding-population Concept.* The concept of the multidimensional species is a collective concept. It considers species as groups of populations that actually or potentially interbreed with each other. Such populations, in order to retain their identity, cannot coexist at the same place and at the same time. The multidimensional-species concept thus deals with allopatric and allochronic populations, populations distributed in the dimensions of space and time, and classifies them on the basis of mutual interbreeding.

This concept has the weakness of all collective concepts, that of practical difficulties of delimitation: which discontinuous populations shall be judged "potentially" interbreeding? Even though the multidimensional concept comes much closer to reality than the nondimensional concept, it is evident that it lacks the latter's objectivity.

SPECIES DEFINITIONS

When the term species is applied to inanimate objects, as in "species of minerals," it is based on the typological species concept. When the term is used in biology, it is based to a greater or lesser degree on the two other concepts, the nondimensional ("reproductive gap") and the multidimensional ("unlimited gene exchange"). Parts of these two concepts have been incorporated into nearly all species definitions in biology in the last 100 years. Most of the definitions proposed in the last 25 years have avoided all reference to morphological distinctness. For instance, I defined species (Mayr 1940) as "groups of actually or potentially interbreeding natural populations which are reproductively isolated from other such groups," and Dobzhansky (1950) defined the species as "the largest and most inclusive . . . reproductive community of sexual and cross-fertilizing individuals which share in a common gene pool."

Definitions that stress this dual biological significance of species, reproductive isolation and community of gene pools, are usually referred to as "biological" species definitions. This designation has been questioned for instance by Simpson (1961), on the grounds that this is not an exclusive terminology, since many of the other species definitions also refer to living species and their biological attributes. This cannot be denied and yet the designation "biological species" would seem best for the modern concept for three reasons. First, it has never been used for any other species concept or definition; no confusion can arise as to the intent of an author who uses this terminology. Second, this terminology emphasizes that the underlying concept is based on the biological meaning of the species, that is, to serve as a protective device for a well-integrated, co-adapted set of gene complexes. Third, alternative terminologies are even more ambiguous. This is true, for instance, of the term "genetical concept," preferred by Simpson (1961). This term is full of ambiguity, since it has been applied to the most diverse species definitions. The "genetic species concept" of De Vries, Lotsy, Shull, Bateson, and other Mendelians was a strictly typological species concept that had nothing to do with the biological concept. A later generation of geneticists adopted a genetic species concept based on the sterility criterion, and the result was the recognition of cenospecies and other entities having little to do with the biological species. Nothing in a genetic species concept requires it to stress those particular genetic differences that are essential for species status, those that provide reproductive isolation. The term "biological species concept" for a concept emphasizing interbreeding within the population system and reproductive isolation against others is now so widely adopted and so uniformly used that it could hardly lead to misunderstanding. A history of the development of the biological species concept is given by Mayr (1957a).

If we wanted to single out the aspects most frequently stressed in recent discussions of the biological species concept, we would list these three:

(1) Species are defined by distinctness rather than by difference;

(2) Species consist of populations rather than of unconnected individuals; and

(3) Species are more unequivocally defined by their relation to non-conspecific populations ("isolation") than by the relation of conspecific individuals to each other. The decisive criterion is not the fertility of individuals but the reproductive isolation of populations.

The typological species concept treats species as random aggregates of individuals that have in common "the essential properties of the type of the species" and that "agree with the diagnosis." This static concept ignores the fact that species are reproductive communities. The individuals of a species of animals recognize each other as potential mates and seek each other for the purpose of reproduction. A multitude of devices insure intraspecific reproduction in all organisms (Spurway 1955). The species is also an ecological unit that, regardless of the individuals composing it, interacts as a unit with other species with which it shares the environment. The species, finally, is a genetic unit consisting of a large, intercommunicating gene pool, whereas the individual is merely a temporary vessel holding a small portion of the contents of the gene pool for a short period of time.

These three properties raise the species above the typological interpretation of a "class of objects." The nonarbitrariness of the biological species is the result of this internal cohesion of the gene pool (Chapter 10) and of the biological causation of the discontinuities between species (Chapters 5 and 17).

DIFFICULTIES IN THE APPLICATION OF THE BIOLOGICAL SPECIES CONCEPT

The general adoption of the biological species concept has done away with a bewildering variety of "standards" followed by the taxonomists of the past. One taxonomist would call every polymorph variant a species, a second would call every morphologically different population a species, and a third would call every geographically isolated population a species. This lack of a universally accepted standard confused not only the general biologists who wanted to use the work of the taxonomist, but the taxonomists themselves. Agreement on a single yardstick, the biologically defined category species, to be applied by everybody, has been a great advance toward mutual understanding.

Yet not all difficulties were eliminated by the discovery of this yardstick. Some taxonomists confused themselves and the issue by failing to understand that there is a difference between the species as a category and the species as a taxon. The species as category is characterized by the biological species concept. The practicing taxonomist, however, deals with taxa, with populations and groups of populations, which he has to assign to one category or another, for instance either to the category species or to the category subspecies. The nonarbitrary criterion of the category species, biologically defined, is that of the interbreeding or noninterbreeding. When confronted with the task of having to assign a taxon to the correct category, the occurrence or potentiality of interbreeding is usually only inferred. This, as Simpson (1961) has stressed, poses in most cases only a pseudo problem. Whether a given taxon deserves to be placed in the category species is a matter of the total available evidence.

The evidence that the definition is met in a given case with a sufficient degree of probability is a different matter [from the validity of the concept]. The evidence is usually morphological, but to conclude that one therefore is using or should use a morphological concept of the category (not taxon) species is either a confusion in thought or an unjustified relapse into typology. The evidence is to be judged in the light of known consequences of the genetical situation stated in the definition [of the category] (Simpson 1961:150).

Taxonomy is not alone in encountering difficulties when trying to assign concrete phenomena to categories. Most of the universally accepted concepts of our daily life encounter similar difficulties. The transition in category from subspecies to species is paralleled by the transitions from child to adult, from spring to summer, from day to night. Do we abandon these categories because there are borderline cases and transitions? Do we abandon the concept tree because there are dwarf willows, giant cactuses, and strangler figs? Such conflicts are encountered whenever one is confronted with the task of assigning phenomena to categories.

There are several classes of difficulties one may encounter when trying to apply this yardstick to concrete situations. These have recently been discussed by Mayr (1957b) and Simpson (1961), and for plants by Grant (1957). Such difficulties are caused by lack of information, by evolutionary intermediacy (speciation incomplete), or by genuine inapplicability of the concept (owing to asexuality, that is, uniparental reproduction).

Lack of Information

The Ranking of Variant Individuals. Whether certain morphologically rather distinct individuals belong to the same species or not is a routine problem of taxonomy. The types of variation that are particularly bothersome have been discussed by Mayr, Linsley, and Usinger (1953), who also give hints on how to resolve some of the difficulties. It is important to emphasize the difficulties, caused by sexual dimorphism, age differences, genetic polymorphism, and nongenetic habitus differences, which face the student of insects, of parasites, and indeed of any group of living animals, because some paleontologists seem to believe that it is only in work with fossils that one has to cope with the difficulty of having to draw inferences from morphological types.

No one will deny that the application of the biological species concept to fossil specimens is a difficult task. Yet, in principle, it does not differ from the task of the neontologist who only rarely can study natural populations but is usually forced to classify preserved specimens. The task of the paleontologist is clarified if he remembers that fossils are the remains of formerly living organisms that, when they were alive, were members of genetically defined populations exactly as the species living today are. Morphological criteria are used by the paleontologist as inferences on the natural populations that left the fossil remains. There is no justification for abandoning the biological approach merely because it is sometimes difficult to decide whether or not several morphological types in a sample are conspecific. No one makes the absurd demand that the paleontologist test the reproductive isolation of the species he recognizes. Yet by proper consideration of all the available morphological, ecological, stratigraphic, and distributional evidence it can usually be inferred with high probability whether certain specimens when living were or were not members of the same population. The problems in the application of the species concept to fossils have been discussed by Simpson (1951b, 1961) and Imbrie (1957), and by several contributors to a symposium on the species concept in paleontology (Sylvester-Bradley 1956).

The Ranking of Populations. The criterion of species status, "sympatric coexistence without interbreeding," raises practical problems also where two populations occur in contiguous geographic areas but in very different habitats. Where the evergreen rain forest of central Africa comes in contact with open-country vegetation, one may find the forest drongo *Dicrurus ludwigii* within 50 meters of the very similar savanna drongo *D. adsimilis,* but not on the same tree. Indeed they never interbreed. The same is true of other closely related species pairs wherever habitats meet along a sharp border. Even though such species replace each other spatially they must nevertheless be considered sympatric. The potential mates are within cruising range of each other during the breeding season, and could freely interbreed if they were not kept apart by specific isolating mechanisms. The terms "sympatric" or "coexistence" in species definitions must be conceived broadly, to include populations the individuals of which are within cruising range of each other during the breeding season, even though the habitats in which they occur do not overlap in space (Cain 1953).

Incompleteness of Speciation

Evolution is a gradual process and, in general, so is the multiplication of species (except by polyploidy). As a consequence one finds many populations in nature that have progressed only part of the way toward species status. They may have acquired some of the attributes of distinct species and lack others. One or another of the three most characteristic properties of species—reproductive isolation, ecological difference, and morphological distinguishability—is in such cases only incompletely developed. The application of the species concept to such incompletely speciated populations raises considerable difficulties. The various situations usually encountered can be classified under six headings.

(1) *Evolutionary continuity in space and time.* Species that are widespread in space or time may have terminal populations that behave toward each other like distinct species even though they are connected by an unbroken chain of interbreeding populations. Cases of reproductive isolation among geographically distant populations of a single species are discussed in Chapter 16. For instance, when Leopard Frogs (*Rana pipiens*) from the northern United States are crossed with frogs from southern Florida or from Texas most of the embryos die during development (Moore 1949).

Intermediacy of populations between successive species would be the normal situation in paleontology if all populations had left a fossil record. Actually the breaks in the fossil record are so frequent that it has been possible in only a few cases to piece together unbroken lineages connecting good species. The evolution from *Micraster leskei* through *M. cortestudinarium* to *M. coranguinum* (Fig. 2–1) is one such case (Kermack 1954; see also Nichols 1959). In other cases, cited in the literature, the differences in the lineages are so slight that neontologists would be inclined to consider the consecutive forms merely subspecies of a single polytypic species. Even though the number of cases causing real difficulties to the taxonomist is very small, it cannot be denied that an objective delimitation of species in a multidimensional system is an impossibility.

(2) *Acquisition of reproductive isolation without equivalent morphological change.* This group of cases raises a difficulty more practical than fundamental. When the reconstruction of the genotype in an isolated population has resulted in the acquisition of reproductive isolation, such a population must be considered a biological species, regardless of how

little it may have changed morphologically. Such sibling species are discussed in Chapter 3.

(3) *Morphological differentiation without acquisition of reproductive isolation.* The acquisition of isolating mechanisms in isolated populations

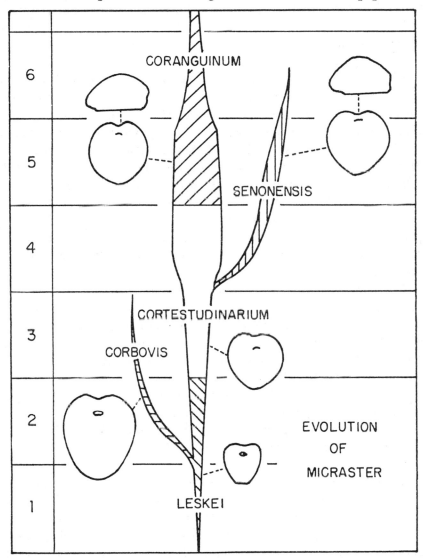

Fig. 2–1. Gradual speciation in time. The echinoid genus *Micraster* changes in the Cretaceous of southern England through six successive geological levels from *M. leskei* through *M. cortestudinarium* to *M. coranguinum.* (From Imbrie 1957.)

sometimes lags far behind morphological divergence (Grant 1957, Mayr 1957a). Such populations will be as different morphologically as good species and yet interbreed indiscriminately where they come in contact. The West Indian snail genus *Cerion* illustrates this situation particularly well. Whenever reproductive isolation and morphological differentiation do not coincide, the decision as to species status must be based on a broad evaluation of the particular case. The solution is generally a rather unsatisfactory compromise.

(4) *Reproductive isolation based on habitat isolation.* Numerous cases have been described in the literature in which natural populations acted toward each other like good species (in areas of contact) as long as their habitats were undisturbed. Yet the reproductive isolation broke down as soon as the characteristics of these habitats were changed, usually by the interference of man. The toads *Bufo americanus* and *B. fowleri* in North America (Blair 1941), and the flycatchers *Terpsiphone rufiventer* and *T. viridis* (Chapin 1948) are well-known examples. Prior to the habitat disturbance no one would have questioned the status of these species, but afterward they behaved like conspecific populations. Such cases of secondary breakdown of isolation will be further discussed in Chapter 6.

(5) *The incompleteness of isolating mechanisms.* Very few isolating mechanisms are all-or-none devices (see Chapter 5). They are built up step by step (except in polyploidy) and most isolating mechanisms of an incipient species will be imperfect and incomplete. Species level is reached when the process of speciation has become irreversible, even if some of the (component) isolating mechanisms have not yet reached perfection (see Chapter 17). To determine whether or not an incipient species has reached the point of irreversibility is often impossible.

(6) *Attainment of different levels of speciation in different local populations.* The perfecting of isolating mechanisms may proceed at different rates in different populations of a polytypic species. Two widely overlapping species may, as a consequence, be completely distinct at certain localities but may freely hybridize at others. Lorković (1953) has described such cases in butterflies. Many cases of sympatric hybridization discussed in Chapter 6 fit this characterization. The compromise solution that the practicing taxonomist often adopts, other things being equal, is to compare the sizes of the areas of undisturbed sympatry and of hybridization. Whichever is the larger determines species status.

The species is a population separated from others by a discontinuity, but not every discontinuity entitles the isolated population to species rank. If we designate as an *isolate* any more or less isolated population or array of populations, we can distinguish in sexually reproducing organisms between geographical, ecological, and reproductive isolates, of which only the last are species. The unspoken assumption made by certain authors, that the three kinds of isolates coincide, is not supported by the known facts and has led to unwarranted conclusions regarding the pathways of speciation.

The six types of phenomena described in the preceding paragraphs are consequences of the gradual nature of the ordinary process of speciation. Determination of species status of a given population is difficult or impossible in many of these cases.

The Difficulties Posed by Asexuality

The criterion of interbreeding among natural populations, the ultimate test of conspecificity in the higher animals, is unavailable in uniparentally reproducing organisms. It is evident that the absence of this criterion provides the most formidable and most fundamental obstacle to the application of the biological species concept. What should the evolutionist consider the "unit of evolution" in such organisms?

Asexuality in existing organisms is almost certainly a secondary phenomenon (Dougherty 1955; Stebbins 1960). All existing asexual organisms seem to be derived from sexual forms. Asexually reproducing lines have, sooner or later, one of three fates: they are lost by extinction, or they mutate, or they exchange genes with some other line by some process of recombination (Pontecorvo 1958). Indeed clandestine sexuality appears to be rather common among so-called asexual organisms. The expression "uniparental reproduction" is being used increasingly, instead of "asexual reproduction," to overcome this and other difficulties. Many biologists, for instance, are reluctant to refer to parthenogenesis as asexual reproduction.

It is too early for a definitive proposal concerning the application of the species concept to asexually or uniparentally reproducing organisms. If mutation and survival were random among the descendants of an asexual individual, one would expect a complete morphological (and genetic) continuum. Yet discontinuities have been found in most carefully studied groups of asexual organisms and this has made taxonomic subdivision possible. For this phenomenon I have advanced the explanation (Mayr 1957b) "that the existing types are the survivors among a great number of produced forms, that the surviving types are clustered around a limited number of adaptive peaks, and that ecological factors have given the former continuum a taxonomic structure." Each adaptive peak is occupied by a different "kind" of organisms and if each "kind" is sufficiently different from other kinds it will be legitimate to call such a cluster of genotypes a species (see also Chapter 15).

Various proposals have been made to resolve the difficulty that asexuality raises for the biological species concept. Some authors have gone so far as to abandon the biological species concept altogether and return to the morphological species for sexual and asexual organisms. I can see nothing that would recommend this solution. It exaggerates the importance of asexuality, which is both secondary and limited in its extent, and reintroduces the subjectivity and arbitrariness of the morphological species (see Hairston 1958 and Simpson 1961 for illuminating discussions of this problem).

A second solution, that of using a neutral term ("binom") for kinds of asexual organisms and restricting the term species to biological species of sexual organisms (Grant 1957), is logical and consistent. It minimizes, however, the fact that the word "species" signifies not only the biological unit of a reproductively isolated population, but also the classifying unit of a kind of organism. It is perhaps this consideration that has induced most practicing taxonomists to be frankly dualistic: they define the term species biologically in sexual organisms and morphologically in asexual ones. The fact that degree of genetic difference is on the whole responsible both for reproductive isolation and for morphological difference gives this procedure biological justification. To draw conclusions from the degree of morphological difference on the probable degree of reproductive isolation is a method of inference that has long been applied successfully to isolated populations in sexual organisms. There is no reason not to extend its application to asexual types. It results in the combining in a single species of those asexual individuals that display no greater morphological difference from each other than do conspecific individuals or populations in related sexual species. Subjectivity and arbitrariness cannot be avoided in such situations, particularly when there are no related sexual species.

Cain (1954a), Meglitsch (1954), and Simpson (1961) speak of asexual or uniparental "populations." However, every individual and its descendants are reproductively isolated in asexual organisms, and I am at a loss to define the term "population" in such organisms. Simpson

(1961) states that uniparentally reproducing individuals can be treated as species because their unity "is maintained by community of inheritance, by the capacity for genes to spread throughout the population (which therefore has a gene pool), and by the inhibition of their spread to other populations." It seems to me that these statements are contradicted by the known facts. Cases like that of the bdelloid rotifers show that community of inheritance does not help to define uniparental species and I know of no evidence that genes in uniparental species have a capacity to spread to individuals other than descendants (excepting phenomena of parasexuality). I feel, for these reasons, that introducing the term "population" into the definition of asexual or uniparental species is not legitimate.

Is the biological species concept invalidated by the difficulties in its application that have been listed?

One can confidently answer this question: "No!" Almost any concept is occasionally difficult to apply, without thereby being invalidated. The advantages of the biological species are far greater than its shortcomings. Difficulties are rather infrequent in most groups of animals and are well circumscribed where they do occur. Such difficulties are least frequent in nondimensional situations where (except in paleontology) most species studies are done. Indeed the biological species concept, even where it has to be based on inference, nearly always permits the delimitation of a sounder taxonomic species than does the morphological concept.

THE IMPORTANCE OF A NONARBITRARY DEFINITION OF SPECIES

Whoever, like Darwin, denies that species are nonarbitrarily defined units of nature not only evades the issue, but fails to find and solve some of the most interesting problems of biology. These problems will be apparent only to the student who attempts to determine species status of natural populations. The correct classification of the many different kinds of varieties, of polymorphism (Chapter 7), of polytypic species (Chapter 12), of biological races (Chapter 15), would all be meaningless, indeed would be ignored, but for an interest in arranging natural populations and phenotypes into biological species. Application of the concept has led to advances in the sorting of fossil specimens (Sylvester-Bradley 1956). Even though the evidence is largely morphological, an interpretation of fossil specimens based on biological concepts forces the paleontologist to make clear-cut decisions: morphologically different specimens found in the same exposure (the same sample) must be either different species or intrapopulation variants (excepting the relatively rare instances of secondary deposits).

It was not possible to state the problem of the multiplication of species with precision until the biological species concept had been developed. Only after the naturalists had insisted on the sharp definition of local species was there a problem of the bridging of the gap between species. And only then did the problem arise whether or not the species is a unit of evolution, and what sort of unit (Simpson 1951b; Thoday 1951).

It should be evident from these comments that the species problem is of great importance in evolutionary biology and that the growing agreement on the concept of the biological species has resulted in a uniformity of standards and a precision that have been beneficial for practical as well as theoretical reasons.

3 ~ Morphological Species Characters and Sibling Species

The morphological species concept, which dominated animal taxonomy during the 19th and early 20th centuries, is steadily losing ground. Yet much contemporary thinking about species still rests on it. In view of the historical importance of this concept and the correctness of some of its elements, it deserves more detailed discussion than was possible in the over-all treatment of species concepts in the preceding chapter.

The argument of proponents of the morphological species concept runs about as follows: "Natural populations considered by general consent to be species are morphologically distinct. Morphological distinctness is thus the decisive criterion of species rank. Consequently, any natural population that is morphologically distinct must be recognized as a separate species." The conclusion is fallacious, even though based on the correct observation of a general correlation between reproductive isolation and morphological difference. It is fallacious because it overlooks the strictly secondary role of morphological differences. The primary criterion of species rank of a natural population is reproductive isolation. The degree of morphological difference displayed by a natural population is a by-product of the genetic discontinuity resulting from reproductive isolation. This consideration necessitates a reevaluation of morphological characters as species criteria. Since sympatry of natural populations indicates reproductive isolation, we can use the amount of morphological difference among sympatric species as a yardstick in the evaluation of the taxonomic status of related isolated populations. The application of the biological species concept is thus facilitated by the proper evaluation of morphological differences.

The vulnerability of a purely morphological species concept in sexually reproducing species can be demonstrated primarily by two lines of evidence: (1) the presence of conspicuous morphological differences among conspecific individuals and populations (intraspecific variation); (2) the virtual absence of morphological differences among certain sympatric populations ("sibling species") that otherwise have all the characteristics of good species (genetic difference and reproductive isolation).

INTRASPECIFIC VARIATION

There is often greater morphological difference between individuals of a single population or between conspecific populations than between related species. In the well-known river duck, the mallard, Linnaeus originally described the male as *Anas boschas* and the female as *Anas platyrhynchos*. In many other groups of birds (birds of paradise, hummingbirds, tanagers, wood warblers, and so forth) the females may appear more different from males of their own species than from females of related species. Even greater are such sex differences among fishes (for example, female deep-sea fishes with attached dwarf males), insects (for example, army ants, mutillid wasps), and lower invertebrates (for example, the dwarf male of the echiurid worm *Bonellia viridis*).

The difference between immature stages and adults in many kinds of animals is astonishingly great. This is true of insects, lower invertebrates (free-swimming larvae of crustaceans, mollusks, echinoderms), and particularly parasites. To discover that "Cercaria" is a stage in the life cycle of trematodes, and "Cysticercus" a stage in the life cycle of cestodes, required real ingenuity.

However, one does not have to take such extreme examples. In many cases, a morphological definition of the species is made difficult by ordinary individual variability. Such variability may be entirely genetic or may, to some extent, be due to nongenetic modification of the phenotype caused by local environmental conditions. Some groups of animals seem particularly subject to such modification. This appears to be the case with fresh-water bivalves (*Anodonta*), fresh-water snails, and some marine snails. More than 250 described "species" of the bivalve *Anodonta* are merely local variants of a single species (Schnitter 1922). In the genus *Melania*, Riech (1937) found 114 "species" to be merely variants of previously described ones, and in the pond snails of the genus *Lymnaea*, more than 1000 names were reduced by Hubendick (1951) to about 40 species (see Chapter 7, under "Individual Variation").

What attitude does the supporter of the "morphological species concept" take toward this variation? Curiously, he takes precisely the same attitude as the proponent of the biological species concept. No matter how different a morphological variant may be, as soon as it is revealed as a member of the same breeding population (sex and age differences, polymorphism), or as a local variant (genetic or nongenetic), he sinks it into synonymy. It is quite evident, then, that even those who profess to hold a morphological species concept base their taxonomic decisions ultimately on the biological criterion of interbreeding. Degree of morphological difference is completely useless as a yardstick for species status unless it is applied in conjunction with such biological criteria as "population," "interbreeding," and "reproductive isolation."

This conclusion is reinforced by a consideration of the reverse situation, involving types that are exceedingly similar morphologically, but belong biologically to different species, the so-called sibling species. Again the vulnerability of a purely morphological criterion is evident.

SIBLING SPECIES

The naturalist occasionally encounters sympatric populations that are morphologically exceedingly similar, if not identical, but are reproductively isolated. What shall he do with such populations? Adherents of a purely morphological species concept will not classify them as species because for them, as formulated by Sturtevant (1942), "distinct species must be separable on the basis of ordinary preserved material." Natural populations that are not readily distinguishable but are nevertheless reproductively isolated have caused considerable difficulties in the biological and taxonomic literature. Such populations have sometimes been called "biological races," a term under which many unrelated problems and phenomena have been grouped (Chapter 15). Adoption of the biological species concept makes it evident that most "biological races" are indistinguishable from other valid species, except by the slightness of the morphological difference. For such exceedingly similar species, the term "sibling species" was introduced (Mayr 1942) as a translation of the equivalent terms in other languages: *espèces jumelles* (Cuénot 1936) and *Geschwisterarten* (Ramme 1930). Numerous authors, beginning with Pryer (1886), have discussed (under various names) pairs or larger groups of very similar species (see Ramme 1951:313 for a history of the terminology). The contention of some entomologists that sibling species differ from other species in origin and biogeography has not been substantiated. Sibling species may be defined as "morphologically similar or identical natural populations that are reproductively isolated."

Sibling species are of threefold importance in biology: (a) they permit us to test the validity of the biological versus the morphological species concept; (b) they are of great practical importance in applied biology, in agricultural pest control, and in medical entomology; (c) they are of historical importance in the study of speciation (Chapter 15), having been cited by some authors as evidence for a separate type of speciation.

Well-Studied Groups of Sibling Species

The characteristics of sibling species are best revealed by describing in detail some of the better-known cases. In the genus *Drosophila* most of the species complexes contain groups of sibling species (Patterson and Stone 1952). The kinds of differences that may exist between two sibling species are typified by the extensively studied pair *Drosophila pseudoobscura* Frolova and *D. persimilis* Dobzhansky and Epling. Lancefield (1929) originally discovered differences between two kinds of flies, which he designated *D. pseudoobscura*, race A, and *D. pseudoobscura*, race B. Crosses between the "races" produced F_1 hybrids of which the females were fertile, the males sterile. Lancefield found the Y-chromosome of "race A" to be J-shaped; that of "race B," V-shaped. From this modest beginning, the number of known differences has increased steadily, and, when it was discovered that the two "races" coexist over wide areas without interbreeding, they were raised to the rank of full species (Dobzhansky and Epling 1944). The salivary-gland chromosomes of the two species are quite different and completely diagnostic, even though the gene arrangements are variable within either species (Dobzhansky 1944). At first it was thought that the two species were identical morphologically, but then it was found that the average number of teeth in the sex combs of the males is greater in *D. pseudoobscura* than in *D. persimilis*, while the wings of *D. persimilis* are on the average larger than those of *D. pseudoobscura*. Reed, Williams, and Chadwick (1942) calculated a special index from several wing measurements, which ranged from 45.7 to 62.8 (55.7) in different strains of *D. pseudoobscura* and from 68.8 to 76.2 (72.6) in *D. persimilis*. Rizki (1951) showed a clear-cut difference between the species in the shape of the male genitalia. Additional differences were found by Spassky (1957).

What were once considered two morphologically indistinguishable "biological races" are now accepted as two similar species, distinguished by diagnostic characters in salivary-gland chromosomes, male genitalia, sex combs, and relative wing size. This development, reported here for the sake of illustration in considerable detail, is typical of our knowledge of most sibling species. When first discovered, they are believed to be morphologically identical or nearly so, but on closer study one morphological difference after another is discovered, and these are fortified by ecological differences. *Drosophila persimilis*, for instance, has a more northerly distribution than does *D. pseudoobscura*, and is more frequently found at higher altitudes, showing a preference for lower temperatures. The two species also differ in their diurnal activity rhythm, in their reaction to light (Lewontin 1959), and in other ecological-physiological characteristics (Pittendrigh 1958). Females of *D. pseudoobscura* reach sexual maturity at an age of 32–36 hours after hatching; those of *D. persimilis*, at an age of 44–48 hours (Spieth 1958). There may be differences in their scents, as manifested by interspecific mating preferences (Mayr 1950b). In short, under scrutiny almost any property of these species is found to differ, slightly or conspicuously. In spite of their superficial similarity, the two species represent two very different gene complexes. For notes on other sibling species in the genus *Drosophila* see Dobzhansky (1951), Patterson and Stone (1952), Carson (1954), and other recent literature.

Perhaps the most celebrated case of sibling species is that of the malaria-mosquito complex in Europe. According to the older literature, malaria in Europe is caused by *the* malaria mosquito, *Anopheles maculipennis*. A study of the distribution and ecology of this mosquito revealed all sorts of puzzling irregularities. *Anopheles* mosquitoes were found to be quite common in certain parts of Europe where malaria was absent. In some districts they fed only on domestic animals; in others they preferred man. In some districts they were associated with fresh water; in others, with brackish water. Understanding dawned only when Falleroni discovered constant differences in the eggs of mosquitoes differing in biological characteristics. Finally it was proved by Missiroli, Hackett, Bates, Swellengrebel, de Buck, and van Thiel that *the* malaria mosquito of Europe was actually a group of six sibling species (Table 3–1).

The differences in some of the biological characteristics of these mosquitoes are quite conspicuous. For instance, they differ in their requirements for space during mating: *atroparvus* will mate in a small cage without swarm formation; *labranchiae* will mate in a small cage (1 to 0.5 m); *sacharovi* in a room-sized cage; *maculipennis* in a large outdoor cage;

Table 3–1. Biological differences among members of *Anopheles maculipennis* group of mosquitoes.

Characteristic compared	A. melanoon and subalpinus	A. messeae	A. maculipennis	A. atroparvus	A. labranchiae	A. sacharovi (elutus)
Egg color	All black or (*subalpinus*) with dark cross bars	Transverse bars part of a diffuse dark pattern	Two black cross bars on light background	Dappled or with wedge-shaped black spots	Similar to *atroparvus* but paler, dark spots smaller	Gray without pattern
Egg float	Large and smooth	Large and rough	Large and rough	Small and smooth	Very small and rough	None
X-chromosome	Standard	Extensive rearrangement	Standard	Standard	Standard	Small inversion
Third chromosome	Inversion in right arm	Inversion in right arm	Inversion in right arm	Standard	Standard	Inversion in left arm
Habitat	Often rice fields	Cool, standing fresh water	Cool, running fresh water	Cool, slightly brackish water	Mostly warm, brackish water	Shallow, standing water, often brackish
Hibernation	No	Yes	Complete	No	No	No
Feeding on man	?	Rarely	No	Yes	Yes, with preference	Almost exclusively
Malaria carrier	No	No (rarely)	No	Slightly	Very dangerous	Very dangerous
Range	Mediterranean	Continental and northern Europe	Mountains of Europe	Northern Europe	Chiefly southern Europe	Eastern Mediterranean and Near East

while it has been impossible so far to induce either *messeae* or *melanoon* to mate in captivity. All crosses have to be made with *atroparvus* males because only in this species will the males mate under laboratory conditions. Different combinations show different degrees of sterility (Table 3–2).

After 30 years of search the best morphological differences to be found among these species of *Anopheles* are still those of the eggs. How-

Table 3–2. Fertility of interspecific crosses in the
Anopheles maculipennis group.[a]

Hybrid	*A. atroparvus* ♂ with—			
	labranchiae ♀	*melanoon* ♀	*maculipennis* ♀	*sacharowi* ♀
F_1 ♂	Sterile	Sterile	Sterile	Sterile
F_1 ♀	Fertile	Fertile	Sterile	Lethal
Backcross (with *atroparvus*)	20% fertile	Fertile		

[a] In the cross with *messeae*, all eggs are lethal.

ever, additional differences have been found. Number and branching of the hairs in larval appendages are usually diagnostic for each species at a given locality. The shape of the external spines (pointed *versus* blunt or rounded) on one of the segments of the male external genitalic armatures may possibly be diagnostic. There are differences among some of the species in size and shape of the wing scales. Frizzi (1952) has shown that there are constant differences among the species in gene arrangements, as indicated by the banding pattern of the salivary chromosomes (Table 3–1).

Correlated with geographical distribution (temperate or subtropical) and habitat preference are a considerable number of physiological differences between the species (Büttiker 1948). Eggs of the cold-water species *A. maculipennis* develop much more rapidly than do eggs of various warm-water species (Kettle and Sellick 1947). *Anopheles messeae* and *maculipennis* form large fat bodies in the fall, and hibernate in a cold shelter. The other species may be induced throughout the winter to feed or oviposit under suitable conditions.

OCCURRENCE OF SIBLING SPECIES

Sibling species occur in all groups of animals, yet they seem to be much more common in some groups, such as insects, than in others, such as vertebrates. It would be interesting to have exact numerical data on these differences, bt t this will not be possible until the taxonomy of these groups is better known. Surveys of sibling species have been published previously (Mayr 1942, 1948; Brown 1959). Many examples will be listed and discussed in the following pages, but they have been chosen almost at random from the enormous number of such cases in the taxonomic literature. There is hardly a taxonomic monograph or revision that does not give new instances of sibling species.

Vertebrates

Among mammals, sibling species are most common in the order of the rodents. In birds they are comparatively rare, constituting fewer than 5 percent of all species. Stresemann (1948) records twelve differences between the two species of nightingales (*Luscinia megarhyncha* and *L. luscinia*), of which only three are morphological. Sibling species are common among the tyrant flycatchers, particularly the genera *Empidonax*, *Elaenia*, and *Myiarchus*. The cave swiftlets of the genus *Collocalia* in the East Indies, with at least a dozen species, form the largest and morphologically most uniform group of sibling species known among birds. Nest structure may be the best diagnostic character in this genus. Sibling species occur in snakes (for example, *Thamnophis*) and lizards (for example, *Lacerta*, Kramer and Mertens 1938; *Emoia*, Brown and Marshall 1953). They are widespread among frogs with their comparatively simple external morphology. *Rana brevipoda* differs from *R. nigromaculata* in ecology, call notes, and general behavior, and no hybrids are found in the parts of Japan where they are sympatric, even though fertile hybrids can be produced in the laboratory (Moriya 1951). Other examples are *Leptodactylus ocellatus* and *reticulatus* in Argentina (Cei 1949a), *Crinia signifera* and *insignifera* (Moore 1954) and other species of frogs in Australia (Main 1957; Main et al. 1958), and many North American frogs.

Sibling species seem to be widespread among fishes. Minamori (1952) described from western Honshu two "races" of the loach, *Cobitis taenia*, which are evidently good species since they do not seem to interbreed in the streams where they coexist. The smaller of these two species is adapted to warmer waters and muddier bottoms, the larger to cooler waters and sandier bottoms. In many streams only the larger species is found. Later analysis revealed at least one additional sibling species and proved that crosses among these species are largely sterile (Minamori 1956). Three very similar species of the lesser sand eel complex (*Am-*

modytes lancea) differ in ecology, breeding season, and minor morphological characters (Jensen 1941).

In the salmonids, sibling species appear to be common but, for a number of reasons, are **particularly** difficult to define. Most species of this family live in salt water but spawn in fresh water. Often a population becomes landlocked; that is, it becomes cut off or voluntarily refrains from return to the ocean and becomes a straight fresh-water population. If the same body of water is subsequently reinvaded by colonists of the original parental population, interbreeding may not take place because breeding seasons or ecological niches have become different. The rainbow trout and steelhead forms of *Salmo gairdneri* on Vancouver Island seem to be such a case (Neave 1944), as are the kokanee and sockeye forms of the salmon *Oncorhynchus nerka* in British Columbia (Ricker 1938, 1940). Many members of the genus *Coregonus* have become strictly fresh-water inhabitants even though they ultimately derive from anadromous ancestors. In many rivers and lakes of Europe, particularly in the Alpine region and in northern Europe, there are up to four or five species of this genus. They are not only very similar, but are so plastic under different environmental conditions that it becomes difficult to define their real differences. The number of gill rakers seems least affected by environmental changes (Svärdson 1949a, 1950, 1952). Where more than one species occurs in a lake, there is usually a clear-cut difference in ecology. One species may be surface pelagic, a second deep-water pelagic, a third bottom-living in shallow water, and a fourth bottom-living in the depth of the lake. Intermediate individuals are found in some lakes, to make things even more complicated. These are apparently hybrids between two species (Dottrens 1953), although the low viability of the hybrids usually prevents the development of hybrid populations (Svärdson 1952). It is unknown what controls hybridization, since in many lakes no hybrids are found (Dottrens and Quartier 1949).

The "*Coregonus* problem" is deprived of much of its mystery by the demonstration (Wagler, Svärdson, Berg, and Dottrens) that the *Coregonus* "races," in spite of their phenotypic plasticity, are perfectly good species, resulting from different invasions and subsequent isolation. The evidence for this conclusion has now been summarized by Svärdson (1957). He recognizes five Palearctic species of whitefishes (*Coregonus, sensu stricto*), a few of which also occur in North America. Some of these invaded their present area of distribution early in the late Pleistocene or post-Pleistocene, others later. In certain lakes there has been considerable hybridization between early and late colonists. Dottrens (1959) has reached very similar conclusions, but believes that Svärdson's *lavaretus* consists of two species (*lavaretus* and *wartmanni*). A parallel situation is found among the ciscoes (subgenus *Leucichthys*), which have two Palearctic and several additional North American species (Walters 1955; Svärdson 1957). Four exceedingly similar species which differ in growth curve, ultimate size, and susceptibility to parasitism coexist in Lake Winnipeg (Keleher 1952). The origin of sibling species in the fish genera *Myoxocephalus, Osmerus, Clupea, Coregonus,* and *Gasterosteus* was analyzed by Svärdson (1961) (Table 15–3).

Insects

Sibling species appear to be especially common among insects. Their discovery has been relatively slow, since in most families, even of highly conspicuous insects, many species remain undescribed. On the whole, sibling species are discovered only when the respective group is of economic or medical importance.

Lepidoptera. Sibling species are particularly frequent in the order Lepidoptera. A list of 40 pairs of European sibling species was published by Heydemann (1943), and additional cases in a subsequent publication (1944). Even in the exceedingly well-known butterfly fauna of eastern North America, sibling species are still being discovered, as in the genera *Mitoura* (Rawson and Ziegler 1950) and *Strymon* (Klots and Clench 1952). Nearly all moths that infest stored food products are groups of sibling species (Corbet 1943), usually differing in food preference. Only two of four species of the *Tinaea granella* complex infest grain, and of these *T. granella* prefers wheat, *T. infimella* rye. In the family Coleophoridae, there are numerous groups of sibling species, in some of which the male genitalia are diagnostic, in others not (McDunnough 1946). Sibling species of *Solenobia* from Switzerland were described by Sauter (1956) and Galliker (1958); their reproductive isolation was studied by Seiler and Puchta (1956). A particularly well-analyzed case is that of two budworms of the genus *Choristoneura* (Tortricidae), one of which (*fumiferana*) feeds on spruce and fir, and the other (*pinus*) on jackpine. The two species, although exceedingly similar, differ not only in food plant preference, but also in average size, coloration, wing expanse, wing pattern, male genitalia, and season of flight (Freeman *et al.* 1953:121–151). Most of the sibling species of Lepidoptera were discovered through differences in their preferred food plants and through morphological differences in the male genital armatures. Other characters were usually found as soon as the "host races" became recognized as distinct species.

Diptera. Sibling species are very widespread among the Diptera, and on the whole are well studied, because many of the species are important medically (*Anopheles, Aedes, Simulium*), genetically (*Drosophila*), or cytologically (*Sciara, Chironomus*). Sibling species in *Drosophila* and in the *Anopheles maculipennis* complex have been discussed above. Numerous other groups of sibling species have been described in the Culicidae (Bates 1949). The North American *Anopheles maculipennis* complex consists of four species (Aitken 1945; Kitzmiller 1959). The Oriental "species" *A. hyrcanus* is actually a complex of at least eight sibling species (Reid 1953). *Anopheles punctulatus* and the dreaded *A. gambiae* (with *melas*) are other groups of sibling species. There are also several groups in the genus *Aedes*, such as the *Aedes scutellaris* complex (Marks 1954).

That not all "biological races" in the Culicidae are necessarily sibling species is indicated by the forms of the *Culex pipiens* complex. They were at one time assigned to three species: cold-climate, rural *pipiens;* Mediterranean or urban *molestus;* and subtropical *fatigans* (= *quinquefasciatus*). Physiological characters ascribed to the three species were as follows: (A) *pipiens:* (1) lay eggs only after a blood meal, (2) require large spaces for mating, (3) have a period of inactivity in winter, (4) feed almost entirely on birds and amphibians, not on mammals, (5) are restricted to rural habitats; (B) *molestus* and *fatigans:* (1) frequently lay eggs without a blood meal (autogeny), (2) are able to mate in small spaces (stenogamy), (3) are reproductively active throughout the year, (4) feed readily on mammals, (5) are often found in urban environments. Unfortunately this simple, diagrammatic diagnosis does not always hold true. Strains with combinations such as $B_1A_2A_3B_4A_5$ or $B_1A_2B_3A_4B_5$ are common, although all A_1 seem also to be A_2. The belief that these are species is further undermined by the frequency of mixed broods (for example, A_1 and B_1) produced by single, fertilized, wild-caught females (for details see Mattingly *et al.* 1951, 1953; Kitzmiller 1953). This variation is not yet fully understood, but it is possible to advance a tentative interpretation. There is reason to believe that each of the forms of *C. pipiens* was originally a polymorphic geographic race. Even today *fatigans* and *pipiens* are largely allopatric and merge in their zone of contact, forming intermediate types ("*pallens,*" "*comitatus*"). These races have strongly ecotypic characteristics, anautogeny and hibernation being special developments of cold-climate populations. As pointed out by several authors, most populations are highly polymorphic, but toward the north the percentage of autogenous females decreases steadily. Two major factors complicate investigation of the *Culex* problem. One is that through human transport new colonies were established by mosquitoes originally from far-distant areas. For instance, shortly after the first large-scale entry of American troops into Australia in about 1942, *molestus* was discovered there, side by side with the native *fatigans*. At that time the two were separated by a clear-cut morphological discontinuity. Within 10 years the gap between the forms had been filled through interbreeding (Drummond 1951). It seems that many of the putative morphological characters of the physiological *Culex* types are actually the incidental morpho-, logical characteristics of local populations and by no means necessarily correlated with stenogamy or autogeny. The cold-climate rural *pipiens* seems least subject to human transport, and this has permitted the evolution of many differences between European and North American *pipiens*. It has also permitted the evolution of an endemic Australian race. The sedentary habits of the nondomestic *Culex* have resulted in the formation of comparatively uniform and well-defined geographic races, while the intermixing, through human transport, of the domestic forms explains in part why these forms are so highly polymorphic.

The second complication is more disturbing. In *pipiens* Laven (1953, 1959) found cytoplasmic factors that produce sterility barriers between certain local races. Five groups of strains have been found so far: one in western Europe (London, Paris, and elsewhere), one in northern Germany, one in southern Germany, and two in the Mediterranean area (Fig. 3–1). The first two groups of strains are cross-sterile in either direction, the embryos dying. Males of the southern German strain have normal fertility with females of the other two European strains. The reciprocal crosses, however, are sterile (except for a fraction of 1 percent). American *pipiens* and *fatigans* strains are fully fertile with the northern German strain, but more or less sterile with the others. The mode of inheritance indicates the presence of a cytoplasmic factor.

These groups of strains answer the definition of sibling species, yet there is serious doubt whether it would be legitimate to label as "species" allopatric strains that may differ only by a single genetic factor. With *pipiens* and *fatigans* behaving like subspecies, and *molestus* and *pipiens* also like subspecies or polymorphs, it would seem paradoxical to consider *pipiens* itself a complex of sibling species. Further research will undoubtedly clarify this problem, but for the present the *Culex pipiens* complex is best interpreted as a single polytypic species with many ecologically polymorphic geographic races, some of which have acquired imperfect isolating mechanisms. By the nature of these mechanisms no two

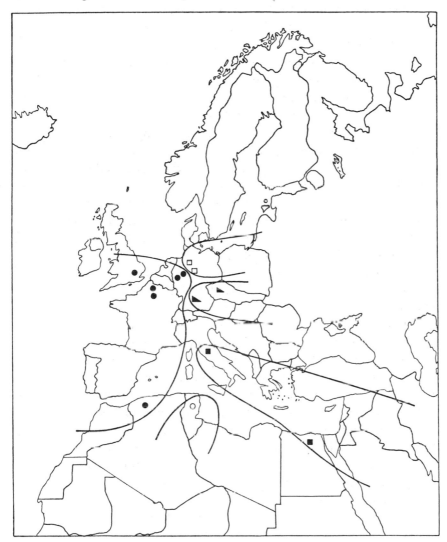

Fig. 3–1. Strict geographical replacement of five cytoplasmic races of *Culex pipiens* in Europe. (From Laven 1959.)

such strains can successfully coexist permanently. Cytoplasmic sterility barriers between geographic races exist also in *Aedes* and in other culicids. (See also Chapter 15.)

The classification of the numerous sibling species in the genus *Chironomus* is gradually being elucidated with the help of morphological and ecological information and particularly through an analysis of the salivary chromosomes (Strenzke 1959, 1960, with literature). On the basis of the available data there is little doubt that *C. tentans* and *pallidivittatus*, as well as *C. thummi* and *piger*, are good species.

Beetles. Groups of sibling species occur in almost every family of beetles, but are particularly common in Curculionidae, Chrysomelidae (Brown 1958), and Lamellicornia. Brown (1945) described an interesting group of about 25 species of leaf beetles of the genus *Calligrapha* (Chrysomelidae), many of which cannot be separated by morphological characters. Feeding and breeding tests, and the study of the beetles in nature, indicate that these are valid species. The analysis is impeded by the fact that some of these "species" are parthenogenetic (in at least five "species" no males are known). The genus *Chrysomela* is also rich in sibling species (Brown 1956). The weevil "*Calandra oryzae* L." actually consists of a large-sized species that infests corn and a small-sized species that infests wheat. The two do not interbreed (Birch 1954).

Orthoptera. Sibling species are very common among the Orthoptera. In a series of papers Fulton (1931, 1933, 1937) established that three kinds of crickets, hiding under the name *Nemobius fasciatus*, differ not only in habitat preference and song, but also in a number of minor morphological characters. This was fully confirmed by Cantrall (1943) and there is now no doubt that *fasciatus, tinnulus,* and *allardi* are sibling species (Alexander and Thomas 1959). In view of the known differences in their songs, it is not surprising that differences are found in the sound files of the males (Table 3–3). *Nemobius fasciatus* cannot be crossed, while *allardi* and *tinnulus* produce fertile hybrids with a characteristic song. The absence of this song where the habitats of the two species overlap indicates that the two species do not hybridize in nature. Song differences among sibling species of the *Nemobius carolinus* group were described by Alexander (1957b).

Fulton (1925) found two "races" of *Oecanthus fultoni* in Oregon: one ("race A") agrees essentially with *O. niveus* of eastern North America in song and in its habit of depositing a single egg on the bark of a tree; the other ("race B"; *O. rileyi*) has a much slower song (90 compared to 160 stridulations per minute at 70°F) and lays its eggs in rows on twigs of wild rose and berry bushes. Morphologically the two forms are indistinguishable, but, even though A is not truly different from eastern *niveus*, in many respects B agrees more closely with another eastern species, *O. nigricornis,* although the two differ morphologically. It would be interesting to know more about songs and egg-laying habits of so-called *O. niveus* from Cuba and Mexico. There is no doubt, however, that race B of Oregon is a distinct sibling species. The same is evident for the "physiological races" in *Gryllus* and *Anaxipha,* described by Allard (1929).

Table 3–3. Morphological differences of the sound file and characteristics of the song in the *Nemobius fasciatus* group of crickets (from Pierce 1948).

Measurement	*N. allardi*	*N. fasciatus*	*N. tinnulus*
Average number of teeth in file (right wing)	192 (165–220)	118 (101–126)	214 (196–218)
Average file length (mm)	1.438 (1.32–1.50)	0.992 (0.81–1.12)	1.600 (1.5–1.74)
Duration of pulse (sec)	0.002	0.006–0.010	0.02
Number of teeth struck per pulse	162 (81%)	56 (47.5%)	120 (58%)
Frequency (cy/sec)	7500	7740	6300
Number of pulses or chirps per second	14–20 pulses per second, lasting 8 sec	4–12 pulses per chirp, 1.4–5.0 chirps per second	5–10 single-pulse chirps per second
Nature of song	Series of separate and distinct pulses	Series of discrete chirps or trains of pulses	High-pitched bell-like note

The field crickets, *Gryllus* (*Acheta*), are a notoriously difficult group and, although 17 "species" had been described in North America, a detailed morphological analysis induced leading specialists for a long time to accept only a single species. Yet there are obviously biologically different types (Allard 1929), and an analysis of the field crickets in the restricted area of North Carolina permitted Fulton (1952) to distinguish four different kinds, tentatively given vernacular names: triller, woods cricket, mountain cricket, and beach cricket. They differ in song (in part), annual cycle (Fig. 3–2), choice of habitat, and minor morphological characters. That these are good species has been confirmed by Alexander (1957) and Bigelow (1960). It should be interesting to compare their cytology.

Hymenoptera. Sibling species seem common among bees, wasps, and other groups of hymenopterans. Ants are particularly rich in sibling species. This has been the cause of much confusion, and is in part responsible for a peculiar polynomial nomenclature that has long characterized ant taxonomy. The application of the modern species concept to North

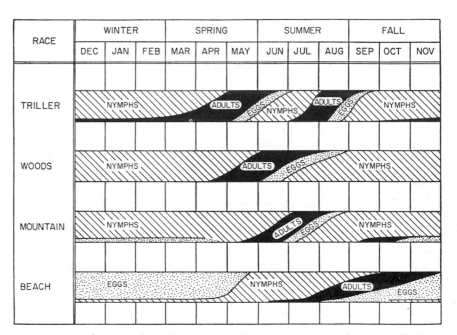

Fig. 3–2. The annual cycle of four sibling species of crickets (*Gryllus*) in North Carolina. Each species has a distinctive seasonal history. (From Fulton 1952.)

American ants by Creighton (1950) led to a dramatic simplification of this literature. Sibling species in the genus *Lasius* have recently been described by Wilson (1955), and occur also in *Myrmica* (Brian and Brian 1949) and in *Camponotus* (Brown 1950). The situation seems particularly difficult in the genus *Formica*.

Scattered cases of sibling species have been reported for many groups of insects, such as aphids and collembola (Gisin 1947). Only rarely has the classification matured sufficiently to permit quantitative analysis of the frequency of sibling species. Kontkanen (1953) found that about 40 percent of the 292 species of leafhoppers (Homoptera) of Finland are sibling species, with as many as 13 in a single genus. In some cases no ecological differences have so far been found between sympatric pairs of sibling species, such as *Criomorphus affinis* and *bicarinatus,* and *Calligypona flaveola* and *straminea.*

Other arthropods. Sibling species are undoubtedly common among mites, but the study of them has hardly begun. There are three very similar mites on the honey bee, one of which (*Acarapis woodi*) is pathogenic. *Acarapis externus* lives in a ventral fold between the head and the propleura of the bee, at the base of the tentorium; *A. dorsalis* in a dorsal fold between the mesoscutum and the mesoscutellum; and *A. woodi* in the tracheae (Morgenthaler 1934). Tarsal length and distance between stigmas help to distinguish the three species. Sibling species occur in many genera of spiders, for instance in the poisonous black widow spiders (*Latrodectus;* Levi 1959).

Several genera of crustaceans are noted for the occurrence of sibling-species complexes. In the genus *Gammarus* a number of similar forms were described as "races" or "subspecies," but Kinne (1954) showed that they are full species: *G. zaddachi, G. salinus,* and *G. oceanicus.* Diagnostic morphological criteria are now known for all these species, some of which were first discovered through their preferences for habitats with different salinities. A similar situation occurs in the isopod genus *Jaera;* Forsman (1949) and Bocquet (1953) have shown that a number of sibling species exist, each characterized ecologically and chromosomally (Staiger and Bocquet 1954). A sibling species has recently been described in the copepod genus *Calanus* (Bowman 1955). In fresh-water crustaceans the situation is frequently more complex. In *Daphnia* there has been much debate as to the number of species in North America, estimates ranging as low as four, with numerous "subspecies" and varieties. Brooks (1957a) finds that no fewer than 13 species can be distinguished quite clearly, but nongenetic variation (cyclomorphosis and so forth) and occasional hybridization may make delimitation difficult.

Other Invertebrates

In mollusks, likewise, sibling species are common, and some are pointed out in almost every major revision. The difficulty in recognizing them is increased by the great individual variation in many genera. Intraspecific variants are frequently more different than are separate species. Very similar species occur in land snails, in slugs (where they are sometimes best recognized by differences in courtship patterns), in fresh-water snails, and in marine mollusks. Four sibling species of limpets of the genus *Patella* are currently recognized in Europe and, now that the differences are understood, the species seem to be well defined in most localities. In France and England, however, intermediate specimens have been found that may be hybrids, even though this possibility has been denied (Evans 1953). The sibling species in *Lymnaea* have been excellently monographed (Hubendick 1951); in other genera (for example, *Pisidium*) the analysis is not yet complete.

In view of the relative frequency of sibling species in such morphologically elaborate organisms as diptera, hymenoptera, coleoptera, lepidoptera, and crustaceans, one would expect an even greater frequency in the lower invertebrates with their simplified external morphology. However, the difficulty of the discovery of sibling species is correspondingly increased in such groups. This is particularly true for groups like the nemerteans, turbellarians, and nematodes, in which most taxonomic characters are internal. Yet even in these groups a close study of ecological and physiological characteristics has led to the recognition of previously overlooked complexes of sibling species. Once recognized, such species often are found to be well defined morphologically. For instance, the nemertean *Lineus ruber*, believed to be a single species with two "varieties," was found to consist of four sibling species which differ in color, size, proportions, size and position of the eyes, and color of the cerebral ganglia. Some species can contract, while others roll up in a bundle; two species reproduce sexually, the others reproduce asexually; gamete formation and larval development are species specific. The asexual species regenerate extremely well, but the sexual ones do not (Gontcharoff 1951). The nematode genus *Rhabditis* appeared to be rich in sibling species, but Osche's (1952) painstaking analysis indicates that there are more morphological differences than was formerly suspected.

Sessile marine invertebrates are believed to be subject to much nongenetic variation, depending on local water conditions. In the coral genus *Millepora* finely branched and coarse-stemmed individuals grow adjacent to each other in the surf zone and in quiet waters. Yet Boschma (1948) showed that the genus contains ten species rather than the single variable one recognized by some recent authors. The taxonomy of most lower invertebrates is, on the whole, too unsettled, and they are too devoid of medical or genetic interest, to permit determination of the frequency of sibling species. However, sibling species have been recorded among sponges (Lévi 1956; Table 3–4).

The simpler a group of organisms is morphologically, the more difficult it should be to distinguish species. On the basis of this consideration.

one would expect the highest number of sibling species to occur in the morphologically simple protozoans. They were first discovered there by Sonneborn (1938), in *Paramecium aurelia* and by Jennings (1938) in *P. bursaria*. Although usually recorded as "varieties," they are reproductively isolated and correspond to the species of higher organisms, as is evident from Sonneborn's (1957) analysis. Such sibling species have also been found in other species groups of *Paramecium* and *Euplotes*. The

Table 3–4. Some differences between two sibling species of sponges, *Halisarca* (after Lévi 1956).

Characteristic compared	H. dujardini	H. metschnikovi
Structure of spherical cells	Rosette-like	Globular
Reproductive period	June to September or later	April to June
Ecological zone	Medium and deeper water, attached to pebbles and rocks	Shallow waters in estuaries, attached to bases of marine algae
Spermatogenesis	Simultaneous	Successive
Spermatozoon	Aberrant, discoid	Normal
Oocytes	Small, multinucleolate	Large, simple
Large cells at posterior pole of larvae	Always with flagella	Bare
Rhagon	Without folds, asconoid	Folded, syconoid

recognition of sibling species in the class Protozoa depends mainly on experiment (but see Hairston 1958). In the most intensively studied species group, *P. aurelia*, no fewer than 16 "varieties" are now known, some of which are rather local. This number might be increased substantially if a sufficient number of local populations from all areas of the globe were tested. In *P. caudatum*, similarly, 16 "varieties" are known, and in *P. bursaria*, five (Sonneborn 1957).

Even more difficult is the problem of determining sibling species in parasitic and partially asexual protozoans. In many of them, so-called biological races are known (Hoare 1943), but it is not known to what extent they represent reproductively isolated species. Here we are dealing partly with the eternal problem of defining species in asexual organisms (Sonneborn 1957), partly with the impossibility of testing the crossability of "strains" that cannot be grown on artificial media. The widespread occurrence of sibling species in these organisms is highly probable on the basis of much indirect evidence.

Because of their superficial morphological similarity, sibling species are normally discovered through various differences in habits, ecology, or physiology. Some examples of biological attributes that distinguish sibling species and that may aid in their discovery are recorded in the following paragraphs. Others were mentioned in the preceding section.

Biometric Differences. Although qualitative structural differences between sibling species may be absent, their distinctiveness can sometimes be substantiated by biometric studies. This has been done, for example, for *Drosophila pseudoobscura* and *persimilis* (Reed and Reed 1948) and for the darters *Etheostoma nigrum* and *olmstedi* (Stone 1947). Such an analysis need not be based exclusively on metric characters, but may involve any type of multiple-character analysis, as was demonstrated by Petersen (1947b) for butterflies of the genus *Boloria*, or by Lorković (1942) for lycaenid butterflies of the genus *Everes*.

Breeding Tests. In many cases sibling species were first discovered when strains were crossed in the laboratory. Patterson and Stone (1952) gave a detailed account of this in the case of *Drosophila*. Other instances are the grain beetle *Calandra oryzae* (Birch 1954) and the sibling species of *Paramecium* (Sonneborn 1957). Crossing is particularly important when sibling species are not sympatric. Moore (1954), when crossing frogs of the genus *Crinia* of eastern and western Australia (previously classified as *C. signifera* and morphologically indistinguishable), found that these crosses were sterile. The western Australian population was actually a sibling species, *C. insignifera*. Such breeding tests are the only way in which the existence of allopatric sibling species can be substantiated.

Habits. Nest structure is helpful in species recognition among cave swiftlets (*Collocalia*), in termites (Desneux 1948; Emerson 1956), in ants (Gösswald 1941), and in certain wasps. In the wasp *Polistes fuscatus*, 18 "varieties" were traditionally recognized, differing in color pattern but not in structure. Rau (1942) found that near Kirkwood, Missouri, three of these so-called "varieties" differed in many biological attributes (Table 3–5). No interbreeding occurs; in fact there is active hostility among them. Rau (1946) concludes: "I, for my part, am only too happy to meet and greet *pallipes* [= *metricus*] and *variatus* as two distinct species," *rubi-*

Table 3–5. Characteristics of three sibling species of the *Polistes fuscatus* group (after Rau 1942, 1946).

Characteristic	P. metricus	P. variatus	P. rubiginosus
Color	Somber black-brown body	Yellow bands on brown	Solid bright brick red
Nest	In well-lighted parts of man-made structures or in dense vegetation	In hollow places in the ground, such as old mouse holes	In total darkness in hollow trees or between walls of buildings, under the roof
Colony founding	One queen	One queen	Many queens
Average size of colony at end of summer	70–85 cells	120–140 cells	≫ 140
Hibernation	In cracks of buildings	In cracks of buildings	In hollow logs
Guards at nest entrance	Absent	Absent	Present, also ventilating nest

ginosus being the third. The three forms are now universally recognized as full species.

In a study of the behavior of predatory wasps, Adriaanse (1947) found that individuals of *Ammophila* "*campestris*" fell into two distinct groups. Among 69 observed individuals, 11 conformed to the *campestris* pattern of the literature and 58 had a previously undescribed pattern. The behavior differences between the two groups are listed in Table 3–6. A specialist confirmed that specimens of the new type of wasps were a different species (*adriaansei* Wilke = *pubescens* Curtis).

The two bee species *Trigona braunsi* and *T. araujoi* were recognized as distinct only owing to biological differences. *T. braunsi* has the brood cells in clusters, not vertical layers; it defends its colony by pouring honey into the entrance, not by fighting; and its nests are commonly robbed by *Lestrimelitta cubiceps*, a predation to which *T. araujoi* is relatively immune (Portugal-Araujo and Kerr 1959).

A study of the light flashes produced by eastern North American fireflies led Barber (1951) to recognize 18 species of *Photuris* instead of the traditional two or three species. Once recognized, these species were found to differ not only in the frequency, pattern, and color (yellow, green, or reddish) of their flashes, but also in breeding season, in preferred habitat, and in minor color differences. Even though there are no

Table 3–6. Behavior differences between *Ammophila campestris* and *pubescens* (after Adriaanse 1947).

Ethological character	A. campestris	A. pubescens
Source of material used to fill nest hole	A quarry	Flown in
Choice of food	Sawflies	Caterpillars
Sequence of egg laying and provisioning	First egg, then prey	First prey, then egg
Breeding season	Earlier, ends in August	Later, until middle of September

differences in the male genitalia, there is little doubt that good species are involved (Fig. 3–3). A study of courtship patterns permitted Gerhardt (1939) to discriminate between species and color variants among slugs.

The breeding seasons are often different in sibling species, such as those of the frog *Leptodactylus* (Cei 1949a), *Coregonus* (Svärdson 1957), *Gryllus* (Fulton 1952), *Photuris* (Barber 1951), the moth *Choristoneura* (Smith 1953), and the sponge *Halisarca* (Lévi 1956), to mention only a few.

Vocalization. From the very beginning of natural history, differences in songs and call notes were used to discover morphologically similar species. This is how Gilbert White, the vicar of Selborne, recognized the Chiffchaff (*Phylloscopus collybita*) and Willow Warbler (*P. trochilus*) as distinct species 30 years before the formal description by Vieillot. Differences in voice have led to the discrimination of sibling species in grasshoppers (Faber 1929) and crickets (Fulton 1925, 1937, 1952). They often permit rapid field identification, even in mammals such as chipmunks of the genus *Eutamias* (Miller 1944a). The recent advances in electronic sound-recording instruments have led to a rapid expansion of this area. By the new techniques, species discrimination has been advanced in the avian genera *Sturnella* (Lanyon 1957), *Empidonax* (Stein 1958), and *Myiarchus* (Lanyon 1960), as well as in the amphibian genera *Hyla* (Johnson 1959), *Crinia* (Littlejohn 1959), *Bufo* (Bogert 1960), and *Acris* (Blair 1958), in the orthopteran genera *Nemobius* (Alexander 1957b) and *Gryllus* (Alexander 1957a), and in the homopteran genus *Magicicada* (Alexander and Moore 1962).

Host Preference. Sibling species that feed on plants or are parasitic are often discovered through differences in host specificity. Among insects

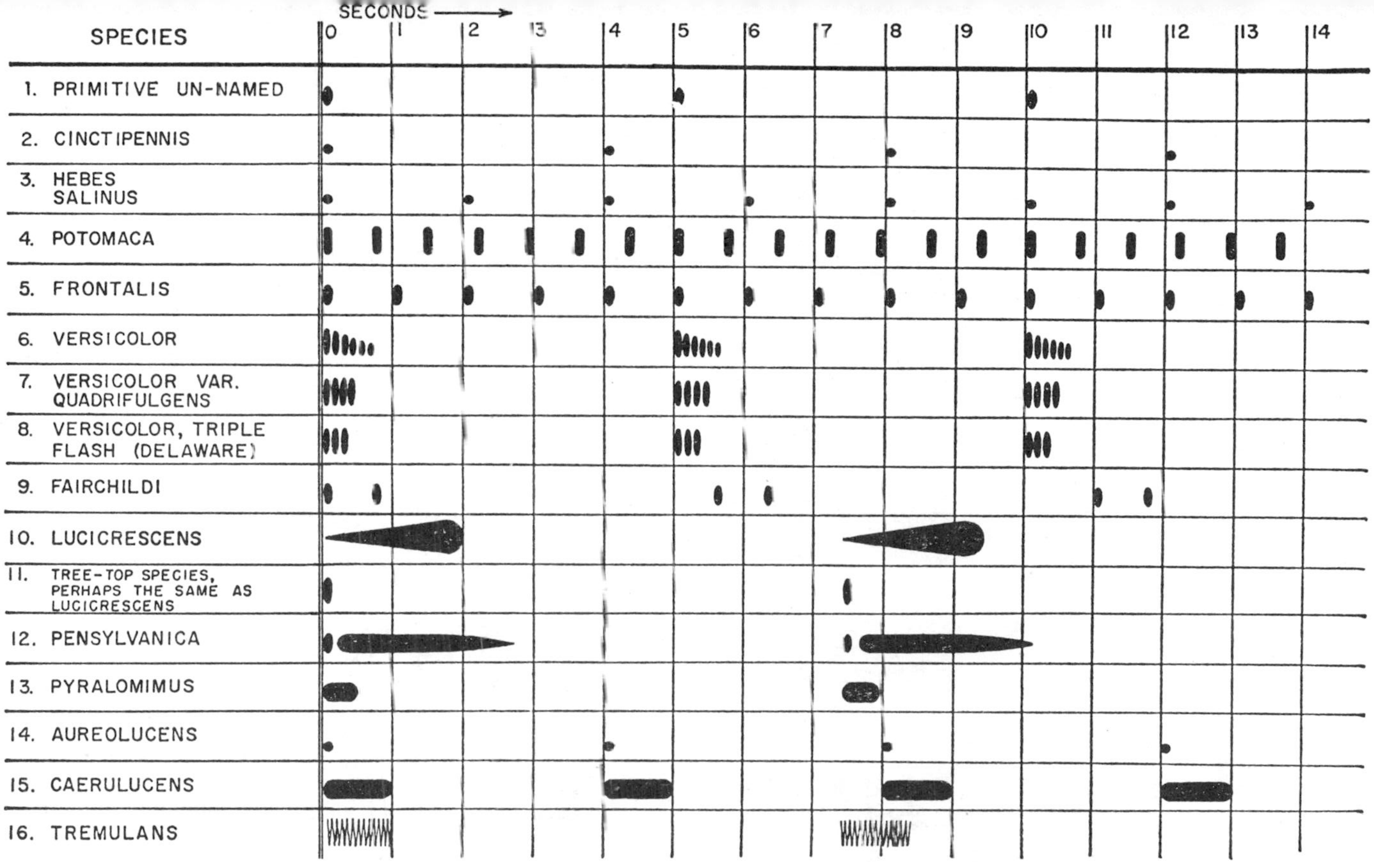

Fig. 3–3. Pattern of light flashes in North American fireflies (*Photuris*). The time scale gives the usual frequency and duration of the flashes. The height and length of the marks indicate the intensity and pattern of the flashes. (From Barber 1951.)

this is particularly true of moths, beetles, and leafhoppers. Sibling species have been found even in the much-studied butterflies when individuals raised from different host plants were compared (see Chapter 15, under biological races).

Pathogenicity. The first indication that *"Anopheles maculipennis"* of authors is a complex of sibling species was given by the spotty distribution of malaria in areas supposedly occupied by this species. The same clue has led to the scrutiny of many tropical species groups of *Anopheles* and to the discovery of numerous sibling species. In supposedly well-known Malaya the number of species of *Anopheles* increased from about 30 in 1935 to about 50 in 1960, mainly owing to the recognition of sibling species.

Parasites, Commensals, and Symbionts. Sibling species often differ in the number or the kind of parasites they carry. A pair of sibling species of *Octopus* in California was distinguished (Pickford and McConnaughey 1949) when it was found that some octopuses were parasitized by the mesozoan *Dicyemennea abelis* and others by *D. californica.* The *Octopus* individuals parasitized by *D. abelis* live in deep water on rocky bottoms, and have minute eggs (1.8 to 4 mm long) with long stalks, attached in festoons. In the adult the arms are relatively longer, the suckers larger, and the hectocotylized arms relatively shorter. These were distinguished as *Octopus bimaculatus.* The individuals parasitized by *D. californica* live in shallower water where rocks rest on soft bottom, and have large eggs (9.5 to 17.5 mm) with shorter stalks, attached in small clusters. This species was given the name *O. bimaculoides.* Males of one species will not mate with females of the other.

Emerson (1935) discovered a pair of sibling species of termites (*Nasutitermes*) through differences in the faunas of staphylinid beetles in their nests. The previously overlooked turbellarian species *Polycelis tenuis* was discovered in England by Reynoldson (1948) when he found that some *Polycelis* had an average of 22 individuals of the peritrichan *Urceolaria mitra* on their surfaces, while others (*P. nigra*) had an average of only 0.7 individual. However, the peritrichans on various hosts did not differ (Reynoldson 1956). Differences between sibling species regarding susceptibility to parasites have also been reported for fish (Keleher 1952), for wasps (Bohart 1942), and for Homoptera (Lal 1934; Maramorosch 1958).

Cytology. A study of chromosomal patterns has led to the discovery of numerous sibling species, or at least it has established differences between stocks and strains that were suspected of belonging to different species because they were difficult to cross in the laboratory. Among a number of such cases described for the genus *Drosophila* (Patterson and Stone 1952; White 1954), the case of *D. pseudoobscura* and *persimilis* is best known (Dobzhansky and Epling 1944). Cytological analysis showed that *Sciara "fenestralis"* consists of two species which differ in the banding patterns of the salivary-gland chromosomes, in the sex ratio of their offspring (offspring in one species are unisexual, either all males or all females), and, finally, in the failure to produce hybrids when crossed (McCarthy 1945). A study of the banding patterns of salivary-gland chromosomes has helped to remove difficulties also in the taxonomically difficult genus *Chironomus*. The same is true for black flies (simuliids). Specimens that key out to *Prosimulium hirtipes* were revealed by cytological analysis to include two sibling species (Rothfels 1956) differing in the gene arrangements on the long arm of the third chromosome. There are apparently numerous such sibling species in the Simuliidae (Syme and Davis 1958; Dunbar 1959).

A comparison of the chromosomes of the buprestid beetle *Agrilus anxius* from birch and poplar (Smith 1949) showed that two sibling species are involved. The birch borer has 22 chromosomes, while the poplar borer has lost the original Y-chromosome, and the X-chromosome has fused with an autosome, resulting in a new chromosome number of 20. A subsequent taxonomic analysis showed that the birch species tends to have more coppery reflections on pronotum and head and broader lateral lobes of the male genitalia than does the poplar borer. The poplar species emerges 6 to 16 days earlier than the birch species. There is no evidence of hybridization in nature (Barter and Brown 1949).

Cytological analysis revealed a previously unrecognized species of grasshopper very similar to *Austroicetes pusillus* (White and Key 1957) and several sibling species of mammals (Matthey 1959). Good chromosomal differences between sibling species exist in a number of genera of crustaceans, such as *Cyclops* (Harding 1950) and *Jaera* (Staiger and Bocquet 1954). These are merely a few selected examples of conspicuous chromosomal differences between species that are morphologically exceedingly similar.

Biochemical Analysis. It is to be expected that sensitive biochemical methods, such as electrophoresis, chromatography, and other methods of protein analysis will be employed to an increasing extent to confirm suspected species differences. Eichhorn (1958) could substantiate by paper chromatography the differences of three species of fir aphids, which had been originally discovered through biometric analysis.

SPECIAL KINDS OF SIBLING SPECIES

While most sibling species differ from other species in no way except the slightness of their morphological differences, there are a few exceptions. For instance, autopolyploids, not infrequent among plants, often are hardly distinguishable from their diploid ancestors. Yet in cases where they do not backcross with the diploids, or where backcrossing produces inviable triploids, they undoubtedly represent independent gene pools and thus separate species, biologically and genetically speaking. Whether or not autopolyploids occur in sexually reproducing animals is not yet established (Chapter 15).

Another special class is represented by the "parthenogenetic species" that occur in many insects and lower invertebrates. Among the sibling species of chrysomelid beetles described by Brown (1945), many are known only in the female sex and apparently reproduce strictly parthenogenetically. The same is true of the so-called species of white-fringed weevils (Buchanan 1947), of psychid moths of the genus *Solenobia* (Sauter 1956), and of isopods of the genus *Trichoniscus* (Vandel 1940). Some of the "biological races" of *Trichogramma minutum* also seem to be, at least in part, parthenogenetic and reproductively isolated from each other (Harland and Atteck 1933). Whether or not to list such clones as sibling species depends on the criteria adopted for "species" in asexual organisms (Chapter 15). Parthenogenetic animals tend to develop polyploidy and this adds another level of complexity. The polyploid "races" of weevils (Suomalainen 1950) and *Solenobia* (Seiler 1961) are well-analyzed cases; others are found in *Trichoniscus* and additional genera discussed by White (1954). Most of these polyploids are reproductively isolated from the parental diploid species even in the cases in which no morphological differences are visible. Since they are definitely not "biological races," they are best considered sibling species. Too little is known about the chromosomal nature of parthenogenetically reproducing nematodes to permit a definitive decision as to their taxonomic rank (Osche 1954).

Those who adhere to a purely morphological species concept usually refer to sibling species as "biological races." By far the majority of the so-called biological races of the literature (for instance, Thorpe 1930, 1940) are now acknowledged to be sibling species. In Thorpe's words, they are forms "which on every biological ground should be classified as distinct species." With the replacement of the morphological species definition by a genetic-biological definition, there is no longer any reason why such cryptic species should be designated "races." See Chapter 15 for a detailed discussion of biological races.

THE SIGNIFICANCE OF SIBLING SPECIES

Speciation among sibling species is in no way different from that in other species. Aside from a few cases of autopolyploidy and of partheno-genesis (see above), geographic speciation is the normal process by which sibling species originate. This is very evident in the taxonomically better-analyzed groups. Among the anophelines, for instance, subspecies are frequent and many species are still essentially allopatric: *Anopheles atroparvus* and *labranchiae*, or *occidentalis*, *freeborni*, and *aztecus* (Bates 1949). Reid (1953) concluded that the A. *hyrcanus* group of sibling species "seems to conform to the classical pattern of speciation by geographical isolation." The most distinct forms of this group occur in the Philippines, which constitute the most isolated portion of its range. Patterson and Stone (1952) came to the same conclusion with respect to speciation in the sibling species of *Drosophila*. Entirely or essentially allopatric sibling species occur in many genera, strengthening the evidence for geographic speciation.

A much more difficult question concerns the evolutionary significance of sibling species. Why are some closely related species very different, others morphologically indistinguishable? Sibling species are definitely not species *in statu nascendi*, which have acquired only the "biological" properties of species but not yet the genetic ones. All the available evidence indicates that sibling species show the same number of genetic differences as do other closely related species.

It seems that sibling species are a problem of developmental genetics. Recent work in this field has found much evidence (see Chapter 10) that there is a selective premium on the maintenance of the phenotype. Any disturbance of the developmental process by a gene mutation will result in a selection pressure in favor of other genes that restore development along the normal, time-tested channels (Lerner 1954; Waddington 1956a).

CONCLUSIONS CONCERNING SIBLING SPECIES

1. There is no sharp division between ordinary species and sibling species. The latter are merely near the invisible end of a broad spectrum

of increasingly diminishing morphological differences between species. The occurrence of natural populations with all the genetic and biological attributes of good species but with little or no morphological difference reveals the vulnerability of a purely morphological species concept.

2. Sibling species, when subjected to a thorough analysis, usually are shown to differ in a whole series of minor morphological characters. Like ordinary species, they are separated from each other by distinct gaps.

3. Sibling species are apparently particularly common in those kinds of animals in which chemical senses (olfactory and so on) are more highly developed than the sense of vision. Although indistinguishable to the eye of man, these sibling species are evidently dissimilar to each other, as is shown by cross-mating experiments. Sibling species are apparently rarest in organisms such as birds that are most dependent on vision in the recognition of epigamic characters.

4. There is no indication that sibling species arise by a process of speciation different from that which gives rise to other species.

5. Degree of morphological similarity in sibling species is an indication not of genetic similarity, but rather of developmental homeostasis. A reconstruction of the genotype, resulting in the reproductive isolation of two species, can take place without visible effect on the morphology of the phenotype.

6. Evolutionary changes in the genetic constitution seem to occur in groups of sibling species at the same rate as in groups of morphologically very distinct species.

4 ~ Biological Properties of Species

When first introduced into biology, the term "species" designated primarily a morphological-systematic unit, or, worse, merely a Latin binomen. As the study of species was taken up by field naturalists on one hand, and by laboratory biologists on the other, it became increasingly clear that the morphological distinctness of each species is an indication for the presence of a distinct biological system.

BIOLOGICAL SPECIES CHARACTERISTICS

The taxonomic literature stresses "species characters." This term in general refers to any attribute of a species that differentiates it from other species (and is therefore "diagnostic"), and that is reasonably constant (invariable), so that a species can be recognized by it at once. Morphological characters are most useful for diagnostic purposes and with preserved material; this is the reason for their high esteem in the taxonomic literature. More detailed comparisons, however, always show that species may differ from each other not only in aspects of external morphology, but also in size, color, internal structure, physiological characters, cell structures, chemical constituents (particularly proteins and nucleic acids), ecological requirements, and behavior. Summaries of such species differences have been given previously (Mayr 1948; Mayr, Linsley, and Usinger 1953). Whenever a particular aspect of animals receives special attention, numerous previously unexpected species differences are discovered, for instance in comparative physiology (Prosser 1955, 1957) or in comparative studies of animal behavior (Mayr 1958).

Darwin and the majority of naturalists were convinced that species differences are, in the last analysis, always adaptive. Others have maintained the opposite: "A survey of the characters which differentiate species (and to a less extent genera) reveals that in the vast majority of cases the specific characters have no known adaptive significance" (Robson and Richards 1936:314). A renewed analysis of species characters reveals that this conflict of opinions can be largely resolved. Every species is the product of a long history of selection and is thus well adapted to the environment in which it lives. There is no doubt that the phenotype as a whole, including its physiological properties, is adaptive and is produced by a genotype that is the result of natural selection. This is not contradicted by the fact that an occasional component of the phenotype is adaptively irrelevant. The reasons for this—particularly the possibility that the pleiotropy of genes is responsible for selectively neutral species characters—will be discussed later (Chapter 7). This chapter and the next (which constitute a digest of several unpublished manuscript chapters) will be devoted to a discussion of three sets of biological attributes of species, those that (1) adapt species to their physical environment, (2) enable species to coexist with potential competitors, and (3) permit species to maintain reproductive isolation from other species (Chapter 5).

ADAPTATION TO THE PHYSICAL ENVIRONMENT

The statement that every species is adapted to its environment is a self-evident platitude. In continental areas without physical barriers the border of the species range indicates the line beyond which the species is no longer adapted, and the very existence of such borders is tangible proof of the limitations of this adaptation. Some factors that contribute to the adaptation are obvious, particularly those expressed in the visible phenotype. The white color of many Arctic birds and mammals or the sandy coloration of desert species are among these evident adaptive characters. More elusive, yet far more important, are various physiological regulatory mechanisms that not only permit survival in the breeding range but also secure adequate reproduction to maintain the size of the breeding population at a more or less steady level. We have shown in the last chapter that even very closely related and similar species may differ from each other in various nonmorphological characters relating to physiology, ecology, and behavior. Numerous such species differences are cited in the vast physiological literature, for instance by Prosser (1955, 1957), and in the ecological literature (for example Allee *et al.* 1949; Hesse, Allee, and Schmidt 1951). All these comparisons of closely related species indicate that each species is a separate biological system with species-specific tolerances to heat, cold, humidity, and other factors of the physical environment, habitat preference, productivity, rate of population turnover, and numerous other biological constants. Related species may overlap in these properties, yet each species is characterized by well-defined mean values. Steady physiological states are maintained by rather complex homeostatic systems that are shared in their essential properties not only by members of a single local population but by all members of the species, a certain amount of geographic variation notwithstanding. The dispersal of individuals through smaller or larger portions of the species range, resulting in the mixing of genotypes, places a premium on the existence of species-wide homeostatic mechanisms. It is probable that generalized adaptive genes, particularly those adding to fitness in heterozygous condition, are involved in the optimal functioning of these species-specific homeostatic mechanisms. They serve on one hand as the element that gives a species its unity (Chapter 10) but are at the same time largely responsible for the existence and location of species borders. Such homeostatic devices have definite limits of tolerance beyond which they can no longer adjust to external conditions. The result is that every species has an optimal environment, presumably somewhere near the center of its range, and definite limits of tolerance with respect to latitude and altitude.

A few examples may illustrate such physiological differences between species. Wallgren (1954) has shown that two northern European species of buntings differ in temperature preference and tolerance (Fig. 4-1). The more northerly species, *Emberiza citrinella*, is less resistant to high temperatures and particularly to prolonged spells of heat. *Emberiza hortulana*, a more southerly species, is tolerant of much greater heat, but shows greater heat loss at low temperatures. It has a very different fat metabolism, correlated with a strong migratory drive. That the environmental temperature might be very important even in warm-blooded vertebrates had previously been inferred by Kendeigh (1934) on the basis of a study of temperature tolerance and distribution pattern of the house wren (*Troglodytes aedon*). Far more conspicuous is the temperature dependence of aquatic cold-blooded vertebrates, for instance as shown by Moore (1949) for frogs of the genus *Rana* (Table 4-1). In *Telmatobius schreiteri*, the high-altitude frog of the Argentinian Andes, Cei (1949b) found continuous sexual activity and spermatogenesis throughout the year even though freezing spells occurred in winter. Spe-

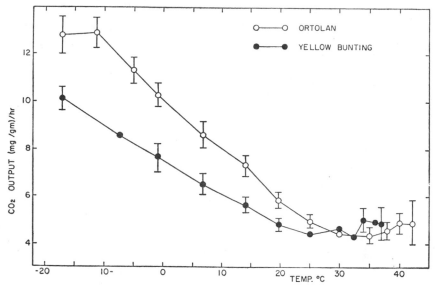

Fig. 4–1. Carbon dioxide output of two species of buntings at various environmental temperatures. Ortolan = *Emberiza hortulana*, Yellow Bunting = *E. citrinella*. (From Wallgren 1954.)

cies of marine invertebrates living in the Arctic Oceans are likewise adapted to go through the full reproductive cycle at temperatures hardly if at all above 0°C (Thorson 1936, 1950). The rate of larval development in Arctic marine invertebrates is about as rapid as in their temperate or tropical relatives, the compensation being achieved by higher oxygen consumption and rate of cell division at equal temperatures. The oxygen consumption in adult lamellibranchs is about the same in Arctic species at 0°C as it is in Boreal species at 8°C, in Mediterranean species at 12°C, and in tropical species at about 25° to 30°C.

Among the limiting factors for terrestrial animals, temperature tolerance and humidity requirements show up most frequently as species differences. Pittendrigh, for instance, has demonstrated differences in the humidity requirements of closely related species of *Drosophila* (1958) and of *Anopheles* (1950). *Drosophila persimilis* is positively phototactic when undisturbed, while *D. pseudoobscura* is negatively phototactic under conditions of low excitement (Pittendrigh 1958; Lewontin 1959). Six species of ants of the genus *Formica* occurring in the Chicago region differ markedly in their temperature and humidity tolerance. At high humidity *incerta* is superior to *montana* and *subsericea* to *ulkei*, at low humidity the situation is reversed (Talbot 1934). The differences are

Table 4–1. The relation between geographic distribution, breeding habits, and adaptive embryological characteristics in five species of North American frogs, *Rana* (from Moore 1949).

Characteristic	Species				
	R. sylvatica	R. pipiens	P. palustris	R. clamitans	R. catesbeiana
Limiting latitude:					
Northern	67°	60°	51–55°	50°	47°
Southern	34°	9°	32°	28°	23°
Beginning of breeding season	Mid-March	Early April	Mid-April	May	June
Limiting embryonic temperature [°C]:					
Lower	2.5	5	7	12	15
Upper	24	28	30	32	32
Time (hr) between stages 3 and 20 at 20°C	72	96	105	114	134
Egg diameter (mm)	1.9	1.7	1.8	1.4	1.3
Type of jelly mass	Globular, submerged	Globular, submerged	Globular, submerged	Film, at surface	Film, at surface

correlated with habitat preferences and geographic range (Table 4–2). Of two closely related species of European snails (*Helicella*), the eastern species (*obvia*) has a preference for higher temperature and is more resistant to extreme temperatures. The western species (*ericetorum*) has a preference for much higher humidity and is negatively phototactic. The adaptation to an oceanic climate for the western species and to a continental climate for the eastern species is evident (Melchinger 1955). Five species of Cuban *Anolis* lizards differ in temperature and perch preference (Ruibal 1961).

Table 4–2. Average survival times (hours) of ants (*Formica*) at different temperatures and relative humidities (from Talbot 1934).

Species	Temperature (°C) and relative humidity (percent)		
	$30.4 \pm 0.34°$, 44.3 ± 1.2	$30.6 \pm 0.3°$, 12.1 ± 2.7	$41.2 \pm 0.8°$, 8.3 ± 0.1
F. *obscuriventris*	—	14.06	1.93
montana	25.35	9.50	1.57
incerta	54.49	8.12	1.43
subintegra	—	6.62	1.04
ulkei	18.16	6.18	0.84
subsericea	41.64	2.81	0.92

Different species reach different degrees of adaptation to such extreme environments as the desert. Certain species of birds, for instance, are independent of drinking water and can live on metabolic water and perhaps occasional dew (Hoesch 1953), while all other birds need to drink water. Certain desert rodents, for instance pocket mice (*Perognathus*) and kangaroo rats (*Dipodomys*), likewise can subsist without drinking water (Lindeborg 1952).

Students of population dynamics have found that many components of productivity, such as fertility, fecundity, life expectancy, and, in social animals, colony size, have a strong genetic component and may differ from species to species. Schneirla (1957) gives some of the differences in colony size and reproductivity between two common American species of army ants, *Eciton hamatum* and *E. burchelli* (Table 4–3). Differences among closely related species with respect to longevity and fecundity have been recorded not only for insects but for many other animals, for instance mice of the genus *Peromyscus* (Table 4–4). How much difference there can be for all sorts of biological characteristics has been demonstrated by Sonneborn (1957) and his co-workers for sixteen sibling

Table 4-3. Species comparison in army ants, *Eciton*
(from Schneirla 1957).

Characteristic	E. hamatum	E. burchelli
Functional cycle	Nomadic phase 16–18 days (mode, 17) Statary phase 18–21 days (mode, 20)	Nomadic phase 11–16 days (mode, 13) Statary phase 19–22 days (mode, 21)
Raiding pattern	Branching-column pattern, small terminal groups	Swarm pattern: large advance group, basal fan, consolidation column
Brood	Worker brood, 60,000 + Sexual brood (at maturity), *ca.* 1500 males, 6 queens	Worker brood, 120,000 + Sexual brood (at maturity), *ca.* 3000 males, 6 queens
Colony population	150,000–250,000	300,000–1,500,000
Duration of egg-laying (days)	*ca.* 7 (peak, 2–3)	*ca.* 10–12 (peak, 3–4)
Number of eggs in each brood	60,000 +	120,000 +

Table 4-4. Biological differences (average values) of three species
of *Peromyscus* (from McCabe and Blanchard 1950).

Characteristic	P. maniculatus	P. truei	P. californicus
Number of litters per season	4.00	3.40	3.25
Number of young per litter	5.00	3.43	1.91
Number of offspring per breeding female per season	20.00	11.66	6.21
Known life span (days)	152	190	275

species of the *Paramecium aurelia* complex. These species may differ not only in temperature preference, size, and rate of maturation, but also in mating-type system, length of mature period, serotype system, and the form of fertilization in senility (autogamy *versus* selfing).

These examples could be extended *ad libitum*. It should be clear, however, that every recent investigation has merely confirmed the widespread conviction that each species is a unique biological system. No matter how similar two species may be morphologically, they will be found to differ in physiological preferences and tolerances and in many aspects of their life cycle from the egg stage to death. The precise description of such differences and their relation to the particular niche of a species, its geographic range and its evolutionary history, however, has hardly been begun.

ADAPTATIONS THAT MAKE THE COEXISTENCE OF SPECIES POSSIBLE

Among the numerous biological characteristics of species, two sets are of particular significance in the process of speciation (Mayr 1948). They are those special properties that permit sympatry between closely related species, the properties that an incipient species must acquire in order to complete the process of speciation: (1) mechanisms that guarantee reproductive isolation (see Chapter 5), and (2) the ability to resist competition from other species that utilize the same or similar resources of the environment.

The remainder of this chapter will be devoted to a study of these ecological interactions among closely related species, the nature of such interaction, and its evolutionary effects.

Competition

Darwin (1859) greatly stressed the evolutionary significance of competition and its contribution to natural selection:

As the species of the same genus usually have, though by no means invariably, much similarity in habits and constitution and always in structure, the struggle will generally be more severe between them, if they come into competition with each other, than between the species of distinct genera.

Unfortunately, he referred to competition in such drastic terms as "severe struggle" and "one species has been victorious over another in the great battle of life." To certain authors ever since then, competition has meant physical combat and, conversely, the absence of physical combat has been taken as an indication of the absence of competition. Such a view is erroneous. To be sure, actual combat or aggression among competitors does occur, particularly among territory-holding higher vertebrates (see p. 87). Among sessile marine invertebrates, where the amount of available substrate may be the limiting factor, the fastest-growing species may overwhelm and literally kill slower-growing competitors. Such a situation has been described for competing species of intertidal barnacles (Connell 1959). In most cases, however, dramatic competition occurs only where two species come newly into contact or where a radical change of the environment has upset the previously existing dynamic balance. As Brown and Wilson (1956) state correctly, such an acute phase of competition is often "a relatively evanescent stage in the relationship of animal species" and will be replaced by a new balance in which severe competition is avoided. Thus the relative rarity of overt manifestations of competition in nature is proof not of the insignificance of competition, as asserted by some authors, but, on the contrary, of the high premium natural selection pays for the development of habits or preferences that reduce the severity of competition.

No unanimity has yet been achieved concerning the precise definition of competition, yet it always means that two species seek simultaneously an essential resource of the environment (such as food, or a place to live, to hide, or to breed) that is in limited supply. Consequently competition becomes more acute as the population of either species increases. Any factor, the effect of which becomes more severe as the density of the population increases (and this may be true for all causes of mortality and fecundity) is called a density-dependent (= controlling) factor. Williamson (1957) defines a controlling factor as "a factor which acts more severely against the individuals of a population when the population increases . . . and so tends to control the population size . . . One can then say that two species are in competition when they have a controlling factor in common" (but see Milne 1961).

Part of the study of competition is the study of population dynamics, a field that has both benefited and suffered from the early development of a mathematical theory, by Volterra and Lotka, with its simplifying and sometimes unrealistic assumptions. Here we shall steer clear of the controversial aspects of population dynamics and concentrate on the actual observation of the occurrence or avoidance of competition (as defined above) by species. For arguments on the definition of "density-dependent," on the relative importance of density-dependent and density-independent factors, and on the relative importance of intraspecies and interspecies competition, the reader is referred to the ecological literature (for example, Crombie 1947; Andrewartha and Birch 1954; Gilbert *et al.* 1952; Nicholson 1954, 1957; Park 1954; Philip 1955; Birch 1957; Andrewartha 1961:168).

The logical consequence of competition is that the potential coexistence of two ecologically similar species allows three alternatives: (1) the two species are sufficiently similar in their needs and their ability to fulfill these needs so that one of the two species becomes extinct, either (a) because it is "competitively inferior" or has a smaller capacity to increase or (b) because it had an initial numerical disadvantage; (2) there is a sufficiently large zone of ecological nonoverlap (area of reduced or absent competition) to permit the two species to coexist indefinitely. The consequence of these alternatives is summarized in the statement: "No two species can coexist at the same locality if they have identical ecological requirements." This theorem is sometimes referred to as the Gause principle, after the Russian biologist Gause (1934), who was the first to substantiate it experimentally. Yet, as Hardin (1960) and others have pointed out, the principle was known long before Gause. Darwin discussed it at length in his *Origin of Species,* and Grinnell and other naturalists have referred to it frequently in the ensuing 100 years. Instead of associating the principle with the name of one of its many independent discoverers, Hardin suggests calling it the "competitive exclusion principle."

The exclusion principle has great heuristic value. Attempts to prove or disprove it have stimulated numerous comparisons of the ecologies of closely related species and analyses of the factors permitting them to coexist. Yet a complete analysis of the process of competition has proved difficult.

The operation of competition between individuals of two species is best understood if one compares it with the fate of two "competing" genes in a single gene pool. Let us assume that gene A is superior to its allele a in a population and that the heterozygote Aa is intermediate in viability between the two homozygotes. Then, even if the difference between A and a is very slight, a will inevitably disappear from the population, as calculated by Fisher (1930) and first demonstrated experimentally by L'Héritier and Teissier (1937). Into a population cage containing a pure population of more than 3000 *Drosophila melanogaster* with the Bar gene they introduced a few wild-type flies. The frequency of the Bar gene dropped rapidly at first, then more slowly, until, 600 days later, the gene had a frequency of less than 1 percent. Many similar experiments have since been conducted with different genes and gene arrangements; the outcome has always been the same, except in cases of heterozygote superiority. Two species are rarely as similar as genotypes of a single species. In cases, however, in which they depend entirely on the same resources of the environment, one or the other species will prove superior in the utilization of these resources. This will lead to an exclusion of the other species from the zone of potential ecological overlap. To avoid such fatal competition, the two species must utilize the resources of the environment in a somewhat different way. This is why ecological compatibility with potential competitors is one of the most important species characteristics. In order to survive, each species must be supreme master in its own niche.

Competition is not the only ecological interaction between species. Of almost equal evolutionary importance are various kinds of direct interference and modifications of the environment. A flock of sheep that converts a meadow into a pasture affects the living conditions not only of the meadow inhabitants, but also of the soil inhabitants. A predator that reduces the density of the sheep population may reverse the direction of these changes. Every predator and every herbivore changes the environment, as does the termite that eliminates wood and organic debris, the mole or gopher that tunnels under the soil, or the woodpecker that makes holes in trees. They all affect the needs of other species positively or negatively. Any change of the phenotype that mitigates the ill effects of the impact made by coinhabitants of the habitat will be favored by natural selection. Differences in these biotic interactions are among the reasons for the diversity of evolutionary changes in isolated populations.

Observed Exclusion

The avoidance of competition is achieved by either conspicuous or subtle factors. As conspicuous factors, I classify geographical separation, occurrence in different habitats, and temporal isolation of potentially competing stages of the life cycle. Under subtle factors, I include all niche differences in the same habitat, that is, all differences in the utilization of the habitat. Lack (1944) found, in an analysis of the British songbird fauna, that, where a genus had more than one British species, in 21 cases there was conspicuous exclusion (geographical 3, habitat 18), while in 11 to 13 cases exclusion was more subtle. Whenever two species share the same habitat, a more refined analysis usually reveals differences. For instance, the cormorant (*Phalacrocorax carbo*) and the shag (*P. aristotelis*) seem to have identical ecologies. However, the cormorant feeds in shallow water and takes fish that live on or near the bottom, such as pleuronectids and gobies, while the shag feeds in the open sea on free-swimming fish (clupeoids, *Ammodytes*) (Lack 1945).

Moreau (1948) analyzed the birds of a tropical country, the province of Usambara (3000 square miles) in East Africa. The analysis deals with 172 species in 92 genera. There are 173 possible ecological overlaps between species of the same genus and 1474 between other species of the same family. In only 16 percent of these possible overlaps is there an actual overlap of habitat, and in only a third of these an overlap in diet. However, in some of these cases, very rare species are involved, apparently without true competition. Of the congeneric species 94 percent are ecologically isolated, and of the species of the same family, 98 percent. The one apparent exception is weavers of the subfamily Ploceinae, but these birds show much movement and normally feed cn locally superabundant grass seeds. Factors other than food seem to control their population numbers. Skutch (1951) found that all sympatric species of 23 genera of birds of the American tropics have differing habitat preferences, and Hamilton (1962) found the same for the genus *Vireo*, with two possible exceptions. He presents a detailed discussion of the factors that facilitate or preclude sympatry of related species. Of the 12 species of hedge sparrows (Prunellidae) only one pair is not separated geographically or by habitat (Marien 1951). The North American titmice (*Parus*), in contrast to their European congeners, co-occupy the same habitat only rarely, as Dixon (1961) has shown in an excellent analysis. Where co-occupancy occurs, one of the species is considerably (1.6–2×) larger than the other species. The less diverse the habitat, the more rigid the exclusion. The most detailed analysis of niche exclusion is that made by MacArthur (1958) of North American wood warblers (*Dendroica*) in coniferous forests. Several species that at first sight seemed to have identical niches actually occupied different parts of a tree or fed on the outer or inner parts of a branch. Coexistence in the same habitat is facilitated by differences in vocalization that reduce interspecific antagonism.

Numerous other cases of exclusion are recorded by Lack (1954a:148–151) and Andrewartha and Birch (1954:456–465). In studies of reptiles it was found that there is either a preference for different foods, for instance earthworms or amphibians where two species of snakes (*Thamnophis*) share the same habitat (Carpenter 1952), or a separation of habitat where several species share the same food, as in the case of the four species of termite-feeding whiptail lizards (*Cnemidophorus*) of Texas (Milstead 1957, 1961). Specialization for different prey species has been reported for many species of fresh-water fish, for instance by Nilsson (1955, 1960) for trout (*Salmo trutta*), char (S. *alpinus*), and whitefish (*Coregonus*) in northern Swedish lakes. Numerous cases of exclusion have been described for invertebrates. Dobzhansky and Pavan (1950) report seasonal and habitat differences for sympatric species of *Drosophila,* and da Cunha, Shehata, and de Oliveira (1957) show how these *Drosophila* differ from each other in their preference for and utilization of 43 species of yeasts occurring on fermenting fruits in the area. Tretzel (1955) shows that congeneric sympatric species of spiders are in most cases isolated by habitat or season. Of ten species coexisting in a habitat, eight or nine (84 percent) belong to different genera. At the other end of the scale is the molluscan genus *Conus* in which more than 20 species coexist in the littoral zone of Hawaii. Deleterious competition is reduced or prevented by specialization on different foods (polychaetes, gastropods, fishes), superabundance of a source of food (the polychaete *Perinereis helleri,* the chief food of the three most abundant *Conus*), or difference in habitat preference. In spite of the sympatry of so many species, there is little evidence of serious competition (Kohn 1958). The subtlety of exclusion is well illustrated by the fauna of cow dung. No less than 51 species of nematodes have been found in cow dung on European pastures. For 22 species it is the only habitat, for 10 it is a typical habitat, for 19 an occasional habitat. As the microflora (bacteria and fungi) of the dung on which the nematodes feed changes successively, so does the composition of the nematode fauna, which is further specialized into "surface" and "interior" species (Sachs 1950).

In the cases mentioned the taxonomic approach was employed, that is, the ecological relations of closely related or congeneric species were studied. A different approach is to take the species of a habitat and examine them for the occurrence of possible competition. The first systematic survey of this type is that of the South Haven peninsula in England (Diver 1940). This survey confirms that potential competitors differ in habitat, breeding season, or preferred food. Brian (1956) shows that two species of ants (*Myrmica*) that appear to coexist are actually segregated on a micro scale, either species being competitively superior in part of the mosaic of soil and vegetation types. Elton's (1946) analysis of community structure comes to the same conclusion as the South Haven study. In mountainous areas, altitudinal segregation is the rule, with each species occupying a different vertical zone although adjacent species may overlap widely. Predators of an area usually specialize in different prey species (Lack 1946; Fitch 1947). In stream organisms one may find that each of a series of potentially competing species occupies a definite zone determined by water temperature, oxygen and mineral content, current, and other chemical and physical factors. The crayfish of the genus *Cambarus* illustrate this situation well (Hobbs and Marchand 1943). According to Blasing (1953) and Dahm (1958), the adaptation of the turbellarian *Planaria alpina* to the cool oxygen-rich headwaters of streams and of *P. gonocephala* to the warmer courses has reached a stage in which there would be no niche expansion even if the related species were not present. At the seashore, likewise, related species occupy different levels in the intertidal zone, with or without overlap. Among 17 more or less sympatric species of Californian limpets (*Acmaea*), Test (1945) found only two instances of possible interspecific competition. These concern three related species that live high in the intertidal zone and another group of two species coexisting under similar circumstances in a restricted area of the coast. In all other cases, there is strict exclusion either through food specialization or through restriction of habitat.

The study of exclusion in the oceans is only beginning. I know of no good evidence, for instance, demonstrating competition or exclusion among plankton-feeding pelagic fishes. The more specialized a species is, the less the probability that it would share its niche with another species. On these premises one would expect rather rigorous exclusion among parasites. This is not entirely borne out by the parasitological literature, where frequently three or four species, for instance of cestodes, are reported from the intestinal tract of a single host. Further studies are needed to decide, on the basis of statistically significant data, how often true coexistence occurs and how often the potential competitors are segregated in different portions of the intestinal tract or in different individuals of the host species.

EVIDENCE FOR COMPETITION

The mere fact of exclusion does not prove that it is the result of selection in favor of avoiding competition. Other explanations may apply to certain cases. There are, however, situations that are difficult to explain except as the result of competition.

Exclusion in Impoverished Habitats. Species that habitually coexist in a rich, diversified habitat may exclude each other in more homogeneous or ecologically marginal habitats. Islands are generally far less diversified ecologically than are the neighboring mainlands. Stresemann (1939), Mayr (1942), and Lack (1942) have shown that related species of birds often replace each other on islands while they are able to live side by side on the mainland. Other recently published illustrations of this phenomenon for birds and other animals concern the islands on the west coast of Australia (Serventy 1951), the hawk eagles (*Spizaetus*) on the western Sumatran islands (Amadon 1953), the babblers of the genus *Malacopteron* in Indonesia (Voous 1951), honey eaters (*Myzomela*) and sunbirds (*Nectarinia*) in Indonesia (Ripley 1961), the skinks of the genus *Emoia* in the Marshall Islands (Brown and Marshall 1953), and other island lizards. Lizards of the genus *Lacerta* were found by Radovanović (1959) on 46 small islands off the Dalmatian coast. On 28 islands, the genus was represented by *L. melisellensis* and on 18 islands by *L. sicula*. On no island did both species of lizards coexist. Which species reaches a given island first is presumably a matter of accident in many cases; yet once established it prevents colonization by newcomers. The island principle is also illustrated by isolated mountain forests in the African savannas. Each mountain island is occupied by only one species of woodpeckers, but it may be a *Dendropicos*, a *Campethera*, a *Mesopicos*, or a *Yungipicus* (Moreau 1948). There is apparently neither sufficient niche diversity nor total available habitat to permit occupation by a second species of woodpeckers. The numerical advantage of the earlier colonist is decisive.

Exclusion Limited to Area of Overlap. Widespread species sometimes display a sharp reduction of their ecological tolerance in areas where their range is overlapped by related species, or conversely a marked broadening of tolerance in a zone of nonoverlap. In the Philippines, where only three species of the *Anopheles hyrcanus* group occur, the two species *A. lesteri* and *A. peditaeniatus* are found both on the coast and inland. On the Malay peninsula, however, where there are seven species of this group, *lesteri* is on the whole restricted to the coast near or in brackish water, while *peditaeniatus* is an inland species (Reid 1953). The rotifer *Polyarthra dolichoptera* coexists with the more successful species *P. vulgaris* only in waters poor in oxygen, but occurs in all types of water from which *vulgaris* is absent (Pejler 1957a:49).

In a narrow area of overlap of the jays *Cissilopha beechei* and *C. sanblasiana* in Nayarit, Mexico, one species is restricted to the mangrove, the other to evergreen thickets along the fields in the more arid interior; north and south of the area of overlap, either species occurs in both types of habitat (Selander and Giller 1959a). At the south end of Baja California where both species of *Cissilopha* are absent, the mangrove habitat is occupied by still another species of jay (*Aphelocoma coerulescens*), which nowhere else in its wide range enters this habitat (Pitelka 1951a). The Chaffinch (*Fringilla coelebs*) occurs throughout its European range in both broad-leaved and coniferous woods. On the islands of Gran Canaria and Tenerife where a second species occurs, the Chaffinch is restricted to deciduous woods, being replaced in the pine woods by the Blue Chaffinch (*F. teydea*) (Lack and Southern 1949). Ground Squirrels (*Citellus*) have sharply reduced habitat tolerance in areas of overlap of two species (Durrant and Hansen 1954). The swallowtail *Papilio glaucus* feeds on plants of several families in eastern North America, where it is the only representative of its species group. In western North America each of the three species of the group is restricted in its food plants, *P. multicaudatus* to the Rosaceae, Oleaceae, and Rutaceae, *P. rutulus* to the Salicaceae, Corylacaceae, and Platanaceae, and *P. eurymedon* to the Rhamnaceae (Brower 1958).

The Impact of Invasions. The effectiveness of competition in nature is best demonstrated by the impact of an invading species on species already established in the area. The starling (*Sturnus vulgaris*), a European species introduced into North America, competes with American birds not only for food but also for nesting sites. During a one-hour visit, a flock of starlings can eliminate from a meadow an amount of food that would sustain a single meadowlark (*Sturnella*) for the better part of a winter. Sharp reduction in the number of meadowlarks in areas visited by starlings is well documented and yet the numerical decrease of the native species is small in comparison to the increase in the invader. The history of invasions has been reported by Elton (1958) but he says little about the impact of the new competition.

Island faunas are particularly vulnerable to new competitors. Most of the species of birds that have become extinct during the past 200 years have been island birds (Greenway 1958). The Moth Skink (*Lygosoma noctuum*) disappeared from the Hawaiian Islands, where it was once common, soon after its close relative *Lygosoma metallicum* was introduced (Oliver and Shaw 1953). The Red Squirrel (*Sciurus vulgaris*) disappeared from many parts of the British Isles after colonization of the district by the introduced American Gray Squirrel (*S. carolinensis.*) Although much of the British countryside is suboptimal habitat for the Red Squirrel, essentially an inhabitant of coniferous woodlands, its disappearance from long-occupied districts was in most cases clearly correlated with competition by the invader (Shorten 1954). The many cases of double invasions, old and recent, prove that a new invader may be sufficiently different from the related resident in food and habitat preference to avoid drastic competition.

Much indirect evidence suggests that many of the birds that became extinct in New Zealand and the Hawaiian Islands succumbed to the ravages of diseases carried by introduced species. For instance, the New Zealand quail (*Coturnix*) became extinct within a few years after the European quail was released on New Zealand. In none of these cases has it been possible to study the effect of disease independently of the other ecological consequences of the invasion. The role of disease as a weapon of competition has been discussed by Haldane (1949c).

Particularly dramatic are the competitive encounters of entire faunas. Simpson (1940, 1950) has described graphically what happened when, at the end of the upper Pliocene, North America became attached to South America across the Isthmus of Panama. Some 15 families of North American mammals spread into South America and seven South American families spread in the reverse direction. North American types turned out to be competitively superior, with a few exceptions, and in the end many of the most characteristic types of South American mammals, like the notoungulates and marsupial carnivores, became extinct. The cause of the extinction of a few North American types, like the horses and mastodons, is still uncertain. Grinnell (1925) proposed an empirical rule: "When a species native to a large area is successfully introduced into a new small area the related species which is native in this area and with which the former comes into competition is soon supplanted." Darwin (1859) favored a similar generalization, according to which natives of larger and more diversified areas are in general superior to natives of small and more uniform areas, and Matthew (1915) made this observation the basis of a sweeping zoogeographic theory. It is substantiated by the recent encounters of members of the Australian fauna with the fauna of the northern continents (Storr 1958). There is little doubt that species with certain types of genetic population structure and adaptation for a diversified biotic and physical environment have a competitive advantage over species lacking these properties. Yet some species from large diversified areas lack competitive superiority and some species from insular areas or southern peninsulas are conspicuously successful invaders. There are thus many exceptions to any conceivable generalization. More research in this area is badly needed.

There are essentially two methods by which enrichment of a fauna can be achieved. The newcomer (let us say a termite-feeding lizard) may utilize the same niche ("termites") as an incumbent species, but in a somewhat different habitat. There will be a local exclusion of the two species paralleling the replacement of the respective habitats. Or else the newcomer succeeds in entering the same habitat as the incumbent and forces him to yield part of a heterogeneous niche in a well-diversified habitat. The coexistence of five or six species of Eurasian *Parus* or North American *Dendroica* in the same piece of woodland is made possible by such a specialization of their niche requirements. The richer and more diversified a habitat, the more easily such minor niche differences can develop.

Islands are greatly impoverished both as to the number of available habitats and the diversity of niches in the habitats. This is compensated by the vacancy of major niches, owing to the difficulties of transoceanic colonization, and this, in turn, permits adaptive radiation.

As far as the over-all evolutionary picture is concerned, it should be mentioned that every new arrival in an area tends to add to the total diversity and to enrich thereby the opportunities of other organisms except the most immediate competitors. The new food pyramid that the evolution of the angiosperm flora made possible is a striking illustration of this principle.

Experimental Competition. The validity of the exclusion principle has been tested in numerous recent experiments in which mixed populations of two species were established in a uniform environment. In virtually every case, one of the two species was eliminated sooner or later. *Paramecium aurelia* eliminates *P. bursaria* (bacterial food) (Gause 1934); among flour beetles *Rhizopertha* exterminates *Sitotroga* and *Tribolium* eliminates *Oryzaephilus* (Crombie 1946); the large sibling species of *Calandra oryzae* (grain weevil) eliminates the small one in maize while in wheat the small species eliminates the large one (Birch 1953). *Daphnia pulicaria* eliminates *D. magna* invariably whether the food consists of yeast or algae (Frank 1957); *D. pulicaria* is equally superior to *Simocephalus vetulus* (Frank 1952). In all these experiments the outcome was invariable. In competition experiments involving species of *Drosophila*, the outcome is often a matter of specified conditions. In a mixed population of *Drosophila melanogaster* and *D. simulans*, the former species is always superior at 25°C and the latter species usually at 15°C. A temperature shift of only 3 C deg reverses the competitive superiority of the fruit flies *Dacus tryoni* and *D. neohumeralis* (Birch 1961). Introgression occurs, however, under certain conditions in crowded laboratory cultures, and precludes a further analysis of the competitive situation. The most thorough and best-controlled competition experiments are those of Park (1948, 1954). When *Tribolium confusum* and *T. castaneum* were placed in competition in pure wheat flour, one or the other species was invariably eliminated, the outcome being partly indeterminate, partly depending on the constellation of environmental factors (Table 4–5). More recent experiments by Lerner and Dempster (1962) have shown that much, if not all, of the indeterminacy can be eliminated by removing the genetic heterogeneity of the founding populations. In properly matched cultures *T. castaneum* is always victorious.

Table 4–5. Two species of *Tribolium* in competition (from Park 1954).

Condition		Number of replicas	Victorious species (number of trials)	
Temperature (°C)	Humidity (percent)		T. confusum	T. castaneum
34	70	30	—	30
29	70	66	11	55
24	70	30	21	9
34,29	30	60	53	7
24	30	20	20	—

The situation is completely different in experiments where the environment is not homogeneous. A culture bottle of *Drosophila* may have enough differential between the dry crust and the moist interior to allow a long-continued coexistence of two species (L'Héritier and Teissier 1935; Moore 1952a). When two competing species of grain beetles are raised in whole wheat kernels rather than in sifted wheat flour, the two species can usually coexist indefinitely (Crombie 1947). These experiments in heterogeneous environments indicate how subtle the differences between the ecological niches of two closely related species can be.

DIFFICULTIES OF THE EXCLUSION PRINCIPLE

Naturalists have described numerous cases in which two or more related species appear to occupy the same niche. Ross (1957) describes the coexistence in the Mississippi Valley of six species of leafhoppers (*Erythroneura*) on the same food plant (*Platanus occidentalis*). Cooper (1953) finds that the three most common species of eumenine wasps in New York state frequent the same flowers, hunt the same caterpillars on the same trees, and nest in 6-mm burrows in the same places at the same time. They appear to be competing for all the important elements of their existence. Sokoloff (1955, 1957) believes that *Drosophila pseudoobscura* and *D. persimilis* do not reach a population density in nature that would induce actual competition. Congeneric species of saw flies often infest the same host species, as described for the genera *Periclista* (Beer 1955a) and *Strongylogaster* (Beer 1955b).

In all these cases (and they are only a small selection from an extensive literature), it has been impossible so far to determine the factor or factors that permit coexistence of several species in the "same" niche. The real question is, of course, whether these species really share the identical niche. Bagenal (1951), Savage (1958), and Udvardy (1959) point out that niche is not the same as habitat, and that the mere occurrence of six species of insects on the same species of host plant does not necessarily mean that they coexist in the same niche. Much of the argument depends on the chosen definition of niche (Hutchinson 1957). The niche of a species is the outward projection of its needs, and identity of the niches of two species cannot be proved until it is shown that the needs of the two species are identical. How careful one must be in judging the "identity" of needs is nicely illustrated by the case of *Drosophila mulleri* and *D. aldrichi* (Wagner 1944). The larvae of these two closely related species occur simultaneously in the ripe fruit of the cactus *Opuntia lindheimeri*, feeding on the microflora of the fermenting pulp. Competition between the larvae of the two species appears to be complete. Yet analysis of the intestinal contents showed that the two species of *Drosophila* tended to feed on different species of yeast and bacteria with a certain amount of nonoverlap in the requirements. Similar differences in food selection have been established for grasshoppers feeding together in a meadow (Isely 1946) and other coexisting groups of species. Two exceedingly similar mosquito species on Trinidad, *Anopheles bellator* and *A. homunculus*, appear at first sight to share the same niche. A detailed analysis showed, however, that *bellator* is more tolerant of lower humidities while *homunculus* is superior in utilizing small bromeliads as breeding places (Pittendrigh 1950). The ecological overlap is only partial. In all these cases a careful distinction must be made between habitat and niche. The difference in food utilization of the two species of *Drosophila* in *Opuntia* proves that their niche is not the same. Nor is niche identity proved for the various species of nematodes, flies, or dung beetles that live in cow dung.

There has been a tendency in recent years to exaggerate food as the controlling factor in competitive situations. As a matter of fact, many other controlling factors must be considered. In the tropics, for instance, food appears to be less important than in the temperate zone, and in all zones competition may occur even where food is superabundant. In some cases competition is reduced not because of differences in the kind of preferred food but because the competing species differ in the place where they search for food. This has been demonstrated for Hawaiian honey creepers (Baldwin 1953), for European titmice (*Parus*) (Gibb 1954; Betts 1955), for American wood warblers (Parulidae) (MacArthur 1958), and for European psocids, of which six species feeding by preference on *Pleurococcus* coexist on the same species of larch (*Larix*). Near relatives may differ by predominating on living or dead twigs, preferred altitude in the mountains, and slight seasonal differences. In spite of these niche differences, there are indications that competition among these psocids might occur under deteriorating conditions (Broadhead 1958). The role of food may be different even among close relatives. Carnivorous cyclopoid copepods, for instance, appear to compete for food and differ markedly in prey preference, while there is no significant difference of diet among those that feed on algae, a superabundant source of food (Fryer 1957).

These studies indicate that natural environments are far more heterogeneous than appears at first sight and that competition may be severe only during infrequent adverse seasons. The respective superiority of the competing types may shift with environmental conditions and with population density. Competition is further mitigated by differences in habitat selection resulting from differences in the physiology of the species. The two salamanders *Plethodon dunni* and *P. vehiculum* take almost the same food in their area of coexistence in Oregon but tolerate different temperature and humidity limits. Differences in the secondary food of the species contribute toward reducing competition (Dumas 1956). The diets of two European species of sticklebacks (*Gasterosteus aculeatus* and *G. pungitius*) appear to be virtually identical. Competition is apparently mitigated through a preference for different breeding habitats in which the level of population density is regulated by territorial habits (Hynes 1950). Two species of parasitic wasps continued to coexist in a population of the azuki bean weevil (*Callosobruchus chinensis*) for 70 generations, without either being able to eliminate the other. It appears that one species of parasites is more successful at low, the other at high densities of the host species (Utida 1957). Lack (1946) showed that there is no competition between ten species of hawks and owls in Europe in years of abundance of their primary prey, the meadow vole (*Microtus*). In years of vole scarcity, each of these predators falls back on a different species of secondary prey and thus escapes competition with the other vole predators.

POSSIBLE ABSENCE OF EXCLUSION

The frequency of demonstrable exclusion is so great that its importance can no longer be questioned. We are now ready to ask, without denying the principle as such, whether there are factors or constellations of factors that may reduce or altogether negate the operation of exclusion. There are indeed factors that reduce the impact of interspecific competition to a greater or lesser extent. The density of a species may be fixed owing to intraspecific intolerance at so low a level that its individuals are not in competition with a potential competitor whose populations are equally widely spaced. Marshall's (1960) case of niche co-occupancy by two mutually tolerant species of towhees (*Pipilo aberti* and *P. fuscus*), may be an illustration of such a situation. There is no apparent difference between the two species with respect to habitat utilization for breeding or feeding. Pontin (1961) similarly believes that the coexistence of two species of ants (*Lasius flavus* and *L. niger*) is the result of intraspecific density-dependent pressures, as a result of which an increase of either species is prevented and coexistence permitted. The considerable differences in food utilization of the two species indicates, however, that contrary to Pontin's belief there is only partial niche overlap and thus exclusion.

Two situations seem to be exceptions to the exclusion principle. One is that of generalized herbivores, which are seldom food-limited and are not likely to compete for common resources (Hairston, Smith, and Slobodkin 1960). Their numbers are often controlled by predators, diseases, or other density-dependent factors that have no obvious connection with exclusion. The other situation occurs in areas, for instance near the periphery of the species range, where the severity of the environment prevents the populations from building up to the limits of the niche capacity, that is, where population size is kept low by density-independent factors (Andrewartha and Birch 1954; Savile 1960). This is particularly true where a peripheral population is replenished in each generation by a steady stream of immigrants from a more favorable portion of the species range. Competition, of course, is not eliminated in either case, but it will never culminate in exclusion.

Other disruptive factors in nature reduce the relative importance of competition. These factors include rapid changes of the environment and selective advantage of rarity; they are not, of course, alternatives to competition, but, being likewise able to influence the density of species populations, they are in their effect superimposed on competition.

EVOLUTIONARY RESULTS OF COMPETITION

No matter how slight or intermittent, any kind of competition is always an important selection pressure. If two species overlap in part in their requirements and if any one of these requirements is a controlling factor, selection will favor those members of either population that are least dependent on this factor. The observed avoidance of competition is the result of such selection according to the theory of exclusion. To test the validity of this assumption, let us look more closely at various alternative forms of ecological relation among species. Whenever the ranges of two closely related species overlap, any of the following five alternatives may take place (Lack 1949):

(1) One species is superior to such a degree that it eliminates the other one. This is the situation we discussed above under introductions and invasions. It can be proved only during the actual period of displacement.

(2) In part of the geographic area one species is competitively superior; in another part, the other species is superior. As a consequence the two species will replace each other along a sharp but frequently fluctuating line of balance. Range expansion by either species is prevented by its competitor.

(3) Where two species come into contact along a similarly well-defined border, expansion is prevented not by the competing species but by the ecological unsuitability of the terrain across the borderline.

The choice between (2) and (3) is not always simple. Allopatric species pairs are very common, and Mayr (1951a) and Vaurie (1955) have listed many such pairs among birds. In most cases they meet along a sharp climatic or vegetational break, and it appears probable that differences in adaptation rather than competition are responsible for the geographic replacement. This is evidently the case where such conspicuously different habitats come into contact as savannas and the Congo rain forest in Africa. The interpretation must, however, remain tentative where more elusive climatic or biotic factors are involved. The fact that the ranges of two widespread Asiatic bee-eaters, *Merops superciliosus* and *M. philippinus*, are largely allopatric has, for instance, been ascribed to competition (Marien 1950). Yet these two species coexist in some border districts and there is a strong probability that the essentially allopatric distribution pattern is due to differences in physiological adaptation. *Merops superciliosus* appears to be superior where the rainfall is less than 20 inches per year, *M. philippinus* where it exceeds this amount. Whether either species could invade the range of the other species if the latter were not present is the big question. I do not believe, for instance, that competition has been involved in the many cases of a northward retreat of northern animals during the recent amelioration of the climate and the corresponding northward expansion of southern species.

It is highly probable that most cases of allopatric species would have to be listed under (3) rather than under (2). However, there are also some clear-cut instances of (2). The essential geographic replacement of *Gammarus pulex* and *G. duebeni* in the British Isles is due to the reproductive superiority of *pulex* in fresh water and of *duebeni* in brackish waters (Hynes 1954). The heterogeneity of factors will in most cases make an unequivocal decision between (2) and (3) impossible.

(4) One species is superior in some habitats, the other in other habitats. The result will be geographic overlap combined with habitat exclusion. The case of the Mexican jays (*Cissilopha*) described above or that of the chaffinches (*Fringilla*) on the Canary Islands are typical illustrations. The vertical ranges of mountain species are often very much compressed where they encounter a competitor. This has been described, for instance, by Mayr and Gilliard (1952b) and Gilliard (1959) for two species of New Guinea honey eaters.

(5) Both species enter the same habitat but occupy different niches. Most of the previous discussion was devoted to such cases.

The meeting of two similar or closely related species along a line of contact or in a zone of overlap may thus have very different consequences. The study of this group of phenomena has merely begun, but students of evolution are to an ever-increasing extent concerned with the study of such interactions.

SYMPATRIC CHARACTER DIVERGENCE

One further consequence of the exclusion principle is that competition in an area of geographic overlap should exert strong selection pressure. Those individuals of two overlapping species should be most favored by selection that have the least need for the resources jointly utilized by both species. Darwin (1859) devoted an entire section in his chapter on "Natural Selection" to the "Divergence of Character," saying, "The principle which I have designated by this term is of high importance . . . Natural Selection . . . leads to divergence of character; for more living beings can be supported on the same area the more they diverge in structure, habits, and constitution." Although such divergence will at first be strictly ecological, consisting of a different utilization of the environment (see above, p. 69), it will subsequently be reinforced by the selection of such morphological differences as facilitate the ecological divergence. Among modern authors, Lack (1947a) was the first to give an example of such character divergence (Table 4–6). The ecological niches on the outlying islands of the Galapagos may be filled by two or three finches of the genus *Geospiza* whose bill sizes are adjusted to the available food niche. On Tower Island three strongly differentiated species divide the habitat among themselves. On each of three other islands only two species are present, one of which occupies two potential niches and is able to adjust its bill size owing to the lowered selection pressure. Amadon (1947) showed for the Hawaiian genera of *Loxops* and *Phaeornis* that intermediate bill size is found on most of those islands on which only a single species of the genus is present. On Kauai where, owing to double invasion, both genera are represented by two species each, centrifugal selection has led to strong divergence in bill size. Other cases of character divergence owing to double invasion are discussed by Nørrevang (1959). A particularly instructive case was described by Vaurie (1950, 1951b). Two species of rock nuthatches (*Sitta*) largely replace each other in eastern and western Eurasia. The two species are very similar in bill size and general coloration wherever only one of them occurs. However, in Iran, in an area of broad overlap of the two species, the bill of one species is strikingly enlarged while that of the other is decreased (Fig. 4–2). Character divergence between Indian species of nuthatches has been described by Ripley (1959). Where the two ant species *Lasius flavus* and *L. nearcticus* coexist in the eastern United States, they differ by at least eight independent characters. In the West, where *nearcticus* is rare or absent, the variability of *flavus* has increased so much that it obliterates all but two of the diagnostic differences between the species (Wilson 1955). Brown and Wilson (1956) have properly stressed the evolutionary importance of Darwin's character divergence (which they call character displacement) and have cited additional examples (although some of their cases seem to illustrate other

Table 4–6. Character divergence in *Geospiza* on Galapagos: bill depth (mm) and niche occupation (from Lack 1947a).

Niche	Tower Island	Hood Island	Wenman Island	Culpepper Island
Large ground finch	*magnirostris* 21.2	*conirostris* 16.0	*magnirostris* 20.4	*conirostris* 16.5
Cactus feeder	*conirostris* 13.0			
Small ground finch	*difficilis* 7.9	*fuliginosa* 8.3	*difficilis* 9.0	*difficilis* 9.0

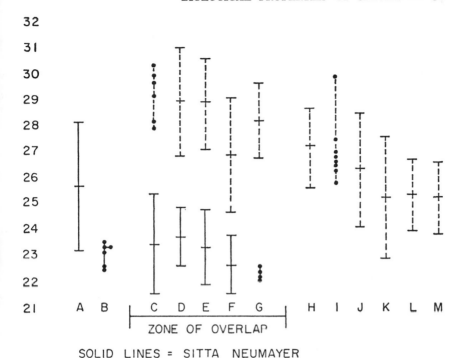

SOLID LINES = SITTA NEUMAYER
BROKEN LINES = SITTA TEPHRONOTA

Fig. 4–2. Geographic variation of bill length in two partially sympatric species of rock nuthatches, *Sitta neumayer* (solid lines) and *S. tephronota* (broken lines). Westernmost population (Dalmatia, Greece) on left (A), easternmost population (Tian Shan) on right (M). Complete character divergence in the zone of sympatry in Iran (C–G). (From Vaurie 1951b.)

phenomena, such as hybridization). Numerous further illustrations of character divergence have since been brought to light, for instance by Hutchinson (1959) and Stein (1960). The time seems to be ripe for a more quantitative analysis than is employed in the studies cited. This is particularly important since there are also cases known in which overlapping species are more similar in their area of overlap (owing, for instance, to Gloger's rule) than elsewhere, seemingly in contradiction to the rule of character divergence. Finally, there is a large class of cases where the overlap seems to have had no effect on the phenotype of the sympatric populations. Only a statistical analysis can bring out the facts needed for valid generalizations.

Another frequent consequence of competition (in addition to character divergence) is reduction in the morphological and ecological variability of one or both species in the zone of overlap. Dobzhansky, Burla, and da Cunha (1950) have demonstrated such a reduction of variability for a genetic character, the occurrence of gene arrangements in populations. They found for *Drosophila willistoni* that their frequency was reduced, other things being equal, in those portions of the range where the greatest number of sibling species coexisted.

A very special case of sympatric character divergence is the reinforcement of ethological (and possible other) isolating mechanisms in zones of overlap of closely related species. A more detailed discussion of this phenomenon is presented in Chapter 17.

OPEN PROBLEMS

Our discussions have clearly demonstrated that there are numerous interactions among ecologically similar species that coexist in the same area. Selection tends to increase the ecological difference between such species and to reduce the zone of ecological overlap. As a consquence, competition has many and important evolutionary consequences. Yet our ignorance of the nature, the mechanisms, and the amount of competition is vast. In recent years the study of competition has become one of the most active branches of ecology, and it promises to yield large returns for many years to come. Some of the problems that invite further study are the following:

Which observed cases of geographical exclusion (allopatry) of species are due to ecological incompatibility and which other cases are simply the result of physiological adaptations that would make the unoccupied area unsuitable even if it were not inhabited by the potential competitor?

Which of the differences among closely related, sympatric species are the result of character divergence after establishment of sympatry and which others had already been acquired as by-products of the preceding geographic speciation?

Are there cases of species that do not compete with each other even though their essential needs are the same? If so, what other mechanisms permit coexistence? In answering this question it is particularly important to avoid circular reasoning. It is very tempting to say that the mere coexistence of species A and B proves ecological exclusion and, therefore, that whatever differences are discovered between these species are responsible for the coexistence. Such a conclusion is not necessarily valid.

What differences, if any, exist between the central and peripheral parts of the species range with respect to competition? Haldane (1953, 1956) suggests that population size near the periphery of the species range is largely controlled by density-independent factors. Competition with other species would be less drastic under these circumstances, and a shift into a new niche facilitated. There is much evidence, as pointed out by Mayr (1954a), that shifts into new niches are particularly frequent in peripherally isolated populations.

To what extent is there aggressive interference among species? We know that the phenomenon exists. Mutual territorial aggression lowers the population density of two species of hummingbirds in California (Pitelka 1951b; Legg and Pitelka 1956). A one-sided intolerance between two species of titmice (*Parus*) has been described for California (Dixon 1954) and territory exclusion is known for other birds (Simmons 1951; Emlen 1957; Selander and Giller 1959b; Ripley 1961). Partial territory exclusion has been found in *Quiscalus* (Selander and Giller 1961). Non-competitive interactions among invertebrates are usually of the nature of predation or interference rather than of interspecific aggression (Birch 1957). What role external metabolites (Lucas 1947) play in the ecological interactions of species is still a complete mystery.

Why are so many apparently empty niches not filled? There are 27 or 28 species of woodpeckers in Borneo and Sumatra and none in the similar forests of the New Guinea region. The important niche filled by woodpeckers in the Oriental region seems to be almost entirely vacant east of Weber's Line. The opening of the Suez Canal permitted more than a dozen Red Sea fishes to invade the Mediterranean, apparently without displacing any native species. The invaders must have found partly unoccupied niches. Invading species only rarely displace a native species from its niche. In most cases they occupy at least in part what appears to have been a previously unfilled niche. This is certainly the case with the European Starling (*Sturnus vulgaris*) in North America, the tremendous increase of which in habitats largely created by man caused some but certainly not an equivalent decrease in other species. Kusnezow (1956) shows that different niches are filled by desert ants in the Old World and in the New World. The picture of a vacant niche is not particularly popular among ecologists. They will point out that insect larvae on New Guinea that are not eaten by woodpeckers may fall victim to parasites or may be eaten as adults by birds or mammals. Any left over are utilized by reducers and decomposers. In my opinion, such utilization does not fill the woodpecker niche. Whenever a local biota can be enriched this proves to me *de facto* that vacant niches had existed. We must leave it with this dogmatic assertion; a deeper analysis would lead us into the difficult problem of the relative efficiency of utilization of natural resources, and this would take us far outside the subject matter of this volume.

Why can certain tropical habitats maintain such an inordinately high number of species? A temperate-zone forest may be inhabited by 60 species of birds, the Amazonian forest by 600. No fewer than 780 species of fish have been found in the harbor of Amboina (Moluccas), a figure not approached anywhere in the temperate zone. MacArthur (1957), Hutchinson and MacArthur (1959), Klopfer and MacArthur (1960), and Fischer (1960) have discussed some of the factors responsible for this richness of tropical faunas. Neither increased complexity of habitat nor, solely, increased specialization, seems to account for this increase in numbers. It is possible that the rarity or short duration of weather-induced catastrophes in the tropics permits a contraction in niche size. The existence of large numbers of rare, yet comparatively unspecialized, species in tropical areas is a particularly puzzling aspect of this problem (Chapter 14).

CONCLUSIONS

Many of the resources of the environment are limited, and competition for them, wherever it occurs, has various evolutionary consequences. It makes speciation more difficult, because an incipient species cannot invade the range of a sister or parent species until it has acquired ecological compatibility with it, that is, until its niche has become sufficiently distinct to permit exclusion (rather than extinction!).

The premium placed by natural selection on increased ecological differences between competing species is a powerful centrifugal force. It favors the entry into new niches and, more generally, adaptive radiation. Competition thus is an element in speciation and is an important cause of evolutionary divergence.

5 ~ Isolating Mechanisms

The mechanisms that isolate one species reproductively from others are perhaps the most important set of attributes a species has, because they are, by definition, the species criteria. That there is a whole set of special devices by which the gap between species is maintained was not realized until rather recently. Previously only the sterility barrier had been recognized. Darwin, thinking that species were something arbitrarily delimited, neglected the problem of the nature and the origin of the species gap. The typologists, on the other hand, accepted the gap as an inherent property of organic nature. Eventually, however, naturalists discovered that many devices other than sterility may prevent the interbreeding of closely related sympatric species. In 1905 Peterson wrote, "It is a fact that each species of animals has devices which permit the recognition and the bringing together of conspecific individuals of opposite sex with such a degree of certainty that hybridization occurs only as an abnormal exception, and in this we see a fundamental difference from the situation in the plant kingdom with its wind and insect pollination." The first systematic treatment of the subject was given by DuRietz (1930), who made a fairly complete listing of the various kinds of barriers, but it was not until Dobzhansky coined the term "isolating mechanisms" and devoted a whole chapter to them in his classic *Genetics and the Origin of Species* (1937) that the great importance of these devices was recognized among evolutionists. Since that time a number of symposia on the subject have been held (Patterson 1942, 1947; Blair 1961) and a truly staggering amount of literature has accumulated. I can deal here only with the highlights and must refer for further details to the works listed in the bibliography (Muller 1942; Mayr 1942, 1948; Emerson 1949; Stebbins 1950; Dobzhansky 1951; Tinbergen 1951; Patterson and Stone 1952; Riley 1952; Thorpe 1956; Spieth 1958; Hinde 1959).

The active study of isolating mechanisms has led to the development of an entirely new branch of biology which combines studies of behavior, ecology, and genetics. These studies have led to a deeper insight into the quality and quantity of species differences and to precise experimental testing of various assumptions.

Before we can begin the discussion of the various mechanisms, two points require clarification: sterility and geographic isolation.

Sterility. In spite of everything that has been written in the last 25 years there are still some authors who seem to think that reproductive isolation and sterility are synonymous terms. Nothing could be further from the truth. An ever-increasing number of very distinct, reproductively isolated sympatric species are known that are not isolated from each other by a sterility barrier. For instance, the mallard, *Anas platyrhynchos*, and the pintail, *Anas acuta*, are perhaps the two most common fresh-water ducks of the Northern Hemisphere. The total world population of these two species may well exceed 100,000,000 individuals. Phillips (1915) showed that in captivity these two species are fully fertile with each other and that there is no reduction of fertility in the F_1, F_2, or F_3, or in any of the backcrosses. One would therefore expect a complete interbreeding of these species in nature, as their world breeding ranges largely coincide. In northern Europe, Asia, and North America they nest side by side on literally millions of ponds, sloughs, or creeks; yet the number of hybrids found among the many birds shot every year is on the order of one in several thousand. Nor is there evidence of backcrossing between these hybrids and the parent species. Obviously, then, the two species are being kept apart not by a sterility barrier but by some other factors.

High, if not complete, fertility is known for many species crosses, not only among ducks and other families of birds, but throughout the animal kingdom. The idea that hybrids always are sterile "mules" is quite erroneous. From the scores of cases of fully cross-fertile species published in the recent literature a few may be cited: the fishes *Notropis lutrensis* and *N. venustus* (Hubbs and Strawn 1956), the midges *Chironomus tentans* and *C. pallidivittatus* (Beermann 1952), the frogs *Rana nigromaculata* and *R. brevipoda* (Moriya 1951), the newts *Taricha torosa* and *T. rivularis* (Twitty 1959, 1961), the isopods of the *Jaera albifrons* group (Bocquet 1953), *Drosophila bocainensis* and *D. parabocainensis* (Carson 1954), the moths *Celerio gallii* and *C. euphorbiae* (Bytinsky-Salz and Günther 1930) and many other cases in Sphingidae, Saturniidae, and other families of moths. Arctic and red foxes (several investigators), and many others. In plants, likewise, sympatric species may have considerable infertility without losing their genetic independence (Stebbins 1950; Epling 1947a). Patterson and Stone (1952) point out that groups of closely related and morphologically similar species occur in many families of animals and plants and that there may be either considerable fertility or complete sterility between the species of which such groups are composed. Cross-fertility does not prove conspecificity.

More decisive evidence for the importance of other isolating mechanisms in animals is the observation by field naturalists that males and females of a species are brought together by sensory stimuli and that it is extremely rare for the male of one species to copulate with the female of another species. The sterility barrier, even when present, is only rarely tested.

Geographic Isolation. It is quite impossible to discuss intelligently the nature and the origin of isolating mechanisms if one includes with them so extraneous an element as geographic isolation. The term isolation refers to two very different phenomena, spatial isolation and reproductive isolation (Mayr 1959a). The term "isolating mechanisms" refers only to those properties of species populations that serve to safeguard reproductive isolation. In order to exclude extraneous phenomena from the category of isolating mechanisms, it is important to define them precisely: *isolating mechanisms are biological properties of individuals which prevent the interbreeding of populations that are actually or potentially sympatric.* This definition clearly excludes geographic isolation. San Francisco Bay, which keeps the prisoners of Alcatraz isolated from the other inhabitants of California, is not an isolating mechanism, nor is a mountain or a stream that separates two populations that are otherwise able to interbreed. Isolating mechanisms always have a partially genetic basis, although some components of behavioral isolation may be reinforced by conditioning.

CLASSIFICATION OF ISOLATING MECHANISMS

Virtually every author who has written about isolating mechanisms in the past 25 years has proposed a classification. All are, on the whole, rather similar. My own classification (Table 5–1) arranges the mechanisms in the sequence in which the barriers have to be overcome. I avoid the terms "genetic" or "physiological" because genetic and physiological factors play a role in every one of the seven major subdivisions.

For the vast majority of animals, it is still not known which particular

Table 5-1. Classification of isolating mechanisms.

1. Mechanisms that prevent interspecific crosses (premating mechanisms)
   (a) Potential mates do not meet (seasonal and habitat isolation)
   (b) Potential mates meet but do not mate (ethological isolation)
   (c) Copulation attempted but no transfer of sperm takes place (mechanical isolation)
2. Mechanisms that reduce full success of interspecific crosses (postmating mechanisms)
   (a) Sperm transfer takes place but egg is not fertilized (gametic mortality)
   (b) Egg is fertilized but zygote dies (zygote mortality)
   (c) Zygote produces an F_1 hybrid of reduced viability (hybrid inviability)
   (d) F_1 hybrid zygote is fully viable but partially or completely sterile, or produces deficient F_2 (hybrid sterility)

isolating mechanisms prevent closely related species from interbreeding. This is especially true of most lower invertebrates and protozoans. The naturalist, the cytologist, the geneticist, all have unlimited opportunity for discoveries in this area.

The numerous mechanisms clearly fall into two very different classes (Mayr 1948): on one hand there are those that prevent interspecific matings and are called "premating mechanisms" (Mecham 1961); on the other hand are all those that reduce the success of interspecific crosses, "postmating mechanisms" (Mecham 1961). There is a fundamental difference between the two types: premating mechanisms prevent wastage of gametes and so are highly susceptible to improvement by natural selection; postmating mechanisms do not prevent wastage of gametes, and their improvement by natural selection is difficult and indirect. A clear understanding of this difference is of the greatest importance for an understanding of the origin of isolating mechanisms (Chapter 17).

INTERSPECIFIC CROSSES PREVENTED (PREMATING MECHANISMS)

The category includes seasonal and habitat isolation, ethological isolation, and mechanical isolation.

Habitat Isolation

The less often two potential mates in breeding condition come into contact with each other, the less likely they are to interbreed. Habitat selection is thus a very effective isolating mechanism. Where closely related sympatric species show habitat exclusion, it enhances other isolating mechanisms effectively. Every naturalist could cite hundreds of illustrations. Various closely related species of chipmunks (*Eutamias*) in California are segregated in different habitats in the zone of contact (Johnson 1943). Soil type seems to be a major isolating mechanism between two species of spadefoot toads (*Scaphiopus;* Wasserman 1957). In the toads *Bufo americanus* and *B. fowleri* Blair (1942) finds that where the two species overlap there is often a differential preference for breeding places:

> *Bufo fowleri* utilizes ponds, sloughs, and large rainpools as breeding sites, but does not breed in the shallow puddles and brook pools that are so extensively used by *B. americanus*. *B. fowleri* breeds in the quieter waters of streams but *B. americanus* rarely or never does so. *Bufo fowleri*, collected near Wilberton, Oklahoma, May 26, 1939, were all found breeding in a creek. *B. americanus* at the same locality, April 8, 1940, were all breeding in shallow rainpools, and not a single toad was found along the creek which was only a few hundred feet from the pools.

Habitat exclusion among close relatives has been known to naturalists ever since Darwin's days (for example, Steere 1894).

The habitat segregation is often very slight in terms of distance. The situation of the spawning grounds in river fish is determined not only by the nature of the substrate (gravel, sand, mud), but also by the rate of water flow (and other factors). A sudden change of gradients in a creek bed may bring the normally well-separated spawning sites of two species so close together that sperm of one species is washed into the nests of the other species, leading to interspecific hybridization (Trautmann *in litt.*)

Two closely related species of dragonflies now overlap in north central Florida, the isolating barrier having broken down in the late Pleistocene. In the zone of overlap, the northern species, *Progomphus obscurus*, is restricted to rivers and streams; the southern species, *P. alachuensis*, inhabits lakes (Byers 1940). In mountainous areas there is often a separation of closely related species by altitude (see Chapter 4).

Habitat isolation is not a very effective isolating mechanism in mobile animals. Consequently, breakdown of sharp borders between habitats, owing to human or other disturbances, has been the cause of most recorded cases of sympatric hybridization in animals (Chapter 6). Habitat segregation plays a much greater role among plants (Stebbins 1950). Its role in isolating very sedentary animals, or animals with highly specific edaphic requirements (for example, soil animals), or completely sessile animals, such as some marine invertebrates, has not yet been investigated.

Seasonal Isolation

Differences in the breeding season can effectively prevent the meeting of individuals of different species. Seasonal isolation is common in plants (Stebbins 1950) and occurs frequently among insects and other invertebrates (Emerson 1949; see also Chapter 15). In the temperate zone related species of birds usually have largely overlapping breeding seasons. It is probable, however, that the two terminal populations in the famous case of circular overlap of the *Larus argentatus* group (Chapter 16), the Herring Gull (*argentatus*) and the Lesser Black-Backed Gull (*fuscus*), show some seasonal isolation. In a colony studied by Paludan (1951), *argentatus* reaches the height of its egg-laying period during the last third of April, but *fuscus* not until the middle of May. Other cases in birds in which the seasonal barrier seems effective have been quoted by Mayr (1942:251).

Seasonal barriers seem particularly frequent among water animals. This is due to two factors: water temperatures are more stable than air temperatures, and embryonic development is more closely attuned to definite temperatures. The two factors combined permit a close regulation of the breeding season. Mertens (1928) lists season of breeding as the principal isolating factor between *Rana ridibunda* and *R. esculenta* and between the toads *Bombina variegata* and *B. bombina*. Blair (1941) found that the toad *Bufo americanus* breeds early in the spring but that *B. fowleri* breeds late. There is, however, an overlap in midseason during which considerable hybridization occurs at many localities. The five species of *Rana* in eastern North America (Moore 1949) likewise have largely overlapping breeding seasons. The time of spawning in fish, particularly fresh-water fish, is often regulated by water temperatures, and the spawning seasons of closely related species may be effectively separated by adaptation to different spawning temperatures.

The actual contribution of seasonal isolation to the maintenance of reproductive isolation between species is largely unknown. In some cases in which the seasonal isolation broke down owing to unusual weather conditions, reproductive isolation was maintained by other factors. No natural hybridization occurred in western Canada between the spruce budworm (*Choristoneura fumiferana*) and the pine budworm (*C. pinus*) when unseasonably high temperatures one year resulted in simultaneous hatching. A combination of different host preferences and behavioral barriers maintained the isolation (Smith 1953, 1954). Similar observations have been made in other cases and suggest that the function of differences in breeding season is chiefly to reduce competition and permit optimal adaptation to the respective niche, rather than to prevent interbreeding.

No case is known to me of two species that are isolated by restriction of their courtship period to different parts of the day, that is, in which preference for a definite light intensity functions as an isolating mechanism. There is evidence, however, that *Drosophila pseudoobscura* and *D. persimilis* differ in their preferred mating hour (Spieth 1958), and similar differences are known for other insects and for closely related sympatric species of fish.

Ethological Barriers (Restriction of Random Mating)

Ethological barriers to random mating constitute the largest and most important class of isolating mechanisms in animals. The word "ethological" (derived from the Greek *ethos* = habit, custom) refers to behavior patterns. Ethological isolating mechanisms, then, are barriers to mating due to incompatibilities in behavior. The males of every species have specific courtships or displays to which, on the whole, only females of the same species are receptive. The specific reaction of males and females toward each other is often referred to loosely as "species recognition." This term is somewhat misleading since it implies consciousness, a higher level of brain function than is found in lower animals (see also Spurway 1955). Rather, these isolating mechanisms function about as follows.

In most animals it is the male that actively searches for a mate. He is usually somewhat easily stimulated to display to objects, sometimes quite inappropriately. When he does not receive adequate responses from his display partner, or is actively repulsed, his display drive soon becomes exhausted. Consequently, if such a displaying male encounters an individual of a different species, or a male of his own species, he will break off his courtship sooner or later. If the male is displaying to a nonreceptive female, the same will happen, perhaps after a longer interval. However, if the male encounters a receptive female of his own species, he will be sufficiently stimulated by her to continue his displays until the female has passed the threshold of mating readiness. This threshold is on the average much higher in females than in males. "Species recognition," then, is simply the exchange of appropriate stimuli between male and female, to insure the mating of conspecific individuals and to prevent hybridization of individuals belonging to different species.

Ethological isolation, therefore, is based on the production and reception of stimuli by the sex partners. What are these stimuli that assure the mating of conspecific individuals? There has been some literature on this subject ever since there have been folk tales and observing naturalists. The song of the nightingale belongs here and so does the strutting of the peacock. It is convenient to classify these ethological devices on the basis of the principal sense organs involved.

Visual Stimuli. The importance of color and form in the reproductive life of visual animals has long been known to naturalists, but their function as ethological isolating mechanisms has been recognized only recently. They play a role in all animals with well-developed eyes, such as birds, certain fishes, many insects and spiders, and even squids. Most of our information in this field is based on observation and inference, only recently supplemented by experiment.

Much current literature in the field of animal behavior (ethology) deals with the analysis of courtship patterns, all of which ultimately serve, directly or indirectly, as isolating mechanisms. There are numerous articles on this subject in every volume of *Behaviour* and *Zeitschrift für Tierpsychologie*. Aspects of behavior that function as isolating mechanisms have recently been summarized by Lorenz (1950), Tinbergen (1948, 1951, 1960), Hinde and Tinbergen (1958), Hinde (1959), and Spieth (1958). The visually perceivable differences between closely related species may be a matter of color, color pattern, form, movement, or a combination of several of these differences. It is not easy to determine the contribution to reproductive isolation of each of these species differences. If the reproductive isolation between *Drosophila melanogaster* and *D. simulans* is the same in the light and in the dark (Manning 1959a; Fig. 5–1), one tentatively concludes that optical stimuli are of minor importance. There are, however, several species of *Drosophila* in which the males will not court in the dark (Spieth and Hsu 1950). If two sympatric species of cichlid fishes have identical display movements, one assumes that the marked difference in color is responsible for the reproductive isolation, unless chemical or other differences are shown to be important.

The analysis is particularly difficult because the signals that function as isolating mechanisms also serve various other functions (Tinbergen 1954): (1) to advertise the presence of a potential mate; (2) to synchronize mating activities; and (3) to suppress fleeing or attacking tendencies in the sex partner. The isolating function cannot be separated from the other three functions. Indeed, one can go so far as to say that only the appropriate, species-specific signals lead to full success of functions (1),

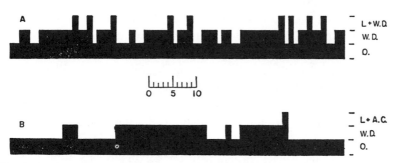

Fig. 5–1. Quantitative differences in the major courtship components of two sibling species of *Drosophila*. A = *D. melanogaster*, B = *D. simulans*. The sequence reads from left to right, the scale showing time units of 1½ seconds. The height of the black columns indicates the courtship element being performed. The lowest level is O (orientation), the middle level W.D. (wing display), and the highest level L + A.C. (licking and attempted copulation). (From Manning 1959a.)

(2), and (3) and that these signals thereby automatically function as isolating mechanisms. Some recent studies on visual stimuli exchanged by potential mates are those of Lorenz (1941) on ducks, Hinde (1955) on finches, Tinbergen (1959) on gulls, Seitz (1951) and Baerends and Baerends (1950) on cichlid fish, Crane (1948–1950) and Drees (1952) on salticid spiders, Crane (1957) and Altevogt (1957) on fiddler crabs, and L. Tinbergen (1939) on squid. Perhaps the most careful and detailed analysis of all is that of Clark, Aronson, and Gordon (1954) on xiphophorin fishes. For further literature see Mayr (1958), Hinde (1959), and Hinde and Tinbergen (1958).

Visual stimuli usually work in conjunction with auditory, tactile, or chemical stimuli. It is true, however, that other stimuli tend to be less well developed when the visual stimuli are predominant. Naturalists have long stated that the most beautiful birds have insignificant songs and vice versa. The number of exceptions, however, is too great for this to be accepted as a general rule. One of the most insignificant songs uttered by any North American bird is that of the Henslow's Sparrow (*Passerherbulus henslowi*), a species with equally insignificant coloration. The members of the cardinal family (Cardinalinae), the Cardinal, the Rose-breasted Grosbeak, the Indigo Bunting, and their relatives, are not only beautiful but also excellent singers. The same is true of both families of orioles (Oriolidae, Icteridae). Most birds of paradise have not only exquisite plumage but also loud, and in some cases very melodious, voices. The bird that is perhaps the most gifted vocalist in the world, the Australian Lyrebird (*Menura superba*), has also conspicuous displays and wonderful plumes. The correlation between strong development of one kind of stimuli and weak development or absence of other stimuli is thus incomplete.

The light signals sent out by male fireflies (beetles of the family Lampyridae) are perhaps the most fascinating visual isolating mechanisms. They differ in timing (Fig. 3–3) and color, which may be clear white, bluish, greenish, yellow, orange, or reddish. Some species emit exceedingly bright flashes, others dull ones. Numerous sibling species in the genus *Photuris* were first discovered through differences in their light flashes (Barber 1951). The stimuli produced by the females are equally specific, as indicated by the behavior of the males (Schaller and Schwalb 1961).

Auditory Stimuli. Songs, calls, and other acoustic signals have a precision and specificity that might well make them even more suitable as isolating mechanisms than visual stimuli. Recent work indicates that this is certainly true of the anurans (frogs and toads) and of most orthoptera. Species-specific sounds play an important role also in the courtship of birds (Marler 1960) and certain diptera. Females may not be ready to mate until stimulated by the appropriate wing vibration of conspecific males in mosquitoes (Roth 1948), *Drosophila* (Spieth 1952), and *Glossina* (Vanderplank 1948). Either song or call notes are species specific in all birds, except perhaps in those with rudimentary vocalization (certain vultures, cave swiftlets, storks, and so on). It is often easier to identify a bird by its song than by the visual characters of its plumage. The female, during pair formation, is attracted to the stations of the singing males of her species. Where the habitats of two very similar species of Californian chipmunks (*Eutamias sonomae* and *E. townsendi siskiyou*) come into contact, one species is readily distinguishable from the other by its call note (Miller 1944a). On the whole, we would not think of calls as important isolating mechanisms in mammals, and yet the recently discovered elaborate "language" of aquatic mammals (whales, porpoises) may, in part, serve this function. Species discrimination by sound may well be widespread in the animal kingdom. It has recently been demonstrated for two species of the fish genus *Notropis* (Delco 1960).

Technological progress in sound recording has greatly stimulated research in this field (Lanyon and Tavolga 1960). The possibility of translating variable sound into visible sound "spectra" permits measurement and precise description. These are now available for nearly all North

American anurans (Blair 1958a; Bogert 1960) and many Australian species (Main, Lee, and Littlejohn 1958). The calls of all species differ in one or several characters from those of sympatric relatives, and observational evidence is accumulating that females move toward conspecific males (for instance, Blair 1958b; Littlejohn and Michaud 1959; Mecham 1961). Yet this does not prevent extensive hybridization between *Bufo woodhousei* and species of the *B. americanus* complex, for instance, which differ considerably in their calls (Chapter 6).

Auditory stimuli are of particular importance among the Orthoptera (Alexander 1960). Nearly every species has a species-specific song. Jacobs (1950) shows that these songs, and the courtship performances into which the songs are integrated, are sometimes a better clue to relationships than are morphological characters. Zippelius (1949) has made a detailed analysis of the function of song in grasshoppers. In the cricket *Gryllus campestris* females search out stridulating males. Females without their hearing (tympanic) organ ignore stridulating males. That additional sense organs play a role in the males is indicated by the fact that males without antennae react toward females as if they were rival males, while blinded males court and copulate normally (Alexander 1962). In the genus *Metrioptera* three sympatric species differ strikingly in their songs. Females do not react to the stridulation of a male of a different species. Receptive females approach males during the stridulating performance but stop as soon as the male interrupts his song. Several cases of sibling species in Orthoptera and Homoptera were discovered because they differed in their song, for example in the genera *Magicicada, Nemobius, Oecanthus,* and *Gryllus* (see Chapter 3). The excellent inventory by Faber (1953) of acoustic and optical signals of Orthoptera is the most complete survey of auditory isolating mechanisms for any group of animals. The importance of sound in the reproductive isolation of two sibling species of cicadas (*Magicicada*) was established by Alexander and Moore (1962).

Chemical Stimuli. Chemical stimuli, which act at a distance ("olfactory") or on contact, often serve as isolating mechanisms. They seem to be particularly important among mammals (except the higher primates) and in some groups of insects. Every naturalist knows that a freshly hatched female moth may attract males for distances of many hundred yards, if not miles. Since the human nose is notoriously insensitive, the study of chemical isolating mechanisms is rather difficult. The experimental approach consists of eliminating visual, auditory, and other nonchemical stimuli and of testing individuals that have been deprived of one sense organ after the other. Amputation of the antennae, for instance, greatly reduces the ability of females of *Drosophila pseudoobscura* and *D. persimilis* to discriminate between their own and alien males (Mayr 1950b). The same operation has little effect on females of *D. melanogaster* and *D. simulans* (Manning 1959a,b), which apparently discriminate through contact chemoreceptors, as do many other species of *Drosophila* (Spieth 1952). Males of the parasitic wasp *Habrobracon* with amputated antennae fail to become excited in the presence of females. However, copulation will proceed normally if such males accidentally touch females (Grosch 1947), and this is also true of some species of roaches. Larger wasps and bees are apparently also attracted to females and are excited by various scents but there is no good evidence yet for species specificity (Kullenberg 1956, 1960). Much work has been done on the chemical nature of sex-attractant scents in insects (Karlson and Butenandt 1959; Jacobson, Beroza, and Jones 1960), but the evidence indicates that the males of other species are frequently attracted as well and that cross-fertilization is prevented by additional isolating mechanisms (Lepidoptera, Götz 1951; Saturniidae, Schneider 1962; Blattidae, Barth 1961). In some roaches (Blattidae) sex stimulation and recognition seem to depend on contact (Roth and Willis 1952, 1954), and this contact chemoreception apparently serves as an isolating mechanism. Individuals are able to distinguish between stimuli received from their own species and those received from others. In certain families of spiders, also, chemical stimulation plays a role (Kaston 1936). Contact stimuli are important in many species groups of *Drosophila*, in which males "tap" the females with their forelegs and the continuation and success of courtship depend on chemical stimuli exchanged during this tapping (Spieth 1952, 1958).

Chemical stimuli are apparently of paramount importance in the spawning of many marine organisms. In the polychaete *Grubea clavata* there is an interesting mutual stimulation. The mature females lose egg protein into the water, which induces some sperm ejection by males. In the sperm liquid is a sex stuff (which can be separated from the spermatozoa by centrifugation) that induces the females to spawn. This, in turn, is followed a minute later by massive sperm ejaculation. The sex stuff is genus specific and probably also species specific (Hauenschild 1951). Such species-specific sex stuffs are probably the most important isolating mechanisms in marine invertebrates with external fertilization, although there are also numerous devices guaranteeing closest contact of conspecific males and females during spawning (Thorson 1950). Internal fertilization, however, is widespread in marine invertebrates, including all the higher prosobranchs, all opisthobranchs and cephalopods, most crustaceans, and some polychaetes (*Capitella*), and places a selective premium on the development of ethological isolating mechanisms.

The Functioning of Ethological Isolating Mechanisms. Analysis of animal courtship by students of behavior (Tinbergen 1951; Spieth 1958) reveals courtship to be an exchange of stimuli between male and female until both have reached a state of physiological readiness in which successful copulation can occur. The male is almost invariably more active in courtship and, in virtually every case, less discriminating. This difference between the sexes is favored by natural selection because a large supply of sperm permits one male (in species without pair formation) to inseminate many females, while the female has a limited supply of eggs and it is of great selective value that they be fertilized in such a manner that zygotes of optimal fitness are produced. The slower a female is in accepting a male, the greater the opportunity for discrimination and the smaller the danger of producing inviable offspring (Richards 1927; Bateman 1948). An extreme case occurs in flies of the genus *Sarcophaga*, in which males mount, almost indiscriminately, individuals (male or female) of their own species, of other species of the genus, and even of other genera gathered on the same source of food. It is apparently entirely the female that determines the success of the mating (Thomas 1950).

The lack of discrimination by males is by no means always as great as in *Sarcophaga*. Males of many species of *Drosophila* display to females of related species only perfunctorily, as was observed by Spieth (1952) for the *Drosophila virilis* group, for instance. Courtship is broken off either at the very beginning or later, depending on the discrepancy of the reciprocal stimulation. In only three, or at best five, of the 30 possible heterogamic crosses between six species of the *Drosophila willistoni* group did the males court the females consistently. In the other 25 to 27 combinations the male broke off the courtship well before reaching the stage of attempting copulation (Spieth 1949). In wild butterflies one can observe particularly often that males court females of other species. Yet these displays almost invariably break off sooner or later, before copulation occurs.

The incompatibilities of courtship can be observed especially well when, in captivity, the male of one species is brought together with the female of a different species, neither having any choice of sex partner. In such cases it can sometimes be observed that the courtship is broken off again and again, in spite of a strong sex drive of both display partners, because one partner (usually the female) does not respond properly to a stimulus given by the other partner. Under natural conditions this would be the end of the courtship.

An isolating mechanism is rarely an all-or-none affair. It has been remarked by several students of *Drosophila* courtship, for instance, that the basic pattern of courtship is the same in closely related species. The differences are quantitative rather than qualitative and yet such quantitative differences are apparently sufficient to prevent the successful synchronization of readiness for mating. The quantitative nature of court-

Table 5–2. Average bout length (seconds) in *Drosophila melanogaster* (from Bastock 1956).

Activities	Yellow males	Wild-type males
High-motivation (vibration and licking)	4.35	5.85
Low-motivation (orientation)	10.35	8.25

ship differences has been recorded also for mallard (*Anas platyrhynchos*) and black duck (*A. rubripes*) (Johnsgard 1960b), and seems to be the rule for many related species (Hinde 1959). It has been particularly well analyzed for *Drosophila*. Males of the mutant yellow (*y*) of *Drosophila melanogaster*, which are strongly discriminated against by females, display at a tempo different from that of normal flies (Table 5–2). The "low-motivation" components of courtship occupy a much larger fraction of the courtship time than in normal males (Bastock 1956). Yellow males have very little mating success in the presence of normal males, but in their absence have high insemination percentages. Bösiger (1960), in a series of investigations, likewise found that among competing males of *Drosophila melanogaster* those with the highest sexual drive succeeded most frequently in inseminating females regardless of their own genetic constitution and that of the females. Heterozygotes between inbred laboratory stocks are as successful as wild-type stocks. Mutants like vermillion and cinnabar, when made otherwise isogenic with wild-type stocks, are as successful as the wild-type males. There is no evidence of homogamy. All this confirms the conclusion that ethological isolation is the result of an interaction between external stimuli and the totality of internal drives. If no appropriate sex partner is available, the internal drive continues to build up until readiness for mating can be induced even by a highly inadequate stimulus. This is the reason for the increased frequency of hybrids near the periphery of the species range, where certain individuals (particularly females) may be unable to find a conspecific sex partner (see Chapter 6).

The relative importance of ethological isolating mechanisms in preventing hybridization is documented by the genera of animals in which hybridization is comparatively frequent. It has been shown by Blair (1941) for *Bufo,* by Hovanitz (1948a) for the butterfly species *Colias philodice* and *eurytheme,* and by Sweadner (1937) for the moth *Platysamia,* that ethological isolating mechanisms are weak in these three genera. The isolation of these closely related species is normally maintained by other mechanisms. These, being less efficient than ethological isolation, break down rather easily, resulting in an unusual frequency of hybrids.

Mechanical Isolation. Soon after the discovery of the manifold structural differences in the genital armatures of different species of insects, it was asserted by Dufour (1844) that these armatures act like lock and key, preventing hybridization between individuals of different species. There is occasional observational evidence to support this claim. Sturtevant, Stalker, and Spieth have shown that interspecific crosses in *Drosophila* may cause injury or even death to the participants and the same is true in *Glossina* (tsetse fly). Other isolating mechanisms presumably prevent such interspecific crosses in nature.

The mechanical barrier formed by incompatibilities of the copulatory apparatus may well be an important isolating mechanism in the pulmonate snails. For instance, Webb (1947) has shown that in the subfamily Polygyrinae of the polygyrid snails the complicated structure of the genitalia prevents interspecific mating. In the closely related subfamily Triodopsinae, the genitalia are of a simpler structure, and in this group interspecific crosses have been reported. In *Cepaea* a combination of ethological and mechanical barriers usually prevents copulation between the species *hortensis* and *nemoralis.*

Mechanical isolation was for a long time considered a most effective isolating mechanism, particularly when it was found how widespread genitalic differences are among various orders of insects. Karl Jordan (1905) concluded, however, on the basis of comprehensive detailed studies, that Dufour's hypothesis is not valid. Among 698 species of Sphingidae that he examined, 48 were not different in their genitalia from other species of the family, while in about 50 percent of the species with geographic variation in color there was also geographic variation in the structure of the genitalic armatures. Since that time much additional information has accumulated indicating the slight importance of the genitalic armatures as isolating mechanisms (see Goldschmidt 1940 and Dobzhansky 1951). For instance, Beheim (1942) showed that in the carabid beetles hybrids can be produced even by species with very different forceps, as, for instance, *Carabus coriaceus* and *C. violaceus.* In the lady beetles of the genus *Hippodamia,* hybrids can be produced easily between species that show pronounced differences in their genitalia (Shull 1946). Rosen and Gordon (1953) came to the same conclusion for poeciliid fishes, Kullenberg (1947) for hemiptera, Kunze (1959) for homoptera, Gering (1953) for agelenid spiders, and Edgren (1953) for the hemipenis in snakes. In some species of insects strong polymorphism in the structure of the genital armatures of the males does not result in incompatibilities.

The most complete experimental study of mechanical isolation was made by Sengün (1944). He removed or altered by operation many parts of the male genital armatures of *Bombyx mori* and found that copulation and fertilization were normal except when part of the penis itself was amputated. He lists crosses between different species, genera, and even families of Lepidoptera that were successful even though the sclerotic structures of the genitalia were sometimes very different. The conclusion of all these studies is that mechanical isolation plays a very minor role as an isolating mechanism in most groups of animals.

The true significance of the differences in the genitalia is presumably the following. The genitalic apparatus is a highly complicated structure, the pleiotropic by-product of very many genes of the species. Any change in the genetic constitution of the species may result in an incidental change in the structure of the genitalia. As internal structures, they are less subject to the corrective influences of natural selection than are components of the external phenotype, provided the basic function of gamete transfer is not impaired.

INTERSPECIFIC CROSSES UNSUCCESSFUL (POSTMATING MECHANISMS)

The ecological and ethological barriers in animals are very efficient and, in most cases, prevent interspecific crosses. If, however, this first set of barriers should fail, a second set of barriers may prevent successful hybridization, although it cannot prevent wastage of gametes. The potential mates complete copulation but no offspring are produced, or the off-spring have reduced viability or fertility. This second group of isolating mechanisms can be classified (in part after Patterson and Stone 1952) into four categories.

Gametic Mortality. The sperm may encounter an antigenic reaction in the genital tract of the female and be immobilized and killed before it has a chance to reach the eggs. Patterson found that an "insemination reaction" occurs in many species of *Drosophila* which leads to an enor-mous swelling of the walls of the vagina and the subsequent killing of the spermatozoa. This insemination reaction is very widespread in the genus

Table 5–3. Number of offspring after a single insemination in *Drosophila* (from Dobzhansky 1947).

Male	Female	
	pseudoobscura ♀	*persimilis* ♀
pseudoobscura ♂	269.5 ± 33.4	99.5 ± 10.6
persimilis ♂	156.3 ± 27.6	235.2 + 24.9

Drosophila (Patterson and Stone 1952:361–383), but it is not known to what extent it occurs in other genera. Antigenic reactions may occur even when there is no evident insemination reaction. Kosswig and Sengün (1945) found that sperm in interspecific crosses was less successful than conspecific sperm in fertilizing female cyprinodonts (*Xiphophorus*). Dobzhansky (1947b) found likewise that crosses between *D. pseudo-obscura* and *D. persimilis* were less fecund than intraspecific crosses. Fe-male *pseudoobscura*, after a single insemination by a male *pseudoobscura*, stay fertile for 12.6 ± 1.9 days; mated with a male *persimilis*, for only 9.2 ± 1.4 days; reciprocally, the figures are 9.7 and 5.4 days. The number of offspring is correspondingly reduced (Table 5–3). In other cases the sperm dies because it cannot penetrate the egg membrane of the alien species.

A very different type of gametic mortality occurs in species with ex-ternal fertilization. A male with nonfunctioning gonads but normal sex behavior (such as a sterile F_1 hybrid) may induce a female to spawn. Her (unfertilized) eggs will, of course, die. As Volpe (1960) has made plausible for the toad species *Bufo fowleri* and *B. valliceps*, this mortality will eventually lead to an increase in the reproductive isolation of the two species (see also Chapter 17).

Zygotic Mortality. The development of the fertilized hybrid egg is often irregular, and development may cease at any stage between fertilization and adulthood. The details and causes of this incompatibility are thoroughly described in the textbooks of embryology, cytology, and genetics (for instance, Dobzhansky 1937, 1951).

Hybrid Inferiority. Many naturally occurring animal hybrids have been observed to leave no offspring, even though they seem to have somatic hybrid vigor and are fully fertile (normal eggs or spermatozoa). The explanation seems to be that ecologically and in their behavior they are intermediate between the two parental species and are therefore discriminated against during pair formation. Hybrid inferiority has been very little studied (it is difficult to study!), but may well be the reason for the small amount of introgression in groups (like the ducks) in which hybrids often are fully fertile.

Hybrid Sterility. Species hybrids have considerable or complete fertility in some groups of animals, but are more or less sterile in others. The cytological and genetic reasons for this are excellently discussed by White (1954) and Dobzhansky (1951). There is a detailed survey of hybrid sterility in *Drosophila* (Patterson and Stone 1952), but few investigators have systematically crossed all the available species of a genus and determined their cross fertility. This has been done for various families of birds (Gray 1958) and for the American toads *Bufo* (Blair 1959). Certain other aspects of hybridization are discussed in Chapters 6 and 13.

THE COACTION OF ISOLATING MECHANISMS

The interbreeding of closely related sympatric species of animals is usually prevented by a whole series of ecological, behavioral, and cytogenetic factors, usually several for each species pair. One factor is often dominant, such as acoustic stimuli in certain grasshoppers, cicadas, and mosquitoes, chemical contact stimuli in some insects, sex stuffs in Lepidoptera and certain marine organisms, and visual stimuli in certain birds, fishes, and insects. The sterility barrier may be strong or weak, but it is rarely tested except when the other isolating mechanisms break down.

That each of the categories of Table 5–1 is, furthermore, composite is evident in the case of ethological isolation; it has been shown also for sterility factors, and is indicated for habitat and temporal isolation. As a consequence, the total isolation between two species is normally due to a great multitude of different isolating factors, each more or less independent of the others. If we were to rank the various isolating mechanisms of animals according to their importance we would have to place behavioral isolation far ahead of all others. In plants, on the other hand, sterility is the most important factor. The ability of animals to search actively for potential mates is undoubtedly the reason for this striking difference between the two kingdoms.

In only a few cases has an attempt been made to analyze all the factors that isolate two species from each other. The most detailed information available is for the genus *Drosophila* and is summarized by Dobzhansky (1951), Patterson and Stone (1952), Spieth (1952, 1958), Bastock and Manning (1955), Pittendrigh (1958), and Manning (1959a,b). Each analysis has added some new factors previously overlooked, as we have described for *Drosophila pseudoobscura* and *D. persimilis* (Chapter 3). The relative importance of the various factors changes from one species group to another. Most of the differences are quantitative rather than qualitative. Petersen and Tenow (1954) analyzed the isolation between two species of butterflies of the genus *Pieris* and found that differences in habitat selection and considerable hybrid inviability contributed most of the isolation. Brower (1959b) analyzed the relative importance of various isolating mechanisms in butterflies of the *Papilio glaucus* complex. Partial seasonal and altitudinal isolation seems to have evolved between the two species with the weakest ethological barriers. Fabricius (1950) described the complex constellation of internal and external factors necessary to induce spawning in fresh-water fish. Such a complexity greatly reduces the probability of two species coming into spawning condition exactly at the same time.

The isolating mechanisms are arranged like a series of hurdles; if one breaks down, another must be overcome. If the habitat barrier is broken, for example, individuals of the two species may still be separated from each other by behavior patterns. If these also fail, the mates may be unable to produce viable hybrids, or if hybrids are produced they will be sterile. A clear demonstration of this principle is provided by Smith (1954), who showed that, when the usual seasonal barrier between the spruce budworm and the pine budworm (*Choristoneura*) broke down, hybridization still did not occur, owing to sexual isolation. Cases of a breakdown of all isolating mechanisms are discussed in Chapter 6.

THE GENETICS OF ISOLATING MECHANISMS

From the multiplicity of isolating mechanisms by which every species is protected, one can conclude that a considerable number of genes are involved. Almost any gene that changes the adaptation of a population may have an incidental effect on the interaction between male and female. Conversely, a mating advantage of a male will become established in a population provided it does not seriously lower the fitness of his offspring. Elaborate plumes and other types of striking sexual dimorphism have evolved in birds usually only in species in which the male does not participate in the raising of the young, and consequently does not endanger his brood by his conspicuous presence.

The genetic basis of behavior has been reviewed by Caspari (1958). Detailed analyses of the genetics of isolating mechanisms other than sterility exist only for the genus *Drosophila*. Tan (1946) found that numerous factors influence the ethological isolation between *Drosophila pseudoobscura* and *D. persimilis,* but the main factors appear to be located on the X chromosome and the second chromosome. That sexual isolation is a product of the entire genotype is demonstrated particularly well by experiments on the relative discrimination of females of *Drosophila melanogaster* against males of certain mutant genotypes, such as yellow (Bastock 1956), white and bar (Petit 1958, 1959), or other mutants (Bösiger 1958). Constant levels of sexual isolation changed when the mutants were outcrossed to different genetic backgrounds, or when new genotypes originated in the stocks owing to mutation or recombination. A complete analysis of the genetic basis of isolating mechanisms would seem rather futile in view of this complex basis. Indeed if, as it appears, behavioral isolation is the result of differences in activity rhythms, levels of activity, and other generalized factors that influence the coordination of the drives of males and females, it would be very difficult to express the isolation in terms of specific genes. It might be remarked, incidentally, that selection continuously adjusts the mating propensity of females to the prevailing level of the mating drive of the males. In *Drosophila americana,* for instance, Spieth (1951) found that "eastern" males had a high sexual drive and the females low readiness to mate, "western" males a low sexual drive and females a correspondingly high readiness to mate. Mixing eastern males and western females resulted in a much higher percentage of copulation (55.8 percent) and mixing western males and eastern females a much lower percentage of copulation (21.5 percent) than intrastrain eastern (30.7 percent) or western (50.8 percent) crosses.

There is evidence that the genetic isolating mechanisms are occasionally modified or reinforced by conditioning. Young birds can sometimes be imprinted on a foster species when raised by foster parents.

There is no danger that such conditioning will lead to hybridization because in nature young birds are invariably raised by parents of their own species. In parasitic birds—cuckoos and cowbirds, for example—such conditioning is absent, and species recognition is rigid. Young cowbirds, for instance, leave the company of their foster parents and flock together as soon as they are independent. Conditioning, however, is an important isolating mechanism in the human species. The free interbreeding of individuals coexisting in a geographical region is strongly influenced by religious, economic, and cultural barriers. The origin of isolating mechanisms will be discussed in Chapter 17.

THE ROLE OF ISOLATING MECHANISMS

It is one of the unquestioned functions of the isolating mechanisms to increase the efficiency of mating. Where no other closely related species occur, all courtship signals can "afford" to be general, nonspecific, and variable. Where other related species coexist, however, nonspecificity of signals may lead to wasteful courtship and delays, even where no heterospecific hybridization occurs. Under these circumstances there will be a selective premium on precision and distinctiveness of signals.

This is perhaps best documented by situations in which a single species moves into an isolated area and is no longer exposed to stabilizing selection for precision and distinctiveness. As a result, the high specificity of the isolating mechanisms may be lost. Such a loss has been recorded not only for visual characters (Mayr 1942) but also for the song of birds on islands, for instance *Parus* on Tenerife and *Regulus* on the Azores (Marler 1957).

Each species is a delicately integrated genetic system that has been selected through many generations to fit into a definite niche in its environment. Hybridization would lead to a breakdown of this system and would result in the production of disharmonious types. It is the function of the isolating mechanisms to prevent such a breakdown and to protect the integrity of the genetic system of species. Any attribute of a species that would favor the production of hybrids is selected against, since it results in wastage of gametes. Such selection maintains the efficiency of the isolating mechanisms and may indeed add to their perfection. Isolating mechanisms are among the most important biological properties of species.

6 ~ The Breakdown of Isolating Mechanisms (Hybridization)

Not all isolating mechanisms are perfect at all times. Occasionally they break down and permit the crossing of individuals that differ from each other genetically and taxonomically. Such interbreeding is called *hybridization*. This term is difficult to define precisely and has therefore been applied to very different phenomena. Lotsy, following De Vries, designated as hybridization any cross-mating of genetically different individuals. Since nearly all individuals in sexually reproducing species, such as man, are genetically different, all members of such species would be hybrids according to this definition. Others have defined a hybrid as "a zygote produced by the union of dissimilar gametes," which shifts the difficulty of definition to the word "dissimilar."

In many cases there is no doubt of the propriety of using the term "hybrid." Individuals that are the offspring of a cross between two good species are unquestionably hybrids. Both the concept and the term "hybrid" were taken from the realm of animal and plant breeders. They referred originally to the crossing of two unlike individuals, usually members of two different species. Difficulties arise when one attempts to extend this typological concept from individuals to populations. Is it legitimate, for instance, to refer to the interbreeding of conspecific populations as hybridization? If so, under what circumstances? In most cases it would seem highly misleading to apply the term hybridization to ordinary gene exchange among conspecific populations. Yet when there is a secondary contact between previously long-isolated populations, the term "hybridization" is sometimes appropriate. Perhaps hybridization may be defined most conveniently as "the crossing of individuals belonging to two unlike natural populations that have secondarily come into contact." The disadvantage of a subjective determination of "unlike" in this definition is counterbalanced by the emphasis on populations and the avoidance of circular reasoning. The definition of hybridization as "the crossing of individuals belonging to two different species" results in circular argument because the decision whether or not to include two populations in the same or in two different species may depend on the occurrence of hybridization.

At this point it is necessary to discuss one more matter of terminology. Anderson and Hubricht (1938) coined the convenient word *introgression* for the incorporation of genes of one species into the gene complex of another species as a result of successful hybridization. This term is especially useful as a contrast to the term "gene exchange." Gene exchange results from gene flow among different populations of a single species. It would lead to confusion and imprecision to use the terms "introgression" and "gene exchange" synonymously. To be sure, the phenomena designated by these terms merge into each other in the cases of gene exchange between allopatric populations that are on the borderline between subspecies and species level. Yet these intermediate situations do not seriously undermine the usefulness of the terminological distinction. One cannot discuss profitably the evolutionary importance of introgression if entirely different phenomena are included under this one term. Those who consider introgression and gene exchange synonymous would do better to adopt the older and more familiar term, "gene exchange."

GENETIC ASPECTS OF HYBRIDIZATION

In treatises on hybridization the emphasis is placed primarily on hybrid sterility and its causes, secondarily on other genetic aspects of hybridization. For a detailed discussion of such genetic and cytological aspects of hybridization we refer to the authoritative treatments by Dobzhansky (1951), Stebbins (1950), and White (1954). Only a few facts will be summarized here.

Hybrids differ from individuals of the parental species not only in morphology but usually also in fertility and viability. The inability of most hybrids to produce the normal number of viable gametes is called "hybrid sterility." It may range from slight to complete and may be caused by a variety of genetic factors. Such a reduction in fertility is not necessarily correlated with a reduction in viability. Indeed, hybrids often show a marked phenotypic "luxuriance," the so-called "hybrid vigor." The hybrid vigor of the mule, sterile product of the cross of horse and donkey, is proverbial, yet it does *not* constitute Darwinian fitness (Dobzhansky 1955a).

The lack of correlation between fertility and viability works both ways. Fully fertile hybrids are known that nevertheless fail to reproduce in nature. They may be less well adjusted to existing ecological niches, and, where definite behavior patterns and species-specific stimuli play an important role, the hybrids are usually less successful in courtship than individuals of pure species. Ecological and ethological inferiority reduces their chances of leaving offspring. Backcross individuals are likely to be even more strongly inferior owing to various imbalances of their gene complexes. Incompatibility of the genes of the parental species may lead (even in the F_1) to severe or lethal physiological disturbances. A well-known case is that of the melanotic tumors that may arise in certain crosses between the fish species *Xiphophorus maculatus* and *X. helleri* (Gordon 1937). Similar incompatibilities of the parental genomes have been described for species crosses in a number of animal and plant genera.

First-generation (F_1) hybrids are generally intermediate between the parental species and tend to be uniform in most characters. Occasionally a character crops up in the hybrid that does not occur in either parental species. The F_2, when it occurs in nature, generally shows increased variability. Yet, for various reasons, only some of the possible recombinations of the parental characters are found (Anderson's "recombination spindle"; Dempster 1949). Chromosome doubling in a hybrid may lead to allopolyploidy, a mode of species formation important in plants (Stebbins 1950) but rare in animals (Chapter 15).

Artificial Hybrids. It is well known that many species can be crossed in captivity, but do not produce hybrids where their ranges overlap in nature. For instance, many sympatric species of *Drosophila* can be hybridized in the laboratory (Patterson and Stone 1952). The same is true of birds and many species of fishes (for instance, in the genus *Xiphophorus*). The production of such hybrids has been cited as the basis of two kinds of wrong conclusions: first, that hybridization is frequent in the animal kingdom; and second, that the possibility of such hybridization indicates conspecificity, particularly when the hybrids produced are fertile. It must be emphasized once more (see Chapter 5), that among animals sterility is only one of many isolating mechanisms, and that other mechanisms are much more important in maintaining the distinctness of species. The mere possibility of hybridization in captivity proves nothing as far as species status is concerned.

HYBRIDIZATION AS A POPULATION PHENOMENON

The present treatment will emphasize the natural history and particularly the population aspects of hybridization, a subject in which the literature is enormous and widely scattered. No attempt will be made to present an exhaustive survey, but all the major problems will be touched upon.

When hybridization is considered as a population phenomenon, various aspects ignored by the breeders become important. Does a given case of hybridization involve otherwise valid species, or allopatric populations of the same species? Does it consist of the occasional production of a hybrid individual, or is it a massive phenomenon resulting in a more or less complete breakdown of the barrier between species? These various aspects of hybridization differ in their evolutionary significance and it would be misleading to confuse them.

On the basis of the stated criteria, five kinds of hybridization might be distinguished, as far as the naturalist and taxonomist are concerned:

(*a*) The occasional crossing of sympatric species resulting in the production of hybrid individuals that are ecologically or behaviorally inviable or sterile and therefore do not backcross with the parental species.

(*b*) The occasional or frequent production of more or less fertile hybrids between sympatric species, some of which backcross to one or both of the parental species.

(*c*) The formation of a secondary zone of contact and of partial interbreeding between two formerly isolated populations that failed to acquire complete reproductive isolation during the preceding period of geographic isolation. (Cases in which isolating mechanisms had not yet been acquired at the time of the secondary contact will not be discussed in this chapter. Here the emphasis is on the breakdown of isolating mechanisms. The secondary hybrid belts that develop when geographic isolates reestablish contact with the parental species will be treated in Chapter 13.)

(*d*) The complete local breakdown of reproductive isolation between two sympatric species, resulting in the production of hybrid swarms that may include the total range of variability of the parental species.

(*e*) The production of a new specific entity as the result of hybridization and subsequent doubling of the chromosomes (allopolyploidy) (virtually restricted to plants).

The first four categories (*a–d*) grade into each other, and it is sometimes difficult to decide where a given case should be listed. Cases of hybridization involving "good species" are here arbitrarily included under (*a*) and (*b*) above, even when the distribution is largely allopatric. They do behave like sympatric species in the area of overlap.

In connection with a systematic survey of the occurrence of these five clases of hybridization in nature, it will be possible to consider some more general questions, such as the following:

(1) How common is hybridization in nature among different groups of animals?

(2) Is there a striking difference in the incidence of hybridization between plants and higher animals and, if so, why?

(3) What effect does hybridization have on species structure and intraspecific variability?

(4) What is the role of hybridization in speciation and evolution?

Occasional Hybridization

Hybridization is common among plants. In most groups of animals it is sufficiently exceptional to justify a report in the literature whenever a hybrid is discovered. The suggestion is sometimes made that this rarity of animal hybrids is more apparent than real, but, as we shall see below, the frequency of hybrids is as low in groups where hybrids can be discovered easily (by cytological or other methods) as where the recognition of hybrids is difficult. This is confirmed by numerical estimates that can be made in some groups. On the basis of my examination of random collections, I estimate that perhaps 1 out of 60,000 wild birds is a hybrid. Hall (1943) found 6 hybrids in 100,000 randomly collected mammal specimens, but his ratio of hybrids is essentially the same as in birds since only two pairs of species were involved. Suchetet (1897) has given the most complete summary of the earlier literature on hybridization in birds, while Gray (1958) and Sibley (1961) have summarized the more recent work. Johnsgard (1960a) analyzes the occurrence of hybridization among the Anatidae. Cockrum (1952) has recorded some 75 hybrid crosses among North American birds, of which 59 are cases of occasional hybridizations of good sympatric species (mostly in the duck family), 11 are crosses between subspecies or allopatric semispecies, 2 are cases of polymorphism, and only 3 are indicative of introgression. These 3 cases involve less than 1 percent of the species of North American birds. Additional cases of occasional hybridization are described annually, lately for the genera *Calcarius, Spizella,* and *Dendrocopos.* Contrary to the assertions of earlier authors, hybridization is exceptional even in the case of species morphologically so similar as Eastern and Western Meadowlark

(*Sturnella;* Lanyon 1957). Although there are many closely related sympatric species of mammals, such as those among the squirrels (*Sciurus*), only very few genuine hybrids have ever been reported. Even species that can be hybridized easily in captivity do not produce hybrids in the wild where their ranges overlap.

Hybrids appear to be even rarer among reptiles, although Bailey (1942) reported a natural hybrid between two different genera of rattlesnakes, *Crotalus horridus* and *Sistrurus catenatus*. Mertens (1950) discusses 40 reports of reptile hybrids. Of these most were produced in captivity, several were based on misidentifications, and others were intergrades from the border belt between subspecies. Only about half a dozen were good species hybrids and, except for a few cases in *Thamnophis*, I am unaware of any indications of backcrossing.

In contrast to the evident rarity of hybridization among reptiles is its comparative frequency in the amphibians. Here its significance seems to change from genus to genus. Hybridization is rare in true frogs of the genus *Rana*, but widespread in toads of the genus *Bufo* (see below). Where *Bufo woodhousei* and *B. valliceps* coexist in southeast Texas, intermediate specimens are occasionally found that appear to be hybrids. An analysis of the hybrids and of artificial backcrosses indicates that cross-sterility is sufficiently great to make introgression in nature highly improbable (Thornton 1955). The same situation exists for *B. fowleri* and *B. valliceps* (Volpe 1960). Fertile hybrids between *Hyla cinerea* and *H. gratiosa* are not uncommon in the region of overlap. For a further discussion of various aspects of anuran hybrid viability see Moore (1955). Hill (1954) found evidence for occasional hybridization between two species of *Amphiuma* in Louisiana, and occasional hybrids (*"Triturus blasii"*) are produced in France where the ranges of *Triturus cristatus* and *T. marmoratus* overlap (Vallée 1959).

The situation in fishes is in strong contrast to that in nearly all land vertebrates. Fertilization is usually external and for this reason (among others) hybridization is frequent. Excellent reviews by Hubbs (1955, 1961) summarize the previous literature and record the known data, particularly for the fresh-water fishes of North America (mostly based on the work of Hubbs and his students). Occasional or extensive hybridization was found among lampreys (Petromyzontidae), chars, trout, and salmon (Salmonidae), whitefish (Coregonidae), Cyprinidae, catfish (Ameiuridae), pike (Esocidae), goodeid killifishes (Goodeidae), livebearers (Poeciliidae), silversides (Atherinidae), perch (Percidae), sunfishes (Centrarchidae), and a few other families. In most cases only sterile F_1 individuals were found and there is little or no evidence of backcrossing with the parental species. For instance, where the coastal Californian chub genus *Gila* was introduced into the range of *Siphateles* in the Mohave River, 442 hybrid specimens were found in a total collection of 5604 specimens, the rest being clear-cut representatives of the parental species. The characters of a few specimens indicated the possibility of introgression, yet there was no blurring of the species border. Similar observations were made in the family of suckers (Catostomidae; Hubbs *et al.* 1943). Among 2000 *Catostomus commersoni* and *C. catostomus* caught in the Platte River there were 5 hybrids (0.25 percent); among 2379 specimens of *C. macrocheilus* and *C. syncheilus* caught in the Columbia River system in Washington and Oregon there were 69 hybrids (3 percent). Every collected specimen was either clearly an F_1 hybrid or a specimen of the parental species. The same is true of hybrid specimens found among marine fishes. There are, however, a few cases of evident introgression in fishes which will be discussed below.

The frequency and distribution of hybridization in other groups of animals are difficult to determine because the information in the taxonomic literature is exceedingly scattered, and the primitive condition of the taxonomy of most groups of animals does not permit comprehensive reviews by specialists. The general impression one gets from a perusal of the literature is that hybrids are rare. Swan (1953) considers one in several thousand sea urchins of the genus *Strongylocentrotus* at Friday Harbor (Pacific coast) to be a hybrid. He found individuals indicating the following hybrid pairings: *S. purpuratus* × *S. droebachiensis; S. purpuratus* × *S. franciscanus; S. droebachiensis* × *S. franciscanus.* Those among the presumed hybrids that were carefully examined had small gonads and did not spawn. Although this does not prove sterility, the fact that these sea urchins are "old" (Tertiary) and widespread species indicates that such occasional hybridization does not necessarily lead to visible introgression. Sympatric species as close as these three species of *Strongylocentrotus* are rare among the sea urchins.

Good quantitative data on the frequency of hybrids among invertebrates are available for some groups of Lepidoptera and for *Drosophila* (see below).

Introgression in Animals

The incorporation of genes of one species into the gene complex of another species, introgressive hybridization, occurs frequently in the plant kingdom (Anderson 1949). Among animals rather few cases have been found so far, primarily because only a small fraction of hybrids will backcross to either of the parental species. The better-known cases are the following. Where the Golden-wing Warbler (*Vermivora chrysoptera*) and the Blue-wing Warbler (*V. pinus*) meet in eastern North America, hybrids are not uncommon. They are distinctive in color and song ("Brewster's Warbler," "Lawrence's Warbler") and numerous backcrosses have been studied by field ornithologists (Parkes 1951). This hybridization has been known for a century and presumably began some 200 years ago when the natural habitat barrier between the species was obliterated by deforestation and farming. Yet the delimitation of the two parental species is still quite sharp in most areas. There is no evidence of a blurring of the species border except in the zone of overlap. In another case, the so-called "Potomac Warbler," an isolated population of the Yellow-throated Warbler (*Dendroica dominica*) in West Virginia has introgressed with the Parula Warbler (*Parula americana*). Pair formation was apparently facilitated by similarity in the nesting behavior of the two parental species.

Prairie Chickens (*Tympanuchus cupido*) and Sharp-tailed Grouse (*T. phasianellus*) hybridize frequently in Wisconsin and Ontario. Hybrids seem fully fertile and form 5–25 percent of many populations, showing every degree of intermediacy between the parental species (Lumsden MS). Yet hybrid males seem to be less successful in their courtship than are males of the pure species (Hamerstrom MS).

Two yellow butterflies in North America, *Colias philodice* and *C. eurytheme*, hybridize regularly (Hovanitz 1948a, 1953; Remington 1954). This breakdown of reproductive isolation is in part caused by the spread of *eurytheme* over wide areas previously occupied only by *philodice*, following the increased planting of alfalfa, its food plant. Where the overlap is recent, as in New Hampshire, only 1.0–1.5 percent of the population are hybrids (Gerould 1946). Elsewhere the percentage of hybrids seems to have risen to a level of about 10–12 percent, where it has stayed ever since this hybridization was first observed nearly 50 years ago. This stabilization is somewhat surprising since the mating between the two species is reported to be almost random, and the F_1 not only are fertile but have been shown to backcross to some extent with the two parental species. Presumably the hybrids are sufficiently inferior in general viability to prevent the development of a genuine hybrid population. Hybrid individuals seem to be more common in some other species crosses in butterflies (Hovanitz 1949). Occasional hybridization has been reported in snails, but gastropod systematics is still far from having reached a level of maturity that would permit determining how often introgression is involved.

In the majority of cases where introgressive hybridization in animals has been reported, two species are involved that had been conspecific until recently and are still largely allopatric. They are *semispecies* (in the emended definition of Lorković 1953), showing some of the characteristics of species and some of subspecies. The fact that they hybridize to a greater or lesser extent proves that they had not acquired complete reproductive isolation during their geographic isolation. Such cases have been described for the hedgehogs *Erinaceus europaeus* and *E. roumanicus* in Europe (Herter 1934), for the white-footed mice *Peromyscus leucopus* and *P. gossypinus* in the southern United States (McCarley 1954), for the buntings *Passerina cyanea* and *P. lazuli* (Sibley and Short 1959), for the frogs *Microhyla carolinensis* and *M. olivacea* in Oklahoma (Hecht and Metalas 1946; Blair 1955), for swallowtail butterflies of the *Papilio glaucus* group in British Columbia (Brower 1959a), and for the cabbage whites *Pieris napi* and *P. bryoniae* in the southern and eastern Alps (Petersen 1955).

These cases differ from others recorded in Chapter 13 in that the integrity of the species is in no case broken down completely, even though there may be a high percentage of hybrids in the area of overlap or even a complete breakdown of isolation locally. The zone of hybridization may be either localized (*Peromyscus, Microhyla*) or widespread (*Erinaceus, Pieris*). There is definite opportunity for introgression in these cases, and backcrossing with the parental species has indeed been recorded in all the stated cases. The separation of these cases from those in the next section and those recorded in Chapter 13 is somewhat arbitrary.

Sympatric Hybrid Swarms

The barrier between two sympatric species sometimes breaks down so completely, locally or over wide areas, that the two parental species are replaced by a hybrid swarm that serves as a continuous bridge between the two parental extremes. A thorough knowledge of the taxonomy of the respective groups is a prerequisite for a sound analysis of such situations. This is presumably the reason why so few such cases have so far been described, and none among mammals. Groups that are easily observed, like birds, or easily caught, like butterflies and fishes, supply the best-substantiated instances. In view of their great evolutionary interest, some of these cases will be described in detail.

Birds. In southern Europe and western Asia there are two closely related and widespread species of sparrows, the House Sparrow (*Passer domesticus*) and the Willow Sparrow (*Passer hispaniolensis*). In most areas the two species coexist side by side without any signs of interbreeding (Fig. 6–1). In such areas the House Sparrow is associated with human habitations while the Willow Sparrow lives in willow groves or other kinds of woods in river bottoms. Such sharp separation occurs in Spain, Morocco, the Balkans, Asia Minor, Iran, and Turkestan. However, in a few areas the barrier between the two species has broken down and more or less unrestricted hybridization between them is taking place (Meise 1936b). One of these areas is Tunisia, another is Italy and adjacent islands (Sicily, Corsica), a third is Crete. Conditions are different from place to place. Sometimes one species prevails, while the other is rare and occasional hybrids appear. More frequently the hybrid population includes phenotypically the two parental species and all conceivable combinations of the parental characters with indication of complete random mating. Finally, there are some areas, like Italy and the oases of southern Tunisia, where an intermediate hybrid type has become stabilized and the parental extremes have disappeared. A clue to the origin of the hybridization is provided by the fact that the Willow Sparrow occupies the ecological niche of the House Sparrow (human habitations!) in certain areas, such as Sardinia, eastern Tunisia, and Cyrenaica. Where the aggressive House Sparrow invades such areas, the conditions are present, in the absence of ecological separation, for a breakdown of reproductive isolation.

Chapin (1948) describes a particularly interesting case from tropical Africa. There are three African species of Paradise Flycatchers (*Terpsiphone*) of which two, *rufiventer* and *rufocinerea*, live in the rain forest, while the third, *viridis*, lives in second-growth woods and in savanna forest. In most areas where these species come into contact with each other they live side by side without any signs of intergradation or hybridization, each one restricted to its own habitat. However, along the edge of the African rain forest there are many areas where parts of the forest have been partially or completely cleared in recent years, and in such areas *viridis* interbreeds with the two forest species, particularly with *rufiventer*. As a result there are now three areas with hybrid populations: one in northwestern Angola (*T. rufocinerea rufocinerea* × *T. viridis*

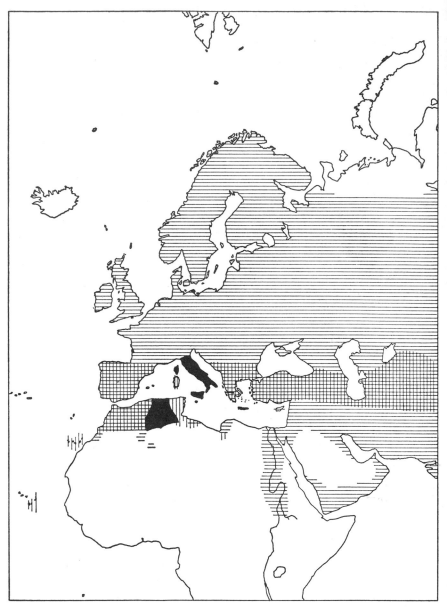

Fig. 6–1. Largely sympatric distribution of the House Sparrow, *Passer domesticus* (horizontal hatching) and the Willow Sparrow, *P. hispaniolensis* (vertical hatching). Hybridization and introgression (black) in various Mediterranean and North African areas. (After Meise 1936b.)

plumbeiceps); one in West Africa (*T. rufiventer nigriceps* × *T. v. viridis*); and one in Uganda (*T. rufiventer somereni* × *T. viridis ferreti*). In each case the hybrid population has settled down to a reasonable constancy, so that the new stabilized hybrid populations were at first considered separate species or subspecies: *bannermani* in Angola, nominate *rufiventer* at the Gambia River, and *emini, poliothorax,* and *albiventris* in Uganda. Where much of the original rain forest is left, hybrids are sporadic; where it has been destroyed and the remaining stands have been invaded by *T. viridis,* complete hybrid populations have evolved. The available evidence suggests to Chapin that the hybridization is quite recent, being in all cases due to the clearing of the forest by the African natives.

Perhaps the most thoroughly analyzed case of the breakdown of isolation between two species of birds is that of two members of the genus *Pipilo* in Mexico (Sibley 1950, 1954a; Sibley and West 1958). The Red-eyed Towhee (*P. erythrophthalmus*) and the Collared Towhee (*P. ocai*) are more or less widespread as "pure" species (Fig. 6–2). *Pipilo ocai* occurs from Oaxaca to Jalisco. *Pipilo erythrophthalmus* is widespread in North America and extends south as far as Chiapas and Guatemala. In Oaxaca the two species live side by side without intermixing. In Puebla 16 percent of the 117 known specimens show evidence of hybridization. In the other states of the Mexican plateau from northern Puebla through Nayarit and Michoacan to Jalisco a series of introgressed hybrid populations is found, which in the east and north are similar to *P. erythrophthalmus* and toward the south and west are similar to *P. ocai.* If a hybrid index is designed which gives pure *erythrophthalmus* the value of 24 and pure *ocai* the value 0, an east-west chain of populations is found with the mean values 22.4–19.8–16.9–15.8–13.5–7.8–4.0, and a north-south chain with the values 23.5–22.8–22.6–13.7–8.0–2.8–0.17. The variation within a local population is great but does not span the total range. In a population with a mean index of 13.7 it varied from 6 to 20 in 76 specimens; in another with a mean of 8.0 it varied from 3 to 16 among 58 specimens. If it were not for the pronounced differences between the species and their sympatry in Oaxaca, one might be tempted to consider them conspecific. Sibley's original papers must be consulted for many other interesting aspects of this hybridization. Again, it is apparently a very recent event, caused by man's agricultural activities, and not dating back further than 300–500 years.

A parallel, and in many ways very similar, situation exists among the honey eaters of New Guinea (Gilliard 1959). The gray-billed *Melidectes*

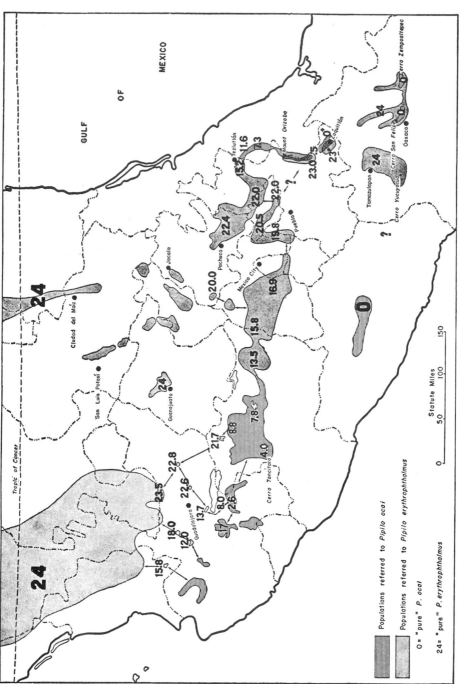

Fig. 6–2. Distribution of the Red-eyed Towhees (*Pipilo*) in Mexico. Pure *erythrophthalmus* (24) in the north and southeast. Pure *ocai* (0) in the south and southeast. The numbers (from 0 to 24) indicate the mean character indices of various hybrid populations. Note the sympatry of the two species at several localities in the southeast. (From Sibley 1954a.)

rufocrissalis and the black-billed *M. belfordi* differ in numerous characters from each other and occur as pure, unmixed, allopatric species on various mountains of New Guinea. Wherever they come into contact, however, they form either highly variable hybrid swarms (Wahgi Mountains, Mount Hagen, Mount Wilhelm, and so on) or stabilized hybrid races (Mount Goliath, Herzog Mountains). Again, it seems that this hybridization is very recent, caused by the ecological disturbance resulting from the destruction of the mountain forest by man. Another case in which two species are well defined in parts of their range but freely interbreed in other parts is that of the bulbuls *Criniger calurus* and *C. udussumensis* (Rand 1958). In the case of the Malayan Kingfishers *Ceyx erithacus* and *C. rufidorsus* (Ripley 1942; Voous 1951) the hybridization is now so extensive that Sims (1959) considers them to have merged into a single species.

An especially interesting case in the genus *Bufo* was analyzed in detail by Blair (1941). Fowler's Toad (*B. fowleri*) and the American Toad (*B. americanus*) interbreed extensively in many areas of their nearly completely coincident geographic ranges. The most interesting aspect of this hybridization is that the toads at the beginning of the breeding season (April) are almost pure *americanus* and at the end of the breeding season (June) almost pure *fowleri*. However, the samples captured at the middle of the breeding season (May) bridge the two species. A study of rate of development and temperature tolerance of the embryos (Volpe 1952) has confirmed the hybrid nature of some populations. The amount of hybridizing is different from locality to locality, and there is much evidence that it is caused by the breakdown of the ecological barriers between the species through agricultural activities. *Bufo americanus* is primarily a toad of wooded areas, *B. fowleri* of grasslands, fields, and other open areas (Corey and Manion 1955). Interestingly there are apparently "pure" populations of *americanus* and *fowleri* at localities near the hybrid areas. Various other species of *Bufo* show evidence of introgressive hybridization. For instance, where the ranges of *B. microscaphus* and *B. woodhousei* overlap in the watershed of the Virgin River (southwestern Utah), there were 99 hybrids in addition to 376 *microscaphus* and 196 *woodhousei*. The gonads of the hybrids were of normal size and there was some evidence for introgression in the parental species populations (Blair 1955). The same was found by Volpe (1959) for the cross *B. terrestris* and *B. fowleri*.

Two species of salamanders, *Batrachoseps attenuatus* and *B. pacificus*, still coexist without mixing at one or two localities in southern California and on Santa Cruz Island, but form a single hybrid population wherever else in southern California they have come in contact (Hendrickson 1954).

Fishes. In spite of the high frequency of hybridization in fishes, particularly fresh-water fishes, the situation is only rarely auspicious for the occurrence of introgression. Hubbs (1955, 1961), summarizes the relevant facts. Although sunfishes (Centrarchidae) hybridize commonly, only three places have been found so far where the characters of the especially abundant hybrids grade into those of each parental species. These three hybrid swarms are further exceptional in that they do not consist predominantly of males. Another case occurs in the cyprinid fishes. In a stretch of the San Juan River in California two sympatric species of the genera *Hesperoleucus* and *Lavinia* have crossed to the extent of forming a hybrid swarm, with the borders between the species breaking down completely locally. Hybrids between *Notropis rubella* and *N. cornuta* have a normal sex ratio and seem to bridge the gap between the two species completely, thereby providing opportunity for introgression. The parentage of the hybrid cyprinodont *Anatolichthys* in Asia Minor is still uncertain (Villwock 1958). Hybridization seems widespread in the genus *Coregonus*.

Invertebrates. If introgressing hybrid swarms are not more frequently reported among insects and other invertebrates, it is because taxonomic analysis has rarely reached the point at which hybridity of a variable or intermediate population can be proved unequivocally. Sailer (1954) reports a probable hybrid swarm between several species of stink bugs (*Chlorochroa*), and Pejler (1956), apparent introgression in several genera of rotifers. Hybridization seems widespread in *Daphnia*. Brooks (1957b) ascribes the difficulty of identifying many populations of the *D. pulex* group to the fact that they are introgressing hybrid populations between *D. middendorfiana, D. schoedleri,* and *D. pulex*. In many areas, however, pure populations of these species can be found.

It must be emphasized that not all so-called "hybrids" recorded in the literature are really the product of hybridization. Many species are exceedingly variable and extreme types may deviate very far from the normal appearance of the species. In fact, such extreme types may often be quite similar to closely related species, a phenomenon that is perhaps due to the inherent potentialities within a genus so often commented upon by naturalists. For instance, Alberti (1954) showed that the reputed frequency of hybridization in European species of *Colias* could not be confirmed but was simply a misinterpretation of natural variability. For similar studies in *Etheostoma* (Stone 1947) see Chapter 3. Cytological analysis proved (White and Key 1957) that all the individuals of a population of grasshoppers (*Austroicetes*) superficially appearing to be a hybrid swarm clearly belonged to one or the other of two well-defined species.

CAUSES OF THE BREAKDOWN OF ISOLATING MECHANISMS

A study of the occurrence of natural hybridization shows that it is not a random phenomenon. There are certain factors that clearly facilitate the breakdown of isolating mechanisms. We shall endeavor to point out some of these factors.

Methods of Fertilization

Internal fertilization is normally preceded by a more or less extended courtship between the potential mates, and, as described in Chapter 5, this normally prevents interbreeding between individuals that are not conspecific. Ethological factors have only limited value as isolating mechanisms in aquatic animals with external fertilization. The eggs and sperm of many fishes are freely discharged into the water, and, if there is any intermingling of species, the eggs of one species might easily be fertilized by sperm of another species. For instance, in many ponds that are largely filled with aquatic vegetation, spawning fishes of several species are crowded into a few gravel areas, where occasional hybridization is inevitable (Hubbs 1955). In small brooks, sperm may be washed out of the "nest" of one species into the "nest" of another species (Trautmann *in litt.*). External fertilization is unquestionably the reason for the relatively high frequency of hybridization among fishes as compared with mammals, birds, and reptiles. In fishes with internal fertilization, as in the viviparous cyprinodonts, hybridization is very rare, but, as with birds, there are a few exceptions (Hubbs 1955), apparently none of them leading to a breakdown of the species border.

However, external *versus* internal fertilization does not provide the entire answer. Frogs (*Rana*) and toads (*Bufo*) have an identical mode of fertilization, yet differ considerably in frequency of hybridization. Newts (*Triturus*) are aquatic, and fertilization is internal, even though the spermatophore is not directly inserted into the female genital opening. Yet the courtship is sufficiently complex to permit the functioning of ethological isolating mechanisms. The more precisely the reproductive activity of an aquatic species is limited to a specific temperature, the less danger of hybridization there is. In addition, there seems to be a considerable sterility barrier in the genera (such as *Rana*) in which hybridization is rarely or never found (Moore 1955). It seems evident that the various isolating mechanisms (see Chapter 5) differ in the efficiency with which they prevent gene exchange between species: sterility ranks first, ethological barriers are a close second, while ecological barriers are easily disturbed and hence least reliable. A comparative study of the isolating mechanisms in groups with low frequency of hybridization (birds, reptiles, spiders) and in groups with more frequent hybridization (mollusks) would seem highly desirable.

Nature of the Mating Bond

Even in species with internal fertilization, there are great differences in the frequency of hybridization, according to the nature of the mating bond. In most species of birds, for instance, males and females form a definite pair during the mating season and share the duties of incubation and raising of the young. In such pair-forming species there is usually a more or less lengthy "engagement" period before copulation takes place. There are innumerable displays between the two mates during this period, and pairs not composed of conspecific individuals apparently break up at this stage. This is presumably the reason why hybrids are so rare in birds with this type of mating bond. Hybrids are more frequent among those groups of birds in which copulation is not preceded by pair formation and an "engagement" period. In most birds of paradise (Paradisaeidae), for example, one or several males perform on a display ground and a female appears there only when she is ready for fertilization. After this has been accomplished, she alone builds the nest, incubates the eggs, and raises the young. On rare occasions it happens that a female is attracted to the display ground of the wrong species and is fertilized. At least 14 kinds of such hybrids, involving 10 genera, are now known. The total number of known hybrid specimens of birds of paradise is some 30 to 50, which is indeed a small number considering that more than 100,000 skins of birds of paradise were exported from New Guinea between 1870 and 1924. No hybrids are known from seven New Guinea genera of birds of paradise, perhaps primarily because the species of these genera are exceedingly rare. The only exception is the genus *Manucodia*, with three common species between which no hybrids are known. The puzzle as to why this genus is not involved in intergeneric or interspecific crosses was solved by Rand (1938), who found that the birds of this genus form pairs and the males take part in raising the young. This is in contrast to all other common birds of paradise, with the apparent exception of the closely related genus *Phonygammus* and of *Macgregoria*. These three genera have failed to develop conspicuous sexual dimorphism and are not subject to hybridism. A relatively high frequency of hybrids, many of them intergeneric, has been described in all the other families of birds in which pair formation is absent (at least in certain genera): humming-birds (Trochilidae; Banks and Johnson 1961), grouse (Tetraonidae), and manakins (Pipridae; Parkes 1961). Even where such hybridization is common, as between Capercaillie (*Tetrao*) and Black Cock (*Lyrurus*), it has not led to introgression.

The frequency of hybrids in these families is easy to understand. The contact between males and females is apparently very short, leaving much room for error and little chance for correcting it. The absence of pair formation explains another peculiarity in these families, the enormous development of secondary sex characters in the males. Apparently, the male with the most stimulating plumage or display is favored by prob-ability to leave the greatest number of offspring and lowest number of hybrids, and this gives great selective advantage to those possessing special plumes and performing elaborate courtships (Sibley 1957).

Rarity of One Parental Species

The individuals that occur beyond the solid range of their species often have difficulty in finding a conspecific mate. In the absence of ade-quate stimuli, that is, stimuli from conspecific individuals, they are apt to respond to inadequate stimuli, that is, to individuals belonging to a dif-ferent species. Many of the known hybrids of animal species are found at the margin of the normal geographic range of one of the two parental species or even beyond it. The "Cincinnati Warbler," which appears to be a hybrid between the Blue-wing Warbler (*Vermivora pinus*) and the Mourning Warbler (*Oporornis formosa*), was found in an area south of the range of the Mourning Warbler. Likewise, the "Potomac Warbler" comes from an area where the Parula Warbler (*Parula americana*), one of the parental species, is rare. The same phenomenon has been reported for flycatchers, woodpeckers, bulbuls, and other birds, and for fishes and many other animals.

This rule holds not only for geographic distribution in the narrow sense of the word but also for vertical and for habitat distribution. The factor common to all these situations is that an individual may find itself in an area without an adequate number of suitable mates belonging to the same species. This is not different in principle from situations in which a species hybridizes in captivity in the absence of conspecific mates. Sometimes one of the parental species is rarer at a locality than the hybrids it has produced, or one of the species is so rare that its presence can be determined only because hybrids are found.

Disturbance of Habitat

By far the most frequent cause of hybridization in animals is the breakdown of habitat barriers, mostly as a result of human interference. Hybridization between species that are normally separated by habitat preference can often be clearly shown to be due to agricultural activities. Hybridization in the African Paradise Flycatchers (*Terpsiphone*), in the American toads (Blair 1941), and in the yellow butterflies (*Colias*) is apparently due to man's disturbance of the natural habitats. The same is true of the hybridization between Blue-wing and Golden-wing Warblers (*Vermivora*), of the towhees of Mexico and the honey eaters of New Guinea, and of some of the cases of hybridization among fishes. Stebbins (1950) lists this as the most important single cause of hybridization in plants.

Not all habitat disturbances that lead to hybridization are man-made. Hubbs and Miller (1943) have shown that the desiccation in the Western deserts forced certain species of fish together in the confined waters of springs, leading to a breakdown of the isolating mechanisms between these species. The same situation was found by Kosswig (1953) and Aksiray (1952) for the cyprinodonts of Asia Minor.

Hubbs (1955) emphasizes the great contrast between marine and fresh-water fish faunas, and, within the latter, between tropical- and temperate-zone faunas with respect to the frequency of hybridization. Most fresh-water fishes live in somewhat isolated, somewhat temporary bodies of water where extensive changes have taken place in the Northern Hemisphere since the peak of glaciation. Under these circumstances a rigid adaptation to a fixed set of ecological conditions could not occur, and occasional introgression, an adverse factor under stable ecological conditions, was not selected against. Yet it appears that even under these favorable conditions introgression occurred only rarely.

DIFFERENCE IN HYBRIDIZATION BETWEEN PLANTS AND ANIMALS

It is evident that there is a considerable difference between higher animals and plants with regard to the frequency of hybridization. Contrary to the belief of some botanists, the low frequency of hybridization among animals, already emphasized by Weismann (1902), is not an artifact produced by insufficient analysis of animal species. Rather, the striking difference between these two kingdoms is correlated with basic differences in their physiology, population structure, genetic constitution, and ecological needs.

Animals have mobility and can actively search both for a suitable mate and for a suitable habitat. This allows much greater specificity and specialization than is normally found among plants. Speciation is made easy and occurs with great profusion. Even though the plant kingdom is much older than that of animals, there exist perhaps five times as many species of animals as of plants. An increase in variability, such as is produced by hybridization, normally results in a lowered efficiency in the chosen niche, especially if it is a narrow one.

Plants cannot move. A seed germinates where it drops and must succeed or die. Therefore, in plants natural selection favors great nongenetic plasticity as well as genetic variability. Hybridization replenishes such variability and is therefore favored. Frequent hybridization is inevitable because the transfer of pollen from one plant to another is done by extrinsic agents such as wind or insects. This permits many "mistakes" that lead to hybridization. Thus, both the ecological needs of plants and their reproductive mechanisms favor hybridization. Because there is much hybridization and because hybrids are as viable under many conditions as the parental types, there will be selection pressure in favor of an increased ability to cope with the consequence of hybridization. This leads to further increase in hybridism, and, conversely, reduces the frequency of speciation, counteracts too narrow a specialization, and places a premium on instantaneous speciation. Finally, as Stebbins (1950) points out, among perennial plants there is not nearly so high a selection pressure in favor of high fertility as in short-lived animals. A reduction of fertility due to hybridization will not be selected against as severely in these plants as in animals that produce a new generation, or even several generations, each year.

The observed difference in frequency of hybridization between animals and plants is thus the logical consequence of the known differences in ecology and reproduction prevalent in the two kingdoms. These differences become apparent if one compares the data presented in this chapter with some recent summaries on hybridization in plants (Stebbins 1950:chap. 7, 1959; Baker 1951; Heiser 1949; Anderson 1953; Grant 1952b, 1957). To utilize the evidence of only one of the two kingdoms—let us say plants—and derive from it sweeping generalizations on the role of hybridization among all organisms would be neither scientific nor justified by the available evidence.

THE EVOLUTIONARY ROLE OF HYBRIDIZATION

Hybridization has played an important role in the evolutionary theories of some authors, while others have virtually ignored this factor. The backward state of the taxonomy of lower animals precludes at present a balanced evaluation, as the possibility cannot be ruled out that hybridization is more common among the invertebrates than among the better-known vertebrates. On the whole, the various aspects of the evolutionary significance of hybridization that have been stressed in the past can be classified under three headings.

1. *Perfection of isolating mechanisms.* It has been suggested that the occurrence of hybridization may occasionally lead to the perfecting of isolating mechanisms, provided hybrids are sufficiently rare and inviable. The validity of this suggestion will be investigated in Chapter 17.

2. *Source of new species.* If hybrids between two species were to form a third species coexisting with the two parental species, this would be a process of speciation through hybridization. Such a process has often been postulated, but has never been unequivocally established. The difficulty, except in the case of allopolyploidy, consists in keeping such a population of hybrids segregated until it has acquired reproductive isolation. Whether or not this is possible will be discussed in Chapter 15.

3. *Increase of genetic variability.* The claim has been made (Anderson 1953) that species owe much of their genetic variability to introgressive hybridization. This claim is based on several assumptions that must be discussed first before the validity of the general thesis can be considered.

The first assumption is that hybridization is widespread and frequent but often overlooked. Two steps are involved in the process of introgressive hybridization: the production of the original F_1 hybrid and the backcrossing of this hybrid to one or both of the parental species. In the preceding account we have stressed how rare the production of F_1 hybrids is in the better-known groups of animals. The few cases quoted were taken from literature involving thousands of species. The recognition of hybrids is a relatively easy matter in the better-known groups of animals (birds, butterflies, grasshoppers, *Drosophila,* and others), and in the absence of contrary evidence it would seem legitimate to extrapolate the findings of the better-known to the less-known groups. In birds, for instance, most species differ conspicuously from each other either in coloration or in song, or both, and hybrids arouse the attention of ornithologists by their peculiar songs (McCamey 1950). Laboratory-bred hybrids in the cricket genus *Nemobius* have very distinctive songs. These have not been heard in nature in the area of contact of these species (Cantrall 1943). The same is true of many other orthopterans and anurans.

The actual frequency of hybrids in nature can be determined beyond question wherever they can be diagnosed unequivocally, for instance on the basis of their gene arrangements. This is the case in the genus *Drosophila,* where, in spite of the similarity of the species, hybridization in nature is exceedingly rare. Although the gene arrangements of more than 100,000 wild-caught *Drosophila* flies have been examined by various investigators, wild hybrids have been found only a few times, so far involving only three pairs of species. In 31 collections of *D. mulleri* and *D. aldrichi* made in the years 1940 and 1941, Patterson obtained 3244 normal males of these species and 30 hybrid males. Backcrossing cannot take place because the hybrids are completely sterile (Patterson and Stone 1952). Dobzhansky has collected *D. pseudoobscura* and *D. persimilis* for over 20 years. No hybrids were found during this period, including 9 years of intensive collecting at Mather, California, until two females of *D. persimilis* were captured in 1954 that had been inseminated by *D. pseudoobscura* males. A third case involves the species *D. montana* and *D. flavomontana* (Patterson 1953). In spite of much searching and easy detection (with the help of cytological criteria), no case of introgression has yet been discovered in the genus *Drosophila.*

The rarity of introgression in animals manifests itself not only in the original rarity of F_1 hybrids, but also in the absence of an increased variability of the parental species in the areas of geographic overlap. Although hybridization does not necessarily lead to an appreciable increase of phenotypic variability, as shown, for instance, in the case of such highly multifactorial characters as size (Mayr and Rosen 1956), there will be an increase in variability for characters with an oligogenic basis. This has in fact been shown in all known cases of hybridization. One can state with assurance that the presence or absence of hybridism can be demonstrated fairly conclusively if a careful investigation is made.

A second assumption is that introgression would not lead to a breakdown of the isolation of species. This assumption overlooks the fact that successful leakage of genes from one species into another is normally a self-accelerating process. Each successful case of introgression weakens the genetic isolating mechanisms of the introgressed species and leads to increased frequency of hybridization, until ultimately the two species are connected by a continuous hybrid swarm. The fact that such hybrid swarms develop only rarely forces one to the conclusion that introgressed individuals are normally eliminated by natural selection.

A third assumption is that the genetic variability added by introgression is beneficial and will be preserved by natural selection. This assumption is in conflict with much that has been discovered in recent years about the coadaptation of gene complexes. Genotypes containing disharmonious combinations will be discriminated against and in animals a mixture of genes from two species is apparently nearly always disharmonious. A hybrid or backcross individual has lost its normal chromosomal balance, which is replaced by a more or less incompatible mixture of genes. If two hybridizing species could exchange single genes, one could imagine that occasionally an introgressed gene might be superior to its "native" homologue. Actually, in hybridization the units exchanged are internally balanced chromosome sections that are not in relational balance with the remainder of the genotype. Recent experimental evidence indicates that such genotypes usually have a distinctly lowered viability. When the hybrids are fertile, laboratory populations can be started that consist exclusively of F_1 hybrids. In such a closed hybrid population between the semispecies Drosophila mojavensis and D. arizonensis, the mojavensis chromosomes proved on the whole superior to the arizonensis chromosomes, but heterosis of the heterozygotes was sufficiently great to preserve most arizonensis chromosomes or gene arrangements at appreciable frequencies (Mettler 1957). Birch (1961) likewise produced stabilized hybrid populations between two species of fruit flies (Dacus).

In natural populations there is usually severe selection against introgression. The failure of most zones of intraspecific hybridization to broaden (Chapter 13) shows that there is already a great deal of genetic unbalance between differentiated populations within a species. How much greater will be the difficulties of introgression, caused by genic unbalance and lack of coadaptation, in the case of interspecific hybridization! This unbalance leads to an observable lowering of viability. Epling (1947b) and others have pointed out that the integrity of the parental species is not necessarily broken down even in many cases where hybrids are frequent and fertile, for example in the plant genera *Quercus, Salvia,* and *Arctostaphylos.* In animals, hybridity usually leads to incompatibilities in mating behavior, combined with ecological inviability.

The total weight of the available evidence contradicts the assumption that hybridization plays a major evolutionary role among higher animals. To begin with, hybrids are very rare among such animals, except in a few groups with external fertilization. The majority of such hybrids are totally sterile, even where they display "hybrid vigor." Even those hybrids that produce normal gametes in one or both sexes are nevertheless unsuccessful in most cases and do not participate in reproduction. Finally, when they do backcross to the parental species, they normally produce genotypes of inferior viability that are eliminated by natural selection. Successful hybridization is indeed a rare phenomenon among animals.

Hybridization and the Origin of Domestic Animals

In view of the well-known hybrid origin of many kinds of crop plants, it has been suggested frequently that introgression is responsible for the great diversity of most of our domestic animals. This claim can be refuted for most of the species involved: all the breeds of the domestic fowl are derived from *Gallus gallus,* and sheep, goats, pigs, and donkeys each from a single ancestral species. In a few cases (cat, cattle, horse) there is the possibility that several subspecies are involved or that the original domestic stock interbred locally with various other wild populations of the same species. In the case of cattle there seem to have been repeated independent domestications, one starting from the (now extinct) wild *Bos taurus* of the western Palearctic, another starting from the wild *Bos banteng* and leading to the Indian cattle. Although these have been crossed recently, this hybridization has nothing to do with variability of the domestic cattle. Another domestic animal for which claims have been made for a diphyletic origin is the dog. Its evident ancestor is the wolf (*Canis lupus*), but there are claims that the jackal (*Canis aureus*) has also been domesticated and has contributed genes to some dog races, particularly those of the subtropical and tropical Old World (including the dingo). This double ancestry of the dog is very doubtful. Indeed Matthey (1954) insists (on the basis of an analysis of dog chromosomes) that the wolf is clearly the exclusive ancestor of the dog. It is probable, however, that different geographic races of the wolf (particularly *lupus* and *pallipes*) have been domesticated independently.

Invoking hybridization to account for the astonishing variety of races in some domestic animals hardly seems necessary if one recalls the racial differences that have evolved in some species within less than a hundred years (budgerigar, turkey, fox, mink) or in the course of a few hundred years (canary, guinea pig, rabbit). Darwin rightly emphasized the extraordinary variation of the domestic pigeons, which, without question, derived exclusively from the Rock Dove, *Columba livia*. If such variability could evolve in so short a time, one need not be surprised about the variability of the dog, which has been with man for many thousand years. The dogs of the American Indians no doubt have a single root, with the wolf certainly the ancestor. Yet what a difference between an ordinary Indian dog and a hairless Chihuahua!

The high degree of inbreeding and the unbalancing selection connected with domestication often have interesting results. A disturbance of the genetic and developmental homeostasis is apt to reveal unexpected components of the genetic heritage of a species (Chapter 10). Phenotypic characters may suddenly appear that are unknown in the directly ancestral species, but are found in other species of the genus or even in different genera. A potentiality in related species for homologous characters has been pointed out by students of mutations in *Drosophila*, and particularly by Vavilov ("law of homologous series"). There is no need to invoke hybridization to account for the emergence of such character traits.

SUMMARY

Occasional breakdown of isolating mechanisms has been found to occur in most taxonomically well-known groups of animals. Relatively most frequent among the various forms of hybridism is the occurrence of occasional sterile, or at least nonreproducing, species hybrids. Evidence of backcrossing with one or both of the parental species is found much more rarely, and rarer still is the complete breakdown of the barrier between species resulting in hybrid swarms.

The evolutionary importance of hybridization seems small in the better-known groups of animals. Even when fertile hybrids are produced, genetic unbalance of the hybrids results in strongly lowered ecological and ethological adaptedness, and there is little or no introgression. The contribution to the genetic variability of a population made by non-eliminated genes, remaining as a residue of introgression, can be considered negligible in comparison with the contribution made by mutation and regular gene flow from adjacent conspecific populations.

In view of the rarity of allopolyploidy and successful hybridization, it is evident that reticulate evolution above the species level plays virtually no role in the higher animals. The systematics of lower animals is too little known to permit generalizations on the evolutionary role of hybridization. The possibility that they are more similar to plants than to higher animals can be neither supported nor excluded by the available evidence.

7 ~ The Population, Its Variation and Genetics

Between the individual and the species is a level of integration of particular importance to the evolutionist, the level indicated by the word *population*. It is proper that the study of natural populations has become a major preoccupation of several branches of biology: genetics, ecology, and systematics. The term "population" is used in several different ways and recourse to a dictionary is of little help. Ecologists may speak of the plankton population of a lake, including in it the individuals of several species. It is customary to speak of the human population, referring to the totality of individuals of a single species, the human species. These are legitimate uses of the term. Under the impact of modern systematics and population genetics, a usage is spreading in biology that restricts the term "population" to the *local population*, the community of potentially inter-breeding individuals at a given locality. All members of a local population share in a single gene pool, and such a population may be defined also as "a group of individuals so situated that any two of them have equal probability of mating with each other and producing offspring," provided, of course, that they are sexually mature, of opposite sexes, and equivalent with respect to sexual selection. The local population is by definition and ideally a panmictic unit. An actual local population will, of course, always deviate more or less from the stated ideal. A species in time and space is composed of numerous such local populations, each one intercommunicating and intergrading with the others (Wright 1931a, 1943a, 1949a).

In view of the diverse meanings of the word "population," it would be useful to have a technical term for the local population, as defined above. The term *natio* used by Semenov-Tianshansky (1910) is a little too broad, but two other terms may be considered eligible: *deme* (Gilmour and Gregor 1939) and *ethnos* (Vogt 1947). Unfortunately neither is rigidly defined nor clearly restricted to the local population. Since the term "deme" in its original publication is nothing but a needless synonym for the prior term "population," it may be given a more specific meaning. This was done by several zoologists, for instance Simpson (1953a), and Wright (1955), who adopted the term "deme" for the local population, the interbreeding community, as defined above. Since there is no other technical term for this evolutionary unit, the term "deme" has been widely accepted for it, even though this was not the meaning of the term when originally proposed (nor that in the more recent use of the term by Gilmour and Heslop-Harrison; see Chapter 12).

An understanding of the evolutionary role of the population is facilitated by discussing its relation to the two adjacent levels of integration, those of the individual and of the species. In no other species is the individual so important as in the human species with its highly developed consciousness and social tradition. This has been rightly emphasized by Simpson (1941), Dobzhansky (1955a), and other writers on the subject, who stress the lasting impact of certain individuals on human society. In all other sexually reproducing organisms, however, the individual is only a temporary vessel, holding a small portion of the gene pool for a short time. It may, through mutation, contribute one or two new genes. It may, if it has a particularly viable and productive combination of genes, somewhat increase the frequency of certain genes in the gene pool, yet in sum its contribution will be very small indeed compared to the total contents of the gene pool. It is the entire effective population that is the temporary incarnation and visible manifestation of the gene pool. It is in the population that the genes interact in numerous combinations, genotypes. Here is the proving ground of new genes and of novel gene combinations. The continued interaction of the genes in a gene pool provides a degree of integration that permits the population to act as a major unit of evolution.

The deme contributes importantly to an understanding of each of the three species concepts (Chapter 2). Its variability discloses the invalidity of the typological-morphological species concept. The multidimensional species is an aggregate of demes. The nondimensional species is, in each concrete case, a single deme of a multidimensional species. The interaction of one species with another, in a nondimensional situation, is that of a deme of one species with a deme of another species.

The population, in addition to its static role as local representative of the species, has the capacity to change in time. This dynamic aspect of the population is of even greater biological significance than the static role. Evolution is sometimes defined as "a change in the genetic composition of populations" (Dobzhansky 1951). Such genetic changes manifest themselves phenotypically in various ways and the study of these phenotypic changes has been the basis of most evolutionary theories and hypotheses of speciation. A study of the variation of populations is a prerequisite for the understanding of these theories.

KINDS OF VARIATION

From the evolutionary point of view one can distinguish two kinds of biological variation: "group variation," referring to differences among populations (which will be discussed in Chapters 11–13), and "individual variation," referring to differences among individuals of a single population (which will be the main theme of Chapters 7–10). The nature, source, maintenance, and biological functions of this individual variation and the factors that account for the genetic changes of populations deserve detailed consideration.

INDIVIDUAL VARIATION

No two individuals of a sexual species are completely alike. To say that The Robin or The Wolf has such and such characters is a generalization, and not necessarily always a correct one. An oversimplified treatment of the species as a unit phenomenon was adopted in the early chapters of this book only for its obvious didactic advantages. However, in much of the taxonomic literature authors speak of *the* species for reasons that have a historical and philosophical background. Adopting a typological approach, they actually assume that all members of a species conform to the "type." Anyone who makes that assumption is forced to minimize variability. Conscious or unconscious neglect of variation is at the bottom of most of the difficulties encountered by evolutionists and has led to the establishment of most of the erroneous theories of evolution.

Individual variation has long been of practical interest to the descriptive morphologist, and even more so to the taxonomist who wants to find out which of the described "species" are merely variants of previously known ones. A detailed discussion of this aspect of variation can be found in volumes on taxonomy (Hennig 1950; Mayr, Linsley, and Usinger 1953; Simpson 1961). To the typologist, any variant is merely an imperfect copy of the *eidos*. Indeed, many early taxonomists discarded variants from their collections as disturbing and confusing elements. Only such specimens were kept as "conform to the type." When the study of evolution began to come to the fore, there was a tendency to fall into the opposite extreme. The existence of species was denied: "only variable individuals exist." We now combine the correct aspects of these opposing viewpoints into a more realistic interpretation.

What interests the evolutionist is not the mere occurrence of variation, but rather its significance. On the basis of the criterion of heritability, all manifestations of intrapopulation variation can be divided into nongenetic and genetic variation. Broadly speaking, it can be generalized that nongenetic variation adapts the individual, while genetic variation adapts the population. Actually, they are alternate strategies of population adaptation, one of them sacrificing the individual, the other not.

Nongenetic Variation

Modifications of the phenotype that do not involve genetic changes have long been considered, as a reaction to Lamarckian concepts, to lack evolutionary significance. This view is not correct. Nongenetic * variation is usually adaptive and controlled by natural selection, since genetic factors determine the amount and the direction of the permissible flexibility of the phenotype. A few examples may be helpful in bringing out the essential difference between genetic and nongenetic variation (Table 7–1).

Age Variation. The young animal at birth or on hatching from the egg may be either very different from or very similar to the adult. A young whale or snake resembles the adult closely except for size. A caterpillar is very different from a butterfly and the same is true of the larvae of all holometabolic insects. Even more different are the larval stages of most animal parasites and the free-swimming larvae (nauplius, tornaria, and so forth) of more or less sessile marine invertebrates.

Age variation is of considerable practical importance to the taxonomist, since age variants have often been described as separate species. In addition, this variation has broad biological significance. The evolutionary importance of morphogenetic changes through the life cycle of an individual has been the subject of much argument. A balanced evaluation of

---

* It must not be overlooked that the capacity of a genotype to produce several phenotypes is very much under genetic control. "Nongenetic" in the present discussion means simply that the differences as such between the modified phenotypes are not caused by genetic differences.

Table 7–1. Noninherited variation.

1. Individual variation in time
   (a) Age variation
   (b) Seasonal variation of an individual
   (c) Seasonal variation of generations
2. Social variation (insect castes)
3. Ecological variation
   (a) Habitat variation (ecophenotypic)
   (b) Variation induced by temporary environmental conditions
   (c) Host-determined variation [a]
   (d) Density-dependent variation [a]
   (e) Allometric variation [a]
   (f) Neurogenic color variation
4. Traumatic variation
   (a) Parasite induced [a]
   (b) Accidental and teratological variation [a]

[a] For details, see Mayr, Linsley, and Usinger (1953).

these phenomena was impossible until it was understood that an individual is exposed to selection at every stage of its life cycle. Age variation must be interpreted as the product of interaction between selection and genic endowment. The observed changes of morphology during the life cycle are the result of two somewhat antagonistic tendencies. One is a tendency toward ever greater differentiation from birth to maturity which may result in a greater dissimilarity of the adults of related species than of the immatures. This divergence of the adult phenotype may also be due in part to the greater ease of incorporating new genes that affect late rather than early stages of the developmental process. Furthermore, where there is parental care or heavy competition for mates or both, much selection will affect the mature stage. The second trend is toward adaptation at every stage of the life cycle. This may lead to greater differences between larval stages of related species than between the adults, particularly in those species in which the larval stages are exposed to high selection pressures. It is accentuated where intraspecific competition between adults and immatures is high. Instead of discussion of these phenomena in simple biological terms that are open to analysis, there has been an unfortunate tendency among comparative embryologists, beginning with Haeckel, and among certain paleontologists, to obscure the problems by an elaborate set of terms, such as anaboly, archallaxis, caenogenesis, gerontomorphosis, hypermorphosis, paedogenesis, paedomorphosis, palingenesis, proterogenesis, tachygenesis, and the like. For an interpretation of the phenomena underlying these terms see de Beer (1951) and Rensch (1954).

Seasonal Variation. Adult individuals of certain species of animals are subject to seasonal changes of the phenotype. For instance, mammals in the temperate and cold regions may molt in fall into a winter fur. The snow-white winter dress of certain species of weasels (*Mustela;* Hall 1951) and hares (*Lepus*) is well known. A seasonal change of plumage is very frequent in birds. The winter plumage of ptarmigans (*Lagopus*) is white like the winter fur of ermines. Other kinds of birds have, during part of the year, a simple eclipse plumage which is replaced by a bright nuptial plumage before the beginning of the breeding season. A brightening of colors and a change of certain epidermal structures during the breeding season occurs in many species of fishes and in some invertebrates. The significance of this seasonal variation is to adapt individuals to the changes of the environment through the seasons and of their own life cycles.

Generations. In organisms with a rapid sequence of generations, seasonal variation may be found to involve generations rather than individuals. In well-known cases in insects, summer generations differ from spring generations, or dry-season individuals from those of the rainy season. The most frequent difference is one of coloration, but in some insects (for instance, *Gerris paludum*) a long-winged generation may alternate with a wingless or short-winged one (Brinkhurst 1959). The differences, particularly where only two generations per year are involved, can usually be shown not to have a genetic basis.

One form of seasonal variation deserves special discussion; this is *cyclomorphosis,* the cyclic change of form in a series of genetically identical generations. It occurs in planktonic fresh-water organisms, dinoflagellates, cladocerans, and rotifers, which reproduce by asexual or parthenogenetic processes. In some species of *Daphnia,* for instance, spring populations are made up of round-headed individuals, while a definite pointed "helmet" distinguishes the generations born during the summer. Since the summer individuals are parthenogenetically produced, the difference cannot be due to genetic change. Brooks (1946, 1947) has shown that changes in the temperature of the water and in its turbulence are involved in the production of this cyclomorphosis. The morphological changes are completely reversible and have, of course, nothing to do with speciation. Higher temperatures and (at least in *Daphnia galeata*) increased turbulence seem to result in increased metabolism and growth rate, and these, in turn, in the growth of the helmet. There is no evidence that this phenotype has any adaptive significance (Brooks 1957b). The equivalent variability of rotifers was studied by Buchner *et al.* (1957) and by Buchner and Mulzer (1961).

Habitat Variation. Everyone is familiar with the difference between two plants, one planted in good, the other in poor soil. The direct effect of the physical environment on the phenotype is rarely as pronounced in animals as in plants. It is perhaps most evident in sessile marine invertebrates such as sponges and corals and in some mollusks (for example, oysters); also in some fresh-water bivalves. Whether such animals grow in quiet waters or in the surf, whether they grow in clear water or in water rich in plankton or silt, whether they grow in an environment rich or poor in calcium (lime), all these factors may greatly affect the appearance of an individual. Phenotypes resulting from modification by edaphic or other ecological conditions, and not the product of genetic differences, are sometimes called *ecophenotypes.*

Nongenetic Variation in Fresh-water Fish. Certain groups of fresh-water fish (for instance, *Coregonus, Leucichthys*) display a plasticity of the phenotype that far exceeds what is usual in animals. This has led to erroneous interpretations. Some years ago an author described many "subspecies" of the whitefish *Leucichthys artedi* (Lesueur) from lakes in Michigan and Wisconsin that are only a few miles apart, but differ in water temperature, pH, bound CO_2, and other chemical and physical properties. When the fish populations were arranged according to the physical and chemical characteristics of the lakes they inhabit, they fell into the same series as when arranged according to their morphological characters. Even in a single lake, morphological changes occurred with changing water conditions. For instance, the 1928 population of Muskellunge Lake was very different from the 1929 population. Ultimately it was established (Hile 1937) that the adult characters of these fishes depend largely on their rate of growth. The populations with the most rapid growth had shorter heads, maxillaries, paired fins, and dorsal fins, but wider bodies and a smaller eye diameter. The growth rate, in turn, depended largely on the available food supply.

Parallel findings were made by Svärdson (1950) on European *Coregonus.* Two species that agree in growth rate and body proportions but differ in number of scales and gill rakers, spawning season, and choice of spawning ground were transplanted separately, from the same northern Swedish lake, into a lake previously unoccupied by *Coregonus.* In the new environment one of the species showed accelerated growth, while growth

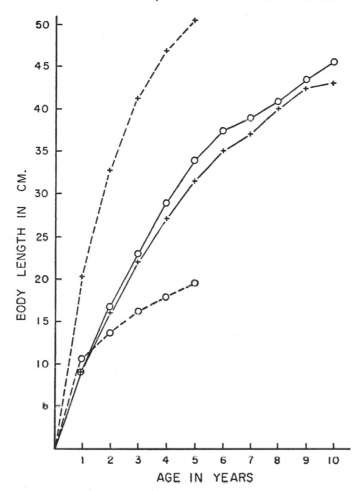

Fig. 7–1. Growth rates of two *Coregonus* species, "storsik" (circles) and "aspsik" (crosses) in different Swedish lakes. In Lake Uddjaur (solid line) the two species have almost identical rates; in Lake Kälarne (broken line) the growth of aspsik is accelerated, that of storsik retarded. (From Svärdson 1950.)

was retarded in the other species (Fig. 7–1). Ultimate body proportions differed greatly between the two groups. On the other hand, the difference in scale numbers disappeared in fingerlings raised under identical conditions. It is evident that in this genus the phenotype is unusually responsive to environmental conditions. The situation in *Leucichthys* and *Coregonus* cannot be generalized. The phenotypic differences between most local populations of fishes seem to be genetically determined, as shown by the work of Gordon and others.

Fresh-water invertebrates, on the whole, have an unusually plastic phenotype (Brooks 1957b). Special caution must therefore be exercised in the interpretation of their morphological variation. For instance, the existence of statistically significant differences between samples of the copepod (*Mixodiaptomus laciniatus*) taken from different parts of the same lake (Lago Maggiore, Italy) does not necessarily prove the existence of genetically determined geographic variation, as suggested by the original investigators (Baldi *et al.* 1945). There is so much movement of water masses in a small lake that its pelagic population must be considered essentially panmictic. The differences between the phenotypes are probably due to the nature of the water masses in which crucial growth stages took place.

Other Nongenetic Variation. The production of genetically identical but morphologically different castes in social insects, regulated by the nutrition of the larvae, or the morphological (and physiological!) modifications produced by crowding in army worms (*Lepidoptera*) and grasshoppers (Brett 1947; Key 1950), are other forms of nongenetic adaptive variation, ultimately steered by natural selection.

Very impressive color changes occur in individual animals in response to environmental influences (neurogenic variation), for instance change of substrate color. This phenomenon, typified by the proverbial color changes of the chameleon, occurs particularly commonly among marine organisms such as crustaceans, cephalopods, and fishes. Slugs of the genus *Arion* seem able to change slowly from blackish brown to reddish brown, in accordance with environmental factors (Albonico 1948), a modifiability superimposed on the genetic polymorphism in this genus. Various kinds of insects associated with distinctly colored substrates acquire during their last larval instar a coloration that conceals the adult on this substrate. This was shown by Popham (1941) for waterstriders, and by Ramme (1951), Burtt (1951), and Ergene (1952, 1957) for grasshoppers. If such adults are passively transported to a contrasting background, they tend to become restless and move about until they chance upon another harmonizing background.

Genetic or Nongenetic Variability?

Of phenotypic variability observed in nature, it is never possible to tell, except by careful breeding experiments, what part should be ascribed to nongenetic modification and what part to genetic factors. One reason for this difficulty is that the genetically determined phenotypic change favored by natural selection in a particular environment may be evoked somatically (without change of the genotype) by the same environmental condition. The short mammal tail favored by selection in a cool climate occurs also as a developmental response: mice raised at 26.3°C had an average tail length of 93.1 mm, whereas those raised at 6.2°C had tails that averaged only 75.9 mm long (Sumner 1909). Harrison (1959) reviews the role of the environment in the determination of the phenotype.

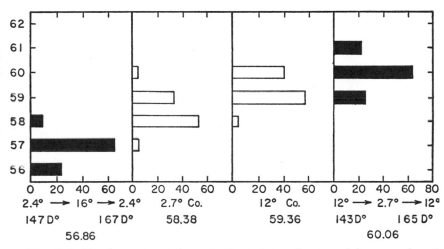

Fig. 7–2. Numbers of vertebrae (ordinate) in offspring of four samples of sea trout. Experimentals black, controls white. Heat shock (16°) during the supersensitive period at left, cold shock (2.7°) at right; $D°$, day-degrees since fertilization. Untreated controls at constant temperatures of 2.7° or 12°. (From Tåning 1952.)

The respective contributions of environment and genotype to the final phenotype have been studied particularly well for fishes. Raising fish at cool temperatures leads to an increase in the number of vertebrae, but populations from cool waters will have a genetic tendency toward an increased number of vertebrae (Hubbs 1922; Gabriel 1944). We owe to Tåning's (1952) work on Danish populations of the Sea Trout (*Salmo trutta*) the most detailed information on the effect of environmental conditions (temperature, oxygen and CO_2 pressure) on the phenotypic expression of meristic characters. Fish that were heat-treated during the temperature-sensitive stage had 56–58 vertebrae, cold-treated fish had 59–61 (Fig. 7-2). Cold- and heat-shock treatments of Danish Sea Trout produced phenocopies of Scandinavian and Mediterranean races of this species (as far as meristic characters are concerned). The results of these and similar studies are of considerable practical importance, since different "races" or "stocks" of commercially important food fishes are often characterized by differences in meristic characters such as the number of vertebrae or fin rays. The recent studies suggest that many differences have no genetic basis but are due to the different temperatures of the waters in which the fish larvae develop. Yet the choice of the spawning grounds and various physiological properties of these local races may well be genetically determined.

In snakes, similarly, there is evidence of nongenetic modification of taxonomically important phenotypic characters (Fox 1948b; Fox et al. 1961). Garter Snakes (*Thamnophis elegans terrestris*) kept in a cool room bore young with significantly fewer longitudinal scale rows, supralabials, ventrals, subcaudals, postoculars, and lateral blotches than snakes born in a warm room. For instance, in females, at cool temperatures the number of ventrals was 130–154 (150), at warm temperatures, 147–167 (155).

Determination of the precise contribution of inheritance and of environment to a given character or character complex is of immense importance in the field of animal breeding. It has led to the development of elaborate statistical methods (Kempthorne 1957; Lerner 1950; Mather 1949).

The Significance of Nongenetic Variation

The competence of the phenotype to be modified by environmental factors has been interpreted in many different ways. Some Lamarckian naturalists thought that such nongenetic modifications could be transformed into genetic changes. This led to the experiments of Standfuss and others in which the coloration of butterflies was changed by exposing the pupae to cold or heat. These phenomena are not puzzling when considered in the light of physiological genetics. After all, the genotype is not a mold into which the characters are cast, but rather a "reaction norm" interacting with the environment in the production of the phenotype (Kühn 1955). Since the rates of chemical processes are changed by temperature, it is understandable that temperature changes may be accompanied by changes in the phenotype. This does not suggest that such modifications become hereditary.

An inquiry into nongenetic variation in *Daphnia* and *Coregonus* was particularly important because both genera had been cited by Lamarckians as illustrations of rapid evolution under the direct influence of the environment. This argument collapsed when the nongenetic nature of most of the changes was established. In fact there is doubt even as to the amount of actual change (as compared to the original claims), as shown by Wagler (1951) for the *Coregonus* transplanted into the Laacher See and indicated by D'Ancona and D'Ancona (1949) for the *Daphnia* of Lake Nemi.

Why some organisms have a very stable phenotype, while others are easily modifiable, is not understood (Dobzhansky 1956a). A stable internal environment, as in warm-blooded vertebrates, undoubtedly contributes to the stabilization of developmental processes and thus to phenotypic uniformity. Yet this does not explain, for instance, the generally lower morphological variability of birds compared to mammals. The coefficient of variability in linear measurements of birds is often as low as 1–2.5, while in mammals 3–5 are low values. Nor does it explain why some insects are very stable, others very plastic. In plants, even under very uniform conditions, the coefficient of variability for linear measurements rarely falls as low as 5 (Went 1953). (For stability of the phenotype, see also Chapters 9 and 10.)

It is sometimes claimed that nongenetic changes pave the way for equivalent mutations, but there is no evidence for this claim. The ability of the phenotype to respond to demands of the environment without mutation greatly reduces selection pressure. One must assume therefore that easy modifiability of the phenotype has a retarding effect on evolution, contrary to the claims of the adherents of the "Baldwin effect" (see Chapter 19).

Phenotypic flexibility is classified by Thoday (1953) into two types, developmental flexibility and behavioral flexibility. Developmental flexibility, as usually understood, leads to the development of different phenotypes under different environmental conditions. Strictly speaking, as Thoday points out, such inability of the phenotype to escape from the dictates of a given set of environmental conditions indicates an actual lack of developmental flexibility in a higher sense. Developmental canalization, on the other hand, leads to formation of the same phenotype under the most variable external conditions. This, then, is much more truly an indication of plasticity of the developmental pathways.

Behavioral flexibility includes all behavioral elements that permit temporary adaptation to temporary environmental conditions, for instance habitat selection and various measures for controlling the habitat (beaver dams, termite and bee nests) or providing protection against the environment. Man has this behavioral flexibility to a vastly higher degree than any other animal, and animals as a whole, a higher developmental canalization than plants.

It is evident that phenotypic modifiability, phenotypic stability (owing to developmental canalization), and behavioral flexibility are biological phenomena of considerable evolutionary importance. In the past these phenomena were erroneously considered the exclusive domain of the developmental physiologist. They deserve much greater attention by evolutionists than they have so far received. Waddington (1957) has rightly emphasized this point and has stated various evolutionary problems in terms of developmental physiology.

Genetic Variation

No two individuals are genetically identical in sexually reproducing species, presumably not even most monozygotic twins (allowing for somatic mutations). Much of this genetic variation contributes to the variation of the phenotype. Consequently almost any feature of an animal may vary individually, be it a morphological character, a physiological attribute, a cytological structure (such as number, pattern, and form of the chromosomes), or whatever else. Morphological variation may be classified in various ways. It may concern *meristic* characters that can be counted, like numbers of vertebrae or scales, *quantitative* characters that can be measured, like dimensions or weight, or *qualitative* characters, like the presence or absence of spots. This is merely a distinction of convenience; the genetic basis for these classes of characters is the same. Reptiles, fishes, and arthropods are usually rich in meristic characters, while birds, mammals, and mollusks vary mainly in nonmeristic characters. Variation can also be classified as to whether it is *continuous* (larger *vs.* smaller, darker *vs.* lighter) or *discontinuous* (blue eyes *vs.* brown eyes, with white spots *vs.* without white spots). The occurrence of several discontinuous types within the same population will be discussed below under the heading *Polymorphism*.

It cannot be stressed too strongly that, so far as we know, all these types of variation have essentially the same genetic basis. During the latter part of the last century and the first decades of this century, a bitter fight raged between the biometricians (Pearson and associates) and the early Mendelians (De Vries, Bateson, and associates) with respect to continuous and discontinuous variation. The Mendelians thought that a simple study of frequencies would solve all difficulties, while the bio-metricians considered discontinuous variation an inconsequential excep-tion. They accused the Mendelians, at that time not without reason, of not having solved the genetics of continuous variation, the type of varia-tion that seemed especially important to evolutionists in general, and in particular to those interested in selection and adaptation.

The quandary was resolved when it was shown that multiple factors having small, similar effects would produce the picture of continuous variation and that the apparent continuity of the variation was even more difficult to resolve if the effect of these genes was small in relation to non-heritable modifications of the same character (Mather 1943). Although the multiple-factor basis of continuous variation had been suspected rather early in the history of genetics (Johannsen, Nilsson-Ehle, East, Baur), the early attempts to analyze it (Castle, MacDowell, Payne) did not contribute much to our thinking. It was not until new mathematical tools had been devised that significant advances were made in this field. While Mendelian genetics deals as far as possible with the analysis of isolated, single genetic factors, the genetics of continuous variation deals with the entire variation of a character but cannot sort out the respective contributions of individual genetic factors (Mather 1943, 1949, 1953; Lerner 1950). The available genetic evidence is consistent with the as-sumption that continuous and discontinuous variation have in principle the same genetic basis, regardless of the superficial difference of the visible phenomena. And yet there is a subtle difference. In the study of discontinuous variation (polymorphism) we seem able to discover the immediate action of an individual gene on the phenotype. The multifac-torial genetic effects displayed by continuous variation, on the other hand, seem to consist more in the interaction of genes than in their indi-vidualistic actions. This, then, represents a level of complexity higher than single-gene effects. Although this separation is artificial (there are no genes that do not interact), the two aspects of gene action will be used as arbitrary categories by which to separate the material on the genetics of population into two different chapters. In the present chapter emphasis will be placed on the viewpoint that might be called that of classical genetics, dealing with the atomistic, particulate, and additive aspects of genes. Where the evolutionary change of whole populations is concerned, as in speciation, the consequences of interaction of genes is of special importance. This subject matter will be dealt with in a separate chapter (Chapter 10).

A character with strictly continuous variation, such as body size in most animals, will have a highly complex genetic basis, while the variation of a character with discontinuous variation will be controlled by merely a few genes, in fact usually by alleles at a single locus. It will be convenient for didactic reasons to begin the study of the genetics of natural populations with characters the genetic determination of which is controlled by one or only few loci. Characters showing polymorphism are particularly suitable for this purpose.

Polymorphism

The occurrence of discontinuous phenotypes within a species is not at all uncommon in the animal kingdom. They may occur as evident freaks such as albinos, or they may be a normal component of a species population, such as redheads in the white human population.

Polymorphism always refers to variability *within* a population. It was originally defined as "the occurrence of several strikingly different discontinuous types within a single interbreeding population." The term "polymorphic" must be strictly distinguished from "polytypic," a term applied to composite categories. A species is polytypic if it is composed of several subspecies, a genus is polytypic if it is composed of several species. Any human population, for instance, is polymorphic, but the human species as a whole is polytypic (Haldane 1949a).

When the term "polymorphic" was first coined, it was used rather vaguely for any kind of phenotypic variation, regardless of its genetic basis. Ants and termites were called polymorphic because their castes differ morphologically, even though they are genetically identical in nearly all cases. A species like the Little Blue Heron (*Florida caerulea*) was called polymorphic because the young are white and the adults blue. Certain staghorn beetles were called polymorphic because, owing to allometric growth, small individuals have very short horns and large ones have very large horns (Mayr, Linsley, and Usinger 1953). In order to make the term "polymorphism" more useful and precise, there is now a tendency to restrict it to genetic polymorphism. Since this would leave nongenetic variation of the phenotype without a designation, the term "polyphenism" is here proposed for it. Polyphenism is discontinuous when definite castes are present (certain social insects) or definite stages in the life cycle (larvae *vs.* adults; sexual *vs.* parthenogenetic) or definite seasonal forms (dry *vs.* wet; spring *vs.* summer). Polyphenism may be continuous, as in the cyclomorphosis of fresh-water organisms and some other seasonal variation. Huxley (1955b) has proposed the term "morphism" to replace "genetic polymorphism." However, since invariable species also have at least one "morph," the normal wild type, the term "morphism" fails to bring out what the *poly*-component indicates, namely the occurrence of several variants. Acceptance of the term "morphism" for discontinuous genetic variability might therefore be misleading. The term "morph" (Huxley), however, is a suitable designation for the variants that contribute to polymorphism, and that have in the past been referred to variously as varieties, mutants, forms, or phases.

Polymorph variants ("morphs") are sometimes strikingly different from the "normal" type of the population, so much so that many morphs were originally described as separate species. Indeed, the phenomenon of polymorphism has been quite confusing to the taxonomist, particularly to adherents of a strictly morphological species concept. Yet it is evident that, no matter how different such intrapopulation variants may be superficially, they are not different species, nor are they subspecies or "races." The banded Florida land snail (*Liguus fasciatus*) has very numerous morphs. One of the standard treatises of this group recognizes as many as eight different "subspecies" from a single Florida key (Lower Matecumbe Key) and as many as four or five "subspecies" in a single Florida hammock. Needless to say, these intrapopulation color variants are neither subspecies nor races. In order to eradicate such erroneous terminology, it should be emphasized strongly that a phenotype within a population does not constitute a race. The silver foxes, the black hamsters, the blondes of a local European population, and the Rh-negative individuals are not races. One must be aware that polymorphism is an intrapopulation phenomenon, while race is an interpopulation phenomenon.

Polymorphism received special attention by taxonomists during the De Vriesian period of genetics when it was believed that species originate by macromutation. Each morph was considered an incipient species. The importance of polymorphism today, after this earlier interpretation was found to be erroneous, is that the study of it is a convenient approach to an understanding of the genetics of populations. The genes involved in polymorphism have, in general, conspicuous discontinuous effects, and different genotypes (except some heterozygotes) can be distinguished phenotypically. Such genes are therefore much easier to analyze than the vast majority of the genes in a population that are more or less cryptic. Thus the working method of the study of polymorphism corresponds to that of classical genetics where, likewise, genes with conspicuous effects and full penetrance were selected as study material. A study of polymorphic genes is a particularly suitable introduction into population genetics.

The Occurrence of Polymorphism. Polymorphism is exceedingly widespread. It has been recorded in virtually every class of animals from protozoans to vertebrates. In birds more than 100 cases are known in which a morph was originally described as a separate species. Such cases have been summarized by Stresemann (1926), Mayr (1942, 1951a), and Huxley (1955a). The literature on polymorphism has become so enormous that we must refer to recent summaries (da Cunha 1955; Dobzhansky 1951; Ford 1940, 1945, 1953; Huxley 1942, 1955a, 1955b; Kennedy 1961; Remington 1958); some additional papers on polymorphism are Zimmermann (1961) on *Clethrionomys,* Hrubant (1955) on *Otus,* Kramer (1941) on *Lacerta,* Volpe (1955, 1961) and Pyburn (1961) on anurans, Haskins *et al.* (1961) on *Lebistes reticulatus,* Fryer (1959) on cichlids, and de Lattin (1951) on terrestrial isopods. Particularly well-analyzed cases are those of the hamster *Cricetus cricetus* (Gershenson 1945), the fish *Xiphophorus maculatus* (Gordon 1947; Gordon and Gordon 1950, 1957), the butterfly *Colias* (Hovanitz 1953; Remington 1954), and the moth *Panaxia* (Ford 1953; Sheppard 1953a, 1961). Certain aspects of these cases will be discussed below. Detailed analyses of polymorphism have also been made for several species of marine animals: *Tisbe reticulata* (copepods) (Bocquet 1951), *Sphaeroma* (isopods) (Bocquet, Lévi, and Teissier 1951), and *Jaera marina* (isopods) (Bocquet 1953).

The Recognition of Polymorphism. The genetic nature of a doubtful case of polymorphism is most easily established by breeding. Many species, particularly vertebrates, cannot be bred in the laboratory and a conventional genetic analysis is therefore impossible. However, a tentative analysis can be made by analyzing broods in nature or the offspring of captured pregnant females, or by hatching in the laboratory freshly laid clutches or egg masses. Two otherwise similar kinds of Australian hawks were long considered separate species, because one (*Accipiter cinereus*) is gray, the other (*A. novaehollandiae*) snow-white. The hypothesis that they are polymorph variants of a single species was not accepted until nests were found with both gray and white young. Another interesting case is that of the King Snakes (*Lampropeltis getulus*) of San Diego County, California. There are two color patterns, a striped one, named *californiae* in 1835, and a ringed one, named *boylii* in 1853. These two types were considered full species until Klauber (1939, 1944) found 44 ringed and 13 striped young in the offspring of ringed mothers, and 12 ringed and 52 striped young in the offspring of striped mothers. These frequencies are close to expectancy if the two types are considered morphs of a single population (Dunn, in Mayr 1944), and this interpretation is now universally adopted. How such broods can be analyzed simply by application of the Hardy-Weinberg formula is shown by L. C. Dunn (in Mayr 1944), by Goin (1947, 1950), and by Hrubant (1955).

Polymorph Characters. The suffix "-morphism" suggests a limitation of the phenomenon to structural characters. Actually the term "polymorphism" covers any phenotypic character, be it morphological, physiological, or behavioral, provided it is genetically controlled and more or less discontinuous in its phenotypic expression. Polymorphism in color, being so conspicuous a character, is most often described, but presence or absence of certain tooth structures in mammals, wing veins or entire wings in insects, winding in snails (dextral *vs.* sinistral), and asymmetry in flat fishes are other well-known cases of polymorphism. A very large and increasingly important group of characters are the blood group genes in vertebrates, demonstrated to be polymorphic not only in man (Race and Sanger 1954), but also in other mammals (Mourant 1954) and in birds (Irwin 1947).

The Genetic Basis. Polymorphism results from the simultaneous occurrence in a population of several genetic factors (alleles or gene arrangements) with discontinuous phenotypic effects. Very often there are merely two alternative types ("dimorphism"), such as male and female phenotypes, or white *vs.* colored in *Colias* butterflies. In other cases more than two morphs occur, sometimes as many as a dozen, a score, or more.

The genetic analysis has been completed only in relatively few cases. For instance, the genetics of banding in the snail *Cepaea nemoralis,* studied for more than 50 years, is still by no means fully understood (Cain and Sheppard 1957). In a number of cases, such as spotting genes in coccinellid beetles, pattern genes in grasshoppers, and blood groups in cattle, polymorphism is controlled by a large series of multiple alleles (or by pseudoalleles at closely linked loci). The reason for such large series of alleles is not at all clear. Reduction of the frequency of homozygotes has been suggested as one possible selective advantage.

The most frequent phenotype in a polymorph population is by no means necessarily caused by the "dominant" gene. Genetic dominance does not equal numerical predominance. The recessive gene in the dimorphic moth *Leucodontia bicoloria,* for instance, is, in most districts, far more common than its dominant allele (Suomalainen 1941). The same is true of *Cepaea nemoralis,* where recessive genes outnumber the dominant alleles at many localities. Why the most common allele in a series of multiple alleles is so often the universal recessive is not at all clear. A selective advantage of numerical equivalence of various phenotypes may be responsible in part. It must also be remembered that morph genes are selected in most cases for their physiological effect, and that a gene that is recessive for its morphological contribution to the phenotype may be dominant with respect to its physiological phenotype. Such dominance would permit the rapid spreading of a morphologically recessive gene.

Various changes in the structure of chromosomes constitute a special class of polymorph characters. This includes supernumeraries, fusions, translocations, and inversions. Genotype and phenotype coincide in this type of polymorphism; indeed, the genetic material itself is polymorphic in its arrangement. In such cases, the revelation of the genotype by the phenotype permits a complete genetic analysis, while in most cases of genic polymorphism the recognition of heterozygotes is prevented by dominance. The various forms of cytological polymorphism in animals have been treated by White (1954) in masterly fashion. Polymorphism in chromosome number is most often produced by a (sub)terminal fusion of two rod-shaped (acrocentric) chromosomes into a single metacentric (V-shaped) chromosome. Fragmentation of chromosomes is more difficult to attain successfully, since this requires the acquisition of new centromeres and presumably telomeres, and it has been suggested by White (1957b) that this can be achieved only by some sort of translocation. Polymorphism with respect to chromosomal fusion or fragmentation has been recorded in various insects, a mollusk, and a species of mammals, *Sorex araneus* (Ford, Hamerton, and Sharman 1957; Meylan 1961). These same events are the reason why subspecies or semispecies often have distinctly different chromosome numbers (for example, *Gerbillus pyramidum,* Wahrman and Zahavi 1955).

Gene Arrangements. A particularly revealing type of chromosomal polymorphism is that caused by the inversion of chromosome sections. In view of the outstanding interest of this phenomenon, a few words must be said about it, although for a full discussion we must refer to genetic and cytological texts (Dobzhansky 1951; White 1954) and to the special literature (Patterson and Stone 1952; White 1957a; Wallace 1954a; da Cunha 1955; Stone 1956; Stone *et al.* 1960; Dobzhansky 1961). Genes are arranged along the chromosomes in essentially linear sequence. If a chromosome breaks in two places and the middle piece is turned around ("inverted"), the gene sequence *ABCDEF* may be changed to *AEDCBF* (Fig. 7–3). This is called an "inversion." In *Drosophila* and some other genera of dipterans, such as *Chironomus, Simulium,* and *Anopheles,* the giant salivary-gland chromosomes permit the direct study of gene sequences. Different parts of these chromosomes stain differently and give a highly individualized pattern of light and dark bands, permitting identification of the various gene arrangements. Dubinin, Sturtevant, and particularly Dobzhansky, Patterson, and their associates have shown that several alternative gene arrangements may coexist in a population, and that such polymorphism of gene arrangements is very frequent in natural

Fig. 7–3. Change of gene sequence by single inversion of the chromosome segment *B C D E*. The looping permits the pairing of the original and the inverted chromosomes. (From Dobzhansky 1951.)

populations of most species of *Drosophila*. The third chromosome of *Drosophila pseudoobscura,* for instance, has at least 16 different known gene arrangements, and *D. willistoni* has more than 50. These are not randomly distributed through the range of the species, but each arrangement reaches maximum frequency in a certain area and may be entirely absent elsewhere, particularly in peripheral populations. This distribution pattern was at first ascribed to historical accidents of sampling, but it is now evident that it is controlled by selection (Mayr 1945; Dobzhansky 1951). Recombination is greatly reduced in the inversion heterozygotes in the inverted chromosome section. The gene complex in the inverted section will therefore tend to function as a unit, as a "supergene." It will be selected particularly for its viability in heterozygous condition (Dobzhansky 1951; Wallace 1954a). The experimental study of these gene arrangements in population cages has contributed greatly to our understanding of polymorphism and of the variability of populations in general.

Discontinuity of the Phenotype and Selection. Polymorphism differs from other kinds of genetic variability of populations in only one respect, the production of several discontinuous phenotypes coexisting in the same population. If the visible phenotype itself has a definite selective value, as in organisms that occur on differently colored substrates (*Cepaea,*

Littorina), a selection model can be constructed. But there is no apparent selective advantage in the many possible combinations of deletion of one or several of the five bands of *Cepaea nemoralis,* or in the number of spots in coccinellid beetles. Why is the tail spot round in some specimens of *Xiphophorus maculatus,* but shaped like a half-moon or a comet in others (Fig. 7–4)? The answer is, in part, that the genotype is the result

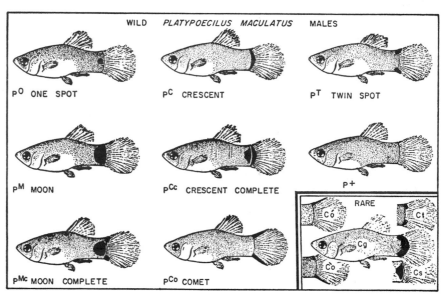

Fig. 7–4. Eight tail-pattern morphs in the fish *Xiphophorus maculatus,* produced by a single gene locus with multiple alleles. The five additional morphs in the insert are rare. (From Gordon 1947.)

of selection for all aspects of the phenotype, not only the visible ones (Dobzhansky 1956a). The particular aspect of the visible phenotype is one problem, the sharp discontinuity between different phenotypes, another.

In the days of early Mendelism there was a tendency to disregard and belittle the occurrence of continuous variation. The pendulum now has swung to the opposite extreme. Every case of phenotypic polymorphism has been claimed by some recent authors to be the result of "disruptive selection." It seems to me that this is true only in a small minority of cases. How could it apply to the spots in coccinellids, the bands in snails, the blood groups or the color phases in birds? Indeed, I know of only two kinds of discontinuous phenotypes in animals, those that occur in mimetic and sexual polymorphism, for which disruptive selection seems to be responsible. In all other cases, phenotypic polymorphism seems to have been the direct result of the original mutation and is maintained either because it contributes directly to fitness (substrate polymorphism) or because the same gene has physiological side effects that contribute to fitness, or both. This statement does not ignore the fact that natural selection may modify the dominance of such morphs and other properties that affect fitness.

Homologous Polymorph Series. Certain types of polymorphism are characteristic of whole genera or families of animals. Polymorphism in banding is exceedingly widespread among gastropod mollusks (snails), occurring in terrestrial (for instance, *Cepaea, Liguus, Hemitrochus*) as well as in marine genera (for instance, *Thais, Littorina*). Spotting polymorphism is widespread in coccinellid beetles, albinism in pierid butterflies, blood groups in mammals and birds, chromosomal inversions in *Drosophila* and other dipterans, and supernumerary chromosomes in grasshoppers, to mention merely a few conspicuous examples. The wide distribution of a single type of polymorphism in entire families and orders suggests that it not only has an important selective significance, but also is of considerable phylogenetic antiquity. Presumably at some time in the history of the phyletic branch a particularly valuable heterotic mechanism became tied up with phenotypic polymorphism and has subsequently been maintained in the entire group. The visible phenotype, such as spotting or banding, is maintained not by selection, but because it is a by-product of a genetic physiological mechanism that is favored by selection, and because the phenotype is not selected against.

In polymorphism, as in so many other areas of evolutionary research, the major open problem concerns the relation between the genotypes favored by selection and the development of the visible phenotype. For instance, it is known that polymorphism is due to the penetrance of an allelic series into the visible phenotype. Yet every population has allelic series at scores, if not hundreds or even thousands, of loci. Nearly all of these are cryptic; they are not expressed in the visible phenotype. Why is visible polymorphism restricted to so few loci?

Furthermore, among two common sympatric genera, why should one (such as *Adalia*) be highly polymorphic while a related and rather similar genus (such as *Coccinella*) contains largely monomorphic species? Likewise, chromosomal inversions are abundant in some species of *Drosophila*, but rare or absent in others. The common and successful species *D. virilis* has no chromosomal inversions, in contrast to all of its close relatives.

There are fewer inversion differences between the full species *mulleri,* *aldrichi, arizonensis,* and *mojavensis,* or between *repleta, neorepleta,* *melanopalpa, canapalpa,* and *limensis,* than occur in heterozygous state in many populations of *pseudoobscura, melanica, subobscura,* and *willistoni* (Patterson and Stone 1952; da Cunha 1955). Notwithstanding the well-established genetic diversity of widespread, successful species (Chapter 13), there is no real explanation, so far, for the phenotypic uniformity of some species and genera and the polymorphism of others. Many forms of phenotypic polymorphism are, however, favored by selection (Chapter 9).

The Genetic Basis of Characters

The existence of polymorphism raises numerous questions. For instance: Why are not all individuals of a population identical in appearance? Is it because diversity is of selective advantage to the population, or is the polymorphism tolerated for exactly the opposite reason—that the phenotypic differences are neutral as far as selection is concerned?

The relative merits of the two alternatives have been argued ever since anyone began to think about the possible evolutionary significance of polymorphism, and the controversy has by no means died down. At one time I considered most polymorphism neutral (Mayr 1942). I asked, why should a snail with a single dark band on its shell be favored by selection over an individual with three or five bands (or vice versa)? Or, why should the number of spots on the elytra of a ladybird beetle (coccinellid) or the shape of the black spot on the tail of a fish affect the viability of its carrier? Such questions reveal a serious misunderstanding, since they are based on the unspoken assumption that a gene does nothing but produce a single character, and that the fate of a gene in a population depends on the "value" of this visible character it produces. Nothing could be more misleading, for the "one gene—one character" hypothesis of early Mendelism fails to take into account the fundamental difference between genotype and phenotype. An understanding of the relation between gene and character is the basis of an understanding not only of polymorphism, but of all genetic variation of populations. A discussion of this problem, thus, must precede a discussion of the factors that are responsible for the maintenance of polymorphism and other forms of genetic variation.

The Physiology of the Gene. The visible phenotype is only a limited part of all the manifestations of the genotype. A pattern of coloration is the end product of a complex physiological process of differentiation and it is highly improbable that any gene which affects the physiology of an organism should be without effect on viability. A gene elaborates a "gene product," presumably an enzyme or part of an enzyme, which interacts with the respective products of other genes. These products diffuse into other cells, stimulate or inhibit growth, and affect the total physiology in an unexpectedly large number of different ways. That any of these gene actions should be without effect on the probability of survival in any of the almost unlimited number of environmental situations in which an organism may find itself appears altogether unlikely. Indeed, as we shall presently see, there is no evidence for the existence of genes that remain selectively "neutral" in all physical, biotic, and genetic environments.

Multiple Gene Effects (Pleiotropy). The capacity of a gene to affect several different aspects of the phenotype is called pleiotropy. Every gene that has been studied intensively has been found to be pleiotropic to a greater or lesser extent (Caspari 1952; Grüneberg 1952; Kühn 1955; Hadorn 1956). It is obviously naïve to regard a gene as a mold into which a character is poured. There is no such mold for the antlers of a stag or the coloration of a bird of paradise. The diversity of manifestations of a gene is often surprising. Nearly every known "coat-color" gene of the house mouse seems to have some effect on body size. Of 17 X-ray-induced eye-color mutations in *Drosophila melanogaster*, 14 showed definite effects on such an apparently altogether unrelated character as the shape of the sclerotic spermatheca of the female (Dobzhansky and Holz 1943).

More important is the fact that pleiotropy often affects the very characters that are of the greatest evolutionary importance, such as fertility, fecundity, sexual vigor, longevity, and tolerance of environmental extremes. In well-studied organisms, such as *Drosophila* and *Ephestia*, it has been found that nearly every known mutation affects one or another component of fitness, usually several simultaneously. Among 33 randomly chosen mutant strains of *Drosophila melanogaster* tested for their ability to fly, it was found that it was unimpaired in less than a third, while 24 percent had completely or nearly completely lost the ability to fly even though in many of them the wings were structurally completely normal (Williams and Reed 1944). Larvae of a DDT-resistant strain of houseflies (*Musca domestica*) developed faster and survived better at high population densities than larvae of a susceptible strain; both strains descended from a single parental strain through divergent selection (Bøggild and Keiding 1958).

A gene (*a*) affecting the pigmentation of the eye in the flour moth

Ephestia kuehniella has ten other known morphological and physiological effects (Caspari 1949; Table 7–2). Further analysis showed that the basic effect of this gene is its inability to synthesize kynurenin, a precursor of tryptophane. The pleiotropic action of this gene appears to be due to its effects on every biochemical process in the body in which the kynurenin-tryptophane reaction plays a role. This illustration may be used as a

Table 7–2. Pleiotropic effects of the genes a^+, a^k, and a in
Ephestia kuehniella (from Caspari 1949).

Characteristic	a^+a^+	a^ka^k	aa
Pigmentation characters			
Eyes	Black	Brown	Red
Brain	Brown		Light pink
Testes	Pigmented	Weakly pigmented–colorless	Colorless
Larval ocelli	Strong pigmentation	Intermediate pigmentation	Weak pigmentation
Larval hypodermis	Pigmented	Colorless	Colorless
Proteins	Pink		White
Chemical constitution			
Kynurenin	Present	Reduced?	Strongly reduced or absent
Protein tryptophane	Normal	?	Increased
Ether-extractable substances	Normal	?	Reduced
General biological characteristics			
Viability	Normal	Reduced	Reduced
Speed of development	Normal	Strongly reduced	Reduced
O₂ consumption	Normal	?	Reduced?

general model of the action of genes, although every level of differentiation may be involved.

The butterfly *Colias eurytheme* in North America has among females a normal orange and a white morph. Observations by Hovanitz (1948b, 1953) suggest that the orange morph is more tolerant of heat and the white morph of cold. This is indicated by geographic distribution as well as by the seasonal fluctuations in frequency. In northern California 74 percent of the butterflies belong to the white morph; in southern California, 13 percent. At Mono Lake, California, in each of two years, the white morph dropped from an early-season (May–June) frequency of 60 percent to a low of 15–30 percent in August, followed by a rise to the spring level in September and October. This indicates that the white morph differs in its survival at different seasons and localities. For details on the genetics of butterflies see Ford (1946, 1953), Remington (1954, 1958), and Sheppard (1961).

In both species of banded snails of Europe, *Cepaea nemoralis* and *C. hortensis,* Sedlmair (1956) and Lamotte (1959) found differences in temperature preference and tolerance between different morphs. There was, however, an interaction between the gene for polymorphism and the residual genotype, so that identical phenotypes from different colonies did not always show identical physiological reactions. On the whole, the yellow and unbanded snails are better adapted for dry and warm conditions, the red and banded snails for cool and humid environments (see also Chapter 9).

Such physiological effects have been found not only for visible morphs, but also for cryptic genotypic differences. Dobzhansky (1947c, 1951) and his associates have analyzed the multiple ways in which the different gene arrangements in *Drosophila* differ in rate of survival under different conditions of temperature, moisture, kind of food, crowding, and so forth.

The human blood-group genes have in the past been held up as an exemplary case of "neutral genes," that is, genes of no selective significance. This assumption has now been thoroughly disproved, lending weight to E. B. Ford's contention (1945 and earlier) that the type of polymorphism displayed by the human blood groups was indicative of a selective balance. The first unequivocal proof of a correlation between blood-group constitution and susceptibility to disease was supplied by Aird *et al.* (1953), who showed that the frequency of blood group A among patients with stomach carcinoma was about 10 percent higher than in the remainder of the population. Evidence of viability effects of blood-group genes has increased rapidly since then (Fraser Roberts 1957). In duodenal ulcer there is an excess of O (about 17 percent), likewise in gastric ulcer (about 10 percent), while there is an excess of A in pernicious anemia (about 13 percent) and in diabetes (8 percent). Other diseases for which there seems to be evidence of an interaction with blood-group genes are pituitary adenoma (increased O), hip fractures (increased A), fatal bronchopneumonia in infants (increased A), and portal cirrhosis (increased A). The frequency of births of certain genotypes deviates from expectancy among mothers of certain genetic constitutions (Levine 1958). Although there are also conflicting data, indicating interaction with local genetic backgrounds, the pleiotropic effects of blood-group genes can no longer be questioned.

An understanding of pleiotropic gene action is the key to the formerly puzzling problem of "neutral polymorphism." By this term had been designated those cases of polymorphism, such as the degree of banding in snails or of spotting in coccinellid beetles, in which there is no apparent selective difference of the alternative visible phenotypes. Such neutral polymorphism, it was claimed, was maintained by "accident." Now that the cryptic physiological effects of "neutral" genes have been discovered, it is evident that such genes are anything but selectively neutral. It is altogether unlikely that two genes would have identical selective values under all the conditions in which they may coexist in a population. The strong geographic variation of most cases of polymorphism (Chapter 11), often closely paralleling climatic gradients, is further evidence of the existence of correlated physiological effects of polymorphic genes. The explanation of balanced polymorphism (Chapter 9) is likewise based on cryptic physiological differences of genotypes. Cases of neutral polymorphism do not exist, as has long been maintained by Ford (1945).

Pleiotropic gene action is the key to the solution of many other puzzling phenomena. The survival of a gene in a gene pool depends on its total contribution to "fitness" and not on the contribution to fitness made by the visible phenotype. Color, pattern, or some structural detail may be merely an incidental by-product of a gene maintained in the gene pool for other physiological properties. The curious evolutionary success of seemingly insignificant characters now appears in a new light.

Some Remaining Puzzles. Recent studies in population genetics have revealed the universal occurrence and extraordinary frequency of genetic polymorphism in animal populations. New cases are found whenever a new approach or technique is applied, such as the study of blood groups, of metabolic genes (Allison 1959), of gene arrangements in the salivary-gland chromosomes and other chromosomal variations, of lethal and sublethal genes, or of isoalleles, to mention only a few. If we consider physiological characters as part of the phenotype, it is legitimate to state that apparently all species are polymorphic and each species is polymorphic simultaneously at several loci. The reason why this genetic polymorphism is usually overlooked is that most of it is cryptic. One may ask, why is only such a minute fraction of the total genetic variation of a population expressed in the visible phenotype?

The occurrence of phenotypic polymorphism is easily explained where the variability of the phenotype itself is being selected for, as in the mimetic polymorphism of butterflies and of the syrphid fly *Volucella,* or in the substrate-adapted component of the polymorphism of *Cepaea nemoralis.* Yet there is no explanation so far for the polymorphism of elytral patterns in coccinellid beetles or for the infinite variety of spotting patterns in the fishes *Lebistes* and *Xiphophorus.* Why this penetrance into the visible phenotype in some species when cryptic polymorphism is just as effective in the majority of species?

It is equally difficult to explain the phenotypic discontinuity in cases of polymorphism. Such polymorphism evolved, says Huxley (1955b), to permit species to cope with the variability of the environment: "all its major components—temperature, humidity, availability of food, abundance of enemies, shelter, and many other factors—[vary] both in space . . . and time." Yes, but most of this variation of the environment is continuous. What is the object or advantage of coping with it by developing a discontinuous polymorphism of the phenotype that in itself is not of selective value? With so much variation continuous, what is the significance of the discontinuities between spotting patterns on the elytra of lady beetles or the wings of grouse locusts? No completely satisfactory answer to these questions is, as yet, available.

The demonstration of the selective differences of the various morphs in a polymorph population leads to still another question: Why does selection not transform inexorably all polymorphism into monomorphism? Since, as Fisher (1930) showed, even exceedingly slight differences in viability eventually lead to the virtual elimination from a population of "inferior" by "better" genes, one would expect polymorphism to be a rare and transient condition. Since this is not the case, it is obvious that mechanisms must exist that maintain a balance of the various competing genes within the population. These mechanisms will be discussed in Chapter 9, together with the entire broad problem of the maintenance of genetic variability in populations. In that context some further aspects of polymorphism will be considered.

THE GENETIC VARIABILITY OF WILD POPULATIONS

The relative phenotypic uniformity of the individuals of most species led to the assumption of early Mendelism that all such individuals are "homozygous wild types," except for a few visible aberrant individuals, the "mutants." As is now realized, this typological description of natural populations is completely erroneous. All populations are variable and some of this variation, as we have seen in the preceding section, is visible, resulting in polymorphism or in polygenically controlled continuous variation. An even greater part of genetic variation in populations is concealed owing to various genetic and developmental devices and does not become apparent until it is revealed by inbreeding or other experimental procedures. Spencer (1947a) and Dobzhansky (1951, chap. 3) have given illuminating surveys of such concealed variation. For instance, Spencer (1947a,b) found that in two populations of *Drosophila immigrans*, sampled in two different years, 110 flies carried 47 visible mutants in 1944 and 51 in 1946, of which 33 (1944) or 24 (1946) were at three loci. Similar frequencies were found in *D. hydei* and other species. Bösiger (1953), in two wild populations of *D. melanogaster* in western Europe, found 274 visibles in the offspring of 128 females and 628 visibles among the offspring of 316 females respectively. Some investigators have concentrated on lethals and other genetic factors affecting viability and have found that these occur in wild populations at astonishingly high frequencies (Dobzhansky 1951, 1959a). None of this genetic variability would have been suspected if it had not been revealed by inbreeding or special techniques. With adequate tests it can be shown that indeed no two individuals of a sexual species are genetically identical. This, of course, has long been known for the human species, and extension of this finding to all species does not really come as a surprise. The failure of tissue grafts among individuals of species of mammals has long been considered an indication of the amount of genetic difference among individuals (Haldane 1954a:111; Medawar 1957).

There is no precise information available on the average genetic difference between individuals of the same deme. The estimate that they may differ in 1 percent of the mutable loci is probably not exaggerated, considering, for instance, in how many traits any two human individuals differ. Yet this might mean an average difference in well over 100 genes. If one could count all the alleles and isoalleles that coexist in a single population, one might get a staggering figure. But even with more modest assumptions, one must still admit the presence of an enormous genetic variability in every natural population. This raises a number of questions: Where does the variation come from? Is it maintained by or in spite of natural selection? What is its function? How can this variation be reconciled with the need of populations for adaptedness? The answers to these questions are of the utmost importance for the understanding of evolution. The remainder of this chapter and Chapters 8 and 9 will be devoted to their discussion.

SOURCES AND MAINTENANCE OF INTRAPOPULATION VARIATION

Every local population is adapted, through natural selection, to the specific environment in which it lives. It is sometimes concluded from this fact that a particular genotype will have optimum survival value at a given locality and that therefore natural selection should cause every local population to become genetically uniform for this genotype. This at first sight logical consideration is in conflict with the observed variation of populations as well as with the obvious "need" of populations for genetic variability to serve as material for an evolutionary response to changing conditions. This contradiction worried Darwin and the early Darwinians considerably. Since they considered inheritance to be blending, they had to assume that, owing to blending, one-half of the total variability would be lost in each generation (Chetverikov 1926; Fisher 1930). At the same time, the availability of an inexhaustible supply of genetic variation was one of the cornerstones of Darwin's theory of evolution. These two assumptions were completely in opposition to each other. There seemed to be only one answer to this conundrum, a truly colossal rate of mutation. Furthermore, in order to maintain adaptedness at all times in spite of this avalanche of mutations, genetic changes had to be appropriate. This, in turn, seemed possible only if the genetic changes were induced as an appropriate answer to a particular set of environmental demands, that is, if they were adaptive. Darwinian considerations of natural selection on the basis of the stated genetic premises thus led "logically" to a Lamarckian interpretation of the induction of genetic changes. It is not surprising then that Darwin himself became converted to Lamarckism (*Origin of Species*, 6th edition) and that some of the best-informed students of evolution around the turn of the century were Lamarckians.

It is one of the many achievements of the science of genetics to have shown that several of the major premises of this "logical" chain of arguments were wrong (Weismann had already undermined others). To be sure, it is true that ultimately all genetic variation is due to mutation (by definition!), but it is not true that a uniform genotype is the highest pinnacle of possible adaptation. Rather, a specific amount and very definite type of genetic variation may actually enhance the adaptedness and adaptability of a population.

The advantages of increased genetic variation are evident: the greater the number of genetic types within a population, the greater the probability that the population will include genotypes that will survive seasonal and other temporal changes, particularly those of a violent nature. If there are drought-resistant genotypes in a population that normally lives in a humid environment, the population will have a chance to survive an abnormal period of drought during which the humidity-dependent genotypes are lost. Genetic variability will also permit a greater utilization of the environment because it will make it possible to colonize marginal habitats and various subniches. It will counteract too great a specialization and will give elasticity.

Yet too much genetic variability will inevitably result in the wasteful production of many locally inferior genotypes. Inviable hybrids and balanced lethals illustrate this extreme. Extreme genetic variability is as undesirable as extreme genetic uniformity. How then does the population avoid either extreme? As so often in the realm of evolution, the solu-

Table 7-3. Factors influencing the amount of genetic
variation in a population.

A. Sources of genetic variation (Chapter 7)
    1. Integrity of genetic factors
        (a) Particulate inheritance (Hardy-Weinberg)
    2. Occurrence of new genetic factors
        (a) Mutation
        (b) Gene flow from other populations
    3. Occurrence of new genotypes through recombination
B. Factors eroding variation (Chapter 8)
    1. Natural selection
    2. Chance and accident
C. Protection of genetic variation against elimination by selection (see Table 9-4)
    1. Cytophysiological devices
    2. Ecological factors

tion is a dynamic equilibrium between two opposing forces. Since adaptation is maintained through natural selection, the most important question is: how can a population retain an adequate supply of genetic variability in the face of the relentless ravages of natural selection? This question can be answered only after a careful analysis of all the factors influencing the amount of genetic variation in a population.

A glance at Table 7-3 shows that one can group these factors essentially into three classes. The first class (A) consists of the mechanisms that produce new variation directly (mutation) or indirectly (recombination, gene flow) and that make it possible to maintain the integrity of these factors from generation to generation (particulate inheritance). The second class (B) consists of the forces that tend to reduce the amount of genetic variation either by natural selection or by chance elimination (Chapter 8). The third class (C) consists of the numerous phenomena and devices that protect genetic variation against the eroding effects of natural selection and of chance elimination (Chapter 9). The relative importance of these factors and their interplay deserve detailed discussion (see also Wright 1949a; Haldane 1954a).

Integrity of Genetic Factors

Particulate Inheritance. The naïve view of inheritance assumes that the genetic potencies of father and mother merge in reproduction and appear in the offspring as a new blend of the paternal and maternal characters, analogous to the mixing of two colored fluids. Virtually all genetic theories of the 19th century, with the spectacular exception of that devised by Mendel, were based on this concept. Curiously, this is true even of those genetic theories that postulated particles as the carriers of inheritance. It was Mendel's great contribution to realize from his experiments that the genetic factors of father and mother, when they combine in the zygote, do not lose their identity, but reassort in the next generation. The full story of the Mendelian theory of particulate inheritance is so well told in various textbooks of genetics (for example, Srb and Owen 1952; Sinnott, Dunn, and Dobzhansky 1958; Dobzhansky 1955a) that we shall not repeat it here.

The hereditary material remains unchanged from generation to generation, unless altered by mutation. A white-eyed *Drosophila melanogaster* may be backcrossed for hundreds of generations to heterozygous red-eyed flies without either the white becoming pinkish or the red becoming diluted by white. This fact, now accepted in genetics as axiomatic, has the most profound evolutionary consequences. Particulate inheritance insures that the variability of a (large) population remains the same from generation to generation under conditions of random mating if the genes are of equal selective value, an enormous contrast to the theory of blending inheritance.

This basic law of particulate inheritance can be expressed in simple mathematical terms. Let us begin with the hypothetical case of two alleles A and a, of equal selective value, occurring together in a population. If the frequency of gene A is q, then that of a is $1 - q$, if A and a are the only alleles at this locus. If we assume random mating, then male gametes $qA + (1 - q)a$ fertilize eggs $qA + (1 - q)a$, producing offspring (zygotes)

$$[qA + (1 - q)a] \times [qA + (1 - q)a]$$
$$= q^2AA + 2q(1 - q)Aa + (1 - q)^2aa \qquad (1)$$

according to the well-known binomial law. Formula (1) gives the total frequency of zygotes (genotypes) in the population. How to calculate the frequency of the genes from the observed frequency of the genotypes is explained in the textbooks of genetics mentioned above.

In a polymorphic species in which more than two alleles coexist or several genes interact, the calculation is somewhat more complex. Such calculations were made, for instance, by Bocquet, Lévi, and Teissier (1951) for a highly variable population of the isopod *Sphaeroma serratum* and by Gordon and Gordon (1957) for *Xiphophorus maculatus;* see also Li (1955a) and Stern (1960).

The law expressed in formula (1) is called the Hardy-Weinberg law, expressing in mathematical terms the fact that owing to particulate inheritance *the frequency of genes in a population remains constant in the absence of selection, of nonrandom mating, and of accidents of sampling.*

Particulate inheritance thus explains how genetic factors may remain indefinitely in a population. What the original source of these genetic variants is, is the next question that needs to be answered.

Occurrence of New Genetic Factors

Mutation. Evolution means change. To be more precise, it means the replacement of genetic factors by others. The ultimate source of genetic novelties was the great unsolved problem in Darwin's theory of evolution. It has become customary in genetics to refer to the origin of a new genetic factor as a *mutation,* which may, with convenient vagueness, be defined as "a discontinuous change with a genetic effect" (Mayr 1942).

The term "mutation" has had a tortuous history, which has not yet been properly presented. In the 17th century the term was repeatedly used for any drastic change of form, whether in insects (Hartlib 1655, *fide* Zirkle *in litt.*) or in paleontology. Robert Hooke, when discussing the use of fossils for the establishment of geological chronology, wrote (about 1675): "And tho' it must be granted, that it is very difficult to read them [the "characters" presented by fossils], and to raise a Chronology out of them, and to state the intervalls of the times wherein such or such Catastrophies and Mutations have happened; yet 'tis not impossible."

The two usages continued in a parallel manner for the next 200 years, with Spring (1838) and others referring to mutation as a change from the "type," in a more or less genetic sense. In paleontology the term was proposed in a more formal manner by Waagen (1869) to denote sudden changes in phylogenetic lineages. Although Waagen's usage became quite general in paleontology during the latter third of the 19th century, this was ignored by De Vries (1901) when he adopted the term "mutation" for sudden genetic changes. In this genetic meaning the term has become so important and of such universal application in biology that it has been virtually abandoned by the paleontologists. However, even within genetics the term "mutation" has changed its meaning repeatedly. The mutations in *Oenothera* on which De Vries based his term were a heterogeneous medley of all sorts of drastic deviations from the "type." Most of them were segregates from balanced heterozygous complexes; others were cases of polyploidy. Only a few were genic changes involving a single locus. Nothing could have been more confusing than De Vries' application of the term "mutation" to all sorts of diverse phenomena with only one feature in common, a drastic variation of the genotype. The violent objection of a large segment of contemporary biologists to De Vries' mutations and to the far-reaching speculations he based on them was justified and not surprising. De Vries concluded that a single mutation could create a new species, while the more experienced naturalists recognized that new species arose from isolated populations by the gradual accumulation of small differences (Jordan 1905). As a result of De Vries' exaggerated claims there was an almost absolute cleavage in biology during the period from 1900 nearly to 1930 between the mutationists and the naturalists. "The mutation theory was often looked upon [by the Mendelians] as a modern substitute for the largely outmoded hypothesis of Darwin" (Stebbins 1950).

In due time it was realized that the spectacular De Vriesian mutations were exceptional phenomena and that the normal genetic changes are "small" mutations (Baur, East, Johannsen, Morgan) which have the same mode of chromosomal inheritance and Mendelian segregation as the De Vriesian mutations but have only slight or even invisible effects on the phenotype. The term "mutation" was consequently redefined to include these slight genetic changes, with the result that the argument of the naturalists ("evolution is due to small variations rather than to spectacular saltations") was no longer in conflict with the findings of genetics. The complete shift in the meaning of the word "mutation" since De Vries has unfortunately escaped the attention of some of the older zoologists, who still attack the evolutionary role of mutations with arguments that have no validity whatsoever when applied to small and cryptic mutations.

The physiological interpretation of the nature of mutations is still somewhat controversial. Customarily (for instance, Dobzhansky 1951:28) mutations are divided into gene mutations, "presumably caused by chemical alterations of individual genes," and chromosomal mutations, "changes of a grosser structural kind, which involve destruction, multiplication, or spatial rearrangement of the genes in the chromosome." Since there is no recognizable difference between the phenotypic effects of these two classes of mutation, gene mutations have also been characterized "as the residuum left after the elimination of all classes of hereditary changes for which a mechanical basis has been detected." Such a highly unsatisfactory definition of gene mutation was largely due to the even more difficult problem of defining the gene. Fortunately, in this endeavor considerable progress has been made in recent years.

The Nature of the Gene. To define the gene as the unit of inheritance is no longer sufficient. Attempts to replace this definition by a more precise one or by an operational definition have led to contradictions and controversies. Until recently the approach to the gene was very indirect and had to be very speculative for this reason. The majority of older authors considered the gene to be a corpuscle on the chromosome with three characteristics: that of definite function, that of capacity to mutate, and that of being the smallest unit of recombination. An increasing amount of evidence has accumulated during recent decades indicating that these three identifying characteristics of the gene do not necessarily coincide (Pontecorvo 1958).

The new concepts of the complex gene locus, such as *cis-trans* effects, and the difference between the units of mutation, recombination, and physiological function were first worked out in *Drosophila* and maize (Stephens *et al.* 1955). They were substantiated and worked out in far greater detail with the help of microorganisms (bacteria, phages) in which the resolving power of the genetic analysis is much greater than in higher organisms. This not only permits the recording of very rare events (of mutation as well as of recombination) but permits also much more precise quantification. On the other hand, it is possible that the process of recombination in higher organisms (with structured chromosomes and meiosis) is somewhat different from that in bacteria and phages.

In spite of these doubts and difficulties, it is possible to sketch out the essentials of our present knowledge. Fortunately, most of the controversial questions are of such a nature that, from the point of view of evolutionary biology, it is immaterial which of various alternatives is correct. Natural selection, after all, deals with phenotypes, while mutation and recombination merely replenish the reservoir of genetic variation.

We owe to Avery, MacLeod, and McCarty (1944) the fundamental discovery that nucleic acid rather than protein (as formerly believed) is the genetic material. A double-stranded coil of deoxyribonucleic acid (DNA) is, in most organisms (including all animals), the carrier of the genetic information (Watson and Crick 1953). A definite sequence of two kinds of nucleotide pairs (adenine-thymine, guanine-cytosine) permits the formation of a linear code—for instance, AT, GC, CG, TA, TA, CG, AT, and so on. A given sequence is believed to determine the formation of a given enzyme. An enzyme, being a protein, is composed of amino acids, and it appears possible that individual sites within a functional gene affect the synthesis or "coupling" (into polypeptide chains) of different amino acids (Ingram 1956). This is also suggested by findings such as those of P. E. Hartman (1956) on the gene loci affecting histidine synthesis in the bacterium *Salmonella typhimurium*. A change at any one of 35 known sites of the *his C* locus will affect production of the enzyme imidazole-acetol-phosphate ester transaminase, whereas a change at any of 25 known sites of the *his B* locus, which is adjacent to *C*, will affect the production of another enzyme, imidazole-glycerole-phosphate ester dehydrase. Most "loci" in higher organisms seem to be more complex aggregates. Various alleles of the *T*-locus in the mouse affect different developmental systems (mesoderm, neural tube) and differ drastically in their pleiotropic effects. In spite of the differences, it has so far been impossible to obtain crossing over between these alleles (Bennett *et al.* 1959). The same is true for many loci in *Drosophila*.

It is now evident why it has become so difficult, if not impossible, to define the gene. A change in a single nucleotide pair already causes a mutation, yet many sites together form a single functional unit, and may be reflected in the phenotype as a unit character. And as far as the unit of recombination is concerned, it seems that, at least in microorganisms, crossing over can be demonstrated between alleles, provided sufficiently sensitive techniques are utilized. The situation in higher organisms may, however, be different.

The new insight into the chemical nature of the genetic material permits a more meaningful discussion of gene mutation. A gene mutation can be considered an error in the replication of genetic material prior to chromosomal duplication. Such an error might involve a large part of the DNA molecule, or it might be confined to a single nucleotide pair. Not all gene mutations are errors of replication; some seem to be failures of replication, resulting in small deficiencies in the chromosomes. The relative frequency of the two types of error has not yet been determined, but deficiencies are probably far fewer than one in ten mutations of normal organisms. Still other mutations appear to be alterations of genes between stages of replication (Muller, Carlson, and Schalet 1961; Ryan 1956).

This account has so far deliberately ignored the presence of genetic material ("plasmagenes") in the cytoplasm. The little that is known about this material does not suggest that it plays a major evolutionary role, important though it may be for the physiology of the organism. Like the chromosomal genes, plasmagenes seem to consist of nucleic acid molecules (including possibly RNA). The cytoplasm is the scene of most of the chemical activity of the cell and the DNA of the nucleus interacts actively, through RNA, with the protein of the cell. Biologists, recognizing that most biological interactions are reciprocal or dynamic equilibria, have long wondered (for example, Mayr 1942:68) whether the activity of the gene loci could not have an effect on their own mutability. There is some oblique and tenuous evidence (Moser 1958; Jacob and Monod 1961b) suggesting such a possibility. In the higher organisms, however, there is such a long metabolic pathway from the gene to the terminal organ of the phenotype which is exposed to the environment that it appears a priori improbable that the phenotype could modify the genotype in a meaningful way.

To summarize, it is now evident that the classical concept of the gene as an entity that serves simultaneously as unit of mutation, unit of recombination, and unit of function is not correct. The unit of function, the cistron (Benzer 1957), is the largest entity. A single cistron may contain several, if not many, units of recombination. The unit of mutation is highly variable in size; it may be an entire chromosome section or a single nucleotide pair. Pontecorvo (1958) has given a brilliant survey of the problems of the gene, and I refer to Demerec and Hartman (1959), Glass (1957a), and the 1958 Cold Spring Harbor Symposium, *Recombination,* for further details.

Curiously, the improved understanding of the nature of gene and mutation has not added, so far, to the understanding of evolutionary phenomena. For most problems of evolutionary biology, particularly in higher organisms, it is legitimate to continue using the classical terminology of genes (loci) and alleles.

Mutability. The capacity of the genetic material to change, its *mutability,* is different from locus (site) to locus (site), some loci being apparently exceedingly stable. It is not clear whether such stability is an inherent characteristic of certain loci or whether mutability is a result of the location of a gene on the chromosome or both. It seems, however, that those loci tend to be more mutable that are located near "heterochromatin," a mysterious component of the chromosome, differing from the remainder of the chromosome (euchromatin) in its staining properties and in being largely inert genetically. The clear proof (Demerec 1954; Stadler 1954) that different alleles may differ strikingly from each other in mutability indicates that mutability has a large intrinsic component. On the other hand, there are now numerous cases known of mutability genes, perhaps more safely called mutability factors, that raise the mutation rate of other loci. For a discussion of such mutability-controlling factors see Ives (1950), Dobzhansky (1951), McClintock (1950), and the 1951 Cold Spring Harbor Symposium, *Genes and Mutations.*

It is exceedingly difficult to arrive at meaningful estimates of the frequency of mutations. Not only is there a difference between loci and between alleles, but there is also a difference between organisms. There is evidence (Russell 1951) that the induced mutation rate in the mouse (*Mus musculus*) is about ten times as high as in *Drosophila melanogaster*. The rate of spontaneous mutation in *Drosophila* may average about 100 or 1000 times as high as that in the bacterium *Escherichia coli*. One may estimate that in the higher vertebrates the average mutation rate per individual per generation is somewhere between 1 in 50,000 and 1 in 200,-000 per locus, although lower or higher estimates can be defended equally well. There are numerous reasons for the uncertainty of these estimates. How can one average mutable and highly stable loci? How can one estimate the frequency of mutation to isoalleles that are neither deleterious nor phenotypically visible? And how can one distinguish between the segregations of concealed recessives, which permit calculation of mutation rates, and of heterotic heterozygotes, which do not?

Even if we take the lowest estimates we come to the conclusion that a species consisting of several million individuals is bound to have a couple of mutations per locus in every generation except at the most inert loci. This conclusion differs radically from the opinion of early authors, who considered mutation a dramatic but rare phenomenon. Admittedly, mutations producing drastic changes of the phenotype are rare and any tendency to produce such mutations would certainly be selected against very strongly. The mutations that do occur on the thousands of mutable loci usually have very slight effects, which can be discovered only by special methods. East (1935) found a steadily increasing variability, generation by generation, in the offspring of a completely homozygous diploid plant of *Nicotiana rustica*. An isogenic strain of "eyeless" *Drosophila melanogaster* "gradually accumulated an abundant supply of modifiers for eye size" (Spofford 1956). Dobzhansky and Spassky (1947) demonstrated the rapid accumulation of viability-enhancing modifiers in stocks with initially deleterious chromosomes. These are merely a few examples from a large literature demonstrating the frequency of mutation. Estimates vary as to the number of loci in a higher organism such as *Drosophila* or man. Calculations based on crossing-over tests suggest a minimum of 5000 to 10,000 loci; calculations based on the chemical structure of DNA suggest the possibility of several hundred thousand mutational sites. Taking this in conjunction with the data on mutation rates, one comes to the conclusion that every individual presumably differs from every other individual on the average by at least one newly mutated site. Even if this estimate were ten times too high, nevertheless it indicates that mutation provides an enormous and steadily recurring source of genetic variation.

The Effect of Mutation. One of the objections frequently raised against the evolutionary importance of mutations is the deleterious nature of virtually all phenotypically conspicuous mutations. This is essentially correct, as far as conspicuous mutations are concerned, and there are two reasons for it. The first is that the number of possible changes at each gene locus is limited and it is therefore probable that a mutation giving the greatest fitness in the particular physical, biotic, and genetic environment has already occurred previously and is now the prevailing allele. A second reason is that any mutation so drastic as to short-circuit the normal developmental feedbacks and greatly affect the visible phenotype is likely to be of selective disadvantage. It can hardly be questioned that most visible mutations are deleterious.

On the other hand, several lines of evidence indicate that many mutations are for diverse reasons beneficial. Plant breeders, for instance, have established a considerable number of cases of the occurrence of "useful" mutations even after such drastic treatment as exposure of seeds to x-rays. Gustafsson (1954) has described mutants of this type that increase the yield of cultivated plants, may make them more resistant to fungus attacks, improve the quality of their stems or fibers, and otherwise alter their qualities. Most of these improvements are correlated, on the respective genetic background, with undesirable qualities (disturbances in pigmentation, sterility, and so on) that require intensive outbreeding and selection before the improvement can be fully utilized. New mutations that are beneficial even in homozygous condition are presumably rare in nature and restricted to such as have only small effects on the phenotype or occur after a change in the environment. Presumably more frequent is the class of mutations that are beneficial in heterozygous condition. Wallace (1958) showed by X-raying a homozygous strain of *Drosophila melanogaster* that the increased heterozygosity caused by the newly induced mutations resulted in an increase in the average viability of the otherwise homozygous individuals.

Whenever there is a drastic change of the environment there is a particularly high premium on mutations that permit coping with the new circumstances. This is best known for microorganisms, disease vectors, and agricultural pests. Whenever bacteria are exposed to antibiotics, resistant strains are rapidly established by mutation. Such resistant strains develop equally quickly among insects (flies, lice, mosquitoes, scale insects) whenever they are exposed to DDT or other insect killers. No one can deny that such mutations are "useful" to the organisms in which they occur.

The selective value of a gene, its "goodness," is determined by a complex constellation of factors in the external and internal environment. Some mutations lead to a breakdown of important metabolic processes and so are always deleterious. Others are deleterious under certain conditions but beneficial under others. In *Drosophila*, for instance, many genes are superior to the normal allele at high temperatures but inferior at low temperatures. A gene may add to viability on one genetic background but have the characteristics of a lethal on another genetic background. Such considerations (discussed in more detail in Chapter 10) make it evident why mutation and its effects on fitness have such a large element of unpredictability.

Mutation as an Evolutionary Force. We are now prepared to consider the question of the relative importance of mutation as an evolutionary force, compared with other factors such as recombination, isolation, selection, and so forth. In the early days of genetics it was believed that evolutionary trends are directed by mutation, or, as Dobzhansky (1959a) recently phrased this earlier view, "that evolution is due to occasional lucky mutants which happen to be useful rather than harmful." In contrast, it is held by contemporary geneticists that mutation pressure as such is of small immediate evolutionary consequence in sexual organisms, in view of the relatively far greater contribution of recombination and gene flow to the production of new genotypes and of the overwhelming role of selection and the prevailing epigenotype in determining the change in the genetic composition of populations from generation to generation (see also Chapters 8, 9, and 17).

A second question concerns the contribution of new mutations to the observed phenotypic variation of a population as compared to the release of potential variability "locked up" in the chromosomes (see Chapter 9). Again, the contribution of mutation appears quite small. Mather (1956) and collaborators found that only about one-thousandth of the phenotypic variation in abdominal and sternopleural bristles in a population of *Drosophila melanogaster* was due to new mutation.

Finally, how important is mutation in maintaining the genetic variation of a given population? Buzzati-Traverso (1954) pointed out the usefulness of radiation in selection experiments to supplement the available genetic variation. Scossiroli (1954) was able to raise the number of sternopleural bristles through selection to well over 40 in an irradiated line, while a control line did not respond to selection and stayed at a bristle number of about 26–28. It is doubtful, however, whether these considerations on the importance of mutations can be extended to natural populations that have, through gene flow, a vastly larger genetic input than laboratory populations. Indeed, gene flow is the principal source of genetic variation for natural populations.

Various exaggerated claims of the immediate evolutionary importance of mutations are thus not valid. Yet it must not be forgotten that mutation is the ultimate source of all genetic variation found in natural populations and the only raw material available for natural selection to work on.

Randomness of Mutations. To sharpen the contrast with Lamarckian ideas of the induction of evolutionary changes by the environment, recent evolutionists, from the Mendelians on, have stressed the "randomness" of mutations. Since this term has often been misunderstood, it must be emphasized that it merely means (*a*) that the locus of the next mutation cannot be predicted, and (*b*) that there is no known correlation between a particular set of environmental conditions and a given mutation. It does not bring into question the facts that the probability of mutation is much higher at some loci than at others and that the number of possible mutations at any given locus is severely limited by the other mutational sites of the cistron and indeed by the total epigenotype. The unity of the genotype places well-defined limits on the potential for variation (Chapter 10).

Gene Flow. All the populations studied in the genetic laboratory are closed populations. There is no input and inbreeding is high. Whenever a genetic novelty occurs in such a population it is due to a new mutation. The average wild population is an open population and thus a totally different system. Among the individuals composing a deme, many are immigrants from the outside. Depending on the species, as many as 30 to 50 percent of the members of a deme may be such newcomers in every generation. Even though data on such immigration are rapidly accumulating in the ecological literature, they are strictly measures of dispersal (Chapter 18). We may be able to determine for a deme what percentage of the individuals has newly arrived but we are not informed how different these individuals are genetically from the native population. It is to be assumed that most of the new arrivals come from adjacent populations and differ only slightly in their genetic content.

The really crucial question for the evolutionist is what percentage of the genetic novelties in a local population is contributed by new mutation and what percentage by immigration. No precise information is available, but one may guess that immigration contributes at least 90 percent, if not more than 99 percent, of the "new" genes in every local population. Several lines of evidence support such high estimates of the magnitude of gene flow. One is the phenotypic uniformity of contiguous populations, often over very wide areas. Such uniformity indicates (Mayr 1954a) that there is a high premium on stabilizing genes to counteract the disturbing effect of alien genes. A second indication is provided by the observation that dispersal curves are usually strongly skewed (Bateman 1950). Dispersal is not a purely passive phenomenon, like the scattering of ashes by a volcano, but one in which the individuals concerned participate actively, various opposing tendencies balancing each other (Chapter 18). Obviously this is an exceedingly difficult subject to study and the available evidence is meager and none too reliable. Cook (1961) has set up mathematical models that permit some first estimates.

The effect of gene flow can also be demonstrated in the cases in which it overrides the effects of local selection. Blair (1947) collected reddish deer mice (*Peromyscus maniculatus*) at a station with pinkish-gray soils 4 miles away from the red soils of the Tularosa area. At all other stations the color of the mice agreed essentially with that of the substrate. In the exceptional case reddish individuals from the densely populated Tularosa area had apparently swamped the adjacent pinkish-gray local population. Another case concerns the black races of mammals on lava flows. Well-defined blackish races have evolved only on those rocky lava flows that are completely isolated by sandy or otherwise rockless areas (Dice and Blossom 1937). In the Valencia, New Mexico, badlands, where 0.1 to 0.6 of the margin of the lava flow is in contact with desert-colored rocky areas, there has been no development of well-defined lava races, although in four of eight species the lava populations average darker than the desert populations (Hooper 1941). Taxonomists find clear evidence of gene flow in virtually every geographically variable species. For instance, the plumed Pigeon (*Lophophaps plumifera*) is invariably grayish through northern Australia from Queensland to the East Kimberley District. In the Kimberleys, however, a series of populations (*mungi, proxima*) shows the increasing effect of gene flow from the red West Australian race *ferruginea* (Mayr 1951b). More or less extensive gene exchange takes place whenever there is a secondary contact of two previously isolated populations (Chapter 13). Such gene flow usually comes from other populations of the same species. Occasionally, particularly in plants, it is supplemented by introgression, that is, gene exchange with different species (Chapter 6).

Gene flow, as will be shown in more detail in Chapter 10, is important also for another reason. Intrinsically it is a disturbing element and threatens to counteract local ecotypic adaptation by breaking up well-integrated gene complexes. As a consequence, there will be selection for those alleles that are able to produce harmonious phenotypes with the greatest number of different alleles. And, in general, there will be a selection pressure to establish all sorts of feedbacks, developmental regulators, and canalizations which, in spite of the shifts in gene contents, will tend to produce a uniform phenotype. Essentially all such mechanisms are stabilizing and tend to shield variations of the genotype against natural selection. Gene flow and its consequences are essentially a retarding element as far as evolution is concerned. Gene flow thus has far-reaching effects on geographic variation, ecotypic adaptation, speciation, and long-time evolution.

The Occurrence of New Genotypes through Recombination

It is not the "naked gene" that is exposed to natural selection, but rather the phenotype, the manifestation of the entire genotype. Genotypes rather than genes are what the student of evolution must consider. Genotypes are combinations of genes and the number of possible combinations of even a small number of genes is staggering. An organism with 1000 loci, each with 4 alleles, could produce 4^{1000} gametic types, resulting in 10^{1000}

diploid genotypes. Even a single pair of parents, differing from each other, let us say, in 100 loci, could in the course of time give rise to 3^{100} genetically different descendants. Since a given gene on different genetic backgrounds is likely to have a wide range of contributions to viability, potentially ranging from lethal to highly fit, genotypic variation is of crucial importance in evolution. Recombination, thus, is by far the most important source of genetic variation.

A haploid, asexual organism normally lacks the capacity for recombination (for exceptions, see below). It depends on mutation for its variability. By the simple device of permitting exchange of genetic material among sexual individuals the production of different genotypes can be vastly increased. This is the biological significance of sexual reproduction, as pointed out by Weismann long ago. Genuine sexual reproduction always involves the fusion of two haploid gametes resulting in the production of a diploid zygote. At some time between fertilization and gamete formation, *meiosis* takes place, a sequence of two cell divisions during which the chromosome number is halved. In the higher animals this reduction of chromosome number immediately precedes gamete formation. In many lower organisms it immediately follows fertilization. The complex and not yet fully understood details of meiosis are described in the cytological literature. For our purposes it is sufficient to point out that the paternal and maternal chromosomes tend to pair with each other during one stage of meiosis, break in several places, and exchange pieces. This is called crossing over (Fig. 7–5). The chromosomes of the resulting gametes thus are a recombination of the homologous chromosomes of the parents. The amount of genetic variability released by such crossing over is enormous. Through recombination a population can generate ample genotypic variability for many generations without any genetic input (by mutation or gene flow) whatsoever. The experiments of Spassky et al. (1958) and Spiess (1959) are convincing demonstrations of this fact.

Sexuality is, in certain situations, difficult to define. One might think that the occurrence of sexual dimorphism, the existence of sexual organs (ovaries, oviducts), and the production of eggs would be abundant evidence of sexuality. Yet a female insect with the stated properties that reproduces by permanent ameiotic parthenogenesis lacks the essential component of sexual reproduction, genetic recombination. On the other hand, among microorganisms there have been discovered in recent years a number of processes of genetic recombination that have little in common with the diploid sexuality of higher organisms, such as transforma-

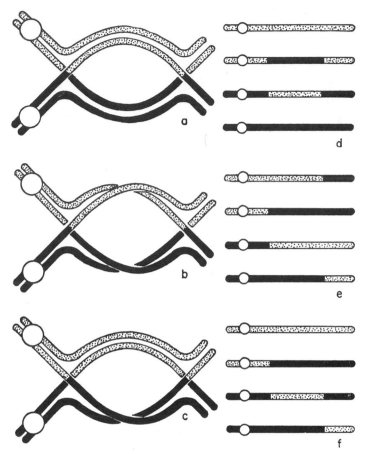

Fig. 7–5. Reciprocal (a, d), complementary (b, e), and diagonal (c, f) pairs of chiasmata. (From White 1954.)

tion, transduction, and heterokaryosis. Since these processes may serve the same function as sexuality among higher animals, they cannot be ignored in a biological definition of sexual reproduction (see Chapter 14).

There seems to be an inverse correlation between the number of generations per unit of time and the relative importance of recombination (as compared to mutation) as a source of new genotypes. In an organism with a slow sequence of generations, a new mutation can be tested on a different genetic background only at intervals of many years. In man, for instance, a new mutation may be tested once every 25 years; in a large tree in a climax forest, a new mutation may be tested only once every 100–200 years. Mutation pressure as a determinant of evolutionary change is of negligible importance in such organisms. Such types rely for evolutionary plasticity on the storage of genetic variation and the production of a great diversity of genotypes by sexual recombination.

At the other extreme are microorganisms. They owe their evolutionary plasticity to an ability to keep up with their changing environment largely by mutation. A bacterium that divides every 20 minutes would have given rise, theoretically, to 2^{72} ($\simeq 10^{22}$) individuals by the end of the day. A favorable mutation, which in these haploid organisms affects the phenotype almost immediately, will spread with extreme rapidity. Considering the enormous population size in microorganisms and the high number of generations per time unit, even a low mutation rate can provide an amount of variability that might offer all the needed material for selection in a slowly changing environment. It is possible that mutation may in part take over in microorganisms the function performed by recombination in higher organisms. I make this statement in full awareness of the fact that various forms of recombination (haploid sexuality, transduction, transformation, heterokaryotic fusion) are far more frequent in these microorganisms than is generally acknowledged. Indeed, there are obvious limits to the adaptation by mutation. Yet the rapidity of the genetic shift, compared with the changes of the environment, and the shortness of the pathway from gene to phenotype, permit microorganisms to do a great deal of adapting merely by mutating. Like Dougherty (1955) and Stebbins (1960), I believe that most currently existing forms of asexuality are secondary. Organisms would not have relinquished sexuality to such a large extent if it had not been advantageous to do so, and thus favored by natural selection. Experimental proof of the efficiency of the mutational system of adaptation has been produced by Ryan (1953), and its dynamics has been investigated by Moser (1958).

Phenotypes are produced by genotypes interacting with the environment, and genotypes are the result of the recombination of genes found in the gene pool of a local population. This is why the Mendelian population is such a crucial link in the evolutionary chain. The study of variation is the study of populations.

Summary. Most of the genotypic variation found in a population is due to gene flow and recombination. All of it, however, ultimately originated by mutation. The diversifying power of these factors is sufficiently great that no two individuals in a sexually reproducing population are identical.

8 ~ Factors Reducing the Genetic Variation of Populations

If mutation, gene flow, and recombination were to operate unchecked, the genetic diversity of populations would soon be far greater than is found in nature. Actually, a continued increase in the genetic diversity of populations is prevented by factors which, generation after generation, erode away part of the accumulated genetic diversity. These eroding factors can be grouped under two headings, natural selection and accidents of sampling. Aspects of natural selection that serve to increase genetic variation will be discussed in Chapters 9 and 10.

NATURAL SELECTION

The elimination of excess genetic variability, which is one of the functions of natural selection, is in a way a negative process and certainly a conservative one. This stabilizing power of selection, well known to Blyth and other pre-Darwinians, had long been neglected in the post-Darwinian period owing to the other equally important role of natural selection, that of the direction-giving factor in evolutionary change. The few authors who were aware of the importance of normalizing selection, like McAtee (1937: "Survival of the ordinary"), thought that this property of selection disqualified it as an evolutionary agent. These two aspects of selection cannot be neatly separated from each other, and both will be considered in our discussion of the influence of selection on the genetic variation of populations. We will discuss the basic aspects of natural selection in this chapter, but we will refer again to the evolutionary role of selection in all of the later chapters.

The concept of natural selection was the cornerstone of Darwin's theory of evolution, and anyone speaking today of Darwinism or Neo-Darwinism has in mind a theory of evolution in which natural selection plays a decisive role. However, the concept that differently endowed individuals vary in the probability of survival and reproductive success is much older. Zirkle (1941) brought together an impressive list of pre-Darwinian authors, most of whom had independently arrived at the same conclusion and had expressed it more or less definitely, and Eiseley (1959) added Blyth's name to this list. Nevertheless it was unquestionably Darwin who gave this concept general recognition. Its subsequent adoption was closely, but inversely, correlated with the prevalence of typological thinking. History shows that the typologist cannot and does not have any appreciation of natural selection. The more widely statistical or population thinking has spread in biology, the more apparent has become the tremendous significance of natural selection.

What do we mean by "natural selection"? Darwin had a perfectly clear concept of it. He emphasized again and again that various individuals of a population differ from each other in countless ways and that the nature of these differences had a decisive influence on the evolutionary potential of their bearers. An individual that may "vary however slightly in any manner profitable to itself under the complex and sometimes varying conditions of life, will have a better chance of surviving, and thus be naturally selected." Unfortunately, Darwin sometimes also used Spencer's slogan, "survival of the fittest," and has therefore been accused of tautological (circular) reasoning: "What will survive? The fittest. What are the fittest? Those that survive." To say that this is the essence of natural selection is nonsense! To be sure, those individuals that have the most offspring are by definition (Lerner 1959) the fittest ones. However, this fitness is determined (statistically) by their genetic constitution. Let it be clearly understood that what really counts in evolution is not survival but the contribution made by a genotype to the gene pool of the next and subsequent generations. Reproductive success rather than survival is stressed in the modern definition of natural selection. A superior genotype has a greater probability of leaving offspring than has an inferior one. Natural selection, simply, is the differential perpetuation of genotypes. Most of the objections raised against natural selection and its role in evolution become invalid and irrelevant as soon as the typological formulation of natural selection is replaced by one based on the probability of reproductive success.

When Darwin spoke of selection, differential mortality was foremost in his mind, even though he specifically mentioned also "success in leaving offspring." Since much of this mortality affects juveniles (occurring at prereproductive age), it leads inevitably to differential reproduction. Some differential mortality is postreproductive, however, and therefore without selective significance. At the same time, much differential reproduction occurs without mortality, yet leads to a shift of gene frequencies from generation to generation. Therefore, modern evolutionists include in natural selection any factor that contributes to differential reproduction.

Two refinements in our thinking have encouraged recognition of the importance of natural selection. The typologist interprets natural selection as an all-or-none phenomenon. He assumes that one type is better and therefore survives, while the other type is inferior and is therefore wiped out. Natural selection in this interpretation is immediate, absolute, and final. Under these premises one would expect every population to consist uniformly of perfect individuals. Since natural populations do not conform to this expectation, one must reject the importance of natural selection. If, however, natural selection is regarded as a statistical phenomenon, it means merely that the better genotype has "a better chance of surviving," as Darwin said so rightly. A light-colored individual in a species of moth with industrial melanism may survive in a sooty area and reproduce, but its chances of doing so are far less than those of a blackish, cryptically colored individual. It happens not infrequently in nature that, for one reason or another, a superior individual fails to reproduce while an inferior one does so abundantly. But the statistical probability of such an unlikely event is the smaller, the greater the difference in viability of the two genotypes. Natural selection, being a statistical phenomenon, is not deterministic; its effects are not rigidly predictable, particularly in a changeable environment.

The second improvement in our thinking is the realization that natural selection favors (or discriminates against) phenotypes, not genes or genotypes. Where genotypic differences do not express themselves in the phenotype (for instance, in the case of concealed recessives), such differences are inaccessible to selection and consequently irrelevant. Most of the phenotypic variation on which natural selection works (in sexual species) is the result of recombination and not of new mutations. The fact that the phenotype determines fitness is the reason for the extraordinary evolutionary importance of the developmental processes that shape the phenotype, as is so eloquently emphasized by Waddington (1953a, 1957). Any improvements in the "epigenotype," any genes that buffer development better against fluctuations of the environment or metabolic errors will contribute to fitness. And, as we shall see in Chapter 10, the phenotype is not a mosaic of independent characters, but the integrated manifestation of the total genotype.

The modern attitude toward natural selection has two roots. One is mathematical analysis (R. A. Fisher, Haldane, Sewall Wright, and others), demonstrating conclusively that even very minute selective advantages eventually lead to an accumulation in the population of the genes responsible for these advantages. The other root is the overwhelming mass of material gathered by naturalists on the effect of the environment. This evidence was given a largely Lamarckian interpretation in the days when mutations were believed to be saltational and cataclysmic. It became a powerful source of documentation for the selectionist viewpoint when small mutations were discovered, indeed when it was realized that all variation had ultimately a mutational origin.

Natural selection is not only the keystone of evolution, but one of the most stimulating phenomena to have challenged the human mind. Most leading evolutionists have devoted chapters of their books to it, and some entire symposia have dealt specifically with this subject. Reading these thoughtful discussions will facilitate comprehension of the intricate working of natural selection. I would like to mention especially Fisher (1930, 1954), Haldane (1932), H. J. Muller et al. (1949), Stebbins (1950, chap. 4), Dobzhansky (1951, chap. 4), Simpson (1953, chap. 5), Ludwig (1954), Wright (1956), and Lerner (1959). The selective factors of the environment are treated by Allee et al. (1949) and by Andrewartha and Birch (1954). The most systematic account in the older literature is that of Plate (1913), in some respects still readable today, even though Plate, like Darwin previously, somehow managed to combine a selectionist interpretation with Lamarckism!

The various objections raised against natural selection have been considered, point by point, by Plate (1913), Zimmermann (1938), and Ludwig (1940). Most of the criticisms appear irrelevant to the evolutionist familiar with modern genetics. Yet a discussion of some of these objections may help to further clarify the issues.

Basically, the arguments of the antiselectionists rest on an inability to appreciate the statistical nature of selection. Consequently all those objections are irrelevant that are based on imperfections of adaptations or on conflicting selection pressures. For instance, antiselectionists will point out that much of the mortality of young animals is purely accidental rather than selective, as in the case of plankton, scooped up indiscriminately by a large fish or a whale. This observation overlooks the fact that among the remaining individuals (and it is immaterial whether there remains 50-percent, 5 percent, or 0.01 percent of the population), selective factors largely determine reproductive success. A second type of false argument stems from the observation that even among protectively colored animals some will fall prey to predators, or that an insect protected by its cryptic coloration against a vertebrate predator may succumb to a hymenopteran parasite. Such arguments reveal typological thinking at its worst. No one claims that natural selection gives immortality! It merely determines the probability of survival and of relative reproductive success among the members of a population. Whatever increases this probability will be selected for, and it is immaterial in this selection what the residual mortality factors are.

Two other well-known objections are (1) that other species survive under similar circumstances, without the stated adaptation, and (2) that the species already has a device that works quite well, so why should it get another that may not even work so well?

For instance, if a desert animal has various behavior adaptations, such as nocturnal habits and secretiveness, why does it need protective coloration (Hoesch 1956)? Why should natural selection favor large body size in the coolest portions of the range of warm-blooded species, if heat can be conserved so efficiently by increasing the density of fur or plumage (Scholander 1955)? These questions ignore the fact that different ways of adaptation to the same condition of the environment are not strict alternatives, but may be superimposed on each other. No adaptation is so perfect that it would completely deprive parallel adaptive mechanisms of all selective advantage.

Antiselectionists often question the selective value of genes that add only very slightly to fitness, overlooking the fact that selection works not on individual genes but on phenotypes. A lot of very small advantages when combined into a single genotype may add appreciably to fitness. Selective values are cumulative. To illustrate this, let us imagine an arbitrary model of a population with 1000 unfixed polyallelic loci, each allele having an exceedingly slight individual effect on viability. Depending on the fluctuating environment and other factors, an allele may have a positive or a negative selective value. Owing to recombination and the particular local and contemporary constellation of environmental factors, most individuals will have an average mixture of positive and negative factors. But some individuals are bound to be in the tails of the curve of variation. Those that have mostly positive factors are bound to have a considerably increased chance to survive and to reproduce more successfully. Those that have mostly negative factors will almost certainly die without leaving offspring. Thus in every generation the frequency of the negative genes will decline and that of the positive genes will rise. This model, although crude, helps to illustrate the additive effect of viability factors.

Selection pressure is often treated as if it were a uniform condition at all times and in all places. Actually it fluctuates enormously, being most potent in times of a crisis. A severe drought, an exceptionally hard winter, a food shortage, the appearance of a new predator or disease, all such occurrences will eliminate genotypes that would "get by" under average conditions (Taylor 1934). The total effect of natural selection can be calculated only if one takes into account the crises and castatrophes experienced by every population at regular intervals. Central Europe, for instance, in 1946–47 had one of the hardest and longest winters in decades. The ground was frozen for 111 days at a locality east of Berlin, leading to an almost complete local extermination of moles (*Talpa europaea*). They died of starvation, being unable to reach their normal food (insect larvae, earthworms). Since lowered food consumption was a more decisive selection factor under these conditions than improved heat conservation, it resulted in a high selective premium on small body size. As a consequence there was a dramatic reduction in body size in the population of survivors (Stein 1951; Fig. 8–1).

"Improbable" events and constellations of genes play a role in selection difficult for the typologist to understand. Mathematicians have pointed out that evolution deals with numbers of such astronomical dimensions that even "improbable" events may occur. Most species have millions of genetically unique individuals in every generation, each producing thousands or millions of different gametes. There are thousands or millions of generations during the geological life span of each species. Under these conditions an event may become a certainty even if the chance of its occurrence is only one in a billion. Yet the total number of possible genotypes in a species is infinitely greater than the possible number of individuals.

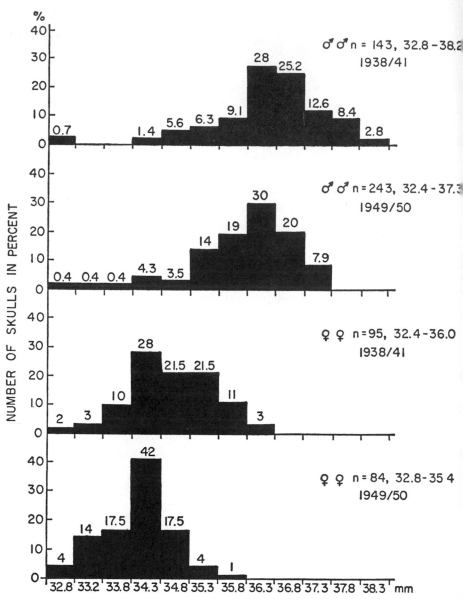

Fig. 8–1. Skull length in adult male and female moles (*Talpa europaea*) before (1938–1941) and after (1949–1950) a catastrophic winter kill. (From Stein 1951.)

Reaction Norm and Selection. The phenotype never reveals the total potential of the genotype. The same weasel that is brown in summer has the potential to be a white ermine in winter. The difference between the office worker with a soft palm and the laborer with heavily callused hands is not genetic. This interaction between genetic potential and developmental response is overlooked in a frequently cited antiselectionist argument. Many vertebrates are born with calluses where the bare skin touches the ground, as on the knees in the warthog or on the breast in the ostrich. These calluses first appear in embryos, long before they are used. On this basis it is argued that, if calluses are the reaction to friction on the skin and if there was no such friction in the embryos, the appearance of the callus in the embryo must be a case of "inheritance of acquired characters." This argument overlooks a number of points. First of all, most organs must be laid down long before they are used, outstanding examples being the eye and many parts of the central nervous system. If the presence of calluses is advantageous for the young animal, selection will certainly favor their formation at an early stage so that they are available when needed soon after birth. In the absence of preformed calluses, friction will produce blisters and inflammation, possibly leading to serious infection or feeding inefficiency. Ease of callus formation is under partial genetic control, like all other components of the phenotype. The shift of a species into a new niche, in which there is suddenly a "demand" for calluses, will set up strong selection pressure in favor of callus formation. How a given individual responds depends on the frequency in its genotype of genes facilitating callus formation. Even though the presence of such genes may be revealed only by a modification of the phenotype under exceptional conditions, it is nevertheless these genotypes that will be favored by selection. As callus-facilitating genes accumulate in the gene pool, the probability of their early penetrance into the phenotype increases. Finally calluses appear at the embryonic stage. This is an evolutionary model that has been well established for other characters and organisms.

Phenotypic Response (Phenocopy) and Selection. The same genotype may produce different phenotypes under different environmental conditions. An extreme environment may bring out developmental potencies that are not expressed under more normal conditions; it permits genetic factors to manifest themselves that do not normally reach the threshold of phenotypic expression. This threshold effect explains recent selection experiments by Waddington (1953a). When pupae of *Dro-*

sophila melanogaster, aged 21–23 hours, are given a temperature shock (4 hours at 40°C), about 40 percent of the hatching flies are "crossveinless." Stocks formed from such selected crossveinless flies respond more readily to the treatment with each generation, until crossveinless flies emerge even from untreated pupae. The simplest interpretation is based on the observation that about 60 percent of the flies do *not* respond to the shock treatment, owing to a lack or insufficient number of genes capable of responding to the treatment. Only those flies respond that have a sufficient number of the many genes contributing to the crossveinless condition (Stern 1958; Milkman 1960a,b, 1961). The presence of crossveinless genes at low frequency is not enough under normal conditions to lift the phenotype above the threshold of visibility. The treatment, however, reveals the carriers of such genes, and their continued selection permits an increasing accumulation in the gene pool of genes contributing to crossveinlessness until they express themselves phenotypically even without the shock treatment. Milkman (1960a,b, 1961) and Bateman (1959) have further analyzed the genetic and developmental aspects of this selection and Waddington (1957) has described another similar case (bithorax). The term "genetic assimilation" which Waddington uses for such situations seems to me poorly chosen, because it fails to bring out the essential point that the treatment merely reveals which among a number of individuals already carry polygenes or modifiers of the desired phenotype. What we really have is threshold selection. Many observations, previously interpreted in Lamarckian terms or as the Baldwin effect, are presumably due to the same threshold-selection effect.

Deleterious Structures. One further antiselectionist argument approximates the following statement: "Natural selection cannot be very efficient, because it permits the establishment of nonadaptive or even deleterious structures." This claim is founded on several unproved assumptions or evident misinterpretations that need to be discussed.

First of all, one can never assert with confidence that a given structure does not have selective significance. The peculiar tarsal combs of the males in certain species of *Drosophila* turned out to have an important function during copulation; the color patterns of *Cepaea* snails have cryptic significance, mitigating predator pressure (Cain and Sheppard 1950). Many aspects of the phenotype may have functions that are not yet fully understood. The visible phenotype is to a considerable extent merely the incidental by-product of a pleiotropic genotype selected for its over-all fitness. As long as selection favors certain genes and gene combinations, it matters little whether some manifestations of the phenotype are not directly adaptive, provided that they are not deleterious. There are numerous such "permissive" features in every phenotype, as we have already mentioned in the discussion of polymorphism (Chapter 7).

Many structures have different selective values at different stages of the life cycle. A genotype may be favored by selection because it is superior at an early stage in the life cycle even though it produces a disadvantageous phenotype in late adulthood or during postreproductive life. It is quite possible, for instance, that the large antlers in the Irish Elk (*Cervus megaceras*) were selected because stags with the relatively largest antlers were most successful in the first age class that participated in reproduction. This would do more to increase the frequency of genes for large size of antlers than a certain amount of counterselection against all-too-large antlers in the higher age classes (assuming throughout that large body size was of selective advantage). What is selected is the genotype as a whole. Another category of characters contains those which although not of evident selective advantage are at least not disadvantageous. Selectively neutral components of the phenotype can become established in a population or species if they are the pleiotropic by-product of a selectively superior genotype (see Chapter 10).

One final group of objections raised against the potency of natural selection concerns the assertion that such complex characters as eye or brain, which display perfect adaptation, could not have evolved gradually by chance mutation and subsequent selection. This problem will be considered in Chapter 19.

The Force of Selection

The classical calculations of the power of natural selection (Fisher 1930, Haldane 1932) were deliberately based on very slight differences in the selective value of competing genes, in order to demonstrate that evolutionary changes would occur even when one gene was superior to its allele by as little as 0.1 percent. Much evidence is now accumulating to indicate that the differences in selective values among genotypes occurring in natural populations may be as high as 30 or 50 percent. This has been shown by Gershenson (1946) for the black morph of the hamster (*Cricetus*), by Goldschmidt (1947), Haldane (1957), and Kettlewell (1961) for industrial melanism (Table 8–1), and by Dobzhansky (1951) and associates for gene arrangements in *Drosophila* (see also Chapter 9).

The rapidity with which DDT-resistant strains of the housefly (*Musca*)

Table 8-1. Differential mortality of *Biston betularia* released
in different woodlands (from Kettlewell 1961).

No. released		Type of woodland	No. eaten by birds		Percent recaptured alive	
Melanic	Pale		Melanic	Pale	Melanic	Pale
Equal		Grayish	164	26	—	—
Equal		Sooty	15	43	—	—
473	496	Grayish	—	—	6.3	12.5
447	137	Sooty	—	—	27.5	13.0

and other insects have become established or with which various pathogens have become resistant to antibiotics likewise proves the magnitude of some selection coefficients.

A quick response to the selective factors of the environment is also revealed by the sensitivity of response of the genotype to changes of the environment. This has been established for seasonal cycles in gene frequencies, as in the case of inversions in certain populations of *Drosophila pseudoobscura* (Dobzhansky 1951) and of *D. funebris* (Dubinin and Tiniakov 1945) and of color forms in the housefly (Sacca 1956), in the beetle *Adalia* (Timoféeff-Ressovsky 1940b), and in the hamster (Gershenson 1946). The magnitude of the selection coefficients permits an experimental study of such genes in the laboratory, and the magnificent work of Dobzhansky and his group on the selective significance of gene arrangements in *Drosophila* has greatly added to our understanding of the operation of natural selection.

Particularly impressive is the extraordinary sensitivity of the selective response to slight changes in the environment. The literature contains numerous records of genes or gene arrangements that are highly advantageous in one environment, say at 25°C, but neutral or deleterious in another environment, say at 16°C (Dobzhansky 1951). One typical example may be cited. Experimental populations of *Drosophila pseudoobscura* were fed on either of two species of yeasts (*Zygosaccharomyces* and *Kloeckera*). The populations contained two gene arrangements (Standard = ST and Chiricahua = CH), the heterozygotes of which were distinctly superior at 25°C. This heterosis of ST/CH disappears for the *Zygosaccharomyces* stocks at 21°C, when ST/CH and ST/ST are equal in fitness and superior to CH/CH. At 16°C no appreciable difference seems to exist in the fitness of any of the three genotypes under the two culture conditions (Dobzhansky and Spassky 1954b). The extreme sensitivity of the genotype to environmental conditions cannot be doubted even in the cases where the viability differences are not expressed in the visible phenotype. Drastic changes in a population may occur within a few generations.

In the exploitation of natural resources, great care must be taken to avoid undesirable selection. For instance, hunters who consistently shoot the stags with the finest sets of antlers set up a selection pressure in favor of inferior conformation of the antlers. The reduction in the frequency of the silver fox and the spread of undesirable qualities in several game species (Voipio 1950) may in part be due to such negative selection. Commercial fishing also sometimes sets up such undesirable selection pressures. The Pink Salmon (*Oncorhynchus gorbuscha*) of southeastern Alaska consists of many local populations, which differ in the time of spawning. Fishing is permitted only during the open season, but the opening date precedes the beginning of the salmon migration and only the closing date is effective in protecting salmon runs. As a consequence, the heaviest fishing pressure has been on the early runs of this species and has resulted in their virtual extinction. In 20 of 22 localities, there has been a consistent retardation of the date by which 75 percent of the catch has been made. In one locality, for instance, this 75-percent point was reached on July 28 in 1921, on August 2 in 1926, on August 9 in 1930, on August 16 in 1936, and on August 18 in 1944. If the open season had been arranged to protect not only the late runs, but also the early ones, fishing could not have resulted in this virtually complete elimination of the populations with genes for early spawning (Vaughan 1947).

The efficiency of natural selection bedevils man in many other ways. One is the rapid development of resistant strains of insects whenever insecticides are applied. In agriculture and horticulture new insecticides have to be introduced continuously to cope with the problem of resistance. Within 2 years after DDT had been used for the control of houseflies, resistant strains had developed independently in different parts of the world. The literature on this one subject alone is so voluminous that monographs and special bibliographies have been published (for example, Babers and Pratt 1951; Brown 1958). Microorganisms, similarly, have the capacity to develop strains resistant to antibiotics and other drugs. This resistance results from the selection of a few resistant mutations or gene combinations, exactly as in the higher organisms. In view of its practical importance, the development of resistant strains in microorganisms and insects has been studied in various laboratories. The outstanding finding of these studies is the multiplicity of pathways by which the same goal can be reached. Resistance to DDT in *Drosophila melanogaster* was reached in three very different ways in three different laboratories (Crow 1954; Sokal and Hunter 1954; King 1956). For more recent summaries see Milani (1956), Crow (1957), King and Sømme (1958), and Oshima (1958).

Disease is another selective factor of utmost importance, as brilliantly discussed by Haldane (1949c). Resistance to disease may give a decisive advantage over a competitor, individual or species, that is not immune. Different genotypes will differ in their immunological properties. Highly host-specific diseases (and types of parasitism) will have very different ecological effects (for example, making food resources available for competitors) than less specific diseases.

The Phenotype a Compromise

One of the reasons why the role of natural selection in evolution is sometimes questioned is, curiously, that it is considered not sufficiently effective. Why does a species not acquire greater running speed if this would aid in escape from predators? Why does it not produce more young? Such questions overlook the fact that the phenotype is a compromise among all selection pressures and that some of these are opposed to one another. The breeder knows how much the production of eggs in the chicken or of milk in the cow has been increased by strong selection, but he also knows that this success was achieved at a price. The high-yielding strains of domestic animals would not be able to survive in nature, exposed to the elements and in competition with other species. Fitness is a property of the total genotype. The phenotype is the product of a compromise necessitated by the need for balance. Many illustrations of this compromise have been reported in the recent literature. Lack (1954b) showed that the clutch size in birds is regulated to produce a maximum number of offspring. If too many eggs are laid, the feeding efficiency of the parent birds drops and the number of birds ultimately raised becomes smaller (Table 8–2). Kramer (1946) and Stebbins and Robinson (1946) have shown that in lizards, likewise, two opposing selection pressures determine clutch size or sizes of newborn. Where the adults have high mortality there is a premium on large clutch size, to fill empty territories. When there is little predation, as on small rocky islets or in the mountains, the adults will have great longevity and will occupy all available

Table 8–2. Survival of Swiss Starlings (*Sturnus vulgaris*) after leaving nest (from Lack 1954).

Number in brood	Number of young ringed	Number of recoveries more than 3 months old:	
		Per 100 young ringed	Per 100 broods ringed
		Early broods	
1	65	—	—
2	328	1.8	3.7
3	1,278	2.0	6.1
4	3,956	2.1	8.3
5	6,175	2.1	10.4
6	3,156	1.7	10.1
7	651	1.5	} 10.2
8	120	0.8	
9, 10	28	0.0	—
Total	15,757	1.94	—
		Late broods	
1	44	}	
2	192	} 2.3	5.8
3	762	}	
4	1,564	2.2	8.9
5	1,425	1.8	8.8
6	438	1.4	8.2
7	49	0.0	—
Total	4,474	1.99	—

territories. Young born at an advanced state will have the greatest chance for survival under these conditions which results in a selection for few but large young. Various exceptions to the ecogeographical rules are apparently due to a conflict of selection pressures (Snow 1954a; Mayr 1956; Hamilton 1961; see Chapter 11). Several cases are known in *Peromyscus* (Blair 1947; Hayne 1950) in which the border of a "substrate race" does not coincide with the border of the substrate. When a pale beach mouse extends into the area of dark soils, or a race from reddish soils extends several miles into the range of gray soils, evidently the general (physiological) superiority of the more "aggressive" race outweighs the disadvantage of insufficient blending with the substrate. Since we are dealing in these cases with recent shifts in distribution, it is to be expected that selection will correct this discrepancy in due time.

These cases illustrate very nicely the lag between a new situation and the response to selection. A similar lag is sometimes found when a mimic colonizes an area where the model is absent. This has happened, for instance, with some of the snakes that mimic the highly poisonous coral snake. The mimicry of coral snakes is indeed particularly suitable to demonstrate the manifold compromises made by natural selection. Mimicry in these snakes generally starts out from a generalized disruptive pattern of bands and may go on from there to Batesian mimicry, Mullerian mimicry, or both (Hecht and Marien 1956). The phenomenon of the so-called "superoptimal stimuli" of the ethologists is one final proof of the many compromises made by natural selection. Animals have innate tendencies to react to certain stimuli presented by the environment and par-

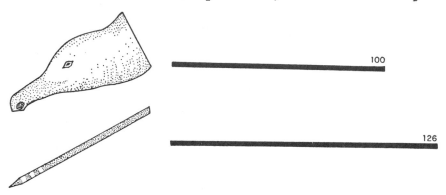

Fig. 8–2. Superoptimal bill model elicits 126 begging responses in *Larus* chicks to 100 for facsimile model of head of feeding adult. (From Tinbergen and Perdeck 1950.)

ticularly by other, conspecific individuals. The hungry chick of the herring gull (*Larus argentatus*), for instance, gives a begging response when it sees the yellow bill of the parent with a red spot near the tip. Experiments have shown, however, that a longer and thinner object with more red and more contrast will elicit a far stronger begging response (Tinbergen and Perdeck 1950; Fig. 8–2). A bill incorporating the stimuli for maximal feeding reactions of the young has not evolved, in spite of the evident selection pressure for it, because it would be a very inefficient tool for a scavenging predator like the herring gull.

The phenotype is determined in all cases of conflict by the relative force of the two opposing selection pressures. This seems obvious enough but is too often forgotten in discussions of selection. Let us take, for example, the famous case of the *fremddienliche Zweckmässigkeit* of gall-making insects and their hosts. Why, it was asked (Becher 1917), should a plant make the gall such a perfect domicile for an insect that is its enemy? Actually we are dealing here with two selection pressures. On the one hand, selection works on a population of gall insects and favors those whose gall-inducing chemicals stimulate the production of galls giving maximum protection to the young larva. This, obviously, is a matter of life or death for the gall insect and thus constitutes a very high selection pressure. The opposing selection pressure on the plant is in most cases quite small because having a few galls will depress viability of the plant host only very slightly. The "compromise" in this case is all in favor of the gall insect. Too high a density of the gall insect is usually prevented by density-dependent factors not related to the plant host.

Evidence for natural selection is so universal in nature and experimental proof of it is so abundant that it seems curious, in restrospect, that its importance should ever have been denied. Though even today there are still a few biologists who continue to deny the importance of natural selection, it hardly seems worth while to try to accumulate more evidence for its widespread occurrence. Rather, the time has come to take up several of the puzzling and partly unsolved problems associated with the phenomenon of natural selection. The two problems I would like to single out are of very different nature: one raises the question how selection operates, the other, how it affects the further course of evolution.

The Population as the Unit of Selective Advantage

Populations have a number of attributes that are of strong selective advantage to the population as a whole but seemingly not to any one individual or genotype within the population. In some of these cases it is difficult to see how selection of individuals will establish the particular balance of genotypes that produces optimum results for the population.

Aberrant Sex Ratio. Fisher (1930), Bodmer and Edwards (1960), Kolman (1960), and Shaw (1961) have demonstrated that natural selection will normally tend to equalize the parental expenditure devoted to the production of the two sexes. There are species in which more or less aberrant sex ratios occur (for many different reasons) and where the mode of selection (intra- or interpopulation) is still obscure. Deviations from the normal sex ratio (50:50) are counteracted in *Asellus aquaticus* by reduced fertility of crosses between stocks that deviate from normal in the same direction ("high female" or "high male") and heterosis in crosses between stocks that deviate in opposite directions. This tends to maintain the polygenic balance for normal sex ratio (Tadini 1958).

Two types of sex-ratio factors have been described in *Drosophila*. In one type infection of the females by spirochetes kills the eggs fertilized by Y-bearing sperms so that only daughters are produced (Poulson and Sakaguchi 1961). In the other type, represented by *D. pseudoobscura,* the condition is carried by the males. A sex-linked genetic factor causes the elimination of the Y-chromosome during spermatogenesis and an additional division of the X-chromosome. Such males produce only daughters (Wallace 1948). The opposing selection pressures by which these sex-ratio factors are maintained in populations are discussed by Shaw (1959) and Watson (1960).

Mutation Rate. It has been pointed out by many geneticists that there is presumably an optimal mutation rate for any given population. In a population that is already highly variable genetically, owing to various devices discussed in Chapters 7 and 9, additional mutation may not be "necessary," indeed it may be detrimental. On the other hand, in a genetically rather homogeneous population an increase in the mutation rate might be of selective advantage even though this may be detrimental to the specific zygotes involved, in view of the deleterious nature of the majority of mutations. How can the optimal number of "mutator genes" be established in the face of the strong counterselection?

Dispersal Rate. The opposing tendencies for sedentariness and for its converse condition, a flow of individuals from population to population, have an optimum balance. Too much dispersion destroys ecotypic adaptation, too little induces inbreeding. Again the advantage of the proper balance seems to be for the population as a whole rather than for the individual. And this is true of the whole intricate balance of factors that determine the degree of inbreeding or outbreeding of a population. Only a few of these factors benefit an individual *per se,* much as their proper balance benefits the population as a whole. How then is this balance regulated by natural selection?

There are numerous other properties of populations in which a similar conflict seems to exist between the welfare of the individual and that of the population. A priori reasoning convinces me that the conflict must be spurious. If a given condition exists, it must be the result of selection, that is, the unequal success of genotypes. Yet a more detailed analysis of such cases would seem desirable. Haldane (1932:207) was perhaps the first author to call attention to the problem of what he called "socially valuable but individually disadvantageous characters." He was particularly concerned with "altruistic traits," such as would be of importance to mankind. He found that in large populations "the biological advantages of altruistic conduct outweigh the disadvantages only if a substantial proportion of the tribe behave altruistically." However, parental care, which depends on altruistic traits, is clearly of selective advantage in man and other species.

The solution usually proposed for the difficulty raised by the conflict between a benefit for the individual and one for the population is to make the population rather than the individual the unit of selection. The species, then, is envisioned as an aggregate of competing populations, each with a fortuitously differing mixture of traits. Those with lucky combinations will prosper, those with "losing tickets" will die out. This interpretation would be sufficient if species were such imaginary aggregates of complete isolates. In reality there is extensive gene exchange among populations, generation after generation, but the effect of this factor on interpopulation selection has, to my knowledge, never been evaluated properly. It would seem preferable to search for solutions based on the selective advantage of individual genotypes, such as Fisher's explanation of an even sex ratio.

Selection for Reproductive Success

Even selection, with its almost unbelievable efficiency and sensitivity, has a weakness in its armor. Fitness is measured in terms of the contribution made to the gene pool of the next generation, that is, in terms of reproductive success. Normally, there is no better way of determining overall fitness. Yet this places a premium not only on all-around viability but also on "mere" reproductive success. Within recent years an increasing number of cases have been described in which a genotype had spread in a population not for any reasons of general superiority but merely because it was a superior reproducer. Sometimes this is comparatively innocuous, as in various forms of sexual dimorphism. Among male birds of paradise that individual which is most stimulating sexually will inseminate the greatest number of females and contribute the greatest number of genes to the gene pool of the next generation. This is the reason for the almost absurd ornaments of the males in many of those bird groups in which a single male may fertilize many females. One may assume that the most vigorously displaying males have, on the average, superior viability, yet the extreme conspicuousness of such males may maneuver the species into an exceedingly precarious position, and the appearance of a new and more efficient predator might be the doom of such a species. In nonpairing species of animals there is a high premium on the sexual vigor of males. The most active male will leave the most offspring, other things being equal. Many genes seem to have an effect on the level of sexual activity (see Chapter 5 for a more detailed discussion).

There is reason to believe that many mutant genes in *Drosophila* are deleterious primarily through being responsible for a reduction of sexual success. Reed and Reed (1950) found no difference in viability between otherwise largely isogenic stocks of wild-type and white-eyed flies, but white-eyed males were only 75 percent as successful in mating as were wild-type males. However, as shown by Morpurgo and Nicoletti (1956), factors other than female selectivity also affect the frequency of these genes in populations. That genes may have pleiotropic effects on sexual vigor has been proved also for other organisms, for instance by Caspari (1950) for a pigment gene in *Ephestia*. Male sexual vigor may vary geographically within a single species. In such cases there is a fine adjustment in each region between the sexual vigor of the males and the threshold of readiness of the females.

Selection for high reproductive success may be one of the contributory factors in the enormous fluctuations in population size among certain species of rodents or indeed any kind of animal subject to catastrophic population declines. The individuals surviving such a crash find themselves in a virtually vacant niche which places a high premium on the genotypes with the highest reproductive potential. Aphids or *Daphnia* in spring solve this problem by abandoning sexual reproduction, voles of the genus *Microtus* by early maturity and rapid succession of generations. A female field mouse, *Microtus arvalis*, may be fertilized at an age of about 13 days, before or immediately after weaning, and produce her first litter 20 days later. She copulates again almost immediately after parturition and may produce litters every 3 weeks under favorable circumstances. A female in captivity produced 24 litters in 20 months (Frank 1956). Even though this runaway type of reproduction often leads to a partial or almost complete destruction of the habitat, natural selection has evidently been unable to incorporate factors that would damp the unhealthy fluctuations in the face of the high premium for sheer reproductive success. There may be an inverse correlation between the importance of density-dependent factors for a species and the premium on reproductive success. The reproductive capacity of pelagic animals supports this hypothesis.

In all the cases mentioned genes are favored because they give reproductive advantage, but there is no clear evidence on the nature of the residual contribution of these genes to general fitness. If such genes are highly deleterious, they will no doubt bring about the extermination of the respective populations. The occurrence of genes with high reproductive advantage but with a viability so low as to lead to extermination is a distinct possibility.

Crosby (1949) has postulated that in two English populations of the primrose (*Primula vulgaris*) an unfavorable gene complex is spreading because it leads to increased pollination of its bearers. The result would be the replacement of heterostyly, normal for the Primulaceae, by homostyly. Ford (1957), however, believes that the homostyly is not spreading steadily and that the balance in the polymorphism for the two conditions shifts whenever local environmental conditions demand an increase or decrease in the amount of inbreeding (through homostyly). Meiotic drive (Chapter 9) is another condition maintained in populations merely because it adds to reproductive success, but not by improving viability or competitive ability (Hiraizumi *et al.* 1960). In all these cases (see also Fisher 1941) selection favors genotypes that are merely successful reproducers but do not add to the survival value of the species as a whole. This essential weakness of natural selection is a potential danger to every species, including mankind (Chapter 20).

Darwin had an inkling of the importance of reproductive success and discussed it in part under the heading of "Sexual Selection." Yet selection in favor of reproductive success has in many instances nothing to do with sexual selection. How much of Darwin's sexual selection falls under the heading of "mere reproductive success" can be determined only after a complete reevaluation of his material.

Is Selection Destructive or Creative?

In the days when the validity of "Darwinism" was still vigorously disputed, the crucial question was whether or not selection is "creative." If evolutionary change were due to the sudden appearance of entirely new types, the only possible function of selection could be the a posteriori one of acceptance or rejection. Likewise, thinking in early Mendelian terms, if one were to consider mutations the principal actors on the evolutionary stage, one would have little use for selection. As long as selection is defined typologically, as something that eliminates all deviations from the type or "anything inferior," we must assign to it a purely destructive role. He who thinks typologically must answer the question, "Can selection be creative?" with an emphatic "No!"

A different interpretation of the process of selection leads to the opposite answer. Is not a sculptor creative, even though he discards chips of marble? As soon as selection is defined as differential reproduction, its creative aspects become evident. Characters are the developmental product of an intricate interaction of genes and since it is selection that "supervises" the bringing together of these genes, one is justified in asserting that selection creates superior new gene combinations. This viewpoint has been ably presented by Muller (1929), Simpson (1947), Fisher (1954), Dobzhansky (1954), Lerner (1959), and virtually every recent writer familiar with the newer findings of population genetics. Simpson gives a very pertinent description of the patient way in which natural selection brings together favorable combinations of genes:

How natural selection works as a creative process can perhaps best be explained by a very much oversimplified analogy. Suppose that from a pool of all the letters of the alphabet in large, equal abundance you tried to draw simultaneously the letters *c*, *a*, and *t*, in order to achieve a purposeful combination of these into the word "cat." Drawing out three letters at a time and then discarding them if they did not form this useful combination, you obviously would have very little chance of achieving your purpose. You might spend days, weeks, or even years at your task before you finally succeeded. The possible number of combinations of three letters is very large and only one of these is suitable for your purpose. Indeed, you might well never succeed, because you might have drawn all the *c*'s, *a*'s, or *t*'s in wrong combinations and have discarded them before you succeeded in drawing all three together. But now suppose that every time you draw a *c*, an *a*, or a *t* in a wrong combination, you are allowed to put these desirable letters back in the pool and to discard the undesirable letters. Now you are sure of obtaining your result, and your chances of obtaining it quickly are much improved. In time there will be only *c*'s, *a*'s, and *t*'s in the pool, but you probably will have succeeded long before that. Now suppose that in addition to returning *c*'s, *a*'s, and *t*'s to the pool and discarding all other letters, you are allowed to clip together any two of the desirable letters when you happen to draw them at the same time. You will shortly have in the pool a large number of clipped *ca*, *ct*, and *at* combinations plus an also large number of the *t*'s, *a*'s, and *c*'s needed to complete one of these if it is drawn again. Your chances of quickly obtaining the desired result are improved still more, and by these processes you have "generated a high degree of improbability"—you have made it probable that you will quickly achieve the combination *cat*, which was so improbable at the outset. Moreover, you have created something. You did not create the letters *c*, *a*, and *t*, but you have created the word "*cat*," which did not exist when you started.

The validity of this analogy is particularly evident when applied to polygenic characters. Any gene will accumulate in a population that contributes to the favored phenotype.

Natural selection sets severe standards and constitutes a sieve through which only a minority can pass. There is nothing accidental, nothing blind about its outcome. Since it means differential reproduction, it is something that tests a zygote again and again until the end of its reproductive cycle, in good times and in bad. The theory of natural selection escapes the fatal weakness of all vitalistic theories whose "improvements of the type" factors lead the type into dead ends whenever the environment changes. Natural selection, on the other hand, by always being strongly opportunistic and by selecting simultaneously for the preservation of genetic variability as such, is ready in each generation to jump off in new directions.

CHANCE AND ACCIDENT

Selection is the most important of the factors that induce evolutionary changes by affecting the frequency of genes in populations. However, it is not the only one, as was pointed out at an early stage in the history of evolutionary research. Among the other possible factors is one, namely chance, the significance of which is still greatly disputed. Let me illustrate, on the basis of a single example, how important chance is. The human male produces many billions of gametes during his lifetime, and the human female many hundreds during hers. Still, one human couple can produce, at best, only about a score of children. It is largely a matter of chance which among the countless gametes will form the few successful zygotes. Since virtually all gametes differ from each other genetically, owing to the almost unlimited number of possible combinations of the parental genes, it is obvious that accident plays an important role in determining the genetic constitution of the F_1 of a set of parents. Chance affects every step in the life cycle of an individual. Mutation is largely governed by chance. So is crossing over and the distribution of the chromosomes during meiosis. Success of the gametes is largely a matter of chance, as is the difference in the genetic constitution of the two gametes which form a new zygote. R. A. Fisher (1923) has mentioned accidents of sampling, and Chetverikov (1926) and particularly Sewall Wright (1931b, 1951a, 1960) have calculated their possible effects in combination with other factors, such as mutation pressure, size of the population, selective values of the respective genes, and so on. Sewall Wright has shown that of two alleles in a population one might be entirely lost under certain calculable circumstances with the other reaching a frequency of 100 percent, and has termed this event "random fixation."

Definitions of Genetic Drift

Instead of restricting itself to rigidly definable terms, like "random fixation," the literature on random events in evolution has increasingly employed such vague terms as "genetic drift," a term with which each author seems to equate a different genetic phenomenon. If one were to restrict the term to random fixation, one could rather easily prove that genetic drift has no evolutionary significance whatsoever; if one defines as genetic drift any change in gene frequencies, as was recently done by Wright (1955), one can demonstrate convincingly that all evolution is due to genetic drift. Dobzhansky (1951) takes an intermediate position and defines genetic drift as "random fluctuations in gene frequencies in effectively small populations" (p. 156). It seems to me that the term "drift" has been rather discredited by its application to utterly different and otherwise inexplicable phenomena. During the period from about 1935 to 1955 it was fashionable to attribute puzzling evolutionary changes to "drift" or to the "Sewall Wright effect" in the same manner in which the preceding generation of evolutionists had explained similar changes as due to "mutation." Applying a technical term such as "mutation" or "drift" to an unexplained phenomenon has a peculiarly soothing effect on the human mind! Among the many inappropriate uses of "drift" as an "explanation" of evolutionary phenomena, none is so farfetched as its use to interpret gaps in the fossil record. To make saltations by drift is no better than to make them by macromutations or by creation.

These excesses are particularly regrettable since random phenomena deserve a thorough and objective evaluation in view of their enormous frequency in natural populations. The first modern author to attribute population differences to accident seems to have been Gulick (1873, 1894, 1905) in his endeavor to explain the characters of *Achatinella* snails in different valleys on Oahu, Hawaiian Islands. Lloyd (1912) thought that it accounted for the differences between rat populations in different houses in India. The Hagedoorns (1921), elaborating an earlier study (1917), gave great weight to the role of accident in the evolutionary process and Fisher (1922:328) has therefore designated random survival the "Hagedoorn effect."

That gene frequencies will be affected by random factors is evident and not questioned by any author. What is disputed is the evolutionary significance of such random fluctuations of genes in local populations. The more frequently the label "genetic drift" is attached to evolutionary changes by various authors, the more important it has become to determine exactly what genetic drift is and to evaluate its significance. It would seem that there is no hope for a real analysis of the concept drift until an attempt is made to classify the exceedingly diverse phenomena that have been recorded under this term. A preliminary survey indicates that at least five such categories exist, some very different from each other, some partially overlapping. Confusing ambiguity of terminology has been the cause of much of the recent controversy on drift.

(1) *Random Fluctuations Leading to Fixation.* Accidents of sampling are of universal occurrence in natural populations. The evolutionary significance of such random fluctuations depends on the contribution to fitness by the genes involved and on the size of the effective population, its isolation, and its permanence. The best available evidence comes from the study of human populations. Birdsell (1950) found striking gene differences among neighboring Australian tribes that live in essentially identical environments. Suggestive as this evidence is, it permits also other interpretations, since each tribe has had a somewhat different history and has been subject to somewhat different admixtures. This potential objection can be met if one compares the gene frequencies of a religious isolate with that of the remainder of the population living in the same environment. Two such studies are available. Glass *et al.* (1952) made a study of the Dunkers in Pennsylvania, an isolated religious sect of German descent, with an inflow of only 10 to 15 percent of intermarrying individuals per generation. In the absence of drift the gene frequencies in the Dunker isolate should fall between those of the United States and of the country of origin (Rhineland). However, in several of seven characters investigated there was a pronounced deviation from both of these populations. The frequency of blood group A was expected to lie between 39.5 and 44.6 percent, but had risen to 59.3 percent. The gene B had almost disappeared from the population. Blood groups M and N are of almost identical frequency in the Rhineland and in the United States, yet among the Dunkers type M had increased from the expected 30 percent to 44.5 percent while N had decreased from 20 percent to 13.5 percent. Several other characters (for instance, mid-digital hair, ear lobes) also differed strikingly from the frequencies recorded for the American white population.

An even more striking deviation from the neighboring American population was found by Steinberg (MS) for the Hutterites. They are a religious sect that had originally migrated from Germany to Russia and from there in the 1880's to North America where they now form a series of colonies in the Dakotas, Montana, and adjacent parts of Canada. Blood group O occurred in about 29 percent of the individuals investigated as compared to well over 40 percent in most European and American populations. Blood group A had a frequency of about 43 percent, well above the normal 30–40-percent level of European and American populations. Blood group B was reduced in frequency and had completely disappeared from two of the colonies. Even more significant is the amount of difference between various colonies of Hutterites, even though all of them have descended from closely related founders. For instance, gene A varied from 32 percent (colony 85) to 52 percent (colony 80). Among the Rh genes, R^1 varied from 27 percent to 68 percent, R^2 from 4 percent to 32 percent, r from 27 percent to 64 percent, to cite merely a few examples (from Steinberg MS). It is interesting that many deviations were in the same direction and almost of the same magnitude as those found by Glass for the Dunkers. The MN frequencies of all but one of the colonies differ from those of nearly all known white populations but agree closely with those of the Dunkers. The data for the Kell blood group differ, so far as I know, from those of any other known human population. The frequency of the K gene is zero in Negroes, Mongoloids, and Indonesians, and varies between 2 and 6 percent in all known white populations. In six colonies of Hutterites it is 13, 20, 21, 22, 23, and 34 (21.2) percent (Steinberg unpublished). The occurrence of accidents of sampling in these human isolates can hardly be questioned without denying that selection may have contributed to some of the shifts in gene frequencies (see below).

We can now ask the much more precise question: "Is a fixation of genes owing to gametic randomness occasionally responsible for lasting changes of evolutionary significance?" Those who in the past have answered this question affirmatively (most of them not being geneticists) have tended to make one or several of these assumptions: (a) that one can determine the selective value of genes on the basis of the visible phenotype, (b) that there are genes which are essentially or effectively neutral, and (c) that the selective value of a gene is absolute, remaining the same on various genetic backgrounds. Valid objections can be raised to each of these assumptions (for instance, Cain 1951a,b). It is evident from Wright's calculations that three factors contribute most effectively to the possibility of fixation. These factors are the smallness of the population, the completeness of isolation, and the selective neutrality of the allele involved. Completely isolated, small populations are a very special kind of population, insular populations, which will be discussed separately (Chapter 17). Wright (1951b) has emphasized correctly that "fluctuations in gene frequencies in small, completely isolated communities rarely if ever contribute to evolutionary advance, but merely to trivial differentiation, or in extreme cases to degeneration and extinction." Of greater pertinence to our problem is the question of the existence of neutral or near neutral genes.

The order of magnitude of selective difference between alleles coexisting in a population is difficult to determine. It is quite evident that a difference in selective value between two alleles amounting to, let us say, 5 percent or less will be difficult to demonstrate in view of nongenetic variation and experimental errors. However, attempts made in recent years to determine differences in the selective values of alleles found in wild populations often yielded values as high as 10 percent, 30 percent, or even 70 percent. Entirely neutral genes are improbable for physiological reasons. Every gene elaborates a "gene product," a chemical that enters the developmental stream. It seems unrealistic to me to assume that the nature of the particular chemical (enzyme or other product) should be without any effect whatsoever on the fitness of the ultimate phenotype. A gene may be selectively neutral when placed on a particular genetic background in a particular temporary physical and biotic environment. However, genetic background as well as environment change continually in natural populations and I consider it therefore exceedingly unlikely that any gene will remain selectively neutral for any length of time.

It must be repeated once more (see Chapter 7) that a gene is not necessarily selectively neutral merely because it does not seem to make an adaptive contribution to the visible phenotype. Studies of cases of balanced polymorphism have shown again and again that a gene with a most irrelevant visible manifestation (for example, pigmentation of the sheath of the testis in a flour moth; Caspari 1950) may have profound effects on viability. The claims of Gulick and other naturalists that certain phenotypic differences among natural populations are "obviously" due to accidents of sampling are unproved and based on a confusion of gene and character, of genotype and phenotype.

Selective neutrality can be excluded almost automatically wherever polymorphism or character clines are found in natural populations. This clue was used to predict the adaptive significance (previously denied) of the distribution pattern of the gene arrangements in *Drosophila pseudoobscura* (Mayr 1945) and of the human blood groups (Ford 1945). Virtually every case quoted in the past as caused by genetic drift due to errors of sampling has more recently been reinterpreted in terms of selection pressures. Elton (1930), for instance, believed that the gradual decline in the relative frequency of the silver fox was due to drift. Actually there are numerous indications that selection is involved, such as the simultaneous amelioration of the climate and the enormous extent of the area in which the decline has occurred (randomness should have caused a strong increase in some areas), the regularity with which it is correlated with population densities, and the fact that similar genes in other mammals, for instance the black gene in the hamster, are known to have high selective values. The haphazard distribution of white and blue flowers in *Linanthus parryae* in the Mojave Desert was interpreted as evidence of drift by Epling and Dobzhansky (1942). Wright (1943b), however, showed by an analysis of variance that systematic pressures were involved. The stability of the distribution pattern during the past 20 years (Epling, Lewis, and Ball 1960) also indicates the presence of highly localized selective factors, even though it is not yet known whether these consist of the chemical or the physical properties of the soil or are related to ground-water level, exposure, or some unknown biotic factor. Hartmann (1953) ascribed the occurrence of melanistic races of lizards (*Lacerta*) on small islets in the Mediterranean to drift. Kramer (*in litt.*), however, showed that all melanistic races were restricted to islands of a single type, namely small and rocky ones, and that wherever endemic races had formed on such islands the trend was invariably in the direction toward more pigment deposition. If random fluctuations had been involved, about half of these races should have been paler and only half of them darker than those of the adjacent large islands. The black color helps the lizards to warm up during the cool seasons and in the morning. Black races do not occur on similar islands in areas where it is continuously hot (as in the Red Sea) (Mertens 1952). The blackness has a polygenic basis and the oldest islands, among islands of equal size, tend to have the blackest races.

One of the classical cases of variation that has often been attributed to drift is the variation in the color patterns of the banded snails *Cepaea nemoralis* and *C. hortensis*. These snails are unusually polymorphic, with the ground color either yellow, pink, or brown and with various banding patterns ranging from unbanded to five-banded with intermediate band numbers in various combinations. It is now known that these phenotypic differences have a measurable selective significance on various substrates

(Cain and Sheppard 1950) and that the genotypes producing the different color patterns have various cryptic physiological properties (Schnetter 1951; Sedlmair 1956; Lamotte 1959) that affect viability differentially under different microclimatic conditions. Lamotte (1952) ascribes to drift the fact that differences between smaller colonies are greater than between large colonies. He bases this conclusion on the assumption that, "if the gene frequencies were in all instances rigorously determined by the interplay of selective forces, their distribution would be independent of the number of individuals in the colonies." This assumption ignores sev-

Table 8–3. Temperature and polymorphism of *Cepaea nemoralis* in different districts of France (from Lamotte 1959)

Mean July temperature (°C)	16–18	18–19	19–20	20–21	21
Average frequency of unbanded (percent)	22	24	26	29	30
Mean January temperature (°C)	4	4–2	2–0	0	
Average frequency of yellow (percent)	57.9	61.0	73.4	78.8	

eral facts. Small colonies will occupy, on the average, more uniform habitats. Furthermore, the effects of selection will be more immediately apparent owing to the reduced heterozygosity of small colonies. The differences in the amount of variation of small and large colonies is presumably, in part, due to selection. This does not contradict the probability that many colonies may have been originated by a single fertilized individual (Mayr 1942:32) and may thus also demonstrate the founder principle (see below).

All recent researches indicate numerous differences between genotypes of *Cepaea* with respect to various components of fitness. Unbanded yellow snails are more resistant to heat and cold than banded ones. There is a positive climatic correlation, in France, between the frequency of unbanded snails and increasing July temperatures, and between the frequency of yellow and decreasing January temperatures (Lamotte 1959; Table 8–3). De Ruiter (1958) found that five-banded snails laid almost twice as many eggs annually in the laboratory as unbanded or single-banded snails.

The relative frequency of certain banding patterns has not changed in some local areas from the Pleistocene to the present, as was first demonstrated by Diver (1929) for the south of England and later by Lamotte (1952) for the country north of the Pyrenees. This, together with the constancy in the spotting pattern of the Mexican fish *Xiphophorus maculatus*, was erroneously interpreted (Mayr 1942) as an indication of an absence of selection pressure. Actually, of course, such constancy proves stabilizing selection since random fluctuations of neutral genes would have resulted, no doubt, in a great change in gene frequencies during the vast elapsed span of time. A long-term stability of polymorphism has also been demonstrated for the isopod *Sphaeroma serratum* (Bocquet and Teissier 1960).

The distribution of the human blood groups has been frequently cited in the past as a classical demonstration of drift, on the basis of the former assumption by leading geneticists that these genes are selectively "neutral." This interpretation has now been largely abandoned since numerous effects on viability have been established since 1953 for blood-group genes (mainly the ABO system). The selective significance of the blood groups is further substantiated by the fact that only a very few of the possible frequencies have become materialized among human races (Brues 1954; Fig. 8–3). Selection may have played a role even in the changes found in the religious isolates discussed above. If the population of the Dunkers is divided into three generations (Glass 1956), the frequency of LM grows from 0.55 for the oldest (56 years plus) through 0.66 to 0.735 for the youngest (3–27 years). To be sure, such steady trends can be produced also by random fluctuations, but the fact that they are paralleled by similar trends for the same genes in some of the Hutterite colonies strengthens the case for selection. It is probable that selection is operating strongly even in these small isolates which are so subject to errors of sampling. These human isolates are smaller and even more completely isolated than are most isolates in animals about which we have information. In most human populations there is a considerable amount of intermixing with other semi-isolates (Lasker 1952).

Now that the occurrence of errors of sampling has been firmly established by calculation, observation, and experiment, the question has shifted to the evolutionary significance of this phenomenon. Obviously, the shift in frequency of an "effectively neutral gene" from, let us say, the 50-percent to the 60-percent level would hardly qualify as an event of evolutionary significance. It is evident from Wright's (1931b, 1951a) calculations that complete fixation is the only fluctuation which is unquestionably "of evolutionary significance." Such fixation does indeed occur, as in the loss of blood group B in some colonies of Hutterites or of pattern genes in colonies of *Cepaea nemoralis*. Yet this is not enough. Such small

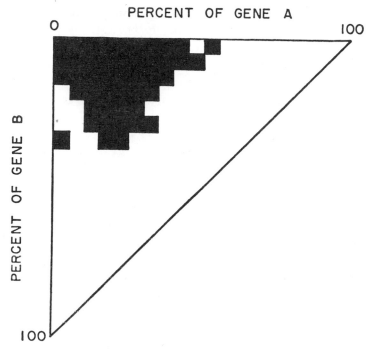

Fig. 8–3. ABO blood-group frequencies actually recorded in various populations throughout the world in relation to the complete possible range. (From Brues 1954.)

isolated colonies are only temporarily withdrawn from the free gene exchange of the species. As soon as contact with the parental species is reestablished, the lost gene will be restored. For all these reasons, it appears probable that random fixation is of negligible evolutionary importance. Sewall Wright (1949b) has come to a similar conclusion: "Accidents of sampling . . . may be responsible for nonadaptive differentiation of small island populations but are more likely to lead to ultimate extinction . . . than to evolutionary advance."

(2) *Founder Principle*. This term (Mayr 1942:237) designates the establishment of a new population by a few original founders (in an extreme case, by a single fertilized female) which carry only a small fraction of the total genetic variation of the parental population. The descendant population contains only the relatively few genes that the founders had brought with them until replenished by subsequent mutation or by immigration.

The founder principle is often responsible for the genetic and also the phenotypic uniformity of animal colonies, or peripherally isolated populations, and of colonies in temporary bodies of water (Hubendick 1951), in short, of any population established by one or a few founders. While in the case of random fixation there is a secondary decay of the originally present variation owing to gametic randomness, in the case of the founder principle there is a primary poverty of genetic variability owing to zygotic randomness. The evolutionary role of such founder populations will be discussed in Chapter 17.

(3) *Rapid Changes in Population Size and Concomitant Shifts in Selection Pressures.* A rapid change in population size will set up various changes in selection pressure (Chapter 9). During the period of increase in the size of a population, total selection pressure decreases, and, what is more important, there will be an increased premium on those alleles that as heterozygotes make the greatest contribution to viability. During a rapid decline of population size, there will not only be a sharp rise in total selection pressure, but also a shift to alleles favored under the conditions of increased inbreeding. The actual size of natural populations (large *vs.* small) is perhaps of far less evolutionary consequence than sudden drastic changes in population size. The rapid changes that occur in small isolated populations immediately after their establishment, owing to shifts in selection pressure and to the integration of new gene complexes, may lead to drastic differences which mimic random fixation and have often been erroneously thus recorded. The evolutionary consequences of these sudden shifts in selection pressure will be discussed in Chapter 17.

(4) *Correlated Effects of Cryptic Selection.* Various evolutionary trends have been mislabeled in the recent paleontological literature as being due to "slow genetic drift." This label is usually applied to slow phylogenetic changes in a phenotypic character for which no selective significance has been established. Such an interpretation confuses the genotypic and the phenotypic levels. There is every reason to believe that such a slow but steady change of the phenotype in geological time is the correlated response of a systematic selection pressure on the genotype. This is not drift. The existence of such correlated effects, known even before Darwin, has found its interpretation through the study of pleiotropic gene action. Furthermore, in some of the cases cited it has not even been demonstrated unequivocally that the phenotypic character itself is without selective significance.

(5) *Selective Equivalence of Genotypes.* Perhaps the greatest single source of randomness and indeterminacy in evolution is the selective

equality of genotypes. Since phenotypes are the product of the interaction and collaboration of many genes, it happens not infrequently that different assortments of genes may produce phenotypes that react in an identical manner to a given selection pressure. Various models have been suggested to help us visualize the phenotypic equivalence of different genotypes. Wright (1956) has made a special study of such models and has discussed their evolutionary consequences. When four dominant genes affect a single selectively important character, which genotypes will be favored in a given population will depend upon the interaction between these four genes, their favorable or unfavorable effect on other characters, and finally their interaction with the residual genetic background. An optimum may exist, for instance, for those genotypes that contain only two dominants; that is, *ABcd, AbCd, AbcD, aBCd, aBcD,* and *abCD.* In another model one peak may be *ABCD* and another of equal selective value *abcd,* with the more heterozygous genotypes inferior. These simple models do not even begin to describe the complexity of the actual relation between genotype and phenotype. The genetic analysis of certain characters (for example, pseudo tumors in *Drosophila;* Barigozzi 1957) was impeded because many very different genotypes produce identical phenotypes. It is obvious that chance may play a considerable role in determining which of various equivalent genotypes reaches the greatest frequency in a population and is available to interact with shifts in the residual genetic background and the changing environment. It seems to me—and this has likewise been stressed by Wright—that it is this indeterminacy of the selective aspects of genotypic recombination that introduces the greatest element of chance into evolution. The truth of this assertion has been abundantly established in recent selection experiments. Variation in response was found in almost every case in which several daughter populations of a single parental population were exposed to the identical selection pressure. The uncertainty produced by the indeterminacy of genetic recombination and by the phenotypic equivalence of different genotypes is the most important random factor in evolution.

Random Phenomena and Genetic Drift

It is now apparent that at least five different phenomena have been included by various authors under the heading "genetic drift." Some of these are essentially selective phenomena, others include a random component. To apply the term "drift" to nondirectional random fluctuations is unfortunate, since in the daily language we generally use the term "drift" for passive movements in a more or less unidirectional manner.

To speak of the drift of icebergs or clouds conforms to this usage. The indeterminacy of the sampling of genes by recombination and by the founder principle is precisely the opposite of what is colloquially understood by "drift."

It is now evident that the term "drift" was ill-chosen and that all or virtually all of the cases listed in the literature as "evolutionary change due to genetic drift" are to be interpreted in terms of selection (see above). Random phenomena, like recombination (with selective equivalence of different genotypes) and the founder principle, introduce a considerable degree of indeterminacy into evolution. They may even temporarily override selection in completely isolated and at least initially small populations. How important such populations are for speciation and ultimate evolution is still rather obscure (Chapter 17). For the time being it seems to me that it would clarify evolutionary discussions if authors would refrain from invoking "genetic drift" as a cause of evolution.

Interaction of Chance and Selection

The effect of chance on genetic variation is ambivalent. On one hand, it may lead to a depletion of genetic variation through random fixation. On the other hand, by counteracting selection, it may delay the elimination of temporarily disadvantageous genes. It certainly enhances the indeterministic aspect of evolution. Consequently it would be entirely misleading to say that chance directs the course of evolution. Nor does chance cause "drift" in the sense of a steady movement in one direction. On the contrary, if chance has any influence on the direction of evolution, it is that it "jars" it at frequent intervals and may occasionally be responsible for a jump to another track.

Let us remember that evolutionary change is a two-factor process. One stage consists in the generation of genetic variation. It is on this level that chance reigns supreme. The second stage is concerned in the choosing of the genotypes that will produce the next generation. On this level natural selection reigns supreme and chance plays a far less important (although not altogether negligible) role. Chance causes disorder, selection causes order. Chance is disoriented, selection is directional (including stabilizing selection). Chance is often destructive, selection is frequently creative. Yet both chance and selection are statistical phenomena and this permits them not only to coexist but, one might even say, to collaborate harmoniously.

9 ~ Storage and Protection of Genetic Variation

Rating all genes and new mutations as clearly either superior or inferior, the classical theory of genetics took it for granted that superior mutations would be incorporated into the genotype of the species while the inferior ones would be eliminated. As a result, uniformity of species would be guaranteed and genetic variation kept at a minimum. Natural selection would see to it, at all times, that the genetic purity of species would be preserved. Studies of the genetic variation of natural populations have, however, raised doubts concerning this assumption: not only do natural populations contain a richer store of genetic variation (Dobzhansky 1959a) than would be expected on the basis of the classical theory, but also numerous devices have actually been discovered that protect variation against the relentless ravages of natural selection. This chapter is devoted to a discussion of the known mechanisms for the storage and protection of genetic variation in natural populations.

The contradiction between the postulated efficiency of natural selection and the observed genetic variation has worried evolutionists from Darwin and the early geneticists on. Their answer to the puzzle—continued high production of genetic novelties—we now know to have been wrong. Only a small fraction of the observed genetic variation of populations is the immediate result of new mutations. Most of it instead is due to the release of variation from a storage reservoir where it is more or less protected against natural selection. The questions, then, that we would like to see answered are: what devices permit this storage, and how large a store of variation can a population maintain? Some tentative answers to these questions have been given by Haldane (1954, 1957), Lerner (1954), Dobzhansky (1955b, 1959a), Crow (1955), Wright (1960), and others.

The relative importance of some of these factors has been calculated by Dempster (1955) and Kimura (1955), who should be consulted for the mathematical aspects of the interacting forces. In my own treatment I shall concentrate on those aspects that are of special interest to zoologists and general naturalists.

In his discussion of the factors responsible for the genetic variation of populations, Haldane (1954) lists ten. Several additional ones have been mentioned by other authors. Yet they all fall into three major categories:

(1) *Cytophysiological or developmental devices,* which thwart selection by combining the effects of a gene on the phenotype with that of other genes in such a way that different genotypes in which a given gene may participate have unequal selective values. There are five or six of these devices, such as recessiveness, modification of penetrance, heterosis, linkage, and epistasis.

(2) *Ecological devices,* consisting of the diversity of the environment in space and time, which often lead to a neutralization of opposing selection pressures and thereby reduce the effectiveness of selection.

These various devices are summarized in Table 9–4, p. 255.

(3) *Regulation of the size of the gene pool,* that is, an increase or decrease in the amount of outbreeding.

The occurrence, relative significance, and functioning of the cytophysiological and ecological devices will be discussed in the present chapter. The control of outbreeding, which affects species structure fundamentally, will be treated in Chapter 14.

CYTOPHYSIOLOGICAL AND DEVELOPMENTAL DEVICES

The phenotype is the point of attack of natural selection. Consequently any developmental process which reduces the phenotypic expression of genes that might otherwise be deleterious will protect these genes against selection, at least until changes in the environment improve the selective value of these genes. There are many aspects to the relation between genotype and phenotype. For instance, the heritable component of many characters is quite low, as has been well established by animal breeders (Fig. 9–1). A selective equivalence of different genotypes (*Ab* and *Ba,* or *AB* and *ab*) is another phenomenon that favors, at least temporarily, the maintenance of genetic variation in the face of selection. Diploidy and the organization of the genetic material into chromosomes provides various opportunities to reduce the exposure of genes to selection. Let us consider some of them.

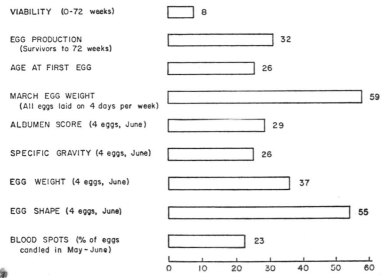

Fig. 9–1. Approximate degree of heritability (percent) of variation in nine phenotypic characters in a flock of Leghorn fowl. (From Dickerson 1955.)

Complete Recessiveness

The less a gene affects the phenotype, the less exposed to selection it will be and the more slowly it will be eliminated from the population. The completely recessive allele is fully protected while it is in heterozygous condition. The physiological theory of complete recessiveness is based on the assumption that at most loci a single dose is sufficient for the normal phenotypic expression of the gene. According to the classical theory of genetics, the heterozygous genotype Aa and the homozygous dominant genotype AA will have identical phenotypic properties. The presence of the recessive gene a in the heterozygote will be completely concealed and heterozygous a will escape all attacks by natural selection even when it is greatly inferior to A. Such an inferior gene will be exposed to selection only when it becomes homozygous (aa) through recombination. The frequency of a completely recessive allele in a population is thus determined by the balance between its rate of mutation and its elimination in homozygous condition. For a long time most genetic variation in natural populations was ascribed to the segregation (becoming homozygous) of previously concealed recessives.

Various findings in recent years have undermined belief in the frequency of completely recessive alleles. Every gene produces a gene product and there is no reason why the product of gene *a* should not have an effect on the heterozygote *Aa*. That there is a difference between homozygotes and heterozygotes even when there is no visible difference has been established by paper chromatography (Hadorn 1951), by immunochemistry, and by comparing the viabilities of heterozygotes and homozygotes. The difference in fitness between heterozygote and dominant homozygote is not necessarily expressed on all genetic backgrounds or in all environments, and where it occurs it may be in either a plus or a minus direction. Deviations in a plus direction (resulting in a superiority of heterozygotes) will be discussed in the next section. Deviations in a minus direction are often due to incompletely recessive genes that are lethal or semilethal in homozygous condition. In *Drosophila* such lethals have a semidominance of about 4–5 percent according to some studies, while others have established at least a 2.6-percent reduction in preadult viability. With this high selection pressure against them, they will be eliminated within 20–50 generations in spite of their classification as "recessives." But even if they were completely recessive, the equilibrium level between mutation and elimination by selection would keep these genes at a low frequency. A gene *a* with a mutation rate of 1 in 50,000 ($u = 0.00002$) and a selective disadvantage of 10 percent ($s = 0.1$) would reach this balance at the frequency $q = \sqrt{u/s} = 0.014$ or about 1 heterozygous individual in 35. At this frequency it would manifest itself in the phenotype of 1 in 5000 individuals (as recessive homozygote). A gene with a disadvantage of only 1 percent would occur in 1 of 12 individuals and express itself in 1 of 500 individuals. These would be maximum figures, since only a few genes are completely recessive and the average deficiency in fitness of deleterious genes is presumably a good deal higher than 1 percent.

The elimination of deleterious recessives from natural populations is retarded by the variability of the environment in space and time. A gene that is inferior on the average may be favored at certain times or in certain subniches, and thereby gain a temporary respite. Yet, if the input of genetic variation could be prevented, selection would in due time make the population homozygous at all loci. We might call this part of the stored variation the "input variability." It may be defined as "the presence, owing to mutation and immigration, of inferior alleles in a population, and the delayed elimination of such inferior alleles owing to various protective devices." This is the variation that Muller has described in a masterly way in his essay "Our Load of Mutations" (1950a).

Suppression of Phenotypic Variability

Recessiveness is only one of many ways by which the pressure of selection is mitigated through the reduction of phenotypic variation. One is tempted to think of "dominance" and "recessiveness" as inherent qualities of genes, and, indeed, new mutations are often strongly dominant or recessive from the very beginning. Fisher (1930), however, has quite rightly pointed out that natural selection will favor any modifying genes that make the heterozygote more similar to the phenotype of the superior homozygote. There is much evidence in favor of this interpretation, particularly in cases of polymorphism (Sheppard 1958), although it is apparent that additional factors affect the degree of dominance and overdominance.

Recessiveness, however, is only one way by which a gene may fail to express itself in the phenotype. Even dominant alleles may have no visible effect in the case of "incomplete penetrance." Most of the homeotic mutants in *Drosophila*, for instance, have low penetrance, sometimes not even reaching the 10-percent level. Incomplete penetrance may be of major significance in human genetics, and may greatly delay the analysis of the mode of inheritance of important pathological conditions. A gene is incompletely penetrant if it fails to express itself in the phenotype when placed on certain genetic backgrounds or when living in certain environments. There will be natural selection in favor of background genes ("suppressors") that make deleterious genes innocuous in this manner (Chapter 10). These penetrance modifiers are only one set in a large class of genes, all of which contribute to prevent deviations from the "normal" phenotype of the species. The greater the number of such homeostatic devices in a gene pool, the greater the amount of genetic variation that can accumulate in the population under standard conditions without being exposed to selection. Such phenotype-stabilizing genes presumably play a dual role. They may suppress environmentally caused variations of the phenotype, but also buffer against the effects of mutation. Their ability to canalize developmental processes presumably gives them high selective values.

The amazing phenotypic uniformity of many species, in spite of the great concealed genetic variation, suggests that there must be a high selective premium on the standardization of the phenotype (Dobzhansky 1956). The morphology of all individuals is essentially identical in these species even though produced in each case by a different combination of genes and with each genotype possessing different physiological properties. The high frequency of sibling species (Chapter 3) proves that genotypes can be entirely reconstructed, and that new isolating mechanisms and an entirely new ecology can be acquired, without leaving a visible impact on the morphology. The same is demonstrated by the abundance of isoalleles, which differ in their physiology but do not affect the visible phenotype (Stern and Schaeffer 1943; Spencer 1944).

We must ask what mechanisms permit preservation of the favored phenotype in spite of the extensive reconstruction of the genotype? Waddington (1942, 1948, 1953a, 1957) has discussed the nature of this developmental stability in a series of papers. He postulates that the genotype is buffered in such a way that the development is "canalized." Regardless of the genes of the gene pool that come together, a series of developmental tracks will guarantee a standardized end product. This developmental stability can cope not only with variation of the gene pool but also with that of the environment. As long as the environment varies within "normal" limits, it will not be able to make the phenotype jump out of its normal developmental track. The term *canalization* stresses the inevitability of the developmental pathway, in spite of disturbances and temporary roadblocks caused by the external and genetic environment. The term *developmental homeostasis,* used by other authors for essentially the same phenomenon, stresses the dynamic aspect provided by the homeostatic mechanisms that restore development to its normal course after deviations, no matter how they were caused. In the case of canalization, the end product is stressed; in the case of homeostasis, the process of producing it (see also Lewontin 1957). The terms "canalization" and "developmental homeostasis" merely verbalize conceptual models. Actually nothing is known except rather indirectly about the developmental mechanisms responsible, or about the genes that regulate them. The genetic and developmental problems raised by the developmental flexibility have been discussed by Muller (1950b), Thoday (1953), Lerner (1954), Mather (1954), and Waddington (1957).

Canalization of development is best illustrated by genetic switch mechanisms. Here a single gene, as in some cases of sex determination or mimetic polymorphism, may switch development into a very distinct pathway with specified interactions with numerous other genes. Such a switch gene, of course, does not "produce" these exceedingly different phenotypes; it merely determines a developmental pathway, the final phenotype being the product of a great many different genes. In some cases it is possible to demonstrate the switching properties of a gene by transferring it to a new genetic background, for instance by hybridization. Canalization is a developmental phenomenon, but does not differ in its genetics from other cases of polygenic inheritance. To assume that each dominant and overdominant gene, each isoallele, and developmental feedback has its own *ad hoc* modifiers and polygenes would lead to the postulation of a number of genes far in excess of what can be accommodated on the chromosomes. The other alternative, the one in which I firmly believe, is that each gene serves as a modifier for many, if not most, other gene loci. Any interference with such a system will lead to a chain reaction. The multiplicity of interactions, including multiple pathways, gives such a system great plasticity as well as stability.

Superiority of Heterozygotes

Although much of the genetic variation in populations is due to this lag between the input of genetic variation in populations and its elimination by natural selection, there is evidence that some of the genetic variation is directly maintained by natural selection. We may call this portion of the genetic variance "selected heterozygosity." Several devices have been found in recent years that fall in this category, foremost among them being heterozygote superiority.

In our prior treatment of polymorphism (Chapter 7) we merely stressed its widespread occurrence but did not try to solve the puzzle of its maintenance. The solution was provided by Muller (1918) and R. A. Fisher (1922, 1930), who showed that two alleles can be maintained in a population at high frequencies if the fitness of the heterozygotes (*Aa*) is higher than that of either homozygote (*AA* or *aa*). Even if *aa* should be considerably inferior to *AA*, and possibly even lethal, the gene *a* would be retained in the population, provided that heterozygote *Aa* has a higher selective value than has homozygote *AA*. The reservoir of gene *a* will be replenished continuously by segregation of the favored heterozygous genotype *Aa*. Polymorphism maintained by such "overdominance" of the heterozygotes has been designated by Ford (1945) *balanced polymorphism*.

The potency of overdominance may be illustrated by an example. If a gene *a* is completely lethal as homozygote (*aa*), it will nevertheless be retained in the population with an equilibrium frequency of 0.01 if it raises the fitness of the heterozygote (*Aa*) by 1 percent over that of the homozygote (*AA*). The frequencies of the genes *A* and *a* will reach the equilibrium, when the frequency of the lethal allele has reached $s/(s+t)$, where t is the coefficient of selection of the lethal and s that of the non-lethal allele.

Among the first to demonstrate balanced polymorphism in natural populations was R. A. Fisher (1939), who calculated that in the grasshopper *Paratettix* the selective superiority of the heterozygotes exceeded that of the homozygotes in the three largest samples by 6.6, 10.4, and 14.2 percent in each generation. The first case of selective superiority of heterozygotes in a laboratory population was found by L'Héritier and Teissier (1937) for the gene "bar" in *Drosophila melanogaster*. Since then many similar cases have been found in *Drosophila* and other animals. The heterosis is sometimes caused by the interaction of a pair of alleles at a single locus, in other cases by the interaction of chromosome sections (gene arrangements) acting as supergenes. Dobzhansky (1951) and his school have made a particularly extensive contribution to our understanding of the working of such balanced chromosomal polymorphism. It occurs not only in *Drosophila* but in many other organisms (White 1954, 1957a; da Cunha 1955; da Cunha *et al.* 1959). In the Australian grasshopper *Moraba scurra* inversion polymorphism occurs in two chromosomes. The widespread occurrence of polymorphism in far-distant and highly isolated colonies of this species indicates not only the antiquity of this polymorphism, but also its strong heterosis in the face of sampling errors in often very small populations (White 1957c). The same has been demonstrated for laboratory populations. Of 27 polymorph strains of *Drosophila* maintained at a population size of only 20–40 individuals, 24 were still polymorph after 130–211 generations of random transfers (Levene and Dobzhansky 1958). Indeed, it is becoming more apparent daily that most cases of genuine polymorphism in natural populations are maintained by the superiority of the heterozygotes, regardless of any additional factors (such as mosaicism of the environment) that may contribute to genetic diversity. Nearly all of the literature on polymorphism, referred to in Chapter 7, can be included under balanced polymorphism (Ford 1945; Dobzhansky 1951; Huxley 1955a,b).

By far the most astonishing aspect of balanced polymorphism, to me at least, is the magnitude of the differences in the selective values of the different genotypes. Almost all tested highly inbred lines of chickens (in which selection strictly favors high productivity, and thus indirectly high viability) were polymorph at the B locus of blood-group genes. The selective advantage of the heterozygotes was thus so great that it was able to override the tremendous pressure in favor of genetic uniformity (homozygosity) (Shultz and Briles 1953; Briles, Allen, and Millen 1957). The exact amount of the heterozygote superiority has been calculated in a number of cases, for instance for the three genotypes (*EE, Ee, ee*) that determine body color in *Drosophila polymorpha*. All three genotypes are phenotypically distinguishable, *EE* being darkest, *Ee* intermediate, and *ee* lightest. The viability of the three genotypes in a laboratory experiment was as follows. If the viability of the heterozygote *Ee* was taken as 100 percent, homozygous *EE* had a viability of only 56 percent, while homozygous *ee* was semilethal with a viability of only 23 percent (da Cunha 1949, 1953). Similarly great viability differences among genotypes were found by Gershenson (1945) for a melanistic gene in the hamster (*Cricetus cricetus*). Calculated coefficients of selection were remarkably high, ranging from 0.79, 0.83, 0.89, and 0.94, and 0.98 down to −0.82, and −0.86. The melanics were favored (positive values) in some years and districts and discriminated against (negative values) in other years or areas. Dobzhansky (1947a) kept a polymorph population of *Drosophila pseudoobscura* from Piñon Flats, San Jacinto Mountains, California, in a population cage under specified conditions, and found that flies homozygous for one of the gene arrangements (Standard) had only 76.2 percent of the fitness of the heterozygotes, those homozygous for the other gene arrangement (Chiricahua) only 37.9 percent. Similar differences have been found for other gene arrangements in the same and other species of *Drosophila*.

The fitness of each genotype as compared to the others is not absolute but is exceedingly sensitive to shifts in environmental conditions, including changes in the genetic background. This explains the many apparently conflicting observations made on such polymorph species. For instance, in both natural and laboratory populations, Epling et al. (1953, 1957) found apparent deviations from the results of Dobzhansky and co-workers on chromosomal polymorphism in *Drosophila pseudoobscura*. There are a number of ways of accounting for these discrepancies. For instance, fitness includes not only survival into the adult stage, but also the totality of factors (sexual vigor, fertility, fecundity, and so forth) that contribute to "reproductive success" (Levene and Dobzhansky 1958; Spiess and Langer 1961).

In a number of cases clear superiority of the heterozygotes was found, in spite of lethality or near lethality of one of the homozygotes. It is not permissible in such cases to base the calculation of mutation rate on the frequency of the lethal gene in the population, as is possible in the case of strictly recessive genes. Rosin *et al.* (1958) have shown, for instance, that the widely accepted mutation rate for the hemophilia gene in man must be drastically revised in view of the greatly increased fertility of the female carriers of this gene (about 1.15 compared to 1.00 normally). A fitness for females of 1.22 would be sufficient to maintain the hemophilia gene in the population without additional mutation, even though the observed fitness of the males is only 0.64.

Heterozygosity in Natural Populations

Visible polymorphism is only a small fraction of the total genetic polymorphism found in natural populations. Indeed, the amount of genetic variation present in wild populations is staggeringly large. The steadily accumulating evidence for this assertion consists primarily of two sets of data. One is the discovery of numerous very similar alleles—isoalleles—at most loci whenever they were looked for seriously. There may be an over-all "wild-type" phenotype, but the early Mendelian idea of one wild-type gene for every locus has been rather thoroughly refuted. The second set of data, long known under the designation of "hybrid vigor," has been subject to causal analysis only since genetics has become a science. It is an almost universal observation that severe inbreeding leads to "inbreeding depression," a serious reduction of fitness in its various components. Loss of fertility, increased susceptibility to disease, growth anomalies, and metabolic disturbances are among the manifestations of inbreeding depression (see Lerner 1954:22–27 for numerous examples). Countless laboratory stocks have been lost owing to inbreeding. There is much evidence to indicate that part of this inbreeding depression is due to a loss of overdominance. Fitness declines as more and more loci become homozygous, but can be restored dramatically when two such depressed inbred lines are crossed. Hybrid corn is one of the many practical results of this new stress on the benefits of heterozygosity. I refer to Lerner (1954) for a detailed discussion of this important topic.

Of the many recent experiments demonstrating a superiority of heterozygotes, I will cite only one. Carson (1958a) introduced a single wild-type Oregon-R third chromosome into a large *Drosophila* population homozygous for five third-chromosome recessives (se, ss, k, e^s, and ro). After about 15 generations (1 generation = 14 days) the three recessives that were closely followed had stabilized at the following frequencies: $ro = 53.4$ percent, $se = 25.3$ percent, and $ss = 12.3$ percent. The heterozygosity due to the single introduced chromosome had resulted in more than tripling the productivity of the parental population (Table 9–1). Indeed, the performance of the new populations (E–1 and E–2) was in all respects superior even to the wild-type Oregon-R laboratory population, which, with its long history of inbreeding, was presumably far more homozygous than the new experimental population.

The Causes of Heterozygote Superiority

The reasons for the selective superiority of heterozygotes, where it occurs, are manifold, ultimately physiological, and not yet fully understood, in spite of several symposia and conferences specifically devoted to this subject (for example, Gowen 1952; Cold Spring Harbor 1955; see also Dobzhansky 1951, chap. 5).

(1) *Heterosis Due to Dominance.* The standard interpretation of heterozygote superiority used to be that the increase in viability is the result of an increase in dominant genes in the hybrids at loci where more or less deleterious recessives had become homozygous in the parental strains. This interpretation is founded on the probability that numerous deleterious recessives are concealed in all lines and that different inbred lines would become homozygous for different, nonallelic recessives. If one line is *AAbb,* the other *aaBB,* the hybrid would be *AaBb,* with the dominant gene suppressing the deleterious recessive, in the heterozygote, at both loci.

This interpretation has been proved valid in many situations observed by animal and plant breeders, but is not sufficient to account for all of the superiority of heterozygotes in polymorph populations (Crow 1948, 1952). In particular, the dominance theory fails to account for single-locus heterosis, which has been reasonably well established not only in plants (Stubbe and Pirschle 1940) but also for a number of cases in animals. Buzzati-Traverso (1947, 1952) found it for the "light-eye" gene of *Drosophila melanogaster,* da Cunha (1953) for the pigmentation gene E in *Drosophila polymorpha,* and Rosin et al. (1958) in the hemophilia gene of man, to mention a few cases.

(2) *Heterosis Due to Overdominance.* According to this theory a locus is overdominant if the heterozygote has higher fitness than the homozygotes.

There are several possible explanations of such heterozygote superiority, where it exists. One is that heterozygosity gives greater biochemical versatility. Consequently, the heterozygote, having a combination of different gene products available, is able either to cope with a greater di-

Table 9-1. Size and production of experimental populations of *Drosophila melanogaster* (from Carson 1958a).

Population	Number of weeks at equilibrium	Population size		Production	
		Mean number of individuals (weekly count)	*Mean wet weight measured weekly (mg)*	*Mean number of individuals per week*	*Mean wet weight (mg/week)*
Controls:					
C-1 se ss k e^s ro	32	161.6 ± 6.4	90.3 ± 3.0	100.8 ± 4.4	48.4 ± 1.9
C-3 se ss k e^s ro	20	154.4 ± 4.4	88.7 ± 2.0	61.0 ± 3.3	27.6 ± 1.5
Experimentals:					
E-1 se ss k e^s ro with *n* Oregon autosomes	20	457.4 ± 13.7	292.5 ± 9.1	171.0 ± 10.6	91.0 ± 5.9
E-2 se ss k e^s ro with *n* Oregon autosomes	20	502.6 ± 14.9	318.6 ± 8.9	201.1 ± 14.4	105.2 ± 8.0

versity of developmental needs than the homozygote, which has only a single gene product, or to operate more efficiently on a diversified genetic background. In view of the ability of a single allele at many loci to produce in a "single dosage" all the gene product necessary for the development of a normal phenotype, it would appear that the presence of a second allele at the same locus, with an optimal activity at different temperatures or under other conditions, might result in a combination that would operate as a physiological mosaic.

Another possibility is that the heterozygote is less exposed to selection pressures than either homozygote. The prototype of such a situation is presented by balanced lethals, a less extreme case by the gene for sickle-cell anemia (Allison 1955). The anemia caused by the sickle gene results in at least a 90-percent reduction in fitness among the homozygotes. Indeed, few of them reach even the age of 5 years. On this basis one would expect a rapid elimination of the gene until it reached the level of maintenance by recurrent mutation. Yet there are large areas in tropical Africa where some 20 to 40 percent of the natives are heterozygous for the gene. This would require either a fantastically high mutation rate of this gene or a high selective advantage of the heterozygote. Allison clearly established the latter. He found that the areas of the highest gene frequency coincide, on the whole, with areas with an excessively high rate of morbidity due to subtertian malaria caused by *Plasmodium falciparum*. Children heterozygous for sickle-cell anemia have a greatly lowered rate of infection by *P. falciparum* and heterozygous adults are less susceptible to artificial infection. Although additional factors enter the picture (Livingston 1958), there can no longer be any doubt that the clinically non-apparent inferiority of the blood cells of heterozygotes is a protection against subtertian malaria and thus gives the heterozygotes superior fitness. Italian doctors had previously suggested the same explanation for the high frequency of thalassemia (another hemoglobin disease) in Italy in districts with a high rate of endemic malaria.

A balance of opposing selection pressures has been found also for the sex-ratio gene in *Drosophila* (Wallace 1948), for the *T*-locus in the house mouse (Dunn 1956), and for the bleeder (hemophilia) gene in man (Rosin *et al.* 1958). The balancing selection pressures for the human blood-group genes were discussed in Chapter 7. Spotting polymorphism in the butterfly *Maniola jurtina* also seems to be in part maintained by opposing selection pressures. A selection pressure of about 70 percent (caused by a hymenopteran parasite) operates during the last seven weeks of pre-imaginal life against larvae destined to give rise to females with two or more spots (Dowdeswell 1961). It is not yet known what favors these genotypes at other seasons.

A heterozygote that owes its superiority to a balance of disadvantages of the homozygotes may easily lose it when conditions change. This may well be the cause of some cases of transient polymorphism and of secular fluctuations, discussed below.

The continued persistence of polymorphism has often been ascribed to a fluctuating balance of opposing selection pressures against homozygotes. This is possible for rapid fluctuations but unlikely in the long run since one or the other allele will be lost sooner or later owing to errors of sampling (Kimura 1955). A permanently balanced polymorphism can be maintained only if the opposing selection pressures add up to a superiority of the heterozygote or of the rarer homozygote (Dempster 1955).

(3) *Double-dose Disadvantage.* Huxley (1955b) has suggested that one should look at the superiority of the heterozygotes from the viewpoint of the homozygote that "suffers" from the disadvantage of a double dose of the same gene product. I do not think that the evidence favors this interpretation. As Muller (1950b) has shown, most phenotypes are remarkably immune to excess doses of the same gene product. The homozygote clearly suffers from a lack of biochemical or ecological versatility and not from an overdose of an enzyme and its products.

The presence of heterozygous loci in a population, guaranteed by heterosis (heterozygote superiority), is of double advantage. First, it produces highly viable individuals that are buffered against environmental fluctuations. More importantly, it gives the population a great and much-needed diversity, because the number of genotypes in the population is more than simply proportional to the number of alleles: if the number of alleles is n, the number of genotypes is $\frac{1}{2}(n^2 + n)$. A locus with three alleles (A, a, a') produces six genotypes (AA, Aa, Aa', aa, aa', $a'a'$), each optimal at somewhat different environmental conditions. Such versatility makes genetic variability available at all times for an immediate evolutionary response to a change in the environment. Thoday (1953) and Lerner (1954) have rightly stressed the dual aspect of heterosis, by which the advantage to genetically diverse individuals (the heterozygotes) gives an advantage of evolutionary plasticity to the population as a whole. Being composed of several genotypes, a population that is polymorphic owing to heterosis can better utilize different components of the environment (different subniches). Populations of *Drosophila pseudoobscura*, polymorphic for two gene arrangements, produce more individuals than equivalent monomorphic populations; they produce a greater total biomass, and show less phenotypic variation (Beardmore, Dobzhansky, and Pavlovsky 1960).

Heterozygotes may have still another advantage. A study of phenotypic variance showed unexpectedly that it is lower in heterozygotes than in homozygotes (inbred lines) (Wigan 1944; Mather 1950; Robertson and Reeve 1952a; Dobzhansky and Wallace 1953; Grüneberg 1954; summarized by Falconer 1960). Animal and plant breeders have long been familiar with this phenomenon. It is evident that the increased phenotypic variation in homozygotes or inbreds is largely nongenetic. They are more strongly affected by variation of the environment, they are less able to compensate physiologically for the unbalancing impact of environmental factors, they are less well "buffered." Why? Robertson and Reeve (1952a), following Haldane, suggest that the developmental superiority of the heterozygotes is due to their greater biochemical versatility. In a homozygote each gene (locus) produces a double dose of the same gene product, while a heterozygote has a single dose of each of two different (though closely similar) gene products available. At each set of environmental conditions either one or the other allele will have superior developmental efficiency, resulting in optimal viability.

Heterozygotes would be specially favored, according to this theory, in a fluctuating environment. If allele A performs optimally from 18° to 26°C and allele a from 14° to 22°, then each homozygote has an environmental "resistance" of 8 C deg, the heterozygotes of 12 C deg (from 14° to 26°). Dobzhansky and Levene (1955) found that strains of *Drosophila pseudoobscura* homozygous for certain chromosomes suffered far more drastic changes of viability when transferred to new environments than heterozygous strains. Conversely, making the environment completely uniform deprives heterozygotes of much of their superiority, while increasing the relative superiority of one of the homozygotes over the other. In a closed population this may lead (but does so only rarely) to a loss of heterosis of the heterozygotes and thus of the balanced polymorphism (Lewontin 1958). Heterozygotes are, however, favored not only in a fluctuating environment, but more generally in most adverse environments. Under optimal conditions homozygotes and heterozygotes may show similar viabilities, while the heterozygotes may be definitely superior under unfavorable conditions, for instance in the copepod *Tisbe reticulata* (Battaglia 1958).

The superior homeostasis of the heterozygotes under varying conditions is one of the reasons for their selective superiority. Heterozygote superiority thus leads in this case to the production of a uniform, standardized visible phenotype, while in other cases it leads to a conspicuously discontinuous polymorphism. A completely uniform phenotype is only one extreme in a rather broad spectrum of phenotypic variability. At the other extreme are such cases as that of the Ruff (*Philomachus pugnax*), a bird in which hardly any two adult males look alike, the color varying from uniform to barred, and from white to black or rufous, and with every possible combination of these colors. It is probable that the method of male selection in this species places a high premium on phenotypic uniqueness. Other cases of high individual variation of phenotypes were discussed in Chapter 7.

The ultimate effect of a tight system of developmental homeostasis is to minimize the effect of natural selection. No matter how much genetic variation there is in a gene pool, the less it penetrates into the phenotype, the smaller the point of attack it offers to selection. The developmental mechanisms that guarantee a uniform standard phenotype regardless of the amount of underlying genetic variation are thus another device contributing to the maintenance of genetic variability in populations. How this potential variability can be made available in time of need is a problem that will occupy us elsewhere (Chapters 10, 17, 19).

The Origin of Heterosis

The role of natural selection in the production of heterosis is still rather obscure. Two extreme viewpoints are possible, the relative merits of which I have discussed previously (Mayr 1955a). According to one view, heterozygous loci are always potentially superior to homozygous loci. The other extreme view is that heterozygotes are superior to homozygotes only in those cases in which there has been selection for such superiority, that is, where the genetic background has been selected to give the heterozygotes maximal fitness for all components of fitness affected pleiotropically by the locus.

The best evidence against the extreme selectionist viewpoint comes from population crosses. Starting from the consideration that the genes of a gene pool are the product of a long history of selection for optimal interaction, one would always expect to find a drastic loss of fitness in interpopulation crosses. Surprisingly, the exact opposite is sometimes the case. Stone (1942) was the first to find heterosis when crossing strains of *Drosophila* from different localities. This occurred both in *hydei*, a world-wide species building up very large populations, and in *virilis,* with its more scattered distribution. First-generation hybrids (F_1) were superior both in fertility (number of fertile pairs) and in fecundity (number of offspring per pair per day). Wallace and Vetukhiv (1955) and Vetukhiv (1956) confirmed this for *D. pseudoobscura* (Fig. 9–2). Selection cannot account for these cases of interpopulation heterosis (see also Chapter 10).

Yet there is also much evidence that selection is important (Mather 1943; Dobzhansky 1951; Thoday 1955). For instance, the relative superior-

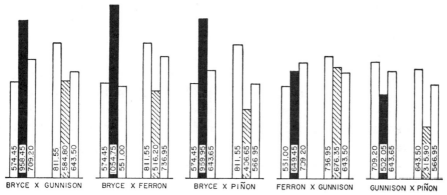

Fig. 9–2. Fecundity of geographic populations of *Drosophila pseudoobscura* and of hybrids between them. The height of the column gives the mean number of eggs deposited per life per female. The white columns represent the parental populations, the black columns the F_1 hybrids, and the crosshatched columns the F_2 hybrids of these populations. (From Vetukhiv 1956.)

ity of heterozygotes often increases sharply in experimental populations. Also, heterozygotes may lack heterosis in inbreeding species, as has been demonstrated for many plant species. Finally, nothing could be more heterozygous than species hybrids, yet most hybrids are decidedly inferior with respect to most components of fitness. Heterosis therefore cannot be an automatic by-product of heterozygosity. Wallace (1954a, 1955) tested the relative effects of heterozygosity and chromosomal hybridity and demonstrated that heterosis was due not to heterozygosity as such but to a heterozygosity of selected ("balanced") chromosomes and chromosome sections.

Fisher's theory of dominance, according to which dominance evolves as a result of selection for modifying genes ("background genes") that give the heterozygote the phenotype of the dominant homozygote, can be modified to explain the origin of overdominance. If the dominance for each phenotypic character controlled by a single pleiotropic gene locus can be modified independently, a balanced polymorphism will result, provided that each of the homozygotes has some advantageous phenotypic manifestations (Sheppard 1953a). Selection will cause the disadvantageous phenotypic expressions of the gene to become recessive and the advantageous ones to become dominant. As a consequence, the heterozygote will exhibit the more advantageous phenotype of each of the pleiotropic effects. Selection then will clearly favor the heterozygote. A gene that answers this postulate has been found by Caspari (1950) in *Ephestia kuehniella,* a flour moth. The only visible effect of this gene seems to be that the dominant allele *Rt* causes the testes to be brown while *rt* testes are red. When three components of fitness were investigated it was found "that the heterozygote is in every case at least equal to and possibly superior to the more favored homozygote." The allele *Rt* was superior and dominant as far as speed of development and copulation behavior are concerned; *rt* is superior and dominant with respect to viability.

The understanding of the origin of such a system is greatly facilitated if we remember that new genes become incorporated into populations in heterozygous condition. Homozygotes become frequent only long after the heterozygotes have already become very frequent. It is during this initial period, as Parsons and Bodmer (1961) point out correctly, that an overdominance of the heterozygotes will be favored by selection. As also shown by Mayr (1954a), such modifiers will be selected under those conditions that have the greatest fitness-enhancing effect on heterozygotes.

In the long run heterozygote superiority does not seem a particularly efficient mechanism for the maintenance of fitness in populations. It cannot avoid in each generation the wasteful production of inferior homozygotes through segregation. In view of this wastefulness of balanced polymorphism Muller (1950a) expressed the opinion that "a mutant gene . . . advantageous in its heterozygous degree of expression but deleterious homozygously . . . would usually have become replaced after a while by mutant genes of less deviant expression . . . which give an equivalent advantage when they were present homozygously." Lerner (1954) and Wallace (1956) have argued that this is not necessarily the case and that it is doubtful whether a homozygote is in principle superior to a heterozygote. The latter may have greater developmental plasticity and capacity for developmental homeostasis but also, by producing three different genotypes in each generation, give the gene pool a greater evolutionary plasticity, permitting a more rapid shift in case of changing environmental conditions. To what extent this is achieved by interpopulation selec-

tion remains to be studied (Lewontin 1958). Optimum fitness of a genotype is presumably achieved by a proper ratio between homozygous and
heterozygous loci, the particular ratio depending on breeding system,
population size, selection pressure, and other factors.

One way in which it would be possible to decrease the number of
homozygotes is to increase the number of alleles (presumably mostly isoalleles) in a population, since the proportion of homozygotes is $1/n$ if the
number of alleles is n (and if all alleles are equally frequent). This is presumably the situation for many loci in open natural populations. It is, however, usually not a feasible solution for gene arrangements where, for a
given chromosome section, often only two alternatives can effectively coexist in a single population (Wallace 1953).

With only two alleles or alternative supergenes maintained in a population by heterozygote superiority, even the rarer homozygote usually
occurs at an appreciable frequency. One would therefore expect strong
selection pressure in favor of improving the fitness of the homozygotes. Yet
analysis of several natural populations has revealed such large deficiencies
of homozygotes that they must be presumed to be near lethal. In a population of *Drosophila tropicalis* from Honduras nearly one-quarter of the
zygotes die in each generation on account of homozygosis for an unfavorable gene complex in an arm of the second chromosome (Dobzhansky and Pavlovsky 1955). Because of this differential mortality, 70
percent of the population consists of the heterotic inversion heterozygotes.
Yet at this particular locality *D. tropicalis* seems to be more successful
than in other areas where the second chromosome lacks this homozygote
lethality. Other species of *Drosophila*, particularly *D. subobscura*, have
sometimes also significant homozygote deficiencies. It seems that species
with the high fecundity of *Drosophila* can afford to sacrifice many of their
potential offspring for the sake of the increased adaptive advantages
gained by the surviving heterozygotes. Kitzmiller and French (1961) find
in *Anopheles quadrimaculatus* that 95 percent of the individuals are
heterozygous for an inversion.

It is probable that a given gene pool can accommodate only a limited
number of overdominant loci. One reason is that otherwise too many genotypes would be produced that are deleterious homozygotes at one locus
or another. The second reason is that there is interference among several
balanced heterozygous systems, at least among systems of balanced gene
arrangements (Chapter 10).

Epistatic Interactions

Single-locus heterosis is only one of many forms of interaction among genes. Epistatic interactions among nonallelic genes in a variable environment may have the same effect of preserving genetic variation in populations. These will be discussed in Chapter 10 (see also the discussion of multiple peaks, p. 213).

Prevention of Free Recombination

If there were no chromosomes, there would be no limit to the possible assortment of genes. However, with the genes linked together on chromosomes and with the amount of crossing over in every generation restricted, recombination between the parental genomes is severely limited. It is important to find out to what extent and by what mechanisms recombination can be reduced or prevented altogether and how this affects the storage of genetic variation in populations. The mixing of the gene contents of a species or population is determined at two levels, the gametic and the chromosomal. It is advisable to restrict the term *recombination* rigidly to the factors controlling the degree of mixing of the gene contents of parental chromosomes. The various phenomena, such as hybridization, isolating mechanisms, dispersal, and population size, that determine the degree of genetic difference of the parental gametes are of a very different nature and will be discussed in Chapter 14 under outbreeding.

On the chromosomal level, there are mainly two sets of factors that determine the amount of recombination: the number of chromosomes and the frequency of crossing over (Darlington 1939; White 1954).

Chromosome Number. The various nonhomologous chromosomes assort independently during meiosis. The greater the number of chromosomes on which the genes of a gamete are distributed, the greater the possible number of combinations. The increase is exponential, so that, for instance, for a haploid number of chromosomes $n = 7$ the number of possible gametic chromosome combinations is 128; for $n = 14$ it is 16,384. For plants some definite regularities have been worked out (Stebbins 1950; Grant 1958), indicating that annuals have on the average lower chromosome numbers, perennials and woody plants higher chromosome numbers. For most groups of animals, as for plants, there exists a "typical" chromosome number, which happens to be high in birds and low in most dipterans. But the reasons for these differences are not yet certain (White 1954, 1958), nor has chromosome number or total amount of DNA been correlated with the breeding system (see Chapter 14).

Frequency of Crossing Over. The recombination of genes linked on the same chromosome is favored by a high chiasma frequency, a random distribution of chiasmata, and structural homozygosity of chromosomes. Conversely, the amount of crossing over is reduced by a lowering of the number of chiasmata, by localization of chiasmata, and by structural hybridity of chromosome sections (particularly through inversions) which prevents crossing over in the particular section. A chromosome section that is protected from crossing over acts, for the purposes of recombination, as a single gene and has therefore been called a *supergene* (Darlington and Mather 1949). Such supergenes may have two advantages. They permit indefinite preservation of a particularly valuable assortment of genes, and they permit collection in the gene pool of several such supergenes, which are heterotic when combined. The advantages of single-locus heterosis are thus potentially expanded to an entire chromosome section. Furthermore, the genetic loads caused by deleterious homozygotes are thereby made to coincide.

Linkage. The term "linkage" has been applied to two different phenomena, the linkage of genes and the linkage of loci. If two loci occur on the same chromosome they will be permanently "linked" unless they are separated by chromosome breakage. Crossing over does not normally change the distance between loci. Because the same phenotypic character is often controlled by closely linked loci, Fisher (1930:102) and Sheppard (1953a) have postulated that such loci were brought together by natural selection through translocation or other forms of chromosomal rearrangement. Duplication of chromosome sections (perhaps by translocation between homologous chromosomes) is an alternative explanation for proximity of functionally related loci. Once established, such linkage of loci ought to be a fairly permanent condition in view of the relative rarity of translocation.

A very different phenomenon is the linkage of definite alleles at different loci, let us say allele a^1 on locus A and allele b^1 on locus B. Such linkage is quite temporary, owing to crossing over, except in the presence of cross-over inhibitors. Usually, "the frequencies of combinations among loci are in the long run those of random combinations, in a random breeding population, irrespective of linkage, unless the latter is complete" (Wright 1949a). Linkage may, however, prevent complete equilibrium under special circumstances (Lewontin and Kojima 1960). Certain experiments, likewise, seem to support a greater importance of linkage. "In one experiment the breakdown by recombination of balanced combinations in *Drosophila* necessary to achieve a certain response to selection took twenty generations" (Mather 1953). Perhaps the linkage was very close (adjacent loci) or a cross-over inhibitor was involved. There are many ways and means of reducing crossing over or eliminating it altogether. Inverted gene arrangements show very little crossing over in heterozygous condition. Chiasma localization is another mechanism for preserving a chromosome segment with a particular assortment of genes. At present not much is known about the occurrence and operation of these cytogenetic devices (White 1954, 1958). Reduction of crossing over in one chromosome (or part of one) is often compensated by an increase (up to thirtyfold) elsewhere (Carson 1953), so that the total amount of recombination remains approximately constant. The mechanism of this compensation is still obscure (Oksala 1958).

One must conclude from these observations that, on the whole, linkage is a rather inefficient mechanism for the preservation of genetic variation.

The term *linkage* is sometimes used incorrectly in the discussion of pleiotropic effects of genes. If left-handed coiling of the shell in a mollusk or foraminifer (Ericson 1959) is correlated with a cold-water environment and right-handed coiling with warm water, this is not due to linkage of the coiling allele and the gene giving superior viability at certain water temperatures (such linkage would soon be terminated by crossing over), but due to the versatile pleiotropic manifestations of a single gene or supergene.

Unequal Segregation

This term refers to very peculiar mechanisms by which a deleterious gene may escape elimination by natural selection. If such a gene determines simultaneously that the chromosome which carries it is favored during meiosis, a higher than expected percentage of the gametes will carry the gene. Sandler and Novitski (1957) have applied the term *meiotic drive* to a force able to alter the mechanics of meiotic cell division in such a manner that the gametes produced by a heterozygote do not occur with an equal frequency of 50 percent. A number of genes with this effect have been recorded in the literature.

In several species of the *obscura* group in *Drosophila*, a sex-ratio gene leads to the degeneration of the Y-bearing spermatocytes, compensated by a double division of the X-chromosomes. As a result the normal number of spermatozoa is produced, but all produce females (since they carry an X-chromosome) and all carry the sex-ratio gene. This in due time would lead to extermination of the males (and hence of the species) if it were not for a pronounced superiority of the heterozygotes. The resulting balanced polymorphism has been carefully analyzed by Wallace (1948). A similar gene (or supergene) in *Drosophila paramelanica* was analyzed by Stalker (1961). A gene in *Drosophila melanogaster* ("segregation distorter," SD) tends to make the gametes with the normal allele nonfunctional (Sandler *et al.* 1959).

An analogous situation has been found for the *T*-locus in the house mouse (*Mus musculus*). Many populations from different parts of the United States contain heterozygotes for *t*-alleles at considerable frequencies, even though most of these alleles are lethal as homozygotes. It is found that 90–99 percent of the gametes produced by these heterozygotes contain the *t*-gene (instead of the expected 50 percent *t* and 50 percent normal). There seems to be a prezygotic advantage of the *t*-gene-carrying sperm. Matsunaga and Hiraizumi (1962) found a similar (but much smaller) advantage of O-bearing spermatozoa in the sperm of fathers heterozygous for the A and B blood groups (AO, BO). The 4.5-percent advantage of O found by them would lead to a rapid elimination of A and B, if it were not otherwise compensated (Chapter 7). The frequency of the *T*-alleles in the mouse is, likewise, determined by several factors, not only the lethality of the homozygotes. The recorded frequency of the *t*-genes (25–30 percent) in wild populations is well below the expected frequency of nearly 50 percent. This appears to be due to the fact that the *t*-genes tend to be lost by random elimination in many of the small and semi-isolated populations in which the house mouse occurs (Lewontin and Dunn 1960).

It is very unlikely that unequal segregation or sperm success is of any appreciable evolutionary significance. Yet it may serve occasionally to keep otherwise deleterious genes in populations for a length of time sufficient to permit the accumulation of modifiers that cover the deleterious qualities.

ECOLOGICAL PROTECTION OF GENETIC VARIATION AGAINST ELIMINATION BY SELECTION

A second factor that reduces the efficiency of natural selection is the diversity of the environment. The contribution to fitness made by a given phenotype changes in time and space and this causes fluctuations in the selective values of genes. A gene that is inferior under some conditions is superior under others and the chance that it will be preserved in the gene pool is thereby greatly increased. Another conservative factor is the possibility that a gene may be retained in the population if its fitness is inversely correlated with its frequency. Finally, the complex integration of the epigenotype prevents too precipitous a submission to every environmental change.

Opposing Selection Pressures

Since the genotype as a whole is an interacting, integrated system, virtually all aspects of the phenotype are the result of a compromise between opposing selection pressures. An increase in body size, for instance, might be an over-all advantage for an organism, but, since it would necessitate changing the relative size of numerous organs, opposing selection pressures would be set up until the various proportions had been reconstructed. An increase in tooth length (hypsodonty) was of selective advantage to primitive horses shifting from browsing to grazing in an increasingly arid environment. However, such a change in feeding habits required a larger jaw and stronger jaw muscles, hence a bigger and heavier skull supported by heavier neck muscles, as well as shifts in the intestinal tract. Too rapid an increase in tooth length was consequently opposed by selection, and indeed the increase averaged only about 1 mm per million years (Simpson 1944). Haldane (1949c) showed that, even in rapidly evolving lines, changes of length and proportion are usually on the order of only 1–10 percent per million years. Genetic variation will not be rapidly depleted when selection pressures are so nearly in balance with each other.

Finally, many selection pressures are severe only during periods of crisis—epidemics, exceptionally severe winters, droughts, and the like. Such crises, however, tend to be localized and the genes lost during a crisis may be subsequently restored by gene flow. All these factors combine to soften the impact of natural selection and to prevent it from depleting genetic variation too rapidly.

Diversity of the Environment in Time

Temporary and Secular Changes in Selection Pressure. The storage of genetic variation in populations is greatly affected by shifts in selection pressures. These shifts may involve either the intensity of selection or the direction of the selection pressure. A shift in the intensity of selection pressure nearly always accompanies (or results from) a change in population size. Any sudden change in population size will affect genetic variability. Sudden contraction of a population not only may increase homozygosity, but as a consequence may also permit the emergence of deviant phenotypes (Lerner 1954) that have a selective advantage under special conditions. Increased selection pressure and reduced population size unquestionably lead to an increased rate of selective elimination of genes from a population. Conversely, temporary relaxation of selection pressure will permit survival of otherwise inferior genotypes, facilitating rare genetic recombinations that would be impossible during periods of severely adverse selection. Certain of these "improbable" genotypes may represent new adaptive peaks, and may be retained in the population when the period of higher selection intensity is reestablished. There is observational evidence that cycles in population size affect the genetic variability of populations according to these theoretical expectations (Ford and Ford 1930; Tetley 1947).

Shifts in the Direction of Selection Pressure. All environments are changing, which means, in terms of evolution, that the selection pressures shift continuously. Every such shift might ease the selection pressure on those genes and genotypes that had previously been under the heaviest pressure or even favor them selectively. Detailed samplings of polymorph or otherwise genetically variable populations have indeed established in many cases a parallelism between environmental and genetic fluctuations. Here we can really watch evolution at work. The underlying assumption is that genes may have very different selective values under different environmental conditions. This has now been well established by population geneticists. Dobzhansky *et al.* (1955) have given a particularly clear demonstration of the effect of temperature and food on the selective value of different genotypes in *Drosophila pseudoobscura*.

Temporal changes in selection pressure can be classified into seasonal changes, secular fluctuations and cycles, and long-term trends, depending on the time interval involved in the change.

Seasonal Changes. One of the first to be described in detail concerns a population of ladybird beetles (*Adalia bipunctata*) near Berlin (Timoféeff-Ressovsky 1940a). The population occurs in two phenotypes which may be designated "black" and "red." The frequency of "blacks" among adults dropped from 55–70 percent (in different years) at the beginning of the winter to 30–45 percent at the end of the winter, owing to the higher mortality of "blacks" during hibernation. However, the superior viability of the "blacks" during the hot season permitted it to return by the end of the summer to the original frequency of 55–70 percent. Contrary to Timoféeff's original interpretation, such a fluctuating polymorphism cannot be maintained by a balance of selective advantages of the two phenotypes. Such a system would sooner or later inevitably lead to fixation of one or the other type owing to random fluctuations (Kimura 1955) if it were not maintained by a superimposed superiority of the heterozygotes.

In the snail *Cepaea nemoralis*, the same generation is exposed throughout the seasons to shifting selection pressures. Sheppard (1951a) showed for southern England that in early spring, when the woodland floor is

Table 9–2. Seasonal change in frequency of yellow morphs among snails killed by thrushes (after Sheppard 1951).

Place	Date	Percent killed	Place	Date	Percent killed
Marley	April 11	43	Ten Acre	April 19–26	42
Woods	23	41	Copse	April 28–May 2	22
	30	34		May 4–8	11
	May 7	26		11–16	12
	19	16		17–22	0
	22	14		28–June 5	11
	26	14			

brown with leaf litter and exposed earth, thrushes (*Turdus philomelos*) destroy a higher percentage of contrasting yellow-shelled snails than of brown or pink ones, but that the situation is reversed in late spring after the woodland floor has become green (Table 9–2). (See also De Ruiter 1958 and Lamotte 1959.) In the marine isopod *Sphaeroma serratum*, the frequency of a spotted gene at Morgat, France, rose from about 23 percent at the beginning of the summer to about 28 percent at the end (Bocquet, Lévi, and Teissier 1951). Similar seasonal changes in *Colias* were discussed in Chapter 7.

By far the best evidence for such cyclical fluctuations comes from *Drosophila*. Seasonal changes in the relative frequency of gene arrangements have been found in *D. funebris* (Dubinin and Tiniakov 1945), *D. pseudoobscura* (Dobzhansky 1943, 1956b), and *D. robusta* (Levitan 1952). For instance, in a population of *D. pseudoobscura* occurring at Piñon Flats, Mount San Jacinto, California, the gene arrangement Standard decreased regularly in the years 1939 to 1946 from March to June while the frequency of another arrangement (Chiricahua) increased correspondingly. The reverse change took place between June and November (Dobzhansky 1951, 1956). If an artificial population with material from this locality is established in the laboratory, an equilibrium is always reached at 25°C with a frequency of about 70 percent Standard and 30 percent Chiricahua chromosomes. The fact that Standard is of superior viability at high population densities while Chiricahua is superior at low densities (Birch 1955) may be in part responsible for the annual cycle. At nearby Keen Camp no such seasonal cycle was found and interestingly enough it disappeared from the Piñon Flats population after 1946 (Epling, Mitchell, and Mattoni 1953; Epling and Lower 1957). These cycles were discussed by Dobzhansky in several publications (1952, 1956, 1958).

Secular Changes. In the *Drosophila* case just mentioned, more than seasonal change is involved. Indeed, in nearly every case in which polymorphic populations were studied carefully over a series of years, shifts in the frequencies of morphs have been found. A melanistic-plumage type of the Rock Dove (*Columba livia*) had a frequency of at least 70 percent in the isolated feral pigeon population of St. Mark's Square in Venice in 1951 (personal observation), while ten years earlier this blackish type occurred only at low frequency (Buzzati-Traverso *in litt.*). In a population of *Cepaea hortensis,* Schnetter (1951) found an increase of 16.2 percent in the frequency of yellow unbanded snails during a series of years (1942–1950) that were warmer and drier than average. In many of these secular changes there is a fairly well-defined correlation between climatic cycles and shifts in morph ratios, and in both *Drosophila* and *Cepaea* there was a reversal when the climatic trend changed.

Long-Term Trends. Some shifts in gene frequency are long-term trends extending over a considerable number of years. When such a trend leads to the fixation of one allele, it is called *transient polymorphism* (Ford 1940), and is defined as the condition in which an advantageous gene in a natural population displaces a previously frequent allele to the extent that the latter is preserved only by recurrent mutation. Such a replacement may be due to one of two conditions, either the origin by mutation (or immigration) of a new, superior gene (which then spreads through the population), or a rapid change in the environment resulting in a drastic change in selective values of the two competing alleles. The second alternative seems valid in most cases of long-term trends. Transient polymorphism rarely leads to complete fixation of a new favorable gene in a previously polymorph species. Most recorded cases of apparent transient polymorphism are merely shifts from one to another level of balanced polymorphism.

The selective factor leading to such a shift in gene frequency is only rarely understood. Perhaps the best-known case is that of the so-called industrial melanism (Hasebroek 1934; Ford 1955; Sheppard 1961). In a number of species of moths the normal grayish or whitish morph is being replaced in the manufacturing districts of England and western Europe by black or dark forms. While the black morphs were rare or absent 100 years ago (before the period of industrialization), it is now the light forms that are rare. The shift, in some cases, was amazingly rapid, requiring less than 50 years. This indicates a selective advantage of the dark gene amounting to somewhere between 10 and 30 percent (Goldschmidt 1947; Haldane 1957). The dark gene in most of these cases is dominant and appears slightly more viable under laboratory conditions. Yet this does not explain the magnitude of its sudden selective advantage. The puzzle was solved by Kettlewell (1956, 1961), who found strong predation by birds on the adult moths resting on the bark of trees. In unpolluted areas with lichen-covered tree trunks, the light-colored moths are cryptically colored; in sooty areas it is the black moths. Even this is, however, not yet the whole answer. The melanistic form has invaded also strictly rural areas in which the tree trunks are still grayish and covered with lichens (for example, Rensch 1960b). It is possible that a physiologically superior genotype was built up in the industrial areas which was then able to spread into rural areas in spite of a lack of cryptic coloration, or else the phenotype is maintained by strong gene flow. Occasional melanism is of widespread occurrence in many species and genera of moths, indicating that there may be a basic superiority of heterozygotes. Yet the particular level at which the ratio of the gene becomes stabilized depends largely on the predator pressures on the two phenotypes. Observed shifts in dominance modifiers strengthen this argument.

Climatic trends seem to be responsible for other shifts of polymorph ratios observed in recent years. The rapid spread of the black hamster (*Cricetus*) in Russia seems to have coincided with an amelioration of the climate and certain other ecological changes (Gershenson 1945). The blue morph ("Blue Goose") of *Anser caerulescens* in the American Arctic seems to have greatly increased in frequency in the past 30 years at the expense of the white morph or "Snow Goose." Interestingly enough, the southernmost portion of the range of this dimorphic species seems to be occupied by populations in which the blue gene has reached fixation. In the Red Fox (*Vulpes*) two genes ("Canadian" and "Alaskan") produce the coloration "Silver Fox." Both have been declining steadily in frequency in historical times (Butler 1947; Calhoun 1950; Oksala 1954). The decrease has been ascribed to selective hunting, but this is highly improbable. Since most of the hunting is done by trapping and the red furs also have considerable commercial value, it is more likely that the decrease of the silver genes is somehow correlated with the general amelioration of the climate in the American north. The fact that the strongest shifts in

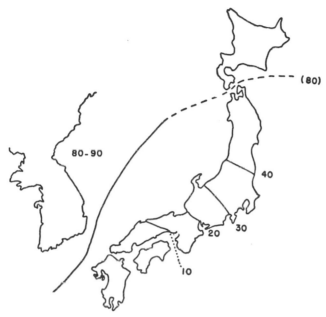

Fig. 9–3. Frequency (percent) of the *succinea* morph of *Harmonia axyridis* on the Japanese islands and the adjacent Asiatic mainland. Note the regular decrease of the morph from north to south. On Hokkaido Island, north of the dotted line, the frequency changed drastically between 1923 and 1944 (see text). (From Komai, Chino, and Hosino 1950.)

frequency of the black gene are correlated with cycles in the abundance of the foxes strengthens this assumption. Much evidence indicates that there is an inverse relation between population density and the frequency of the silver genes. The evidence for changes in the relative frequency of morphs in various game animals has been summarized by Voipio (1950), particularly for the Fenno-Scandian area.

The causes of shifting morph ratios in many species of insects are even less understood. In the ladybird beetle *Harmonia axyridis*, the frequency of the *succinea* morph dropped at Sapporo (Hokkaido) from 84 percent in 1923 to 43 percent in 1944 (Komai, Chino, and Hosino 1950; Fig. 9–3).

In *Drosophila pseudoobscura* there was a spectacular increase in the frequency of the Pikes Peak gene arrangement in California between 1945 and 1961, resulting at Yosemite in an increase from near zero to more than 10 percent at the expense of other gene arrangements (Dobzhansky 1958), fully confirming the ephemeral nature of the distribution pattern of these gene arrangements predicted by Mayr (1945). Equivalent changes in *Drosophila melanogaster* were described by Ives (1954). The best-studied case of shift in morph ratio concerns the tiger moth *Panaxia dominula*, of which several small and isolated colonies in the south of England were studied by Sheppard (1951b, 1953a). A rare gene producing the *"medionigra"* phenotype in heterozygous condition increased in frequency from the 1920's to 1940 at one of these stations until it reached a frequency of 11.1 percent and since then has steadily decreased to a level of about 3 percent. It is not known why the currently disadvantageous gene should have been so highly advantageous before 1940.

The cited cases of strong shifts in morph ratios are selected from an extensive literature. They are more than compensated for by the evidence of high stability in gene ratios of human blood groups, Neolithic *Cepaea*, *Xiphophorus maculatus*, *Sphaeroma serratum*, and many other cases (p. 211). I have pointed out previously (Mayr 1942) that some of the reports in the ornithological literature on transient polymorphism in the genera *Rhipidura*, *Coereba*, and *Lybius* (see also Benson 1946) are not supported by the available data.

Stabilizing Effect of Environmental Fluctuations. The observed fluctuations in the gene ratios of many species indicate frequent reversals of selection pressures owing to changes in environmental conditions. These changes contribute to the preservation of genetic variation in natural populations because they greatly delay total elimination of a gene from a population. Indeed, even daily fluctuations of the environment may lead to an increase of the genetic variance of a population (Beardmore 1961). The composition of a gene pool at any one time rarely reflects the current selection pressures owing to the inevitable lag in response to environmental fluctuations. Yet eventual elimination is inevitable unless prevented by heterozygote superiority or other factors mentioned in this chapter.

Diversity of the Local Environment

Ecological Mosaicism. In discussions of fitness the simplifying assumption is usually made that the environment is uniform. This, of course, is never the case. In the preceding section we examined evidence for the heterogeneity of the environment in time, and in this section we will consider heterogeneity in space. It would evidently be of great selective advantage for a species to be able to utilize simultaneously a number of different aspects of the environment. Since there is a limit to the ecological tolerance and efficiency of a given genotype, a population will be more successful that is able to diversify genetically in order to broaden the utilization of its environment and to expand into various subniches. This poses the problem of maintaining the harmonious coadaptation of the gene pool while producing a number of different phenotypes, each specializing in a particular aspect of the environment.

Ludwig's Theorem. As far back as Darwin and his predecessors there has been a feeling among naturalists that it would be advantageous for a species to have "varieties" utilizing different subniches. The more widespread and numerous a species, the more likely it would be to have such ecological variants. This thesis was placed on a firmer basis by Ludwig's (1950) calculations that a genotype utilizing a novel subniche could be added to the population even if it were of inferior viability in the normal niche of the species. Levene (1953), Dempster (1955), and Li (1955b) have shown under what circumstances such a system of ecological polymorphism can be maintained without superiority of heterozygotes.

A possible case of such ecological polymorphism is that of the gene arrangements in *Drosophila*. Mayr (1945) suggested that new inversion types might be "able to occupy ecological niches that are inaccessible to other members of the ancestral population." Dobzhansky found much evidence (summarized in 1951) to support the thesis that each gene arrangement is adapted to a different subniche within the general habitat of the species. Hence the greatest amount of this polymorphism is found in the most favorable areas of the species range (Chapter 13). However, this is only one of several possible interpretations. Carson (1958b) believes that the frequency of gene arrangements is regulated by selection for free recombination. White (1958) calls attention to the striking differences between species, some highly euryecous species being monomorphic, and some highly polymorphic species showing no reduction of the polymorphism in peripheral populations. Evidently the last word on the ecological significance of polymorphism has not yet been said.

Substrate Polymorphism. A comparatively simple situation is presented where phenotypic polymorphism is correlated with a "polymorphism of the substrate." The marine snail *Littorina obtusata* is strongly polymorphic, having uniform yellow, blackish olive, and banded morphs.

On light gravels or on rocks covered with whitish barnacles the yellow type predominates. On blackish gravels or on black rocks covered with the seaweed *Fucus,* the yellow type may be rare to the point of virtual absence (Mayr MS). Not much is known about predators, but the absence of a pelagic larval stage favors local maintenance of a phenotype once established by selection. Better analyzed is the celebrated case of the banded snails (*Cepaea*) of Europe which formerly was often cited as a case of

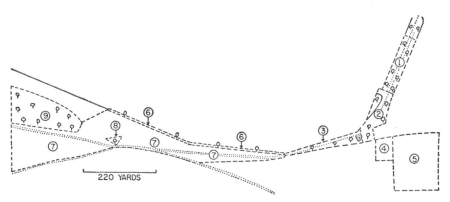

Fig. 9–4. Composition of *Cepaea nemoralis* populations in neighboring habitats at Wiltshire, England. Areas 1, 2, 8, 9 are beechwoods, 5 and 7 shortgrass down, 3, 4, 6 intermediate. The percentages of yellow snails in the nine areas are: 1, 28.5; 2, 24.5; 3, 35.7; 4, 35.5; 5, 47.7; 6, 15.7; 7, 45.1; 8, 20.0; 9, 16.0. (From Sheppard 1952a.)

neutral polymorphism (Mayr 1942; Lamotte 1951). Cain and Sheppard (1950) found a close correlation between certain morphs and definite habitats, for instance unbanded yellow and shortgrass downs, unbanded reddish and beechwoods, or banded yellow and hedgerows. The distance between colonies of strikingly different compositions is often a matter of only a few meters (Fig. 9–4).

Substrate polymorphism entirely controlled by predator pressure would seem a rather wasteful method of adaptation. There is evidence, however, that several mechanisms greatly mitigate the effect of the predators. One mechanism is differential survival of the different genotypes owing to physiological differences; another is habitat selection.

Habitat Selection by Genotypes. If individuals of substrate polymorphic species dispersed at random, all those would be in great danger that happened to get onto a contrastingly colored substrate. An ability to choose the "right" background would tremendously increase the probability of survival. Evidence of such habitat selection by genotypes is beginning to accumulate. A few cases may be mentioned.

In a polymorphic population of the isopod *Asellus aquaticus* the least-pigmented individuals have the greatest preference for darkness and conversely the darkest individuals the greatest reluctance to seek dark places (Janzer 1950). The relevance of this observation for the origin of the adaptations of cave species is evident (Janzer and Ludwig 1952), although one should not overrate the importance of this factor (Kosswig 1944). In *Cepaea hortensis* a new habitat, a grassy ditch, was colonized almost exclusively by yellow unbanded individuals from an adjacent mixed population (Schnetter 1951). Preference tests in the laboratory yielded considerable differences among the phenotypes of *Cepaea*. Unbanded yellow *hortensis* preferred a temperature of 20°C, five-banded ones preferred 17°C. Yellow three-banded *nemoralis* preferred a mean temperature of 20°C, red three-banded ones preferred 14°C. There were considerable differences in activity between the various genotypes when exposed to different conditions of temperature, humidity, and light (Sedlmair 1956). Cumber (1949) showed that bumblebees, *Bombus agrorum*, visiting the deep flowers of *Symphytum officinale* had a long mentum, 345 to 413 (376) micrometer units, while others taken at the same locality and season visiting the shallower flowers of *Epilobium hirsutum* had a short mentum, 268 to 365 (311) micrometer units. Conditioning may be involved in this case. Differences in habitat selection of different mutants of *Drosophila melanogaster* has been established by Waddington, Woolf, and Perry (1954). A selection of the suitable substrate occurs even when the phenotype is not determined by genetic difference as in the case of certain hemiptera (Popham 1942) and grasshoppers (Ergene 1955), where the color of the adult is determined during the last larval stage.

When several morphs are found in different subniches of the same area at different frequencies, it is not always possible to determine whether this is due to habitat selection, selective mortality, or both. For instance, in the lady beetle *Harmonia axyridis* in southern Japan, the *conspicua* morph occurs on pine at a frequency of 12 to 31 percent, and on wheat at 40 to 62 percent, while the *axyridis* morph was more common on pine (20 to 33 percent) than on wheat (7 to 11 percent) (Komai and Hosino 1951).

The more closely a species is studied, the more likely it is that some evidence will be found for ecological polymorphism or gradual ecological variation. This variation is still largely ignored by ecologists, most of whom discuss the ecological requirements of species in a strictly typological manner.

Sexual, Social, or Other Special Forms of Differential Niche Utilization. Many forms of polymorphism, such as sexual dimorphism, have a primary selective significance that is not related to the diversity of the niche. Yet secondarily it may serve to improve niche utilization. Rand (1952) has collected numerous instances of bird species in which male and female differ in size, morphology of the bill, and feeding or migratory habits, in such a way as to minimize competition between the sexes, thereby providing the species with an expanded ecological universe. Amadon (1959) has added further illustrations. Among the insects much of this ecological diversity is achieved without the help of genetic variation. The larvae of most insects, for instance, live in a different niche from the adults. The caste specialization among the social insects (worker, soldier, sex form) is likewise nongenetically achieved in nearly all cases. An exception occurs in the stingless bees (*Melipona*) in which caste is determined by genetic factors (Kerr 1950). In certain species of marine organisms it is advantageous under some ecological conditions to shift from the normal separation of sexes (gonochorism) to hermaphroditism. Both types may occur polymorphically in the same population (Bacci 1949, 1955a). The ratio of these two genetically determined types varies geographically and seems to be adjusted by natural selection. The same is true of the ratio of neotenic to metamorphosing individuals in some populations of *Ambystoma tigrinum* (Dunn 1940). In the pyralid moth *Acentropus niveus,* both flying and flightless females occur. The flightless female is aquatic and protrudes its genital opening through the surface film of the water to elicit copulation. The larvae feed on submerged plants (Beirne 1952). In all these cases a different utilization of the environment is permitted by the different adaptation of the genotypes.

Mimetic Polymorphism. In many species of (mostly) tropical butterflies several color types occur within a single population, each mimicking a different, distasteful species (its "model") and thereby providing protection against insect-eating birds. Such mimicry works best when the mimic is sufficiently rarer than the model to prevent the development of counterconditioning. The frequency of butterflies as food of birds and the speed of conditioning of birds against distasteful prey are now sufficiently well established by observation and experiment to dispel all doubts as to the selective basis of mimetic polymorphism. There have been various illuminating discussions of this phenomenon in the recent literature (Cott 1940; Ford 1946, 1953; Sheppard 1958; Clarke and Sheppard 1960b). In the genus *Papilio* only the females are mimics; in some other genera both sexes may mimic. A good example is the African Nymphalid butterfly *Pseudacraea eurytus,* in which Carpenter (1949) recognizes thirty-three color forms, each mimicking some species of the acraeid genus *Bematistes.* Male and female *Pseudacraea* usually mimic different models. Twelve or more forms of *Pseudacraea* may coexist at a single locality. If under special circumstances the model becomes rarer than the mimic, the latter may "go to pieces." As a result of relaxed selec-

Table 9-3. Inverse relation between frequency of mimic and precision of mimicry in *Pseudacraea eurytus* on islands in Lake Victoria (from Carpenter 1949).

Location	Frequency of mimic in model-mimic association (percent)	Frequency of deviant variants among mimics (percent)
Mainland	18	4
Damba Island	62	35
Bugalla Island	73	56
Kome Island, 1914	23	51
1918	68	54
Buvuma Island	43	28

tion the precision of the mimicry breaks down and a considerable part of the population consists of intermediates between the elsewhere sharply discontinuous types. This has happened in *Pseudacraea* on some islands on Lake Victoria (Table 9-3). This proves, contrary to Goldschmidt's (1945) contention, that selection is continuously active in building up the precision of the mimicking types and in maintaining discontinuity between them. Even though originally established through small mutational steps, the polymorphism in any local population of *Papilio dardanus* is now controlled by a few major switch genes. However, when one crosses different geographic races one gets arrays of intermediates between the morphs (Clarke and Sheppard 1959, 1960a,b). This proves that the strict discontinuity of the morphs within a population is the result of selection of the appropriate genetic background. The sharp definition is lost when two such polygenic complexes from different regions are mixed.

In the Holarctic region we have the interesting case of flies of the species *Volucella bombylans* that resemble various species of bumblebees (*Bombus*). This polymorphism was shown by Gabritschevsky (1924) to be caused by three alleles, dominant to each other in the sequence b, p, h. The phenotype of b is like that of B. *lapidarius,* that of p is like B. *horto-rum,* and that of h is like B. *agrorum, montanus,* and others. In the Caucasus Mountains the various morphs of *Volucella* are colored exactly like the local species of *Bombus* and the corresponding situation is true in North America (Johnson 1916, 1925), where species like B. *pennsylvanicus* and B. *ternarius* are imitated.

The polymorphism in the plumage of female cuckoos (*Cuculus*) is believed by some authors (Voipio 1953) to be mimetic because one plumage type resembles the Kestrel (*Falco tinnunculus*), the other the Sparrow Hawk (*Accipiter nisus*). The so-called "host races" in cuckoo eggs are, however, not a case of polymorphism (Southern 1954).

All cases discussed in this section have two aspects in common. The visible phenotypes are directly adaptive, and they coexist in the same population. To what extent superiority of the heterozygotes contributes to the maintenance of the polymorphism is unknown in nearly all these cases. It is conceivable that the relative frequency of the several morphs is in some cases controlled by density-dependent factors operating separately on the various morphs, in such a way that the rare morphs are favored by selection.

Adaptive Nature of Polymorphism. There is great diversity in the forms of polymorphism discussed in the preceding section and no consistent classification of types of polymorphism has yet been proposed (Williamson 1958). This raises a number of terminological difficulties. Dobzhansky (1951) has remarked correctly that in most cases of observed polymorphism it is quite impossible without experimental evidence to determine what portion is due to superiority of the heterozygotes (balanced polymorphism *sensu stricto*), what portion is due to mutation and immigration pressure, and what portion is due to a selective balance of the phenotypes (Ludwig effect). Yet, regardless of the mechanism, "polymorphism . . . or any other kind of diversity of sympatric forms increases the efficiency of the exploitation of the resources of the environment by the living matter." Dobzhansky refers therefore to polymorphism as "adaptive." Cain and Sheppard (1954) have questioned whether a species with polymorphism is "better adapted" than a species without it, and whether it is legitimate to refer to such polymorphism as "adaptive." There are, as stated previously, many possible definitions of "adaptive," but species with ecological polymorphism have two attributes that contribute to their adaptedness. First, as stated by Dobzhansky, their greater genetic diversity permits them to utilize the environment better and more completely. This statement is well supported by observational as well as by experimental evidence. Chromosomally polymorphic populations of *Drosophila pseudoobscura* produced more flies and more biomass under standard conditions than monomorphic populations. This was equally true when the larvae were crowded and the adults uncrowded (Beardmore, Dobzhansky, and Pavlovsky 1960) and when the larvae were uncrowded and the adults crowded (Dobzhansky and Pavlovsky 1961). Furthermore, a polymorph population, being composed of a number of different genotypes, each specialized in a slightly different way, is better buffered against possible shifts of environmental conditions. Finally, genetic and ecological diversity is of advantage in both intra- and interspecific competition. The more the genotypes differ in their requirements, the less they will compete with each other. Nor should it be forgotten that an increased diversity of genotypes in a population improves the ability of this population to compete with other species. On the basis of all these arguments, it seems completely legitimate to consider ecological polymorphism as adaptive. Polymorphism is based on and produced by definite genetic mechanisms, such as genes for differential niche selection and the heterosis of heterozygotes. A population that has not responded to selection for such mechanisms and therefore lacks polymorphic diversity is more narrowly adapted, more specialized, and therefore more vulnerable to extermination. The widespread occurrence of genetic mechanisms that produce and maintain polymorphism is directly due to selection and is in itself a component of adaptiveness. It seems appropriate, therefore, to speak of "adaptive polymorphism."

Considering the advantages of all forms of polymorphism that lead to ecological diversity, it is surprising that it is not more widespread. In many cases it can perhaps not be incorporated into the gene pool, on account of the added genetic load (p. 253). In other cases the right mutations may not have turned up at the right time. Furthermore, it is not the only way to cope with the challenge of the environment, since many of the ecological specialists seem also very well adapted (Chapter 14). Even though there are thus a number of unsolved questions concerning ecological polymorphism, one point is clearly established: diversity of the environment favors the genetic diversity of populations and thus counteracts the depletion of genetic variation.

Diversity of the Species Range

For the sake of simplicity we have acted up to this point as if each population were a completely closed system. This is, of course, not the case. Indeed, as will be shown in Chapters 14 and 17, the most important source of genetic variability in any but the most isolated populations is the immigration of genes from other locally adapted populations, each of which has a somewhat different assortment of genes. Such gene flow not only introduces new genes but also restores genes that had been temporarily lost during a catastrophe or owing to accidents of sampling. The amount of gene flow is determined by the breeding system of the particular species, a topic that will be discussed in Chapter 14.

Assortative or Random Mating

The existence of definite preferences in the choice of mates may affect gene frequencies. Homogamy, the preference for a phenotypically similar mate, will tend to favor inbreeding and the production of homozygotes for the genes controlling that part of the phenotype involved in the homogamy. Other things being equal, this will facilitate the exposure of genes in homozygous condition to natural selection, which in turn may affect gene frequency.

Inverse Relation Between Fitness and Gene Frequency

The rarer allele is usually at a selective disadvantage not only because it is "intrinsically" inferior but also because the remainder of the genotype (the genetic background) is continuously selected to maximize the fitness of the most common allele. Most deleterious recessives illustrate this rule. There are, however, situations where the rare gene is favored for no other reason than that it is rare. Such a property would automatically help to maintain genetic variation in a population where it occurs. Haldane (1949c, 1954a) has pointed out that this might well be the case with "disease-resistance genes." Pathogens are selected to overcome antibodies and those mutations will be favored in pathogens that can cope with the most frequent antibody-forming genes in the host organism. Those host individuals will have superior survival in an epidemic that have rare antibody-forming genes to which the pathogen has not yet become "adapted."

A rarer gene will also be favored if several genotypes differ somewhat in their ecological requirements and each obeys density-dependent factors independently of the other genotypes. As a model we may postulate two genotypes in *Drosophila* of which one is favored in the dry and the other in the moist part of the food medium. Mortality will be increased in the more crowded part of the food medium. The effect of relative frequency on survival has been stressed by Teissier (1954), who points out that shifts in fitness with changes in frequency permit a polymorphism not maintained by a superiority of heterozygotes. The occurrence of a gene equilibrium independent of heterozygote superiority has been likewise demonstrated by Lewontin (1958). It had already been known that a mixing of genotypes leads to increased yield in plants (Gustafsson 1953). The relative importance of this mechanism for the maintenance of genetic diversity of populations is entirely unknown, yet it is evident that wherever it occurs it will contribute to the diversity of the gene pool.

Genetic Load

Our survey of mechanisms to prevent the depletion of genetic variability has revealed that there are indeed numerous devices that thwart the homogenizing tendencies of natural selection to a greater or lesser degree. However, the evolutionary plasticity that these mechanisms give to the population is bought at a price! The storage of genes results, through recombination in every generation, in the inevitable segregation of an appreciable number of inferior genotypes. The difference between the actual mean fitness of the population and the postulated value of a genotype with maximal fitness is referred to as the genetic load of the population. One can also call this load "the cost of the evolutionary potential," since the mechanisms that generate genetic variation (like mutation or gene flow) and protect currently deleterious genes against elimination are responsible for this load.

Haldane (1937) seems to have been the first to call attention to the problem of the genetic load and to estimate its size. Muller (1950a), Dobzhansky (1955b, 1957a), Crow (1958, 1960), Kimura (1960), and other recent authors have added to our understanding of the problem. Some six or seven kinds of genetic load have been distinguished by these authors, but they may be classified in three groups, A, B, C (see below). First, one must distinguish the cost of possessing an evolutionary potential from the cost of evolutionary change. The genetic load caused by the potential is the result of stabilizing selection which eliminates all inferior genotypes. For instance, in the case of a deleterious recessive gene *a*, the *aa* homozygotes are to some extent a reproductive waste and reduce the over-all fitness of the population. If *a* is a lethal, *aa* homozygotes are a total loss. Such a genetic load exists even in a completely stable environment. This cost of the evolutionary potential is composed of the "input load" (A) and the "balanced load" (B). In contrast is the cost of evolutionary change, called by Kimura (1960) the "substitutional load" (C).

A. *Input Load.* This load is produced by the presence of inferior alleles in a gene pool owing to mutation and immigration. Natural selection would reduce this load very rapidly if it were not continually restored and if many of these genes were not temporarily protected (see earlier in this chapter).

(1) *Mutational load.* This consists of recurrent harmful mutations and has been described by Muller (1950a) in a masterly way in his essay "Our Load of Mutations." Haldane (1937) has shown that the population fitness is depressed by an amount roughly equal to the sum of all mutation rates.

(2) *Immigration load.* This is the incorporation into a gene pool of alien genes that depress fitness in their new genetic environment. It is irrelevant whether these genes had been advantageous or deleterious in their former environment. The compensation load in man (Li 1953; Lewontin 1953) and similar rare phenomena result in the renewed input of deleterious genes into the gene pool of the next generation and are perhaps best included with the input load.

B. *Balanced Load.* The frequency of inferior genotypes produced by some loci is too high to be accounted for by the input load. Crow (1948), Lerner (1954), Dobzhansky (1955b, 1959a), and Wallace (1958) have called attention to another source of such variants, called by Dobzhansky the "balanced load." This load is due to the fact that selection favors an allelic or epistatic balance of genes which by recombination and segregation produces inferior genotypes in every generation. Here again different subtypes can be distinguished (Crow 1958). A balanced load is created by many of the protective devices discussed in this chapter, such as adaptation to the heterogeneity of the environment (Ludwig effect), selective advantage of rare genotypes, genes for strong dispersal, genes for negative assortative mating, and so forth. (Genes responsible for distorted segregation ratios should probably, by definition, also be included here.) Indeed, each one of the devices for storing genetic variability (Table 9-4) causes a genetic load of its own. Best known are:

(1) *The homozygous-disadvantage load.* This load is caused by the segregation * of inferior homozygotes at loci where the heterozygote has higher fitness than either homozygote. Every case of balanced polymor-

---

* The mutational load is likewise exposed to selection through the segregation of recessive homozygotes. Hence it is best not to call any of the loads "segregational."

Table 9–4. Protection of genetic variation against
elimination by selection.

1. Cytophysiological and developmental devices
   (a) Complete recessiveness
   (b) Control of penetrance and expressivity
   (c) Superiority of heterozygotes
   (d) Prevention of free recombination
2. Ecological factors
   (a) Inefficiency of natural selection
   (b) Change of selection pressure in time
   (c) Mosaicism of the local environment (Ludwig effect)
   (d) Geographic variation of the environment and gene flow
   (e) Heterogamy (inverse assortative mating)
   (f) Selective advantage of rare genes

phism creates such a segregation load. We have mentioned above (p. 233) cases like that of *Drosophila tropicalis* or *Anopheles quadrimaculatus* where the genetic load of a single gene or supergene amounts to 20–50 percent.

(2) *The incompatibility load.* In mammals, this load is caused by a deleterious antigenic interaction between the embryo and its mother, owing to incompatibilities of the genotypes. A human embryo of blood group A or B has apparently a 10-percent greater risk of dying if its mother is of blood group O than if she is of its own blood group. The amount of prenatal death due to incompatibility at the ABO locus is estimated as 2.4 percent (Crow and Morton 1960). This lethality would rapidly lead to an elimination of the rarer genes if it were not compensated by other factors, presumably a selective superiority of the heterozygotes.

(3) *The environmental-heterogeneity load.* See above (pp. 237–250).

The Proportion of Input and Balanced Loads. The problem of the relative importance of the two kinds of load (A and B) for the total genetic load is still unsolved. The difference of opinion is particularly wide concerning the respective contributions made by the mutational and balanced loads. Three viewpoints are represented in the literature:

(1) Virtually all genetic variation in populations is due to deleterious recessives revealed through homozygosity. The balanced loads are negligible.

(2) The contribution to the genetic variance made by the vast majority of the loci is due to the mutational load. There are, however, a few overdominant genes or supergenes that may make a major contribution to genetic variance. Crow (1952) stated this in the following words:

Inbreeding depression and recovery on crossing are mainly the result of loci at which the favorable allele is dominant and the recessives are at low frequency. On the other hand, the variance of heterozygous populations and the differences between different hybrids are due mainly to loci with intermediate gene frequencies. It appears likely that such loci are due to selectively superior heterozygotes, but there are several other possibilities.

The importance of overdominant supergenes has been demonstrated particularly well by Dobzhansky and Pavlovsky (1960). It is further supported by the great frequency of species in nature with some sort of polymorphism.

(3) A large proportion of loci are overdominant, particularly those with polygenes or sets of isoalleles. As a consequence, most of the observed genetic variance is due to balanced polymorphism.

This viewpoint has been advanced by Lerner (1954), Dobzhansky (1955b, 1959a), and Wallace (1958). An extreme formulation of this view was presented by Wallace:

We feel that at every locus there are heterozygous combinations of alleles which, on the average, give rise to individuals of higher viabilities or greater fitnesses than do homozygous combinations of the same alleles. Subject to the limitations imposed by chance elimination of alleles, by mating of close relatives, and by the finite number of alleles at a locus, we feel that the proportion of heterozygosis among gene loci of representative individuals of a population tends toward 100 percent.

The discussions in the recent literature reveal that it is very difficult to come to a decision between the three alternatives (or at least between 2 and 3), the difficulties being of a conceptual as well as of a technical kind. The main conceptual difficulty is a result of the definition of genetic load as "the amount by which the average fitness in the population is lower than [that of an individual with] the optimal genotypic composition." Does not this definition (and variants of it) go back to the early Mendelian concept of an ideal population consisting of a uniform homozygous genotype? Could not the fitness of a highly heterotic population be maximal in spite of the production of a large number of deleterious homozygotes? To me, at least, it seems as if the stated definition of genetic load is meaningless as far as the balanced load is concerned. The two aspects of overdominance, increased fitness of heterozygotes and deleteriousness of homozygotes, are two aspects of a single phenomenon and one cannot be segregated as *the* genetic load.

There is a second difficulty. The concept of the genetic load is based on the situation in the human species where the preservation of every individual is of ethical value. In animal and plant species, where an overwhelming percentage of the zygotes is dispensable (in all but a few cases), the production of an optimal genotype for the few survivors may well be favored by selection even if this is at the expense of the losers. Is it not sound strategy for the gene pool to gamble, considering its huge stack of chips? There is every reason to believe that the composition of the genetic load is different in species with different reproductive potentials. A species like man evidently cannot afford as high a balanced load as *Drosophila*. A situation such as that in *Anopheles quadrimaculatus*, where the homozygotes of a given gene arrangement appear to be lethal, would be intolerable in man. This single factor leads to a 50-percent mortality. Even the sickling gene in the tropical-malaria districts of Africa does not cause anywhere near that high a mortality. A species with as small a number of offspring as man can evidently afford only a far smaller balanced load than as fecund a species as *Drosophila melanogaster*. This does not mean that only one or a few loci can be heterozygous. It merely means that comparatively few truly deleterious or lethal homozygotes can be segregated.

Technical difficulties add to the complexity of the problem. Those laboratory geneticists who minimize the role of the balanced load tend to make a number of subconscious assumptions, such as:

(*a*) That populations are closed and input is to be measured in terms of mutations;

(*b*) That the average mutation is adequately represented by the highly penetrant, conspicuous, and deleterious mutations used by the laboratory geneticist (for instance, it seems doubtful to me that one can use the frequencies of deaf-mutism, muscular dystrophy, and mental defect as clues to the frequency of heterotic loci in man);

(*c*) That genes and genotypes have relatively constant or absolute selection coefficients;

(*d*) That current gene frequencies reflect steady states rather than unbalanced conditions, caused by continuous changes in the physical, biotic, and genetic environments, changes with which selection has not been able to catch up owing to the inertia of the gene complex.

It is too often forgotten in these discussions that the contribution to the genetic load made by supergenes may be very different from that of genes. Most calculations demonstrating a high segregation load are based on a study of supergenes known to be overdominant. Here, by definition, the homozygotes are of lower fitness. Most calculations demonstrating a high load of mutation (and a small or absent balanced load) are based on the study of loci carrying recessive lethals or recessive detrimentals. Most of these calculations (not in man) are based on the study of small closed populations under constant environmental conditions. That monomorphism is occasionally favored under these conditions (for example, Lewontin 1958) does not tell us much about optimal conditions in open populations. Finally, it is often forgotten that the numerical contribution to genetic variance (and hence to the genetic load) made by a single overdominant locus may be several hundred times higher than the contribution made by the average recessive or semidominant locus. The ratio of mutational to balanced load is not given by the ratio of recessive to overdominant loci.

Time will ultimately provide an answer. It is evident that no population can maintain simultaneously too many overdominant genes or supergenes. This may well be the reason why double heterozygotes are often of lowered fitness. Fisher (1939) found that double heterozygotes for two polymorph loci in the grasshopper *Paratettix texanus* were at a 40-percent disadvantage. Lewontin and White (1960) likewise found negative interaction of heterozygous gene arrangements on two different chromosomes in the grasshopper *Moraba scurra*. Evidence of interference among gene arrangements was also found by Wallace (1954a), and yet in *Drosophila subobscura* and *D. willistoni* individuals have been found that were heterozygous for a remarkably high number of different gene arrangements (up to 16). There are other ways of minimizing the load due to deleterious homozygotes. One is to accumulate a large number n of isoalleles at many loci, so that the number of homozygotes $1/n$ is kept at a minimum. Another way is to select for an increase in the viability of the homozygotes, to reduce the contrast in fitness. In view of the evident load caused by drastically overdominant loci, it is highly probable that most heterozygosity in populations that is not due to the mutational load is due either to the Ludwig effect or to gene immigration involving highly polyallelic loci.

The readiness of populations (even closed ones) to respond to selection, the large amount of genetic variance remaining when the phenotypic response is exhausted, and the much greater genetic variance of open than of closed populations, all these factors indicate to me a far richer reservoir of genetic variation than would be available if the genetic load were essentially only a mutational load.

C. The Cost of Evolution

The genetic load discussed so far results from the action of stabilizing selection in a stable environment. A new load is added wherever directional selection favors the replacement of an existing gene by a new allele. Every evolutionary change imposes such a substitutional load. We owe it to Haldane (1957) to have recognized this problem and to have calculated its consequences. Kimura (1960) has confirmed and extended Haldane's findings.

The cost of replacing an allele in a species (or population) by a new one is very considerable. The number of deaths required to carry through this process is (over a wide range) independent of the intensity of selection. Totaled up for all generations, this number is about five to fifteen times the total size of the population in each generation. Haldane and Kimura agree in two somewhat unexpected conclusions:

(1) The replacement of one gene by another is, owing to the high number of genetic deaths, a slow process. Haldane estimates that it takes an average of 300 generations per substitution.

(2) The number of loci at which genes can be substituted simultaneously is low. Kimura estimates that not more than about a dozen loci can be involved at any one time, or else the survival of the population is jeopardized.

These estimates of the cost of evolution lead to the conclusion that evolutionary change is normally an exceedingly slow process. If two species differ at 1000 loci, Haldane estimates that it may have taken at least 300,000 generations to complete speciation. With many organisms having several generations per year, this would certainly be a minimal estimate. But Haldane also believes "that the rate of evolution is set by [and inverse to] the number of loci in a genome and the number of stages through which they can mutate." This does not necessarily follow. Elephants and other slow-breeding mammals seem to have evolved much faster than *Drosophila* with its numerous annual generations. Other factors seem to be more important than the number of loci (Chapter 14).

Even greater difficulties are raised by organisms that are known to have evolved and speciated with extreme rapidity, like some fishes in fresh-water lakes (Myers 1960; Fryer 1959). Naturalists have repeatedly called attention to situations of extremely rapid change. Complete resistance to various insecticides has developed in houseflies (*Musca*), *Anopheles,* and *Culex* in periods of 2 years or less (Reid 1960). Even if the resistance were due to a single gene, this is exceedingly rapid evolution for a wild population. Zimmerman (1960) has pointed out that five endemic Hawaiian species of the pyraustid moth genus *Hedylepta* are restricted to the banana plant which was introduced by the Polynesians to Hawaii only 1000 years ago. These five allopatric species seem to have evolved from a parental palm *Hedylepta* in this extraordinarily short period.

The work of Haldane and Kimura is only a beginning, as both these authors emphasize. A few factors may be suggested that may have to be incorporated in the calculations of the cost of evolution, to give them wider applicability and greater precision.

(1) The contributions to genetic death made by each gene substitution are treated as if they were independent of each other. It is quite likely, however, that they are partially synergistic, when coming together in a genotype. Intuitively I would assume that an individual with three deleterious genes, each with a 1-percent loss of fitness as compared to the normal allele, would have a loss of fitness far in excess of 3 percent, perhaps amounting to as much as 10 percent. But even if fitness was decreased only by 3 percent, the probability of the elimination of this genotype is greatly increased. To be sure, if each of these genes has a frequency of 1 in 1000, the probability that all three genes would come together in a single individual would be only 1 in 1 billion. This, evidently, would be negligible. However, since populations contain numerous genes with minor deviations (+ or −) from mean fitness, there always will be some genotypes with several poor genes. A single death of one such genotype will remove several deleterious genes simultaneously.

(2) Selection coefficients are usually treated as constants. Yet they presumably change their values from generation to generation, according to the total pressure of other mortality factors. It is probable that the over-all "fitness" of a population over long periods is essentially constant (with the limits set by the capacity of the niche) and that either the elimination of deleterious genes or the addition of superior new genes simultaneously changes the selective values of the remaining genes. This effect will mitigate the impact of gene substitution and permit an increase in the rate of substitution. If the total fitness remains constant and likewise the number of zygotes eliminated (not reproducing) in each generation, then the concept of "load" becomes rather nebulous. Under these premises gene substitution could be accelerated or retarded, regardless of "load," as long as much of the mortality is density dependent and does not result in lowering the ability of the species to compete with other species.

(3) Haldane quite rightly mentions the possible importance of density-dependent factors, but it is difficult to calculate their effect. In a species where one pair normally produces 100 juveniles, 98 die on the average prior to reproduction. Mean population size will remain fairly constant; consequently it is irrelevant what factors are responsible for the mortality of the 98. This part of the offspring is expendable and will be eliminated regardless. If an allele is converted from advantageous to deleterious, it may simply be charged to this expendable surplus. The earlier in the life cycle such mortality occurs, the less effect it will have on the fitness of the population as a whole, by reducing intrapopulation competition at later stages in the life cycle. The higher a percentage is killed by gene substitution, the lighter will be the "ecological load" on the remainder of the population. The survival of populations is guaranteed by this feedback mechanism, in spite of the increased substitution load.

(4) Haldane shows that the cost to a population of making a gene substitution depends only on the natural logarithm of its initial frequency (as long as selection is not too intense). If one assumes the most general case, that evolution deals with large populations, that selection intensities are low, and that the initial frequency is determined by mutations, one arrives at extremely slow evolutionary rates. It can be shown, however, that there are numerous situations in nature where none of these three assumptions is correct. Most importantly, the input of new genes into natural populations is largely through immigration, and a new founder population may start off with a relatively high initial frequency of a gene that was rare in the parental population.

None of these considerations invalidate or even weaken Haldane's and Kimura's basic approach. It is probable, however, that the numerical values suggested by them for rates of gene substitution and, hence, of evolution will require considerable revision. This is especially true of those populations that are of the greatest importance in speciation, peripherally isolated populations (Chapter 17).

CONCLUSIONS

Genetic variation in a population is controlled by three sets of factors: (1) the input of new genetic material through mutation and immigration, (2) the erosion of this variation by selection and errors of sampling, and (3) the protection of the stored variability by cytophysiological devices and ecological factors. The diversity of these protective devices indicates that they are strongly favored by natural selection. Too uniform an "optimal" genotype would evidently favor too uniform a survival, and would result in the double disadvantage of density-controlled mortality in favorable periods and of extinction under adverse conditions.

The genetic variation of a population is not merely a "mutational load," because heterozygote superiority and selection for ecological diversity greatly add to it. Indeed, heterozygote superiority may well be responsible for a greater portion of the genetic variation of wild populations than all other factors combined.

It is not yet known how much variation a population can cope with. It has been estimated (Wallace 1958) that 50 percent or more of the loci of a given individual in an open, natural population may be occupied by dissimilar alleles. Others calculate that 12–20 variable loci already place a severe strain on a population, owing to the segregation of inferior genotypes. Far more research and a clarification of the basic concepts is needed to narrow the gap between these widely diverging estimates.

This much is well established, however, that all natural populations contain abundant genetic variation which serves as potential raw material for evolutionary change. The problems for adaptation and speciation posed by this variation will be considered in the ensuing chapters.

10 ~ The Unity of the Genotype

The procedure of the classical Mendelian genetics, of studying each gene locus separately and independently, was a simplification necessary to permit the determination of the laws of inheritance and to obtain basic information on the physiology of the gene. When dealing with several genes, the geneticist was inclined to think in terms of their relative frequencies in the population. The Mendelian was apt to compare the genetic contents of a population to a bag full of colored beans. Mutation was the exchange of one kind of bean for another. This conceptualization has been referred to as "beanbag genetics." We are all familiar with the atomistic concepts of that period: genes were believed to be clearly either recessive or dominant; mutation was thought to lead to a steady increase in variation until an equilibrium was reached, owing to the elimination of deleterious recessive homozygotes; genes were given constant selective values; and there was a tendency to equate genes and characters, as if there were a one-to-one relation.

Work in population and developmental genetics has shown, however, that the thinking of beanbag genetics is in many ways quite misleading. To consider genes as independent units is meaningless from the physiological as well as the evolutionary viewpoint. Genes not only *act* (with respect to certain aspects of the phenotype) but also *interact*. It had long been known that a gene which adds to fitness on its normal genetic background may be deleterious or even lethal when placed on a different genetic background, for instance through hybridization (Dobzhansky 1937). Such a shift in the selective value of a gene is not an isolated phenomenon, since every gene interacts with others even in its usual genetic environment. This interaction has been described, in an obviously exaggerated form, in the statement: *every character of an organism is affected by all genes and every gene affects all characters*. The result is a closely knit functional integration of the genotype. With recombination producing in every generation new assortments of genes (new genotypes), which in turn have to form well-balanced and fully viable phenotypes, it is evident that the integration has to extend beyond the level of the individual. There must be harmony among all the genes of which a local gene pool is composed. This gives the local population its cohesion and makes it a significant level of integration. Whatever phenotypic variation may be observed in a local population is not in conflict with this basic unity, because all the genotypes in a population are products of the same gene pool (Dobzhansky 1951, 1955b).

This internal cohesion of the gene pool is an immensely conservative force. It serves as a powerful brake on all forces that attempt to change the contents of the gene pool and makes it well-nigh impossible to create discontinuities within the gene pool. A thorough discussion of the cohesion of the gene pool is a prerequisite for understanding the genetics of speciation (Chapter 17).

GENE AND CHARACTER

To understand the cohesive factors in a gene pool, we must begin with the discussion of the relation between gene and character. This relation is not preformistic. No white eye is encapsuled in a white-eye gene, nor a clutch of four little eggs boxed into the chromosome of a bird laying a four-egg clutch. Development is epigenetic: genes merely give the potentiality to produce or to contribute to the production of a given phenotype, a "tendency" to produce it, "all other things being equal." These "other things" include the external as well as the genic environment. The precise details of gene action are still unknown and currently subject to intense research. Yet it is possible to suggest a simplified model that is sufficiently diagrammatic to be equally valid for most of the suggested theories of gene action (Fig. 10–1). The diagram indicates that a gene elaborates a gene product, which may be utilized in the differentiation of several organs (pleiotropy), and, conversely, that any one character may be affected by many genes (polygeny). Though the model indicates an intricate network of interrelations and interactions, we shall presently see that even this is a gross oversimplification.

Pleiotropy. The property of a gene of affecting several different aspects of the phenotype has been discussed in Chapter 7. This phenomenon is now so well substantiated that we are beginning to wonder whether any genes exist, in higher organisms, that are not pleiotropic. Since the pri-

GENE   GENE PRODUCT   CHARACTER

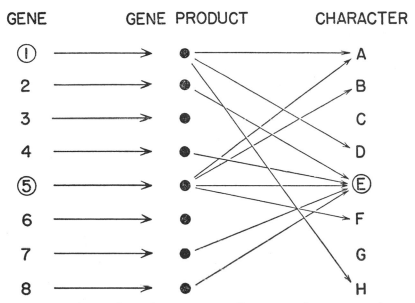

Fig. 10–1. The product of a gene may affect many characters; a character may be affected by the products of many genes.

mary gene action in multicellular organisms is usually several steps removed from the peripheral phenotypic character, one can safely state that nonpleiotropic genes must be rare, if they exist at all. Their contribution to fitness can be assumed to be negligible.

The importance of pleiotropy for evolution was emphasized by Chetverikov as early as 1926 (Dobzhansky 1959b:30):

> The concept of a pleiotropic action of genes . . . is very important for understanding the action of selection. This concept leads us to the idea of the genotypic milieu which acts from the inside on the manifestation of every gene in its character. An individual is indivisible not only in its soma but also in the manifestation of every gene it has.

Polygeny. We have pointed out the deep conflict between evolutionists and Mendelians in the early days of genetics. The sharply discontinuous characters with which geneticists then experimented seemed to have little evolutionary significance; continuous variation, which appeared to be the basis of all evolutionary change, did not seem to obey Mendel's laws. Attempts to divide variation into two classes go back at least to Darwin, who spoke of "sports," yet placed his chief emphasis on small variational differences which he thought could be compounded into the greater distinctions between varieties and species. The large discontinuous genetic differences clearly obeyed Mendel's laws; the continuous variation of the minor variants seemingly did not. The early Mendelians were not able to resolve this conflict with the techniques available to them. The solution became obvious when it was realized that several genes can contribute to the phenotype of a single character.

Multifactorial inheritance (polygeny), then, is the cause of continuous variation. Let us study this with the help of a very simple model. Assume that in an organism large body size is determined by contributions from two loci A and B, each with two alleles (A, a, B, b). Let the heterozygotes (single dose) be intermediate in size between the homozygotes and let either locus make a roughly equal contribution to increased size. We then have five classes of phenotypes: four size genes ($AABB$), three size genes ($AABb, AaBB$), two size genes ($AAbb, AaBb, aaBB$), one size gene ($Aabb, aaBb$), and no size genes ($aabb$). With a third locus (with two alleles) affecting size, the number of classes of phenotypes rises to seven, with a fourth, to nine, and so on. Since it is unavoidable that each member of each class is subject to a certain amount of nonheritable environmental modification, it becomes obvious that, as the number of genes rises, sharp classes are increasingly difficult to distinguish since the genetic classes will overlap phenotypically with each other. The final result will be a more or less smooth curve of continuous variation even though the genetic basis consists of discrete loci and alleles exactly as in the case of completely discontinuous polymorphism. Since well-defined classes of phenotypes cannot be distinguished in the progeny of crosses, it requires special biometric techniques (Falconer 1960), to separate the heritable from the nonheritable portion of variation and to determine the various genetic components (multiple factors).

It had been known since the earliest days of Mendelism that several genetic factors may affect the same character and the same component of the phenotype. The pioneer studies of Castle on hooded rats and of Mac-Dowell and Payne on bristle number in *Drosophila* showed that such characters respond to natural selection, but these early students were unable to discriminate the genetic components. The first genetic analysis of multiple factors was that of Nilsson-Ehle (1909) on color factors in wheat, but the field did not become really active until Mather (1941, 1943, 1953) began his revolutionary studies in the polygenic inheritance of bristle number in *Drosophila*. Studies conducted since this pioneer work are too numerous to be discussed or enumerated in detail. They include the work of MacArthur (1949) on body size in the house mouse, of Robertson and Reeve (1952b), Scossiroli (1954), Dubinin (1948), Breese and Mather (1957), and Milkman (1960a,b, 1961) on various characteristics in *Drosophila,* and much other work summarized by Falconer (1960). In addition, there is an enormous body of data on polygenic characters in the literature on animal and plant breeding. Some of this, including much of his own work, is ably summarized by Lerner (1950, 1954, 1958). All these studies agree in their conclusion that "continuous variation" (see Chapter 7) differs from discontinuous variation only in the number of genetic factors affecting the particular aspect of the phenotype. Both kinds of variation obey the same laws of particulate inheritance. Quantitative characters are controlled by multiple genetic factors, some with larger, some with smaller contributions to the phenotype, and are thus, by definition, polygenic. The methods of genetic analysis have to be different for the two kinds of variation.

THE GENETIC BASIS OF POLYGENIC INHERITANCE

The terminology of multifactorial inheritance has tended to change as the concepts of gene complex and gene action matured. In the days of classical genetics, "major genes" and "modifiers" were the terms for multiple factors. No clear difference seemed to exist between the two kinds of genes except for the amount of their contribution to the shaping of the phenotype. In his earlier papers (for instance, 1943), Mather thought that there was a difference in principle between genes with conspicuous discontinuous contributions to the phenotype and genes with slight contributions resulting in continuous variation. He thought that the latter, which he called *polygenes,*[*] were located in the heterochromatic portions of the chromosomes. The available evidence does not support such a distinction, since the phenotype is the product of the genotype as a whole. Every character is affected by numerous genes and is therefore polygenic; virtually every gene is a polygene. Although it is useful to speak of polygenic characters and polygenic inheritance and although genes located in the "heterochromatic" portion of the chromosome may have mostly modifying effects, there is no evidence for a separate class of polygenes.

One should not go to the other extreme and believe that all polygenic characters are determined by large numbers of polygenes with very small effects. Regardless of the number of genes that may affect a polygenic

[*] The term had first been introduced into the literature, with the same meaning, by Plate (1913); see Fabergé (1943).

character, there are nearly always a few genes that make the major contribution to the phenotype. The fewer these major genes and the larger their individual contributions, the more discontinuous the inheritance will be. The extreme is the "switch gene," which shifts the phenotype into an entirely new direction, such as the sex-determining genes in certain organisms. Yet even such a switch gene interacts with numerous other genes when producing the phenotype, the final product being polygenic. The switch gene on one hand and the large array of small polygenic factors on the other are therefore only the extremes of a broad spectrum of possible genetic determination, not two separate classes. Another extreme is lethal genes which by their action (most often by their failure to produce an essential gene product) block normal development and prevent the production of a viable individual.

Polygenic inheritance confers two great advantages. First, it provides a storage system that not only has an enormous capacity but can readily respond to the slightest shift in selection pressure. Second, in view of the similar action of the factors in a polygenic set, the same phenotype can be produced by many different combinations, for instance, Abcd, aBcd, abCd, and abcD. Each of these genotypes may have somewhat different correlated phenotypic responses (owing to pleiotropy). Selection pressure on the correlated responses thus can produce a shift in the polygenic system without disturbing the major phenotypic character. For instance, if selection pressure for increased body size should develop in a population, all genes in the genotype will be favored that have a pleiotropic effect on size increase. As long as the selection pressure is mild, only those genes will respond whose other phenotypic manifestations will not be selected against. Indeed, a mild selection for a polygenic character often has no other visible effect on the phenotype. A more drastic selection pressure generally affects other characteristics and produces the so-called "correlated responses" (see p. 290).

Both pleiotropy and polygeny contribute to the genetic cohesion of the population. Pleiotropy contributes because each gene through its manifold effects on development enters into a teamwork with numerous other genes. The replacement of one gene by another in a gene complex may affect numerous interactions of the genotype and lead to an upset of stability which can be compensated only gradually. Polygeny contributes because selection pressure for or against any aspect of the phenotype will affect all the genes that polygenically add to this character. A chain reaction of gene substitutions may result, the resistance to which is deter-

mined by the magnitude and nature of the correlated pleiotropic manifestations of the genes involved.

Consideration of the interactions between pleiotropy and polygeny
may help to solve a frequently posed question. Are the small phenotypic
effects of quantitative inheritance caused by genes whose only function is
to be specific modifiers, or are they the pleiotropic by-product of genes
having other essential functions in the organism? It is of considerable evolutionary importance which alternative is right. Mather's original concept
of polygenes tended to ascribe to each polygene a specific and singular
function. This view is similar to the modifier concept of the early geneticists to which Goldschmidt (1953) curiously still subscribed. These concepts are ultimately based on the "one gene–one character" hypothesis. It
is not surprising that the adherents of this concept are forced to reject the
idea of the cohesive gene pool, as developed here, because it would require an impossibly large number of modifying genes. This imaginary
difficulty disappears (as already pointed out by Chetverikov in 1926), as
soon as one realizes that the product which a given gene contributes to
the developmental "soup" may be partaken of by numerous other genes.
Thus a gene that may be a major gene in its own right, say in the development of the eye, may serve also as a modifier of other genes, hence as
polygene for other characters such as body size or rate of development.
There may be a few narrowly specialized modifiers, but the large majority of them are surely pleiotropic genes with multiple contributions to the
phenotype. The fact that numerous modifiers await every new mutation
is fully consistent with this interpretation. To assume that these genes had
been neutral or "dormant" up to that time would be absurd.

INTERACTIONS OF THE GENOTYPE

Pleiotropy and polygeny are concepts that stress the two end points in
the developmental pathway, the gene and the character. In reality, genes
interact with each other all along the epigenetic pathway in numerous
direct and indirect ways. This abundant and continuous interaction leads
to an even greater strengthening of the internal cohesion of the gene pool.

Allelic Interactions

The various modes of interactions among alleles at the same locus are
indicated by such terms as dominance, recessiveness, and overdominance
(superiority of the heterozygote). "The variance traceable to the differences between the average values of the different alleles in all genetic

combinations in which they appear is termed the additive portion of the total genotypic variance" (Lerner 1958). This concept has been discussed in Chapter 9. A particularly illuminating review of the role of allelic factors in the organization of the gene pool was given by Dobzhansky (1955b). The available evidence indicates that the selective superiority of many heterozygote gene combinations (heterosis) not only serves as a major source of genetic variation but also functions as a strongly cohesive factor.

Epistatic Interactions

It was discovered in the earliest days of Mendelism that different loci may interact with each other in their effect on the ultimate phenotype. Any elementary textbook of genetics will tell us that two loci with a dominant and a recessive allele each, segregating independently, will produce ratios of 9:3:3:1. The frequent occurrence of aberrant ratios such as 15:1 or 9:7 or 9:3:4 indicates other modes of interaction, for which, always in terms of two loci, an elaborate terminology was adopted. This terminology became the less applicable the more it was realized that usually more than two loci are involved in such interactions. The collective term "epistatic interactions" is now generally applied to interactions between different loci on the same or on different chromosomes. An extreme case of epistasis exists, for instance, if an albino gene precludes the manifestation of all other pigmentation genes. More often the effect is less absolute and consists only in a modification of the degree to which a character is shown, such as the interruption of a wing vein in *Drosophila*. Many genes have "incomplete penetrance," which means that in certain physical or genetic environments they do not express themselves at all in the phenotype even though by definition they are otherwise dominant. The genetic literature abounds in analyses of the epistatic effect of one gene on the phenotypic expression of another gene. Some of this literature is summarized in Wagner and Mitchell (1955).

Suppressor genes are one of several special types of interacting genes. Glass (1957b) has analyzed such a gene, situated on the second chromosome of *Drosophila melanogaster,* which suppresses the effects of the *erupt* gene on the third chromosome. The erupt gene, when present without the suppressor, produces a wartlike eruption on the eye. Both the erupt and the suppressor genes were found to be widespread in natural populations. The evidence indicates that not merely a single suppressor allele and a single erupt allele exist, but that there is on either locus a whole set of isoalleles of differing strengths. In no case was the erupt gene found without the suppressor being present in the same population, but in a number of populations suppressor genes were present without the erupt genes.

What may be the evolutionary interpretation of this curious system of interactions? Glass quite rightly suggests that the erupt gene is maintained in the populations because it adds to fitness through some other unknown pleiotropic manifestations. This maintenance sets up a selection pressure in favor of suppressor genes that remove from the phenotype the undesirable effect of the erupt gene on the morphology of the eye. On the other hand, the frequency, if not the universality, of the suppressor gene indicates that it also has pleiotropic manifestations for which it is maintained in the gene pool even in the absence of the erupt gene.

Another well-analyzed suppressor system is that which controls the color of the hemolymph in certain moths (Stehr 1959). The details of the biochemical pathways that permit the suppressor effects do not concern us here, but rather the effect on the phenotype which ultimately determines the fitness of the genotype.

Epistatic Interactions and Fitness

The following picture emerges from recent work on gene interactions. Each gene has numerous pleiotropic manifestations which add to or detract from the goodness of the phenotype. Other loci are able to enhance those manifestations that make the greatest positive contribution to fitness and suppress other manifestations having a deleterious effect, such as the erupt phenotype of the erupt gene. The selective value of a gene is thus only partially determined by what we might call its direct immediate effect on the phenotype. An equally or even more important component of the selective value of a gene is its effect on or contribution to the fitness-enhancing qualities of other genes. Plant geneticists were perhaps the first to fully appreciate this multiple function of genes, a point particularly stressed by Harland and his school (Harland 1936; Stephens 1950). The analysis of hybrids and of mutator genes, the work on polygenic characters, on dominance, on sex determination by genic balance, and studies on dosage compensation (Muller 1950b), all indicate the universal occurrence of genic interactions regulated by their contribution to the total fitness of the phenotype.

It is evident that the genes which harmoniously interact in the gene pool of a population were brought together by the action of natural selection. The total gene complex of a population is "coadapted," to use Dob-zhansky's felicitous term. In natural populations exposed to a steady influx of alien genes and to all sorts of environmental disturbances, this process of coadaptation must go on steadily. Dobzhansky and his school have made this area their particular field of research and have greatly advanced our understanding through a series of brilliant experiments.

A good example of this research is the exposure of populations to chronic irradiation (Wallace and King 1951; Wallace 1956). Two experimental populations of *Drosophila melanogaster* which received about 5 r per *hour* rapidly accumulated a great deal of chromosomal radiation damage. About 80 percent of all the second chromosomes tested after 70 generations were lethal when homozygous. One would expect heterozygous combinations of chromosomes from these populations to display evidence of greatly lowered viability. Yet when such tests were made it was found that flies carrying two randomly chosen second chromosomes from one of the irradiated populations had on the average only a slight (1–8-percent) reduction of viability compared to the standard population. Considering the number of accumulated lethals, one would have expected a greater loss of viability. Indeed, in a certain percentage of such crosses viability was above the average of wild flies. It is evident from these results that there must be a high selection pressure in these closed experimental populations for all genes and chromosome sections that combine into harmonious, efficient genotypes. Whenever a chromosome is damaged by radiation, this damage is compensated by the selection of a mutant that restores the balance. Some bypass is found for every interrupted metabolic pathway. The average viability of recombinants remains nearly normal in spite of the steady increase in the frequency of "deleterious" genes.

Wallace's experiments have shed new light on an observation that has long been an enigma to population geneticists. With the help of ingenious techniques, it is possible to study in *Drosophila* the relative viability of individual chromosomes in homozygous and heterozygous condition. When testing wild-type flies from certain localities in Russia, Dubinin *et al.* (1934) discovered that between 10 and 20 percent of the second chromosomes in wild flies are lethal when homozygous. This has since been confirmed for natural populations of five or six species of *Drosophila* (Dobzhansky and Spassky 1954a). Indeed, the figure is around 30 percent in the majority of the populations studied. Nearly all the remaining chromosomes, although not lethal, are at least somewhat deleterious when homozygous. The frequency of the lethals is particularly noteworthy considering that most of the lethals isolated from natural populations are somewhat deleterious even as heterozygotes (Cordeiro 1952). It appears possible that part of the "lethal chromosomes" are the product not of lethal genes, but of lethal gene combinations. Evidence for this comes from experiments in which viable chromosomes were recombined into lethals.

Synthetic Lethals. It was found by Dobzhansky (1946) and confirmed by Wallace *et al.* (1953) and other authors that chromosomes which are of approximately normal viability both as heterozygotes and homozygotes may become lethal chromosomes after crossing over. An experiment by Dobzhansky (1955b) may be used as illustration:

Ten second chromosomes were taken from a population of *Drosophila pseudoobscura* collected near Austin, Texas. These chromosomes produced normally viable to subvital homozygotes. Females heterozygous for all possible (45) combinations of these ten chromosomes were obtained, and ten chromosomes from the progeny of each heterozygote were tested for viability effects in homozygous conditions (450 chromosomes in all). Among these 450 chromosomes 19 were lethal and 57 were semilethal in double dose; 30 of the 45 combinations produced at least one lethal or semilethal among the 10 chromosomes tested. Most remarkable of all, the combinations including the chromosomes which were known to be subvital when homozygous gave actually fewer lethals and semilethals than the normally viable to supervital chromosomes.

It must be remembered that there is no crossing over in *Drosophila* males and there is no question that the lethal chromosomes were derived from the viable chromosomes of the females.

What is the explanation of this atonishing change in fitness of these chromosomes? Normal recessive lethal loci can be excluded because two generations earlier the same chromosomes had been of normal viability when homozygous. Their frequency is far too high to be due to newly arisen lethal mutations. Therefore we must conclude that these lethal chromosomes are not the result of lethal alleles but rather of epistatic interactions due to lethal gene combinations, newly created by recombination. This is confirmed by the fact that such lethal chromosomes can be restored to normal viability through crossing over (Dobzhansky and Spassky 1960). The more genic diversity there is in a species (owing to gene flow among many widely distributed local populations), the greater the risk of disharmonious combinations, that is, the greater the probability of the occurrence of "synthetic" lethals or detrimentals. The fact that lethals are more frequent in common and widespread species, like *Drosophila pseudoobscura* and *D. willistoni*, than in rarer ones, like *D. prosaltans*, is consistent with this hypothesis. Yet Hildreth (1956) and Spiess and Allen (1961) found no evidence of the production of synthetic lethals in *D. melanogaster*.

Incongruity of recombination need not always result in lethality or semilethality; it may affect any component of fitness. There is, for instance, some evidence for "synthetic" sterility (Krimbas 1960). The interest of these "synthetic" chromosomes is that the drastic reduction in fitness, resulting from the recombination of highly fit parental chromosomes, is most telling evidence for the importance of epistatic interactions. They must be epistatic, since the synthetics are tested in homozygous condition, and the original chromosomes (before recombination) were also highly fit when made homozygous.

It is now well established that allelic as well as epistatic interactions decisively affect the fitness of the genotype. What is still uncertain is the relative contribution to fitness made by either type of interaction. So many beneficial effects of heterozygosity have been described (Chapter 9) that one might be tempted to adopt the extreme viewpoint that all heterozygosity is to the good and all increase in viability and developmental stability is due to heterozygosity. "The degree of adaptiveness of individuals and of groups in cross-fertilized species then may well be a function of their degree of heterozygosity," as one author suggested. This viewpoint is often correlated with the assumption that the superiority of heterozygotes is the result of selection for a genetic background on which the heterozygotes would show heterosis. A series of experiments by Vetukhiv (1954, 1956), Brncic (1954), and Wallace (1955) shows that neither assumption is correct. In all three sets of experiments the viability of intrapopulation F_1 and F_2 hybrids is compared with that of F_1 and F_2 interpopulation hybrids (and some other combinations). Viability is measured for various components of fitness, as, for instance, fecundity of females or larval survival when in severe competition with a standard strain of another species of *Drosophila*. The results of these experiments are quite consistent. The F_1 hybrids between populations are significantly more viable than intrapopulation F_1 hybrids. This superiority is lost when the F_2 cross is made (Fig. 9–2).

Two conclusions are obvious. The first is that the superiority of the F_1 interpopulation hybrids cannot be due to a previous selection of a suitable genetic background because the populations tested came from far distant localities. The increase in the fecundity of the females shows that more than mere "luxuriance" of the larvae is involved. Thus we have a case of genuine heterosis that was not the product of selection. The second and even more important conclusion is that this superiority is not due to mere high heterozygosity, because the F_2 lose this superiority in spite of retaining much heterozygosity and indeed a very much higher heterozygosity than the parental populations. The conclusion is inevitable that the loss of viability in the F_2 is due to a loss of epistatic balance among interacting loci and that this overrides the beneficial effects of high heterozygosity. A number of other tests confirm this conclusion, particularly the results of three-way and of double crosses.

The integration of the gene complex, its coadaptation, therefore depends on the presence of two kinds of harmonious balance. One of them, a balance among alleles leading to overdominance and through it to balanced polymorphism, Mather (1953) has called *relational balance*. The other, a balance among different loci, Mather has designated *internal* (= epistatic) *balance*. It is still unknown whether or not the two kinds of balance intergrade (as one would expect from the nature of the gene locus), and how much difference it makes whether two interacting loci are on the same chromosome or not. What evidence there is indicates that in higher organisms it makes little difference whether interacting loci are on the same chromosome or not.

The Physiological Model of Gene Interaction

The standard model of gene action considers the gene as a sort of template on which are formed proteins (enzymes) that are the determinants of the developmental process. This classical picture has been somewhat modified through recent research, since the DNA (deoxyribonucleic acid) of the chromosomes does not form the proteins directly but rather with the help of various kinds of RNA (ribonucleic acid) which control the assembling of the amino acids. The synthesis of the proteins takes place in the cytoplasm. The details are of little known evolutionary significance. The various models of gene action agree in postulating that genes control the production of chemical substances and that in the process they utilize some substrate.

The first level of interaction of genes is within a given cell whenever several genes are simultaneously active in elaborating their gene product, particularly in the production of complex protein molecules. These products will diffuse into other cells and stimulate the growth and division of cells. A second level of interaction is thus initiated, that of development, with all its intricate interactions among tissues and organs. Normally all cells have identical chromosome sets, yet cells and tissues tend to become very different in their potencies in the process of differentiation. It is evident, therefore, that different genes are active in different tissues and organs and that a distinction must be made between the presence of a gene and its activity. It is still largely unknown what controls the activity of a gene and the site of its activity, but contributory evidence is beginning to accumulate rapidly (Beermann 1961; Jacob and Monod 1961a). It is becoming increasingly evident that a gene is characterized not only by the chemical nature of its product but also by the period in the developmental process when it is active and by the quantity of its product. The abandonment of a grossly corpuscular definition of the gene greatly facilitates the understanding of this mode of gene function. Since a gene product may in turn serve as substrate for the activity of other genes and since for its proper functioning a gene requires the presence of the specific substrate at a specific time (period in development) and in fixed amounts, it becomes evident why the interaction of genes is of such vital importance for the viability of the phenotype ultimately produced by the developmental process.

We can use this information to set up a model of gene interaction. Let us say that an alien gene A is introduced into a gene pool where it produces too much of its gene product, with the result of inhibition of gene action at locus B. In order to restore balance this will set up a selection pressure in favor of one of the following three kinds of genes: (1) a gene substitution at the original locus A resulting in a reduction of the amount of the gene product or a retardation in the time of its appearance, (2) a gene substitution at the locus B resulting in resistance to the effect of the superabundance of the enzyme produced by A, or (3) a gene substitution at a third locus C that would utilize the excess amount of enzyme A and thereby neutralize its effect on locus B. The same model, with slight modifications, can be used to describe the effects of the loss of a gene, with the resulting deficiency of its gene product. The essential aspect of this model is that it emphasizes the chemical unbalance that the gene substitution at A has produced and that may be corrected in many different ways by gene substitution at the same or other loci. The result will be a shift of balance and the establishment of a new equilibrium. The best evidence for the occurrence of biochemical imbalances and their correction by subsequent gene substitutions has come from work with microorganisms. We must refer for further information to the relevant literature (for example, Wagner and Mitchell 1955, and numerous more recent symposia).

It is obvious not only that such an interacting system is highly sensitive but also that it permits numerous feedbacks and systems of regulation. The students of development have various terms for these regulatory powers, such as buffering, canalization, and developmental homeostasis. These terms apply to models that help us to visualize the action of genes in the developmental process but should not blind us to our basic ignorance of the exact details by which the universally observed regulation during development is achieved. We refer to various textbooks and essays on epigenetics for further details on the physiology of differentiation of the tissues and organs in relation to gene action (Goldschmidt 1955; Waddington 1956a, 1957; Kühn 1955).

In order to make an optimal contribution to fitness, a gene must elaborate its chemical gene product in the needed quantity and at the time when it is required for normal development. The total genotype can be considered a "physiological team," an analogy that has considerable illustrative value. Some of the best-known athletes are poor team players, or might star as members of one team but not of another. Some musical virtuosos, unexcelled as soloists, are only mediocre in an ensemble. Genes are never soloists, they always play in an ensemble, and their usefulness, their "selective value," depends on their contribution to the goodness of the product of this ensemble, the phenotype.

Position Effects. We have spoken up to now of gene interaction as a physiological phenomenon affecting the course of development. No special role was attributed to the particular location of the interacting genes on the chromosomes. The frequency with which genes interact that are known to be located on different chromosomes indicates that there does not need to be a physical closeness of such genes. This does not preclude, however, the possibility of a more intimate or more frequent interaction of genes that are located on the same chromosome. In addition to the classical position effect (Lewis 1950), there are other interactions of neighboring genes which Dobzhansky (1959b) has designated "organization effects." Levitan (1955, 1958), Levitan and Salzano (1959), and Stalker (1960), for instance, have shown that in several species of *Drosophila* it affects fitness whether two inversions occur both in the same chromosome A and none on the homologous chromosome A' or whether one inversion is on the A chromosome and the other on the homologous A' chromosome. It is probable that some of these phenomena will have to be interpreted in terms of linkage and epistasis (Lewontin and Kojima 1960). That position on the same or opposite chromosomes affects the phenotype had long been known for the bar locus in D. *melanogaster* (Lewis 1950). An even more complex functional interaction of adjacent loci is known for some pseudoallelic sites in *Drosophila,* for instance *bithorax* (Lewis 1951), and a number of similar cases are known in microorganisms (Pontecorvo 1958). Among all the cases of gene interaction, those that are due to a physical relation of these genes (cistrons) on the chromosomes appear to be, at least in the higher organisms, perhaps in the minority. The integration of genotypes is not primarily based on gross chromosomal architecture, as was believed during the early period of cytogenetics. Instead, much of the integration occurs at a different level, with natural selection its agent and the developmental process its locale. If genes that interact with each other are found to be linked, it is often apparent that they have been secondarily brought together on the same chromosome through translocation (Sheppard 1953b).

In addition to the cellular and developmental levels, there is a third level of interaction among genes. This level is the interaction of genes in the total gene pool of a population. The fitness of a gene being a relative value depending on the particular team of other genes with which it is associated in the production of a particular phenotype, it is evident that the goodness of a gene is "statistical." It is averaged from all the different genetic backgrounds in the many genotypes in which the gene is exposed to selection. The selective value of a gene is the mean of the selective values of all the combinations in which it appears in a given population (gene pool). The evolutionary consequences of this and the other aspects of the cohesion of the gene pool are manifold.

The Consequences of the Cohesive Factors

Recent findings and interpretations have fostered a new concept of the genotype. Genes as evolutionary phenomena can no longer be considered disconnected entities to be studied in isolation. Their selective value is no longer considered absolute. On one genetic background a given gene may add to the fitness of the genotype; on another genetic background the same gene may create an unbalance and produce a severely deleterious effect. To the two well-known classes of environment, the physical and the biotic environments, we must add a third, the genetic environment. A given gene has as its genetic environment not only the genetic background (Chetverikov 1926) of the given zygote on which it is temporarily placed, but the entire gene pool of the local population in which it occurs. The evolutionary fate of this gene will depend, in the long run, on how it cooperates with the other genes of this gene pool, how well it is coadapted to them. The student of gene physiology is justified in separating individual genes to study their action and function. The evolutionist, however, who is interested in selective values and adaptation, cannot separate a gene from its genetic environment; he must treat them together as a whole. Not the actions of the individual units are important, but rather their interactions, their joint contribution to the total fitness. As a consequence of this new interpretation of the gene complex, it becomes necessary to revise some of our concepts, particularly those of the relation between genotype and phenotype, and of the operation of natural selection. This, in turn, will affect our interpretation of the genetics of speciation.

The Phenotype. The word "phenotype" is indicative of the typological thinking of the period during which it was coined. It was applied to the appearance of the "normal" individuals of a species, which in turn were believed to be the individuals homozygous (or heterozygous dominant) for the "wild-type" genes of the species. We now know how far this interpretation missed the mark. Yet our new knowledge of the diversity of the genotypes raises a puzzling problem. Why does the phenotype remain so constant in spite of the underlying genetic diversity? The prevailing phenotype of a species is usually remarkably uniform, aside from a few cases of polymorphism and some quantitative variation. We are forced to conclude that there must be a selective premium on phenotypic uniformity and that optimal fitness is closely correlated with one particular phenotype. Animal and plant breeders and students of experimental populations have indeed found evidence in support of this assumption.

We are still far from understanding completely how this stabilization of the phenotype functions, but some of the phenomena that contribute to it can be perceived (Lerner 1954; A. Robertson 1955). Heterozygosity is one factor. Within a harmoniously coadapted population, the more heterozygous an individual is, the greater will be its developmental homeostasis (Chapter 9). Likewise, the more heterozygous an individual of such a population is, the nearer to the population mean it is apt to be for all quantitative characters. This has been ascribed to pleiotropy by some authors and to linkage by others. One can assume that most genes will be pleiotropic and make some contribution to fitness and also some contribution to one of the more trivial aspects of the phenotype (color pattern, number of bristles, and so forth). These genes will show dominance or overdominance for their fitness-producing characters (through selection of the appropriate genetic background) and additive effects (heterozygotes intermediate) for their trivial characters. According to Breese and Mather (1960), however, the observed phenomena are better explained by linkage. At any rate, those individuals of a population that are most "mediocre" and "ordinary" (as far as quantitative morphological characters are concerned) are the fittest! This was expressed by Wright (1951a) in the statement: "The best adapted form in a species is usually one that is close to the average in all quantitatively varying characters." To cite the "survival of the ordinary" (McAtee 1937) as evidence against natural selection fails to understand the normalizing character of most natural selection.

The Constancy of the Phenotype. The classical experiments of genetics had resulted in a widespread impression, particularly among nongeneticists, that every mutation or gene exchange will be reflected in the visible phenotype. Actually, genes of such high expressivity seem to be very much in the minority and much turnover in the gene pool seems to take place without any effect whatsoever on the visible phenotype. There appear to be many developmental mechanisms and canalizations that prevent gene substitutions from expressing themselves in the phenotype (Waddington 1957). This constancy of the phenotype is by no means fully understood, but it is, no doubt, one aspect of the general phenomenon of the unity of the genotype. The discovery of the formerly unexpected frequency of sibling species (Chapter 3) has revealed that even gene substitution so extensive as to lead to the origin of new species is possible without visible effect on the phenotype (Dobzhansky 1956a). Each individual gene substitution during the origin of these species was somehow compensated for and prevented from affecting the end product of development, the phenotype. The degree of difference (and the incompatibility) of the gene complexes of sibling species can be demonstrated when it is possible to cross them. In F_1 hybrids between *Drosophila melanogaster* and *D. simulans*, Sturtevant (1929) found many disturbances in bristle development, even though the bristle patterns of the two parental species are identical. In 167 hybrid females (with 334 bristle sites on both sides of the dorsum) there were 142 anterior dorsocentrals (192 missing), 181 posterior dorsocentrals (153 missing), 192 anterior scutellars (142 missing), and 198 posterior scutellars (136 missing). Evidently the developmental pathways leading to the phenotypically identical bristle patterns of the parental species are very different.

It would lead too far afield to discuss here the genetic and developmental system that maintains the constancy of the phenotype in spite of gene substitution. The best analysis is that of Rendel (1959) for the scutellar bristles in *Drosophila*. The phenotype "normal number of scutellar bristles" is so tightly canalized that even a considerable amount of genetic substitution will not result in a visible change of the phenotype. Only after the character has been moved outside the zone of canalization will it begin to respond again to selection. The normal bristle number is four, but it takes about eight times as much genetic change to move from three bristles to five than it does to move from one bristle to three.

The statement recently made that "*Drosophila* has remained unchanged since the Eocene and is therefore not particularly suitable material for evolutionary studies" fails to take these facts into consideration. The active speciation in the genus (probably 2000 species or more), its radiation in every continent, and its numerous specializations (as on Hawaii) show what an enormous amount of genetic reconstruction has gone on under the mask of phenotypic stability. There are very few evolutionary phenomena that cannot be demonstrated to have occurred in *Drosophila*.

Gene substitution without effect on the morphological phenotype is analogous to the restoration of a Gothic cathedral. In the course of the never-ending repair work, many of the stones have been replaced, sometimes repeatedly, since the original construction some seven centuries ago. Yet the "phenotype" of the medieval edifice has remained unchanged. The tendency of the morphological phenotype not to respond to a far-reaching reconstruction of the genotype is manifested even above the level of sibling species. The same "character" may go through an entire genus, family, or even higher category. In many instances, there is no obvious reason why a particular aspect of the phenotype should be favored by selection since a different phenotype seems to serve equally well in a related taxon that coexists in the same environment. Interpreting these characters as not produced by specific genes, but as the by-product of the entire well-integrated gene complex held together by genetic homeostasis, greatly facilitates an understanding of such "conservative characters."

A well-knit system of canalization tends to narrow down evolutionary potential quite severely. It accounts for parallel evolution which, for instance, induced several separate lines of mammallike reptiles to cross the borderline to the mammals independently (Fig. 19–2). Which gene will mutate and at what time, is "random," but the subsequent fate of such a mutation is strongly controlled by the gene complex in which it occurs. The direction of evolution is, therefore, not random, and yet it is not predictable either.

The Effects of Cohesive Factors on Selection

Selection acts on phenotypes. In view of our new understanding of the production of phenotypes by genes that are intimately welded together into a single closely knit whole through pleiotropy, dominance (and overdominance), epistasis, and polygeny, it is evident that the effects of selection on the genetic composition of gene pools and on the visible variation of populations is vastly different from the conventional picture.

Stabilizing Selection. Let us begin with the simplified assumption of a constant environment. Some earlier authors thought that this would mean a cessation of selection. Of course this is not at all the case. To be sure, there may not be any "directional selection," but even in a normally stable environment there will always be "stabilizing selection" (Schmalhausen 1949) (Chapter 8). This somewhat elusive concept has been discussed, among other authors, by Dobzhansky (1951), Simpson (1953a), Lerner (1954, 1958), and Waddington (1957). The term simply refers to the fact that the processes of recombination and of segregation, acting in a population with high genetic variation, will inevitably produce numerous individuals that deviate from the average phenotype in one respect or another, and are of lower fitness. The discrimination against these phenotypically peripheral individuals of a population by natural selection has been called stabilizing selection, centripetal selection (Simpson), or normalizing selection (Waddington).

There is much concrete information available to show that selection works to maintain an optimal phenotype and that it discriminates against any "phenodeviants" (Lerner 1954) that may segregate in the normal course of recombination. Snakes and lizards, for instance, do not change the number of scales or scutes between birth and adulthood. Yet E. R. Dunn (1942) could show that the total variability in scales of adult samples in reptiles is much smaller than that of juvenile samples, owing to the higher mortality of the extreme types. In the snake *Natrix natrix*, Mertens (1947) found that unusual variants of the postocular scales comprised 26.67 percent of the immature sample, but only 13 percent of the adult sample. Immatures have 38.33 percent of variants among the sublabials, while the adults have only 23.76 percent. In water snakes (*Natrix*) on islands in Lake Erie there was a selective elimination of banded individuals between birth and maturity, apparently by avian predators (Camin and Ehrlich 1958). In the West Indian Gecko, *Aristelliger praesignis*, there is selection against large as well as small extremes of size (Hecht 1952). Variability in wintering wasps (*Vespa vulgaris*) drops markedly during hibernation owing to differential mortality of the extremes (Thompson *et al.* 1911). Naturalists and biometricians have published many examples of greater mortality among extreme types than that found in the phenotypic average. The paper by Bumpus (1896) is classical; others have been cited by Huxley (1942) and Lerner (1954). Haldane (1954b) has given the mathematical treatment of this normalizing selection. Not all the evidence cited in support of such selection is necessarily conclusive. The fact that the milk dentition of mammals is sometimes more variable than the permanent dentition does not necessarily prove that the more extreme juveniles were killed. It is equally possible that the permanent dentition has a better buffered developmental control.

Waddington (1957, 1960c) has pointed out correctly that stabilizing selection has two different aspects. The negative one, the elimination of all individuals that are phenodeviants, he calls normalizing selection. The positive one, a selection in favor of all sorts of feedback mechanisms that would produce the standard phenotype in spite of considerable substitution of genes in the genotype and of environmental fluctuation, he calls canalizing selection. The two processes are, in a way, merely two aspects of a single process, since canalizing selection by necessity operates with the help of normalizing selection.

The effectiveness of stabilizing selection depends on the occurrence of phenodeviants and the genetic reasons for their appearance are thus of considerable evolutionary interest. Two, not necessarily mutually exclusive, explanations for the occurrence of phenodeviants have been advanced.

(a) *Increased homozygosity.* Lerner (1954) stresses the fact that any kind of inbreeding automatically raises the frequency of homozygotes in a population. If inbreeding is sufficiently severe, it leads to a lowering of fitness in a normally outbreeding species ("inbreeding depression") and to the appearance of deviating phenotypes. Increased homozygosity often leads to developmental disturbances in addition to lowering the resistance to environmental changes (developmental homeostasis) as discussed in Chapter 9. Closed populations, such as those studied by breeders and laboratory geneticists, are particularly vulnerable to the depressing effects of homozygosity.

(b) *Disturbances of the internal balance.* In an open population there is a steady high influx of alien genes. These set up all sorts of epistatic effects, some of which result in the production of inferior phenotypes. Synthetic lethals and steriles are extreme examples. The main activity of stabilizing selection in such populations will be the elimination of ill-fitting genes and the selection of those genes that have a maximum fitness in the greatest possible number of combinations (Mayr 1954a). The occurrence of excessive homozygosity in such a gene pool is improbable since many loci are normally occupied by numerous isoalleles.

It must be admitted that it is exceedingly difficult to supply conclusive proof of the high frequency of isoalleles, which seems to have been established by Stern and Schaeffer (1943), Spencer (1944), Glass (1957b), Green (1959), and many other recent authors. In view of the impossibility of making the stocks used in these experiments completely isogenic, it is virtually impossible to discriminate between isoallelism and epistatic effects. For a further discussion of the problem of isoalleles or polygenes, see Reeve and Robertson (1953) and Mather (1954).

These two modes by which phenodeviants may arise in a population must be clearly distinguished in order to analyze the action of stabilizing selection. There are consequently two kinds of stabilizing selection, both active at all times, but one usually more important owing to the genetic structure of the particular population. If the elimination of homozygotes is the main problem, the pressure of stabilizing selection will be low. This pressure will be mainly directed toward the selection of a genetic background that increases the overdominance of the heterozygotes in order to guarantee a high level of heterozygosity. In particular, it will assure the dominance of all those phenotypic expressions of pleiotropic genes that make the greatest contribution to fitness (see Chapter 9).

On the other hand, if the main object of stabilizing selection is to minimize the effects of upsetting epistatic interactions, then such selection will favor certain types of genes. First, genes will be favored that I have dubbed "good mixers" (Mayr 1954a), that is, genes that make a positive contribution to fitness on the greatest number of possible genetic backgrounds. Secondly, supergenes will be favored, that is, well-balanced combinations of genes that are protected against recombination (through suppression of crossing over). The amount of genetic input into a given gene pool will determine which of these two processes of stabilizing selection will be more important. We shall return to this point again.

Selection in a Changing Environment

The model we have discussed in the preceding section has led us to the conclusion that the cohesive factors in a gene pool will result in stabilizing selection as long as the environment itself is essentially stable. But what will happen to the integrated gene pool if a new environmental factor arises which initiates a strongly directional selection and attempts to shift the phenotype away from its present norm? This is an extremely well-explored field of research, chiefly because most plant and animal breeders try to achieve exactly this shift when they breed for an increase in yield per acre in a crop plant or in the annual production of milk in cows or of eggs in hens. For a discussion of much of this work, see, for instance, the 20th Cold Spring Harbor Symposium (1955), Lerner (1958), and Falconer (1960).

In most breeding work, selection is simultaneous for a great many factors, particularly for most of the components of fitness. In such cases it is usually quite impossible to determine what happens during the process of selection. Therefore I have chosen for analysis a case that, though complex, is easier to analyze because it deals with severe selection for a single character in a laboratory animal. I refer to the pioneering and now classical studies of Mather and his associates on bristle number in *Drosophila* (Mather and Harrison 1949). The object of the selection was an increase in the number of bristles (chaetae) on the ventral surface of the fourth and fifth abdominal segments in *Drosophila melanogaster*. Two selection experiments were run, one for increase and one for decrease in bristle number (Fig. 10–2). In the starting stock, the combined average bristle number of males and females on these segments was about 36. Selection for low bristle number was able to lower this average after 30 generations to 25 chaetae, after which the line soon died out owing to sterility. A mass "low" line (maintained without selection) was started with 32 chaetae and remained nearly stable for 95 generations. However, all attempts to derive from this line others with lower bristle numbers proved failures. The lines invariably died out owing to sterility before selection had made much progress. The evidence indicates that the resistance of the low line to further selection was not due to an exhaustion of genetic variability but in part due to a balanced sterility system. Yet the possibility

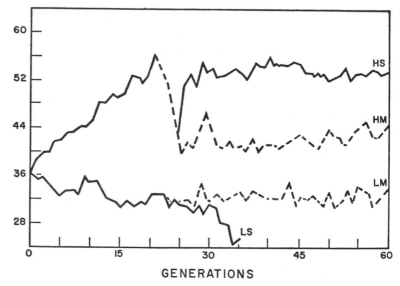

GENERATIONS

Fig. 10–2. Bristle number in four lines of *Drosophila melanogaster,* derived from a single parental stock: *HS,* high selection line; *LS,* low selection line; *HM,* unselected mass culture derived from the high line; *LM,* unselected mass culture derived from the low line. (After Mather and Harrison 1949.)

could not be excluded that the genes of this stock favoring low bristle numbers had a depressing effect on viability when in homozygous condition.

In the "high line" (selection for high bristle number), progress was at first rapid and steady. Within 20 generations bristle number had risen from 36 to an average of 56, without marked spurts or plateaus. At this stage sterility became severe and a mass culture (without selection) was started. Average chaetae number fell sharply and was down to 39 in 5 generations, a loss of over 80 percent of the gain of the preceding 20 generations. Without selection, this line fluctuated somewhat up and down (going at the 29th generation as high as 46), but finally settled approximately at an average of 40. New high selection lines were repeatedly taken from this "high" mass line. The first two (started at the 24th and 27th generation) regained the previous high bristle number as quickly as the line had lost it when selection was stopped. However, viability was now much improved and a viable line could be maintained under constant selection pressure for high bristle number (without, however, much phenotypic response). A mass line taken off this new high line some five or six generations later maintained its high bristle number rather than dropping precipitously as had the earlier mass line. Other reselection lines taken off the original high mass line at later periods were much less successful, indicating that the high mass line had restabilized itself in a new way, with a loss of the combinations leading to higher bristle numbers.

We must refer to the original paper (Mather and Harrison 1949) for a report on the fate of the many additional selection lines, on the crosses between the lines, and on the detailed genetic analysis. We shall concentrate on a number of generalizations that can be derived from this and similar studies (Falconer 1960). The first point is the immense storage capacity of the genotype for variability. Selection was able to double bristle numbers even though this was a closed population and no new genetic material was added after selection had started. The second point is that such a well-defined character as the number of abdominal bristles depends on a high number of genes. The genetic analysis showed that genes affecting bristle number could be found on every chromosome and indeed at many loci on each chromosome (Breese and Mather 1957). The third point is that any intense selection results in various correlated effects. Some of these, like changes in pigmentation and in the number of spermathecae (which increased in some of the stocks), had no particular effect on viability. Others affected fertility, fecundity, and larval survival. The counterpressure of natural selection finally became so strong that a plateau was reached without further response to the artificial selection. This limit (and the same observation has been made by animal breeders) was not due to the exhaustion of genetic variability, since genetic variance remained considerable. A relaxation of selection (it would be more precise to say a shift to selection for high fitness) sometimes resulted in a rapid loss of the "artificially" acquired character. An extreme phenotype (high bristle number) can be produced only by very specialized genotypes. Such genotypes are bound to be unbalanced in more ways than one, as, for instance, with respect to various components of fitness.

One curious finding of these experiments was the occasional occurrence of a great delay in response to selection. One line, for instance, was unable to rise above the level of 53 or 54 chaetae from the 28th to the 82nd generation in spite of continued selection. In the next two generations it rose to 56 or 57 chaetae, where it stayed for the next 50 generations. Another line became stabilized at 54 chaetae for 23 generations and then climbed within a few generations to a new plateau of 63 chaetae where it remained in spite of continued selection. Similar behavior was shown by several other lines. Among several possible explanations, three appear most probable: mutation, replacement of one heterozygous balance by a different one, and recombination in chromosome sections previously immune to crossing over. In the cases studied by him, Mather favors the last of these three explanations, although the other two seemed equally compatible with the known data.

Genetic Homeostasis

One of the most interesting findings of these experiments is the tendency of the phenotypes to return to the original condition when selection is discontinued after a population has been exposed to a severe selection pressure for a specific phenotypic character, whether increased bristle number or body size in *Drosophila* or increased egg number or egg size in the domestic fowl. The many observations of the selective superiority of morphological intermediates is merely another aspect of the same phenomenon. Lerner (1954) has designated this phenomenon *genetic homeostasis,* defining it "as the property of the population to equilibrate its genetic composition and to resist sudden changes." * Stressing the static more than the dynamic aspects of the conservative properties of the gene pool, Darlington and Mather (1949) have referred to this phenomenon as "genetic inertia." The precise definition and interpretation of "genetic homeostasis" is still under consideration (Waddington 1957; Lewontin 1957), but this does not affect the essential point made here.

The reason for genetic homeostasis should be evident from the preceding discussions. A naturally existing phenotype is the product of a genotype that has a long history of selection for maximum fitness. Any selection for a new phenotype will force the abandonment of the previously integrated genotype and will thus lead to lowered fitness, due to either an accumulation of homozygous recessives or a disharmony between the newly favored genes and the remainder of the genotype. Relaxation of the selection for the new phenotype permits at least a partial return by natural selection to the historical combination that had given maximum fitness, particularly heterozygous combinations. As a by-product there will be a partial restoration of the original phenotype. If the return to the original phenotype is only partial and some of the phenotypic gains of the preceding selection are preserved, this indicates either that some homozygous fixations had occurred or, more likely, that an alternate adaptive peak had been climbed. This alternative peak is equivalent to the

---

* The term *homeostasis* (Cannon 1932) was originally applied to the capacity of an organism to hold certain physiological steady states at definite optimum levels.

original genotype as far as general fitness is concerned (see Chapter 8) but superior with regard to the specific phenotypic character that had been under selection pressure. Genetic homeostasis may well provide the solution for many previously unsolved phenomena, such as evolutionary "stagnation." If two related but long-isolated human races retain the same frequency of fingerprint patterns or blood-group genes, in spite of the numerous selection pressures in favor of shifting the frequency, it may well be due to the superiority of this particular frequency on the common genetic background.

Genetic homeostasis determines to what extent a gene pool can respond to selection. The less associated with general fitness a particular aspect of the phenotype is, the greater the probability of a response to *ad hoc* selection. If the character does not contribute to fitness in nature, like high bristle number in *Drosophila* or fancy color in pigeons and parrots, it is understandable why natural selection had not previously taken advantage of the possibilities revealed by artificial selection. The more specific a character and the more monogenic its basis, for example resistance to a specific toxic substance, the more rapid will be the response to selection. "A single dominant or partially dominant factor or additively acting genes if more than one is involved are the kinds of genes that are most readily increased by selection in a sexual population" (Crow 1960). Resistance to insecticides is proof of this.

Fitness itself is not apt to show much response to artificial selection because it is merely a continuation of a constant selection pressure and it is to be expected that a plateau had long since been reached (A. Robertson 1955). However, individual viability factors may respond to artificial selection, particularly in a newly established closed population placed in a new environment. In lines of the chalcid parasite *Microplectron fuscipennis* bred in the laboratory, Wilkes (1947) was able to lower the percentage of sterile males from 35 to 2 percent, to increase the mean number of offspring per mother from 34 to 68, and to achieve other improvements pertaining to rate of development, oviposition, and life span. In such cases one must assume that part of the improvement in the stated characters has been made at the cost of a loss of general viability or of specific resistance to various factors of the physical environment. Obviously any drastic improvement under selection must seriously deplete the store of genetic variability. An open, natural population has to cope with numerous conditions to which the closed and sheltered laboratory population is not exposed. Protected against the effects of immigration and an adverse and variable environment, such a population can afford to develop genotypes that would be of inferior viability in nature. A closed population can respond to special selection pressures in the new uniform environment and on the standardized genetic background of a fixed and limited number of genes (Beardmore 1960).

Correlated Responses. Animal and plant breeders have long known that various "correlated responses" may occur as a consequence of selection for a particular character, that is, changes in seemingly independent aspects of the phenotype. Darwin (1859), for instance, stated that "if man goes on selecting, and thus augmenting any peculiarity, he will almost certainly modify unconsciously other parts of the structure, owing to the mysterious laws of the correlation of growth." The discovery of the cohesion of the gene complex and of the frequency of pleiotropic genes sheds some light on this mysterious correlation. The occurrence of sterility during selection for changed bristle number in *Drosophila* is a typical case of a correlated response. Many other responses have been listed in the recent literature (Haskell 1954; Lerner 1954). Correlated responses occur during virtually every case of selection, since most genes seem to affect simultaneously some components of fitness and some more trivial aspects of the phenotype. The evolutionist is especially interested in cases where selection for a physiological trait has produced a correlated response in morphological characteristics. For instance, DDT-resistant strains of the housefly (*Musca domestica*) and of *Drosophila melanogaster* differ from each other and from nonresistant strains in various morphological characteristics (Sokal 1959; Sokal and Hiroyoshi 1959). The diversity of the responses indicates that it is not the DDT-resistance-giving genes which are responsible for the morphological changes, but rather the reconstruction of the total genotype. Various morphological differences, particularly in pigmentation, occurred in several lines of *Drosophila melanogaster*, all descended from a single stock, that Wallace (1954b) had exposed to different dosages of radium irradiation. Important in both of these cases is not only the occurrence of morphological changes without selection for them, but also the rapidity of the response. The elapsed time is negligible in terms of geological ages.

The most frequent "correlated response" of one-sided selection is a drop in general fitness. This plagues virtually every breeding experiment. An exceptionally well-analyzed case is that of hybrid turkeys (*Meleagris gallopavo*), in part derived from domestic stock, that were released in Missouri to "strengthen" the depleted native stock. It was found that the feral birds were inferior to the wild birds in every aspect of viability studied (Leopold 1944). They also had lower relative brain, pituitary, and adrenal weights. Native turkeys raised larger broods than hybrids and a larger proportion of wild hens was successful in raising broods. The commercial qualities of the domestic birds had been bought at the price of abandoning qualities favoring survival in the wild.

Unrepeatability. The response may be quite different if unidentical genetic stocks are exposed to the same selection pressure. This diversity is the inevitable consequence of the fact that the phenotype is the product of an incredibly intricate interaction of a large number of constituents of the genotype. The rate of response to the selection pressure, the final level of response, and the occurrence of correlated responses will depend not only on the original constitution of the gene pool, but also on the choice of the many alternate multiple pathways that are available during each generation. Dobzhansky (1951), Dobzhansky and Pavlovsky (1957), and Dobzhansky and Spassky (1962) have given a convincing demonstration of this unrepeatability. They have furthermore demonstrated that the predictability increases with the increasing initial homogeneity of the gene pool. As described in Chapters 7 and 9, most populations of *Drosophila pseudoobscura* are polymorphic for gene arrangements in the third chromosome, with the heterozygote being superior. When an experimental population is started with flies from the same locality bearing different locally occurring gene arrangements, sooner or later a predictable balance of their frequencies is reached (Dobzhansky 1951). On the other hand, if the experimental population is started with heterogeneous material, that is, with flies from different localities, or with gene arrangements from different populations, the results become indeterminate. In six parallel experiments in which chromosomes with the *ST* arrangement from California were mixed with chromosomes with the *CH* arrangement from Mexico, there was a rapid elimination of the *CH* chromosomes during the first 100–150 days of the experiment (Fig. 10–3). After that period, each population behaved differently. In four populations, *CH* reached the vanishing point sooner or later, while in two populations heterosis between the *CH* and *ST* evolved at the 69-percent and 80-percent level of *ST* frequency. The foundation stock of these experiments was a mixture of twelve strains from Mexico and fifteen from California. In two out of six cases, recombination between these numerous chromosomes permitted the piecing together of new balanced chromosomes which displayed heterosis in the inversion heterozygotes. In one of

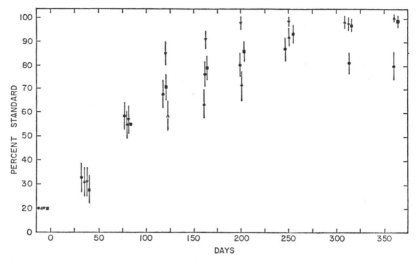

Fig. 10–3. Divergent behavior of four experimental populations of *Drosophila pseudoobscura*, with the *ST* gene arrangement from California and the *CH* arrangement from Mexico. Although all four populations were started with a 20-percent frequency of *ST* chromosomes, each population responded differently to natural selection, as shown by rate of change and final level. (From Dobzhansky 1960.)

the populations, the effects of the coadaptation became apparent already after three generations and reached equilibrium after ten, in the other, after four to nine generations. A similar unpredictability has been recorded in many cases during parallel selection experiments based on more or less heterogeneous material. For instance, among nine lines of *Drosophila pseudoobscura* selected for an increase in the rate of development, five lines had responded strongly, two lines weakly, and two not at all after 25 generations of selection (Marien 1958).

Limits to Genetic Variability. In the early days of genetics Johannsen, on the basis of breeding work with beans, came to the conclusion that the potential variability of populations was quickly exhausted by selection. This finding was vigorously contested by many animal and plant breeders who, under selection, achieved steady progress from generation to generation. The reason for the apparent discrepancy is now evident. Johannsen was dealing with an exceptional organism, a highly inbred, self-fertilizing, and virtually homozygous stock, while the other breeders were dealing with highly variable stocks of cross-breeding organisms. The more recent selection experiments of Mather and others have shown how much concealed variability is available in such populations. This type of work, however, as well as an exaggerated emphasis on heterozygotes and a neglect of epistatic effects, might lead to the extreme assumption that populations are virtually inexhaustible reservoirs for the storage of genetic variation, ever ready for utilization. Muller (1950a) has raised two objections to this view. The first is that most of this variability is not available to selection since the segregation of recessive homozygotes occurs only at a very low rate, and the second, that the rate of accumulation is greatly depressed by the incompleteness of the recessive condition. These arguments are entirely valid, but do not properly take into account the enormous increase in storage capacity owing to genic interaction and environmental heterogeneity.

We are now approaching a more balanced view on this subject. There are many factors that help to increase the storage capacity of the gene pool (see Chapter 9), yet it cannot be overemphasized that this capacity has severe limits. On one hand, developmental homeostasis has limits and all genes producing extreme phenotypes are threatened by normalizing selection. On the other hand, there are differences in fitness (as we have seen above) even among gene combinations that do not produce visibly distinguishable phenotypes. As a result, only those genes will accumulate in a gene pool that add to fitness and help to produce a harmonious and well-adapted phenotype. Epistatic interactions narrow down the capacity of the genetic reservoir more severely than any other factor. The best evidence for the widespread incompatibility of genes and gene complexes comes from crosses between populations of the same species, as described above. Each local population is a separate integrated system, which can accumulate only so much genetic variability as can be combined into a harmonious gene complex. The narrowness of secondary hybrid belts (Chapter 13) is further evidence of this incompatibility of genes. Where an isolated hybrid population has been formed between two dissimilar but fertile parental populations, a period of "coadaptation" may lead to the elimination of the most disharmonious elements. In artificial hybrid populations between the "species" Drosophila arizonensis and mojavensis, the mojavensis chromosomes generally proved superior. The arizonensis X-chromosome was usually lost altogether and the second and third chromosome retained only because of heterosis with the mojavensis chromosomes (Mettler 1957). There was a deficiency of homozygotes of these arizonensis chromosomes.

Genetic variability is limited to such an extent that heterotic systems which coexist in a population may interact adversely when they come together in a single individual. A particularly well-analyzed case is that of gene arrangements in *Moraba scurra* (Lewontin and White 1960). Haldane (1957) and Stone *et al.* (1960) have discussed the limits in the number of heterotic systems that can coexist in a single population. This severe limitation on genetic variability explains in part the sharp gaps that we find between species and it also explains why each population has its own particular gene complex. Finally, this limitation makes it evident why speciation, the creation of new gene pools, is not a great loss of genetic variability, as is sometimes claimed (Heuts 1952).

At the present time, there is no way to determine the upper limit of the genetic storage capacity of a species. There is little doubt that successful, widespread, nearly panmictic species pay for their success in many ways. The high frequency in these species of chromosomes that are lethal in homozygous condition may be one of these penalties, particularly if an appreciable percentage of these chromosomes are synthetic lethals. The frequent occurrence of devices to cut down recombination and to "lock up" favorable gene combinations may be another indication of the drawbacks of gene flow and of excessive genetic variability in such species. The inverted gene arrangements in *Drosophila* are one of these "lock-up" devices. "The principal genetic effect of inversions is suppression of recombination of genes located in the chromosomes involved in inversion heterozygosis. If the high fitness, the heterosis, of the inversion heterozygotes is due to the interaction of gene complexes carried in the chromosomes, then the preservation of these gene complexes as units is advantageous. Crossing over and gene recombination would break up these gene complexes" (Dobzhansky 1951). Wallace (1953) shows that, if three overlapping gene arrangements coexist at a locality, there is an opportunity for a serial transfer of genes and thus for a destruction of the coadapted gene arrangements. He shows that in agreement with this argument not more than two overlapping arrangements are found with high frequency at one locality in most species of *Drosophila*. However, there are exceptions, as pointed out by Levitan, Carson, and Stalker (1954). Wallace (1959) has given consideration to these exceptions in a revised interpretation. Other cytogenetic devices also seem to have the function of preventing the breaking up of coadapted gene complexes by recombination. Any mechanism that prevents the separation of genes by recombination will cause such a block of genes to be inherited as if it were a single gene (supergene).

SUMMARY

Our findings on the genetic cohesion of gene pools can be summarized in the following statements:

(1) A considerable number, perhaps the majority, of loci of a species is represented in each population by several alleles. Many of these are so-called isoalleles and produce indistinguishable phenotypes. Each individual is therefore normally highly heterozygous. Since many genes occur in natural populations primarily in heterozygous condition, that genetic background will be favored by selection which enhances the selective value of these genes in heterozygous condition.

(2) The phenotype is the product of the harmonious interaction of all genes. The genotype is a "physiological team" in which a gene can make a maximum contribution to fitness by elaborating its chemical "gene product" in the needed quantity and at the time when it is needed in development. There is extensive interaction not only among the alleles of a locus, but also between loci. The main locale of these epistatic interactions is the developmental pathway. Natural selection will tend to bring together those genes that constitute a balanced system. The process by which genes are accumulated in the gene pool that collaborate harmoniously is called "integration" or "coadaptation." The result of this selection has been referred to as "internal balance." Each gene will favor the selection of that genetic background on which it can make its maximum contribution to fitness. The fitness of a gene thus depends on and is controlled by the totality of its genetic background.

(3) The result of the coadapting selection is a harmoniously integrated gene complex. The coaction of the genes may occur at many levels, that of the chromosome, nucleus, cell, tissue, organ, and whole organism. The nature of the functional mechanisms of physiological interaction are only of minor interest to the evolutionist, whose main concern is the viability of the ultimate product, the phenotype.

(4) There is a definite upper limit to the amount of genetic diversity that can be incorporated in a single gene pool.

(5) Many devices tend to maintain the *status quo* of gene pools, quantitatively and qualitatively. The lower limit of genetic diversity is determined by the frequent advantage of heterozygosity (and the very pronounced disadvantage of many homozygotes under variable or severe conditions). The upper limit is determined by the fact that only those genes can be incorporated that are able to "coadapt" harmoniously. No

gene has a fixed selective value; the same gene may confer high fitness on one genetic background and be virtually lethal on another (genetic theory of relativity).

(6) The phenotype is the by-product of a long history of selection and is therefore well adapted. The effect of selection will normally be to stabilize or to normalize this phenotype. Since this well-integrated phenotype is adapted to make a maximal contribution to fitness, it will resist change (inertia, genetic homeostasis) in the face of new selection pressures.

(7) The result of the close interdependence of all genes in a gene pool is tight cohesion. No gene frequency can be changed, nor any gene be added to the gene pool, without an effect on the genotype as a whole, and thus indirectly on the selective value of other genes.

(8) A sudden rise in the input of genes into a gene pool, let us say by hybridization, inevitably results in a disturbance of the internal balance and in the production of many genotypes of lowered viability. Disharmonious combinations will be eliminated by natural selection until a new balance has been reached.

(9) The cohesion of the gene pool results in various characteristic responses to new selection pressures. The amount of response to selection is unpredictable because different genotypes have different correlated phenotypic responses, particularly with respect to fitness. The uniqueness of every chromosome, of every individual (in sexual species) and of every population results in an immense diversity of responses to selection.

11 ~ Geographic Variation

The study of individual variation has revealed the invalidity of the concept that considers all individuals of a species to be replicas of the type. This typological concept of the species is further undermined by the findings presented in this chapter. These findings show that variation occurs not only within populations but also between populations. The occurrence of differences among spatially segregated populations of a species is called *geographic variation*. The emphasis on "geographic" has historical reasons since the phenomenon was first discovered when specimens from geographically far-distant populations were compared. We know now that even neighboring populations differ from each other, indeed, that in sexually reproducing organisms, no two demes can ever be identical. Some authors prefer, therefore, a separate term for local variation among neighboring demes, calling it "microgeographic variation" or "spatial variation." In this volume the term geographic variation is used in the broadest possible sense, to include all population differences in the space dimension.

The number of species of animals in which even a rather elementary analysis has failed to establish the occurrence of at least some geographic variation is very small. Geographic variation can therefore be considered a universal phenomenon in the animal kingdom. Its existence was well known to Linnaeus, Pallas, Esper, and other founders of animal systematics. Lamarck (1815) described it as follows:

Let anyone pass slowly over the surface of the earth, especially in the north-south direction, stopping from time to time to give himself leisure to observe; he will invariably see the *species* varying little by little, and more and more as he is farther from his starting point. He will see them follow, in some sense, the variations of the localities themselves, the conditions, exposed or sheltered, and so on. Sometimes he will even see varieties produced, not by habits required by the conditions, but by habits contracted accidentally, or in some other way. Thus man, who is subject by his organization to the laws of nature, himself shows remarkable varieties within his species, and among these some which seem to be due to the causes last mentioned.

The biological significance of the genetic (and usually also phenotypic) differences between populations of a species has been fully appreciated only within recent decades. The study of geographic variation has helped to close the gap between genetics, the study of the dynamics of genetic factors within populations, and paleontology, the study of evolutionary phenomena of the higher categories. It has yielded clues to the solution of some of the most stubborn evolutionary puzzles, particularly that of the multiplication of species. It has contributed toward the development of a new brand of biology, population genetics. In view of its importance it seems justified to give considerable space to a discussion of geographic variation. For previous surveys of the subject see Rensch 1929, 1934, Huxley 1942, and Mayr 1942 (Chapters 3 and 4).

The Genetic Basis of Geographic Variation. Evolution, according to De Vries, Bateson, and other early Mendelians, is due to spectacular mutations. The existence of geographic variation was a source of considerable annoyance to these mutationists. That populations from different portions of the range of a species should differ from each other by "gradual" characters, and that these differences should be greater the greater the distance, was so completely at odds with the "mutationist" interpretation of speciation that it forced the mutationists to deny the genetic nature of this variation. The present generation of evolutionists can hardly appreciate the enormous impact of the demonstration by Schmidt (1918) for the fish *Zoarces,* by Goldschmidt (1912–1932) for the gypsy moth (*Lymantria dispar*), and by Sumner (1915–1930) for the deermouse *Peromyscus* that the slight differences between geographic races have a genetic basis. This finding had been anticipated by naturalists, who had shown that bringing an animal from its native habitat to a zoo or to a new locality only rarely affects its phenotype or that of its descendants. The evidence for the genetic basis of geographic variation, accumulated since the pioneer investigations of Schmidt (1918), Goldschmidt (1934), and Sumner (1932), is too vast to be cited in detail (see also Rensch 1954:28; Dobzhansky 1951). Outstanding in this field is the work of Gordon (1947) on fishes of the genus *Xiphophorus,* of Dice and his school (Blair 1950) on mice of the genus *Peromyscus,* on various genera of lepidoptera (Ford 1953; Remington 1954), and on numerous species of *Drosophila* (Patterson and Stone 1952). The genetic uniqueness of each local population is an inevitable consequence of sexual reproduction. Since no two individuals are genetically identical, one would not expect any two groups of individuals to be identical. If they live in slightly different environments, as most geographically distributed populations do, one would expect such differences to become accentuated.

One must, however, avoid the other extreme, the silent assumption that all phenotypic differences among populations have a genetic basis. This is not necessarily true. We have recorded numerous instances in Chapter 7 of phenotypic differences produced by changes of environmental conditions rather than of the genotype. This has been recorded for fishes, grasshoppers, water bugs (hemiptera), butterflies (seasonal races), and various fresh-water organisms (cyclomorphism). Special care must be exercised when the geographic variation of organisms that are subject to such nongenetic modification of the phenotype is studied.

Descriptive Aspects of Geographic Variation

The study of geographic variation began as a by-product of taxonomic research. The working systematist, when comparing different specimens or population samples, wants to determine whether or not they belong to the same species. In the course of these comparisons he is often forced to compare specimens from different parts of the range of a species and he will record in the literature whatever differences he finds. In view of the simplicity of the method, it is not surprising that we find records of geographic variation as far back as the Linnaean period and earlier. Interest in the phenomenon increased rapidly during the period 1830–1870 as the size of the study collections grew and different parts in the range of species were increasingly well represented. It is not feasible to attempt a historical treatment here, but the names of some of the pioneers may be mentioned: for mammals, J. A. Allen, Osgood, and Sumner; for birds, Gloger, Schlegel, Baird, Coues, Seebohm, Kleinschmidt, and Hartert; for fishes, Heincke, Schmidt, D. S. Jordan, and Hubbs; for insects, Wagner, Eimer, Staudinger, Standfuss, and K. Jordan. The students of snails were particularly influential in promoting interest in the study of geographic variation. Rossmässler and Kobelt studied chains of Mediterranean "species"; Gulick explored the remarkable variability of *Achatinella* in the Hawaiian Islands, and the Sarasins chains of species on Celebes; Crampton selected the Polynesian genus *Partula* for his classical biometrical studies (for *Partula* see also Lundman 1947 and Bailey 1956).

Today no substantial taxonomic work dealing with the better-known groups of animals fails to include information on the geographic variation of the species treated. To prepare a list of such publications would be to prepare a catalogue of the modern taxonomic literature. Review papers on some of this literature have been published, for example, on game animals (Voipio 1950), birds (Mayr 1951a), and insects (Hubbell 1956). The most recent surveys of the fascinating island lizards (*Lacerta*) of the Mediterranean are by Radovanović (1956) and Buchholz (1954). There are numerous excellent papers on geographic variation in North American reptiles. Among the countless papers on insects, the review paper by Hubbell (1956), those of Rensch (1943) and O. Park (1949) on beetles, of Michener (1947b) on the bee *Hoplitis albifrons,* of Lorkovic′ (1953) on *Erebia,* and of Saccà (1953, 1956 ff.) on the housefly *Musca* and its geographic races may be mentioned. Biometric analysis often reveals geographic variation where it is otherwise not apparent (Götz 1959).

The study of geographic variation in fresh-water organisms is made particularly difficult by the great capacity of many species for nongenetic modification of the phenotype. To separate the nongenetic from the genetic contribution has been the endeavor of numerous recent workers. Among studies on geographic variation in fresh-water species one may cite Ernst (1952) on the newt *Triton,* Gordon (1947) on the platyfish *Xiphophorus,* Miller (1948, 1955), Hubbs (1922), and Heuts (1947) on various other genera of fishes, Hubendick (1951) and Forbes and Crampton (1942) on *Lymnaea,* Kiefer (1952) and Tonolli (1949) on copepods, and Brooks (1957a) on *Daphnia.* It occurs even among protozoans (Gause *et al.* 1942).

Except for important food fishes, notably herring and sardine, the study of geographic variation of marine animals has been rather neglected. The pioneer work of Heincke and of Schmidt was followed by Schnakenbeck's (1931) studies on local races of the herring (*Clupea harengus*), studies of still disputed significance (for example, Blaxter 1958, Wielinga 1958).

A great deal of information on the geographic variation of marine animals is scattered through the taxonomic literature, as pointed out, for instance, by Rensch (1947). The well-known echinoderms, with their numerous taxonomic characters, are particularly favorable material (Tortonese 1948, 1950; Mayr 1954b; Vasseur 1952). The widespread occurrence of geographic variation among plants, expressing itself both in local (often ecotypic) and in broadly regional variation, is abundantly substantiated (Stebbins 1950; White 1962).

Microgeographic Variation. One of the rather unexpected results of recent studies is the extreme localization of phenotypically distinct populations in some species. Gulick (1905) had pointed out long ago that every valley or ridge on Oahu (Hawaii) had its own characteristic *Achatinella* (see also Welch 1938, 1942, 1958), and Crampton (1916, 1932) had confirmed this for the *Partula* snails of the Society Islands (Tahiti and others). It is perhaps not altogether surprising to find high localization in animals as sedentary as these snails. Considerable localization has, however, been demonstrated also for highly mobile animals such as mice of the genus *Peromyscus* (Dice 1940; Hayne 1950; Blair 1950), and in *Drosophila* (Dobzhansky 1951; Patterson and Stone 1952). Three Chinese populations of *Chironomus* midges, separated by only 3.5 km in one case and 7.5 km in another, differed significantly in their chromosomes (Hsu and Liu 1948), as do adjacent populations of the Australian grasshopper genus *Moraba* (White 1957b). In extreme cases it is possible to demonstrate differences between populations living only a few meters apart. In the banded snail *Cepaea nemoralis*, for instance, Sheppard (1952a) found that a population in a hedgerow was quite different from the population of an adjacent meadow (Fig. 9–4), and Lamotte (1951) calculated that the diameter of a deme of this species, even in relatively homogeneous terrain is only about 50 meters. In fresh-water fish several races and species are known to be restricted to a single spring (Gordon 1947; Miller 1948; Clark Hubbs and Springer 1957). Habitat selection by genotypes (Chapter 0) may accentuate local differences. On the other hand, there seems to be nothing known in animals corresponding to the extreme localization of genetically determined ecotypes found in some species of plants, where localization is often reenforced by self-fertilization or other forms of extreme inbreeding.

Comparative Amount of Geographic Variation. No two species or characters agree in their pattern of geographic variation. Generalizations on the factors controlling geographic variation can be derived from comparative studies in taxonomically well-known groups of animals. An analysis of the amount of geographic variation within a single family is particularly illuminating.

The small family of drongos (Dicruridae) was quantitatively analyzed by Mayr and Vaurie (1948). The 20 species of this Old World family of birds extend from Africa through India to north China, Australia, and the Solomon Islands. Though all species are black, blackish, or gray, parts of the plumage may be white in some subspecies of some of these species, and the tone of gray, as well as the reflection of the glossy black plumage, may vary geographically. Among 20 species, 13 show pronounced geographic variation in size (wing length of adult males). Of the seven showing no obvious variation, five are restricted each to a single rather small island. Of the two continental species without visible geographic variation, one is strongly migratory. Many different aspects of the phenotype participate in the geographic variation. The color of the iris, for instance, varies in one species, being red, white, or brown in various parts of the range. Proportions or structural characters may vary. Drongos are long-tailed birds, and the length and shape of the tail varies geographically in most of the species. The same is true of the form and size of the crest in the crested species. An even more detailed analysis was made by Peterson (1947a) of 16 species of Fenno-Scandian Lepidoptera. No geographic variation could be discovered in three species, two species varied in size only, while the other 11 species varied in several characters.

Chronological Variation. What has been said of geographic variation applies also to variation in time. Whenever samples of the same species from different time levels (geological horizons) are compared carefully they are found, almost without exception, to be different. Such variation is sometimes irregular, sometimes showing definite trends ("chronoclines"), although usually only for some characters and for limited periods before these trends are modified or reversed.

Absence of Geographic Variation. The population structure of species and the need for local adaptation would lead us to suppose that geographic variation is a universal phenomenon. Yet in all groups of animals some species occur without visible geographic variation. In most cases this is due to insufficient study. Adequate biometric analysis has revealed geographic variation even in the morphologically notoriously uniform genus *Drosophila* (Stalker and Carson 1947; Prevosti 1954; Teissier 1958). In other cases where morphological variation has not yet been definitely established, for instance for *Drosophila pseudoobscura,* it has been demonstrated by genetical and cytological analysis that there is pronounced geographic variation in many cryptic characters.

The comparatively few cases of total absence of geographic variation (also most cases of very slight variation) can be explained by one of the following four possibilities, or a combination of them:

(1) The total range of the species is so small that there is no opportunity for geographic variation. The environment in the species area is essentially uniform. Rothschild's Starling (*Leucopsar rothschildi*), for instance, is restricted to a limited area in northwestern Bali (itself a small

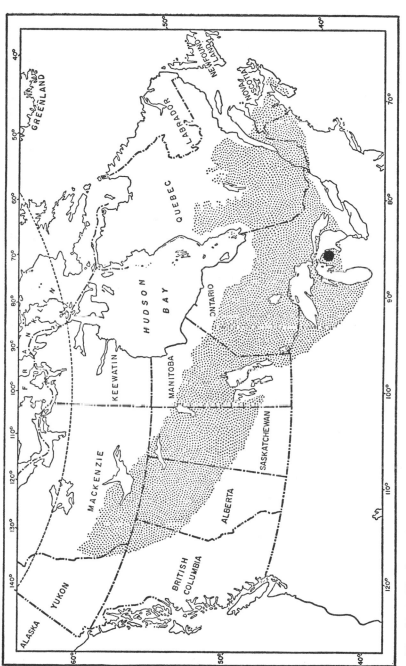

Fig. 11–1. The large dot indicates the nesting area of Kirtland's Warbler (*Dendroica kirtlandi*), comprising only a minute fraction of the range of its habitat, jack-pine forests (dotted). (From Mayfield 1960.)

island). Kirtland's Warbler (*Dendroica kirtlandi*) occurs only in a few counties in Michigan (Mayfield 1960; Fig. 11–1). Several species of fresh-water fish are found only in a single spring.

(2) The means of dispersal are so great that the species is virtually panmictic, regardless of the geographic extent of its range. This may well be the case for many of the very small "cosmopolitan" fresh-water organisms, such as protozoans, rotifers, tardigrades, cladocerans, and the like. The same seems to be true for certain ducks. Most migratory species of the North American warbler genus *Dendroica* are geographically invariable (Fig. 18-2).

(3) The phenotype is stable. The case of sibling species proves that little of the genetic variation penetrates into the phenotype in the presence of strong homeostatic devices. Presumably the greatest part of the apparent absence (or slightness) of geographic variation is due to such developmental homeostasis.

(4) The genotype is stable. It is sometimes postulated that ancient forms, like the horseshoe crab (*Limulus*), have lost all mutability. There is no evidence available to support this frequently proposed "explanation," but also none against it. In the absence of any evidence, it is easier to believe that the uniformity in space and time is due to a highly perfected "buffering system," in other words, to genetic homeostasis.

Lack of geographic variation, then, can be attributed either to an essentially panmictic condition of the species population or to low penetrance of genetic variation, combined with stabilizing selection. Why some species should have pronounced phenotypic variation, while others with similar population structure lack it (owing to factors 3 and 4), is still a complete mystery.

WHAT CHARACTERS ARE GEOGRAPHICALLY VARIABLE?

It can be stated that any character, be it external or internal, morphological or physiological, may vary geographically. Many instances of geographic variation are cited in the literature (Mayr 1942; Huxley 1942).

External Morphology

Size and Proportions. Size is perhaps the character most universally subject to geographic variation. It varies in virtually every species of animal that has an extensive geographic range.

Proportions also are highly variable in the great majority of species of animals. To cite merely one example, the tail of the Asiatic-Australian drongo, *Dicrurus hottentottus,* bears the following ratios to wing length (= 100) in various localities: Samar-Leyte, 66–71; Sumatra, 69–71; Sumbawa, 78–82; continental Asia, 78–85; New Guinea, 84–94; Moluccas, 100–112; Timor, 111–114; Tablas (Philippines), 116–134; and New Ireland, 202–213 (Vaurie 1949). Certain regularities in the variation of proportions will be discussed below under Allen's rule.

Epidermal Structures. All structures of the outer surfaces of animals are known to vary geographically. In mammals this includes the length, thickness, shape, and distribution of hair, and the size and shape of horns or antlers in antelope, deer, and caribou; in reptiles it includes the num-

Table 11-1. Conspicuous geographic variation on an archipelago
(*Zosterops rendovae*). [a]

Subspecies	Eye-ring	Bill	Black on forehead	Belly	Green on breast
(a) *vellalavellae*	Large	Yellow	None	White	Little
(b) *splendida*	Large	Black	Much	Yellow	None
(c) *luteirostris*	Large	Yellow	Little	Yellow	None
(d) *rendovae*	Absent	Black	None	Yellow	Much
(e) *tetiparia*	Absent	Black	None	White	Much
(f) *kulambangrae*	Small	Black	Little	Yellow	Little

[a] This polytypic species is sometimes treated as a superspecies with four species: (a), (b), (c), and (d–f).
See also Table 18–1.

ber and form of scales; in amphibians the texture of the skin; in insects the surface of the sclerotic covering, such as the sculpture of the elytra of beetles or of the wing scales in butterflies; in snails the presence or absence of spines and ribs; and so forth. Very often the differences within a single species are as great as between species, and sometimes even as great as between genera.

Patterns of Coloration. Color pattern is particularly important in those animals in which vision plays an important part, such as birds and certain reptiles, fishes, and insects, but a certain amount of geographic variation in tone of color is found in almost any group of animals. That striking differences may occur within a very limited area is illustrated by the variation within *Zosterops rendovae* (Table 11–1).

The color or color pattern of a given animal consists of many unit elements. These often vary independently in the species range. Geographic variation in *Lalage* (Mayr and Ripley 1941; Fig. 11–2), *Rhipidura* (Mayr and Moynihan 1946), *Pachycephala* (Mayr 1932a,b; Galbraith 1956),

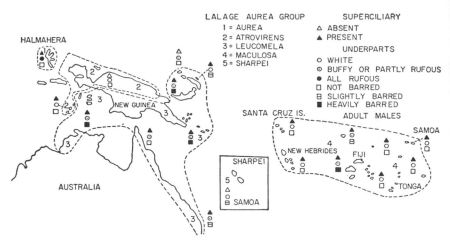

Fig. 11–2. Geographic variation in the taxonomically important characters of the *Lalage aurea* group. Samoa was colonized twice, first by *sharpei*, later by *maculosa*. (From Mayr and Ripley 1941.)

and *Ptilinopus* (Ripley and Birckhead 1942; Cain 1954b) give a rich picture of the type of changes that are induced in birds by geographic variation. Cain has made a detailed analysis of the significance of geographic variation in various genera of parrots (1955). Among insects such variation has been described for butterflies, beetles, and hymenoptera (for example, van der Vecht 1953, 1959; Fig. 11–3).

Internal Morphology

Anatomy. Too little is as yet known about the anatomy of different populations of the same species to present in detail a picture of geographic variation. Perhaps more is known in this respect of the human species than of any other. It has been demonstrated that for many anatomical structures there are differences between Whites, Negroes, and Japanese, particularly in the frequency of rare deviations from the normal type. In mammals the structure of the skull often varies geographically, as well as the structure and the number of teeth. In echinoderms there is geographic variation in the number, the shape, and the location of the subcutaneous plates. Geographic variation of internal structures has been recorded for mollusks (Hubendick 1951) and for earthworms (Lumbricidae). The reason that so little is known about geographic variation of anatomical features is not that such variation is rare, but that it has been investigated so far by only few authors.

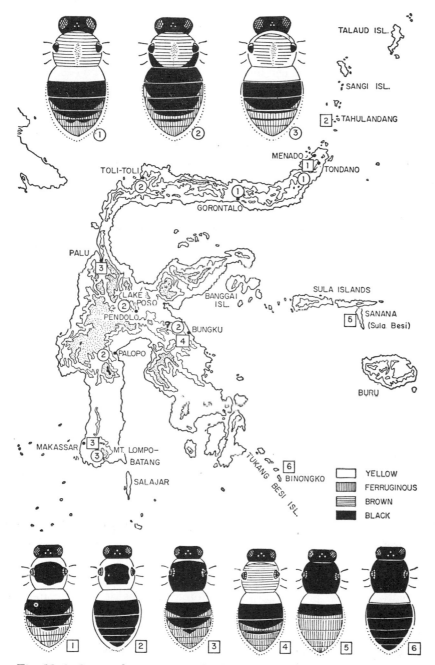

Fig. 11–3. Geographic variation of color in two species of carpenter bees on Celebes and adjacent islands. *Xylocopa diversipes* (above) with 3 subspecies (circles), and *X. nobilis* (below) with 6 subspecies (squares). (From van der Vecht 1953.)

Cytology. Evidence beginning to accumulate indicates that there is much intraspecific variation of chromosome structure. The presence and frequency of inversions, the fusion of chromosomes, and the existence of supernumeraries, all are subject to geographic variation. A detailed presentation of this variation can be found in the books of White (1954), Dobzhansky (1951), Patterson and Stone (1952), and Stebbins (1950). In the early period of cytogenetics, it was believed that chromosomal variation was intimately connected with the origin of new species. It is now known that this is not necessarily true, since many chromosomal changes are without effect on cross fertility. They appear to have, as primary function, either the increase of adaptability and adaptedness of these populations through balanced polymorphism of entire chromosome sections or the regulation of the amount of recombination. These phenomena are discussed in the relevant chapters (Chapters 9, 10, and 17).

Nonmorphological Characters

The study of geographic variation was at first almost exclusively the domain of the taxonomist. As he works mostly with preserved material, he quite naturally emphasizes morphological characters. The study of various physiological, behavioral, and ecological characters is, however, even more interesting, as such characters are often more directly of survival value or form potential isolating mechanisms. (All so-called morphological characters are, of course, ultimately also the result of physiological processes.) Geographic variation of physiological characters is quite universal, whether or not they manifest themselves phenotypically. The distribution of the blood groups in the races of man is an outstanding example of such physiological differences (Mourant 1954). A more detailed treatment will be given in the second half of this chapter, since much, if not all, of this variation is evidently adaptive.

More and more cases are found of geographic variation in behavioral and ecological characters. The data presented by Desneux (1952) and Emerson (1956) indicate geographic variation in nest-building habits of termites (see also Schmidt 1955a,b, 1958). Since it is still customary in termite taxonomy to describe each geographically isolated and morphologically different population as a different species, it is quite possible that *Apicotermes trägårdhi* (Zululand) and *holmgreni* (Nyassa) are conspecific. Conspecificity is similarly possible for *A. arquieri* (northern Congo), *rimulifex* (Katanga), and *occultus* (French Guinea), for *A. desneuxi* and *gurgulifex,* and for *A. angustatus* and *porifex.* The fact that different species and races may differ from each other in behavior has, of course, long been known for various kinds of animals (Huxley 1942; Mayr 1942, 1948), and nothing is gained by introducing a special term for species or geographic races that differ more in behavior than in morphology.

Much of the geographic variation in behavioral and ecological characters is potential raw material for the formation of isolating mechanisms and will be discussed in Chapter 16.

THE SIGNIFICANCE OF GEOGRAPHIC VARIATION

When taxonomists first became aware of geographic variation, they studied it as a phenomenon of purely practical taxonomic interest. Only rarely did they inquire why taxonomic characters so often are geographically variable and why not all populations of a species are identical. The majority opinion was, unquestionably, that most of this variation is irrelevant and without biological significance. It has been asserted as recently as 1936 (Robson and Richards) that variation below the generic level is nonadaptive, yet it was realized surprisingly early that there is close correlation between the environment and much of geographic variability. In 1833, Gloger devoted an entire book of 159 pages to the subject of *The Variation of Birds under the Influence of Climate*. Bergmann (1847), J. A. Allen (1877), and others, in notable contributions to the subject, came to similar conclusions long before 1900.

That specific as well as subspecific characters are, on the whole, adaptive and acquired through natural selection was maintained by many leaders of evolutionary research at an early date. Wallace (1889:142) stated:

It has not even been proved that any truly "specific" characters—those which either singly or in combination distinguish each species from its nearest allies —are entirely unadaptive, useless, and meaningless; while a great body of facts on the one hand, and some weighty arguments on the other, alike prove that specific characters have been, and could only have been, developed and fixed by natural selection because of their utility.

And Darwin (*Life and Letters,* 3:161) wrote to Semper, November 30, 1878:

As our knowledge advances very slight differences considered by systematists as of no importance in structure, are continually found to be functionally important . . . Therefore it seems to me rather rash to consider slight differences between representative species, for instance, those inhabiting the different islands of the same archipelago, as of no functional importance, and as not in any way due to natural selection.

A strong reaction to these adaptationist views developed in the 1880's and dominated evolutionary thought during the ensuing three or four decades. Gulick (1873) concluded that there was no correlation between the phenotype of the *Achatinella* snails from Hawaii and the respective environmental conditions in which the various local varieties occurred. Bateson (1913:131, 248) asserted "that it is useless to expect that such local differentiation can be referred to adaptation in any sense," and "a broad survey of the facts shows beyond question that it is impossible to reconcile the mode of distribution of local forms with any belief that they are on the whole adaptational." Crampton (1932:194), on the basis of his studies of variation in the snail genus *Partula,* likewise comes to the conclusion "that environmental circumstances produce no discernible effects upon the course of organic differentiation" and Diver (1939) echoes these sentiments with respect to the polymorphism in snails of the genus *Cepaea*. Leading systematists continued to maintain the opposite viewpoint throughout this period, vigorously defending the contention that much of variation found in nature was adaptive and directly correlated with the local conditions of the environment (Jordan 1896; Grinnell 1926: 260; Rensch 1929).

In retrospect it is evident that much of the heated discussion was based on semantic misunderstandings and insufficient analysis. The sources of misunderstanding were manifold. The major ones may be listed as follows:

(1) Use of the expression "effect of the environment" does not necessarily connote a Lamarckian interpretation. The selective "effect of the environment" is now recognized as an evolutionary force of primary importance.

(2) There has been considerable confusion over the significance of the terms "character" and "phenotype." Some visible components of the phenotype may in fact not be adaptive, but may be the by-product of a genotype selected for its invisible, cryptic contributions to fitness.

(3) Different assortments of genes may produce populations with similar phenotypes or viabilities.

(4) The genetic composition of populations is, in part, the result of stochastic phenomena and of the interaction of the totality of selective factors (internal and external ones).

A simple analysis of selective factors is therefore virtually impossible. The same selective factors may produce very different phenotypes in different populations.

Each local population is the product of a continuing selection process. By definition, then, the genotype of each local population has been selected for the production of a well-adapted phenotype. It does not follow from this conclusion, however, that every detail of the phenotype is maximally adaptive. If a given subspecies of ladybird beetles has more spots on the elytra than another subspecies, it does not necessarily mean that the extra spots are essential for survival in the range of that subspecies. It merely means that the genotype that has evolved in this area as the result of selection develops additional spots on the elytra. When studying geographic variation in the voice of birds, in the plumes of birds of paradise, or in the color patterns of parrots and pigeons, one must never ignore the possibility that some of the phenotype is merely the incidental by-product of the pleiotropic action of genes selected for other contributions to the viability of the phenotype (Dobzhansky 1956a). Yet close analysis often reveals unsuspected adaptive qualities even in minute details of the phenotype, for instance, in the body proportions of island lizards (Kramer 1951).

EVIDENCE FOR THE ADAPTIVE NATURE OF GEOGRAPHIC VARIATION

The geographic variation of species is the inevitable consequence of the geographic variation of the environment. A species must adapt itself in different parts of its range to the demands of the local environment. Every local population is under continuous selection pressure for maximal fitness in the particular area where it occurs. In the external environment there are principally two groups of factors that may exert a selection pressure on the phenotype: (a) climatic factors and (b) habitat and biotic factors. These two sets of factors usually manifest themselves differently in their effects on geographic variation. Climatic factors generally change rather slowly over wide areas (except where altitude is involved) and this results in a variability expressed in regular gradients. Biotic factors, and, even more, habitat factors, are often very local and irregular. The extreme in this respect is adaptation to the color of the substrate, which often shows a veritable checkerboard type of distribution, utterly different from the clinal variation of characters under climatic control. The evidence for the adaptive nature of geographic variation can be summarized under four headings, (1) geographic variation of physiological characters, (2) ecogeographic rules, (3) substrate adaptation, and (4) geographic variation of balanced polymorphism.

(1) Geographic Variation of Physiological Characters

Each local environment exerts a continuous selection pressure on the localized demes of every species and models them thereby into adaptedness. As a consequence, local populations differ not only by morphological characters, but also by numerous genetically controlled adaptive features of habit, ecology, and physiology. For more than 100 years, the literature of natural history has included occasional references to physiological differences among different populations of the same species. However, only within the last several decades has this problem been studied systematically and its genetic basis been determined. Yet, to the present day, a "comparative physiology of races or subspecies hardly exists" (Heuts 1949). The universal occurrence of an adaptive geographic variation of physiological traits is indicated by the work summarized in the following paragraphs (see also Mayr 1949b).

The pioneer work of Goldschmidt on the gypsy moth (*Lymantria dispar*) was reported in a series of papers, beginning in 1911 and summarized in 1934. He showed that among geographically varying characters are such physiological attributes as rate of growth, size, pigmentation, and strength of sex genes. The most obviously adaptive character he studied is the length of the incubation period. During the winter the young larvae undergo a diapause and remain enclosed in the egg. Even if the diapause is broken artificially and the eggs are exposed to a constant warm temperature, there is a lag of a certain number of days before the larvae hatch. The length of this incubation period depends on a considerable number of factors, such as the experimental temperature and the length of the preceding period of diapause. However, it is also closely correlated with the climatic characteristics of the place of origin of the eggs. In northern Europe, Korea, and the northernmost islands of Japan, where the vegetation develops very rapidly in spring after the long winter, the incubation period is short. In the Mediterranean, where the winters are warm but the vegetation is slow in the spring, there is a very long incubation period, 87 days at 11°C, compared to 55 days for central European eggs. The selective advantage of this different timing is obvious: the larvae would die of starvation if induced by the warm winter weather of a Mediterranean country to hatch prior to the emergence of vegetation. Geographic variation of the diapause or length of lag period has since been found in many other insects, for instance, the Australian plague grasshopper *Austroicetes cruciatus* (Andrewartha and Birch 1954:690).

In many butterflies and moths there is geographic variability of the number of broods per year and this requires the acquisition of many physiological adjustments. In the Polyphemus moth, Dawson (1931) found that in a southern race (reaching north to Nebraska) the moths are larger, chestnut brown; the cocoons are of coarser silk; the egg diameter is larger (2.8 mm); the number of eggs per female is smaller (average 236); and there are two broods per year. In a northern race (characteristic of Minnesota) the moths are smaller, yellowish brown; the cocoons are of finer silk; the egg diameter is smaller (2.2–2.4 mm); the number of eggs per female is larger (average 291); and there is only one brood per year. In the southern population, the larvae of the second brood complete their growth in the middle of September and spin their cocoons by the end of the month when the seasonal isotherm of 60°F reaches that locality. Northern larvae of the single annual brood complete growth and pupate in August, when the seasonal isotherms of 69–63°F pass through the region. Exposing the northern larvae in their last stages to lower temperatures produces dormant pupae. The same temperature drop fails to induce dormancy in the southern larvae. The northern population is prevented by these physiological mechanisms from starting a second generation in the fall, a generation destined not to mature successfully. Petersen (1947a:408) has shown similar genetic differences between southern and northern populations of butterflies in Scandinavia. Climatic races in insects have also been reported for butterflies by Hovanitz (1942), for *Carabus* beetles by Krumbiegel (see Mayr 1942), and for *Drosophila funebris* by Timoféeff (1935). The story of the climatic races of the corn borer (*Pyraustes nubilalis*) and of the spruce sawfly (*Gilpinia polytoma*) has been told by Andrewartha and Birch (1954:687).

Among insects, particularly striking differences often exist between low- and high-altitude races of the same species. In the moth *Nemeophila plantaginis*, Pictet (1938) found that above the altitude of 2000 meters the caterpillars hibernate during two winters rather than one. In crossing experiments a single switch gene seemed primarily responsible for the number of hibernations of the larvae. Aestivation in the butterfly *Gonepteryx rhamni* lasts much longer in the hot parts of China than elsewhere in the range. In the cool mountains there are two generations per year; in the lowlands there is time for only one (Mell 1941). In *Pieris* there is much physiological difference between the alpine *bryoniae* and the lowland *napi*, even where the two forms are not reproductively isolated (Petersen 1955).

The sensitivity of the adjustment to the local climatic conditions is so great that selection may produce genetic changes in succeeding generations in those species that have numerous generations in a single year such as *Drosophila* (Dobzhansky 1951; Prevosti 1955a,b) or *Musca* (Saccà 1952) (see Chapter 9). Warm-blooded animals are better shielded against the vicissitudes of local climates, yet in all cases in which widespread species of birds were carefully studied numerous local adaptations in physiology or ecology were discovered (Nice 1937; Blanchard 1941; Johnston 1954).

Fresh-water animals appear to be even more sensitive to changes in temperature than are land animals. As a consequence they show much geographic variation in temperature tolerance, growth rate, and other physiological constants that adapt them for living in waters of particular temperatures. An instructive example is the development of the Leopard Frog, *Rana pipiens* (Moore 1949; Ruibal 1955). There are pronounced climatic differences in the vast range occupied by this species, extending from Canada to Panama. These climatic differences have resulted in great physiological differences among the populations of this species. The most vulnerable period in the life cycle of the frogs is the embryonic stage between the fertilization of the egg and the metamorphosis of the young frog. Selection is most severe in this stage, and in this period is found the greatest number of adaptive physiological differences among the various geographic races. Northern races (Quebec, Vermont, Wisconsin, New Jersey) and races from Louisiana develop normally at temperatures as low as 5° (all temperatures are given here in degrees Celsius), while the lower limit is 9° in central Florida (Ocala), 12.1° in southern Florida (Englewood), about 10° in Texas, and 12° in Mexico (Axtla). In turn, the southern populations are able to tolerate higher temperatures and still develop normally. The upper temperature limits for normal development in various populations are as follows: Quebec, Vermont, Wisconsin, and New Jersey, 27.5°; Louisiana, 32°; and Mexico, 33°. At low temperatures, northern frogs develop fast; that is, at the temperature of 12° the number of hours required to reach embryological stage 20 (Moore 1949) is: Vermont, 325; Louisiana, 348; Florida (Ocala), 354; Florida (Englewood), 364; Texas, 429; Mexico (Axtla), 396. That an adaptation to local water temperatures is involved is indicated by the study of various altitudinal races in Mexico (Ruibal 1955). Lowland populations have a rapid development, highland populations are slower. The rate of development of a population from the mountains of Costa Rica is sufficiently similar to that of frogs from the northern United States (Vermont) that development of the hybrids is nearly normal (Moore 1950). Yet Costa Rica frogs are not especially cold adapted (Volpe 1957). Other work that demonstrates geographic variation in various physiological properties of fresh-water animals, particularly tolerance to temperature or salinity, is that of Heuts (1947, 1956) on the stickleback *Gasterosteus*, DuShane and Hutchinson (1944) on the salamander *Ambystoma maculatum*, Hart (1952) on various fresh-water fishes, Volpe (1953) on the toads *Bufo fowleri* and *B. americanus*, Forbes and Crampton (1942) on the snail *Lymnaea*, and Johnson (1952) on *Daphnia*.

The occurrence of temperature races in marine animals has been known for a long time, particularly as a result of a comparison of Mediterranean populations (Naples) with northern European populations of the same species. The North Atlantic population of the bivalve *Chlamys opercularis* seems to find optimum conditions between 4° and 13°, while the Mediterranean subspecies thrives between 13° and 26° (Ursin 1956). The pioneer work on this geographic variation of temperature preference and tolerance was done by Runnström (1927, 1929, 1936).

Most marine organisms have a pelagic larval stage and, in connection with it, a considerable ability to adjust to the environmental conditions of the place where the larvae settle (Prosser 1957). The question has therefore been raised: what part of the differences between local races in temperature and salinity tolerance has a genetic basis? This problem was extensively discussed in a recent symposium on this subject (Fage and Drach 1958). As a result there can be no doubt of the widespread occurrence of genetic races. They are particularly well established for oysters, in America *Crassostrea virginica* (Stauber 1950; Loosanoff and Nomejko 1951), and in Europe *Ostrea edulis* (Korringa 1958). Oysters from warm waters, even when transferred as spat, fail to spawn in cold waters. Northern cold-water oysters, taken into warm southern waters, discharge their gametes long before the local populations. Yet the temperature preference of local populations is often irregular, not necessarily paralleling the change in latitude. Along the coast of Spain, for instance, the preferred spawning temperature of *Ostrea edulis* is 12–14°C, while near Bergen (Norway) it is above 25° (Korringa 1958). Other cases of geographic variation in physiological properties of marine animals have been listed by Bullock (1957), Dehnel (1955, 1956), and Schlieper (1957). On the whole, the genetic component of the phenotypic variation with changing water conditions seems smaller in marine animals than the genetic contribution to the phenotype of climatic races in land animals. The demand for phenotypic plasticity is particularly great in sedentary intertidal species and in all species whose pelagic larvae are at the mercy of currents. Successive spawnings of the same local populations may be forced to colonize rather different areas. The very free gene flow will result in a highly panmictic condition. All this counteracts local genetic differentiation and favors developmental flexibility.

The evidence for regularities in geographic variation of marine animals is as yet contradictory. In some cases there seems to be an increase in size toward the north, in others a decrease (Johnsen 1944; Ray 1960). Cold water delays the attainment of sexual maturity and also slows growth. There seems to be a tendency for heavier shells in warm waters (not always, see Rao 1953) and for a higher percentage of pelagic larvae. In fishes there is a tendency for an increase in the number of fin rays and vertebrae in cooler waters (Tåning 1952). Other examples are quoted in Hesse, Allee, and Schmidt (1951) and Allee *et al.* (1949); (see also Chapter 7).

The adaptive significance in the geographic variation of many other physiological characters is not nearly as evident as in the case of temperature tolerance. This is particularly true for many of the changes in reproductive pattern through the range of species.

Sex Races and Developmental Stages. Bacci (1950) has shown that in many marine invertebrates certain geographic races are gonochoristic,* while others are hermaphroditic. For a review of this subject see Montalenti (1958). A similar condition prevails in some species of amphibians. In the frog *Rana temporaria* so-called differentiated sex races occur in the Alps and along the Baltic. In these populations the males acquire their characteristics early in embryological life. In the climatically milder parts of Europe there are undifferentiated races in which all newly metamorphosed frogs are phenotypically females. The genetic males among these acquire their visible male characters by way of a hermaphroditic condition during or after the first year of life (Witschi 1930). In the salamander *Ambystoma tigrinum* some geographic races reproduce in a neotenic larval stage similar to that of the axolotl (*A. mexicanum*) while other races metamorphose into an adult condition (Dunn 1940). *Planaria alpina*, a

* Of individuals, having functional gonads of only one sex; of breeding populations, composed of male and female individuals; bisexual.

well-known European flatworm, has several geographic races, differing in the prevalence of asexual reproduction; the northern Scandinavian populations reproduce entirely asexually. Similar races have been found in several other species of turbellarians (Dahm 1958). The capacity to reproduce sexually or asexually is genetically determined and retained in the laboratory. Sexuality varies geographically also in bivalve mollusks (Bloomer 1939).

Many species of marine snails can produce two kinds of eggs, small ones with little yolk which produce pelagic larvae, and large, yolk-rich eggs which produce young snails (skipping the larval stages). In species that are thus dimorphic the percentage of eggs producing pelagic larvae decreases from the warm to the cold parts of the range and arctic populations produce only yolk-rich eggs. The elimination of the pelagic larval stage is an obvious adaptation to the arctic water conditions (Lemche 1948; Thorson 1950).

Fecundity, that is, egg production per female or per breeding season, has been shown to be geographically variable in vertebrates and invertebrates. The fact that egg number is partially independent of body size (and other evidence) indicates a genetic component in this variation.

Geographic Variation of Coloration. The biological significance of the coloration of an animal is many-sided and subject to a particularly sensitive adjustment to the changing environment. The relation may be very direct, as in the case of those birds (ptarmigans) and mammals (weasels, hares) that molt each autumn into a white winter dress (Hall 1951). A white winter dress may be absent in the populations that live in areas without extensive snow cover. The close parallelism between humidity and coloration will be discussed below (Gloger's rule).

The precise significance of color is less easily determined in other cases, particularly where color merely reflects an internal physiological condition. This is true, for instance, of the geographic variation in the degree of sexual dimorphism in birds (Mayr 1942:48–52). In the Australian Robin (*Petroica multicolor*) the males are brightly colored red, white, and black and the females cryptically brownish in most races, but on some islands the males are hen-feathered and on others the females are cock-feathered. "The loss of sexual dimorphism through feminization of the male plumage seems to develop only in well-isolated and rather small populations . . . It . . . seems to occur only in localities where no other similar species exist, i.e. where a highly specific male plumage is not needed as a biological isolating mechanism between two similar species"

(Mayr 1942:49). In the Golden Whistler (*Pachycephala pectoralis*) this has happened independently at least three times: on Rennell Island near the Solomons (*feminina*), on Norfolk Island near Australia (*xanthoprocta*), and on the islands of Roma (*par*) and Letti (*compar*) in the Banda Sea.

Geographic Variation of Seasonal Adjustments. The seasons, particularly winter and summer, differ in severity in different latitudes, and seasonal adjustments may vary geographically through the range of a species. In many species of birds the northern races are migratory, the southern races more or less sedentary. Such cases have been described by Mayr (1942:55, 1951), Voipio (1950), and Salomonsen (1955). A particularly detailed analysis of the physiological differences between migratory and sedentary races in a species of bird was made by Blanchard (1941) for *Zonotrichia leucophrys*. Miller (1960) has summarized much of the information on the geographic variation of breeding seasons.

The mode and annual number of molts may vary geographically in birds and insects (Mayr 1942:50). The variation of habitat preference and other ecological and behavior characteristics through the range of a species will be discussed in Chapter 16.

Some Practical Consequences. One conclusion emerges from these observations more strongly than any other: every local population is very precisely adjusted in its phenotype to the exacting requirements of the local environment. This adjustment is the result of a selection of genes producing an optimal phenotype. The discovery of this physiological adaptation of local populations is of considerable practical importance, for instance, in wildlife management (Aldrich 1946; Voipio 1950). Populations that are well adapted in their native environments are often very vulnerable when transplanted into different environments. The literature on game animals records many instances in which stocks died out rapidly after introduction into a different region. If they survive long enough to breed, they will contribute to the deterioration of the native stock. Millions of dollars of taxpayers' money spent on raising and releasing ill-adapted game stocks could have been saved if those in charge had been aware of the physiological differences among local populations.

(2) Ecogeographical Rules

One should find much parallel variation in different species if the assumption is valid that every species adjusts to local conditions and if there is a gradual change ("climatic gradient") of these conditions with latitude and longitude. Such parallelism in geographic variation is indeed widespread and has led to the establishment of a series of generalizations, the so-called climatic or ecogeographical rules.

Gloger (1833), Bergmann (1847), Semper (1881), and J. A. Allen (1877) were the first to investigate this phenomenon. The historical significance of the climatic rules is that they focused attention on the importance of the environment in a period during which many biologists denied the environment any evolutionary role. For a more complete survey of the ecogeographical rules we must refer to the papers of Rensch (1934, 1938, 1939). Lukin (1940) devoted an entire book (in Russian) to the subject, and Mayr (1956) discussed the meaning and validity of the rules.

The ecogeographical rules are purely empirical generalizations describing the parallels between morphological variation and features of the physical environment. For instance, Bergmann's rule states that "races from cooler climates tend to be larger in species of warm-blooded vertebrates than races of the same species living in warmer climates." The validity of these rules depends on the statistical validity of the data on which they are based. Bergmann's rule, for instance, is valid only if it is really true that in more than 50 percent of species of warm-blooded vertebrates there is an average increase of size in the cooler portions of the range of the species. The validity of these empirical findings is independent of the physiological interpretation given to the observed regularities. Some recent critics of the ecogeographical rules (Scholander 1955; Irving 1957) have failed to understand the real significance of this population phenomenon (Mayr 1956).

All these rules have only statistical validity; they are not unalterable "laws," even though they may be true for "most" species or races. The degree of validity of these rules varies from one group of animals to another and from one region to another. Rensch, who devoted a long series of investigations (1936, 1938, 1939, 1940, 1948) to a precise analysis of the percentual validity of these rules, found that their application depends on "other things being equal." Hamilton (1959, 1961) has made a particularly careful analysis of the interacting and often conflicting environmental factors that affect the direction of these trends.

In view of occasional misunderstandings in the literature, it is important to emphasize "that the validity of the ecological rules . . . is restricted to intraspecific variation . . . A more northerly species is by no means always larger than its nearest more southerly relative" (Mayr 1956). As we shall presently see, separate species have different means of adaptation available than have open populations within species. The climatic rules are ecotypic, not phylogenetic, phenomena.

Rules for Warm-Blooded Vertebrates. Variation in size. No rule is more widely known than "Bergmann's rule," which states that body size in geographically variable species averages larger in the cooler parts of the range of a species. The amount of difference can be very impressive, the largest races sometimes being more than twice as big (heavy) as the smallest. The percentage of exceptions found by Rensch (1936) are as follows: sedentary Palearctic birds, 8 percent; Malay birds (with Palearctic relatives), 12.5 percent; North American birds, 26 percent; North American mammals, 19 percent; western European mammals, 40 percent. Rensch's percentages are not based on original measurements but on data published in the taxonomic literature. Although subject to correction in detail, they support Bergmann's generalization very convincingly. The rule seems to be true also for marine birds (penguins) and mammals (whales). Even though whales undertake extensive migrations into warmer waters for part of the year, nevertheless whales from the very cold antarctic waters average larger than their counterparts in the warmer waters of the Nothern Hemisphere (Tomilin 1946); since water conducts heat 27 times better than air, the differences might be even greater if it were not for the remarkable heat-regulating mechanisms of whales.

Bergmann's rule tends to be valid also for changes of size with altitude. Among 60 species of New Guinea birds, 16 showed change of size with altitude, 15 increasing in size (Rand 1936). Further cases of altitudinal size increases in birds have been recorded by Mayr (1944) and Traylor (1950) and in mammals by Rümmler (1938).

Changes of size with time have been recorded by paleontologists. The races of several species that occurred in North America, in Europe, and in the East Indies during cold periods at the height of the Pleistocene glaciations averaged larger in body size than the races of the same species now occurring in these regions (Howard 1950; Hooijer 1949; Degerbøl 1940; Wüst 1930). The adaptive nature of these consistent changes cannot be doubted, even though in some of the cases it is not certain whether the changes from the Pleistocene to the present are due to a genetic reconstitution of local populations or to a latitudinal displacement of populations (through migration).

It is sometimes claimed that island populations are of smaller body size than populations on adjacent mainlands. This is by no means necessarily true. To be sure, Ceylon birds are usually smaller than those of Peninsular India, but populations on arctic islands, such as Greenland, usually have very large body size. On the whole there seems to be a close inverse correlation in the temperate zone between body size and the temperature of the surrounding air. In the tropics much of the size variation on islands seems to be unpredictable and dependent on local factors (Mayr and Vaurie 1948), although with remarkable frequency body size on small islands is greater than on adjacent large islands.

The usual physiological explanation of Bergmann's rule is based on the fact that the volume of the body increases as the cube and the surface as the square of a linear dimension. The larger a body, the relatively smaller its surface. In a cool climate there should be a selective advantage in the relative reduction of surface resulting from increased size, since the metabolic rate is more nearly proportional to body surface than to body weight (Kleiber 1947; Hemmingsen 1960). In hot climates the premium should be on small body size and relatively large surface. This interpretation has been attacked (Scholander 1955) on the basis that other devices in warm-blooded animals (feathers, fur, circulatory mechanisms, and so forth) prevent heat loss far more efficiently than slight shifts in surface/volume ratios. Yet Scholander's argument in favor of an all-or-none solution is alien to the facts of variation. Multiple solutions for biological needs are the general rule in evolution. Selective advantages are independent and strictly additive. The fact that a thicker fur or denser plumage reduces heat loss does not eliminate completely the selective advantage of an improved body surface/volume ratio. Harrison (1958) has proved experimentally for mice the heat-regulating value of the tail.

A study of exceptions sheds further light on the significance of the ecogeographical rules. Exceptions show that the phenotype is a compromise between conflicting selection pressures, and that in many species of warm-blooded vertebrates the northernmost populations are exposed to conditions tending to neutralize the advantage of increased body size. For instance, in many temperate-zone bird species the populations with the northernmost breeding ranges spend the winter south of the resident southern populations ("leapfrog" migration). The average annual temperature at which the "northern" birds live is higher, and their body size smaller, than in the "southern" birds (Salomonsen 1955; Fig. 11–4). In many Eurasian and North American species of birds the largest body size is found not in the coldest part of the range (near its northern periphery), but unexpectedly in the highlands of the semiarid subtropics (Iran,

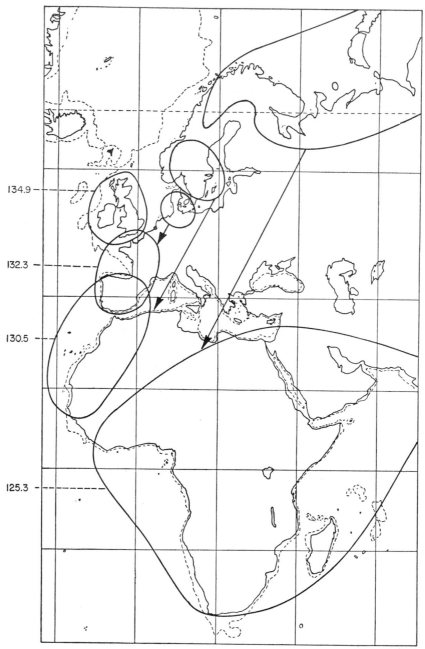

Fig. 11–4. Mean wing length of four populations of the plover *Charadrius hiaticula*. Arrows connect breeding range and winter quarters. The British population is resident. The warmer the wintering area, the shorter the wing length. (From Salomonsen 1955.)

Atlas Mountains, Mexican highlands). Snow (1954a) ascribes this "latitude effect" to the shortness of the arctic winter day, which depresses size by reducing daily food intake. Hamilton (1959) suggests that need for water conservation in an arid area may be part of the selection pressure. The effects of aridity on geographic variation have not yet been fully studied. In the kangaroo rat *Dipodomys ordii* the area of nasal mucosa is decreased in the hotter, more arid parts of the range (Setzer 1949). Other adaptations to desert conditions were discussed by Etchécopar and Hüe (1957).

Burrowing mammals almost consistently fail to obey Bergmann's rule. They are well protected against the cold, particularly in areas with snow cover, and for them the amount of food available in winter seems to be the decisive factor determining body size. This has been shown for rodents (for example, *Thomomys, Microtus*) and by Stein (1950) for the European Mole (*Talpa europaea*). In all these cases of exceptions to Bergmann's rule some environmental selective factor other than temperature has come into conflict with the size trend. What this factor is has not yet been determined for most of the known exceptions.

An animal is not a globe, and the relation of surface to volume is related also to various aspects of form. In man, northern Europeans have a body-weight/body-surface ratio of 37–39 kg/m², while in tropical peoples the ratio frequently drops to less than 36, being as low as 31.5 in Congo pygmies and 30.2 in Bushmen (Schreider 1950). Furthermore, the relative length of the heat-radiating limbs is much greater in warm than in cold climates (Schreider 1957), the extremes being the Eskimos on one hand and the slender Nilotic tribes on the other. The theoretically most heat-resistant physique of man was found to be typical for the hot deserts (Baker 1958). Barnicot (1959) has summarized our knowledge on the climatic adaptations of human races.

It has long been known that in warm-blooded animals protruding body parts like bill, tail, and ears are shorter in cooler than in warmer climates (Allen 1877). "Allen's rule" is actually an extension of Bergmann's rule, dealing likewise with the surface-to-volume relation. Among mammals, Rensch (1936) found 14 percent of exceptions concerning the tail, and 16 percent concerning ears; among birds he found 20 percent of exceptions concerning the bill. In island birds the bill is often exceptionally long (Murphy 1938).

Because wings and tails of birds do not contribute to heat loss, being composed merely of feathers, they naturally do not obey Allen's rule. On the contrary, birds from northern populations of migratory species normally have relatively longer wings than birds from more southerly populations of the same species (Williamson 1958). The same appears to be true of high-altitude populations, except that here an additional factor may play a role, namely, the fact that the thin air of the higher altitudes provides less lift than the denser air of lower altitudes.

The wings of northern birds are often not only longer, but also more "pointed," that is, with the outer wing feathers elongated. This change in the shape of the wing is correlated with an increase in the efficiency of the wing stroke. For further discussion of the "wing rule" see Rensch (1939), Meise (1938), and Kipp (1942). Still unexplained is the increase in relative tail length in the cooler parts of the range of many temperate-zone Passerine birds (Snow 1954a).

A particularly impressive result of studies of ecogeographical rules is the discovery of the extreme sensitivity of body proportions to natural selection. The former belief that proportions are determined by "built-in" allometry factors and change automatically with changes in body size is not supported by these findings (see also Kramer 1960).

Exceptions to Allen's rule involving the length of the bill in birds are usually related to food requirements. In the crossbills (*Loxia*), bill size is determined by the genus of conifers on which a given population feeds most frequently (Kirikov 1940). If it is *Pinus* the bills are large and robust, if it is *Abies* or *Picea* the bills are intermediate, if it is *Larix* or *Tsuga* the bills are small and slender. In some species of titmice (*Parus*) there is a decrease of relative bill length with decreasing environmental temperature. However, no further decrease takes place toward the north, once a certain bill length is reached, as if this represents minimal functional bill length (Snow 1954a).

Pigmentation. Races in warm and humid areas are more heavily pigmented than those in cool and dry areas. Black pigments are reduced in warm dry areas, and brown pigments in cold humid areas. This rule, called Gloger's rule, seems to have comparatively few exceptions, but its physiological basis is not at all clear. The selective advantages of the genes responsible for these pigmentation differences are not evident. Pigmentation in warm-blooded animals is believed to be somehow controlled by the thyroid, but this does not help to explain the pigmentation rule, as Gloger's rule applies also fairly well to animals without a thyroid, such as beetles, flies, and butterflies (Dobzhansky 1933; Ford 1937; Mayr 1942:90; LeGare and Hovanitz 1951). Neither can it be ascribed to substrate-adapted cryptic coloration, as arboreal and even nocturnal animals obey Gloger's rule well. The precise selective factors responsible for Gloger's rule are still a mystery.

Other kinds of ecogeographic variation in birds. Rensch and others have found, in addition to the rules discussed above, several other regularities in the geographic variation of birds.

The clutch-size rule states that the average number of eggs per clutch in a species may increase with latitude (Rensch 1936). This rule has been examined carefully by Lack (1947b) and Moreau (1944a,b, 1947), but is denied by Kipp (1948), and Lack's interpretation is questioned by Skutch (1949). The early interpretation of increased clutch size in cooler climates was that birds lay larger clutches because they "need" larger clutches to compensate for the greater mortality in the higher latitudes. However, birds cannot anticipate the percentage of loss in their offspring, nor control consciously the number of eggs they lay, so this obviously cannot be a valid explanation. Lack has presented much evidence to show that the optimum number of eggs per clutch is regulated by selection on both ends of the curve: broods that are too large are unsuccessful because the individual young are poorly fed; broods that are too small do not produce enough young for complete replacement. Evidence for the validity of this thesis has been found in the starling and support for it among the thrushes, but so far it has been impossible to prove it for titmice. In titmice the upper part of the curve is perhaps exposed to the full selection pressure only in years with highly adverse weather conditions. In tropical localities, where adults have a high life expectancy, there may actually be a selection pressure in favor of small clutches. The smaller the nest and the fewer the feedings per day, the greater the probability that the nest will not be discovered by the exceedingly numerous nest predators. Nonhibernating herbivorous North American mammals show an increase of litter size with latitude (Lord 1960).

Rensch (1936) has discovered a number of geographical regularities in the relative size of internal organs, for instance that the stomach is smaller and the intestines shorter in tropical races of widespread species of birds.

Ecological Rules in Reptiles and Amphibians. The study of the adaptive aspects of geographic variation in reptiles and amphibians is only beginning. Schuster (1950) found few regularities and these are somewhat contradictory: drought-tolerant amphibians are largest, humidity-dependent species smallest, in the warmest parts of their range. Among lizards, the longest legs were found where the maximum substrate temperatures are highest. For various species of the lizard genus *Liolaemus*, Hellmich

(1951) describes regular changes occurring in Chile from the north toward the cooler, more humid, and less sunny south. Mediterranean lizards (*Lacerta serpa*) on flat islands have shortened hind limbs "in adaptation to the reduced demands on locomotion [absence of predators] in the insular environment" (Kramer 1951). There is a rather general trend toward melanization in lizards (*Lacerta*) on small islands (Kramer 1949; see Chapter 8). Certain body proportions in frogs (*Rana*) vary clinally parallel with certain climatic gradients (Ruibal 1957).

Ecological Rules in Insects. Adaptive regularities in the geographic variation of insects have been described by earlier authors and were summarized by Mayr (1942). Hovanitz (1942) found that in the butterfly *Melitaea chalcedona* pupae from coastal Southern California are largest and heaviest, and those from the Mojave Desert smallest and lightest. The change in the pupal characters runs parallel with the climatic conditions. Michener (1947b) found regular geographic variation in the pigmentation of the bee *Hoplitis albifrons,* and Lorković (1942) in butterflies of the genus *Everes.*

Many regularities in geographic variation were described by Petersen (1947a, 1952) in his fine studies of Fenno-Scandian butterflies. For beetles we have detailed studies by Rensch (1943) for Europe and by O. Park (1949) for North America. Rensch found that in two species of *Carabus* the subspecies from the subtropical Mediterranean differ from the races of temperate central Europe by having smaller size, relatively smaller elytra, and relatively narrower abdomen, while the following measurements were larger: relative length of pronotum, relative length of antennae, and relative length of the extremities, particularly the tarsi. Alpatov (1929) had found a similar decrease of size in the honeybee, while the relative length of the extremities increased with decreasing latitude. Lieftinck (1949) found definite regularities in the variation of various species of New Guinea dragonflies with altitude: increase of size, dulling of the metallic body parts, narrowing of the wings, and darkening of the pterostigma. Altitudinal variation in the frequency of gene arrangements has been found in several species of *Drosophila* (Dobzhansky 1948; Stalker and Carson 1948).

The weakness of most of these studies is that it is possible only to correlate the character gradients with certain environmental gradients, not to prove that the variation of the phenotype itself is in all cases adaptive. This problem has been given special attention by Prevosti (1955a,b) for wing length and other characters in *Drosophila subobscura.*

The difficulty in determining the adaptive significance of geographic variation is particularly acute for size variation in invertebrates and cold-blooded (ectothermal) vertebrates. For these, of course, Bergmann's rule, with its implication of conservation of body heat, has no validity. It seems that three factors, to some extent antagonistic, determine the size trend. In species with a single generation every year, the length of the available growing period ("degree days") determines maximum larval size and hence adult size. Maximum size is usually reached in such species in the warmest, most humid portion of the range. In species in which sexual maturity is delayed for several years, as in certain marine animals, largest size may be reached in the coolest portion of the species range, by individuals having had the greatest number of (shortened!) growing seasons. If individuals of a single population are raised at different temperatures, those exposed to the lower temperature grow more slowly, but usually reach ultimately larger size (Ray 1960). Finally, where factors other than temperature (such as humidity, food supply, absence of disease or competitors) affect body size, largest size will be reached in the optimal portion of the species range, independently of temperature (Rensch 1932). The relative importance of these three factors has not yet been determined for any group of invertebrates or cold-blooded vertebrates. In a given genus or species group there is usually a fair degree of regularity (as one would expect in a character determined by natural selection), but, taking ectothermal animals as a whole, one may find the largest body size in the warmest or in the coolest portions of the species range, in a peripheral or in a centrally located population.

Conclusions. The climatic regularity in much of geographic variation proves that different species may react to the same factor of the environment in a similar manner. Since it can usually be shown that these differences have a genetic basis, the ecogeographical rules constitute evidence for the selective role of the environment. The multiplicity of largely independent regularities indicates the multiplicity of selective components of the environment. The regularity, the "smoothness," of the character gradients resulting from the climatic rules indicates that geographical changes of phenotype result from the interaction of numerous genetic factors, each with only a small phenotypic effect.

(3) *Geographic Variation in Substrate Adaptation*

The third set of factors proving the adaptative nature of geographic variation is presented by cryptically colored animals. An amazing agreement between the coloration of animals and the color of the substrate on which they normally live has been recorded not only for species (Cott 1940), but for local races in many animals of open lands, lava flows, and deserts. To cite one specific example: the soils, gravels, and rocks of the Namib Desert in Southwest Africa are of a characteristic reddish yellow, a coloration displayed by most of the inhabitants of this desert:

small succulent rock plants of the genus *Lithops* show the same color adaptation to the substrate as the elephant shrews *Elephantulus intufi namibensis*, the rodents *Gerbillus g. leucanthus* and *G. vallinus*, the larks *Certhilauda curvirostris damarensis*, *Tephrocorys cinerea spleniata*, and *Ammomanes g. grayi*, the bustard *Heterotetrax rueppelli*, the viper *Bitis peringueyi*, the lizards *Eremias undata gaerdesi* and *Meroles suborbitalis*, and finally a diversified group of wingless grasshoppers (Batrachotettiginae). The phenomenon of a "local coloration" which is shared by a large part of a local fauna is found also in other districts of Southwest Africa, e.g. the Usakos district, the Waterberg area, the Etosha Pan, and the Kaoko Veld (Hoesch 1956).

The more open the country and the more contrastingly colored the substrate, the more striking these substrate races; some of the most characteristic are found on lava flows and white gypsum sands. The Tularosa Basin in New Mexico includes both a large lava flow (40 miles long by 1–6 miles wide), and also 270 square miles of dazzling white gypsum sands. According to Blair (1943), of the five species of mammals occurring on the lava flow, four have formed darkened endemic races. The exception, *Peromyscus eremicus*, is found very commonly in suitable locations of the surrounding desert and the active population interchange between desert and lava flow has presumably prevented the formation of an endemic race. Among the six species of mammals regularly occurring in the gypsum dunes only two are restricted to them; the other four are common also in the surrounding terrain. Both ecologically restricted species have developed endemic races: *Peromyscus apache*, as expected, has formed a whitish race; the pocket gopher *Geomys*, however, a blackish form! It is not known to what extent this unexpected contrast is due to recent immigration of a very numerous animal and to what extent to the concealed mode of life of this strictly subterranean mammal. For other discussions of lava races, see Benson (1933), Dice and Blossom (1937), Hooper (1941), Hoffmeister (1956), and Baker (1960). Similar substrate races are found in lizards (Lewis 1949).

The older literature abounds in conflicting "explanations" of such cryptic coloration, many of them with an openly vitalistic flavor. Attempts to explain "desert races" with the help of climatic factors (heat, dryness, solar radiation) are not very convincing in view of the amazing agreement with the color of the substrate and the close proximity, under virtually identical climatic conditions, of black lava and white limestone or gypsum-sand races. The majority opinion now is that substrate races are the result of predator selection. That predators tend to take first the most discordant individuals when given a choice of cryptically and noncryptically colored individuals has been proved by numerous observations and experiments (Faure 1932; Boettger 1931; Sumner 1934, 1935; Isely 1938; Ruiter 1955). Mice of colors that blended well with the background were taken by owls significantly less often than other mice, even at exceedingly low light intensities (Dice 1947). An analysis of snails (*Cepaea nemoralis*) taken by thrushes clearly showed the selective advantage of cryptic coloration (Sheppard 1953a), as has Kettlewell's work (1961) on melanistic moths.

Whether or not predator selection is the complete answer, convincing though it seems, is not certain. It is somewhat disconcerting that some of the naturalists who have the greatest experience with desert animals deny that selection pressure by predators could have anything to do with the formation of concealingly colored substrate races (Heim de Balsac 1936; Kachkarov and Korovine 1942; Meinertzhagen 1954; Hoesch 1956). These authors insist that predators are exceedingly rare in the deserts and account for only a small fraction of mortality, "too small to be of selective significance." Niethammer (1959), in a carefully reasoned analysis, refutes the validity of this assertion. That there are complexities, however, is indicated, for instance, by the difference in the degree of localization of substrate races between African and Asiatic larks (Vaurie 1951a:442–446).

(4) Geographic Variation of Balanced Polymorphism

Most polymorphism found in nature, as we saw in Chapter 9, is due to a delicate balance between three (or more) genotypes, the heterozygote *Aa* being on the average selectively superior to the two classes of homozygotes, *AA* and *aa*. This delicate balance tends to change with the seasons and one would expect it to be somewhat different from locality to locality. In other words, one would expect the relative fitness of the three genotypes *AA, Aa,* and *aa* to be subject to geographic variation. This is indeed the case; geographic variation has been found in every adequately studied case of polymorphism. To list the examples of geographically variable polymorphism would be to list all known cases of polymorphism. Bateson (1913:121) already knew quite a few. Others have been listed in the reviews of Mayr (1942, 1951a), Ford (1945), Huxley (1955a,b), Voipio (1950), and Andrewartha and Birch (1954:681), and in other papers on polymorphism cited in Chapters 7 and 9. Among

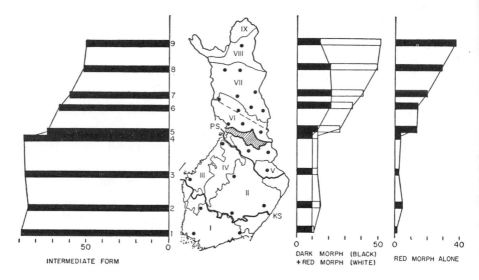

INTERMEDIATE FORM

DARK MORPH (BLACK)
+ RED MORPH (WHITE)

RED MORPH ALONE

Fig. 11–5. Geographic variation of polymorphism in the squirrel *Sciurus vulgaris* in Finland. Areas I–IX are the major vegetation belts. The shaded area separates south Finland, with a low frequency of the dark and the red morphs, from north Finland, with a much higher frequency of these morphs. (For details see Voipio 1957.)

the more detailed studies of geographic variation of polymorphism in individual genera or species one may mention studies on the hamster (Gershenson 1945), the squirrel *Sciurus* in Finland (Voipio 1957; Fig. 11–5), the tooth structure of the vole *Microtus arvalis* (Zimmermann 1952; Stein 1958), the wheatears *Oenanthe* (Mayr and Stresemann 1950), the bridled guillemot *Uria* (Southern 1939), the platyfishes *Xiphophorus* (Gordon 1947), bumblebees *Bombus* (Reinig 1939), the lady beetle *Harmonia axyridis* (Dobzhansky 1933; Komai and Hosino 1951), the gene arrangements in *Drosophila* (Dobzhansky 1951; Patterson and Stone 1952), and the crustacean *Sphaeroma* (Hoestlandt 1958).

A number of generalizations emerge from these studies. There are some species that show a rather even degree of polymorphism throughout their ranges, others that are highly polymorphic in some areas and rather uniform in others. Among the rather evenly polymorphic species are, for instance, *Oenanthe pleschanka, hispanica, picata,* and *monticola* (Mayr and Stresemann 1950); or in the genus *Drosophila, D. willistoni* (Dobzhansky 1957b) and *subobscura* (Stumm-Zollinger 1953). The same four gene arrangements of *D. willistoni* that were most common in southern Brazil were also most common in Cuba and Florida (Townsend 1952; Dobzhansky 1957b). A different pattern occurs in the polymorphic moth *Zygaena ephialtes,* which shows much regional fixation. In Italy and Yugoslavia only the yellow morph with colored hindwings is found, in the western and northern parts of the range (from France to Russia) only the red form with black hindwings. The species is highly polymorphic in the intervening areas (Bovey 1941).

If one singles out an individual morph gene and follows it through the range of a species, one can make the following generalizations: (1) rare genes often have very restricted, local distributions; the same gene may reappear in far distant parts of the species, usually with equal rarity; (2) where the species range is essentially continuous, gene frequencies change usually clinally and the morph clines generally run parallel to climatic gradients (Komai, Chino, and Hosino 1950; Fig. 9–3); (3) at the periphery or in isolated parts of the species range some genes are generally lost while others may reach "fixation" (a frequency of 100 percent).

The adaptive nature of this variation is in most cases merely a hypothesis, being based on the known selective differences among genotypes in the cases of balanced polymorphism. Geographically variable polymorphism was a rather puzzling phenomenon before pleiotropy was fully understood and before it was realized that the visible portion of the phenotype determines only a part of the selective value of a genotype. How important the invisible physiological component of the phenotype is, is demonstrated by the often striking selective differences between heterozygotes and visibly indistinguishable dominant homozygotes (see Chapters 7 and 9 for a fuller discussion of cryptic pleiotropic factors).

There are some cases, however, in which the geographic variation of the visible phenotype itself is clearly adaptive. This is true of some of the microgeographic variation of such substrate-adapted polymorphic species as the snails *Cepaea nemoralis* and *Littorina obtusata.* It is even more true of mimetic polymorphism, where the frequency of the mimetic types in different regions is determined by the local frequencies of the model species. The parallelism in the geographic variation of models and mimics is particularly compelling evidence of the sensitivity of the selection process.

CONCLUSIONS

The findings emerging from the study of geographic variation have had a decisive impact on the development of evolutionary thought. The assertion of the early Mendelians that mutations are something drastic and disruptive and without any relation to selection forced students of geographic variation to adopt a Lamarckian interpretation. The smallness of the geographical differences, the gradualness of the changes from population to population, the obvious correlation with factors of the environment, in short all the findings of the students of geographic variation, negated De Vries' mutation theory. When the geneticists themselves proved the error of the mutation theory, and not only adopted the idea of small mutations, but also began to admit the selective importance of the environment, there was no longer any obstacle to a reconciliation of geneticists and students of geographic variation. The latter gave up their Lamarckian interpretation except for a few who are still unaware of De Vries' refutation by modern genetics.

The human mind finds it difficult to focus simultaneously on two separate phenomena or on two different aspects of a single phenomenon. The result is that we have in evolutionary biology many spurious dichotomies: ecological and geographical races, ecological and geographical barriers, or adaptive and phylogenetic variation, to mention a few. Goldschmidt (1940, 1948a) fell into this error when he considered the role of geographic variation in speciation. He emphasized quite correctly that "most, if not all, differential traits of subspecies and still lower categories are directly or indirectly adaptational," but then concluded that therefore they cannot have anything to do with speciation and phylogenetic evolution. This conclusion was based on the arbitrary assumption that an organism has two sets of genetic properties, one dealing with "existential adaptation," as he called it, and another dealing with genetic changes of evolutionary significance. Actually, all the available evidence indicates that there is no such distinction. An organism has only a single genotype. Adaptation to local conditions and evolutionary change are two aspects of the same genetic phenomenon, the continuous adjustment of an integrated gene complex to a changing environment.

The study of geographic variation has resulted in a number of well-established conclusions.

(1) Every population of a species differs from all others genetically and, if sufficiently sensitive tests are employed, also biometrically and in other ways.

(2) The degree of difference between different populations of a species ranges from almost complete identity to distinctness almost of species level.

(3) The area occupied by superficially identical populations may be extremely small, as in some land snails, or cover the entire species range, as in some sibling species.

(4) The various characters of a species may and usually do vary independently. Neighboring populations agree, therefore, in some characters and differ in others.

(5) All characters employed to distinguish species from each other are known also to be subject to geographic variation.

(6) The characters of a given population have at least in part a genetic basis and tend to remain in most cases rather constant through the years.

(7) Geographic variation as a whole is adaptive. It adapts each population to the locality it occupies. However, not all the phenotypic manifestations of this genotypic adaptation are necessarily adaptive.

(8) The ecotypic adaptation of local populations is a centrifugal evolutionary force. It leads to an increased genetic diversity of the species and results, as a by-product of gene flow, in a continuous readjustment of the local gene complexes.

Geographic variation is a population phenomenon that has greatly contributed to our understanding of the nature of species. It demonstrates the invalidity of the typological concept of species. It permits the conclusion that much, if not all, of the variation is adaptive in response to the varying environmental demands. Some variable components of the phenotype, such as general size, proportions, and general coloration, are usually clearly adaptive. The geographic variation of other components, such as certain color patterns, does not seem directly adaptive, but it can often be shown by an appropriate analysis that these "neutral" phenotypes are the outward manifestations of genotypes that simultaneously control cryptic physiological characters established and maintained by natural selection.

The adaptive response of the different populations of a species affects in many ways the structure of a species as a whole. These various aspects of the population structure of species will be discussed in Chapter 13.

12 ~ The Polytypic Species of the Taxonomist

The concept of the nondimensional species has unqualified validity and usefulness only for the local naturalist. It is only in the local situation that the "species" coincides with the "deme," the local population. Combining the properties of a species and of a single population, the nondimensional species has a simplicity the virtues of which the naturalist untiringly extols. Its strength, however, is also its weakness: the applicability of this species concept is strictly limited to the local situation.

That the nondimensional species was an oversimplification became apparent very soon after its introduction into the biological literature, and was indeed already apparent in the days of Linnaeus. Two developments in particular contributed to the gradual undermining of the concept of the typological, monotypic species. Both were consequences of the extensive explorations in the eighteenth and nineteenth centuries, which converted the local naturalist into a widely traveled explorer. This explorer discovered new local populations that deviated somewhat from the population occurring at the type locality of a previously known species. Linnaeus (and his followers) coped with this discovery of geographic variation by calling such deviating populations "varieties," like any other deviation from the type of the species. He used this same term, then, for geographic races, for races of domestic animals and plants, and for nongenetic variants, as well as for the genetic "sports" of the horticulturists (see below, "The Variety"). As an example we may quote Linnaeus' (1758) treatment of the races of man. He recognizes under *Homo sapiens* six varieties, four of which are geographic races, (A) *Americanus*, (B) *Europaeus*, (C) *Asiaticus*, (D) *Afer*, while the other two are monstrosities or mythical.

Pallas, only a little later, used the term *varietas* even more frequently and in most cases in the sense of geographic race. In due time more and more species of the taxonomists contained "geographic varieties."

Esper, in his 1781 essay, *De varietatibus* (p. 18), was apparently the first to make a deliberate distinction between "accidental varieties," which he called *varietates*, and "essential varieties," which he called *subspecies*. In practice the "essential varieties" were geographic races, the subspecies of modern terminology. The use of geographic "varieties," races, or subspecies in animal taxonomy spread only slowly at first. However, beginning with H. Schlegel (1844), the subspecies was used with increasing frequency in the better-known groups of animals, resulting in the recognition of more and more polytypic species.

The second factor undermining the monotypic species concept was likewise rooted in "geographical collecting." As the fauna of the world became better known, it happened more and more often that two allopatric species, originally thought to be completely distinct, were found to be connected by intermediate, intergrading populations. The honest systematist had no choice but to reduce the two "species" to the rank of subspecies and to combine them, together with the intermediate populations, into a single, widespread polytypic species.

Let us illustrate this with the concrete example of the birds of the *Passerella* (+ *Melospiza*) group. Four similar species were discovered in eastern North America by the early explorers and pioneer ornithologists: the Fox Sparrow (*Passerella iliaca* Merrem 1786), the Swamp Sparrow (*P. georgiana* Latham 1790), the Song Sparrow (*P. melodia* Wilson 1810), and the Lincoln's Sparrow (*P. lincolni* Audubon 1834). During the exploration of the West, in the middle of the nineteenth century, several additional forms of *Passerella* were discovered, for instance on Kodiak Island (*insignis*), in Alaska (*rufina*), in California (*gouldi*), and in Arizona (*fallax*). These forms were described as "species" because to their describers they seemed as different from each other as the four original species of eastern North America. However, as the ornithological exploration of North America continued, additional populations were found that were intermediate between these four western "species," and between them and the Song Sparrow (*melodia*) of eastern North America. As a result, all five "species" were finally reduced to the rank of subspecies and combined into a single polytypic species, the Song Sparrow (*Passerella melodia*), comprising more than 30 subspecies (Fig. 12–1).

A classical example of the fusion of allopatric "species," resulting from

Fig. 12–1. Distribution pattern of a polytypic species of bird. Numbers refer to the breeding ranges of the 34 subspecies of the Song Sparrow, *Passerella melodia*. (From Miller 1956.)

the study of population chains, is *Murella*, the *iberus* land snail of Sicily (formerly included in *"Helix"*). Kobelt (1881) studied these snail populations and described his findings graphically and charmingly:

As a result of the methodical exploration of the mountains encircling the Conca d'Oro [the valley in which Palermo is situated], I came irresistibly to the conclusion that the changes in the snails which live in these mountains take place in a very definite direction and according to definite rules. From San Ciro in the east to Cato Gallo in the west, the height of the snail shells increases regularly, the whorls become more bulbous, and the peristome is turned farther back. *Helix globularis* Ziegler becomes *Helix platychela* Menke *typica*, and this transforms westward into *Helix sicana* by becoming ever higher and more bulbous, while in a different direction *Helix rosaliae* evolves from it. . . .

The anatomical study of living material which Mr. Wiegmann had the kindness to undertake failed to reveal any differences. There can be no question whatsoever that *Helix globularis, platychela*, and *sicana*, and of course also *muralis*, are only forms of a single species [p. 62].

I am sure I can assert confidently that additional investigations in the mountains of Sicily will reveal additional intermediate forms and that finally all these snails from western Sicily will have to be combined into a single species which, however, will be a far cry from the current species concept. There will remain for southern Italy only two Formenkreise of *iberus* (snails), which, however, are well separated according to my experiences: that of *Helix strigata*, in which I also include *umbrica, surrentina, carseolana, Mariannae, signata*, and in Sicily *nobrodensis* and *Huetiana*; and that of *muralis* with *globularis, platychela, sicana, provincialis, Ascherae, Tiberiana* and so on until *scabriuscula* [p. 65].

What should the systematist do with such a Formenkreis? They are found not only in the *iberus* snails but also among the *Macularias*, the *Levantinas*, the *Campyleas*, etc. . . . With sufficient material large groups of species will merge into single collective species which would make the classification exceedingly difficult. For this reason I obey a simple, practical rule, no matter how unscientific it may be. I call a good species what I can diagnose without long and careful comparisons and measurements. That which I can distinguish only by precise measurements I call a variety. This simple rule has so far always served me well [p. 66].

Kobelt reveals clearly the conflict between the morphological and the biological species definitions. Whenever such a conflict appeared in the early days of systematic zoology, it was nearly always decided in favor of the morphological species definition.

The full history of the development of the new species concept has not yet been written. Much relevant material has been gathered by Plate (1913), Rensch (1929, 1934), Stresemann (1951), Mayr (1942, 1957a), and other authors listed in their bibliographies. A full history would fill a volume and the following comments are meant to be merely an introduction to this vast subject. The change from the typological-morphological species to the biologically defined polytypic species occurred gradually and at different rates in different areas of taxonomy. It is not yet clear to what extent various authors arrived at their ideas independently or were influenced by others.

As far as ornithology is concerned, this development can be traced in Gloger (1833), who tried to systematize the rules of geographic variation in coloration; in Schlegel who, beginning in 1844, consistently applied trinomials to allopatric populations; in Cassin, Baird, and Coues, who used trinomials extensively and reduced to the rank of subspecies all intergrading "species"; in Seebohm (1888), who applied the concept of geographic speciation to the study of geographic variation in birds; and finally in Kleinschmidt and Hartert, who added the principle of geographical representation as a criterion of conspecificity. Any study of geographic variation was bound to result in questioning the Linnaean species concept, and thus all the authors mentioned in Chapter 11 as pioneers in the study of the geographic variation of mammals, fishes, snails, and insects are likewise important in the history of the polytypic species concept. Among the students of insects, contributions to the development of the concept were made, directly or indirectly, first by Weismann, Eimer, and Staudinger, and later particularly by Karl Jordan. In a classic series of generic and family revisions, Jordan arrived at a species concept that has served as a model for all later workers (Mayr 1955b).

PROBLEMS OF THE POLYTYPIC SPECIES

The shift from the nondimensional to the multidimensional species poses practical and conceptual problems. The two concepts are radically different (Mayr 1957a): the nondimensional species is objectively (nonarbitrarily) defined by the gap separating it from other (sympatric) species, while the polytypic species is characterized by an actual or potential genetic continuity of allopatric populations. This can be determined only arbitrarily in many cases. The nondimensional species is a singular phenomenon; the multidimensional species is a group phenomenon, a collective phenomenon.

Most taxonomists have closed their eyes to the consequences of this change, and continue to use the term "species" even though it has come to mean something very different from the "species" of Linnaeus. Kleinschmidt (1900) was perhaps the first to acknowledge the change. He argued, justifiably, that the polytypic species was in the nature of a higher category since it was formed very often by the combination of many local species, as, for instance, in the combination of the various "species" of sparrows into the polytypic Song Sparrow. He contended that it would be misleading to use the Linnaean term "species" for this collective "higher" category, and proposed for it the term *Formenkreis*. Rensch (1929) pointed out that this term had been used earlier in paleontology for a different phenomenon and also that Kleinschmidt did not clearly distinguish between true polytypic species and groups of allopatric species. He therefore introduced the term *Rassenkreis* for polytypic species and restricted the term "species" to monotypic species. This proposal of a dualistic nomenclature found few adherents, particularly since it was soon realized that even most so-called monotypic species are aggregates of genetically different populations and, thus, collective categories. The term *Rassenkreis* has the additional disadvantage that its English equivalent ("circle of races") suggests a circular arrangement of races and has therefore often been applied incorrectly to cases of circular overlap. The convenient terminology "monotypic species" and "polytypic species," first proposed by Huxley (1940), has now been almost universally adopted. A polytypic species may be defined as a species that contains two or more subspecies; a monotypic species as a species not divided into several subspecies.

THE OCCURRENCE OF POLYTYPIC SPECIES IN THE ANIMAL KINGDOM

The occurrence of polytypic species has been established for most groups of animals. Abundant examples have been listed by Rensch (1934, 1947), Huxley (1940, 1942), Mayr (1942), Schilder (1952), and other authors. More specific reviews were made for birds (Mayr 1951a), reptiles (Laurent 1952), game animals (Voipio 1950), butterflies (Remington 1951), insects (Hubbell 1956), and fresh-water animals (Brooks 1957b), to mention just a few. Some of the best taxonomic publications of today are devoted to detailed revisions or monographs of polytypic species. To list them all would entail listing a major part of the modern taxonomic literature. Those that have recently come to my attention deal with mammals (Hall 1951; Blair 1950), birds (Miller 1941; Vaurie 1949; Galbraith 1956), reptiles (Klauber 1946, 1947; Rodgers and Fitch 1947), frogs (Moore 1944), salamanders (Hairston and Pope 1948), fishes (Gordon 1947; Miller 1948), almost all groups of insects, terrestrial isopods (Vandel 1951, 1953; Fig. 12–2), crustaceans (d'Ancona 1942; Tonolli 1949;

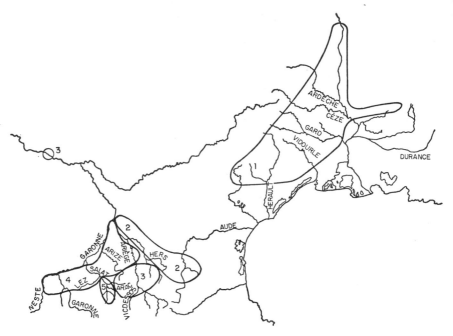

Fig. 12–2. A polytypic species of terrestrial isopod (*Phymatoniscus*) from the south of France. (From Vandel 1953.)

Kiefer 1952; Penn 1957; Johnson 1960), millipedes (Hoffman 1951), scorpions (Vachon 1958), spiders (Holm 1956), and many mollusks. Other instances are cited in Chapter 11.

Rensch (1929, 1947) has repeatedly called attention to the equally widespread occurrence of polytypic species among marine animals. This fact is largely concealed by the current nomenclature in most marine groups. Allopatric populations are usually listed as "varieties" if they are only slightly different, and as full species if they differ sufficiently to be diagnosed by key characters. There is great need to follow up the pioneer work of Döderlein (1902) and unite marine animals into comprehensive polytypic species, although the work of Tortonese (1950) and his associates indicates that some of the "subspecies" of Döderlein are merely individual variants. A cursory survey of the shallow-water echinoids of the West Indies revealed that with one exception all of them were members of polytypic species (Mayr 1954b). A study of Mayer's monograph of the medusae (1910) shows that the same is true for this group of marine organisms.

It is now well established that the occurrence of polytypic species is a universal phenomenon in the animal kingdom. There is no known tribe or family in which polytypic species have not been found when looked for. They are known even in such morphologically uniform animals as *Drosophila* (Patterson and Stone 1952). Claims of some contemporary authors that polytypic species are absent in the families in which they specialize are based either on an insufficient knowledge of these families (not enough samples of peripheral or peripherally isolated populations) or on the tendency of some of these authors to call every morphologically distinct geographical isolate a species.

The phenomenon of polytypic species is not restricted to animals. Most species of plants, unless highly localized, show some geographic variation, and this variation is frequently sufficient to justify nomenclatural recognition of local or regional races. The frequency of polytypic species among plants is sometimes overlooked since it is overshadowed by more conspicuous forms of variation, such as those caused by polyploidy, apomixis, and hybridization. It is therefore only recently that botanists have followed up the pioneer work of Wettstein and Cajander. As in animals, polytypic species are most frequent where ranges are insular, that is, in mountain areas and in archipelagos (see also Stebbins 1950; Grant 1952a, 1957; Lewis 1956).

THE VALUE OF THE CONCEPT OF THE POLYTYPIC SPECIES

Combining numerous more or less isolated and morphologically distinct allopatric "species" into polytypic species had not only the practical result of greatly improving the classification of animals, but also paved the way for a better understanding of the process of speciation (Chapter 16).

Simplification of the System. The most immediate and conspicuous effect of the method is a simplification of the system. This is best illustrated by a few figures. The last complete listing of the birds of the world (Sharpe 1909) gave about 19,000 full species. The arrangement of these species (and many hundreds discovered since 1910) into polytypic species has reduced the total number to about 8600 (Mayr and Amadon 1951). Hall (1951) was able to combine 22 described species of North American weasels into four. Several thousand "species" of Eurasian mammals were combined by Ellerman and Morrison-Scott (1951) into about 700. Some 30 or 40 "species" of Holarctic *Coregonus* were reduced by Svärdson (1957) into five species. Of much greater significance is the restoration of biological meaning and homogeneity to the species category. There are some 28,000 subspecies in the 8600 species of birds, and it is evident that it would represent a rather misleading systematic evaluation if the same rank were given to these subspecies as to full species. Yet such unequal treatment is the rule in many of those groups in which the polytypic species concept has not yet been adopted, and where every isolated population is ranked as a full species.

An additional benefit of the adoption of the polytypic species is that it has contributed to a re-evaluation and simplification of the generic classification. For instance, as soon as the 19,000 nominal species of birds were grouped into polytypic species, it became evident that many of the then accepted genera were monotypic. Each such genus was nothing more than a designation for a morphologically distinct species. The genus, thus defined, coincided with the polytypic species, and failed to perform its true function, namely, that of indicating relationship among species. To restore the balance, a lumping of avian genera on a vast scale became necessary, reducing them from between 6000 and 7000 (the number recognized by several authors around 1920) to about 1700 or 1800 (the number currently recognized). In the course of the re-evaluation of the genus it was possible to show that "morphological" differences distinguish almost any two species (except sibling species), and that mere morphological difference is therefore a useless generic criterion. Indeed, rather striking morphological differences can sometimes be found even among different subspecies of a single polytypic species. Defining the genus as a group of related species permits considerable simplification of the generic classification. It led, for instance, in the opilionid families Phalangodidae and Cosmetidae to a reduction in the number of genera from 108 to 11 (Goodnight and Goodnight 1953). It would not be surprising if the application of the polytypic species concept to fossil animals, particularly to fossil invertebrates, were to lead to a simplification on the level of species and genus similar to that which neontology has experienced.

Heuristic Value. The application of the concept of polytypic species has been particularly fruitful in the elucidation of complex taxonomic situations. In each case it forces an unequivocal decision as to whether or not two forms should be considered conspecific. The need for sorting large numbers of "nominal species" and "varieties" into polytypic, biological species has not only resulted in a great refinement of the taxonomic technique, but also revealed the pathway by which populations reach species level. The criteria by which previously separated species are combined into polytypic species are discussed in the taxonomic literature (Rensch 1934; Mayr, Linsley, and Usinger 1953). Among the questions that need to be considered are the following: whether populations are sympatric or allopatric; whether, when allopatric and in contact with each other, they show indications of gene flow (interbreeding) in the zone of contact; and whether, when completely isolated, they show the degree of morphological distinctness typical of unquestionable sympatric species in this group. With due consideration of these criteria, in most cases the skein of relationship can be disentangled, and those groups of allopatric forms that should rightly be considered polytypic species can be distinguished. In some cases relationships are so complex that a satisfactory arrangement can be reached only slowly, through trial and error. The western North American garter snakes of the *Thamnophis ordinoides* complex illustrate such a case. In a very thorough analysis, Fitch (1940) recognized eleven forms which he classified in three species: *ordinoides* (nine subspecies, most with overlapping ranges), *hammondii*, and *digueti*. Mayr (1942) reanalyzed Fitch's data and, by applying the polytypic species concept, divided the complex very differently into three polytypic species: *ordinoides*, *couchii* (including *hammondii* and *digueti*), and *elegans*. This arrangement eliminated the illogical occurrence of ten areas with sympatric, reproductively isolated "subspecies," but had to allow for two zones of reputed interspecific hybridization. One of these was eliminated by Fox (1948a), who showed that there was no intergradation between *ordinoides* and "*atratus*" of authors (= *terrestris*), and that the latter belongs with *elegans*. This left only an area of hybridization in the Klamath River drainage (Fig. 12–3). The intergradation between *biscutatus* and *elegans* and the free hybridization between *elegans biscutatus* and *couchii hydrophila* prove that the reproductive isolation between *elegans* and *couchii* has broken down completely. The species *couchii* is, in addition, a good example of circular overlap, since the terminal links of a chain or races, *atratus* and *hammondii*, now overlap in Monterey and San Luis Obispo counties without interbreeding (Fox 1951). By studying the behavior of each population toward other allopatric or sympatric populations, it was thus possible to arrive at a much more balanced picture of relationships.

Limitations and Exaggerations. A discussion of the polytypic species would be incomplete if it failed to point out the limitations of this method. Some authors have gone to the extreme of assuming that allopatry by itself is proof of conspecificity. This is not the case. A geo-

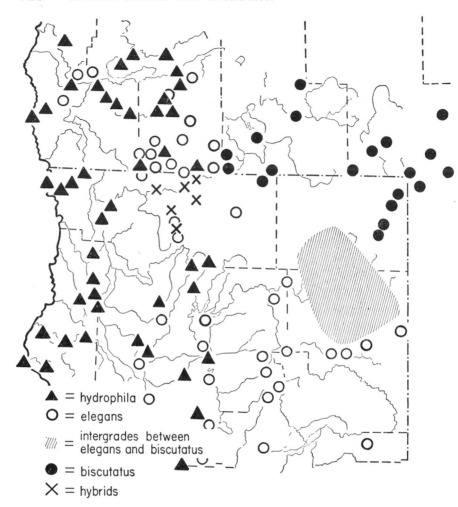

▲ = hydrophila ○ = elegans

○ = elegans

▨ = intergrades between elegans and biscutatus

● = biscutatus

✕ = hybrids

Fig. 12–3. Distribution of two species of garter snakes (*Thamnophis*) in northern California and southern Oregon. The two species are widely sympatric without interbreeding but form hybrid populations (*X*) in the Klamath area. It is possible that *biscutatus* owes some of its characters to introgression from *hydrophila*. (Original, after Fox 1951.)

graphically isolated population may or may not be conspecific with the most closely related allopatric population. The mere fact of geographic isolation supplies only ambiguous information about reproductive isolation, which alone is the criterion of specific distinctness.

There is no doubt that the application of the polytypic species was carried too far in many cases when the concept was consistently applied to a group of animals for the first time. Mayr (1951a) lists about a dozen pairs of species of birds that were at one time considered conspecific; Vaurie (1955) adds several others. In most of these cases the two species are either completely allopatric or overlap only in a narrow border zone without any evidence of intergradation. This complete lack of intergradation in a continental area, *in the absence of geographic barriers*, is now taken as proof of specific distinctness. In his revision of the butterfly *Papilio machaon*, Eller (1939) included a number of North American forms in this polytypic species. However, a genetic analysis by Clarke and Sheppard (1955) showed that several of these forms had already reached species rank. In a magnificent synthesis, Ellerman and Morrison-Scott (1951) have sweepingly arranged the mammals of Eurasia and the Oriental Region into polytypic species. Here again it is rather evident that some of the lumping has gone too far. In island forms, likewise, it can sometimes be shown that forms which on first sight appear to be geographic representatives are actually not so closely related (Myers 1950). It is evident from these instances that the productive working hypothesis that allopatry indicates conspecificity must be tested in every case with the help of all possible corroborative evidence. Though usually reliable, allopatry is not an infallible indicator of conspecificity.

The method of classifying populations into polytypic species is most apt to break down where a species or species group is actively evolving. The *Thamnophis ordinoides* species group, discussed above, is a case in point. In island regions, the decision as to which isolates are to be combined into polytypic species is often quite arbitrary. The difficulties are particularly apparent where subgroups of a species have changed their ecological requirements and have, by renewed range expansion, come into secondary contact with other subgroups of the species. Behavior of the meeting populations is frequently unpredictable. Furthermore, forms that behave toward each other like good, reproductively isolated species are sometimes more similar phenotypically than are intergrading or interbreeding forms. Such a situation exists in many species groups of South Sea Island birds. Examples are *Rhipidura rufifrons* (Mayr and Moynihan 1946), *Edolisoma tenuirostris-morio* (Mayr MS), *Pachycephala pectoralis* (Galbraith 1956), and *Ptilinopus* (Cain 1954b). The cases of circular overlap (Chapter 16) are further instances of the difficulties that evolution occasionally presents to those attempting to apply the polytypic species concept.

THE TERMINOLOGY OF THE SUBDIVISIONS OF THE SPECIES

As implied in its name, the polytypic species has subdivisions. The kinds of subdivisions distinguished by various authors and the terms proposed by them are influenced by the philosophies of these authors. Some have as their ultimate objective the facilitation of the purely pragmatic, formalistic task of classifying specimens; others attempt to find units with specific biological or evolutionary significance. Although we are mainly concerned with the evolutionary aspects of species structure, we cannot entirely avoid a discussion of terms with a purely practical taxonomic significance, since these terms (for instance, the term "subspecies") have been employed a great deal in the evolutionary literature.

The Variety

The variety (*varietas*) was the only subdivision of the species recognized by Linnaeus and the early taxonomists. A variety was anything that deviated from the ideal type of the species. In his *Philosophia Botanica* (1751, no. 158) Linnaeus characterized it as follows: "There are as many varieties as there are different plants produced from the seed of the same species. A variety is a plant changed by an accidental cause: climate, soil, temperature, winds, etc. A variety consequently reverts to its original condition when the soil is changed." A variety thus defined is a nongenetic modification of the phenotype. His discussion of varieties in the animal kingdom (no. 259) indicates that Linnaeus included under the term "variety" not only nongenetic climatic variants but also races of domestic animals and intrapopulation genetic variants. As examples of varieties in the animal kingdom he lists "white and black cows, small and big ones, fat and lean ones, smooth and wooly ones; likewise the races of domestic dogs." An analysis of varieties actually recognized by Linnaeus in his taxonomic writings indicates that they were a highly heterogeneous lot of deviations from the species type (Clausen 1941). Only some were genuine geographic races or subspecies. The term "variety" was plagued with an inherent ambiguity. Much of the argument in the nineteenth century (for example Gloger 1833) over the biological and evolutionary meaning of the variety (see also Darwin's argument with Wagner, Mayr 1959a) was due to the fact that the term concealed two entirely different phenomena: (*a*) individual variants within a polymorphic population; (*b*) distinguishable population in a polytypic species.

It became increasingly evident that this confusion could be resolved only by restricting the term "variety" to one of the two components or by abandoning it altogether. When the three great systematists of the Tring Museum (Rothschild, Hartert, and Jordan) founded a new biological journal, they introduced it with the following editorial preamble (1894, *Novitates Zoologicae*, 1:1):

The term variety, especially among entomologists, has been indiscriminately used to denote an individual variation within a species as well as climatic or geographical races. We therefore, to avoid all possible errors, have determined to discard the term variety altogether. To denote individual variations we shall in this periodical employ the word aberration, and for geographical forms which cannot rank as full species the term subspecies.

The abandonment of the term "variety" for geographic races is nearly complete in zoology. A similar movement in botany has not achieved unanimous acceptance.

The Subspecies

The term "subspecies," when it came into general usage in taxonomy during the nineteenth century, was a replacement for "variety" in its meaning of geographic race. As a consequence the term "subspecies" was from the very beginning endowed with all the typological shortcomings of the term "variety." It was considered a taxonomic unit like the morphological species, with the same objectives but on a lower taxonomic level. Corresponding to the definition of the species as comprising those individuals that conform to the type of the species, the subspecies was defined as those individuals that conform to the type of the subspecies. This definition induced many authors to compare carefully material from every newly established locality with specimens from the type locality of a previously described subspecies. Whenever a thorough biometric-morphological analysis established a mean difference between the samples, this was considered sufficient justification by these authors to describe a new subspecies. In the more intensely studied groups of animals this approach has led to a wild-goose chase for new subspecies, and has seriously impaired the usefulness of the subspecies category.

The method is unsound not only because it leads to practical difficulties but also because its underlying philosophy is fallacious. Species are not composites of uniform subtypes, subspecies, but consist of an almost infinite number of local populations, each in turn (in sexual species) consisting of genetically different individuals. The difficulties of the subspecies concept are intensified by persistent attempts to consider the sub-species not merely as a practical device of the taxonomist, but also as a "unit of evolution." The better the geographic variation of a species is known, the more difficult it becomes to delimit subspecies and the more obvious it becomes that many such delimitations are quite arbitrary. Wilson and Brown (1953) have pointed out that four characteristics of geo-graphic variation contribute to these difficulties in delimitation: (1) the tendency of different characters to show independent trends of geographic variation (see Chapter 13); (2) the independent reoccurrence of similar or phenotypically indistinguishable populations in widely separated areas ("polytopic subspecies"); (3) the occurrence of microgeographic races within formally recognized subspecies; and (4) the arbitrariness of the degree of distinction selected as justifying subspecific separation of slightly differentiated local populations. Each of these phenomena shows how misleading is the excessive splitting of species into subspecies, par-ticularly where contiguous populations are connected by character gradients. They also show that the subspecies, which conceals so much of the inter- and intrapopulation variability, is an altogether unsuitable cate-gory for evolutionary discussions; the subspecies as such is not one of the units of evolution. Yet it is this very oversimplification that makes the subspecies such a useful tool for the classifier and that accounts for the reluctance of the practicing taxonomist to give up this convenient pigeon-holing device.

The modern definition of the subspecies is exceedingly different from that of the Linnaean geographic variety. It attempts to meet the various objections listed above and may be worded as follows: *A subspecies is an aggregate of local populations of a species, inhabiting a geographic sub-division of the range of the species, and differing taxonomically from other populations of the species.*

It is important to emphasize certain aspects of this definition:

(1) A subspecies is a collective category because every subspecies consists of many local populations, all of which are slightly different from each other genetically and phenotypically. (See also Hubbell 1954; Doutt 1955.)

(2) Every subspecies has a formal name (trinominal nomenclature) and it would obviously lead to nomenclatural chaos if every slightly dif-fering local population were to be dignified by a trinomen. Therefore subspecies are to be named only if they differ "taxonomically," that is, by diagnostic morphological characters. How great this taxonomic differ-

ence ought to be can be determined only through agreement among working taxonomists (Mayr, Linsley, and Usinger 1953).

(3) Although it is usually possible to assign populations to subspecies, this is not necessarily possible for individuals, in view of the individual variability of each population and the overlap of the curves of variation of adjacent populations.

(4) A subspecies inhabits a definite geographic subdivision of the range of the species, a necessary consequence of the fact that subspecies are composed of populations and each population occupies part of the range. The distribution of the subspecies will be determined largely by the correlation between the diagnostic characters and the environment; consequently the range of a subspecies may sometimes be discontinuous (polytopic subspecies).

In the case of parasites, in which the principal spatial factor is the separation on different host species, host range takes the place of geographic range, and most subspecies are host races. At the present we know relatively little about these so-called host races. Some seem to be merely nongenetic modifications; others truly correspond to subspecies (Hoare 1952); and still others may actually be sibling species. Their true biological status can be established only by detailed analysis or by experiment.

All attempts either to replace the subspecies by a different terminology or to abandon it altogether have been found unacceptable in taxonomic practice (Mayr 1954c; Inger 1961). The category subspecies continues to be a convenient means of classifying population samples in geographically variable species, in particular in those with phenotypically distinct geographical isolates. It must be realized at all times, however, that in many cases the subspecies is an artifact and that it is not a "unit of evolution." Nor should the subspecies be confused with phenomena of a very different nature, such as character gradients (clines).

The typological history of the subspecies concept is still apparent in much of the current terminology, as, for instance, in the frequently made statement that "subspecies A and B widely overlap in their area of contact." The term "overlap" is used here ambiguously. If the two populations actually coexist sympatrically without interbreeding, then they are not subspecies but species. If on the other hand the two "subspecies" of this statement are merely arbitrary segregates within a single interbreeding population, then they are not two distinct overlapping subspecies. One cannot assign individuals of a single breeding population to two different subspecies. A clear realization of the meaning of such terms as "population," "subspecies," "overlap," and "interbreeding" will lead not only to greater precision of statement but indeed, in some instances, to a more refined analysis.

One should consult the taxonomic literature (for example, Mayr, Linsley, and Usinger 1953) for a discussion of special problems of the subspecies category, such as the taxonomic treatment of slightly differentiated populations and the formal classification of geographical isolates. These questions are of little evolutionary interest.

Other Intraspecific Categories

Various other terms that have been used for intraspecific categories remain to be discussed. Many different meanings have been assigned to the term "race." In the combination "geographic race" it is sometimes synonymous with "subspecies." However, in many branches of taxonomy, as well as in daily life, the term "race" is used colloquially to designate populations or aggregates of populations within formally recognized subspecies. In the combination "ecological race" (see below) it designates ecologically differentiated populations; in the combination "microgeographic race" it applies to local populations. Reference has already been made to various terms (Chapter 7) that have been coined for local populations, such as *deme* (Gilmour and Gregor 1939), *natio* (Semenov-Tianshansky 1910), and *ethnos* (Vogt 1947).

The Temporal Subspecies

Subspecies have been treated in the preceding discussions as subdivisions of the species in the dimensions of longitude and latitude, that is, as spatial units. However, species are as polytypic in time as they are in space, requiring the recognition of temporal subspecies, that is, of subspecies within the time dimension of the multidimensional species. A biological difference between the geographical and the temporal subspecies does not exist.

The history of the adoption of the subspecies category in paleontology is similar to that in neontology, except that it lags behind owing to various practical difficulties. In paleontology, likewise, variety, a purely typological and ambiguous unit, was first used to designate deviations from the species type. More recently, however, paleontologists have found it increasingly useful to take advantage of the greater precision of expression that the subspecies terminology permits. This terminology helps to stem the inflation of purely morphologically defined "species" that threatens to make the classification of certain groups, particularly certain invertebrates, an impenetrable jungle for the nonspecialist. The advantages of this method have been pointed out by Simpson (1943, 1961), Newell (1947, 1956a), and Sylvester-Bradley (1951, 1956). Its application to the living and Tertiary members of the bivalve genus *Pecten,* by Fleming (1957), greatly facilitates understanding of the classification and evolution of this genus. It does not seem advisable to make a terminological distinction between geographical and temporal subspecies because it is usually quite impossible, when different subspecies of a fossil species are found at different localities, to determine whether or not they are precisely contemporary. Even when there is a sequence of subspecies at a single locality, it need not necessarily be purely temporal. Subspecies found in succeeding strata may actually be geographical races that replaced each other owing to climatic or tectonic changes.

It is easy to point out difficulties encountered in the application of the subspecies concept to fossil material. In addition to those mentioned above for neontological subspecies, there is the danger of differential deposition of different age classes and sexes, and the occurrence of nongenetic habitat forms. On the other hand, arranging the multitude of fossil morphospecies into biologically meaningful polytypic species with continuity in time and space will lead to a better understanding of relationships and of the meaning of evolutionary trends. Achievement of this objective justifies the occasional errors inevitably committed in assigning morphological "types" or "samples" to polytypic species. As in living species, it must be kept in mind at all times that the subspecies is merely a classificatory device.

The Ecotype

Rebelling against a purely morphological definition of species and variety, the botanist Turesson (1922) coined a new terminology. He proposed the term *ecospecies* (see Chapter 14) for the "Linnaean species from an ecological point of view" and *ecotype* for "the product arising as a result of the genotypical response of an ecospecies to a particular habitat." Turesson's revolutionary concepts and his experimental analysis of samples of wild plants have had an impact on plant taxonomy that can hardly be exaggerated. He inspired numerous studies of ecotypic characteristics of local populations which brought out the importance of the adaptive nature of much of geographic variation and greatly furthered our understanding of the population structure of species. The vigorous field of experimental taxonomy of plants arose largely out of this work. Turesson's method consisted in transplanting samples of individual plants from various localities within the species range to a uniform garden or greenhouse environment. The localities were selected, on the whole, on the basis of their strikingly different ecology, such as sea cliffs, moist shady lowland forests, sand dunes, and alpine tundra. When the transplanted specimens remained different under the conditions of identical cultivation they were considered to represent different ecotypes.

The picture presented by Turesson in his major 1922 publication brings out to a remarkable degree the strong genetic component of local races (still widely denied at that period) and the variability of each population. The word "ecotype" is not used in the main body of this paper (pp. 211–341). In Turesson's later writings, and even more so in those of some of his followers, there is an increasingly typological interpretation of ecotypes. This has led to a number of implicit or explicit beliefs concerning ecotypes, beliefs not supported by the available evidence:

(1) Turesson later contended that most species are mosaics of "discontinuous adaptational" ecotypes (Turesson 1936:422). This view was strongly influenced by his method of sampling strikingly different habitats. Indeed, in some cases his collecting stations were several hundred miles apart. However, he also sampled intermediate areas and the results strengthened his conviction of the discontinuity of ecotypes: "I continue to believe that the discontinuous variation is the most common type and that continuous variation when it occurs is found only in certain species with a very specific biology" (p. 422).

Other students raised objections: "As habitats do vary continuously, and as ecotypes are, according to the new theory, produced by the selective forces of habitat factors" (Faegri 1937), discontinuous ecotypes should be impossible by definition. Clinal rather than discontinuous variation has indeed been found in most cases where investigators made transects through series of adjacent populations, as Gregor (1946) did for *Plantago maritima* and Ehrendorfer (1953) for *Galium pumilum*. In a transect from waterlogged mud to fertile meadow, plant height in *Plantago* was found to increase gradually from 7.1 ± 0.23 inches up to 22.4 ± 0.42 inches (Gregor 1946). Indeed, such work confirmed the findings previously made by students of land snails and of *Drosophila* that no two populations are identical, and that there are character gradients wherever there are environmental gradients.

(2) In many presentations of the ecotype concept there is an unexpressed assumption that the phenotype responds to the environment as a unit. Observations prove, however, that different phenotypic characters may respond independently to various components of the environment, and that the delimitation of an ecotype may depend on the choice of the character,

(3) The terminology "sea-cliff ecotype" and "alpine ecotype" suggests a unity that does not necessarily exist. If, for instance, the sea-cliff populations of a number of localities are compared, it may be found that each possesses a number of peculiar "local" characters in addition to the ecotypic features of the sea-cliff habitat. There are indications that two modes of origin may be responsible for this mixture of ecotypic and local-geographical phenotypic characters. One kind of ecotype is an invader from far distant areas, and has acquired "local" characters by introgressing with neighboring species populations (see below). The other kind of ecotype is a local product of selection and shares much of its genetic endowment with phenotypically different neighboring populations (in different habitats) from which it was derived. It shares with other more or less distant populations of the "same" ecotype only a few conspicuous adaptive features. To combine such superficially similar and polyphyletically evolved populations into a single polytopic ecotype conceals some of the most interesting aspects in the dynamic history of species. In this the ecotype, as a biological concept, differs from the subspecies.

It seems probable that ecotypic variation is a highly variable phenomenon, the diversity of which has never been properly categorized. As a broad approximation one can distinguish three kinds of ecotypic variation in plants: (1) clinal variation in widespread outbreeders; (2) partially discontinuous variation in inbreeders of various types; (3) partially discontinuous variation in outbreeders as a consequence of secondary contact between populations.

Type 1 is discussed in Chapter 13. Localized ecotypes appear to be rare if not altogether absent in wind-pollinated species. The effect of breeding system on ecotypic variation (type 2) is discussed by Stebbins (1950). One of the reasons for a greater frequency of ecotypic variation in plants than in animals is the more frequent occurrence of extreme inbreeding among plants which favors the development of highly localized types. Phenotypic discontinuity owing to secondary contact (type 3) is well established in animals but has not been studied extensively in plants except recently in connection with introgressive hybridization (Heslop-

Harrison 1958). It appears that specialized habitats such as mountain tops or sea cliffs are not always colonized from adjacent habitats but sometimes from more distant mountain tops and sea cliffs. Many kinds of animals and plants seem to find it easier to overcome distributional barriers by long-distance dispersal than to invade adjacent areas if such invasion would involve a shift into a different ecological niche. As a result of such long-distance dispersal, two different "ecotypes" that have arisen independently and at a considerable distance may meet secondarily at their respective habitat borders and produce a narrow hybrid belt or zone of intergradation. It appears that secondary contact is the most probable interpretation of most cases of sharply discontinuous ecotypes.

Ecotype Versus Subspecies

Those plant taxonomists who divided species into ecotypes believed that this would permit a more precise and more biological description of geographic variation than the division of species into subspecies. Unfortunately, as we have seen, ecotypes are rarely discontinuous, rarely well delimited, often polyphyletic, and always full of intraecotype variability. The ecotype concept, thus, suffers from precisely the same weaknesses as the subspecies concept. However, the situation is even more serious in the case of the ecotype concept because the subspecies is admittedly an arbitrary instrument of the taxonomist, created purely for taxonomic convenience, while the ecotype was established for the very purpose of getting away from the artificiality of taxonomic categories and of replacing them by something more meaningful biologically. It seems doubtful that the ecotype has been successful in this respect. It would appear safer to describe the results of studies in experimental taxonomy in less rigorous terms, such as climatic race or edaphic race, than in terms of the ecotype with its typological implications. However, this terminological conclusion does not in the least deprive study of ecotypic variation of its value. Some botanists define subspecies as "morphologically distinct ecotypes," and suggest the existence of a class of "morphologically indistinguishable ecotypes." Actually, I do not know of a single ecotype described in the literature which, in addition to various physiological characters and potentialities, is not also characterized by morphological attributes.

For a further discussion of the ecotype concept and a review of some of the ecotypic research, one should consult the writings of Baker (1948), Clausen, Keck, and Hiesey (1948), Clausen (1951), Gregor (1947),

Gregor and Lang (1950), Gregor and Watson (1954), Stebbins (1950), and others.

The Ecological Race

This discussion of the biological and evolutionary meaning of the ecotype of botanists will have facilitated understanding of the phenomena to which the animal taxonomist applies the term "ecological race," for it is the exact counterpart of the ecotype.

Local populations that are particularly conspicuously adapted to a local habitat are often referred to as "ecological races." Not all populations thus designated in the literature are truly ecological races; some have recently been unmasked as sibling species, while others are non-genetic modifications of the phenotype (ecophenotypes, Chapter 7). There remains, however, a considerable residue of genuine ecological races, such as those cited by Stresemann (1943) and Mayr (1951a) for birds. Every field ornithologist is familiar with such cases. The Savanna Sparrow (*Passerculus sandwichensis*) of the eastern United States is found in coastal salt marshes and also on dry uplands in the interior. The Redwing (*Agelaius phoeniceus*) normally lives in cattail (*Typha*) swamps, but breeding colonies have been found in alfalfa fields and even in young pine plantations. The Swainson's Warbler (*Limnothlypis swainsoni*) lives in the cane brakes of the coastal marshes in the southern United States but also in the southern Appalachian highlands above 3000 feet in thickets of rhododendron, mountain laurel, hemlock, and American holly.

The question most often debated concerning the ecological race is whether or not it is a category distinct from the geographical race. This question is most easily answered by looking at some specific cases. If this is done, it is found that the two kinds of phenomena cannot be separated. For instance, all lava-adapted ecological races of mammals have well-defined geographic ranges and are thus also geographic races. Entire subspecies or parts of them may have such well-defined habitat preferences that they could just as rightly be called ecological races as geographical races. Among birds some races of the Song Sparrow (*Passerella melodia*), the Herring Gull (*Larus argentatus*) and the Mistle Thrush (*Turdus viscivorus*) may be cited as examples. Other cases were mentioned in Chapter 11.

The study of ecological races in animals has been unaccountably neglected so far. Habitat-restricted populations have been documented by

Stein (1959) for the European mole (*Talpa europaea*). Steinmann (1952) believes that fresh-water turbellarians are particularly prone to vary ecotypically and to form specialized populations under aberrant ecological conditions (caves, changed water chemistry or temperature, and the like). Papi (1954) found a brackish-water race (*breviorispina*) of an otherwise fresh-water species (*Castrada infernalis*) of rhabdocoel turbellarian. However, several so-called brackish-water "races" of marine gammarids have been unmasked as sibling species (Kinne 1954). Ecotypic variation appears to be particularly widespread among fresh-water organisms and has been described not only among planarians and various types of crustaceans, but also among rotifers (Pejler 1957a).

There is a special type of ecotypic variation, the polytopic race, that does not fit too well into the stated picture. Many species of plants and animals have the ability to occupy several specialized habitats and to become adapted to them. The banded snail *Cepaea nemoralis* may develop a prevailing color type in moist beech woods (reddish, unbanded), hedgerows (heavily banded), and short-grass meadows (yellow, unbanded). Yet there is presumably a greater total genetic difference between the reddish unbanded type of England, and those of southern France or eastern Germany than between the three different color types in southern England. No terminology is suitable to cope simultaneously with these two independent systems of variation.

A repeated origin of the same phenotype is found in many fresh-water organisms that invade the same ecological subniche independently in different areas. According to Steinmann (1952), the trout (*Trutta fario*) has developed two distinct ecological races, the Lake Trout and the Brook Trout. These two forms are not isolated sexually and are completely interfertile. Intermediate types occur in larger streams, but are rare because suitable habitats are seldom found in these situations. The differences between Brook Trout and Lake Trout can be tabulated as follows:

Brook Trout	Lake Trout
Early maturing (2–3 years)	Late maturing (5–6 years)
Stationary	Migrates from brooks to lakes
Small food	Fish among its food
Spawning year after year	Spawning only once or twice
Poor growth	Excellent growth
Red spots of juvenile color usually maintained in nuptial dress	Red spots usually lost when silvery nuptial dress is attained

Lake Trout have apparently evolved independently from Brook Trout in many places.

The New Concept of the Geographic Race

These considerations may help to terminate the conflict concerning the terminology of infraspecific units. A species consists neither of an aggregate of strictly morphologically definable subspecies and varieties nor of purely ecologically definable ecotypes or ecological races. A species is actually composed of populations distributed in space and time, which possess morphological as well as physiological and ecological characteristics. The recent work on geographic variation has led to the reinterpretation of the geographic race as a genetic-physiological response to a local environment. There is no antithesis between geographic race and ecological race (or ecotype) since not a single geographic race is known that is not also an ecological race; nor is there an ecological race that is not at the same time at least a microgeographic race. The dual terminology is mainly based on the observation of great differences in the size of the geographic area occupied by phenotypically distinguishable populations. In plants, with their exceedingly strong adaptation to local conditions and well-developed mechanisms for inbreeding, there is a tendency for the development of microgeographic races with strongly ecotypic characteristics, "ecotypes." In animals, particularly warm-blooded animals with much developmental homeostasis and emancipation from the environment, there is a tendency for nonexpression in the visible phenotype of the localized genetic-physiological variation and a considerable phenotypic uniformity over wide areas, "geographic races." That geographic races have physiological-ecological characteristics has long been known to zoologists, as a consequence of the work of Gloger (1833), Bergmann (1847) and Allen (1877). This is one of the reasons why no dualism ever developed in zoology between an ecological and a morphotaxonomic nomenclature of infraspecific categories as the Linnaean and Turessonian terminologies in botany.

Is There a Need for a Dual Terminology?

The species and the subspecific categories have little evolutionary and biological meaning if they are typologically and morphologically defined. This has induced many authors, from Turesson (1922) on, to propose dual terminologies, one for taxonomic and one for evolutionary purposes. Gilmour and Heslop-Harrison (1954) have recently attempted to introduce a whole new terminology for the units of microevolutionary change, based on the term "deme." Unfortunately their definition of the deme, "denoting any group of individuals of a specified taxon," is so vague that it would apply equally well to all members of the taxon Mammalia or even Animalia and thus deprives their elaborate system of any meaningful basis. Their difficulties are in part caused by an apparent belief in the arbitrariness of the species concept and a need for replacing it by something "better." Furthermore, these authors attempt to define demes in terms of individuals instead of accepting the fact that in sexually reproducing species all discernible entities are composed of local populations. The biological species is called a gamodeme and is defined as a deme "composed of individuals which are so situated spatially and temporarily that, within the limits of the breeding system, all can interbreed." Since individuals of many species "can interbreed" with individuals of other species, although the populations to which they belong remain "reproductively isolated," it is obvious that this definition is not a forward step.

It is perhaps no coincidence that nearly all proposals for a dual terminology have been made by botanists, since the tradition of a morphological definition of taxonomic categories seems to have been far stronger in botany than in zoology. To me, as a zoologist, the need for such terminological dualism is not apparent. The term "species" is as useful in purely taxonomic as in evolutionary studies of animals. On the infraspecific level a slight dualism is unavoidable, without calling, however, for an elaborate new terminology. The application of the word "subspecies" is best restricted to taxonomy, while terms like "deme," "race," "geographic isolate," and "cline" belong to the field of evolutionary biology. Finally, terms like "variety" and "biological race" have been such catchalls for heterogeneous phenomena that they would better be discarded altogether.

SUMMARY

In animals as well as in plants, local populations are selected for adaptation to the specific environment in which they live. All races are therefore geographical as well as ecological races. The principal difference between plants and higher animals is the amount of their direct dependence on the environment and the phenotypic expression of the local physiological adaptation. At one extreme are the warm-blooded birds, highly mobile and highly independent of the direct effect of the environment. At the other extreme are certain species of plants and sedentary invertebrates which are highly dependent on their substrate and fully manifest this dependence in their phenotype. The two extremes are differences of degree and are connected by a complete spectrum of intermediate conditions.

13 ~ The Population Structure of Species

As a means of simplification, the practicing taxonomist divides species in a typological manner. He implies in his species catalogues that the subspecies and ecotypes into which he divides his species are well defined, more or less uniform over extensive areas, and separated from other similar units by steep and narrow zones of intergradation. It is now increasingly apparent that this simplified typological picture of the species structure is the exception rather than the rule.

A very different approach, based on the population structure of species, is necessary for a study of the internal variation of species from the ecological and evolutionary point of view. This new approach investigates the degree of difference among neighboring populations, the presence or absence of discontinuities between populations, and the characteristics of those populations that are intermediate between phenotypically distinct populations. It is an objective approach because it does not try to force natural populations into a preconceived framework of artificial taxonomic or ecological units and terms. A new picture of the population structure of species emerges from the new approach. It shows that all populations of a species are involved in one (or several) of the following three structural elements of species: (1) series of gradually changing contiguous populations (*clinal variation*); (2) populations that are geographically separated from the main body of the species range (*geographical isolates*); (3) rather narrow belts, often with sharply increased variability (*hybrid belts*), bordered on either side by stable and rather uniform groups of populations or subspecies.

Nearly every well-studied species, except some of those that for one reason or another are monotypic, shows several of these elements. They will be discussed in the following sections. The only attempt to analyze the species of an entire fauna for these structural components is that of Keast (1961) for the Australian avifauna. The rich findings of this study demonstrate the extraordinary value of this type of analysis.

CLINAL VARIATION

If neighboring populations of a species are compared, one finds that they usually differ from each other, slightly or appreciably, in a number of characteristics. Furthermore, when we trace a character through a series of contiguous populations, the changes usually show a regular progression. Such regular progressions of characters were discussed in Chapter 11 under the ecogeographical rules as an indication of the adjustment of populations to local conditions. Huxley (1939, 1942) has coined the term *cline* for such a character gradient. The study of geographic variation has revealed that much of it is clinal.

There are three chief reasons for the clinal mode of geographic variation. The first is that the environmental selective factors themselves (such as climate) vary along gradients, and, as a consequence, so do those phenotypic characters that respond to this selection. There are only very few features of the environment, such as soil color and other properties of the substrate, that may change abruptly. The second reason for the gradualness of variation is that gene flow among adjacent populations tends to smooth out all existing differences. The potency of such gene flow is particularly apparent where it spills across natural barriers (Galbraith 1956), or where it leads to a discrepancy between coloration of a population and the color of its substrate (Blair 1947). The amount of gene exchange among populations is consistently underestimated in the analysis of continuous variation. Finally, the retarding effect of developmental homeostasis tends to conceal genetic differences among conspecific populations, particularly those with active gene exchange.

Clines are, ultimately, the product of two conflicting forces: selection, which would make every population uniquely adapted to its local environment, and gene flow, which would tend to make all populations of a species identical. This cohesive force of gene flow gives physiological unity to a species but increases the necessity of adjustments to local conditions. Making this compromise between the "typical" physiology of the given species and the needs of the local environments becomes increasingly more difficult toward the periphery of the species range and is ultimately responsible for the phenomenon of the species border that the species cannot transgress.

Haldane (1948) and Fisher (1950) have discussed some of the genetic and biometric aspects of clines.

Clines are widespread and occur in the majority of, if not all, continental species. This is perhaps what one should expect since climatic factors such as temperature, rainfall, evaporation, number of days with frost or snow, and so forth show regular gradients (Huxley 1942). In the drongos of the family Dicruridae, Mayr and Vaurie (1948) found that clines occur in every species that varies geographically. One of the finest analyses of clines in a group of animals is that by Petersen (1947a) of Fenno-Scandian butterflies (Fig. 13–1). He analyzed in 16 species the geographic variation of 59 characters and found that 29 (about 50 percent) varied clinally. The six species that showed no clines were partly migratory. In the other 10 species, 79 percent of the examined characters varied clinally. Other cases of clines in insects have been cited by Hubbell (1956).

Clines and Isophenes. A clear distinction must be made between clines and isophenes. A cline is the total slope from one extreme of the character to the other. Clines, when plotted on a map, are crossed at right angles by isophenes, the lines of equal expression of a character. For instance, in the drongo example mentioned above, the size cline runs from the largest population in the Himalayas south to the smallest populations in Ceylon and the Malay Peninsula. This north-south cline is crossed by the isophenes, lines connecting all populations with the same phenotype, for instance, all those with a mean wing length of 180 mm.

Size, color, or any other kind of morphological and physiological character may vary clinally. Huxley (1942:206–227) has listed a number of such characters. The cline for each character is theoretically independent of the others. Often, however, there is a rather strong correlation among several characters, as, for instance, when general body size in a warm-blooded vertebrate is found to increase toward the north, resulting in a change of proportions and other correlated characteristics. In other cases there is considerable independence among character gradients. Nearly all Australian birds with size variation, for instance, decrease in size from Tasmania northward to Torres Straits along a regular cline, following Bergmann's rule. Intensity of color, however, changes along a very different cline, leading from the most humid periphery of Australia to the most arid interior.

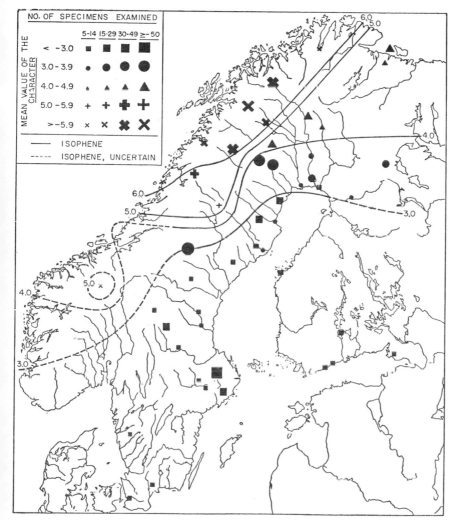

Fig. 13–1. Character gradient concerning pigmentation of the upper side of the wing in females of *Pieris napi* from Fennoscandia. The darkest values occur in the northwest. Size of symbol indicates size of sample. (From Petersen 1947a.)

The term "cline" refers to a specific character, such as size or color, not to a population. A population may belong to as many different clines as it has variable characters. The potential independence of different character gradients makes the cline unsuitable as a taxonomic category (Sibley 1954b; Hubbell 1954). A concordance of the clines for different characters is normally found only where ranges are essentially longitudinal and where the various environmental gradients (temperature and humidity, for example) run by chance more or less parallel.

Character gradients are rarely regular. Areas with gentle slopes may alternate with zones where the clines are very steep. Such clines have been compared with "stepped ramps." The reasons for these steps must be determined separately in each case, but, since many of them are due to the secondary contact of previously isolated populations, the phenomenon as a whole will be discussed below under the heading "Zones of Intergradation."

Clines designate trends of variation. Closer analysis, however, shows that clines are far less regular than is generally supposed. In a study of the beach mouse *Peromyscus polionotus,* Hayne (1950) found that there was a definite gradient from the light coastal beach to the dark soils of the interior, but that the cline did not exactly parallel either the change in soil color or the distance from the beach. Specimens from Crystal Lake were paler and those from Seminole Hills were darker than expected on the basis of their location on the geographical gradient. Similar irregularities are apparent also on the maps published by Petersen for clines in Swedish butterflies. There are two possible interpretations for these irregularities. Either they reflect irregularities in the environmental gradients, or they are the incidental and unpredictable phenotypic by-products of local selective responses of the genotype. It is not known how frequent such irregularities are because detailed careful analyses of clines have been made in only very few species (Womble 1951).

Kinds of Clines. Huxley (1939) proposed a number of prefixes to denote different kinds of clines, such as "ecocline," "genocline" (gradient in genes), "geocline" (geotrend), and others. Actually all clines are simultaneously ecoclines, genoclines, and geoclines, and to distinguish these three kinds of gradients does not seem advantageous. A special term, "chronocline," may be useful for character gradients in geological time. Rate and direction of change in chronoclines may differ for different characters.

A special problem of interpretation arises when a cline develops across a belt of allopatric hybridization. The problem is to determine to what extent the cline reflects selection by the respective environments and to what extent it reflects gene exchange with the population with which contact has recently been made. In the case of the birds of paradise of eastern New Guinea (*Paradisaea apoda*), where a cline exists between birds with red flanks and brown back in the south and birds with orange flanks and yellow back in the north (Fig. 13–2), it seems quite likely that introgression with *Paradisaea minor* is involved, characterized by a strongly differential ability of the introgressing genes to invade the range of *Paradisaea apoda*.

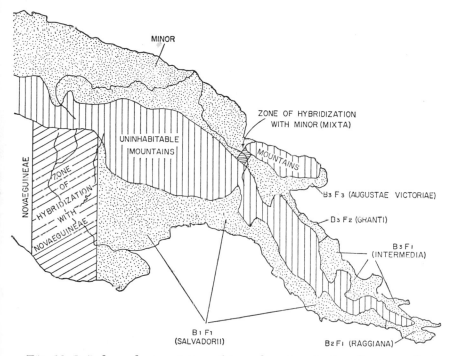

Fig. 13–2. Independent variation of two characters in *Paradisaea apoda* in New Guinea. A cline that is independent of environmental gradients. *B*, coloration of back: *B*1, entire back brown; *B*2, upper back yellow, lower back brown; *B*3, entire back yellow. *F*, coloration of the plumes (flank feathers): *F*1, plumes red; *F*2, plumes orange-red; *F*3, plumes orange. Diagonal hatching indicates hybrid zones. (From Mayr 1940a.)

In many other cases, likewise, gene flow seems responsible for the maintenance of clines to a greater degree than environmental gradients. The importance of gene flow can be demonstrated best by studying isolated populations adjacent to continuous populations and situated on the same environmental gradients. Such studies show that the variation of truly isolated populations is unpredictable and often remarkably independent of the clines found in the adjacent continuous populations (Mayr 1954a). The cohesive effect of gene flow seems to be exceedingly potent. This is shown where there is a slight "spilling" of genes across a geographic barrier, as when different subspecies groups come into close proximity on adjacent islands (for example, *Pachycephala pectoralis*, Galbraith 1956), or across barriers on continents (for example, *Lophophaps plumifera*, Mayr 1951b).

To what extent a given character gradient is the product of gene flow or of an environmental gradient is usually difficult to say. The clines in the frequencies of the human blood groups, formerly ascribed entirely to gene flow, may well have a substantial selective component. The working hypothesis that the presence of a cline indicates a geographically variable selective factor of the environment has proved very productive. It induced Mayr (1945) to postulate that each gene arrangement of *Drosophila pseudoobscura* had a definite selective value at a given locality rather than that this distribution pattern was due to a historical accident as originally postulated by Epling (1944). Subsequent work by Dobzhansky (summarized in 1951) has demonstrated the correctness of the selection hypothesis. Nearly all other known cases of balanced polymorphism show a clinal distribution of frequencies (for example, Mayr and Stresemann 1950).

GEOGRAPHICAL ISOLATE

The typological division of species into subspecies does not give any information on the relation of populations to each other. Whether a subspecies is part of a cline or is isolated completely by geographic barriers is, however, of decisive influence on its evolutionary potential. Huxley (1942:210) has called isolated subspecies "independent," and those in free gene exchange with other conspecific populations "dependent." More informative than this taxonomic approach, when dealing with the evolutionary aspects of infraspecific populations, is a distinction between contiguous populations with clinal variation and "geographical isolates." The term "isolate" has been used in the biological literature with different meanings. In anthropology and human genetics it is usually applied to the inhabitants of a partially isolated area and more broadly to what the naturalists would call any local and usually only very incompletely isolated population. In the present discussion I define the "geographical isolate" as "a population or group of populations prevented by an extrinsic barrier from free gene exchange with other populations of the species." The essential characteristic of the geographical isolate is that it is separated from the rest of the species by a discontinuity. The degree of discontinuity depends on the efficiency of the extrinsic barrier. The isolation is never complete, since a certain amount of gene flow reaches even an isolated oceanic island (or else it could not have been colonized originally). The precise mapping of ranges reveals that virtually every species contains some isolates, particularly near the periphery of the species range. Wherever geographical or ecological conditions produce an insular distribution pattern, the frequency of isolates increases sharply. This is true not only for oceanic islands, but also for all kinds of ecological islands, be they mountains, forest patches in grasslands, or lakes and streams. Kinsey (1937) and Mayr (1942) have discussed the distributional and variational aspects of this insular distribution pattern. The frequency of isolates within a species depends on the structure of the environment and the dispersal facilities of a species. It is well known among taxonomists that species may show great phenotypic uniformity over wide areas where the species ranges are continuous or else an astonishing production of isolates where factors are suitable. Continental lizards have usually few and not strikingly different geographic races; the same species on islands may break up into dozens or scores of highly distinctive isolates (Mertens 1934). The *Lacerta muralis* group has only three races throughout the vast area of the Iberian Peninsula. Yet it has developed more than 50 distinctive races on the small islets around the main islands of the Balearic Islands and Pityusas (Eisentraut 1949; Buchholz 1954). A comparison of the continuous ranges of temperate-zone *Bombus* (Reinig 1939) or planarians (for example, *Dugesia*, Benazzi 1949) with the isolated Mediterranean populations reveals the same phenomenon. Mere mapping is not quite sufficient to establish the occurrence of isolates because in some discontinuously distributed species there is such free interchange of individuals among adjacent colonies that they cannot be considered isolates. The large-scale maps that are customarily found in much of the taxonomic and biogeographical literature are, strictly speaking, misleading because they give geographic ranges as continuities even though species are usually composed of localized colonies.

We know, as yet, little about the frequency of genuine isolates in various groups of animals. This is regrettable, considering the great potential importance of isolates for speciation. The minimum number of isolates in several well-analyzed groups of birds is listed in Table 16–2. Keast (1961) has found that 425 species of Australian birds had developed 211–226 morphologically differentiated isolates. There is, thus, ample incipient speciation even in this continental area. The number of isolates in an equivalent island area is about five times as great.

Geographical isolates may occur throughout the range of a species, wherever barriers occur, but they are most frequent at the periphery. Taxonomists have long been aware of the importance of these peripheral isolates and have pointed out, again and again, that major deviations from the "type" of a species will most likely occur in such populations (Lorković 1943; Mayr and Vaurie 1948; Zimmermann 1950; Mayr 1951a, 1954a). Peripheral isolates have two further, almost universal, characteristics. They are usually comparatively small in area, with a low absolute population size. They differ from the main body of the species population and from each other in numerous, often unique, and sometimes drastic morphological, physiological, behavioral, and other characteristics. For a further discussion of peripheral isolates see Chapter 17.

Geographic isolates have three possible fates (discussed in more detail in Chapter 16). They may become separate species, die out altogether, or reestablish contact with the main body of the species, forming a secondary zone of contact. The taxonomic status and degree of phenotypic difference of isolates depends on the duration of the isolation and on other factors. Some are quite indistinguishable from the remainder of the species population, some are sufficiently different to be regarded as distinct subspecies, and some, finally, are on the borderline between subspecies and species status. Mere geographic isolation of a population on a mountain or an island, in a lake or a cave, is not enough to establish the existence of "reproductive isolation." Corroborating evidence, as discussed in the taxonomic literature (for example, Mayr, Linsley, and Usinger 1953), is required to infer the taxonomic status of phenotypically distinct geographic isolates.

ZONES OF INTERGRADATION

The third phenomenon, in addition to clines and isolates, that is characteristic of the population structure of most species is the existence of contact zones between phenotypically different populations. The terminology, classification, and interpretation of such belts has long been a source of disagreement. Taxonomists have sometimes referred to these belts as "subspecies borders" because widespread and comparatively uniform subspecies often meet in such belts. Yet there is no congruence between the two phenomena. Some subspecies intergrade imperceptibly along a cline while on the other hand distinct steps in clines are sometimes observed within subspecies, separating populations that are not sufficiently distinct to deserve subspecific recognition. It seems probable that these contact zones belong to two phenomena, which, from the evolutionary viewpoint, are rather different (Mayr 1942:99):

Primary intergradation exists if the steepening of the slope developed gradually and took place while all the populations involved were in continuous contact.

Secondary intergradation refers to cases in which the two units now connected by a steeply sloping character gradient were separated completely at one time and have now come into contact again, after a number of differences had evolved.

Cases of primary intergradation are believed to be caused by a corresponding change in environmental conditions, while zones of secondary intergradation are hybrid belts between populations that had become differentiated during a preceding period of isolation. Voipio in a series of thoughtful papers (1950, 1952a,b) has pointed out some of the unsolved problems with respect to zones of intergradation: Is the stated classification exhaustive, or does it omit additional possibilities? How should one classify population shifts, resulting from expanding ecological opportunities, that may bring populations secondarily in contact along habitat borders? Can a zone of intergradation always be classified as primary or secondary merely by a study of the phenotypes of the respective populations? Under what conditions is individual variability increased in a zone of intergradation and what does this signify? The following discussion endeavors to shed some light on these questions.

Allopatric Hybridization

The interbreeding of two previously isolated populations in a zone of contact has been designated *allopatric hybridization* (Mayr 1942). This terminology suffers from the same weakness as any application of the term "hybridization" to the interbreeding of individuals, no matter how unlike phenotypically, that belong to the same species. It is sometimes safer to use the more neutral term *secondary intergradation* in cases where the interbreeding populations have not yet reached species level. Some authors extend the term "hybridization" to any gene flow from population to population, and, by carrying this terminology to a consistent but evidently absurd extreme, they call every population a hybrid population. This terminology goes back to Lotsy, who considered every individual in a sexually reproducing species a *hybrid*. The term "hybridization" loses all usefulness if it is applied in such an indiscriminate manner.

A well-analyzed and carefully described case of allopatric hybridization is that of the Hooded Crow and the Carrion Crow (*Corvus corone* and *C. cornix;* Meise 1928a). The all-black Carrion Crow inhabits western Europe while the Hooded Crow, gray with a black head, wings, and tail, inhabits eastern Europe and most of the Mediterranean region. The two forms come into contact in a narrow zone starting in Scotland, extending through Denmark, central Germany, and Austria to the southern slopes of the Alps, and reaching the Mediterranean somewhere near

Fig. 13–3. Course of the hybrid zone between the Carrion Crow (*Corvus c. corone*) and the Hooded Crow (*Corvus c. cornix*) in western Europe. Note the relative narrowness and unequal width of the zone. (After Meise 1928a.)

Genoa (Fig. 13–3). Pairing within the hybrid belt seems to be random, and there is every conceivable combination of the parental characters as well as all degrees of intermediacy. Beyond the hybrid belt an occasional bird is encountered that does not appear to be quite "pure," such as a Hooded Crow with some Carrion Crow characters, or vice versa, but on the whole the visible effects of the hybridization are rather localized.

Populations meeting in zones of secondary intergradation may show any degree of difference. They range from those as different morphologically as good species (such as the crows) to populations that can be separated only by biometric or genetic tests. Only the more conspicuous cases are usually recorded in the taxonomic or evolutionary literature, and it is therefore difficult to determine the relative frequency of zones of secondary intergradation in different species. In view of the numerous discontinuities in the ranges of most species and the never-ending changes of the environment, one should expect a high frequency of fusions between previously isolated populations. All available evidence indicates that zones of secondary intergradation are indeed very widespread and common. Even the more spectacular cases of allopatric hybridization between very different forms (including semispecies) are remarkably frequent.

Cases of Allopatric Hybridization. Allopatric hybridization usually results from expansion of isolates because of changed environmental conditions. It occurs with particular frequency after periods of climatic change, such as the end of the Pleistocene. During the height of the glaciation, the ranges of many temperate-zone species contracted into small pockets, so-called glacial refuges, which persisted south of the area of glaciation. In Europe, for instance, the Alpine and northern ice caps approached each other to within 300 miles, separated by icy wind-swept steppes. The forest animals retreated into southwestern or southeastern Europe. When conditions improved at the end of glaciation and the populations in the refuges expanded northward, the isolates in southwestern and southeastern Europe had, in many cases, become sufficiently distinct from each other to form hybrid zones in central Europe (Meise 1928b, 1936a; Mayr 1942). Cases similar to that of the crows have been found in Europe in the nuthatches (*Sitta europaea*), the long-tailed tits (*Aegithalos*), the gray shrikes (*Lanius excubitor*) (Salomonsen 1949), and the yellow wagtails (*Motacilla flava*) (Sammalisto 1956; Schwarz 1956), to mention a few examples in birds. Similar hybrid belts in central Europe are known for mammals, amphibians, and invertebrates.

A corresponding great hybrid belt runs through Asia from Iran and Turkestan north to the Siberian tundra (Meise 1928a,b; Johansen 1955). Rand (1948) has postulated that numerous hybrid zones in North American birds likewise owe their origins to the post-Pleistocene expansion of populations that had formerly been isolated in glacial refuges. The grackles (*Quiscalus*) (Huntington 1952), flickers (*Colaptes*) (Short MS), juncos (*Junco*), ruffed grouse (*Bonasa*), towhees (*Pipilo*), white-crowned sparrows (*Zonotrichia*), Canada jays (*Perisoreus*), and myrtle warblers (*Dendroica coronata*) are some avian examples that come readily to mind. Two sets of refuges are usually involved: the various mountain ranges, such as the Appalachians, the Black Hills, the Rocky Mountains, and the Pacific Coast ranges for the more northerly forms, and Florida,

Texas, and southern California for the more southerly forms. There is evidence from many other groups of animals that zones of hybridization are due to expansion from glacial refuges (Deevey 1949; Blair 1951). However, in no case has a whole group of animals been analyzed systematically in order to separate areas of post-glacial temperature-related hybridization from old contact zones that were the result of changes in rainfall either during arid interglacials or prior to the Pleistocene (Mayr 1951a).

A rise in sea level may lead to the isolation of populations on temporary islands and a subsequent drop to a secondary junction. A number of hybrid belts in northern Florida seem to be the result of the joining of Pleistocene Florida islands to the North American mainland (for example, Dickinson 1952; Hubbell 1956).

Most hybrid zones in the temperate region are the result of the fusion of populations expanding into the areas vacated by the retreating ice. In other cases shifts in vegetation zones, indirectly caused by climatic changes, are responsible. The recent hybridization in the northern American plains in the avian genera *Colaptes, Icterus, Passerina, Pipilo,* and *Pheucticus* (Sibley 1959, 1961) is partly due to the planting of trees, which provide avenues of contact across the previously largely treeless prairie. As far as the subtropical and tropical regions are concerned, alternation between arid and humid periods is certainly the main cause of the separation and eventual rejoining of isolates. In Australia, for instance, most birds of the forested areas were squeezed into a number of coastal refuges during a drought period that took place between 4,000 and 20,000 years ago (Gentilli 1949; Mayr 1950d). The present distribution and variation of the tree runners (*Neositta*) indicates that this species group had one such refuge in southwestern Australia, one in northwestern Australia, one in northern Queensland, and two or three in eastern Australia between southern Queensland and Victoria. As the rainfall increased at the end of the dry period, trees and tree runners began to spread, resulting in a secondary contact of the former isolates. There are now five or six zones of hybridization (Fig. 13–4). Keast (1961) has given a detailed treatment of the hybrid zones in the Australian birds. Shifts in the distribution of savannahs and forests, resulting from changes in the amount of rainfall or from deforestation caused by man, are undoubtedly responsible for other well-known cases of hybridization in the tropics, such as the silver pheasants (*Lophura*) (Delacour 1949) and various species of New Guinea birds (Meise 1928b; Mayr 1942), including the Ribbon-tailed Astrapia (*Astrapia mayeri*) and Princess Stephanie Astrapia (*A. stephaniae*)

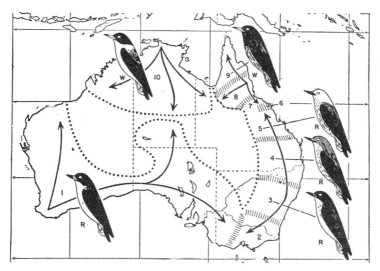

Fig. 13–4. Tree runners (*Neositta*) from Australia. The arrows indicate expansion from post-Pleistocene aridity refuges. Wherever two former isolates have met, they have formed hybrid belts (indicated by hatching). *R*, red wing bar; *W*, white wing bar.

(Mayr and Gilliard 1952a). Hybridization between the various species and subspecies groups in the genus *Paradisaea* is another example (Mayr 1942); still others have been reported from Africa and South America.

Regular hybrid belts occur only when the secondary contact is established along a broad front. In animals with a more insular distribution it happens sometimes that a hybrid population is formed on an intermediate island. For example, on Dampier Island a hybrid population was formed by the interbreeding of *Megapodius freycinet affinis* from the mainland of New Guinea and *M. f. eremita* of the Bismarck Archipelago. *Pachycephala pectoralis whitneyi* in the Solomon Islands is likewise of hybrid origin (Mayr 1942). Similarly, some of the gene flow in California shrews (*Sorex*) between populations of the northern California uplands (*vagrans*) or southern California uplands (*ornatus*) and the endemic populations of the tidal marshes of San Francisco Bay (*sinuosus*) can be best described as insular hybridization of these partially isolated populations (Rudd 1955). Insular hybridization is also observed in mountain species.

Allopatric hybridization has been particularly well studied by ornithologists, because of the advanced state of avian systematics; however, cases of secondary intergradation have been described for many other groups of animals. In addition to those already cited we may list the salamanders *Batrachoseps attenuatus* and *pacificus* (Hendrickson 1954), the toads *Bufo fowleri* and *woodhousei* (Meacham 1962), the snakes *Pituophis catenifer* and *sayi* (Klauber 1947), and the sticklebacks *Gasterosteus aculeatus* and *trachurus* (Münzing 1959).

Various Aspects of Hybrid Belts

When two populations become isolated from each other, their gene pools become independent and they diverge steadily in their genetic composition (Chapter 17). When the geographic isolation breaks down and the two populations reestablish contact, we should find evidence in the zone of contact for the degree of genetic differentiation achieved prior to the secondary contact. An analysis of the characteristics of populations in such hybrid belts reveals indeed much evidence for such prior genetic differentiation.

Random or Selective Mating. If incipient isolating mechanisms had developed prior to fusion, one should expect definite deviations from random mating. In the sapsucker genus *Sphyrapicus,* Howell (1952) found in a northern California hybrid belt 8 apparently pure *nuchalis,* 14 apparently pure *daggetti,* and 20 hybrids. Where the ranges of *nuchalis* and *ruber* meet at Kersley, British Columbia, one form is replaced by the other within the short stretch of 1.5 miles. Among pairs observed in the area, five appeared to be *nuchalis,* three *ruber,* and three pairs either mixed *ruber* × *nuchalis* or *nuchalis* × hybrids, or both hybrids. For additional cases see A. H. Miller (1956) and Sibley (1959). Where the two paradise magpies *Astrapia stephaniae* and *A. mayeri* meet in the Mount Hagen region of New Guinea, there is considerable hybridization, yet there is evidence of maintenance of the parental types, perhaps as a result of mating preference (Mayr and Gilliard 1952a). In such cases of partial breakdown of reproductive isolation a decision on the taxonomic treatment (species or subspecies) is very difficult.

In most cases of secondary intergradation genetic divergence has not yet proceeded to the point where preferential mating is found. Meise (1928a) found no evidence for anything but random mating in *Corvus* nor did Sammalisto (1956) in the polymorphic zone of intergradation in Finland between the central European and northern races of the yellow wagtail (*Motacilla flava*), nor does preferential mating occur between the Tufted and the Black-crested Titmouse (*Parus*) in Texas (Dixon 1955). Usually morphological differentiation seems to take place more rapidly than the acquisition of isolating mechanisms. This sequence is shown to a particularly striking degree in the snail genus *Cerion,* where exceedingly different populations interbreed freely and at random in their contact zones (Mayr and Rosen 1956; Mayr MS; Fig. 13–9).

Ecological Divergence Preceding Hybridization. Isolates often differ in their habitat requirements or in other ecological characteristics. When they come into contact secondarily, without having acquired reproductive isolation, they form a zone of intergradation along a steep ecological gradient or along an ecological discontinuity (Mayr 1942:263). The Song Sparrow (*Passerella melodia*) in the San Francisco Bay area consists of two types of races (Marshall 1948), one group restricted to the tidal marshes, and the other to riparian and other fresh-water habitats of the surrounding uplands. In most areas the two types of habitats are separated by an intervening ecologically unsuitable zone. Wherever a contact exists, however, there is a zone of hybridization or interbreeding. Exactly parallel is the case of the shrews (*Sorex*) of San Francisco Bay described by Rudd (1955). Wherever different isolates of the *Junco hyemalis* group come in contact with each other after the breakdown of previous isolation, there is always a tendency for each of the populations in contact to remain attached to its specific preferred habitat (Miller 1941, 1956, MS). In the many races of Herring Gull (*Larus argentatus*), the geographical isolation is reinforced by differences in habitat requirements. The yellow-footed race *cachinnans* of interior Russia (Volga Basin and so forth) is found mainly inland on bogs and lakes, but within recent decades has spread northward into Finland where it has reached the Baltic and the Arctic Ocean. Only where it has colonized islands in the Baltic has it met the pink-footed race *argentatus* (in this area strictly a saltwater bird) and formed with it highly variable hybrid populations (Voipio 1954). The habitat differences of two races of *Parus bicolor* which meet in an ecological contact zone in Texas were described by Dixon (1955).

The tenacity with which certain ecological preferences remain tied up with gene complexes has been described in a number of cases: for example, the meeting of *lecontei* and *clarus* in the snake *Rhinocheilus* (Klauber 1941). The nominate race of the percid fish *Boleosoma nigrum* occurs in lakes and streams with a firm substrate. In more or less discontinuous pockets of quieter water with considerable aquatic vegetation and somewhat silty or muddy bottoms, one finds the subspecies *eulepis,* which is more extensively scaled on cheeks, nape, and breast. Narrow areas of intergradation are found on the habitat borders (Lagler and

Bailey 1947). The pocket gopher *Thomomys bottae pascalis* of the irrigated lands of the San Joaquin Valley in California meets in a very narrow zone (rarely more than ½ mile wide) the subspecies *T. b. mewa* of the wild uncultivated grasslands. The amount of interbreeding in the zone of contact without visible introgression indicates strong selection against introgressing genes (Ingles and Biglione 1952). Spieth (1947) shows that where four subspecies of the mayfly *Stenonema interpunctatum* meet they maintain their integrity to a considerable degree because each is adapted to different water conditions. The armored (*trachurus*) and the smooth-skinned (*leiurus*) races of the stickleback (*Gasterosteus aculeatus*) hybridize wherever they come into contact, but the preference of *trachurus* for salt water, and of *leiurus* for fresh water limits the extent of the area of hybridization (Münzing 1959).

In the marine snail *Thais lapillus* two chromosomal forms meet in Brittany (Staiger 1954, MS), one with 8 acrocentric and 5 metacentric (= 13) chromosomes, the other with 18 acrocentric chromosomes in haploid condition. The two interfertile forms differ in their ecological requirements; the 13-chromosome type occurs on the rocky coast exposed to surf, while the 18-chromosome type is found in sheltered bays in the shallow-water zone. Heterogeneous colonies with intermediate chromosome frequencies (diploid means of 27–35) are found in intermediate localities. These mixed colonies show "hybrid vigor" in increased shell thickness (but smaller shell size) and high population density, yet also evidence of a certain amount of hybrid sterility and inviability. The different chromosome numbers and habitat preferences had apparently developed during a previous isolation of the 13- and 18-chromosome types without, however, leading to reproductive isolation. Only the 13 X type occurs on the Atlantic coast of North America. The ecological adaptation of the gene complexes meeting in these hybrid zones prevents in all these cases a widening of the zone of intergradation through gene flow. This is paralleled by most cases of ecotypic variation in plants where secondary contacts are involved.

Incompatibility of Gene Complexes. The occasional occurrence of nonrandom mating in hybrid belts and, even more, the ecological stability on either side of the line of contact indicate the amount of genetic difference that must have existed prior to the secondary contact. An analysis of the reasons for the incompatibilities of entire gene complexes would be most interesting, but meets technical difficulties. However, even without experimental analysis much can be inferred from a phenotypic analysis of hybrid populations (Chapter 10). Every analysis undertaken so far has added to the support of the concept of stabilized, well-integrated gene complexes developed by Timoféeff-Ressovsky, Dobzhansky, and Huxley (see Chapter 15). Where such gene complexes come into contact in hybrid belts, various manifestations of incompatibility will become evident and these we shall now discuss.

A few simple generalizations can be made about hybrid zones. The more different the populations coming into contact, the more recent usually is the hybridization; the narrower the hybrid zone, the greater is the probability of unbalance in the recombined genotypes and the greater the probability that the unbalance will express itself as highly increased variability. This is observed in most allopatric hybrid belts. For instance, where the eastern Yellow-shafted Flicker (*Colaptes auratus*) and the western Red-shafted Flicker (*C. cafer*) meet in the western United States (Short MS), individuals may closely resemble either pure parental type or may show any conceivable combination of their characters. Such an increase in variability has been described for nearly every known zone of secondary intergradation: for instance, for *Corvus* (Meise 1928a), *Emberiza* (Paludan 1940), *Icterus* (Sutton 1938), *Junco* (Miller 1941), *Sturnus* (Pateff 1947), *Sphyrapicus* (Howell 1952), *Lanius schach* (Biswas 1950), *Microscelis* (Mayr 1941), *Cerion* (Mayr and Rosen 1956), and *Basilarchia* (Hovanitz 1949), to mention just a few examples. This variability is due partly to the various grades of back crosses of individuals in the zones of intergradation, and partly to a breakdown of developmental homeostasis resulting from mixing of somewhat incompatible genes. Even though both parental phenotypes may be found in the same area, it would be very misleading to say that the parental subspecies overlap in that area. The situation in the zone of contact is not one of two sympatric subspecies, but is, rather, an incorporation of the characters of two parental subspecies into a single population. The term "overlap," used in such cases, conveys a false impression.

In a hybrid belt, some characters may be more variable than others. In some hybrid colonies of *Cerion*, for instance, size and proportions do not show increased variability, while sculpture and pigmentation do (Mayr and Rosen 1956). This difference may be due partly to the much greater number of genes controlling size, and partly to the stabilizing effect of selection, to which size may be more subject than the ornamental shell characters. When two subspecies hybridize in several separate areas, such as the Grey-billed and Black-billed Honeyeaters, *Melidectes*, in New

Guinea (Mayr and Gilliard 1952a), a different character may become stabilized in each area.

It is apparent in these cases that natural selection is unceasingly at work in these hybrid zones, weeding out the most unbalanced combinations. High variability is maintained only by the continued reintroduction of new parental genotypes. A hybrid population may achieve phenotypic stability when subsequent isolation deprives it of gene flow from the two parental populations. Miller (1941) has shown for the genus *Junco* that the race *cismontanus* evidently originated as a hybrid population between *hyemalis* and *oreganus*. Yet its present variability is no greater than that of adjacent "pure" populations. This hybrid population has apparently existed sufficiently long to have discarded all genes producing disharmonious combinations. The same is true for some hybrid populations of *Pachycephala* in the Fiji Islands (Mayr 1932b; Meise 1938). Several human populations that owe their origins to hybridization do not show markedly increased variability (Trevor 1953). Similar stabilization is recorded in certain oases of the northern Sahara for the sparrow populations which originated as hybrids between *Passer domesticus* and *hispaniolensis*, but which have reached phenotypic stability in their isolation (Meise 1936b). Lack of increased phenotypic variability is thus no proof of the absence of former hybridization.

Width of the Zone of Intergradation

One of the least-understood aspects of hybrid belts is their width: some are very wide, others amazingly narrow. In hybrid belts that must have existed for thousands of years this narrowness is a great puzzle. One would expect either that reproductive isolation would be acquired as a result of an inferiority of the hybrids, or that gradual infiltration of the hybridizing genes would steadily widen the hybrid belt until it occupied the greater part of the ranges of the hybridizing populations. Apparently there is a third alternative: a vigorous selection against the infiltration of genes from one balanced gene complex into the other, but without the development of any isolating mechanisms as a by-product of this selection. As examples of narrow hybrid belts I would like to cite *Ctenophthalmus* and *Mus*. Where the *eurous* subspecies group of the flea *Ctenophthalmus agyrtes* meets the *agyrtes* subspecies group the total belt of hybridization is only 6.5 kilometers wide (Peus 1950). In Normandy, where (in the same species) *eurous* meets the western *celticus*, Jordan (1938) found that they approached each other within 100 meters, *eurous* being restricted to a wooded hill, *celticus* to open fields (both occurring on several host species). An unsuitable area separates the two populations. In the house mouse *Mus musculus* in central Denmark, a dark-bellied southern race (*domesticus*) meets a light-bellied northern race (*musculus*). The width of the total area of introgression is only 50 kilometers and the zone of truly intermediate populations is only a few kilometers wide (Ursin 1952). Even narrower is the hybrid belt between the crested titmice *Parus bicolor* and *atricristatus* in Texas (Dixon 1961).

Dobzhansky (1941) believes that the rather narrow hybrid belt (varying between 50 and 250 kilometers in width) between the Hooded Crow and the Carrion Crow (*Corvus*) in central Europe has contracted within historical times. However, the evidence is ambiguous and there are narrow stretches of the hybrid belt both in the south where the belt is oldest and in the far north where it is most recent. It appears that the width of the belt is controlled more by degree of philopatry and by ecological conditions than by duration. This is even more true for the hybrid belt between the Purple and Bronze Grackles (*Quiscalus*) in North America (Mayr 1942; Huntington 1952). This belt is narrow in the south (Louisiana and Alabama) and much wider (500–700 kilometers) in New England at the northern end of the belt. In the south the two kinds of grackles are associated with two different vegetational types and hybridization is confined to the rather narrow border zone between the two stable plant associations. The great width of the hybrid belt in New England is due not only to the lack of any ecological segregation but also to the migratory habit of grackles in this area. As in all migratory birds, there is a more thorough mixing up of populations than in sedentary species. Other cases of narrow hybrid zones were mentioned above under the heading Ecological Divergence.

Narrow but virtually permanent hybrid belts must be interpreted as zones of contact between balanced gene complexes established through selection during isolation. All disharmonious combinations in the hybrid zone will be selected against. Similarly, their penetration into the adjacent populations will be continuously counteracted by selection. This will not entirely eliminate gene flow but will greatly reduce its phenotypic effects. It is to be expected that some genes will be less strongly selected against than others, and that these genes may penetrate beyond the hybrid belt. The more closely related the populations coming into contact and the less disharmonious their gene complements, the more likely it is that such penetration will occur.

The Classification of Zones of Intergradation

In all cases of smooth character clines between contiguous populations there is little doubt that primary intergradation is involved. This includes cases of direct spatial contact of populations as well as those where gene flow between adjacent colonies is essentially unimpeded even though they are spatially separated. A more difficult problem is posed by the phenomenon of a more or less abrupt change of characters, a "step" in a cline, somewhere in a series of contiguous populations. There is uncertainty whether or not a distinct step can ever evolve in a zone of primary intergradation. One gathers from Huxley's (1939, 1942) discussions that he considers most steps in clines as zones of primary intergradation. Indeed Huxley thought that if selection pressures on both sides of an environmental zone of stress were sufficiently different this might lead to the origin of a real discontinuity through semigeographic speciation (see Chapter 17 under "Stabilized Hybrid Belts"). According to this assumption one would have to make a distinction between steep primary and steep secondary zones of intergradation, and numerous authors, using various criteria, have attempted to classify existing zones of intergradation into these two classes.

The most frequently used criterion is that of the phenotypic appearance of the populations in the zone of intergradation. If there is "smooth intergradation" the zone is classified as one of primary intergradation, and if there is "a belt of highly variable hybrid populations" the zone is classified as a zone of secondary intergradation. This phenotypic criterion is unfortunately not reliable, as was pointed out correctly by Miller (1949, 1955). After temporary isolation and the establishment of secondary contact between two populations, either a smooth zone of intergradation or a highly variable hybrid belt may develop, depending on a number of factors, such as the magnitude of the genetic differences acquired during the isolation and the degree of incompatibility of the two gene complexes, the degree of developmental homeostasis in the respective species, and whether or not the hybrid belt is old enough to have permitted secondary phenotypic stabilization. One can summarize the evidence by saying that belts with highly variable populations in a meeting zone between rather uniform populations are almost invariably zones of secondary intergradation (the only exception being a few cases of polymorphism), while zones of smooth intergradation might be either primary or secondary.

Another matter of classification must be discussed at this point. Voipio (1952a:16) proposes to make a distinction between genuine zones of secondary intergradation and "zones of semisecondary intergradation" where a contact "has developed secondarily when a population of foreign origin and thus of different genetic structure has pushed into the neighborhood of another local population differentiated *in situ*." I do not think that this terminological distinction is valid. The cases of the Mistle Thrush (*Turdus viscivorus*) and of the Herring Gull (*Larus argentatus*) show that there is originally just as much spatial separation in these cases of genetically *and* ecologically differentiated populations as in cases of "pure geographical isolation." I employ for this reason the term "secondary intergradation" for all cases of secondary contact, whether or not the meeting populations are ecologically differentiated.

On the basis of all the available evidence it appears to me that all steep zones of intergradation are secondary except for three special cases which are sometimes primary: (*a*) where adaptation to the color of the substrate is involved; (*b*) where habitat selection of genotypes is involved; and (*c*) where the chief phenotypic difference of two adjacent populations is controlled by a polymorph gene locus. This sweeping generalization is offered as a working hypothesis which should stimulate further field work. To the best of my knowledge there is no case of a well-analyzed steep zone of primary intergradation other than those falling under the stated exceptions.

There are several reasons why zones of primary intergradation should be gradual and of gentle slope:

(1) Virtually all climatic and other environmental gradients are gentle except for substrate and habitat borders.

(2) Even where there is a change in substrate, this only rarely has a phenotypic effect on the species living in the area. The major exceptions are cryptically colored local populations of desert animals (and edaphic races in plants). In many cases where a step in the phenotypic characteristics of a substrate-adapted species is observed, it is found that the step affects only a single character. In all other characters a smooth cline may pass through the "stress zone." This independence of the different components of the gene complex indicates that we do not have a break in the populations, but only in a single phenotypic response to the environment. The role of gene flow is emphasized by the shift of the intergradation zone away from the substrate border (Blair 1947) because of population pressures on either side of the border.

(3) It is extremely difficult for an organism to invade a new kind of habitat across an "ecological escarpment." Such colonization is prevented by the same genetic factors that determine the species border (Mayr 1954a). Whenever such an ecological shift does occur, usually under conditions of semi-isolation, the new ecological race will spread rapidly through the newly available habitat and will establish numerous zones of secondary intergradation along preexisting habitat borders.

(4) In polymorph species the frequency of polymorph genes sometimes changes quite drastically in a narrow zone of intergradation. This is due sometimes to substrate selection (*Cepaea*) and sometimes to habitat selection by genotypes (also in *Cepaea*), but is most often due to the secondary intergradation of two populations when interaction with the residual genetic background favors a different frequency of the polymorph genes on either side of the contact zone (Mayr and Stresemann 1950). Even in the case of *Sciurus* in Finland, analyzed by Voipio (1957; Fig. 11–5), the squirrel populations of northern and southern Finland have originated sufficiently far apart that their present zone of contact in central Finland may be considered at least in part a secondary zone of intergradation.

(5) A secondary steepening of originally gentle slopes of primary intergradation is highly unlikely. Environmental factors that would produce a sudden steepening of environmental gradients are rare, and even where this occurs the development of a steep phenotypic gradient would be prevented by gene flow and developmental homeostasis. Even where local races become physiologically adapted, only a small fraction of this adaptation will be reflected in the visible phenotype.

(6) The hypothesis that nearly all steep zones of intergradation are secondary requires the postulate of a very high number of temporary isolates. This postulate has been substantiated whenever the range of a species was mapped accurately.

Every animal is adapted to specific ecological conditions, and its area of distribution consists, therefore, of patches of suitable habitat, surrounded by barriers consisting of unsuitable habitats. The distances between the suitable habitats are, in general, smaller than the normal dispersal potencies of the isolated populations, and in such cases we speak of continuous ranges, even though this may not be strictly true [Fig. 13–5]. In other cases the belt of unsuitable habitat may be wide enough to cause effective isolation (Mayr 1942:230).

The rule seems to be that species, in spite of the general continuity of their ranges, constitute fairly small and relatively independent colonies (Voipio 1952a:8).

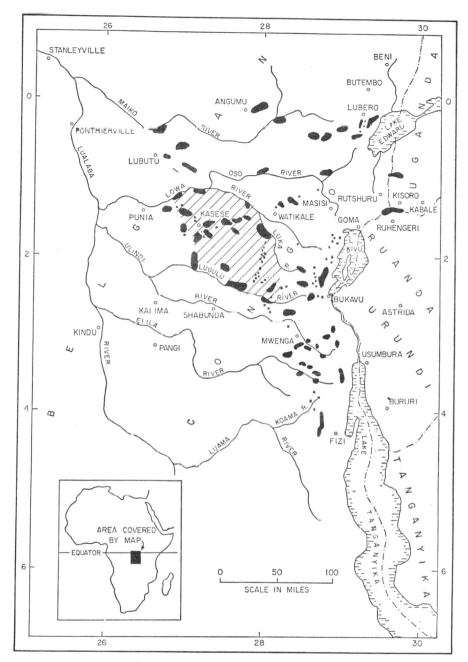

Fig. 13–5. Actual distribution of the Mountain Gorilla in East Africa. Each of the black areas indicates the location and approximate shape of sixty gorilla areas ranging in size from about 10 to about 100 square miles each. The hatching marks a central region of fairly continuous but sparse population. (From Emlen and Schaller 1960.)

The extinction of some and the expansion of others of these isolates leads continuously to the establishment of new zones of secondary intergradation. The farther the genetic, ecological, and morphological differentiation between the isolates had proceeded before contact was reestablished, the more obvious will be the secondary nature of the zone of contact. In the majority of cases the isolation will have been short and the genetic differentiation slight, so that the zone of secondary intergradation will be nothing more than a slight step in a cline.

GEOGRAPHY, ECOLOGY, AND SPECIES STRUCTURE

We conclude from the findings presented in the preceding section that species are not the uniform typological entities envisioned by classical taxonomy. Species actually have a complex population structure, characterized by series of clinal populations, isolates, and zones of intergradation. The relative frequency, importance, and location of these three components of species structure differ from species to species. It is one of the tasks of comparative systematics to determine what kinds of organisms have what kind of species structure. Various geographical and ecological factors, the genetic and developmental potentialities of a given species, and its past history determine species structure. Our knowledge of these factors is still very elementary, but in the better-known groups of animals it is possible to arrive at certain generalizations.

Geography

Different species structures prevail in different geographic regions (Mayr 1942). Populations differ only clinally from each other in continental areas, while in island areas most species consist of strong isolates. For instance, among the passerine species of birds of continental Manchuria, 69 percent show clinal variation and only 3 percent are strong isolates, while in the Solomon Islands only 24 percent show clinal variation and 52 percent are composed of strong isolates (Mayr 1942) (Table 13–1). The superspecies of lizards *Lacerta muralis* in the western Mediterranean consists of three semispecies: *L. bocagei* (Iberian Peninsula), *L. pityuensis* (Pityusas), and *L. lilfordi* (Balearic Islands). The continental species *bocagei* has only three slight subspecies in the immense area of the mainland of Spain, while there are 37 subspecies of *pityuensis* and 13 subspecies of *lilfordi* on the respective islands, in spite of the fact that these lizards are absent on the main islands of the Balearic Islands (Majorca and Minorca), apparently having been exterminated by the lizard

Table 13–1. Species structure in birds of continental and island regions (after Mayr 1942).

Structure	Manchuria (continuous ranges)		Solomon Islands (discontinuous ranges)	
	Number	Percent	Number	Percent
Widespread, uniform species	15 ⎫	69	1 ⎫	24
Minor geographic variation	59 ⎭		11 ⎭	
Species with isolates nearing species level	1 ⎫		17 ⎫	
Groups of semispecies or		3		52
allopatric species	2 ⎭		9 ⎭	
Species with ordinary subspecies		28		24

snake *Macroprotodon*. Yet every island rock nearby has its own race except for a few very small bare rocks (Eisentraut 1949). The same difference between a "continental" and an "insular" pattern of variation (Kinsey 1937) is found wherever natural (geographical-ecological) barriers break up the continuity of populations.

The efficiency of the barriers depends in turn on the ability of the given species to overcome distributional barriers (see Chapter 18). This is particularly evident when we compare the species structure of two closely related species that differ in their ability to spread and to colonize, as, for example, do the two species of pond snails *Lymnaea palustris* and *L. emarginata* (Mozley 1935). *Lymnaea palustris* is very widespread geographically and occurs in ponds and small lakes in several continents. There is evidently much dispersal and little isolation. The species shows a high degree of individual variability and the range of variation at any one locality tends to approximate that which occurs over the entire species range. *Lymnaea emarginata*, on the other hand, is ecologically specialized. It is for the most part found only on the rocky shores of large lakes, a type of habitat that is relatively rare throughout most of its range, so that individual colonies are often separated by several hundred kilometers. There is comparatively little gene flow between colonies and many of the colonies appear to be the progeny of a single founder. The range of variation at any one locality is only a small part of the total variability of the species. Species with similar ecology or with similar histories tend to have similar species structure.

The past history of an area is reflected in the structure of many species. This is the case, for instance, with the location of zones of secondary intergradation (Rand 1948; Meise 1936a; Keast 1961). The effect of local physiographic features on species structure in localized areas has been studied in a considerable number of recent analyses, such as those by Linsdale (1938), A. H. Miller (1951), W. Blair (1950), and R. Miller (1950, 1961). Some of the factors that affect the distribution of populations through the geographic range of a species have been discussed by Mayr (1942:100) and by Schmidt (1950).

Central and Peripheral Populations

Naturalists have long been aware of differences between central and peripheral populations of a species. Discussions of this subject have, however, been almost invariably confused by a failure of the authors to distinguish between various spuriously similar but unrelated phenomena. Matthew (1915) discussed at length the persistence of primitive genera, families, and orders at certain peripheral, isolated localities such as New Zealand, Tasmania, Madagascar, and Ceylon, although the groups to which they belong have become extinct elsewhere. He broadened this observation to the generalization: "At any one time the most advanced stages should be nearest the center of dispersal, the most conservative stages farthest from it" (Matthew 1915). However, the zoogeographic phenomenon of the survival of primitive types has nothing to do with infraspecific geographic variation. Indeed, the generalization one can make concerning infraspecific variation is precisely the opposite of that of Matthew: the "original" phenotype of a species is usually found in the main body or central part of a species range, while the peripheral populations, particularly the peripherally isolated populations, may deviate secondarily in various ways.

A second source of difficulty also involves a confusion of different levels of taxonomy, namely, confusion of genes within a population and strains within a species. Vavilov's (1926, 1951) "centers of diversification" of cultivated plants are not areas in which populations show a maximum genetic variation; rather, they are geographic areas in which the greatest number of distinct cultivated strains are found (or originated). That certain areas (such as Transcaucasia and northeastern Iran, among others) are such outstanding reservoirs of cultivated varieties can in part be explained by the length of time during which the species had been cultivated in these areas, and in part by the abundant opportunities for isolation in the agricultural oases of these mountainous or semiarid regions. That such rich reservoirs of strains exist is largely the result of the many opportunities for isolation, the same cause that is responsible for the evolution of the rich indigenous fauna of the Hawaiian Islands. Vavilov himself was fully aware of this. He emphasized that "extremely interesting" deviates from the average type of the species are found "on the periphery of the areas occupied by a given plant and in places of natural isolation, such as islands and isolated mountain regions" (Vavilov 1951:47), and *not* at the center of the species range.

A third area of confusion surrounds the term "variation." When an author says a species is more variable either in the central part of the range or along the periphery, he should specify whether he means the species as a whole, or a given local population. A species as a whole may be more variable peripherally than centrally, because it has formed many divergent isolates along its periphery, even though each local population in the isolate has far less genetic variability than any local population in the center of the species range (see below). In the following account the terms "high variability" and "low variability" are always used with reference to a single local population, unless specified otherwise.

A fourth area of confusion concerns the "movement" of genes and characters in a species. The following may be taken for granted: all populations of a species actively exchange genes with each other, directly or indirectly, unless such exchange is prevented by dispersal barriers; furthermore, as a result of population surplus the more successful populations will exert a greater "gene pressure" (and will consequently manifest greater population mobility) than the less successful populations; finally, the success of such gene flow depends on the ability of the "alien" genes to compete with the "local" genes. The more deviant the environmental conditions are, as is the case in most peripheral areas, the less likely it is that the alien genes will survive for any length of time. It is not at all simple to translate the events on the level of the gene to the level of the taxonomic character. A curious theory has been proposed according to which "new characters" continuously arise in the center of the range of a species and spread from there outward in all directions, only to be displaced in turn by the next wave of emerging characters. According to this hypothesis the most primitive characters in a species will be found in peripheral populations (as in Matthew's theory for the higher categories). There is no genetic theory that would support this hypothesis, which ignores the two-way movement of genes as well as the local adaptation (through selection) of populations. All the examples used to support it can be explained quite easily along more orthodox lines.

The grasshopper *Ceuthophilus uhleri,* widespread in the lowlands of eastern North America, has also colonized the Appalachian mountains, which are in the center of its area of distribution. In this ecologically divergent area, it has acquired a deviating type of genitalic structure (Hubbell 1954).

A species may evolve a specially adapted population in any ecologically "marginal" area, whether this is in the center of the species range or

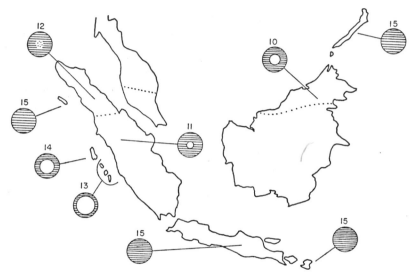

Fig. 13–6. Polytopic subspecies in the drongo *Dicrurus leucophaeus,* the central populations (subspecies 10, 11, 13, 14) with a white mask, the peripheral populations (12, 15) without it. The four populations designated as 15 are indistinguishable. (From Vaurie 1949.)

at its periphery. The polytopic origin of similar populations along the periphery of a species range can be interpreted on the same basis. Vaurie (1949; Fig. 13–6) describes the case of the drongo *Dicrurus leucophaeus* in which comparatively pale populations on Sumatra and Borneo are peripherally surrounded on smaller islands (Palawan, Java, Bali, Lombok, and Simalur) by darker birds with dark lores. Voous (1955) shows that when species of birds from humid Venezuela colonize arid islands in the West Indies (Fig. 13–7) they evolve independently similar phenotypes (paleness and so forth). It would be a great mistake to conclude from the convergent similarity of these populations that they are necessarily derived from each other.

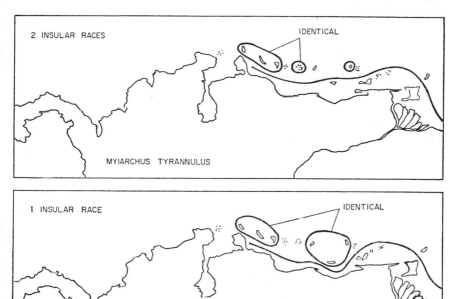

Fig. 13–7. Identical populations may evolve on different islands off the coast of Venezuela, even though independently derived from the mainland subspecies. Top, the tyrant flycatcher *Myiarchus tyrannulus*; bottom, the mockingbird *Mimus gilvus*. (From Voous 1955.)

Characteristics of Central and Peripheral Populations

The populations near the center of the species range are usually completely contiguous; they also show a relatively high population density (per unit area), and greater individual variation than is the average for populations of the species. Peripheral populations tend to have opposite values for each of the three characteristics (frequent isolation, low population density, low individual variation). This broad generalization, long accepted by naturalists, is based on general observations but lacks, so far, detailed quantitative support. It is substantiated by phenotypic variability in polymorph species. A study of such species reveals almost invariably that the degree of polymorphism decreases toward the border of the species, and that the peripheral populations are not infrequently monomorphic, as Reinig (1939) has demonstrated for bumblebees (*Bombus*) and Mayr and Stresemann (1950) for several genera of birds. In most mimetic butterflies there is a similar decrease in the number of mimetic forms per population toward the periphery of the species range. All peripheral populations of the highly polymorphic moth *Zygaena ephialtes* are monomorphic (Bovey 1941).

The best available evidence, however, comes from an analysis of chromosomal polymorphism in *Drosophila*. The widespread and very common tropical American species *Drosophila willistoni* has over 50 different gene arrangements in its variable three pairs of chromosomes. A single individual may be heterozygous for as many as 16 inversions, and 9.4 heterozygous inversions per female is the average found in the most heterozygous population. Only one inversion has a universal distribution from Argentina through Brazil to Central America, the West Indies, and Florida. Most inversions occur throughout the range of the species but are absent in a few peripheral populations. The wide range of most inversions is illustrated by the occurrence in Brazil of all but one of the inversions found in Florida. About a dozen of the inversions are more or less restricted in distribution. The inversional polymorphism is highest in several areas in Brazil, but drops off toward the south (Argentina, Chile), east (easternmost Brazil), and north (northern Central America, Florida, West Indies). At the most isolated point of the range (St. Kitts Island, West Indies) only two inversions are found, with the frequency of 0.2 heterozygous inversion per individual (da Cunha, Burla, and Dobzhansky 1950; Townsend 1952; Dobzhansky 1957b; da Cunha *et al.* 1959).

The picture in *Drosophila robusta* is similar. Some 18 different types of gene arrangements are found in different parts of the range of the species. Again, most arrangements are widespread throughout the range of the species, but the frequency drops toward the periphery of the range. Near the center of the species range (Virginia, Tennessee, southern Missouri) more than 95 percent of the wild females are heterozygous for at least one sizable inversion, and some have as many as one in each of the six major chromosome arms. Peripheral populations have 1–6 inversions on the six major chromosome arms in addition to the standard sequence, while the more central populations have 7–9 such inversions (Carson 1958c). A population from the extreme periphery of the range (northeastern Nebraska), on the other hand, is completely homozygous for all chromosome arms except for two females (among 337 sampled) which had a single inversion in the right arm of the X-chromosome (Carson 1955a,b, 1956, 1958c; Fig. 13–8).

The reason for the reduction of structural heterozygosity in the

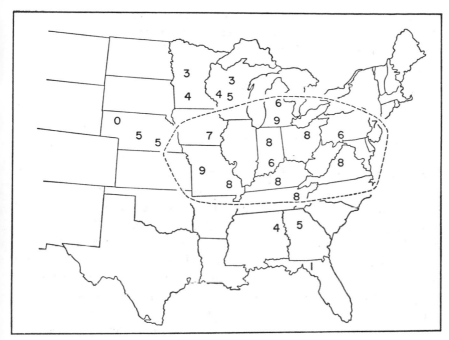

Fig. 13–8. Number of gene arrangements (in addition to standard) in 24 populations of *Drosophila robusta*. The central populations have 6–9 arrangements, the peripheral populations 0–5. (After Carson 1958c.)

peripheral populations of these two species of *Drosophila* is not entirely clear. Two major explanations, which however are not mutually exclusive, have been advanced. According to one hypothesis, the gene arrangements have an ecotypic function (Mayr 1945). This suggestion has been elaborated and broadly documented by Dobzhansky (1951) and his collaborators (da Cunha and Dobzhansky 1954; da Cunha *et al.* 1959). There is a great ecological difference between center and periphery of the species range. A species usually finds itself at its ecological optimum near the center of its range. The physical environment is so favorable that the species can, so to speak, make ecological experiments and occupy various subniches that would be unsuitable under the more adverse conditions at the periphery of the range. A great diversity of gene arrangements is thereby favored according to Ludwig's theorem In peripheral or otherwise ecologically marginal areas the ecological leeway of the species is drastically reduced and only a single ecological variant may be able to survive. This hypothesis is supported by the observation that the greatest number of gene arrangements are, on the whole, found in the ecologically most versatile species. The number of arrangements is reduced in D. *willistoni* in areas with many competing species or with adverse conditions (da Cunha and Dobzhansky 1954).

Carson (1955a,b; 1958c; 1959) generally accepts the above view but has stressed the point that the relative structural homozygosity of the margin will have the result of providing opportunity for increased amounts of recombination. Inversions tie up many genes in the central populations in nonrecombining coadapted groups. This gives two advantages: it prevents the breakup of these coadapted groups by active gene flow and at the same time it utilizes the heterosis of such balanced polymorphisms. Marginal populations that live not only under more severe, but also under more fluctuating conditions are considered to be able to mobilize, through the larger amount of chromosome available for free recombination, genetic adjustment to new conditions. Central populations of D. *robusta* have only 67–71 percent of the chromosomes available for crossing over and near peripheral populations have 84–85 percent, while a truly peripheral population has virtually 100 percent of the chromosomes free. In the central populations situated in the midst of a multidirectional gene flow, there is no danger of an undue lowering of genetic variability even if half of the chromosomes are locked up by inversions. In the peripheral populations, on the other hand, with their small genetic input due to gene flow, anything will be favored by selection that increases genetic variability.

Carson's hypothesis that the geographic variation in the distribution of gene arrangements is the result of selection for free recombination raises a number of as yet unanswered questions. For instance, there are species of *Drosophila,* even widespread and successful ones, that have little chromosomal polymorphism. Still other species are highly polymorphic in the central as well as the peripheral populations. A population of D. *subobscura* from Israel (near the southern periphery of the range of this species) has a richer variety of gene arrangements (26) than any European population studied (14–21) (Goldschmidt 1956). The habitat in this semiarid area is so marginal * that the species can be collected only in spring for some six to nine weeks. Yet this peripheral and ecologically handicapped population of Israel is as polymorphic chromosomally as are ecologically favored populations. The amount of free recombina-

---

* I prefer to use the term *marginal* in its well-known ecological meaning (= near the minimal level of subsistence) and *peripheral* in the strict geographical sense (= near the periphery of the species range).

tion is only slightly larger than in Europe (Stumm-Zollinger and Gold-schmidt 1959). In orthopterans, White (1957a) found no evidence of a consistent decrease of chromosomal polymorphism in peripheral populations. Different species thus appear to have different ways and means by which they cope with the diversities in the genetic needs of central and peripheral populations. Stone *et al.* (1960) call attention to additional considerations. The mere fact that central populations are prosperous and rich in individuals results in the occurrence of large absolute numbers of individuals homozygous for some chromosome sections, even though their relative frequency is drastically lowered. This, and several other factors, will maintain ample recombination in central populations even where its relative frequency is severely curtailed. Furthermore, the arguments presented by Mayr (1954a) (see Chapter 15) and the comparison of species like *Drosophila robusta* and the *D. repleta* group (Wasserman 1960) show that isolation is more important for the genetic composition of populations, owing to the cohesive force of gene flow, than mere peripheral location.

Thus, the genetic differences between central and peripheral populations can be described as follows. The total amount of gene flow is reduced in peripheral populations and the gene flow becomes increasingly a one-way inflow of genes near the periphery. Many of the peripheral populations, particularly the more isolated ones, are established by a single fertilized female or a small group of founders which carry only a fraction of the total genetic variability of the species. Contiguous central populations, on the other hand, are in the midst of a stream of multidirectional gene flow and harbor at all times a large store of freshly added immigrant genes. Environmental conditions are marginal near the species border, selection is severe, and only a limited number of genotypes is able to survive these drastic conditions. Reduction of gene flow and increased selection pressure combined deplete the genetic variability of the peripheral populations. This permits, if it does not favor, a shift into different ecological niches. For, as Haldane (1956) has pointed out, the selection pressure at the periphery is not only more severe, but also different. Central populations, being in the area ecologically most favorable for the species, tend to build up large populations the size of which is mainly controlled by density-dependent factors. Genes adapting for such density-dependent factors are accumulating in such populations. Low-density populations from near the tolerance limits of the species are being selected mainly for adaptation to density-independent factors.

COMPARATIVE SYSTEMATICS AND SPECIES STRUCTURE

There are many ways in which species may differ from each other in their population structure: they may be phenotypically uniform ("monotypic") or they may show geographic variation; the species population may be more or less continuous or it may be fragmented into isolates; there may or may not be a central-peripheral differentiation, to mention only a few of the points made in the preceding sections. Comparison of the patterns of population structure found in different species of mammals, birds, insects, snails, and other organisms is one of the tasks of comparative systematics. This field of research is still very new, and much more needs to be known about the geographic variation in the various groups of animals before such investigations can be placed on a quantitative basis and used for broad biological generalizations.

As a first approach to a study of intraspecific variability one may analyze the presence and frequency of subspecies in various groups of animals. This correlates, by definition, with the degree of geographic variability and depends on a number of previously discussed factors (Chapter 11). Degree of variability may differ quite strongly in families belonging to the same order. For instance, among the North American wood warblers (Parulidae) only 22 (42.3 percent) of the 52 species are polytypic, while among the buntings (Emberizidae) 32 (76.3 percent) of the 43 North American species are polytypic. This difference is real and not an artifact of different taxonomic standards. The extent to which such figures may depend on the standards chosen is illustrated by some data on European mammals. In 1912 only 26.0 percent of the species were considered polytypic, in 1937, 45.3 percent, and currently about 75 percent. Similar shifts have been reported in other groups as their systematics became better known (Table 13–2), and it is therefore difficult to make reliable comparisons. However, that there are some real differences is made apparent by the available information. Of the species of passerine birds in the New Guinea area 79.6 percent are polytypic, while only 67.8 percent of the North American passerines are polytypic. Among the 25 species of *Carabus* beetles from central Europe, 80 percent are polytypic, while in certain well-known genera of buprestid beetles not a single species is considered polytypic.

There are still large groups of animals in which all species are listed under binomials. It would be interesting to know to what extent this is due to a real lack of geographic variation of the phenotype and to what

Table 13-2. Prevalence of polytypic species in a number of groups of animals.

Systematic group	Total number of species	Polytypic species	
		Number	Percent
European mammals: G. Miller 1912	196	51	26.0
European mammals: Ökland 1937	168	76	45.3
Palearctic mammals (nonmarine): Ellerman and Morrison-Scott 1951	362	261	74.9
North American mammals: Miller 1924	1364	369	27.0
North American mammals: Hall and Kelson 1959	922	450	48.8
Palearctic Passeres: Hartert 1923	522	325	62.3
Hartert 1936	516	363	70.0
Vaurie 1959	579	385	66.5
North American breeding birds: AOU Checklist 1957	276	187	67.8
New Guinea Passeres: Mayr 1941	309	245	79.6
New Guinea non-Passeres: Mayr 1941	240	193	80.4
European reptiles: Mertens and Müller 1928	95	47	49.5
Mertens and Müller 1940	104	54	51.9
North American reptiles: Schmidt 1953	143	59	41.2
European amphibia: Mertens and Müller 1928	39	18	46.2
Mertens and Müller 1940	41	20	48.8
North American amphibia: Schmidt 1953	143	59	41.2
Indo-Australian butterflies: Jordan et al. 1927	695	412	59.3
Clausilia snails: Rensch 1933	37	16	43.2
Cypraea sea shells: Schilder and Schilder 1938	165	84	50.9

extent to insufficient taxonomic analysis. Sibling species, of course, are nearly always monotypic.

Number and Area of Subspecies

The classification of species as monotypic or polytypic is only one way of presenting quantitatively the amount of phenotypic variation. Another way is to analyze the subdivisions of polytypic species: What is the average number of subspecies per species in various groups of animals, and what is their average geographic range? There are believed to be about 28,500 subspecies of birds in a total of 8600 species. This comes to an average of 3.3 subspecies per species. It is unlikely that this average will be raised materially (let us say above 3.7) even after further splitting. The average differs from family to family: 79 species of swallows (Hirundinidae) have an average of 2.6 subspecies while 70 species of cuckoo shrikes (Campephagidae) average 4.6 subspecies and 75 species of larks (Alaudidae) 5.1 subspecies. The total number of subspecies is, however, much higher in a few species. The North American Song Sparrow (*Passerella melodia*), for instance, has some 30 subspecies. Species of birds with 20 or 30 well-defined geographic races are not rare in highly insular areas (such as the Indo-Australian region), the extreme apparently being the Golden Whistler (*Pachycephala pectoralis*), with over 70 races. The species of passerine birds that occur in the lowlands of New Guinea display a degree of subspeciation indicated in Table 13–3. Comparatively speaking, subspecies of birds generally have fairly wide ranges. In North American tiger beetles (*Cicindela*) the average subspecies range corresponds in size to that of bird subspecies, and the average number of subspecies per species (4) is also similar. The average range of subspecies is very much smaller in many groups of mammals. Sixty-one species of rodents have an average of 2.75 subspecies in the single state of Utah (Durrant 1952). The Kangaroo Rat (*Dipodomys ordii*) has no less than 30 subspecies in the western states, in an area where no bird has more than 5 or 6 subspecies (Setzer 1949). The pocket gophers (*Thomomys*) are noted for the high number of isolates they form on suitable soils and, consequently, the high number of subspecies distinguished by taxonomists. Of the two species in Utah, one has 11 subspecies in this state, the other 24.

Even smaller appear to be the ranges of recognizably distinct populations in many groups of invertebrates. Nearly every isolated oak grove in Mexico seems to have an endemic population of gall wasps (*Cynips*). Nearly every stream along the Gulf Coast of Florida has its endemic

Table 13–3. Subspeciation among 95 common
lowland species of New Guinea songbirds.

Number of subspecies per species [a]	Number of species
1	30
2	29
3	13
4	17
7	1
8	2
9	1
13	1
15	1
Total 245	95
Mean 2.58	

[a] 30 percent show no geographic variation, 6.3 percent
have 7 or more subspecies.

crayfish (*Cambarus*) (Hobbs 1942, 1945, 1953). Flightless carabid and
tenebrionid beetles tend to form very local subspecies. Even more localized are populations in many genera of tropical land snails. The Hawaiian
species *Achatinella mustelina* has 26 subspecies and 60 additional micro
geographic races in an area 20 by 5 miles in size (Welch 1938). Similar
distribution patterns have been found in *Partula* and *Cerion*. On the
other hand, in a genus with as stabilized a phenotype as *Drosophila,*
special methods (Stalker and Carson 1948; Patterson and Stone 1952)
may be required to discover phenotypic geographic variation. How extensively certain fresh-water crustaceans separate into local populations
has been pointed out by a number of recent authors, in particular by
Kiefer (1952) in his studies of the geographic races of *Thermocyclops*
from tropical Africa and other areas.

Describing intraspecific variation merely in terms of subspecies or
geographic races is a somewhat static approach. It may be useful in studies of species with widespread continuous ranges, but is inadequate in
coping with the intricacies of population structure of insular species.
Here local populations wax and wane. After long periods of distributional stagnation a local isolate may "erupt" and occupy large areas
previously left vacant by other populations of the species. This is sometimes made possible by shifts in habitat preference. Among such cases
are the flycatchers of the *Rhipidura rufifrons* group (Mayr and Moynihan
1946) and the Golden Whistler (*Pachycephala pectoralis*) (Galbraith
1956). In both cases a peculiarly interwoven distribution pattern re-

sults when the "older" isolates persist as islands in a recent "stream" of new colonists. Populations that are quite different in history, gene contents, and phenotype may become immediate neighbors, like white-throated and yellow-throated races in *P. pectoralis*. This situation leads over to other kinds of "borderline cases" between conspecific populations and full species, discussed in Chapter 16.

A somewhat aberrant but highly interesting population structure is found in the land-snail genus *Cerion* in the West Indies. This snail is limited to a narrow strip of coastal vegetation directly above the high-tide line. It rarely occurs more than 100–200 meters inland. An analysis of the populations that occur in northeastern Cuba (Banes district) reveals the following picture. Along a stretch of coast about 50 kilometers in length one finds seven highly distinct types of *Cerion*, all replacing each other geographically (with the single exception of the very different species *lepida*, which coexists with *moralesi* without interbreeding). Of eight potential contact zones, four consist of areas ecologically unsuitable for *Cerion* and thus form complete barriers, effectively preventing gene flow. The other four contact zones are occupied by hybrid populations. The ultimate chain of populations that results from these characteristics of *Cerion* is utterly different from the picture of a geographically variable species of birds such as the Song Sparrow. The species structure of *Cerion* (Fig. 13–9), which at first sight appears altogether irregular, is determined by two factors. One is that morphological divergence seems to take place rapidly but appears not to be correlated with the acquisition of reproductive isolation. As a result, whenever two morphologically divergent populations come into contact the probability is high that they will interbreed indiscriminately. The other factor is the unique dispersal capacity of *Cerion*. On one hand, there is extreme sedentariness, so that a given individual presumably does not move horizontally more than a score of meters in its entire lifetime; on the other, there is a capacity for passive dispersal during hurricanes, permitting colonization hundreds of kilometers away. It is probable that other cases of unusual population structures in animals similarly will find an ecological explanation.

SUMMARY

All species are composed of local populations, and, since in sexual species no two populations are identical, all species show geographic variation, much of it not expressed in the visible phenotype.

Many local populations are completely contiguous and grade into

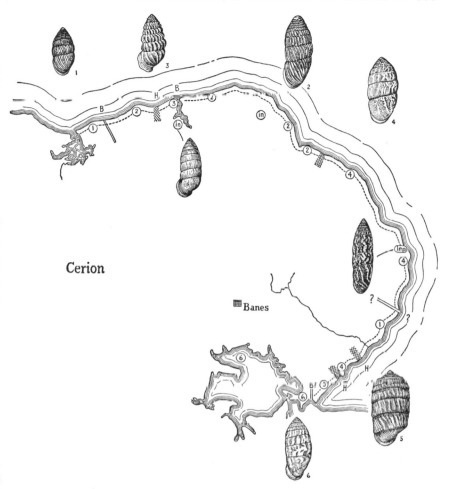

Cerion

▦Banes

Fig. 13–9. The pattern of distribution of populations of the halophilous land snail *Cerion* on the Banes Peninsula in eastern Cuba. Numbers refer to distinctive races or "species." Where two populations come in contact (with one exception) they hybridize (*H*), regardless of difference. In other cases contact is prevented by some barrier (*B*). In = isolated inland population.

each other. Others ("geographical isolates") are more or less isolated by barriers to gene flow. Where the barriers break down, secondary intergradation between previously isolated populations will take place.

Each species has a population structure characterized by a combination of the stated components. Genetic variability and population characters tend to be different in the center and at the periphery of the species range. Comparative systematics attempts to find generalizations concerning the population structure typical of various kinds of animals.

14 ~ Kinds of Species

The unitary aspect of the biological species of animals has been stressed so far in all chapters of this book. While admitting existence of the three species concepts, as described in Chapter 2, we assume that these concepts are merely different ways of looking at a single kind of species in nature. The biological species is usually also a morphological species (except for a few sibling species); the multidimensional polytypic species is nondimensional at any given place or time. Indeed, species of animals on the whole are a singularly uniform phenomenon. Terminologies proposed to differentiate between different kinds of zoological species in most cases were based not on contrasting biological properties of species, but rather on difficulties in the application of the species concept to specimens or samples (see Chapter 2). Paleontologists might speak of paleospecies, fossil species, or chronospecies (Sylvester-Bradley 1956), while practicing taxonomists might distinguish morpho- and biospecies (Cain 1954a), and students of asexual organisms recognize agamospecies. Yet, depending on the criteria applied, a sample of the same population might be called a morphospecies by one author and a biospecies or a paleospecies by another.

The shift from the formerly prevailing morphological species concept to the biological concept has been possible in zoology without much terminological disturbance. One necessary change was the transfer of the majority of "biological races" to the category of sibling species (Chapter 3). Even the shift from the nondimensional to the multidimensional species created no major disturbance. Proposals of dualistic terminologies like those of Kleinschmidt and Rensch (see Chapter 12) were not generally adopted. In contrast with botany, no separate field of "experimental taxonomy" developed in zoology, partly owing to the relative difficulty of breeding animals in captivity (except for *Drosophila* and a few other favored genera), and partly because various attributes of living animals (songs, courtships, and so forth) have long provided an independent check on the findings of museum taxonomy. Finally, the change from typological thinking to population thinking started in zoology at such an early date (in the middle of the nineteenth century) that the transition from alpha taxonomy to gamma taxonomy was gradual and harmonious.

The situation in botany has been different on the whole. Plant taxonomy, in spite of the efforts of a few pioneers, was strongly morphological-typological in its methods up to the twenties or even thirties of this century, resulting in a considerable gap between the herbarium botanists and the plant ecologists. Population thinking did not become popular until the 1930's (Clausen, Anderson, Turrill, Fassett, and others). When Turesson (1922) began his work in experimental taxonomy, he thought that it was impossible to accommodate his new ideas within the framework of the morphological terms of orthodox Linnaean terminology. He distinguished therefore between Linnaean species, which he defined morphologically, and *ecospecies*, which he defined as "Linnaean species from the ecological point of view." As a result, a dualistic terminology developed in botany. In fact, some botanists have declared that the classification of specimens and the (experimental) study of natural populations were two independent branches of biology, one being orthodox taxonomy and the other "genecology" or "biosystematy." This controversial subject has been discussed in a number of recent contributions by Baker, Cain, Camp, Danser, Epling and Catlin, Gregor, Mason, Turrill, and Valentine, reviewed by Heslop-Harrison (1954). Other botanists, for instance Stebbins (1950), have supported a single terminology. Regardless of what happens in botany, it does not seem that an elaboration of terminology is needed in animal systematics. Taxonomically speaking, there is only one kind of species of animals, except for the borderline cases between subspecies and species and the situation in asexually reproducing animals.

The weakness of past proposals for classifying kinds of species is that the choice of the criteria—morphological versus biological, or fossil versus recent—does not touch on any truly significant attributes. Yet there can be little doubt that there are different kinds of species, species characterized by fundamental differences in ecology, population structure, genetic systems, and modes of reproduction. To distinguish kinds of species in a meaningful way one must use such biological attributes as classifying criteria. Perhaps the most elaborate classification so far published is that of Camp and Gilly (1943), who distinguish 12 categories of plant species. Their criteria largely overlap, however, and, because of the overlap and the unwieldy terminology proposed by these authors, their classification has not been adopted by any subsequent author.

CLASSIFYING CRITERIA

At this point it might be useful to consider what kinds of criteria could be utilized in an attempt to classify kinds of species. As soon as one begins to tabulate them, it becomes evident that almost any property of a species might be used but that there is partial overlap between any two sets of criteria. A list with no claim to completeness is given in Table 14–1.

This tabulation could be expanded almost *ad infinitum*. Every species can be classified theoretically into one of the subdivisions of each of the numbered categories. There is thus an enormous amount of overlap between the various classifying criteria. The important question is which of the criteria are most important for the biologist, and particularly the evolutionist. Very little is known so far about the amount of correlation between the different sets of criteria. Certain correlations are obvious, such as that panmictic or sibling species will usually also be monotypic. But rather little is known so far as to the correlation between environmental tolerance, mating system, population structure, and rate of evolution. What ecological properties are correlated with specific genetic systems? It will not be possible to answer such questions until the systematics of animals, particularly of the lower invertebrates, has advanced a great deal further. We badly need to develop a field of comparative systematics that can serve as a firm basis for the study of comparative evolution. Until better data are available, only a tentative discussion can be attempted, to deal particularly with those sets of factors believed to affect the evolutionary potentialities of species. Among these the genetic systems, along with ecological and behavioral factors, seem most important.

GENETIC SYSTEMS

The totality of genetic factors affecting the population structure of a species and its evolutionary potential is sometimes designated the genetic system (Darlington 1939, 1940). White (1954:366), for instance, declares:

Table 14–1. Classifying criteria for kinds of species.

Criterion	Kind of species
1. System of reproduction	Biparental sexual reproduction
	Hermaphroditism
	Parthenogenesis
	Reproduction by fission or vegetatively
2. Degree of intra- and interspecific fertility	Cenospecies
3. Presence or absence of hybridization	Occasional interspecific hybrids
	Occasional introgression
	With allopatric hybrid zones
	Sympatric hybrid swarms
	Amphiploidy
4. Variation in chromosome number or pattern	Variable chromosome numbers (dysploidy)
	Polyploidy
	With more or less extensive structural heterozygosity
5. Difference in origin	Gradual by geographic speciation
	Gradual by sympatric speciation
	Instantaneous by polyploidy or macromutation
	By fusion
6. Structure of species	Monotypic
	Polytypic
7. Size of populations	Constant
	Highly, often cyclically, fluctuating
8. Sequence of generations	Rapid
	Annual
	Slow, a single generation extending over several to many years
9. Amount of gene flow	Essentially panmictic
	With numerous geographical isolates
	Largely inbreeding
10. Pattern of distribution	Cosmopolitan
	Widespread
	Insular
	Relict
11. Environmental tolerance	Euryecous
	Stenecous
12. Rate of evolution	Slow or stagnant
	Rapid
13. Phenotypic plasticity	Sibling species
	Polymorphic species

Under the general term *genetic system* we include the mode of reproduction of the species (bisexual, thelytokous, haplodiploid, etc.), its population dynamics (population size, sex ratio, vagility, extent of panmixia or inbreeding, etc.), its chromosome cycle (meiosis normal in both sexes or anomalous in one or both), its recombination index, presence or absence of various forms of genetic or cytological polymorphism in the natural population, and, in brief, all those characteristics which determine its hereditary behavior over periods of time sufficient for evolutionary changes to occur.

This broad definition embraces most factors listed in Table 14–1 as determining "kind of species," and several additional ones. Most mechanisms listed are merely alternative means to the same end, namely, regulation of the balance between inbreeding and outbreeding. The factors that determine the degree of genetic difference of the zygote-forming gametes have been combined under the term "breeding system" (Darlington and Mather 1949). This controls genetic variability, population structure, and, ultimately, evolutionary change, as emphasized by Darlington (1932, 1939), Dobzhansky (1951), and White (1954). Stebbins (1950, 1960) and other authors have pointed out that the amount of outbreeding depends not only on the usually stressed genetic and chromosomal factors of sexuality, ploidy, and recombination index, but also on such more or less ecological factors as dispersal facility, size of population, stability of population (versus fluctuations), length of life, overlap of generations, number of offspring, differences between larval and adult ecology, and so forth.

We must now consider the relative importance of some of these factors in determining the degree of difference among various kinds of species.

Variation in Chromosome Number

White (1954, 1957a) has given superb surveys of this topic. The occurrence of several different chromosome numbers within a single species may be explained in different ways. It is most frequently caused by the presence of supernumeraries or by Robertsonian fusions or fissions. More drastic differences are caused in certain parthenogenetic species by upsets in the mechanism of maturation and cleavage of the developing egg, as in *Artemia salina* (Goldschmidt 1952). Geographic variation in chromosome number is not infrequent in animals. In the rodent *Gerbillus pyramidum* (Wahrman and Zahavi 1955) chromosome number varies from 40 (Algeria) to 52 (coastal plain of Israel) and 66 (the Negev and adjacent parts of Egypt). The significance of this pronounced variation of closely related populations is not diminished even if one raises these isolates to the rank of full species. Secondary hybridization among such populations might well lead to highly variable chromosome numbers, as demonstrated by Staiger (1954) for *Thais lapillus*. Chromosome number is comparatively constant in most groups of animals but shows tremendous variations in others, for instance in butterflies. The number of chromosomes sets a lower limit to the number of linkage groups, but this number is controlled by additional chromosomal factors, such as chiasma localization.

Fertility

Some botanists recognize a special species category for all populations and species that can hybridize successfully. Clausen, Keck, and Hiesey (1945) have redefined as *cenospecies* "all the ecospecies so related that they may exchange genes among themselves to a limited extent by hybridization." Although it is unquestioned that the establishment of complete cross sterility signifies the point of complete irreversibility of speciation, it must be stated emphatically that sterility is only one of many possible isolating mechanisms. The recognition of cenospecies greatly exaggerates the importance of the sterility barrier and would lead to absurd conclusions in animal systematics. Among animals it is not uncommon that all members of a genus, or even subfamily, can exchange genes with each other even though in nature this occurs rarely or never. To call such a richly diversified group, for instance the river ducks (*Anatini*), a single cenospecies is meaningless. In other groups of animals (*Drosophila*, for instance) considerable barriers to gene exchange may already exist on the subspecific level and most full species are sterile with each other. I can see nothing to favor recognition of a species category for animals corresponding to the cenospecies of Turesson and his school.

MODE OF REPRODUCTION AND KIND OF SPECIES

The biological function of sex, long disputed, is the production of a vast variety of different genotypes (Weismann 1902). Mutation can assume this function only when, as in microorganisms, the generations follow each other much more rapidly than the changes of the environment. The essence of sexual reproduction, thus, is the combining of the genetic factors of two different parent individuals (or cells) in a new zygote. That in animals these gametes are generally produced in special gonads and that gametes (eggs) produced by such gonads can occasionally develop without fertilization is not basic to sexuality. To define sexual reproduction as "reproduction involving gamete formation" is unbiological and misses the point. Any reproduction that does not involve genetic re-combination is biologically speaking asexual, and it does not matter whether the new individual is produced vegetatively (by fission or bud-ding) or from an unfertilized egg cell.

Modes of reproduction other than orthodox sexuality are rare in ani-mals, compared to plants. Vegetative reproduction, such as budding in certain sessile colonial marine and fresh-water organisms or simple fis-sion in certain protozoans, turbellarians, and annelids, is usually only a temporary condition. It normally alternates with sexual stages or genera-tions. However, in certain species no sexual stage has ever been definitely demonstrated. The two most common deviations from bisexual (= gono-choristic) reproduction among animals are hermaphroditism and parthe-nogenesis.

Hermaphroditism

Hermaphroditism, that type of sexuality in which a single individual produces both male and female gametes, is very widespread in the animal kingdom. Its exact significance is still obscure in spite of much discussion (Altenberg 1934). It is certain, however, that hermaphroditism is not equivalent to close inbreeding. In most hermaphroditic species there are numerous mechanisms that reduce or completely eliminate the chances for self-fertilization. Of these, the most important is successive hermaphrodit-ism, termed protandry or proterogyny, and designating a condition in which the gonads at any one time produce only male or only female gam-etes, one set prior to the other. Another such mechanism is reciprocal fer-tilization, best documented in the pulmonate snails. Considering how widespread hermaphroditism is among the lower animals, it is surprising how few cases are known of obligatorily self-fertilizing hermaphrodites. In most reported cases of self-fertilization, it seems to be only an alternate mechanism to cross fertilization. Self-fertilizing hermaphroditism has, for instance, been recorded in the following groups: turbellarians (*Procero-des, Macrostoma, Opistoma*), cestodes (*Anthobothrium musteli* plus many others), trematodes (*Distomum cirrigerum*), cirripedia (*Pollicipes cornucopia*), snails (*Lymnaea auricularis*), and nematodes (*Rhabditis* and so forth).

Hermaphroditism is usually considered a primitive condition, and the loss of one sex, or more precisely of one gonad, the derived condition. However, a good argument can also be made for the opposite evolutionary sequence. Indeed, in nematodes of the genus *Rhabditis* new hermaphroditic "species" may arise suddenly (by unknown chromosomal mechanisms) in cultures of bisexual species (Chapter 15).

It is well established, however, that ambisexual tendencies are present in nearly all groups of animals, including those in which the sexes are normally in different individuals. Except in the few cases of self-fertilizing hermaphroditism, there is no evidence that hermaphroditism affects species structure. It would by no means lead necessarily to closer inbreeding than does the separation of sexes in different individuals. Rather, its significance seems to be that it permits an increase in general productivity. The production of eggs, on the whole, requires much greater metabolic resources, and it is often found that smaller, younger individuals of hermaphroditic species produce male gametes, and larger, older individuals produce female gametes. One might expect that most sessile animals would be hermaphrodites because male individuals would not be able to search for females actively. That this is not true is demonstrated by the corals as well as other coelenterates and by sponges, in which there are numerous cases of separation of the sexes. In many cases hermaphroditism is an exceedingly labile condition. Not only may there be a difference between closely related species, but its expression may vary geographically (Bacci 1955b). Among vertebrates functional hermaphroditism is rare, but it is now well established for several species of fishes (Spurway 1957; Clark 1959; Mead 1960; Schultz 1961).

A condition corresponding to hermaphroditism is widespread among protozoans. Rejuvenation of an asexually reproducing clone depends in *Paramecium* on a sexual process that can be (1) conjugation with an individual of the opposite mating type (and normally a different clone of the same variety), or (2) selfing, that is, conjugation with an individual of the same strain, or (3) autogamy, that is, formation of a fertilization nucleus by union of the male and female gamete nuclei of one and the same individual. Ciliates are characterized by two "ranks" of sex. Although each individual is genetically speaking a hermaphrodite, producing male and female gametes, yet the existence of mating types facilitates and favors outbreeding. Cross-breeding conjugation has first priority and only if there is no opportunity for it will selfing or autogamy take place (Sonneborn 1957).

The technical term in zoology for the separation of the sexes is *gonochorism*. It has been suggested that the botanical terms dioecy and monoecy be substituted for the zoologists' hermaphroditism and gonochorism. Such transfer does not seem justified, since the equivalence is not strict. Dioecy and monoecy refer to sporophytes, while gonochorism and hermaphroditism are phenomena relating to gamete-producing individuals. Sexuality in sessile colonial animals is highly diversified. At one extreme are some hydropolyps (for instance, *Diphyes acuminatus*) in which all individuals are gonochoristic and all individuals of a colony are of the same sex. At the other extreme are colonies in which all individuals are hermaphrodites (ectoprocts, synascidians). Between these extremes are types (most siphonophores, *Plumularia*) which are gonochoristic, but in which a colony may contain individuals of both sexes. Whether or not one includes such cases under hermaphroditism depends on the definition of "individual." All the "individuals" of such a colony are parts of a single genetic individual.

Cross-fertilizing hermaphrodites are not known to differ in species structure from gonochoristic forms. Self-fertilization not only increases the amount of inbreeding, but also permits single individuals to become the founders of new populations. It might therefore lead to a change in evolutionary potential. Unfortunately, however, good comparative taxonomic studies of self-fertilizing hermaphrodites are not available and it is not known to what extent this mode of reproduction affects species structure and speciation.

Close Inbreeding

Self-fertilizing hermaphroditism is not the only mode of reproduction that results in close inbreeding. It results also from the obligatory brother-sister matings observed in some parasitic hymenoptera that mate inside the host, for instance *Telenomus fariai* (Dreyfus and Breuer 1944), in the grass mite *Pediculopsis* = *Siteroptes* (Cooper 1939), and among trematodes (*Didymozoon, Wedlia bipartita*). In such cases all the advantages of sexual reproduction seem to have been abandoned. How such a curious mode of reproduction could have become established in the face of natural selection is still a complete mystery, but it is probable that exceptions occur occasionally in these species and that genetic variability is restored through such exceptional outbreeding. Indeed, it is known in *Siteroptes* that the nymphs will emerge under conditions of high humidity.

The probability of self-fertilization is greatly lessened in plants by the widespread occurrence of self-incompatibility genes (Grant 1958). Such mechanisms are rare in animals and have apparently been reported only in *Ciona intestinalis*. However, the mating types in ciliates (Sonneborn 1957) correspond to the incompatibility factors in plants.

Parthenogenesis

Parthenogenesis denotes the development of offspring from egg cells not fertilized by male gametes. It occurs in two forms, of which only female diploid parthenogenesis (thelytoky) is of interest to us. The other type—the production of haploid males from unfertilized eggs (arrhenotoky)—is a form of sex determination, and its main effect on the genetic system is that it eliminates in each generation all deficiencies, homozygous lethals, and other factors that are inviable in hemizygous condition. This male haploidy has arisen only about seven times in the whole history of the metazoa, five times in the insects, once in the mites, and once in the rotifers (White 1954).

Thelytoky, simply called parthenogenesis in the following account, has arisen repeatedly in most larger phyla of animals. It occurs either as an optional or seasonal condition in otherwise sexually reproducing animals, or is complete, males being entirely unknown. Complete parthenogenesis is usually limited to an occasional species in an otherwise sexually reproducing group. The most recent summaries are that of Suomalainen (1950) and the excellent analysis of the cytological and evolutionary aspects of parthenogenesis by White (1954).

From the cytogenetic viewpoint we may distinguish two types of parthenogenesis which, with White, we may designate the *meiotic* and the *ameiotic* type. Meiosis is entirely suppressed in the ameiotic type, and, since the maturation divisions in the egg are like any mitotic division, the daughters will have a genetic constitution identical with that of the mother, except for an occasional genic or chromosomal mutation. Since there is no recombination, and only dominant mutations will be exposed to selection, there will be an accumulation of recessive mutations and of structural rearrangements, leading to ever-increasing heterozygosity. Furthermore, since the pairing of the chromosomes is eliminated with meiosis, there is no longer any mechanical barrier to the establishment of various chromosomal irregularities, including polyploidy. As a result polyploidy is very widespread in groups of animals with ameiotic parthenogenesis (see Chapter 15).

In the meiotic type of parthenogenesis chromosomal reduction occurs during meiosis but is compensated for by restoration of the diploid chromosome number at some subsequent stage of the life cycle. There are three or four alternate ways of achieving such restoration. Either the first meiotic division is abortive (even though preceded by pairing and crossing over) and the second is a simple mitosis, or meiosis is complete, but two complementary ones of the four pronuclei fuse (automixis). Even though there should be a steady loss of heterozygosity through crossing over, this apparently does not necessarily happen. It is prevented by the fusion of the unlike pronuclei (or the abortion of the reduction division), but more importantly by continued selection in favor of heterozygotes. Among meiotic parthenogenetic groups, polyploidy has similarly arisen several times.

There has been some argument whether or not to consider this uniparental form of reproduction as sexual. Using a genetic definition, in which genetic recombination is the criterion of sexuality, one concludes that meiotic parthenogenesis would qualify as sexual reproduction, but ameiotic parthenogenesis would not. Genetically speaking, there is no difference between reproduction in which new individuals are produced vegetatively and reproduction by ameiotic parthenogenesis.

The phenotypic variation of completely parthenogenetic animals has been critically compared with that of bisexual animals only a few times (for example, Suomalainen 1961). Since each individual and the clone to which it gives rise remain permanently independent of all related clones, one would expect, owing to mutation, a steady genetic divergence between clones, ultimately resulting in high variability in the parthenogenetic "species." Some of the most highly polymorphic species of British sawflies, *Mesoneura opaca* and *Eutomostethus ephippion,* are completely parthenogenetic (Benson 1950). Other parthenogenetic species seem to have no more variation than sexual species. The reason for this might be that the vast majority of mutations is recessive and not able to become homozygous in the absence of recombination. The probability of phenotypic variation is even smaller in tetraploids where each new allele has to compete with three wild-type alleles.

The amount of parthenogenesis in a species quite often varies geographically. In the millipede *Polyxenus lagurus* the fraction of males in population samples decreases from south to north: France, 41.6 percent; Denmark, 8.7 percent; Sweden, 5.6 percent; Finland, 0 percent (Palmén 1949). Parthenogenesis greatly increases productivity because each gamete becomes a new individual. It permits the rapid building up of populations in habitats with highly fluctuating ecological suitability.

Complete parthenogenesis is usually limited to an occasional species or genus in a broad taxonomic group. In the vertebrates, for instance, it has been reported in nature for teleost fishes in the poeciliids (Hubbs and Hubbs 1932; Haskins *et al.* 1960; Miller and Schultz 1959; Schultz 1961) and in *Coregonus* (Melander and Montén 1950), also for reptiles in *Lacerta saxicola* (Darewski and Kulikowa 1961) and in *Cnemidophorus* (Maslin 1962). Studies in nematodes, psychid moths, lumbricids, simuliids, *Drosophila,* and other groups of animals have shown how quickly parthenogenesis can be acquired. However, the only example of a large group of metazoa known to me of which all members show complete, ameiotic parthenogenesis is the rotifer order Bdelloidea (Chapter 15). The absence of a complete continuum of clones between these taxonomic entities in this order can perhaps be explained by secondary elimination of all clones that are competitively inferior. Perhaps this is also the explanation for other cases of discontinuities between closely related parthenogenetic species, such as the white-fringed weevils (Buchanan 1947) or the New Zealand phasmids of the genus *Acanthoxyla* (Salmon 1955).

Complete parthenogenesis poses a taxonomic problem. The orthodox species criterion of interbreeding cannot be applied, because each clone is reproductively isolated not only from the parental species but also from every sister clone. How to treat clones and parthenogenetic species taxonomically must be decided for each case. Where no essential morphological or biological differences exist, such clones should be combined into collective species. Where a parthenogenetic line has originated from a bisexual species by an irreversible chromosomal event (for instance, polyploidy), it is usually advisable to consider it a separate (sibling) species, even though the morphological difference is slight. This practice has proved itself in the classification of polyploid-associated parthenogenesis in earthworms, weevils, and psychid moths.

The evolutionary importance of parthenogenesis is that it permits instantaneous speciation (see Chapter 15). However, by abandoning genetic recombination, it generally gains only a short-term advantage, and, with the apparent exception of the bdelloids, virtually every case of parthenogenesis in the animal kingdom has all the earmarks of recency. Though superficially appearing a "more primitive" type of reproduction, parthenogenesis in recent animals is evidently in all cases secondarily derived from sexual reproduction.

The Occurrence of Asexuality Among Animals

It is sometimes claimed that half or more of the kinds of animals reproduce by self-fertilization or asexually. A study of the literature does not support such claims. Cross fertilization seems to be of regular occurrence even among the protozoans, although it usually alternates with many generations of simple fission or selfing. Mobility, which permits mates to seek each other actively, gives animals a much greater efficiency of fertilization than is possible among plants. This is the reason why gonochorism, combined with normal sexuality and protection against hybridization, is by far the most frequent breeding system in animals. This is also the reason for the absence among animals of many of the elaborate methods known in plants for regulating the degree of outbreeding (Stebbins 1950). Various forms of asexuality therefore play a much smaller role in animals than in plants in producing different kinds of species.

As an aside, it might be mentioned that the widespread occurrence of genetic recombination has been established in recent decades in the procaryota (viruses, bacteria, blue-green algae), organisms without well-defined nuclei and meiosis (Pontecorvo 1958). Such "meromixis" (Wollman, Jacob, and Hayes 1956), in the form of "haploid crossing over," transformation, and transduction, lacks the regularity of genuine sexual recombination, but it may be responsible for the observed cohesion within the "species" of these microorganisms. There is much to commend the view of Dougherty (1955) and Stebbins (1960) that genetic recombination is of such high selective value and of such widespread occurrence among the lowest known organisms that it must have originated very soon after the origin of life itself, and that any now occurring form of asexual reproduction is a derived condition even in the lower organisms. Contrary claims notwithstanding, existing forms of asexuality are not evidence of a primitive condition. The species problem, of course, is far more desperate in these procaryota than anywhere in the animal kingdom.

ECOLOGY AND KIND OF SPECIES

The relation of an animal to its environment is an important factor in the determination of the amount of outbreeding and other components of the species structure. Since every species differs in its ecology from every other species, one might say that every species is a different "kind of species." However, some environmental factors affect population structure more than others and are therefore more important for the evolutionary potential of species.

Frequency. Rare species differ in many ways from common species (Haldane 1932; Huxley 1942). Though every naturalist believes he knows what a rare species is, it is exceedingly difficult to define rarity precisely, because it can be defined only in relation to something else (Andrewartha and Birch 1954). A simple definition is: a rare species is one that has a very much lower frequency per unit area than have other species of the same genus or of related genera (it would be absurd to express on the same scale the rarity of animals as different as whales, birds, and soil nematodes!).

Rare species are usually of one of two types: highly localized or highly specialized. A highly localized species, like Rothschild's Starling on northwestern Bali (*Leucopsar rothschildi*) (Plessen 1926), is sometimes common where found, but is exceedingly restricted. Some species of fresh-water fish are restricted to a single spring or creek and yet may form populations of hundreds or thousands of individuals (Miller 1948; Clark Hubbs and Springer 1957). The evolutionary prognosis of such a localized species cannot be considered very favorable unless the confining barriers break down. In most cases a species is rare because it is highly specialized and finds only few places where its specialized demands are met. The geographical range of such a species is easily affected by distributional barriers. Often the species consists of well-isolated populations that are usually sufficiently distinct from each other to form well-defined geographic races. Common species, on the other hand, show great adaptability. As a consequence, a single population may be able to occupy many different ecological niches, and each of these populations may show such great phenotypic variation that localized subspecies cannot be delimited. Such species tend to show clinal variation. All sorts of intermediate conditions exist between the extremes of very rare and very common species. The most interesting perhaps is the species that is very rare in some years and exceedingly common in others.

It is evident that rarity may have many different causes that, though often called ecological, may ultimately be genetic. Stebbins (1942) believes that on the basis of past history rare species can be separated into two classes of conditions: "One is that the species was once more common, widespread, and richer in biotypes than now, so that its present rarity is due to depletion of the store of genetic variability. The other is that the species never was common, but diverged from a small group of individuals of a widespread ancestral species." He distinguishes therefore two types of rare species, "depleted species" and "insular species," assuming that all rare species are genetically homogeneous. The assumption that rare species that are not insular are ultimately relicts poses numerous problems. What permits them to maintain their population level in the absence of population "density?" Why are they not steadily declining in number?

(Some are, others definitely are not.) How do competitive factors in rare species compare with those in common species? What is "density-dependent" in a rare species?

The greater variability of common and widespread species has been known since the time of Darwin or longer. In the beginning the emphasis was on purely morphological characters, but since such characters have a genetic basis it was extended to genetic variability and indeed, as shown by Fisher, Wright, and others, there is a great accumulation of genetic variability in such species (Chapter 9). The ecological consequences of increased genetic variability are of considerable interest. If each genotype in a population is somewhat specialized ecologically, the more genotypes a species contains, the more versatile and ecologically tolerant it will be. Genetic variability and ecological versatility then form a mutually reinforcing system: the more widespread and common a species is, the more genetic variability it can store, the more tolerant ecologically it can become and the wider it can spread, and so forth. The need for harmonious cooperation among genes during development, however, sets a definite upper limit to this accumulation of genetic variation by a successful species (see Chapter 10).

Habitat Characteristics and Habitat Utilization. The relation between an organism and the environment in which it lives seems sufficiently important to serve as the basis of a classification of kinds of species. One has been proposed by Thoday (1953), who distinguishes three classes of species:

(1) Those that live in a relatively uniform and stable environment and will therefore not be exposed to selection in favor of genetic or phenotypic flexibility; such species will be selected primarily for stability and adaptation to this uniform environment;

(2) Those that live in a heterogeneous environment and will therefore be strongly selected for phenotypic flexibility, especially if the generation time is long;

(3) Those that live in an unstable environment and will therefore be strongly selected for genetic flexibility, although selection for this genetic flexibility will be less, the greater the phenotypic flexibility.

A study of ecological variation in species of animals shows that one can recognize even more categories, categories based more on the reaction of species to the environment than on the type of environment in which they live. Although there are presumably additional ones, I am aware of five major types of ecological specialization:

(1) Specialization for a very narrow niche;

(2) Broad tolerance of the individuals of the species wherever they occur, without the development of genetic polymorphism to facilitate occupation of the extremes of the ecological spectrum occupied;

(3) Polymorphism: the presence of several morphs in the population adapted to particular subniches (Ludwig effect);

(4) Ecotypic variation: the formation within each geographic area of numerous localized populations that specialize in ecologically different subareas, and are not obliterated by gene exchange with adjacent, differently specialized populations;

(5) Geographic polytypicism: the formation of geographic races in response to the environmental variation over the major part of the species range.

It is not certain that these five categories can be sharply delimited. Indeed, it is possible that in certain species some of these categories overlap, for instance (2) and (4), or (3) and (4), or (1) and (5). Except for polymorphism in gene arrangements, nothing at all is known about the genetic systems that permit a species to cope with its ecological needs in such a way that it will be classified in one or another of the above-mentioned categories. Extreme specialization is typical for many insects and virtually unknown among the higher plants. Ecotypic variation is common among plants, particularly in specialized inbreeders, but is rare in animals. The role of physiological-ecological polymorphism, found in *Cepaea* and *Drosophila*, is still very poorly understood. It is possible that a rather simple genetic switch mechanism is involved in producing the manifold differences of the various *Culex* mosquitoes. There is little chance that ecological polymorphism will lead to speciation in sexually reproducing diploid organisms. The genes or chromosomes responsible for the polymorphism cannot be taken out of the gene pools to which they belong without an actual spatial segregation of populations.

BEHAVIOR AND KIND OF SPECIES

Various components of behavior affect the amount of outbreeding and other population properties of species that are of evolutionary significance. The whole field of behavioral isolating mechanisms might be mentioned here as well as behavior-controlled food and habitat selection. Some of these points have been treated in a recent symposium on behavior and evolution (Roe and Simpson 1958). Here I shall single out for discussion merely two behavior elements, parental care and mobility.

Parental Care and Sequence of Generations. The amount of parental care exercised by a species is of great evolutionary importance. An organism that is poorly protected against the physical environment, or in which the young are poorly protected, or that is faced by vast fluctuations in its potential resources will benefit by a vast production of zygotes. Much of the mortality will be nonselective, and intraspecific competition will tend to be limited to certain stages in the life cycle. Conversely, the more an organism is protected against the vicissitudes of the environment, the more independent it becomes of the environment and the greater becomes the role of competition among conspecific individuals. Under these circumstances it is of greater selective advantage to produce well-equipped offspring than many offspring. If there is much competition for females and a long period of parental care, there will be a high selective premium on an increased life span. This again increases intraspecific competition and tends to depress the number of offspring even further. Instead of a survival (to the adult stage) of less than one in a million, as is found in so many species of marine invertebrates, survival may be as high as one out of five or ten. There is a tremendous difference between nematodes and other small invertebrates, with a new generation every few days, and some large mammals and birds, in which only a single offspring is produced every second year and 5 to 10 years are required to reach maturity. Natural selection is considerably curtailed when so few genotypes are available for choice. Even these two extremes of genetic systems among animals are not as distinct as are among plants the long-lived trees with their immense number of seeds and the short-lived microorganisms with a new generation every 10 minutes. Species with a slow turnover of generations and a small number of offspring require special mechanisms for the maintenance of high genetic variability, such as an increase in the number of chromosomes and in the rate of mutation. There are indeed indications of a higher rate of mutation in mammals than in *Drosophila,* and in *Drosophila* than in bacteria.

Mobility. Mobility gives animals an exceedingly flexible means of adjusting the amount of outbreeding. In contradistinction to plants, dispersal in animals is only partially passive. The average distance of dispersal during the dispersal stage in the life cycle of an individual is in part controlled by behavior characteristics. Sedentary habits, territory occupancy, and philopatry (aided by homing ability) tend to reduce the amount of random dispersal. Restlessness, induced by various influences, may lead to long-distance dispersal of a smaller or larger proportion of the population.

In animals mobility and dispersal are the most important devices for the prevention of close inbreeding. In addition there are numerous other devices that reduce the chances of brother-sister matings. Unisexual broods are known in the Sciaridae, Cecidomyidae, and other insects (White 1954). In many insects the males hatch first and have already dispersed before the females of the same brood hatch. In other species sexual maturity of the males may be delayed until after their dispersal flight. In phytophagous insects reproduction often takes place on or near flowers, sometimes at considerable distances from the host plant. Intermediate hosts for larval stages of parasites provide for wide dispersal and for unequal rates of maturation of the offspring of a parent individual. Among higher animals there may be a psychological barrier against brother-sister matings, as reported for the goose genus *Anser*. All these devices reduce the probability of intrafamily matings but do not prevent them altogether.

ANIMAL SYSTEM AND KINDS OF SPECIES

It would seem to be interesting and important to have precise quantitative data on the relative frequency of the various kinds of species in the different phyla and classes of the animal kingdom. Unfortunately such a quantitative analysis is not yet possible, since we know too little about the taxonomy of most groups of animals, and particularly the lower animals where most of the more interesting and more aberrant kinds of species occur.

In some of the better-known groups, such as insects and vertebrates, different patterns of variation have been found in related genera and families. In some families more than 80 percent of the species are polytypic, while in other groups nearly every species is monotypic. The latter seems to be the case, for instance, in *Acmaeodera*, a genus of buprestid beetles (Cazier, verbal communication) and in many Microlepidoptera (Munroe 1951). The genera are large and ill-defined in some families (for example, weevils, solitary bees, some Diptera); other families have large numbers of clear-cut, often monotypic, genera, for instance the longhorn beetles (Cerambycidae), or, among plants, the milkweed family (Asclepiadaceae). It would be interesting to know the reasons for such differences. Different chromosomal characteristics have been suspected, but never demonstrated. Much evidence indicates that ecological factors are very important. For instance, the high incidence of monotypic species among North American warblers (Parulidae) can be explained by their long-distance migrations which churn up the populations every year. Rensch (1933) has pointed out the importance of annual migration in gene dispersal. That migration is not the entire answer is indicated by the large amount of geographic variability shown by the migratory Emberizidae (buntings). However, these are ground-living birds and perhaps are more exposed to selection by predators and microclimates than are species living in tree tops, such as most Parulidae. A species with generalized habits, such as tiger beetles (*Cicindela*), is exposed to more selective factors, including regional differences, than is a species restricted to a single host. The relatively homogeneous environment may explain in part the rarity of geographic variability in many host-specific genera. Another factor contributing to clear definition of genera is the age of the group. An archaic group, such as the gymnosperms, has better-defined genera than has an actively evolving group such as the recent songbirds.

Sonneborn (1957) found that in ciliate protozoans various genetic and reproductive characteristics are correlated with species structure. Outbreeders are characterized by long periods of immaturity and maturity, providing opportunities for dispersal and for finding a mate of a different mating type. Rate of fission is slow and length of generation extends over months or years. The number of mating types may be more than two, thus increasing the opportunity for cross breeding! Outbreeders generally have a wide range of distribution in latitude and longitude, thus indicating high dispersal facility and considerable ecological tolerance. Inbreeders show the converse condition for all these aspects. Mortality of ex-conjugants is high in outbreeders when crossed with sibs, and in inbreeders when crossed with individuals from distant localities. Variety 1 of *Paramecium bursaria* is a typical outbreeder, while varieties 10 and 14 of *P. aurelia* are typical inbreeders. Other species of ciliates are somewhat intermediate (Sonneborn 1957).

DIFFERENCES BETWEEN ANIMAL AND PLANT SPECIES

Some characteristics of animal species become more obvious when compared with species in plants. For the species of plants and animals, though agreeing in many ways, differ in many others. Stebbins (1950:71) quite rightly emphasizes this:

> The generalization emerging from these examples is that we cannot apply uncritically the criteria of species which have been developed in one group to the situations existing in another, particularly if the groups are distantly related to each other and have very different modes of life. This generalization applies with particular force to attempts of zoologists to reinterpret the species which have been recognized by botanists, or of the latter to use their yardsticks of

species on animal material. General principles of systematics as well as of evolution must be based on as broad a knowledge as possible of different groups of plants and animals.

Various authors have attempted to record and classify those differences between animals and plants that might affect their species structure. These comparisons are based on the higher plants and animals, in part because the lower forms are not well known and in part because lower forms somewhat bridge the gap between the two kingdoms. Some of the more important differences between plants and animals, largely based on the discussions of Turrill *et al.* (1942), Anderson (1937), Stebbins (1950), and Grant (1957), are the following:

(1) As animals are mobile and capable of seeking their own environment, habitat selection is among their important characteristics. A plant seedling, on the other hand, must be successful wherever it germinates. From this basic fact follow many consequences, some of which are listed below. It explains why animals can afford to be more specialized, less variable, and less plastic phenotypically. In plants, there is a higher premium on storing genetic variability so that at least some of the high number of seeds will have a genetic combination that will guarantee successful germination.

(2) Individuals among the higher animals are usually unisexual, individuals among the higher plants are usually bisexual.

(3) Polyploidy is common and morphologically differentiated sex chromosomes are rare in the higher plants. The reverse is true for animals.

(4) Animals in reproductive condition seek each other actively, but intermediaries like insects or wind are used by plants to transport the male spores to the female plants. Hence sterility is a much more important isolating mechanism in plants.

(5) Interspecific hybridization, including back crossing to one of the parental species, is much more common in plants than in animals. It may serve as a mechanism for replenishing genetic variability.

(6) Most animals have the ability of active dispersal, whereas dispersal in plants is passive.

(7) There are fewer aberrant kinds of genetic systems among animals than among plants.

(8) Plants have a considerable capacity to alter their phenotype in response to changing conditions of the environment, such as light, moisture, feeding, and so forth. In most animals the genotype has only a very limited range of phenotypical expression.

(9) Reticulate speciation through allopolyploidy and interspecific hybridization is much more common in plants than in animals. In higher animals it is virtually absent.

(10) The life span is very short in most animals, exceeding two years in only a very small number of species. Among plants extremely great ages are known, not only among trees, for instance the famous sequoias, but even in many kinds of plants that reproduce vegetatively. Sexual reproduction in long-lived organisms has a different significance than in the short-lived animals. In the long-lived plants, sexual reproduction serves primarily the purpose of creating new genetic combinations. In animals it serves the additional function of the survival of the species (Stebbins 1950:182–186).

(11) A major evolutionary trend among plants is to become better adapted to the environment; among animals the trend is to become independent of the environment.

In view of all these differences, most of which have an ecological background, it is important to emphasize that the species of animals and plants are nevertheless essentially similar. Plants and animals are virtually identical in their genetic and cytological mechanisms. Geneticists such as Beadle, Demerec, and Mather have switched from the study of a plant to that of an animal and back, and so have some cytologists. Not only the principles but even most of the details of the cellular mechanisms are the same. It is important to emphasize this essential similarity between plants and animals in view of the many differences.

THE MEANING OF "KINDS OF SPECIES"

The survey of the many reproductive, chromosomal, ecological, and behavioral characteristics by which kinds of species differ from each other confirms the opinion of many authors, including Stebbins (1950), Dobzhansky (1951), and White (1954), that they range between two extremes, inbreeding and outbreeding. The average amount of difference of the gametes that produce the zygotes of the next generation depends largely on the amount of outbreeding. Furthermore, this factor is far more easily changed than are the various chromosomal mechanisms (discussed in Chapter 9) that regulate amount of recombination within a gene pool. It appears that the range in the degree of outbreeding is far greater in plants than in animals, at least in the higher animals. Among plants both extremes, complete inbreeding (= selfing) and extreme outbreeding (= hybridization with other species), are common. In the higher animals both extremes are rare, and the normal alternative is between moderate and increased outbreeding. To put it in another way, the size of the deme is the only variable. The larger the deme, other things being equal, the greater the probable genetic difference of fusing gametes. The factors that determine the degree of outbreeding (= the size of the deme) thus determine the average genetic difference of the gametes.

Outbreeders and inbreeders differ from each other in numerous ways. The entire breeding system of outbreeders is so organized as to accumulate and preserve genetic variation in order to have a maximum of ecological plasticity and evolutionary flexibility, but at a price—the production of many inferior recombinants. An outbreeder may also be so well buffered that it stagnates evolutionarily. At the other end is the extreme inbreeder which has found a lucky genotypic combination that permits it to flourish in a specialized environmental situation, but again at a price—inability to cope with a sudden change of the environment. A species thus has the choice between optimal contemporary fitness combined with considerable evolutionary vulnerability and maximal evolutionary flexibility combined with the wasteful production of inferior genotypes. No species can combine the two advantages into a single system. Every species makes its own particular compromise between the two extremes and every species has its own set of devices for achieving this compromise. To provide for more flexibility, devices exist in many evolutionary lines that permit the increase or decrease of the degree of outbreeding according to need. Such a device, for instance, is the highly polygenic sex determination in *Asellus aquaticus*, discovered by Montalenti (1960), which favors outbreeding in this species, one that tends to consist of highly localized, isolated populations. Outbreeding appears to be the original condition in animals and one can assert that any more extreme form of inbreeding (including the various forms of asexuality) is a derived condition.

Outbreeding, that is, genetic flexibility, as stated by Stebbins (1950), is favored by large, structurally complex, slow-growing organisms that have low numbers of offspring and live in a generalized environment. Inbreeding, that is, genetic fixity, is favored by small, structurally simple, fast-growing organisms that have large numbers of offspring and are more or less adapted to special situations. Most animals are essentially outbreeders, most microorganisms essentially inbreeders.

The various chromosomal, reproductive, and ecological factors are not assorted randomly but are adjusted to each other to form a single breeding system. It is difficult to follow the pathway by which the various characteristics are brought together and the success of an outbreeder is achieved. It is much easier to understand the selective advantage of many of the inbreeding devices because, as in the case of parthenogenesis, they nearly always result in a speed-up of reproductive rate. *Daphnia* in a lake in spring, or an aphid when the new foliage grows in spring, or a bacterium successfully invading a new host, all must multiply as rapidly as possible to keep ahead of competitors or deterioration of the environment. A temporary suspension of sexual reproduction in these forms has two advantages: a successful genotype can be perpetuated unchanged, without being broken up by sexual reproduction; but, more importantly, fecundity is doubled. Instead of "wasting" half the zygotes on males, which are not able to reproduce by themselves, all zygotes are fertile egg-producing females. This temporary abandonment of sexuality is characteristic of organisms that invade temporarily vacant niches (lakes and vegetation in spring). The return to sexual reproduction takes place promptly when the habitat is filled and conditions begin to deteriorate.

On the other hand, it is rather difficult to establish a selective model to account for the lengthening of generation time, the lengthening of the immature stage, and many other characteristics of some of the outbreeders. One successful model is Lack's (1954b) explanation of the selective advantage of small clutch size in birds. The advantage in many cases seems to favor the population rather than the individual. Finally, we must not forget that natural selection does not have supernatural powers. The enormous amount of extinction in the organic world indicates that evolution cannot "plan" because natural selection cannot help but favor the currently fittest genotypes, without regard for continued genetic flexibility.

The worst deficiency in our present knowledge is that we have so few studies of the breeding system of a species as a whole. Such items as chromosomal pattern, dispersal facilities, mode of reproduction, life cycles, rates of mutation, and other aspects of the breeding system have been studied individually but rarely or never as components of a single system. The study of breeding systems in animals has hardly begun, and yet it is the real basis of the understanding of "kinds of species."

THE ROLE OF SPECIES

The fact that the organic world is organized into species seems so fundamental that one usually forgets to ask why there are species, what their meaning is in the scheme of things. There is no better way of answering these questions than to try to conceive of a world without species.

Let us think, for instance, of a world in which there are no species, but only individuals, all belonging to a single "connubium." Every individual is different from every other one in varying degrees and every individual is capable of mating with those others that are most similar to it. In such a world every individual would be, so to speak, the center of a series of concentric rings of increasingly more different individuals. Any two mates would be on the average rather different from each other and would produce a vast array of genetically different types among their offspring. Now let us assume that one of these recombinations is particularly well adapted for one of the available niches. It is prosperous in this niche, but when the time comes for mating this superior genetic complex will inevitably be broken up. There is no mechanism that would prevent such a destruction of genetically superior combinations and there is, therefore, no possibility of the gradual improvement of genetic combinations. The significance of the species now becomes evident. The reproductive isolation of a species is a protective device against the breaking up of its well-integrated, co-adapted gene system. Through organizing organic diversity into species, a system has been created that permits genetic diversification and the accumulation of favorable genes and gene combinations without any danger of destruction of the basic gene complex.

15 ~ Multiplication of Species

One of the most spectacular aspects of nature is its diversity. This diversity has the very special property that it is not continuous, but consists of discrete units, the species. To explain the origin of these species is one of the great problems in the field of evolution. The local naturalist knew no answer to it: in fact, the mystery seemed to deepen the more he studied the relation of nondimensional species to each other. Invariably they are separated from each other by bridgeless gaps, as was shown in Chapters 2–5. Not only the naturalists were stumped by this problem; the attempts of early geneticists to solve it were likewise doomed to failure, since dominated by typological concepts they attempted to solve the puzzle with the help of hypotheses of instantaneous speciation involving a single individual. The eventual solution came from a very different direction, namely, from the study of the population structure of species, as shown in Chapters 7–13. This study yielded an abundance of new facts, permitting decisive advances in this field, and, in particular, a precise formulation of the problem.

The true meaning of the term "origin of species" was understood only rather recently. Not only were the pre-Darwinian evolutionists quite vague on this issue, but even Darwin himself seems to have considered "origin of species" the same as "evolution" (Mayr 1959a). Thus he confused two essentially different problems under the single heading, "origin of species." Darwin was primarily interested in demonstrating evolutionary change as such. This process was appropriately called by Romanes (1897) the "transformation of species in time," and by Simpson (1944) "phyletic evolution." It is quite possible to have evolutionary change without any multiplication of species. An isolated population on an island, for instance, might change in the course of time from species *a* through *b* and *c* into species *d* without ever splitting (Fig. 15–1). In the end there

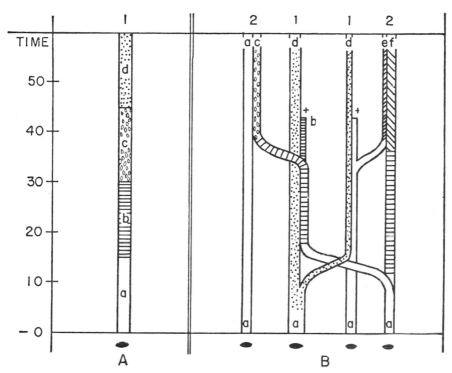

Fig. 15–1. A (on the left) designates a strongly isolated island on which a species *a* changes in the course of geological time through *b* and *c* to *d*. B (on the right) indicates an archipelago with four islands on which an originally monotypic species *a* breaks up into 5 species through geographic speciation and cross colonization. (From Mayr 1949a.)

will be only one species on the island just as at the beginning. This was clearly recognized as early as 1888 by Seebohm:

There is no reason why evolution should not go on indefinitely modifying a species from generation to generation until a preglacial monkey becomes a man, and yet no second contemporary species be originated. The origin of a second species is prevented by interbreeding. So long as the area of distribution of the species is continuous and not too large, the constant intermarriage which takes place between the males of one family and the females of another distributes the inherited and transmittable modifications throughout the race or species; which may advance or retrograde according to circumstances, but is prevented by interbreeding from originating a second species . . . In every species there is a tendency to vary . . . the variations are hereditary and cumulative, so that evolution goes on steadily, though slowly, from generation to generation.

There is some evidence in the paleontological literature that such phyletic evolution without speciation occurs not infrequently in somewhat isolated monotypic genera. The essential aspect of this type of evolution is the continuous genetic and evolutionary change within the populations composing the species, without the development of reproductive isolation between populations of the species, and consequently without its breaking up into several species.

The other evolutionary process is true speciation, or what Romanes in 1897 called "the multiplication of species in space." It is the splitting of an originally uniform species into several daughter species. The normal mode by which such multiplication takes place is the process called geographic speciation, as will be shown in Chapter 16. Its genetic aspects will be discussed in Chapter 17 and its ecological aspects in Chapter 18.

As long as "speciation" was believed to mean simply evolutionary change, not much thought was given to the question whether there is only one way or several ways for species to multiply. Interest in this question developed only after Moriz Wagner raised the issue in 1868. Yet, in spite of much subsequent argument among adherents of theories of speciation by mutation, by ecological specialization, and by geographic isolation, the problem was largely ignored by geneticists and experimental biologists. In all of his writings T. H. Morgan hardly mentioned it, and even in the 1930's the emphasis in discussions on the origin of species was on the origin of genetic differences rather than on the genetics of discontinuity. Even now we are only beginning to understand the speciation process and, at that, in only a few groups of animals. We do not know yet to what extent our findings can be generalized. Do the same modes of speciation occur in all groups of animals? Huxley (1942) asserts that "hymenoptera differ from crustacea in their modes of speciation, corals from higher vertebrates," but does not say how. He may be right, but it is equally probable that new species of vertebrates, hymenoptera, and crustaceans originate in an exactly equivalent manner, as far as population phenomena are concerned. And as far as differences in chromosomal reconstruction are concerned, there will also be considerable independence of the taxonomic hierarchy.

MODES OF SPECIATION

The rapid progress of the last twenty-five years has greatly advanced our understanding and the time has come for a critical discussion of the various possible modes of speciation. Huxley (1942) enumerates three kinds of speciation: geographical, ecological, and genetic. These classifying criteria are not well chosen, since they permit overlapping categories: all geographic speciation, for instance, is simultaneously ecological and genetic; all ecological speciation is also spatial and genetic. A short preliminary classification of four modes of speciation was proposed by Mayr (1942:187), and a more elaborate one, with ten alternatives, by Sewall Wright (1949a). It seems to me that none of these schemes clearly brings out the basic divisions. Sewall Wright's classification is based on two major classifying principles: (1) whether the new species arises from one, or from more than one, ancestral species, and (2) "the size of the population from which the new species takes its origin." His classification includes partially overlapping categories, and the modes of speciation are not listed according to their relative importance.

What, then, would be a more satisfactory classification? Undoubtedly the best would be one based on a careful comparison of the modes of speciation in all kinds of animals from the lowest protozoans up to the higher vertebrates. This cannot be done because we are still quite ignorant about speciation in the lower animals. All we know about speciation (that is, the origin of discontinuities) in any group of animals is ultimately based on taxonomic research. But the taxonomy of the lower animals is still far too insufficiently known to permit generalizations. In fact, the very same groups that might have the most interesting speciation patterns are taxonomically the least known. These include: (a) animals that are ecologically specialized, such as parasites, (monophagous) food specialists, symbionts, marine animals, and (b) animals with unorthodox modes of reproduction, such as forms with parthenogenesis, self-fertilizing hermaphroditism, vegetative reproduction, or complicated alternations of generations (see also Chapter 14). The population structure in such species may be very different from that in species with normal biparental reproduction. Our greatest need then is for more taxonomic work and studies of population structure in these aberrant groups of invertebrates and microorganisms. Until such studies have been completed, it is quite possible that important modes of speciation have been overlooked.

Speciation means the formation of species ("specification," as Darwin called it). In retrospect it is evident that much of the past argument on modes of speciation was due to the application of the single word "species" to three very different concepts (Chapter 2). If we keep in mind the existence of these three species concepts and consider as well most of the ten alternative modes proposed by Sewall Wright (1949a), we come to a revised classification (Table 15–1) which utilizes three major criteria: (1) addition of new, reproductively isolated populations (III), or not (I, II); (2) instantaneous origin through individuals (III A) or gradual origin through populations (III B); and (3) geographic isolation of the speciating populations, or not. Combining these alternatives in various ways results in twelve potential modes. It will be the purpose of this chap-

Table 15–1. Potential modes of origin of species

I. Transformation of species (phyletic speciation)
    1. Autogenous transformation (owing to mutation, selection, etc.)
    2. Allogenous transformation (owing to introgression from other species)
II. Reduction in number of species (fusion of two species)
III. Multiplication of species (true speciation)
    A. Instantaneous speciation (through individuals)
        1. Genetically
            (*a*) Single mutation in asexual "species"
            (*b*) Macrogenesis
        2. Cytologically, in partially or wholly sexual species
            (*a*) Chromosomal mutation (translocation, etc.)
            (*b*) Autopolyploidy
            (*c*) Amphiploidy
    B. Gradual speciation (through populations)
        1. Sympatric speciation
        2. Semigeographic speciation (see Chapter 17, p. 525)
        3. Geographic speciation
            (*a*) Isolation of a colony, followed by acquisition of isolating mechanisms
            (*b*) Extinction of the intermediate links in a chain of populations of which the terminal ones had already acquired reproductive isolation

I2, II, and IIIA2(*c*) may lead to reticulate evolution

ter to discuss whether, and to what extent, these modes of speciation actually occur in the animal kingdom.

Transformation of Species

The evolutionary transformation of a species is referred to by the terms "phyletic evolution," "modification in time," or "modification by descent" (Darwin). Such transformation is quite independent of the origin of discontinuities and does not lead to multiplication of species. It is easy to imagine how a highly isolated species could, in the course of geological time, gradually change into a very different species without budding off any additional species during this period. The occurrence of such temporal changes has been admitted even by some well-known antievolutionists, such as von Baer and Cuvier. It is compatible with the story of creation in the Book of Genesis.

When a paleontologist speaks of speciation, he usually has this type of phyletic evolution in mind. And Darwin, under the somewhat misleading heading *On the Origin of Species,* primarily discussed descent by modification. While Darwin and most early Mendelians did not make a clear distinction between the two processes, that of evolutionary change and that of the branching of evolutionary lineages, it has become increasingly customary in recent times to restrict the term "speciation" to the process that leads to multiplication of species through branching of phyletic lines.

The genetic change that leads to phyletic evolution in a lineage may be *autogenous* (produced by mutation, recombination, selection, and so forth), as discussed in Chapters 8, 9, and 17, or it may be *allogenous,* produced by the incorporation of genes from a different species. Anderson (1953) believes that all important evolutionary changes are due to introgression, but this is evidently not true for the higher animals. In them introgression is rare and probably negligible as an evolutionary factor (Chapter 6). The isolating mechanisms between good species of animals are normally sufficiently efficient to prevent gene leakage. Where introgression occurs, nevertheless, its results are slight owing to the strong counterselection against disharmonious gene combinations.

Fusion of Two Species

The most extreme degree of introgression would be the complete fusion of two species. Such a fusion is, so to speak, the reverse of multiplication of species. Since species are reproductively isolated populations, fusion of two species is, on the whole, a logical contradiction. Yet occasionally a previously existing reproductive isolation breaks down and two previously distinct sympatric species merge. This happens most easily in species in which the isolating mechanisms are primarily ecological.

The frequency of this occurrence is very much in question. Most of the cases cited in the literature allow different interpretations. The secondary fusion of two previously isolated subspecies is, of course, a common phenomenon in birds and other animals (see Allopatric Hybridization, Chapter 13). Indeed, most so-called cases of the fusion of "species" are actually cases of an interbreeding of subspecies, narrowly defined as species. The semantic quandary is obvious. As far as birds are concerned, not a single species is known to me that would be best interpreted as resulting from the fusion of two previously existing species. The Marianas

Mallard (*Anas oustaleti*) has been shown by Yamashina (1948) to be a hybrid population of migratory Mallards (*Anas platyrhynchos*) and of Gray Ducks (*Anas superciliosa*). This fusion is restricted to a single very isolated locality, while the two parental species have retained complete integrity over more than 99.9 percent of their respective geographical ranges. The case of *Pardalotus ornatus* in Australia is suggestive (Hindwood and Mayr 1946) but not sufficiently analyzed (Cooper 1961). Perhaps the nearest approach to a real fusion of two species of birds is the case of the two towhees *Pipilo erythrophthalmus* and *P. ocai* described in Chapter 6. This case and several others discussed in that chapter prove that breakdown of isolation and subsequent fusion is a distinct possibility.

Patterson and Stone (1952) postulate a hybrid origin for *Drosophila americana*, because it combines the gene arrangements of *D. novamexicana* and *D. texana*. But the essential chromosomal similarity of the western *D. americana* populations with *D. novamexicana* and of the eastern *D. americana* with *D. texana* suggests a different interpretation (Mayr 1957c; Fig. 15–2).

Burdur Lake in Anatolia contains a hybrid population of cyprinodonts that may be the product of the fusion of two parental species. The two species *Anatolichthys splendens* and *A. transgrediens*, however, are not of hybrid origin (Villwock 1958), contrary to earlier assertions.

Every taxonomist knows of species that combine certain characteristics of two related species. It is quite inadmissible, in view of the highly polygenic basis of taxonomic characters and the stability of the epigenotype (sibling species!), to consider such intermediacy as evidence for hybrid origin. Ross (1958) utilizes such morphological intermediacy as evidence for a hybrid origin of various species of leafhoppers (*Erythroneura*), but he fails to make a convincing case. All the available evidence indicates that the origin of a new species through the complete fusion of two parental species is an exceedingly rare event among higher animals.

Even less likely is the origin of a new species by fusion of segments of two parental species together with the continued existence of the two parental species. Such a process has been postulated in a number of cases, but its proponents forget that they have to solve the two ever-present problems of speciation, namely, the acquisition of ecological compatibility and that of reproductive isolation. If the two parental species are reproductively so little isolated that they produce hybrids, their reproductive isolation from the resulting hybrids will be even less. The problem then **is** to segregate such a population of hybrids and keep it segregated until

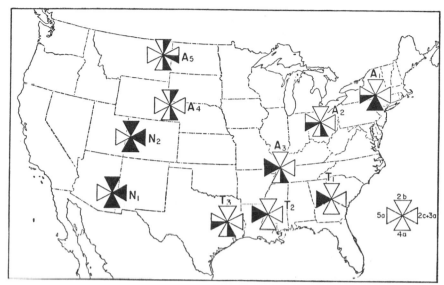

Fig. 15–2. Distribution of some gene arrangements in *Drosophila americana* (A_1–A_5), *D. texana* (T_1–T_3), and *D. novamexicana* (N_1–N_2). Presence of the inversions 2b, 2c + 3a, 4a, and 5a are indicated by black, absence by white arms of the crosses (key on right). A half-filled arm indicates polymorphism. The western populations of *americana* (A_4, A_5) have the gene arrangements of *novamexicana*, the eastern populations (A_1, A_2, A_3) those of *texana*. (From Mayr 1957c.)

it has acquired its reproductive isolation. No mechanism other than geographical isolation is known that could achieve this.

MULTIPLICATION OF SPECIES

The word "speciation," in its restricted modern sense, means the splitting of a single species into several, that is, the multiplication of species. The problem of the origin of discontinuities between species is made more precise by recalling the meaning of the word "species." A species is a reproductively isolated population. The problem of the multiplication of species, then, is to explain how a natural population is divided into several that are reproductively isolated, or, more generally, how to explain the origin of a natural population that is reproductively isolated from preexisting species.

Attempts to find an answer to this problem have led to many heated arguments; the history of some will be discussed in Chapter 16. It appears at present that there are nine potential modes of multiplication of species (Table 15–1). The exact number depends, of course, on the classifying principles employed. The remainder of this chapter will be devoted to a discussion of these various modes of speciation.

Instantaneous Speciation

Instantaneous speciation may be defined as *the production of a single individual* (or the offspring of a single mating) *that is reproductively isolated from the species to which the parental stock belongs and that is reproductively and ecologically capable of establishing a new species population.*

This is what one might call the naive concept of speciation. It was the prevailing concept among the early workers in the field. If one were to ask a lay person how a new species originates, he might say that a species suddenly throws off one individual or several individuals which then form the ancestral stock of a new species. The early Mendelians maintained similar views. De Vries said (1906:vii), "The theory of mutation assumes that new species and varieties are produced from existing forms by certain leaps. The parent type itself remains unchanged throughout this process and may repeatedly give birth to new forms." As recently as 1922 it was stated by Bateson that the problem of speciation was still unsolved: "The production of an indubitably sterile hybrid from completely fertile parents which had arisen under critical observation from a single common origin is the event for which we wait." Obviously he was unable to think of an alternative to instantaneous speciation. Even at present there are still some adherents of instantaneous speciation. On the whole, however, the reasoning has prevailed that species are populations and that new species populations are not normally founded by single individuals. To what extent this might, nevertheless, be possible will be discussed in the succeeding sections.

Instantaneous Speciation Through Ordinary Mutation. Mutation is a phenomenon of such relatively high frequency that in a higher animal with more than 10,000 gene loci almost every individual will be the carrier of a new mutation. Such mutations merely increase the heterozygosity of a population but do not lead to the production of new species, De Vries notwithstanding. Any mutation drastically affecting reproductive behavior or ecology will be selected against if it lowers viability, or will displace the original allele if it is of higher viability. In neither case will there be any origin of discontinuities.

It is evident that ordinary mutations cannot produce new species in sexually reproducing species. The situation is different, however, with uniparental, asexually reproducing animals. Here we face the quandary of having to define species without the help of the criterion of reproductive isolation. A frequent solution, particularly in botany (hawkweed; *Hieracium;* hawthorn, *Crataegus;* and so on), is to consider each morphologically distinct clone a microspecies. If this is done, a single mutation may indeed produce a new "species." Such morphological "speciation" may be even more common among fungi and bacteria than among higher plants. The majority of authors prefer a different solution than calling each clone a species. Even though between strains there are morphological differences that are not bridged by gene flow, there is usually so much morphological agreement among these strains that it is justifiable to treat them as components of collective species (Mayr 1953a).

A similar situation, although unknown among the higher animals, is a distinct possibility among lower animals, where various forms of uniparental reproduction are widespread. Unfortunately not a single case is taxonomically well analyzed. Strains of collective species offer numerous challenging problems: What form of competition is there between strains? How often and under what circumstances do such strains become extinct? If there is continued mutation in such strains, why does it not lead more often to the development of morphologically distinct full species? Why do such strains remain bunched as collective species? From what source do they receive the genetic variability that permits them to cope with natural selection and to continue evolving? Is it possible that a sexual phase does occur in such species under exceptional circumstances and that it provides an equalization of the gene contents of the various diverging strains? For the present we can only confess our complete ignorance of the answers to these questions.

Certain phenomena among rotifers, cladocerans, and nematodes suggest the occasional occurrence of such asexual speciation. Among the most challenging problems is that of speciation in the rotifer order Bdelloidea. In this entire large order not a single male has ever been found with certainty, despite much searching and the analysis of large samples. A secondary extinction of all sexual species appears improbable. It is therefore reasonable to assume that the ancestral species itself was parthenogenetic. This ancestral species evolved into over 200 species, about 20 genera, and 4 families. If our assumption is valid, all this took place without the benefit of increased genetic variation generated by sexual reproduction. One wonders what chromosomal mechanism the bdelloids have invented to permit them to evolve in a way normally available only to sexual species. The ameiotic nature of the parthenogenesis has been established for at least one species (Hsu 1956).

Instantaneous speciation seems to occur also among nematodes. As was first found by Maupas and later confirmed by Osche (1952), there occur occasionally in cultures of the sexual species *Rhabditis papillosa* neotenic hermaphroditic individuals that agree with the description of *R. caussaneli* Maupas, and in cultures of the sexual species *R. producta* individuals of hermaphroditic *R. duthiersi*. In another genus, *Diplogaster*, similar conditions were found by Hirschmann (1951). Here the sexual species *D. l'heritieri* occasionally produces individuals of the hermaphroditic *D. biformis* Hirschmann. No chromosomal analysis has yet been made in any of these cases and nothing can therefore be said about the mode of speciation (Osche 1954). All these cases of spontaneous "speciation" of nematodes, described in the literature, resulted in neotenic hermaphrodites that are able to reproduce without cross fertilization. A spontaneous production of gonochoristic sibling species (which are very common in nematodes) is altogether unlikely and has never been demonstrated. Most known species of free-living nematodes are uniform (monotypic) throughout their geographic range and sometimes cosmopolitan, although much of their wide distribution is presumably an artifact, due to passive dispersal by man. Geographic speciation remains the most probable mode of speciation among nematodes, even though there is not yet any supporting evidence.

Many similar cases of mutative instantaneous speciation, particularly those involving parthenogenetic forms or self-fertilizing hermaphrodites (Chapter 14), are found scattered through the literature. Polyploidy may be involved in many cases. However, I do not know of a single well-analyzed case involving a sexually reproducing, biparental species.

A special situation is presented by the species that alternate between sexual and asexual (parthenogenetic and so forth) generations. In the aphids (plant lice) and cladocerans (*Daphnia*), for instance, a parthenogenetic female will start a clone, all the members of which have the genetic properties of the founding mother. Each clone differs genetically and often also morphologically and physiologically from every other clone. After a series of uniparental generations all these clones return to the common gene pool of the species. Potentially this alternation of modes of reproduction provides a possibility for the sudden origin of a new species during the asexual period. However, there are numerous difficulties. Parthenogenetic organisms are usually not homozygous but are in a state of balanced heterozygosity, maintained even in the cases of meiotic parthenogenesis (Dobzhansky 1951; White 1954). It is improbable that the new type, when returning to sexuality, would be reproductively isolated and ecologically compatible with the parental type, and thus it would be subject to all the difficulties of sympatric speciation (to be discussed below). Kinsey's (1936) studies of gall wasps (*Cynips*) with an alternation of sexual and parthenogenetic generations indicate that they show the same kind of geographic variation as do exclusively sexually reproducing organisms.

Vegetative reproduction should offer as favorable a condition for instantaneous speciation as parthenogenesis. This form of reproduction, however, occurs only in some of the lowest groups of animals, such as sponges, coelenterates (hydroids), turbellarians, and bryozoans, groups in which the taxonomy is as yet too uncertain for a study of speciation. When the form of branching is an important taxonomic character, for instance in hydroids and graptolites, a single somatic mutation may in one step produce a considerable alteration of the phenotype. Such a mutation, of course, does not produce a new species, because the gametes are not reproductively isolated from those produced by the "parental" growth type. Only where a sexual stage is completely eliminated can such a somatic mutation lead to the production of a new morphospecies. A clear distinction must be made between a change of the phenotype and a reorganization of the gene pool. Changes of the phenotype are often saltational, as sometimes in polymorphism and as in the case of chromosomal inversions and translocation. The reorganization of the gene pool, required for successful speciation, is (except in the case of polyploidy) never saltational.

Speciation by Macrogenesis

The sudden origin of new species, new higher categories, or quite generally of new types by some sort of saltation has been termed macrogenesis (Jepsen 1943). Before the nature and structure of species was fully known and before the genetics of speciation was as well understood as it is today, the hypothesis of macrogenesis was one of the most widely accepted theories of speciation and of the origin of higher categories. It is a logical consequence of a typological interpretation of the systematic categories. The proponent of macrogenesis considers, explicitly or implicitly, that the individual is the real unit of evolution, and he believes that such an individual may undergo a major genetic reconstruction.

Every individual is thus potentially an incipient species or higher category. Among modern biologists who have supported theories of macrogenesis to a greater or lesser extent are Goldschmidt (1940, 1948a, 1952a,b), Schindewolf (1936, 1950a), Petrunkevitch (1952), and Willis (1940, 1949). It is inherent in the theory of macrogenesis that it can never be proved, since it is evidently impossible to witness the occurrence of a major jump, particularly one that achieves simultaneously reproductive isolation and ecological compatibility. To support their theory, the adherents of macrogenesis rely primarily on the claim that gradual speciation and the gradual origin of higher categories are impossible. As far as speciation is concerned, this claim will be refuted in Chapter 16, and as far as the origin of higher categories is concerned, in Chapter 19. The present discussion will be limited to pointing out errors in the arguments presented by the believers in macrogenesis and to demonstrating the unworkability of the mechanisms they propose.

The Origin of New Types. The believers in macrogenesis consider the chief support of their thesis to be the "fact" that all new types appear on the evolutionary scene suddenly and abruptly. These types are not connected with the ancestral types by intermediates, they claim, and cannot be derived from them by gradual evolution. Rensch (1960a), Simpson (1953a), and Heberer (1957) have shown how misleading this claim is. As our knowledge of the fossil forms has improved, it has been possible to demonstrate in one case after another how one "type" can be derived from a previously existing one. The fossil record is admittedly very incomplete, particularly among the soft-bodied early invertebrates. Among the organisms with hard parts, however, from the geological periods in which the fossil record is more complete, so-called missing links have been found time after time. Hamann wrote in 1892 that a derivation of the mammals from the reptiles was a clear-cut anatomical impossibility. By now so many intermediate links have been found that a clear-cut separation of the two classes has become an impossibility (Simpson 1959b). The "missing links" between most of the major categories of vertebrates have been found in the 100 years since Darwin. This is quite remarkable in view of the fact that an estimation of the total number of species that must have existed at one time (Simpson 1952b; Cailleux 1954) and of the fossil species already described by paleontologists reveals that only about one out of 5000 species has so far been discovered. As Heberer (1957) has shown, gaps in our phylogenetic lineages are invariably correlated with gaps in the geological record.

Some saltational postulates are based on the assumption of essentially invariant evolutionary rates. If birds, for example, do not show any essential evolutionary change from the Cretaceous to the present, the postulates lead by backward extrapolation to an absurdly early date for the origin of birds. Taking the case of bats, where the first known bat from the Eocene is hardly different from modern bats, Simpson points out that by this system of extrapolation the origin of the mammals would antedate the origin of the world. All we know about evolutionary rates shows clearly how unequal they are and, more specifically, with what rapidity the valley between one adaptive peak and another one is crossed. He who postulates essentially even evolutionary rates cannot help but arrive at absurd conclusions.

The adherents of macrogenesis believe that a new type is created more or less in its final form as a new class or phylum and that it becomes subdivided into families, genera, and species by subsequent evolution. The philosopher Schopenhauer has unwittingly reduced this assumption to absurdity. He remarked that the problem of the origin of new higher categories was easy. Parents, he said, produced normally an individual as offspring. To have new higher categories, parents would simply have to produce a new species, genus, order, class, or phylum. Evolution for this typologist was nothing but a categorical trick. Some of the recent theories of the origin of higher categories through "hopeful monsters" do not differ in principle from this philosophical absurdity. We shall discuss in Chapter 19 how clearly this assumption is refuted by the actual facts. Whenever new types originate, they are virtually identical with the types from which they have branched off. The first artiodactyls are condylarths except for the tarsal joint. When the order of carnivores first appeared on the scene, the genera that subsequently gave rise to several families were more similar to each other than are the extreme genera within any single family of Recent carnivores. Archaeopteryx, the first bird, was essentially a reptile except for its feathers. There are many aberrant forms in every order, class, and phylum of animals. Some of these may some day become the progenitors of new types of animals and will then be transferred to whatever taxonomic category they have given rise to. The delimitation and ranking of the higher systematic categories is essentially subjective and should not be made the basis of evolutionary theory.

The Mechanism of Macrogenesis. Goldschmidt is the only modern adherent of macrogenesis who specifies how he envisions speciation and the origin of higher categories through macrogenesis. It is evident, however, from the writings of other representatives of this school that their thinking is similar. Essentially they all agree that the production of a new type by a complete genetic reconstruction or by a major "systemic mutation" is the crucial event. Such an event will produce a "hopeful monster," as Goldschmidt called it, which will become the ancestor of the new evolutionary lineage. The occurrence of genetic monstrosities by mutation, for instance the homeotic mutants in *Drosophila*, is well substantiated, but they are such evident freaks that these monsters can be designated only as "hopeless." They are so utterly unbalanced that they would not have the slightest chance of escaping stabilizing selection. Giving a thrush the wings of a falcon does not make it a better flier. Indeed, having all the other equipment of a thrush, it would probably hardly be able to fly at all. It is a general rule, of which every geneticist and breeder can give numerous examples, that the more drastically a mutation affects the phenotype, the more likely it will reduce fitness. In his criticism of gradual speciation, Goldschmidt uses the analogue of a marble mosaic with its thousands of individual pieces. He claims that there would be a very small chance of improving the picture merely by changing individual marbles. It seems to me that this simile is far more destructive of the hypothesis of systemic mutation. What Goldschmidt wants us to believe is that one could scatter the thousands of pieces of the mosaic with a giant shovel onto a flat surface and get thereby an improved new picture. One of his critics has remarked that to believe that this might happen is equivalent to a belief in miracles.

Individuals are members of populations. Goldschmidt completely ignores the relation between the freak individual and his parental population. The finding of a suitable mate for the "hopeless monster" and the establishment of reproductive isolation from the normal members of the parental population seems to me an insurmountable difficulty.

Equally great is the problem of frequency. Systemic mutation is bound to be destructive in virtually all its occurrences by producing a disharmonious, inviable genetic combination. In order to produce the occasional ancestor of a new species, a new genus, or a still higher new type, systemic mutations would have to occur at a prodigious rate, so that perhaps one hopeful monster in a million hopeless monsters is produced. On this basis one would expect that a high percentage of the individuals of a species would be monsters. Observation proves, however, that such freaks are comparatively rare, far too rare to be of evolutionary importance. Where a genetic analysis is possible, it shows that systemic mutations in

Goldschmidt's sense are *not* involved. Furthermore, as Heberer (1957) points out, systemic mutations should be equally common at all geological periods and the probability of the production of new phyla should be as great in the Cenozoic as it was in the Pre-Cambrian. The facts, however, are in conflict with this assumption.

If there were a steady production of individuals that are new higher categories, phyla, orders, classes, families, and genera, only later to break up into the lower included categories, one would expect such individuals to occur in nature with considerable frequency. But where are the hopeful-monster phyla that have not yet broken up into classes, where are those classes that have not yet broken up into orders, and so forth? What we actually find in nature is natural populations grouped into species, and species grouped into higher categories.

The objections to the theory of macrogenesis are so numerous, and evidence to support this theory so singularly lacking, that it would be contrary to the scientific principle of parsimony (Occam's razor) to entertain any longer any theory of evolution by saltation except those that are well substantiated by the evidence such as polyploidy, hybridization, shift in sexuality, and chromosomal rearrangements, discussed below and in Chapter 17.

Speciation Through Major Chromosomal Changes

There was a widespread belief among early cytogeneticists that chromosomal rearrangement was the essential step in speciation. Proposed as an alternative to geographic speciation, the chromosomal speciation hypothesis is not valid. However, evidence is accumulating that the building up of isolating mechanisms in geographical isolates may be facilitated by chromosomal reorganization. The problem will be discussed in Chapter 17.

Polyploidy

The only unequivocally established mode of instantaneous speciation is polyploidy. Polyploidy is the multiplication of the normal chromosome number. If, for instance, the normal diploid ($= 2n$) chromosome number of a species is 14, all multiples of 7 higher than 14 are polyploids. Individuals with 3 chromosome sets ($3n$) are called triploids; with $4n$, tetraploids; with $6n$, hexaploids; with $8n$, octoploids; and so forth.

Two types of polyploids are distinguished that have a rather different significance in evolution: autopolyploids and allopolyploids (or amphiploids). An autopolyploid arises when more than two haploid chromosome sets of a single species participate in the formation of the zygote; an allopolyploid has the chromosome sets of two species. Normally an allopolyploid arises through the doubling of the chromosomes in a hybrid. Too little is yet known about animal polyploids to determine which of them are auto- and which allopolyploids. In view of the rarity of hybridization in animals it is probable that allopolyploidy is relatively less frequent in animals than in plants.

Polyploidy is very widespread among plants and is one of the important mechanisms of speciation in the plant kingdom (Stebbins 1950). Probably more than one-third of all species of plants have arisen by polyploidy, although the phenomenon is rare in some groups of plants, as in the conifers and perhaps in the fungi.

The importance of polyploidy among animals, particularly among sexually reproducing animals, is still a highly controversial subject. Most authors agree that it is rare. Nearly all proved cases occur in species that have abandoned sexual reproduction in favor of permanent parthenogenesis or self-fertilizing hermaphroditism. The cause of this rarity of polyploidy among animals is not yet fully understood. Among the reasons that have been cited the following may be the most important.

(1) *Imbalance of Sex Determination.* Sex is determined in most animals by a balance of sex factors, usually distributed over sex chromosomes and autosomes (see White 1954 for an exhaustive treatment). As Muller (1925) pointed out, there will be many unbalanced and hence more or less sterile individuals among the offspring of polyploids with sex chromosomes. If the sex chromosomes of the parents are XXXX ($♀$) and $XXYY$ ($♂$), there will be so many $XXXY$ individuals in the offspring that selection will severely discriminate against the survival of this polyploid. The validity of this arguments has been questioned in view of the occurrence of polyploidy in dioecious species of plants (*Salix* and others). It has been possible to produce experimentally a stable tetraploid strain of the normally dioecious (bisexual) plant *Melandrium album*. Here it was possible to select a stock with sex-chromosomal constitution $XXXY$ ($♂$) and XXXX ($♀$). But this was possible only because the male-determining factors in the single Y chromosome were stronger than the female factors of the combined three X chromosomes. In animals the balance seems to be much more delicate and coupled with sufficient dosage-compensation effects for sex-linked or sex-limited characters (Muller 1950b) to be completely upset by such an inequality in the number of X and Y

chromosomes in males. Potentially it should be easiest to establish polyploid strains where sex is determined not by a balance of factors but by a
single gene. Whenever this gene is present, regardless of dosage it would
produce the heterogamic sex. The only disturbance polyploidy would cause
in such a case is to the sex ratio, since the heterogamic sex would be,
at least in the beginning, much more frequent than the homogamic.
The normal sex ratio could be restored by any factor that would lower the
viability of YY gametes (Darlington 1953). Sex in some groups of fishes
is apparently determined by a single locus (Winge, Gordon, and others)
and it is presumably no coincidence that some evidence for polyploidy
has been found in fishes (see below).

(2) *Gonochorism.* In contradistinction to most plants, in which the
same flower produces male and female gametes, animal individuals are
usually of a single sex. A newly arisen polyploid animal will have difficulties in finding a mate, a handicap which among plants would exist only
for obligatorily cross-fertilizing species. The low frequency of polyploidy
among the digenetic trematodes, pulmonate mollusks, and other groups of
hermaphrodites where self-fertilization occurs (or is at least potential)
may indicate that this difficulty is not the primary reason for the rarity of
polyploidy in animals. However, self-fertilization may actually be very
rare in these groups, which would invalidate this argument.

(3) *Developmental Difficulties.* Stebbins (1950) suggests that polyploidy interferes with cellular differentiation in animals but not in plants
with their much simpler developmental processes. However, the great
frequency of polyploidy among parthenogenetic animals indicates (White
1954) that these developmental factors are presumably not of great importance.

As a consequence of Muller's principle (imbalance of sex determination) one would expect to find polyploidy mostly in groups in which reproduction is hermaphroditic, parthenogenetic, or vegetative (Suomalainen 1958). Unfortunately, most groups of animals in which these forms
of reproduction occur are still very poorly known cytologically and taxonomically. What little is known has been so authoritatively reviewed by
White (1954) that it is unnecessary to go into detail here.

Among hermaphrodites the evidence for polyploidy is this: though
self-fertilization occurs commonly among certain genera of pulmonate
mollusks, particularly *Lymnaea*, the only positive evidence for polyploidy
so far is *Paludestrina jenkinsi*. Neither in the tapeworms (Cestodes) nor
in the digenetic trematodes has polyploidy been found, but it is not in-

frequent among the turbellarians and the annelids. The case of the triclad species *Dendrocoelum infernale* (Steinmann) may serve as a typical example (Aeppli 1952). Morphologically this species ($2n = 32$) resembles very closely the sympatric species *D. lacteum* Müller ($2n = 16$). The only major difference is that *D. infernale,* which has become adapted to caves and alpine brooks, has lost its eyes—an indication that the separation from *D. lacteum* is not very recent. The tetraploid *infernale* has fewer but larger cells than *lacteum* individuals of equal size. There are a number of physiological differences between these species. The temperature preferred by *D. infernale* is 6°C, that by *D. lacteum,* 17°C. The two species also differ in their resistance to temperature shock and to certain chemicals. When both species are kept at their preferred temperatures, *D. infernale* is more resistant to alcohol, *D. lacteum* to glycerin. Analysis of the chromosomes indicated clearly that autopolyploidy, not allopolyploidy, is involved. *Dendrocoelum infernale* satisfies all the demands of a good species but has certainly arisen through polyploidy. However, its reproduction may be through pseudogamy (Benazzi 1957).

Additional cases of polyploidy (mostly in association with parthenogenesis) have been reported among other turbellarians. In earthworms (Lumbricidae), too, polyploidy is widespread among parthenogenetic species. The members of this family are basically cross-fertilizing hermaphrodites, and it was clearly established by several authors, particularly by Omodeo (1951, 1952), that polyploidy occurs also in some sexually reproducing species. It is not yet known whether this polyploidy became established via parthenogenesis or via self-fertilizing hermaphroditism. The usual obstacles to polyploidy are absent in hermaphroditic groups, since there are no sex chromosomes and the difference between the male and female gonad is determined during development. What is most surprising about polyploidy among hermaphroditic groups is its rarity. This is not easily explained, unless one assumes that self-fertilization is very rare.

Polyploidy is much more widespread among groups with parthenogenesis. It has been described in the salt "shrimp" *Artemia salina,* in the terrestrial isopod *Trichoniscus coelebs,* in the grasshopper *Saga pedo,* in psychid moths (*Solenobia* and other genera), in weevils of the subfamilies Otiorrhynchinae and Brachyderinae, in earthworms, and particularly well in simuliids and chironomids. For details and references we refer to White (1954) and Basrur and Rothfels (1959). There has been some question of the extent to which such polyploidy is "speciation." In most cases, as in *Trichoniscus, Solenobia,* weevils, and earthworms, polyploidy produces well-defined forms, variously listed as "races" or "species," depending in the main on the degree of difference. What is more important than the morphological difference is the fact that the polyploid derivative very often differs quite distinctly from the diploid parent in both ecology and distribution. Thus the diploid *Trichoniscus elisabethae* is found in southern France only in humid mountainous habitats, while the derived triploid, *T. coelebs,* lives in arid, hot garigues along the Mediterranean. Even more interesting is the case of *Solenobia* (Seiler 1946, 1961) in which three "races" occur, a sexual diploid, a parthenogenetic diploid, and a parthenogenetic tetraploid. Though these overlap in many areas, they manifest definite regional preferences. The sexual form is found mainly near the edge of the last glaciation and near the nunataks along the northern slope of the Alps. The tetraploids are found in the upper valleys in the northern Alps and along the entire slope of the southern Alps. The Swiss Jura is almost exclusively inhabited by the parthenogenetic diploids. Similar ecological differences have been found among the weevils and earthworms.

Polyploidy has also been reported in that special type of parthenogenesis sometimes referred to as gynogenesis (or pseudogamy). In a number of species of nematodes, planarians, and earthworms the egg develops only if penetrated by a male gamete. Yet the latter degenerates and in the genetic sense no fertilization takes place. No polyploidy has yet been reported in parthenogenetic nematodes, but it does occur in several turbellarian genera, as shown by the work of Benazzi and his school and by Dahm (1958). Autopolyploidy may occur among protozoans (Wenzel 1955).

It is thus evident that polyploidy is clearly established in all groups with permanent parthenogenesis and that it results in the production of species that, by any purely biological or taxonomic test, would be called good species. Indeed, we can say that polyploidy has become the principal method of speciation in a few of these specialized groups. In the lumbricid earthworms, for instance, up to 70 per cent of the species found in a given area may have arisen by polyploidy. In the turbellarians and in certain groups of weevils, the percentage of polyploids is likewise appreciable. There seems to be little doubt (Suomalainen 1950) that in all these cases parthenogenesis has preceded the origin of polyploidy. Most polyploids are autopolyploids, but in some cases allopolyploidy is indicated. However, the situation is quite different from that of amphiploidy in plants (Stebbins 1950). Allopolyploidy in animals is usually produced by the fertilization of a diploid or tetraploid parthenogenetic egg by a haploid sperm, resulting in a $3n$ or $5n$ polyploid. It is still uncertain whether this is the origin of triploid dipterans, or whether they are auto-polyploids that arose as a result of the fusion between three haploid meiotic products in an egg (Basrur and Rothfels 1959).

Pseudopolyploidy in Animals. After botanists discovered that nearly half of the plants are polyploids, active search began for polyploidy in animals. For a time almost any difference in chromosome number be-tween related species was considered indicative of polyploidy. Before listing the well-substantiated cases it would be wise to point out some potential sources of error.

(a) *Chromosome fusion.* It is a common mistake to assume that the lowest chromosome number in a series of related species is the most primi-tive and that the higher numbers are derived. Several investigations sug-gest that precisely the opposite is true for many kinds of animals. Among the copepods Harding (1950) found that the most primitive species have the highest number of chromosomes, and the same is true within subor-ders (for instance, the Cyclopoida). The most primitive isopods have 28 haploid chromosomes, and this number is independently reduced through fusion to 8 in several unconnected lines (*Asellus,* Trichoniscidae, Epicari-dae) (Vandel 1947). The fusion itself is at first sight a rather simple phenomenon, but there are certain cytological difficulties that have not yet been completely solved. In most animals (except those with very con-densed, round metaphase chromosomes) one can distinguish in meta-phase between rod-shaped and V-shaped (or J-shaped) chromosomes. This shape is determined by the position of the centromere ($=$ kine-tochore), which is (sub-)terminal in the rod-shaped ("acrocentric") chromosomes and somewhere along the chromosome ("metacentric") in the V-shaped chromosomes (Fig. 15–3). Through a translocation (and subsequent loss of a centromere) two rod chromosomes can fuse into one single V chromosome. The two arms of the new V chromosome corre-spond exactly to the two former rod chromosomes. The significance of such centric fusions was first clearly recognized by Robertson (1916), who emphasized that the number of major chromosome arms stays nearly constant in many groups ("Robertson's rule"). Matthey (1949) calls this number of chromosome arms the N.F. (*nombre fondamentale*). A par-ticularly instructive case of the effects of centric fusion on chromosome number was described by Staiger (1954). In hybrid populations between

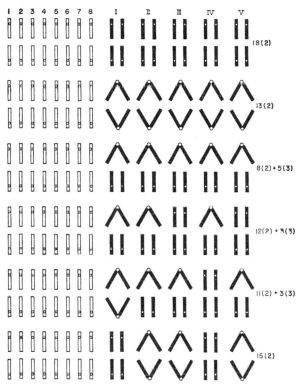

Fig. 15–3. The chromosomes of the marine snail *Thais* (*Purpura*) *lapillus*. Top row, the 18-chromosome race; second row, the 13-chromosome race (5 metacentric); four bottom rows, various combinations in a hybrid population. (From Staiger 1954.)

an $n = 13$ and an $n = 18$ form of the marine prosobranch *Thais lapillus* individuals may be found in which, during meiosis, 5 metacentric chromosomes are paired with 10 acrocentrics, thus clearly demonstrating the mode of origin of the metacentrics (Fig. 15–3). It is thus obvious that chromosome number itself cannot be accepted as absolute proof of polyploidy. A species with $2a$ chromosomes is not necessarily a polyploid of a related species with a chromosomes; rather the a species may have a V chromosomes derived by centric fusion from a species with $2a$ rod chromosomes. Such an interpretation can often be substantiated by the comparison of the N.F. or that of the nuclear DNA content.

A decrease in the N.F. can be interpreted rather easily by translocation and loss of a centromere together with the loss of genetically inert (or dispensible) portions of chromosome arms (White 1954). An increase in the N.F. is more difficult to explain since it involves an increase in the number of centromeres. There is some question whether this can be achieved by the splitting of a centromere, because this process does not account for the origin of a new telomere, apparently also needed. There is need for an additional chromosome fragment to supply a centromere and two telomeres (White 1957b).

(b) *Diffuse or multiple centromere.* Lorković (1949) and others have shown that certain species of Lepidoptera can be arranged in a series of increasing chromosome numbers, suggesting the presence of polyploidy. However, there are many intermediate chromosome numbers and, further-more, Lepidoptera, along with Heteroptera, Homoptera, and some arach-noids, appear to have diffuse (or multiple) centromeres, so that broken-off pieces of chromosomes are not lost (as in species with a localized centromere) but move as "new chromosomes" with the other chromo-somes. Chromosome number is equally ambiguous in *Ascaris* (roundworm) with its compound kinetochore. Differences in chromosome number, fre-quent in related species of all these groups, are not necessarily indicative of polyploidy. Yet the evidence in favor of polyploidy in lepidoptera is fairly suggestive: in *Lysandra*, as in other lycaenids, 24 is the basic chro-mosome number, but *L. bellargus* has $n = 45$ and other species 82–90, *L. argester* about 130–150, and *L. nivescens* ca. 190 (De Lesse 1960). In *Erebia* different species have the following chromosome numbers: 8, 10, 11, 12, 14, 15, 22, 23, 24, 25, 28, 29, 40, 51, 52. Close relatives often have chromosome numbers that strictly correspond to polyploid series, such as 8 and 40, 45 and 90, 51 and 102. Yet the chromosomes of the higher chro-mosome species are correspondingly smaller. De Lesse (1960) therefore suggests that neither polyploidy nor chromosome fragmentation is the answer for the chromosomal multiplications found in the lepidoptera, but the process described under (c).

(c) *Strand segregation.* The chromosomes of higher organisms appear to be highly polytene, that is, composed of many homologous chromatin strands. Schrader and Hughes-Schrader (1956) have suggested that some doubling of chromosome numbers (without increase in the DNA) may be due to reduction of polyteny, that is, due to strand segregation in compound, polytene chromosomes. This suggestion, although consistent with some of the observed facts, remains to be substantiated. The origin of chromosome numbers that are not multiples of the haploid number would be due to fusions or fissions subsequent to the original strand segre-gation.

(*d*) *Supernumeraries and extra chromosomes.* If through mechanisms that are still poorly understood supernumeraries or extra chromosomes are added to the regular chromosome set, this may lead to an increase in chromosome number that simulates polyploidy. In the mole cricket "*Gryllotalpa gryllotalpa*" chromosome number may fluctuate from 12 to 23, and this has been ascribed to polyploidy. However, a whole group of sibling species seem to hide under the name *G. gryllotalpa* and nothing can be said about their chromosome numbers until the taxonomy is clarified.

Polyploidy in Sexual Animals. It has been shown that most alleged cases of polyploidy in animals are based on a misinterpretation of cytological evidence ("pseudopolyploidy") or are limited to forms that lack gametic sex determination (such as hermaphrodites), or that reproduce uniparentally (parthenogenesis). Some other cases, widely cited as proof of polyploidy in sexually reproducing species, remain to be discussed.

(*a*) *Lower arthropods.* E. Goldschmidt (1953b) showed that the euphyllopods (fairy shrimps) of Israel consist of species with either 6 chromosomes (*Lepidurus* spec.), 12 or 12 + 1 chromosomes (*Chirocephalus, Branchinecta*), or 18 chromosomes (*Streptocephalus torvicornis*). No sex chromosomes are distinguishable but reproduction is strictly sexual. In view of the existence of parthenogenesis in related genera (for instance, *Artemia*), it is possible that polyploidy was acquired during a parthenogenetic stage and that sexual reproduction is a reversal.

(*b*) *Insects.* In the earwigs (Dermaptera) three species with low chromosome numbers ($2n = 12$ to 14) are XY in the males, while males in several species with $2n = 25$ are X_1X_2Y and in *Prolabia abachidis* with $2n = 38$ are $X_1X_2X_3Y$ (Bauer 1947). This is interpreted by most authors (for example, Goldschmidt 1953a) as a clear indication of polyploidy, although White (1954, 1957) recommends caution in view of the existence of several XY species with $2n_{\mathcal{J}} = 24$ and an X_1X_2Y species with $2n_{\mathcal{J}} = 21$. Chromosome numbers suggestive of polyploidy have also been found in bisexual phasmids, mantids, and *Gryllotalpa* (for instance, *G. africana* and *G. borealis*), often correlated with an X_1X_2Y sex-chromosome mechanism (Goldschmidt 1953a), but the nonpairing of the various X's in the heterogametic sex is not consistent with polyploidy.

(*c*) *Mammals.* Polyploidy has been often claimed for mammals including man, but all the evidence is opposed to this interpretation. When it was found that the Golden Hamster, *Mesocricetus auratus*, had $2n = 44$ chromosomes, while two other species of hamsters, *Cricetus cricetus* and *Cricetulus griseus*, had $2n = 22$, it was asserted by Sachs (1952) and

Darlington (1953) that this species had arisen as an allopolyploid of the two species with the lower chromosome numbers. This claim has been rejected by Matthey (1953) and White (1957) for various reasons. The primitive chromosome number of cricetids is high (44) both in the Old World (*Mesocricetus*) and in the New World (*Reithrodontomys*) and greatly reduced in the more advanced forms. The following chromosome numbers (2n) are found in hamsters of the Old World: *Mystromys*, 32; *Tscherkia*, 30; *Phopodus*, 28; *Allocricetulus eversmanni*, 26; *Cricetus*, *Cricetulus migratorius*, and *C. griseus*, 22; and *Allocricetulus curtatus* and *Cricetulus barabensis*, 20 (Matthey 1961). It seems that in the cricetine rodents as well as in the sciurids and gerbillids there is great instability of chromosome number, without polyploidy. No case of polyploidy is known among mammals.

(*d*) *Fishes*. Less easily refuted is the evidence for polyploidy among fishes. Svärdson (1945) showed that the chromosome numbers of all investigated species of salmonids are multiples or near multiples of 10, namely $3 \times 10 = 29$, 30; $4 \times 10 = 37$, 40, 40, 40, 40, 42; $5 \times 10 = 51$. White (1946) attributes this to coincidence and Matthey (1949) believes that the number of chromosome arms (N.F.) is in conflict with the polyploid interpretation. But Kupka (1948, 1950) found the evidence rather convincing in *Coregonus*. Most coregonids have 72 chromosomes consisting of four sets of 18. *Coregonus asperi maraenoides* of Zürichsee has only 36. Further evidence for polyploidy in fishes was found by Suomalainen (1958). It seems that sex in fishes is determined by a definite sex gene rather than by a balance of sex-determining factors, and this—if true—would facilitate the occurrence of polyploidy.

Summary. On the whole, speciation through polyploidy is very rare among animals. Only in parthenogenetic groups is its occurrence at all frequent. A great deal of cytological work remains to be done before it can be stated whether and to what extent polyploidy occurs in sexually reproducing groups of animals such as crustaceans, phasmids, mantids, lepidoptera, and fishes. Even if it should be confirmed in some of these groups, one would still be justified in saying that in animals this mechanism is exceptional, in contrast to the situation in plants. There is no evidence that whole species groups or genera owe their origin to polyploidy, as is frequently found in plants.

Polyploids pose a difficult taxonomic problem. An autopolyploid may be virtually indistinguishable, at the time of origin, from the parental diploid. Such a form is often referred to as a "polyploid race," as by

Seiler in *Solenobia*. Yet such a "race" is reproductively isolated from the parental species and is, biologically speaking, a good species. This is another illustration of the frequent conflict between a morphological and a biological species concept.

All gradual speciation is by necessity a population phenomenon. It is the gradual divergence of populations until they have reached the level of specific distinctness. Two modes of gradual speciation have been postulated, those involving geographical separation of the diverging populations (geographic speciation, Chapter 16), and those without geographic separation (sympatric speciation).

Sympatric Speciation

The majority of authors until fairly recently considered sympatric speciation, that is, speciation without geographic isolation, to be the prevailing mode of speciation. Such speciation is based on two postulates: (a) the establishment of new populations of a species in different ecological niches within the normal cruising range of the individuals of the parental population; (b) the reproductive isolation of the founders of the new population from individuals of the parental population. Gene flow between daughter and parental population is postulated to be inhibited by intrinsic rather than extrinsic factors. A rapid process of species formation is implied in most schemes of sympatric speciation. For a more detailed description of this postulated process we may cite one of the adherents of this theory:

> Suppose that on the same oceanic island the original colony has begun to segregate into secondary groups under the influence of natural selection, sexual selection, physiological selection, or any of the other forms of isolation, there will be as many lines of divergent evolution going on at the same time (and here on the same area) as there are forms of isolation affecting the oceanic colony (Romanes 1897:22).

Historical. The concept of sympatric speciation is far older than that of geographic speciation and goes back to pre-Darwinian days. Darwin was rather vague on the subject and made no clear distinction between speciation through individuals and speciation through populations. In some of his statements he seems to give due recognition to the need for geographic isolation while in others he seems to ignore the geographical element altogether. Certain passages in his works as well as in his correspondence with Wagner, Semper, and Weismann indicate a belief in sympatric speciation.

> If a variety were to flourish so as to exceed in numbers the parent species, it would then rank as the species, and the species as the variety; or it might come to supplant and exterminate the parent species; or both might coexist, and both rank as independent species. . .
>
> The small differences distinguishing varieties of the same species speedily tend to increase 'till they come to equal the greater differences between species of the same genus, or even of distinct genera.

He even gave some thought to the question of the mechanisms that might permit sympatric incipient species to coexist without fusion:

> I can bring a considerable catalog of facts, showing that within the same area varieties of the same animals can long remain distinct, from haunting different stations, from breeding at slightly different seasons, or from varieties of the same kind preferring to pair together . . . I believe that many perfectly defined species have been formed on strictly continuous areas.

It would lead us too far afield to give a detailed history of the many attempts to prove sympatric speciation. Part of this history has been presented previously (Mayr 1947, 1955b, 1959a), without, however, any attempt at completeness. Catchpool in England and Dahl in Germany were the first to propose detailed theories. They were joined by Romanes, who studied the subject in a series of individual papers (1886–1889), elaborated and summarized in a book (Romanes 1897). The arguments advanced by these authors were analyzed in detail by Seebohm (1888) and Jordan (1896, 1898) and were shown to be invalid. As Jordan pointed out, all these authors postulated characteristics for the incipient species that automatically made them full species. The subsequent "selection" did not add anything new. None of these schemes explained the origin of the decisive differences between the incipient species. Soon after Jordan's (1896) discussion, Vernon (1897) and Petersen (1903, 1905) followed with new arguments in favor of sympatric speciation which were again shown to be false by Jordan (1897, 1898, 1903, 1905). This back-and-forth interchange between adherents of sympatric and of geographic speciation has continued to our day. The claims of Test (1946) and of Thorpe (1945) were answered by Mayr (1947). Emerson (1949) attempted to revive sympatric speciation but was refuted by Pitelka (1951c).

It is rather discouraging to read this perennial controversy because the same old arguments are cited again and again in favor of sympatric speciation, no matter how decisively they have been disproved previously. None of these authors seems to be aware of the extensive prior literature in the field. In the last analysis, all the various schemes make arbitrary postulates that at once endow the speciating individuals with all the attributes of a full species. They attempt thus to by-pass the real problem of speciation. One would think that it should no longer be necessary to devote much time to this topic, but past experience permits one to predict with confidence that the issue will be raised again at regular intervals. Sympatric speciation is like the Lernaean Hydra which grew two new heads whenever one of its old heads was cut off. There is only one way in which final agreement can be reached and that is to clarify the whole relevant complex of questions to such an extent that disagreement is no longer possible. As a step in this direction an attempt will therefore be made in the subsequent sections to analyze the various premises and postulates of sympatric speciation in considerable detail and to determine whether or not there is evidence suggesting that populations could acquire reproductive isolation without geographic segregation.

Two basic points must be clearly stated before we enter the detailed discussion.

The first is that the theories of sympatric and geographic speciation agree in their emphasis on the importance of ecological factors in speciation. They differ in the sequence in which the steps of the speciation process follow each other. The theory of geographic speciation lets an extrinsic event separate the single into several gene pools, with the ecological factors playing their major role after the populations have become geographically separated. According to the theory of sympatric speciation the splitting of the gene pool itself is caused by ecological factors, and whatever spatial isolation of the thus formed populations ensues is a secondary, later phenomenon.

The second point concerns the definition of sympatric speciation. In order to avoid circular reasoning it must be defined as the origin of isolating mechanisms within the dispersal area of the offspring of a single deme. The size of this area is determined, for instance, in marine organisms by the dispersal of the larval stages. In most insects it is determined in the adult state by the more mobile sex. Since there are normally very numerous ecological niches within the dispersal area of a deme, niche specialization is impossible without continued new pollution in every generation by immigrants.

Reasons for Postulating Sympatric Speciation. The concept of sympatric speciation would not have such widespread support if there were not many phenomena and general biological concepts that seem to support it. Some of these will now be discussed.

Adherents of sympatric speciation are almost invariably students of local faunas who want to explain the bridging of the gap between species within the nondimensional situation. They deal with a number of related sympatric species and make the unconscious assumption that these species originated where they are now found. This assumption is made even in some of the most recent literature (for example, Wieser 1958; Kohn, 1958). Test (1946), for instance, points out that the 17 species of Californian limpets of the genus *Acmaea* have widely overlapping ranges and that some of the most specialized species occur in the midst of the range of the presumably ancestral species. Similar situations are found in nearly every large genus. There are several hundred (perhaps 500) species of bees of the genus *Perdita* in North America. All are oligotrophic, confining their visits to the flowers of a single species or of a group of closely allied species. Many species of *Perdita* may be found at a single locality but never together because they occur on different plants or in some cases at different times of the year. How could all these species have arisen by geographic speciation? This is a question that at first seems difficult to answer.

Another reason for postulating sympatric speciation is based on the observation that all species seem perfectly adapted to the particular environment in which they occur. It is reasoned that such a perfect fitting of species into their ecological niches could not possibly have evolved as an accidental by-product of genetic changes accumulated during geographic isolation. It is this same argument that Lamarckians use to reject natural selection of random mutations as a mechanism that might produce adaptation.

Typological thinking is also opposed to the acceptance of geographic speciation. If species are considered as aggregates of individuals conforming to a type, then speciation is the production of an individual that falls outside of this type. The production of such an individual by sympatric speciation would seem an eminently logical mode of speciation, as was indeed maintained by Schopenhauer.

The observation of greater ecological than morphological difference among many species (for instance, sibling species) has also contributed to the popularity of theories of sympatric speciation. Starting with the unspoken assumption that it is the morphological distinctness which really makes the species, it is argued that such forms acquire their ecological difference while still sympatric and become morphologically distinct species only subsequently.

It is evident, then, that there are quite a few general considerations that at first sight would seem to favor sympatric speciation.

Incipient Sympatric Species. A necessary corollary of any theory of gradual speciation is that there exist in nature some "forms" or "varieties" or "populations" that are "incipient species." If it were possible to list all existing cases of incipient species, one would at once have convincing evidence of the actual frequency of the various possible modes of gradual speciation. Geographical isolates, as we shall see, are such incipient species as far as the process of geographic speciation is concerned (Chapter 16). The question to be solved, then, is whether there are any natural phenomena known that can be interpreted as cases of incipient species in a process of sympatric speciation. The complex of phenomena usually labeled "biological races" has indeed been cited as evidence for such sympatric speciation and must therefore be studied carefully.

Biological Races. Kinds of animals that live in the same area and that, though clearly separable by biological characters, show no structural differences or only slight differences, have been called biological races. The recognition of such a special category is closely associated with a morphological species concept (Chapter 3). Several surveys of biological races have been made, the most important being those of Thorpe (1930, 1940). A close analysis of these cases reveals that a single heading has been applied to a very heterogeneous assemblage of phenomena. They can be sorted, in a preliminary fashion, with the help of a four-square table, using two classifying criteria (Table 15–2).

(1) *Morphs.* Distinct phenotypes of a single population are often erroneously designated races (Chapter 7). Neither the carriers of different human blood-group genes nor the carriers of a given gene arrangement in *Drosophila* belong to a different race from other individuals of the same population. This is self-evident in most cases but must be emphasized for all forms of ecological polymorphism. Genotypes that differ in habitat preference, as established in *Drosophila, Cepaea,* and *Asellus aquaticus,* are not different races. They remain members of a single gene pool showing different preferences. There is a suggestion of genetically determined polymorphism among certain species of insects in their preference for plant hosts. This may be the true explanation of many of the host-

Table 15–2. Phenomena listed as
biological races.

Occurrence	Population	
	Same	Different
Sympatric	Morphs Clones Host "races"	Polyploids Ecological races Host races Sibling species
Allopatric		Geographic races Semispecies Sibling species

transfer experiments reported in the entomological literature. For instance, Thorpe (1928) found that among *Hyponomeuta padella* from hawthorn given a choice between hawthorn and apple leaves 20 percent preferred to oviposit on apple, while 10 percent of the apple form of this species preferred to lay its eggs on hawthorn. There is much suggestion in work of this type that not really different host races are involved, but rather intrapopulation genotypes (Ludwig effect, see Chapter 9). The same is true for the so-called virulent or nonvirulent races in certain insect pests such as the Hessian fly (Cartwright and Noble 1947). Two so-called races of the leafhopper *Cicadulina mbila* differ in that one can easily transmit the virus of streak disease to maize while the other race does not transmit it. Actually, the two types differ apparently only in a single gene which controls the susceptibility of the gut wall to penetration by the virus (Storey 1932). Numerous similar cases have been described. In all of them the application of the term "race" is evidently improper since these are Mendelian differences within populations. Carriers of these genotypes interbreed, usually at random, with other genotypes of the same population.

(2) *Clones.* In forms where sexual reproduction is temporarily or permanently abandoned, parthenogenetically reproducing strains that differ in biological characteristics may be distinguishable. In the Pea Aphid (*Macrosiphum pisi*), strains have been isolated that differ in their virulence toward pea plants. Strains of very different virulence may be isolated from the same local population. After completing its parthenogenetic phase a strain will return to the common gene pool of the local population and therefore cannot be considered a race. Such strains may develop in any species with uniparental reproduction and have been described, for instance, in *Daphnia*, rotifers, nematodes, and protozoans.

(3) *Geographic races*. Geographic races often differ more conspicuously in various physiological characteristics than they do morphologically. Physiological races of this type have, among others, been described for *Fringilla coelebs* (song), *Sturnus vulgaris* (migratory habits and breeding season), *Carabus nemoralis* (temperature preference), *Drosophila funebris* (temperature), *Lymantria dispar* (incubation period and other physiological attributes), and *Cynips* (gall structure). Additional examples are listed in Chapter 11, under the heading "Geographic Variation of Physiological Characters." They differ, of course, in no respect from other geographic races.

(4) *Semispecies*. Geographical isolates sometimes acquire various biological peculiarities and partial reproductive isolation (Chapter 16). When such populations establish secondary contact with each other they interbreed only to a limited extent and retain some of their peculiarities. Animals that are easily transported by man, like the housefly (*Musca domestica*), show such a secondary sympatry of former geographic races particularly well. The races of the human louse (*Pediculus humanis*) may owe their origin to the same process (see p. 464). In all these cases we observe a fusion or partial fusion of former isolates rather than the sympatric origin of incipient species.

(5) *Polyploids*. Polyploid parthenogenetic derivatives of diploid species such as those that occur in the genera *Solenobia, Otiorrhynchus*, and others have also been listed as biological races. This is not correct since they are reproductively isolated. Buchanan (1947) has named several parthenogenetic strains related to the White-fringed Cotton Beetle (*Graphognathus*) which are either polyploids or clones. They must have arisen by instantaneous speciation, and it is preferable to consider them microspecies, as does Buchanan, rather than biological races.

(6) *Ecological races*. This term is often used vaguely and has been applied to many different biological phenomena. It has been used for sibling species, for cryptically colored local geographic races, for seasonal races or generations, and for "habitat races." In order to lend precision to the discussion the term will here be used only for habitat races. All ecological races are intraspecific populations or groups of populations. Such populations occupy a certain area and a certain habitat. As has been emphasized repeatedly in the past (Mayr 1947, 1951a), every race is simultaneously a geographic race and an ecological race. There is no evidence to support the notion of ecological races as a category distinct from that of geographical races. A theory of ecologic speciation would have to prove that a population of a species established itself in a new kind of habitat

within the cruising range of individuals of the parental populations and merely by conditioning and habitat selection segregated itself in the new habitat until it had reached species level. In virtually every case of so-called ecological speciation that has been cited in the past, a reevaluation of the facts has shown that the basic facts have been misinterpreted (Mayr 1947).

An especially interesting case is that of the ecological races of the Song Sparrow (*Passerella melodia*) in the San Francisco Bay area (Marshall 1948). Three subspecies occur in brackish or salt-water marshes and a fourth one near fresh water in the uplands (Fig. 15–4). In most areas the upland subspecies is separated from the salt-marsh subspecies by an ecologically unsuitable dry plain. Wherever this isolation prevails there is a sharply defined taxonomic difference between the upland and the marsh birds, whereas the marsh populations intergrade with each other clinally where they are in contact. Where there is no isolation between marsh and uplands there is also very little or no morphological difference between populations of different habitats: "Without this isolation the Tomales Bay salt-marsh birds are identical in morphological attributes with the birds in fresh-water habitats surrounding them. Wherever it is lacking between bay marsh and upland populations as at Richardsons Bay, Corte Madera, and San Pablo, the respective populations become practically indistinguishable" (Marshall 1948). In a few places the marsh and upland populations have come into secondary contact and the area and amount of intergradation is governed entirely by the configuration of the habitat. At San Francisquito Creek, where two contrasting habitat types meet abruptly, there is little mixing of the neighboring subspecies, which are apparently maintained by habitat selection rather than by nonrandom mating. Marshall's work shows clearly that these so-called ecological races of the Song Sparrow largely owe their origin and maintenance to geographical isolation. They are as much geographical races as they are ecological races.

Petersen and Tenow (1954) believe that the butterfly *Pieris bryoniae* of the Alps might have originated by ecological speciation. They point out, as had been noted by earlier authors, that this species is essentially reproductively isolated from the widespread lowland *P. napi*. However, in Scandinavia there is a northern subspecies, *P. n. adalwinda,* which is phenotypically very similar to *P. bryoniae* but which intergrades with *napi* of southern Sweden through populations with intermediate characters, in contrast to *napi* and *bryoniae* in the Bavarian Alps which are

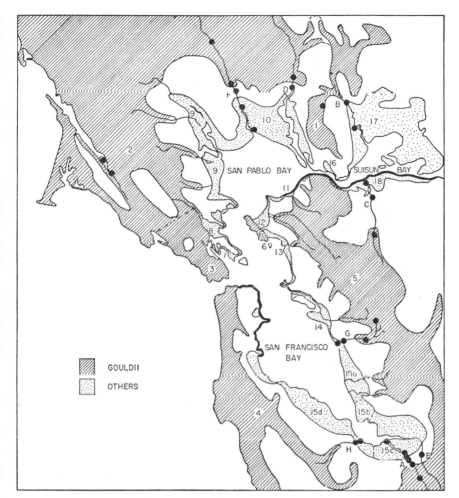

Fig. 15–4. Ecological races of the Song Sparrow in the San Francisco Bay region. The upland race *gouldii* (populations 1–5) is largely separated by unsuitable habitat from the three salt-marsh races *samuelis, pusillula,* and *maxillaris* (6–18), as are the latter from each other. Actual or potential gene flow occurs along certain creeks marked *A–H.* (For details consult Marshall 1948.)

separated by sterility barriers. No evidence is advanced which would contradict the assumption that the contact between the two Alpine species is secondary, and that both the ecological difference and the reproductive isolation were acquired during geographical isolation, presumably during the Pleistocene. The phenotypical similarity between *bryoniae* and *adalwinda* is not necessarily an indication of close relationship and may be a parallel response through selection to similar environmental conditions. The high sterility of the hybrids and low sexual isolation of the adult butterflies in the Alpine species indicate the former existence of an extrinsic barrier to gene flow.

There is no evidence for ecological speciation, as a process distinct from geographical speciation. I have been unable to find a single case of so-called ecological speciation in the literature that did not require the spatial isolation of populations.

(7) *Host races.* In many species of animals, particularly nematodes and insects, temporary strains may develop on specific host plants. In spite of the enormous literature on these so-called "host races," we are still far from understanding this phenomenon and its evolutionary implications. Researchers in this field have been handicapped in the past by their failure to comprehend the basic problems. Some authors, such as Harrison (1927), studied host races "in order to prove" the inheritance of acquired characters; other authors failed to make a distinction between sibling species and host races; and still other authors were intent on proving or disproving sympatric speciation. As an introduction to the literature see the reviews of Brues (1924), Dethier (1947, 1954) and Thorsteinson (1960), and the two excellent papers of Thorpe (1930, 1940), which include abundant references to the earlier literature.

A study of this literature permits the following generalizations:

(*a*) Preference for a given species of host plants may and nearly always does have a double basis—conditioning (including larval conditioning) and a genetic predisposition.

(*b*) Nearly all species with host races concentrate on one host in one given district but have the ability to establish themselves on a variety of other host plants, particularly under crowded conditions, and may have different preferred host plants in different districts. The interesting case described by Armstrong (1945) of a specialized local population of the Apple Codling Moth (*Carpocapsa pomonella*) developing in Ontario in an isolated pear orchard has become questionable since the description by Russian entomologists of a sibling species restricted to the pear. The cerambycid *Xylotrechus colonus*, with which Craighead (1921) did most of his work, primarily attacks oak, but also with almost equal frequency chestnut and hickory. Each of these plant genera contains up to a dozen or more species. It is possible to establish colonies even on ash and maple, although with difficulty.

The degree of host specificity is often overestimated. With Hopkins, the sponsor of the host-selection principle, the wish for new species became the father of the thought and he often described specimens as new species merely because they were from a new host. In the North American Cryphalini alone, 53 of the "species" described by Hopkins turned out to be synonyms (Wood 1954). The worthlessness of these host races and host "species" as evidence for sympatric speciation is obvious.

(c) Much evidence is accumulating which indicates that each local population of insects has much genetic variability with respect to host specificity. Cartwright and Noble (1947) found that it was possible to select from populations of the Hessian Fly (*Phytophaga destructor*) individuals which gave rise to strains that damaged previously resistant varieties of wheat. Any monophagous or oligophagous species of insect will come in contact, during its dispersal phase, with numerous plant species other than its normal host. If it has the appropriate genetic constitution, it will establish itself on the new host and this, according to the Ludwig theorem, leads to an expansion of the food niche of the species. Numerous such cases of an acquisition of new host plants by insect species have been recorded (Andrewartha and Birch 1954).

(d) During the dispersal phase of the species (which in these cases usually coincides with the adult stage and the period of reproduction) a lesser or greater mixing of local strains occurs. This is limited to the males if the females are completely restricted to the plant on which they fed as larvae and if mating takes place on the host species. A dispersal phase is, of course, necessary in every species to permit its spread and to permit utilization of newly opening sites.

(e) If such mixing is prevented artificially and an inbred strain is selected experimentally on a single one of several original hosts, such a strain may become progressively less tolerant ecologically until a stage is reached when it may be difficult to reestablish it on any of the other original host species.

(f) Mortality occurs whenever a strain is established on a new host. The more unusual the host, the heavier the initial mortality. Such cases have been described by Harrison (1927) and Glendenning (1929) and listed by Mayr (1947).

The point on which we have the least information up to now is the degree of reproductive isolation between different host races of the same species coexisting at the same locality. This is a problem not only of great evolutionary but also of practical importance. Apple orchards in California have been heavily infested by the Codling Moth (*Carpocapsa*) since at least 1880. Yet the first damage to the walnut occurred in 1909 and heavy damage did not take place until 1931. Does this mean that the codling moth has established a separate strain on the walnut, or has the apple strain broadened its tolerance to include the walnut? Time of emergence of the apple and the walnut moths is the same in southern districts of California and broadly overlaps in the north. Near Stockton 80 percent of the apple moths had emerged when only 50 percent of the walnut moths had done so. It is possible that microgeographic races are involved, since extensive walnut groves are sometimes isolated from the apple orchards. However, no thorough work on this problem has been done so far. Nor is it known whether there is any incipient sexual isolation between these strains or any increased mortality when larvae are transferred between apple and walnut. One hopes that the interests of economic entomologists have not become so completely chemistry-directed that they continue to overlook this basic biological problem.

In conclusion it may be stated that host races are a challenging biological phenomenon, and constitute the only known case indicating the possible occurrence of incipient sympatric speciation. However, the evidence thus far available indicates that complete stabilization on a new host cannot occur without geographic isolation.

(8) *Sibling species.* By far the majority of the so-called biological races of the literature (Thorpe 1930, 1940) are now acknowledged to be sibling species. As Thorpe said, they are forms "which on every biological ground should be classified as distinct species." This includes among the genera mentioned by Thorpe *Oecanthus, Nemobius, Anopheles, Nasutitermes, Psylla, Paratetranychus, Trichogramma,* and various species groups of *Drosophila.*

Apple, blueberry, cherry, and hawthorn maggots were cited as classical examples of biological races of a single species (*Rhagoletis pomonella*) until Curran (1932) and others showed that a group of sibling species is involved. Numerous similar cases are discussed in more detail in Chapter 3.

This survey of the category "biological races" shows that it contains a highly heterogeneous medley of biological phenomena. Only one of them, the host race, has a good claim on the designation "biological race." Even in this case a process of sympatric speciation is neither established nor even probable. It is not admissible therefore to cite the existence of biological races as evidence for the occurrence of incipient sympatric species.

Reputed Cases of Sympatric Speciation

One of the strongest arguments used in favor of sympatric speciation is the existence of certain situations in nature which supposedly can be explained by sympatric speciation but which, it is claimed, cannot (or only with the greatest difficulty) be explained on the basis of geographic speciation. There are five kinds of phenomena that form the basis of this argument.

The Occurrence of Sibling Species. It is argued that sympatric sibling species could not have originated in any other way except through sympatric speciation. This is claimed particularly in those cases in which the sibling species were listed as biological races until their reproductive isolation was discovered (Chapter 3). This argument overlooks two important considerations. The first is that sympatric sibling species are also common in many groups where host races and host specificity are absent and where (as, for example, in the malaria mosquitoes) speciation could not possibly have happened through host specialization. Equally important is the discovery that isolated populations of a species may acquire sterility factors or other components of reproductive isolation without morphological divergence. This has been described for a number of insects (Chapter 16). Moore (1954) found that New South Wales frogs of the species *Crinia signifera* were reproductively isolated from a morphologically indistinguishable population in Western Australia, which he therefore felt justified in considering specifically distinct (*Crinia insignifera*). These and other similar cases show that reproductive isolation may be acquired in geographic isolation more rapidly than morphological difference and that there is no need to postulate sympatric speciation to explain the speciation of morphologically similar or identical species.

Speciation in Monophagous Species Groups. In many groups of insects there are genera with many species, each of which appears to be limited to a single host. Genera with essentially monophagous species have been described for microlepidoptera (Munroe 1951), solitary bees, buprestid beetles (*Acmaeodera, Agrilus*), chrysomelid beetles (*Calligrapha, Arthrochlamys*) (Brown 1945, 1958), and other groups. Two phenomena, in particular, seem to suggest a mode of speciation in these food specialists that is different from speciation in most other animals. The first is that monophagous insects often, if not usually, belong to large genera. It is argued that sympatric speciation permits more rapid and more frequent speciation than geographic speciation. It would seem even more plausible, however, that the greater number of congeneric species is due to the vastly increased number of available niches and the reduction of competition. Likewise, high food specificity should greatly enhance the efficiency of ecogeographical barriers. This was, indeed, found for flower constancy among desert bees (Linsley and MacSwain 1958).

It is furthermore argued that monophagy and geographic speciation are incompatible concepts anyhow. Indeed, it is asked, how can a monophagous species speciate geographically and thereby become attached to a different host? Without doubt this poses a serious problem. One answer, frequently given, is that all monophagous species are the terminal descendants of an ancestral polyphagous species in which the various host races of the parental species have turned into separate species. The available evidence indicates that this solution is not valid. To begin with, there is good evidence for active speciation in many groups that are essentially monophagous. However, it is equally evident that monophagy is rarely as rigid as is sometimes claimed. There may be subsidiary host species in the entire range or in part of it (for example, Cook 1961b).

As in all speciating animals, the crucial events are apt to take place in peripherally isolated populations and these have not yet been studied adequately in even a single one of the groups of monophagous species. As a working hypothesis one might assume that a subsidiary host may become the primary host in such an isolated population and may offer more favorable conditions under the changed ecological situation of the marginal environment (Fig. 15–5). The shift from one host to another will set up an increased selection pressure that will result in a rapid genetic alteration of the population. The amount of genetic variability appears to be comparatively low in a species that has been selected for life strictly on a single host. Host specificity is thus an ideal prerequisite for rapid speciation. In due time such an isolated population not only may become perfectly adapted to the new host, but as a by-product of the geographical isolation and the genetic changes in the population may acquire reproductive isolation from the parental population. As soon as this is achieved the newly evolved species can reinvade the range of the parental species and live side by side with it. The situation in the leaf beetles described by Brown (1945, 1956) indicates that this is a feasible interpretation of speciation in monophagous groups. The same model has been applied successfully to oligolectic bees (Linsley and MacSwain 1958).

Speciation in Parasites. Among parasites many situations have been described, which, again, appear at first sight difficult to reconcile with

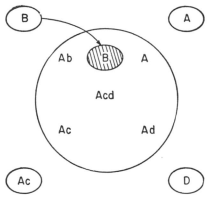

Fig. 15–5. Primary hosts (capital letters) and subsidiary hosts (small letters) of an essentially host-specific species. The large circle indicates the main range of the species; the outlying small circles, the peripheral isolates. New primary hosts (*B*, *D*) are acquired in some peripheral isolates, facilitating subsequent speciation. Reinvasion from the isolate that is host specific for *B* will establish a new species (hatched area) if reproductive isolation had been acquired during the geographic isolation.

geographic speciation. Two types of distribution of parasites, in particular, appear challenging. One is the occurrence of different closely related species of parasites in or on different body parts of the same host, and the second is the coexistence of several related species on different host species in the same geographical area. For instance, four "species" of lice have been described from South American monkeys, all derived from the human louse (*Pediculus*), which is widespread among American Indians. The essential factor in this situation is the spatial isolation of the five host species. There is normally no bodily contact between species of monkeys or between monkeys and humans. Yet occasionally Indians tame a young monkey which is then exposed to infection with lice. If such an individual escapes, it will introduce the infection into its species. The gene flow between the louse population on the monkeys and that on the Indians is so slight that it permitted rapid genetic divergence in populations which are exposed to very different selection pressures. The essential point, then, is not that the habitat of the two populations of lice is different but rather that there is virtually no chance for an interchange of individuals between the two incipient species. The population on each host is a spatial isolate.

Clay (1949) has shown that speciation in mallophaga is quite consistent with the concept of geographic speciation. The essential factor here as always is the interruption of gene flow by extrinsic barriers. Related congeneric species of feather lice often occur in different parts of the plumage (head, body, wing) of the same individual bird. Clay points out that this cannot be interpreted in terms of "ecological speciation" but that these sympatric species are the result of double colonizations with the reproductive isolation acquired either during geographical or host isolation. The zone of contact (and consequently potential gene flow) between lice specialized for distinct body regions is much too great to permit speciation.

The case of the human head and body louse (*Pediculus*) is not yet fully understood. The findings and interpretations of various investigators, such as Busvine (1948) and Levene and Dobzhansky (1959), are in conflict with each other to such an extent that the existence of different genetic strains among head lice, and likewise among the body lice, must be assumed. These races presumably originated as ecotypes in different human races. Well-clothed human races, like the Eskimos, would seem a prerequisite for the development of the body louse; scantily clad races with much head hair, such as the Melanesians, for that of the head louse. It is conceivable, and indeed probable, that the coexistence in many human races of head and body lice is a case of secondary overlap.

Speciation in internal parasites, with their much more complex life cycles, appears to be strictly allopatric in most cases. The same species of parasite may have different hosts or intermediate hosts in different regions, and if the isolation is sufficient there will be the development of a geographical race of parasite in due time which will eventually reach species level. The classification of parasites has not yet reached the stage where a full history of such speciation can be written, but the available data are consistent with the stated hypothesis. Speciation and evolution in parasites is full of fascinating problems, for which we refer the reader to the specialized literature in the field (Lapage 1951; Baer 1951; Rogers 1962; and Caullery 1952).

Species Swarms. Special situations occur in various parts of the world where a considerable number of closely related species, so-called species swarms, are confined to a narrowly circumscribed area, no close relatives occurring elsewhere. Such situations, as far as they concern islands or caves, will be discussed in Chapter 16 under the heading Multiple Invasions. They occur with particular frequency in the so-called ancient freshwater lakes, such as Lake Baikal in Siberia, but also in more recent lakes such as Lake Lanao in the Philippines or Lake Towuti and the neighboring lakes on Celebes. Other lakes with such species swarms are Tanganyika and Nyasa in Africa, Ochrid in the Balkans, and Titicaca in South

America. As early as 1913 Plate cited the Baikal gammarids and Tanganyika cichlids as proof of the occurrence of a mode of speciation different from that of geographic speciation. During the ensuing decades all leading specialists of fresh-water faunas endorsed this viewpoint, such as Woltereck, A. Herre, Myers, Worthington, and others. Rensch (1933:38) dissented from this interpretation and suggested five factors as contributing to the richness of the fauna of these lakes: preservation of relict types, multiple invasions from adjacent river systems, fusion of temporarily separated lakes, fusion of lake basins, and geographic barriers within the lakes. There is little doubt that all these factors are contributory, but the work of Brooks (1950) on the gammarids of Lake Baikal, the analysis by Trewavas (1947) and Fryer (1959) of the fishes of the genus *Haplochromis* of Lake Nyasa, and that by Poll (1950) of the cichlids of Lake Tanganyika all indicate that the localization of populations by extrinsic barriers within the lakes is by far the most important factor. Truly reliable data on the exact distribution of species within a lake are so far available only for the gammarids of Lake Baikal. But enough is now known about the distribution of fish in the east African lakes to permit the assertion that their speciation is consistent with the theory of geographic speciation (Brooks 1950; Fryer 1959, 1960a,b).

The relative importance of the various factors listed by Rensch differs from lake to lake, as was pointed out by Brooks. Sympatric speciation had been postulated by some authors on the basis of the presumed recency of these lakes, such as Lake Victoria in east Africa. However, Greenwood (1951) showed that this lake is much older than was stated in the earlier literature and that it has gone through a series of expansions and contractions during the Pleistocene. There was abundant opportunity for the temporary separation of populations and the acquisition of isolating mechanisms prior to the breakdown of the extrinsic barriers.

The general acceptance of allopatric speciation as explanation for these species swarms still leaves many details to be explained. Lake Lanao on Mindanao Island in the Philippines, for instance, owes its origin to the comparatively recent damming up of a valley by a lava flow. Nevertheless, a swarm of 18 closely related species of cyprinids has evolved in the lake and in the waters connected with it. All are believed to be ultimately derived from *Barbus binotatus*, a widespread Malayan fish. This case, as well as the speciation in Baikal and in the African lakes, demonstrates clearly that the rate of speciation in many of these cases must have been far more rapid than we are accustomed to expect on the basis of calculations derived from terrestrial animals (Myers 1960). Two factors seem to be responsible for this. One is that many of these lake species seem to become specialized for an extremely narrow niche in which they are superior to their competitors. The other reason is that spawning of each species is strictly limited in time and space and that the development of the larvae (fry) is likewise limited to certain water conditions. All this apparently facilitates the rapid establishment of isolating mechanisms, even though the individual steps have not yet been worked out in detail for any of these cases. Lake populations would seem to be a promising material to repeat with fishes the kind of ecological-embryological work done by Moore for *Rana pipiens*.

Now that geographic speciation has been universally accepted as the process by which species flocks in fresh-water lakes develop, it has been possible to shift attention to the ecological situation that has facilitated the explosive speciation (Chapter 18). Sedentariness of local populations and a relative stability of the ecological situation (compared to the annual floods and droughts in rivers) seem to have favored the shift into new niches, some of them remarkably narrow (Fryer 1959, 1960a,b). Hubendick (1960) has attempted to determine the barriers that permitted spatial isolation of ancylid mollusks in Lake Ochrid in Yugoslavia. The Russian zoologists have continued their illuminating work on the Lake Baikal fauna and there is a steady stream of publications on the African lakes. The time will soon be ripe for another review paper on speciation in ancient lakes! Ancient lakes are by no means the only bodies of fresh water where speciation has been rapid and controversial. Complex situations have also been described for the salmon and trout of the temperate zone in both the New World and the Old World, and the complex story of coregonid speciation has been a source of contention among ichthyologists for some 100 years. In spite of their similarity and phenotypic plasticity, these whitefish types are not biological races but local representatives of well-defined, widespread species (Chapters 3 and 7).

The barriers set up by the ice during the various Pleistocene glaciations resulted in the temporary isolation in Eurasia of various lakes or ocean basins and permitted speciation. Svärdson (1961) has attempted to reconstruct the pathway of speciation for pairs of sibling species in the following groups: sculpin (*Myoxocephalus*), smelt (*Osmerus*), cisco (*Coregonus albula*), whitefish (*Coregonus oxyrhynchus*), herring (*Clupea*), and stickleback (*Gasterosteus*) (Table 15–3). Remarkable in all these cases is the relatively short duration of the isolation. The current

Table 15-3. Speciation in Eurasian fish (after Svärdson 1961).

Species pair	Location of isolate	Duration of isolation (years)
Sculpin, smelt	Onega Valley Ice Lake	60,000
Cisco	South Scandinavia	120,000
Herring	Mediterranean	120,000
Whitefish, stickleback	Great Siberian Ice Lake	100,000

coexistence of two sibling species in the same body of water is due to reinvasion after the lifting of the geographical barrier.

The occurrence of two species of lake trout (*lacustris* and *carpio*) in Lake Garda (Nümann 1948), both belonging to the *Salmo trutta* complex, is also best interpreted as a case of double invasion. The widespread species is *S. lacustris*, which spawns from October to December in shallow water in the tributaries of Lake Garda. An earlier colonization produced the endemic species *S. carpio*, which spawns twice yearly, in June and December, in the deep waters of the lake. The lake, in its present form, dates back to the end of the last glaciation. Many other cases of double invasion of lakes are known.

One can summarize the problem of species swarms in lakes by saying that the available evidence is consistent with the hypothesis of a spatial separation of populations corresponding to allopatric speciation in terrestrial animals. However, since such separation in different water masses may result in adaptation to different water temperatures, to different breeding seasons, and to different food niches, it presumably sets up very powerful selection pressures leading to a more rapid divergence of these populations than is usual in terrestrial animals. When secondary contact of such previously isolated populations is established, selection will be intensified by competition among close relatives. Finally, the presence of many previously unoccupied ecological niches, particularly in the older lakes, encourages an amount of adaptive radiation (Brooks 1950; Poll 1950; Hubendick 1952; Fryer 1959) that is quite unparalleled among terrestrial animals except in a few isolated archipelagos such as the Hawaiian Islands. Nothing is thus far known about species swarms in lakes that would necessitate postulating the occurrence of sympatric speciation.

Instantaneous Splitting of Fossil Lineages. Paleontologists have described a considerable number of cases in which, supposedly, a single lineage has suddenly split into two clearly separated species. The Steinheim snails (*Planorbis multiformis*), *Kosmoceras* (Brinkmann 1929), and

Micraster (Kermack 1954) are such cases (Fig. 2–1). The situation of the Steinheim snails has not been reexamined in recent years, but in the case of *Kosmoceras* and *Micraster* it is now evident that the original interpretation was not correct. Whenever a second species appeared suddenly in a horizon, it owed its origin not to sudden speciation but to immigration from elsewhere. Every case of so-called sympatric splitting in *Kosmoceras* is preceded by a break in the strata (as for instance at 680.5 cm, Brinkmann 1929). For *Micraster* Nichols (1959) has shown that *senonensis* invaded the range of *M. cortestudinarium* after the end of the *Holaster planus* zone and did not split off from it. The exposures accessible to the paleontologist are usually localized, and it is in most cases quite impossible to determine from where a species came that had originated elsewhere as a localized geographical isolate.

A reanalysis of all five reputed proofs of sympatric speciation—sibling species, speciation in monophagous groups, speciation in parasites, species swarms in lakes, and the instantaneous splitting of fossil lineages—thus shows that they are consistent with the theory of geographic speciation.

Models for Sympatric Speciation: Assumptions

Some of the supporters of sympatric speciation were not satisfied merely to postulate the occurrence of such a process, but they worked out more or less detailed models of it. An analysis of these models shows that they do not work; they cannot achieve what they are asked to do, in part, because they are based on unsupported and unrealistic assumptions. It is necessary to discuss these often subconscious assumptions before taking up the analysis of the models themselves.

homogamy. According to this concept the most similar individuals of a population tend to mate with each other. Mayr (1947) discussed the concept and its history and showed that the observed evidence contradicts its validity. Homogamy had to be postulated under the pre-Mendelian theory of blending inheritance because without it random mating would soon have led to a complete elimination of all genetic variability. The assumption made by Kosswig (1947) that monogamy would lead to homogamy is contradicted by the facts. Monogamy is the rule among birds, but there is no evidence for homogamy except for some nonrandom mating in the case of the Snow Goose–Blue Goose (Cooch and Beardmore 1959). In the area of contact between blue geese and snow geese (*Anser coerulescens*) there are more pure white or pure blue pairs observed than one would expect if random mating did occur (Table 15–4).

Table 15–4. Nonrandom mating in
Snow Geese (from Cooch,
personal communication).

Cross	Number of pairs	
	Expected	Observed
Snow × Snow	330.4	373
Blue × Blue	21.4	64
Snow × Blue	168.2	83

There are several possible interpretations. Manning (personal communication) believes that the two kinds of geese were formerly isolated geographically and have started to hybridize only recently — a case of secondary intergradation. Another possibility, not excluding the first, is that a bird prefers to mate with an individual colored like its own parents. Even so there are enough mixed pairs of the two kinds of geese to ensure abundant gene flow through the entire population.

A mild form of homogamy occurs in many animals with highly variable adult size. The mates in pairs of such species are on the average more similar in size than expected by chance. The same individual, however, as it continues to grow year after year, passes into larger and larger size classes. Most students of polymorphic species have confirmed the randomness of pairings, in spite of the sometimes striking visible differences of the sex partners. Indeed, in *Panaxia dominula* Sheppard (1952b) found even a slight preference for the unlike partner and the same has been described for morphs in the White-throated Sparrow (*Zonotrichia albicollis*) (Lowther 1961).

Contrary to earlier statements in the literature, homogamy has not been demonstrated for any mutants in *Drosophila*. Reproductive success in these mutants depends largely on activity and there is evidence that this results in discrimination against certain genes in a population. This, however, leads to the eventual elimination of the gene and not (Petit 1958) to a segregation into two separate populations. The only recent case of partial homogamy was described by Popham (1947), who found that in *Coriza distincta* there was a correlation in color of +0.53 between copulating males and females. However, there was also much migration between breeding localities, resulting in a great variety of form and pattern in this species. As Hogben (1946) has mathematically demonstrated, "positive assortative mating can have very little importance as an evolutionary process unless it is exceedingly intense."

Linkage of Mate Preference and Habitat Preference. In nearly all hypotheses of sympatric speciation it is assumed that if a species is found in several niches the individuals in one niche will mate only with individuals adapted for the same niche. And if an individual colonizes a niche that is new for this species, then mating will be strictly limited to the descendants of the founder of this population. Mayr (1947) has shown the fallacies inherent in this assumption. It postulates a much simpler basis for mate and habitat preference than is actually found. It ignores the diploidy of zygotes and the role of heterozygotes as a bridge between the two types of homozygotes. Finally, it ignores the phenomenon of dispersal which maintains gene exchange among populations not separated by isolating mechanisms. In many butterflies, for instance, copulation normally takes place during the dispersal phase and not on the plant on which the caterpillars feed. If the female is sedentary or if she mates immediately after hatching, then the males are active and undertake considerable flights before and during the mating period. And even in the hypothetical case (probably never realized in nature) in which mating always takes place on the host and host selection is entirely rigid, it is doubtful that this would result in sympatric speciation. There are numerous difficulties. If an individual can switch from the parental host species to the new host species, then no doubt the process is reversible, thus preventing a complete separation of the two populations. Also, if a species is so rigidly adapted to a single host species that it occurs nowhere but on that host species, most probably it will have a very low chance for survival on any other host species.

Conditioning. Several recent authors have revived the well-known suggestion that the establishment of a new sympatric species population in a new niche might be achieved through conditioning. Thorpe (1939, 1945), in particular, has presented convincing evidence for the powerful effect of conditioning in shifting insects from one type of food to another. The work of Craighead (1921), referred to above, points in the same direction. Yet Mayr (1947) has shown that complete isolation of the two populations was never achieved in any of these experiments on conditioning. Preference for a type of food may be raised from 35 to 67 percent or lowered from 20 to 8 percent, but this still permits far more gene flow between the two subpopulations than is needed to prevent speciation.

In all carefully conducted experiments of this sort it was possible to demonstrate that there is high mortality on the new host or on the new food. It would seem impossible that a reproductively isolated new host population could become established within the cruising range of the parental population in the face of such high counterselection and with the help of such a slight advantage through conditioning. It is far more likely that the ecological amplitude of the species should be enlarged, according to the Ludwig theorem, by incorporating an additional kind of host into the niche of the species.

Preadaptation and Niche Selection. Most hypotheses of sympatric ecological speciation postulate that dispersing individuals search actively for that particular niche to which they are best adapted on the basis of their particular genotype. There is a nucleus of truth in this assumption, since a given species usually has well-defined species-specific habitat preferences and these are not necessarily identical for all the various genotypes of a species (Chapters 9 and 18). Most habitat selection, is, however, not only species specific in a very generalized way, but also often affected by nongenetic influences (conditioning, and so forth). Only in an exceptional case will an individual search out that particular niche for which it is specifically preadapted by its genetic constitution. Settling down at the end of the dispersal stage in the life cycle of an individual is normally very haphazard, and there is much mortality as a result of settling down at unsuitable places. And even in the few known cases of a limited amount of habitat selection by genotypes, this has merely strengthened an existing polymorphism and has not led to sympatric speciation. The concept of sympatric speciation by preadaptation is strictly typological in its assumption that a single gene preadapts an individual for a new niche. Indeed, it would require a veritable systemic mutation to achieve the simultaneous appearance of a genetic preference for a new niche, a special adaptedness for this niche, and a preference for mates with a similar niche preference. The known facts do not support these assumptions.

Some Proposed Models for Sympatric Speciation

Among the various models of sympatric speciation proposed in recent decades, the following four deserve particular attention.

(*a*) *Speciation by Disruptive Selection.* Fisher (1930) pointed out that the mutationist interpretation of the origin of mimetic polymorphism in *Papilio polytes* is highly improbable, it being far more likely that the striking discontinuities between the mimetic morphs are gradually acquired by natural selection. This assumption has since been largely substantiated (Clarke and Sheppard 1960a,b). Such acquisition through selection of several distinct phenotypes in a population has recently been designated "disruptive selection" (Mather 1955). Muller (1940) was apparently the first to suggest that the accumulation of different sets of specific modifiers might lead to sympatric speciation. Mather (1955), Thoday and Boam (1959), and Millicent and Thoday (1960) have likewise suggested that disruptive selection may be a mechanism of speciation. Actually, all the available evidence indicates that this is not likely to happen in natural populations. Disruptive selection leads to phenotypic polymorphism and in view of changing selection pressures and gene flow from other populations only a relatively oligogenic polymorphism has a chance to become established under these conditions. In such a case the remainder of the genotype (outside the polymorphic locus) does not participate in the discontinuity. Nor would this be favored by selection, since a species would lose all the advantages of improved utilization of the environment through adaptive polymorphism if it were to split into a series of narrowly specialized species. Jordan pointed out as early as 1903, with the clarity so characteristic of his work, that selection of different physiological varieties within a population could have only two outcomes, either polymorphism or extinction of the inferior types, but never sympatric speciation. Mimetic polymorphism, perhaps the most frequent and best-analyzed product of disruptive selection, fully substantiates Jordan's conclusion: in spite of the most powerful selection in favor of a complete discontinuity of phenotypes, it has not resulted in speciation in a single case. It must be remembered that two phenotypes within a single population are not two different races.

Thoday and Gibson (1962) have shown that disruptive selection may lead to incipient speciation under laboratory conditions, if the selection is for two alternate polygene complexes. When simultaneous selection for high and low bristle number was carried on in a population of *Drosophila melanogaster*, almost complete homogamy had developed within 12 generations of selection. Males with high bristle numbers tended to mate with high-number females, and low-number males with similar females. The severity and absoluteness of the selection, the complete prevention of gene flow, and the elimination of mutual competition are, of course, a set of artificial conditions that could scarcely be expected ever to occur in nature.

(b) *Speciation by Cytoplasmic Sterility.* Laven (1959) discovered a cytoplasmic factor in *Culex pipiens* that results in normal fertility when ♀♀ of strain A are crossed with ♂♂ of strain B, but sterility of ♀♀ B when crossed with ♂ ♂ of strain A (for details, see Chapter 3). He postulated that new species could originate by the mutation of a new cytoplasmic sterility factor, and since the "change is already an isolating mechanism in itself, geographical or ecological isolation is not necessary."

A study of Laven's model shows that a new cytoplasmic complex cannot come into independent existence sympatrically with the parental complex. If a new factor B originates by mutation in a male of population A, this ♂ B will be sterile with all A ♀ and will exterminate itself. If B arises in a female, ♀ B will produce only B offspring with A ♂ owing to one-directional incompatibility. This in due time will lead to the extinction of A from the population. A two-directional incompatibility will result in the automatic sterility of its carrier and can never become established.

A cytoplasmic-incompatibility system can arise only allopatrically and, at that, not by a single mutation. It can, however, be built up slowly, step by step, and more rapidly by exterminating the locally preexisting cytoplasmic type. There will be considerable gamete wastage where two populations with incompatible systems meet. The result will be either the extermination of one of the cytoplasmic factors (Caspari and Watson 1959) or the perfection of a preexisting independent system of isolating mechanisms that is the completion of allopatric speciation. In neither case will sympatric speciation be possible.

(c) *Speciation by a Mutation Changing Host Specificity.* Many of the proposals of sympatric speciation by a shift to a new host species make the following genetic assumptions (or similar ones):

Let the animal AA, which is host specific on plant 1, have the recessive mutation a, which in homozygous condition produces host specificity for plant 2. Let us make the following additional postulates:

  (1) Let A live only on plant species 1,
  (2) Let aa live only on plant species 2,
  (3) Let the heterozygotes Aa be exactly like AA,
  (4) Let there be little or no dispersal in the reproductive phase so that A animals do not meet aa animals,
  (5) Let A be ill adapted to plant species 2,
  (6) Let aa be ill adapted to plant species 1,
  (7) Let aa formed on plant species 1 by recombination have difficulties in finding plant species 2, even though the original aa found 2.

Then there will be little mixing between the populations on 1 and 2 and the opportunity is provided for a gradual accumulation of additional genetic differences between these populations and the eventual acquisition of reproductive isolation (Mayr 1947).

Of these seven postulates a minimum of four are necessary to make the model work, yet most of them are quite unlikely in the stated absolute form. The probability is very slight that Aa is exactly like AA or that habitat preference and probability of survival are completely linked. Other objections to this model, such as the neglect of dispersal, have been discussed previously. All in all, it appears exceedingly unlikely that such a typological model could lead to speciation in a natural population.

(*d*) *Speciation by Seasonal Isolation.* It has been postulated by various authors that a species with a very long breeding season might be sympatrically split into two, if the genetic continuity between the earliest and the last breeders of the year could somehow be interrupted. The occurrence of seasonal races is often cited as evidence for such a process of speciation in spite of the grave objections raised by Jordan (1905). The term "seasonal race" is confusing, since at least four different kinds of breeding behavior have been lumped under this heading: (*a*) the succession of generations within a single year, the later generations being direct descendants of the earlier generations; (*b*) in fresh-water and marine animals the occurrence of subpopulations of a species which breed at distinctly different water temperatures but which may interbreed at intermediate temperatures; (*c*) the occurrence of a spring and a summer or fall generation when each year's spring generation is the descendant of the previous year's spring generation and the summer or fall generation a descendant of the previous year's summer or fall generation; (*d*) so-called biological races, which differ in their breeding season, but are actually sympatric sibling species.

As far as (*a*) is concerned, it is of no interest for the problem of sympatric speciation. More interesting is (*b*), a case of which has been described by Spieth (1941). Several formerly geographically isolated races of *Hexagenia munda,* with differences in breeding season and habitat preference, have invaded each other's ranges and begun to interbreed. It is clearly not a case of incipient speciation, but exactly the opposite, a breakdown of incomplete speciation.

More difficult are the cases falling under (*c*), in which different seasonal races of the same species are said not to interbreed. Unfortunately not a single such situation has been well analyzed but what little available evidence there is indicates that we are here dealing with cases of secondary overlap, where the ancestral populations originated as geographic races and had adapted their season of reproduction to local conditions. The various broods and sibling species of the periodical cicada (*Magi-*

cicada) have evidently originated in this manner (Young 1958). Some of the cases of seasonal races recorded in the literature are undoubtedly sibling species. Falla (1946) described as *westlandica* a winter-nesting "subspecies" of the shearwater *Procellaria parkinsoni* from Westland, New Zealand, where the nominate form breeds during the summer. Later on he realized that in reality two sibling species are involved. The indications are the same for the two so-called seasonal races of *Pterodroma neglecta* from the Kermadec Islands (Murphy and Pennoyer 1952). This case had formerly been used by Hutton (1897) as proof of seasonal speciation. Among the many incompletely analyzed cases of seasonal races is that of *Culicoides impunctatus* (Kettle 1950). In Scotland there are two seasonal peaks, one early in June and the other late in July. The two resulting populations differ in their sex ratio and in their vertical and horizontal distribution. In the Liverpool area there is only one peak, which is more extensive, however.

The occurrence of seasonal races and of sibling species that differ primarily in their seasonal activities has led to the proposal of several models of speciation by seasonal isolation. Such a model might postulate the following conditions:

Let there be a species with a single annual generation, but with a prolonged breeding season, which lasts from spring to fall. Let an event take place, like a killing off by climate or a newly invading competitor, that leads to the extinction of the midseason breeders. As a result, only two sets of individuals remain, early-season breeders and late-season breeders. Both kinds are henceforth reproductively isolated and can accumulate genetic differences.

Several of these assumptions are so unrealistic that the operation of this model in nature would seem highly improbable. First, the existence, in an area with pronounced seasons, of a species with such a long breeding season but only one generation per year is altogether unlikely. Prolongation of the breeding season is invariably achieved by having several successive generations or broods per year. Secondly, if there were such a species it would have a wide geographic range and it is improbable that the extermination factor would affect all the populations in a like manner. The midseason gap would then be filled by gene immigration. Segregation by recombination from the gene reservoir of the spring and fall breeders would tend to fill the midseason gap even without immigration. Finally, the assumption that an exterminating factor would single out a segment in the middle of the curve is highly artificial. Competitors as well as climatic factors are far more likely to eliminate the extreme members of the population, that is, the earliest or the latest breeders.

The third of these three objections is successfully met by Bigelow (1958) and Alexander and Bigelow (1960) in their attempt to explain the separation of two field crickets (*Gryllus*) by allochronic speciation. These two very similar sibling species differ most conspicuously in their breeding season and overwintering method. The Northern Spring Field Cricket (*G. veletis*) breeds from May to July and overwinters in a late nymphal stage. The Northern Fall Field Cricket (*G. pennsylvanicus*) breeds from July to October and overwinters in the egg stage. Since the early instars are not cold resistant, Alexander and Bigelow propose that the ancestral species was split into two by the killing off of the early nymphs. Yet their model fails to answer the other two objections raised above. To me, it would seem far more reasonable to assume that the range of the ancestral species was fractionated into several geographical isolates (during one of the glaciations?), in one of which fall breeding proved more adaptive, in the other, spring breeding. Otherwise the two species remained very much the same, so that they could occupy essentially the same region after they embarked on their post-Pleistocene range expansion. The assumption of Alexander and Bigelow that all populations of the parental species would be subjected, in an identical manner, to an elimination of midseason nymphs seems unrealistic. The narrowness of the breeding season in the crickets and grasshoppers precludes the existence of a univoltine continuous breeding season. It is rather probable that the difference in breeding cycle of *G. pennsylvanicus* and *G. veletis* was not yet complete, when they first met, after emerging from their geographical isolation. If so, competition eliminated any tendencies for spring breeding in *pennsylvanicus* and of fall breeding in *veletis*. Related species, not in competition with a close relative, are rarely as narrowly specialized.

The same model of an acquisition of breeding-season differences in geographic isolation (rather than by sympatric speciation) was proposed by Young (1958) for the periodical cicadas. The high improbability of an older model of allochronic speciation by mutation was pointed out by Mayr (1947:277).

It is no coincidence that all the cases of presumed sympatric speciation by temporal isolation concern genera rich in sibling species. The morphological uniformity in these genera conceals the large amount of genetic change induced by or correlated with the shifts in breeding season. Any population will be selected so that the breeding season will coincide with the probability of maximum survival of the offspring. The adaptation of a population to a specific timing of the reproductive cycle will affect many facets of its genetic constitution. It would be altogether unlikely for a population to adapt itself simultaneously to two different breeding cycles. Where a species evolves two such cycles it does so in different, geographically segregated, populations. Seasonal differences in the reproductive cycle may function secondarily as isolating mechanisms.

One can summarize this discussion of models of sympatric speciation by saying that all of them show serious, if not fatal, weaknesses. In not a single case is the sympatric model superior to an explanation of the same natural phenomenon through geographic speciation.

Difficulties of Hypotheses of Sympatric Speciation

Hypotheses of sympatric speciation are usually proposed in order to eliminate so-called difficulties of the theory of geographic speciation. However, the authors of these hypotheses overlook the fact that they create many more difficulties than they remove.

Neglect of Dispersal. Dispersal is one of the basic properties of organic nature, yet it is conspicuously neglected in all schemes of sympatric speciation. The life cycle of every species includes a dispersal phase. In most insects with wingless larvae it is the adult stage. In most marine animals it is the larval stage. The function of dispersal in permitting the species to expand into previously unoccupied areas and insufficiently filled niches is too well known to justify further discussion. What is often overlooked is that a great deal of intermingling of populations is the consequence of this dispersal. This includes individuals not only of different geographical, but also of different ecological origin. Such dispersal also occurs in host-specific species and in those in which mating takes place on the host plant. As shown above, it is this dispersal stage that permits the interchange of individuals between incompletely conditioned populations.

Ecological Plasticity and Polymorphism. Populations and species show much more ecological variability than is consistent with the hypotheses of ecological speciation. This variability exists for individuals belonging to a single population, and for the various populations of a species. Some of the variation may have a genetic basis; much of it, however, is evidently nongenetic. Where a species has a preferred food plant, some individuals are usually also found on other host species, either exceptionally or regularly. If a species is rigidly tied to one host plant in part of the species range, it may be more plastic in other parts. Every population of a species lives in an environment that is different from that of the other populations, and its genotype is continuously adjusted by selection to this specific local environment. Occupation of multiple niches and the shift from one niche to another in a different portion of the species range is the normal situation in most species of animals. If this led to species formation without geographic isolation, we would have a vastly greater number of species than we actually have. To interpret the ecological variability of species as a mechanism of sympatric speciation is to misunderstand the function of this variability. Ecological variation has been misinterpreted in the same way as chromosomal variation. Both are adaptive mechanisms rather than speciation devices. Both, however, may facilitate and accelerate speciation, when superimposed on geographic speciation.

Genetic Difficulties. Serious though all the above-listed difficulties of the hypotheses of sympatric speciation are, the really decisive argument against these hypotheses comes from the field of genetics. The facts that zygotes in sexually reproducing organisms are diploid and that sexual reproduction maintains the genetic cohesion of every local population present serious obstacles to all hypotheses of sympatric speciation.

The bridge effect of heterozygotes. Bateson and other early Mendelians always looked for the singular genetic event that would differentiate a new cross-sterile line from the parental one. However, "one single genic change does not differentiate a new cross-sterile line from the old one" (Stern 1936). This can be illustrated by a simple genetic model. All individuals of sexually reproducing species of animals are diploid (at least in one sex). A new mutation is always heterozygous in the beginning because it occurs on only one of the two equivalent chromosome sets. For instance, if, in a homozygous *aa* population, a dominant mutation *A* were to make its carriers sterile with *a*, every carrier of the gene *A* would automatically be sterile (*aA*). If the production of genotype *aA* were to result not in sterility but merely in reproductive isolation from *aa*, it would still doom the carriers of *A* to extinction, since they would have no potential mates unless *A* were subject to a highly improbable process of mass mutation, or adopted an asexual mode of reproduction. Even then it would segregate 25 percent of *aa* individuals per generation, which would provide for a continuous gene flow from the daughter to the parental species. These are only a few of the most obvious objections to the idea that a single mutation could produce new species. Additional, more general, objections are raised in Chapters 10 and 17.

For successful speciation, a minimum of two complementary genes is necessary (Dobzhansky 1937). This process in its simplest form is presented in Table 15–5. Let us assume that a uniform population, homozygous for *a* and *b*, is separated by geographic isolation into two populations, 1 and 2. Let us postulate that *AA* individuals are reproductively isolated from *BB* individuals. If a mutation *A* occurs in population 1, it will result in the production of an individually variable population with the three genotypes *Aabb*, *aabb*, and *AAbb*. Let us assume that natural selection favors *AAbb*, so that eventually this population becomes homozygous for this genetic constitution. While this happens in population 1,

Table 15–5. Speciation through two complementary factors
(parental population *aabb*).

Population I	Population II
Isolation	
aabb	aabb
Mutation to *A*	Mutation to *B*
Aabb	aabB
aabb	aabb
AAbb	aaBB
Selection for *AA*	Selection for *BB*
AAbb	aaBB

population 2 will go through a parallel transformation, with the *b* locus mutating to *B*, resulting in the genotype *aaBB*. If now the geographic isolation between the two populations should break down, they could intermingle without interbreeding because, according to our initial postulate, *AA* is reproductively isolated from *BB*. From Table 15–5 it is quite evident that in the intermediate stage, when each population consisted of three genotypes, only one of the possible nine combinations of these six genotypes would be reproductively isolated. Such almost complete panmixia would, of course, prevent speciation.

Our model is grossly simplified, since we started with a homozygous population, at least as far as the crucial loci are concerned. Actually, natural populations are highly heterozygous and composed of a genetic environment that favors heterozygosity. A bridge of heterozygosity thus exists not merely at one or two loci but at scores or hundreds of them. There is no reason to doubt that the genes involved in the control of the isolating mechanisms are at the same time part of the total genetic cohesion of the population (Chapter 10). Our second simplification is that we let the reproductive isolation be determined by only two loci. Actually, as we saw in Chapters 5 and 6, reproductive isolation usually has a highly complex basis, and it requires many mutational steps or other genetic re-arrangements to complete reproductive isolation. All these considerations help to make clear why geographic isolation is a prerequisite for successful speciation.

CONCLUSIONS

The discussion of sympatric ecological speciation permits us to conclude that the hypothesis is neither necessary nor supported by irrefutable facts. It overlooks the fact that speciation is a problem of populations, not of individuals, and it minimizes the difficulties raised by dispersal and recombination of genes during sexual reproduction. As was shown by Timoféef-Ressovsky (1943) and Mayr (1947), all populations are geographically, as well as ecologically, defined. The essential component of speciation, that of the genetic repatterning of populations, can take place only if these populations are temporarily protected from the disturbing inflow of alien genes. This can be done best by extrinsic factors, namely, spatial isolation. It appears that such spatial isolation is always maintained by geographical barriers. The possibility is not yet entirely ruled out that forms with exceedingly specialized ecological requirements may diverge genetically without benefit of geographic isolation; the burden of proof rests, however, on supporters of this alternative mode of speciation.

The discussions in this chapter, and in particular the discussions on instantaneous and sympatric speciation, have shown that nearly all the potential modes of speciation (Table 15–1) suggested in the past are improbable and that the reputed instances are quite consistent with the theory of geographic speciation, indeed, are more easily understood through this interpretation than any other. It is now clear that geographic speciation is of overwhelming importance in explaining the problem of multiplication of species. A whole chapter will therefore be devoted to this important process.

16 ~ Geographic Speciation

That geographic speciation is the almost exclusive mode of speciation among animals, and most likely the prevailing mode even in plants,
is now quite generally accepted. And yet this thesis was vigorously contested as recently as 25 years ago and such a distinguished biologist as
Goldschmidt never accepted it. The theory of geographic speciation is one
of the key theories of evolutionary biology and it would seem appropriate
to present in considerable detail both its history and the proofs of its
correctness.

The problem of speciation is one that illustrates particularly well the
basically different mode of thinking of the functional biologist, who is
primarily concerned with mechanisms and their operation, and the evolutionist, who is concerned with natural populations. While evolutionists
have stressed since 1869 that the population structure of species is the key
to the problem of speciation, we still find a complete preoccupation with
the mechanisms in such recent publications as those of Fisher (1930),
Muller (1940), and Babcock (1947a,b). How little the geneticists before
Dobzhansky (1937) understood the problem of speciation is documented
by the fact that R. A. Fisher in 1930 as "motto" of his Chapter 1 quoted
with evident approval W. Bateson's statement, "As Samuel Butler so truly
said: 'To me it seems that the "Origin of Variation," whatever it is, is the
only true "Origin of Species."'" In other words, mutation is the only true
Origin of Species.

Actually, the basic problem of speciation is to explain the origin of the
gaps between sympatric species. These gaps, as we saw in Chapter 5, are
clear-cut, well defined, and maintained by isolating mechanisms. As long
as species were considered static, typological, and nondimensional, it was
virtually impossible to solve the problem of the crossing of these gaps.
Hence, they were termed, with good reason, "bridgeless gaps" by Tures-

son, Goldschmidt, and others. During much of the nineteenth century when typological concepts prevailed, no other mode of the origin of species seemed possible than that by sudden jumps, macromutations. The only apparent alternative was to go to the other extreme, to deny the existence of discontinuities altogether and to believe in the gradual origin of new species populations without geographic isolation by a process that may be called sympatric speciation. How unsatisfactory these two hypotheses are was shown in Chapter 15.

A third, and, as is now known, far more plausible, solution of the seemingly insoluble problem of the bridgeless gaps was made possible by a change in systematic concepts. This consisted in the expansion of the species from the nondimensional and typological to the multidimensional and polytypic species, as described in Chapter 12. The work of numerous taxonomists finally culminated in the theory of geographic speciation which states that in sexually reproducing animals a new species develops when "a population, which is geographically isolated from its parental species, acquires during this period of isolation characters which promote or guarantee reproductive isolation when the external barriers break down" (Mayr 1942).

HISTORY

It will help our understanding of this thesis if we review some of its history. Many naturalists of the late eighteenth and early nineteenth century were already dimly aware of the importance of geographic factors in speciation. Indeed, they were aware of both aspects of the concept of geographic speciation, "climatic variation" and isolation, but failed to unite the two factors into a consistent theory. Climatic variation was described and emphasized by, among others, Buffon, Pallas, von Baer, and Faber (Stresemann 1951:198). Each of these authors, however, denied explicitly or implicitly that this had anything to do with the origin of new species. Gloger, who as early as 1833 devoted a whole volume (159 pages in small print) to the subject of *The Variation of Birds Under the Influence of Climate*, considered this geographic variation apparently a nongenetic adaptation to local conditions. The mere observation of the phenomenon of geographic variation was not enough to lead to a theory of geographic speciation, nor was the observation of the importance of isolation. Wollaston (1856), in a study of the insects (particularly beetles) of the Canary Islands, fully appreciated the importance of isolation. He compared the forms of the various islands with each other or with the nearest relatives of the mainland and pointed out for one character after another how it is affected by isolation: "Isolation, when involving a sufficient period of time, has a direct tendency either to diminish the stature of the insect tribes or else to neutralize their power of flight" (p. 84); "isolation does, in nearly every instance, in the course of time, affect more or less sensibly, external insect form" (p. 452). He illustrated this thesis with numerous examples, yet he remained blind to the obvious consequences of these observations. As a pious creationist he concluded his discussion with the statement: "It does indeed appear strange that naturalists . . . should ever have upheld so monstrous a doctrine as that of the transmission of one species into another" (p. 186). It is remarkable how blind a man can be who is not prepared to see the truth that was so evident to Darwin, Wallace, and Wagner, in their parallel and almost simultaneous studies of insular faunas.

A few early prophets saw things essentially as they are but did not press the point, and we do not know how far-reaching the influence of their early pronouncements was. The most remarkable of these was Leopold von Buch, who wrote in a description of the fauna and flora of the Canary Islands (1825):

> The individuals of a genus strike out over the continents, move to far-distant places, form varieties (on account of the differences of the localities, of the food, and the soil), which owing to their segregation [geographical isolation] cannot interbreed with other varieties and thus be returned to the original main type. Finally these varieties become constant and turn into separate species. Later they may again reach the range of other varieties which have changed in a like manner, and the two will now no longer cross and thus they behave as "two very different species."

This statement has a remarkably modern ring, and, if a few terms were replaced, it could well pass as a modern description of the process of geographic speciation. Two of the points made by von Buch, namely, that geographic isolation is needed to permit the species difference "to become constant" and that proof of the species difference is given by their eventual reproductive isolation, were curiously enough not recognized with the same clarity by later authors. Von Buch's theory, which fitted so well with his observations on the voyage of the *Beagle*, deeply impressed Darwin, as is evident from his notebooks (De Beer 1960, 1961). How highly Darwin rated the role of geographic isolation in his earlier years is evident from his correspondence in 1844 with Hooker: "Do you know of any other case of an archipelago other than the Galapagos, with the separate islands possessing distinct representative species?" and "With respect to original creation or production of new forms . . . isolation appears the chief element." Yet later, under the influence of his work with domestic animals and plants he increasingly abandoned isolation as an important evolutionary factor (Darwin 1859) and his correspondence with Wagner in 1868 shows how hopelessly confused he had become by then: "Nevertheless I cannot doubt that many new species have been simultaneously developed within the same large continental area," and so on. The reasons for Darwin's failure to appreciate the role of geographic isolation have been discussed by me elsewhere (Mayr 1959a). Prominent among them was his confusion of evolutionary change with multiplication of species, and his ambiguous use of the term "variety" (for individuals and for populations). Finally, Darwin's essentially typological species definition made it impossible for him to appreciate the key aspects of the problem of speciation.

Other traveler-naturalists of the period placed far more emphasis on the geographical aspects of speciation. Bates (1863) in his exploration of the Amazon Valley found numerous instances where the Amazon River (or one of its tributaries) separated geographically representative species. "A great number exist only in the form of strongly modified local varieties; indeed many of them are so much transformed that they pass for distinct species . . . We seem to obtain here a glimpse of the manufacture of new species in nature." Specific instances of geographic speciation cited by Bates are *Papilio echelus* at Para, east of the Tocantins, and *P. aeneides* west of the Tocantins (pp. 120–121), and *Papilio lysander* on the upper Amazon and *P. parsodes* at the mouth of the Amazon (pp. 304–305). Yet Bates—and this is equally true of Wallace, his counterpart in the East Indies—though fully aware of the fact of geographical isolation and of the importance of geographical barriers as well as of the aberrant characteristics of isolated populations, never quite welded these observations into a unified theory.

The credit for having done this belongs to the great traveler and naturalist Moritz Wagner, who proposed the theory of geographic speciation in 1868 in his *Migrationsgesetz der Organismen*. A much expanded and somewhat revised version was brought out in 1889 after his death which was more correctly called "the separation theory." Wagner, who had had extensive collecting experiences in Asia, Africa, and America, had observed that the closest relatives of a given species were almost invariably in an adjacent area and normally separated by geographical barriers. Nowhere had he found any evidence for sympatric speciation. He therefore stated that "the formation of a real variety which Mr. Darwin considers as incipient species, can succeed in nature only where some individuals can cross the previous borders of their range and segregate themselves for a long period from the other members of their species." Although Darwin in a letter to Wagner admitted that this was an important point which he had overlooked, it is evident from their correspondence that Darwin by no means appreciated the extent of its importance. This was in part Wagner's own fault. There is no doubt that Wagner presented some powerful observational evidence for the importance of isolation, but we can no longer follow much of his reasoning in connection with the role of isolation. Wagner, like Darwin, neglected the difference between true speciation and phyletic evolution, and insisted that geographic isolation was necessary not only for the multiplication of species but also for evolutionary change. Darwin and Weismann (1872) were quick to point out that such is not the case. The general confusion about the role of isolation that existed in the minds even of the clearest thinkers is best documented by a quotation from Weismann (1902:319):

It would be a grave mistake to assume that every isolation of portions of a species would lead at once to its transformation into a new species, or even, as first maintained by Moritz Wagner and later by Dixon and Gulick, that isolation was an indispensable condition for the transformation of species and that it is not selection, but exclusively isolation which permits the transformation of a species, that is, its splitting into several forms . . . The polymorphism of the social insects proves that a species can split into several forms within a single area through natural selection alone.

Similar statements were made by other contemporaries. To a greater or lesser extent, they all confounded phyletic evolution with speciation, and morphologically distinct individuals with species; they failed to appreciate the importance of populations and thought they had to make a choice between selection and isolation. Even blinder were the early Mendelians, in particular Bateson and De Vries (see p. 517).

The naturalists in the meantime demonstrated ever more conclusively that speciation is a phenomenon involving geographically isolated populations. After Wagner, it was particularly Gulick (1887), in his studies of the Achatinella snails of the Hawaiian Islands, who emphasized the extreme importance of the segregation of populations. In his belief in the importance of isolation he was strongly supported by those malacologists who, like Kobelt and the Sarasins (see Chapter 12), established widespread polytypic species. Among the ornithologists it was particularly Seebohm (1888) who combined the rapidly increasing knowledge of the geographic variation of birds with Wagner's principles. "Evolution may go on from age to age, but without isolation no new species can be produced. However much a species can be changed by Evolution, 'the swamping effects of intercrossing' prevent there being at any given time more than one species." Seebohm criticized with particular vigor the writings of Romanes, who was at great pains to prove that geographical isolation was unnecessary and that "physiological selection" was all that was needed for species formation (see Chapter 15). More and more taxonomists (for example, Dixon, Steere, Staudinger) adopted Wagner's conclusions during the 1880's and 1890's.

Perhaps the most lucid statements on geographic speciation were those of the entomologist K. Jordan * (1896). Even more clearly than Seebohm he recognized the particular role of isolation and the simultaneous need of "transmuting" factors during isolation. "There is only one way possible by which the divarication of a species into two or more can come about; that is the combination of isolation and transmuting factors." Jordan seems to have been the first biologist to have stated clearly that speciation is the product both of mutation and of isolation. Wagner had neglected one of the two factors; the Darwinians and subsequently the Mendelians, the other. Jordan had a clear picture of the process of geographic speciation:

The geographical races thus [by isolation] produced we must assume to be first inconstant, to become more and more constant and divergent by the incessant influence of the transmuting factors, and to develop finally into a form which is so modified that it never will fuse either with the parent form or the sister forms, and that therefore agrees with the definition of the term "species" . . . As this kind of divarication of species is the only possible one [he speaks as a zoologist] . . . the study of localized varieties is of the greatest importance in respect to the theory of evolution: *the study of geographical races or subspecies or incipient species is a study of the origin of species.*

Even though Jordan expressed his conclusions more concisely than any of his contemporaries, a study of the writings of progressive animal

---

* It is somewhat unfortunate and has led to much confusion that three biologists of the name Jordan have written on species and speciation. To prevent further confusion they may be identified as follows: Alexis Jordan, a French botanist (1814–1897), who wrote on elementary species (mostly apomictic clones and aneuploid series) named jordanons by Lotsy; David Starr Jordan (1851–1931), an American ichthyologist, later president of Stanford University, after whom J. A. Allen named "Jordan's rule"; and Karl Jordan (1861–1957), an entomologist born and educated in Germany, who published most of his research as curator of insects at the Rothschild Museum at Tring, England (Mayr 1955b).

systematists at the turn of the century shows clearly that they were quite unanimous in attributing the multiplication of species to the process of geographic speciation. These unanimous and well-substantiated conclusions were either completely ignored or bitterly attacked by the contemporary experimental biologists, including animal and plant breeders. This situation persisted in spite of the strenuous efforts of various Wagnerians (for example, Jacobi 1900; Döderlein 1902; and in America particularly D. S. Jordan) to gain adherents for Wagner's views.

D. S. Jordan (1905) writes that Wagner's conclusion "that geographical isolation is a factor or condition in the formation of every species . . . is accepted as almost self-evident by every competent student of species or of the geographical distribution of species . . . but in the literature of evolution of the present day they [the principles set forth by Wagner] have been almost universally ignored." In particular D. S. Jordan emphasizes the very same point most frequently stressed by Wagner, namely, that closely related forms (species or subspecies) occur in adjacent areas. When J. A. Allen suggested in a review that this observation should be called "Jordan's law," Jordan (1908) modestly pointed out that he was merely reiterating what Wagner had stated 35 years earlier. In this he was right; if the law were to be named at all, which is doubtful, it should be called "Wagner's law." It would be futile to try to give an exhaustive catalogue of the adherents of the theory of geographic speciation during the first four decades of this century. Such a list would include the names of virtually all animal taxonomists writing on species formation. Stresemann (1919) and Rensch (1929) are two more recent representatives of this school of thought.

The universal blindness of the geneticists and experimental evolutionists, of which D. S. Jordan complains, seems hard to understand in retrospect. It had multiple roots. The most important, no doubt, was that the naturalists were working with natural populations and had therefore learned that speciation is a population problem. The laboratory men and the breeders were dealing with individuals and with the offspring of individuals and were quite consistently adherents of typological concepts. The situation was made worse by the compartmentalization of biology during that period. It was a time of greater and greater specialization and no one paid much attention to the work of specialists in other fields. The time had not yet come for the naturalists and the experimentalists to join forces. If the problem of speciation is considered at all in the evolutionary writings of this period, it is almost always solved by some theory of sympatric speciation or by saltation. Among botanists, geographic speciation was almost totally ignored except by a few continental authors (for instance, von Wettstein, Cajander). I believe Baur (1932) and Stebbins (1950) were the first botanists to stress properly geographical isolation among the factors of speciation.

The object of this historical survey is not only to show the source of the different concepts of which the theory of geographic speciation is composed, but more importantly to stress its venerable age. Since it has only recently come to the attention of botanists, paleontologists, and geneticists, it is sometimes erroneously considered a "new" theory. This is quite wrong. The early statements go back 135 years and in its full development by Wagner the theory is more than 90 years old; since then it has been tested and retested by three generations of systematists. It is one of the most interesting examples of a theory that was found strictly empirically and for which the "how?" was answered about 100 years before the "why?"

EVIDENCE FOR GEOGRAPHIC SPECIATION

Every careful modern generic revision provides new evidence for geographic speciation. Yet taxonomic evidence is inferential and a few saltationists still refuse to accept it. Knowing that there are alternative modes of speciation (Chapter 15), the student of evolution is faced by a methodological difficulty. Speciation is a slow historical process and, except in the case of polyploidy, it can never be observed directly by an individual observer. Is there any method by which slow past events can be reconstructed and "proved"?

Yes, there is, and this method is well tested because the evolutionist is not alone in his predicament. Any scientist who must interpret past events, like the archeologist, the historian, or the geologist, or who, like the cytologist, studies dynamic processes that can be observed only at definite fixed stages like the stages of the cell division, faces the same difficulty and solves it with the help of the same method. Obviously it cannot be experiment, but—even though this fact may escape some of the more enthusiastic experimentalists—there are other legitimate scientific methods in addition to experiment. The method most suitable for our purpose consists in the reconstruction of an essentially continuous series by arranging fixed stages in the correct chronological sequence.

Stating our aim more specifically, it should be possible, speciation being a slow process, to find natural populations in all stages of "becoming species." The question then is: "Through what stages does a population pass that is in the process of becoming a separate species?" Darwin and other early evolutionists were quite aware of this approach and looked for "incipient species." A "variety," declared Darwin, is such an incipient species, but his study of varieties was not very fruitful, since he made no distinction between variant individuals and variant populations. It is more rewarding to start with the assumption that speciation is a population phenomenon and look for populations that are incipient species or that appear to have just completed the process of speciation ("new species"). The question to be answered concerning such populations is whether their pattern of distribution and other characteristics are consistent with the theory of geographic speciation. There are three sets of phenomena that give us information on this question: (*a*) levels of speciation, (*b*) geographic variation of species characters, and (*c*) borderline cases and distribution patterns.

Levels of Speciation

Numerous species groups, analyzed by recent authors throughout their area of distribution, have been found to consist of populations that represent every stage of divergence, up to recently completed speciation (for example, Huxley 1942; Mayr 1942; Dobzhansky 1951; R. R. Miller 1948). As an example we may quote findings in the genus *Drosophila* (Patterson and Stone 1952:548):

> We have been particularly intent on demonstrating different degrees of divergence and the correlated difference in factors involved in the several available strains [of *Drosophila*]. These range from partial strain isolation (*repleta, peninsularis*), to subspecific divergence (*texana-americana, fulvimacula-flavorepleta*), to closely related species (*virilis-texana, mojavensis-arizonensis*), to distantly related species in the same species group which would not cross, and to members of different species groups and subgenera which do not ordinarily even show interest or attempt to mate in *Drosophila*. We have been able to show the presence of some isolation between separate subspecies or even strains. We do believe that this evidence favors the view that ordinary species differences have come about through the accumulation of steps, through allopatric strains and subspecies to species.

Similar situations have been described for numerous genera of invertebrates and vertebrates. A particularly well-known case is the platyfishes of Central America (*Xiphophorus maculatus* species group), excellently analyzed by Gordon and associates (Fig. 16–1). The northern-

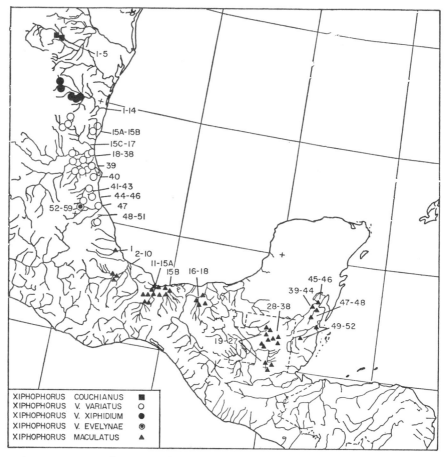

Fig. 16–1. Distribution of the species and subspecies comprising the platyfish superspecies *Xiphophorus maculatus*. (From Rosen 1960.)

most species (*X. couchianus*) is restricted to a single river system. To the south follows *X. variatus*, which has three well-defined subspecies in the states of Tamaulipas, San Luis Potosí, and Vera Cruz, Mexico, among which the subspecies *xiphidium* is so distinct that it has long been considered a full species. Southernmost is *X. maculatus*, famous for the polymorphism of its spotting patterns, which occurs in numerous stream systems from Vera Cruz to British Honduras. The frequencies of the micro- and macromelanophore patterns differ from stream to stream and population crosses have revealed other genetic differences, for instance in the sex-determining mechanisms. Here then we have a series of related, allopatric populations showing every stage from the local genetic race, to the ordinary subspecies, to the almost specifically distinct subspecies (*xiphidium*), to the full species (*couchianus*).

The findings in other recently analyzed species groups fully confirm what we have reported for *Drosophila* and *Xiphophorus*. We find that in every actively evolving genus there are populations that are hardly different from each other, others that are as different as subspecies, others that have almost reached species level, and finally still others that are full species. Sometimes these are still allopatric; in other cases the most distinct ones may already have been able to overlap the ranges of their closest relatives. To find the complete hierarchy of successive levels of speciation, without looking at a single fossil, is most impressive. The fact that these more and more distinct populations remain allopatric until they have reached species level is one of the most convincing proofs of geographic speciation.

Equally impressive is a study of the history of colonization of isolated areas. In the cases of the Australian (Mayr 1944) and North American (Mayr 1946a) bird faunas I have shown that the earliest immigrants from Asia have evolved into new families and genera, later ones into new species or subspecies, while the most recent colonists have not yet started to speciate at all. Again, there is a close correlation between degree (and length) of geographical isolation and level of speciation. Silas (1956) has demonstrated the same for the fresh-water fishes of Ceylon. Of the 56 species indigenous in Ceylon, 2 have reached generic, 10 specific, and 10 subspecific level; 34 species are still taxonomically identical with fishes of the Indian mainland. Similar analyses have been made for many isolated areas.

The Unit of Speciation. The subspecies is considered the "incipient species" in much of the evolutionary literature. Goldschmidt (1952a:84) quite rightly questions whether subspecies, being merely ecotypic responses to the local environment, have much of a potential as incipient species. As we have shown above (Chapter 13) it is the geographical isolate and not the subspecies that is the incipient species. A geographical isolate, if recently established, may still be indistinguishable from the species from which it branched off, or it may consist of a single or of several subspecies. The concept of the incipient species is thus only loosely correlated with the infraspecific categories of the taxonomist. What the student of speciation really must study is geographical isolates.

Two postulates derived from the theory of geographic speciation are that there should be geographic variation in species characters and that the pattern of distribution of natural populations should reflect the pathway of geographic speciation. Any fact that substantiates these postulates would be further proof for geographic speciation. Such facts will be presented in the following sections.

The Geographic Variation of Species Characters

One can summarize the evidence presented in Chapter 11 by saying that any character ever described as distinguishing species is also known to be subject to geographic variation. Of special interest are those species characters (discussed in Chapters 4 and 5) that facilitate sympatry of related species, that is, isolating mechanisms and ecological compatibility factors. If speciation is geographic, these two sets of characters should be as much subject to geographic variation as morphological characters. This conclusion is indeed supported by abundant evidence.

Ecological Properties. Every widespread species that has been carefully studied has been found to contain geographically representative populations that differ from each other to a lesser or greater extent in their ecology. In the Island Thrush (*Turdus poliocephalus*) the populations on Bougainville and Kulambangra in the Solomon Islands live in the high mountain forest above altitudes of 6000 feet. On nearby Rennell Island the species occurs in the lowlands and may feed among coral boulders along the seashore. The Rufous Fantail (*Rhipidura rufifrons*) shows a similar variability of ecological requirements. It occurs on neighboring islands either in the coastal scrub forest, the lowland rain forest, or the mountain forest. Wilson's Warbler (*Wilsonia pusilla*) in North America occurs in Canadian sphagnum bogs in the east, while it is found in chaparral and other exposed hot and rather arid localities in the west. The nominate race of the thrush *Luscinia akahige* lives in the mountains of Japan above 1000 meters and nests on the ground; the race *tanensis* of the Seven Islands occurs at sea level and nests in cavities of trees 1–3 meters above the ground (Jahn 1942). The extensive geographic variation in the habitat utilization of Eurasian tits (*Parus*) has been described in detail by Snow (1954b). Other cases have been discussed by Mayr (1942, 1951a), Stresemann (1943), Stegmann (1935), Huxley (1942), and Hamilton (1962). Geographic variation in habitat occupation is by no means restricted to mammals and birds. Certain species of mosquitoes, as *Anopheles pseudopunctipennis*, *A. bellator*, and *A. labranchiae*, enter houses in part of their geographic range but fail to do so in others (Rozeboom 1952). In most cases the niche requirements of a species remain remarkably constant over wide areas where the populations are contiguous and gene flow is unimpeded. Ecological shifts are characteristic of geographical isolates or zones of secondary contact.

It is easy to see that such shifts preadapt the populations for eventual sympatry. On the basis of available observations I would hazard the guess that ecological exclusion (Chapter 4) of incipient species is in most cases well established before secondary contact is established.

Isolating Mechanisms. The sterility barrier is the isolating mechanism for the geographic variation of which we have the best evidence. Earlier observations of partial sterility among geographically distant races of the same species have been reviewed by Dobzhansky (1951), Ford (1949), Huxley (1942), Mayr (1942), Patterson and Stone (1952), and Rensch (1929). The well-studied genus *Drosophila* supplies many illustrations. In *D. pallidipennis* there is no sexual isolation between a subspecies from Vera Cruz, Mexico, and one from São Paulo, Brazil, and males and females court each other at random. Chromosome pairing in salivary gland chromosomes of hybrid larvae is nearly perfect, yet the F_1 males are as sterile as F_1 males of species crosses (Patterson and Dobzhansky 1943). Other species of *Drosophila* in which sterility barriers between geographic races have been found are *D. peninsularis, D. macrospina, D. athabasca, D. virilis* (Patterson and Stone 1952), *D. tropicalis* (Townsend 1954), *D. cardini* (Heed and Krishnamurthy 1959), and *D. paulistorum* (Ehrman 1960a). Reduction of fertility has also been found in crosses between copepods (*Tisbe reticulata*) from Brittany and the Venetian lagoon (Battaglia 1956) and between various populations of *Tigriopus fulvus* (Božić 1960), to cite cases among marine animals, in a South American freshwater snail (Paraense 1959), and in numerous other terrestrial, freshwater and marine species (for example, Kawamura and Kobayashi 1960, Chabaud 1954). Although the greatest amount of sterility usually occurs among the most distant populations, the amount of sterility is not completely correlated with distance. Rather distant populations are sometimes more fertile with each other than populations less far removed.

Ethological barriers are the most important isolating mechanisms in animals. Unfortunately the evidence for the existence of such barriers between geographical races of a species is scanty. Incipient sexual isolation in the Japanese newt *Triturus pyrrhogaster* has been established for races from the islands Amamioshima, Kyushu, and various stations on Honshu (Kawamura and Sawada 1959). The isolation is nearly complete even for two populations on Honshu Island. The hybrids are, however, perfectly viable. The breakdown of the courtship between males and females belonging to different races seems to be due to the inadequacy of the mutual chemical stimuli. The earliest work on geographical variation in isolating mechanisms among essentially olfactory animals, reacting to chemical stimuli, was done by breeders of moths. Standfuss (1896:107) reports of the Tiger Moth (*Panaxia dominula*) that when freshly hatched females of the Italian subspecies *persona* [*italica*] were exposed near Zurich, Switzerland, only a few males of the native subspecies *dominula* appeared, even though they appeared in large numbers at a nearby place where freshly hatched females of native *dominula* were released. Evidence for the geographical variation of behavioral isolating mechanisms is beginning to be found also in *Drosophila* (for example, Koref-Santibanez and del Solar 1961). According to Goetze and Schmidt (1942), the male honeybee (*Apis mellifica*) has a special scent organ which varies geographically and which may be important in bee courtship. It would be interesting to know whether or not it is due to this difference of scent that it is impossible to cross the European with the Indian bee even though they are believed to be connected by a series of intergrading populations.

The evidence for the geographical variation of ethological isolation based on visual stimuli is almost entirely inferential. The plumes of birds of paradise (Paradisaeidae) vary geographically to such a striking extent that one would expect differences in the courtship poses and in the responses of the females, particularly in the genera *Astrapia*, *Parotia*, and *Paradisaea*. An analysis of such differences would be particularly interesting, since all stages of speciation are represented in these genera. In two species of fiddler crabs (*Uca princeps* and *U. beebei*) the males from Ecuador have more advanced nuptial color patterns than those from Panama (Crane 1944). Behavioral differences between geographic races of birds do occur (Curio 1961; McKinney 1961), but it is unknown to what extent this affects potential isolating mechanisms.

Geographic variation of song, an important isolating mechanism in birds, is widespread. Benson (1948) found distinct geographic variation of song in 33 of 210 species of African birds examined by him. Several cases were reviewed by Mayr (1942). The Pacific warbler (*Acrocephalus*), for instance, has on Guam a song so beautiful that the bird is called the nightingale warbler; on other islands the song is indifferent, and on Pitcairn it seems to have become altogether obsolete. Numerous differences in song of European species of birds between Britain and Tenerife have been described by Lack and Southern (1949). The nuptial song is the most important isolating mechanism in many orthopterans. In this group also, geographic variation of song has been described (for example, Cantrall 1943), but in view of the many as yet undescribed sibling species of grasshoppers none of the recorded cases seems to be unequivocally established. Several cases of geographical variation in the calls of anurans have been reported (Bogert 1960), and completion of speciation by this process is made probable in at least one case (Blair and Littlejohn 1960).

Although the genital armatures of insects are no longer considered important isolating mechanisms, it is interesting to note that there is much geographical variability in these structures (see Chapter 5). Vanderplanck (1948) has shown that the genitalia in tsetse flies (*Glossina*) from different areas vary so much sometimes that a mating of individuals belonging to two different subspecies results in the death of the female, as for instance in the case of *G. palpalis* (see also Machado 1959). Strong geographic variation in the structure of the genitalia has been demonstrated from fishes (Rosen and Gordon 1951) to insects, spiders, crustaceans, and mollusks.

Ultimately, all these differences in fertility, structure, ecology, and behavior between geographical races of a species are nothing but a reflection of an over-all difference in their genetic constitution. The more barriers to gene flow between populations and the stronger their need for local adaptation, the greater will be the genetic rebuilding and the greater the probability of changes in the components of isolating mechanisms. The magnitude of genetic alterations in geographical races is documented by the developmental disturbances in racial crosses in *Rana pipiens* (Moore 1946) and *Rana temporaria* (Kawamura and Kobayashi 1959), by the often-recorded disturbances in sex determination (for example, Goldschmidt 1934 for *Lymantria*), by change in dominance (Ford 1949), and by other aspects of geographic variation discussed in Chapter 11.

Considering the enormous number of existing species, it must be admitted that the study of the geographic variation of the genetic structure of species is still in its very infancy. A few generalizations are, however, possible. Although no two populations of sexual animals are genetically identical, differences among populations that freely exchange genes are usually limited to comparatively inconsequential changes in gene frequencies and chromosomal arrangements. Crosses among such populations, even if they are phenotypically different, will not result in serious developmental disturbances, sterility, aberrant sex ratios, intersexes, or other manifestations of genetic incompatibility. The more extrinsic barriers to gene flow there are and the more distant the populations, the greater is the probability of a profound genetic change. The cohesive effect of gene flow and the importance of geographic breaks for genetic reconstitution (Mayr 1954a) is abundantly reflected in the patterns of geographic variation.

Borderline Cases and Distribution Patterns

The geographic isolate is the key unit in the process of geographic speciation. Under the theory we should be able to find isolates in every stage of speciation. New species, furthermore, should often still be either allopatric to the species from which they have diverged or just barely overlapping their range. It should be possible to find isolates that have some of the attributes of species but lack others ("borderline cases"). Finally, there should be segments of species which consist of former isolates that have failed to acquire fully efficient isolating mechanisms before rejoining the parental species along zones of secondary hybridization. Examples of these various manifestations of geographical speciation will be reviewed in the subsequent section. For fuller details we must refer to the literature cited. We can classify the various kinds of borderline cases into seven categories.

(1) *Peripheral isolates.* The most distinct isolates of a species are nearly always situated along the periphery of the species range (Mayr 1951a). Most polytypic species in well-analyzed groups of animals have such peripheral isolates. They are almost invariably a source of disagreement among taxonomists, some of whom consider them "still" subspecies, others "already" species. An unequivocal decision is possible only through an experimental analysis of sexual isolation, fertility of the hybrids, and developmental compatibility. In the Spangled Drongo (*Dicrurus hottentottus*) there are almost a dozen peripheral subspecies that are considered full species by some authors (Mayr and Vaurie 1948; Fig. 16–2). One of these isolates (*D. megarhynchus* of New Ireland) is so aberrant that until recently it was always considered a separate genus. Other cases in the avian genera *Junco, Ptilinopus, Ducula, Tanysiptera, Dicaeum,* and *Rhipidura* have been listed elsewhere (Mayr 1942, 1951a). The vole *Clethrionomys glareolus* has developed a number of isolated populations on islands to the west of Britain (Skomer, Mull, Raasay) and off France (Jersey) which are considered full species by many authors (Steven

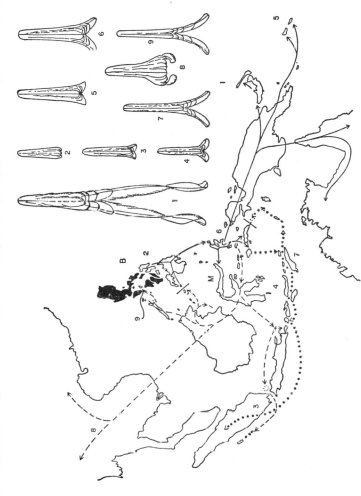

Fig. 16–2. Branches of the polytypic species *Dicrurus hottentottus*. Solid line, oldest group; dotted line, more recent group; broken line, most recent. B = closely related species *D. balicassius*; M = *D. montanus*, the product of a double invasion of Celebes by *hottentottus*. The figures indicate the ranges of the nine forms, the tails of which are shown in the insert. The tails of 4 and 6 are typical for the species; the tails of the peripheral forms 1–3, 5, 7–9 are aberrant and specialized in various directions. (From Mayr and Vaurie 1948.)

1953; Cook 1961). The most peripheral populations of the Golden Whistler (*Pachycephala pectoralis*), those on New Caledonia, Tonga, and Samoa, have rather clearly reached species level, whereas those on island groups (like the Moluccas, the Tenimber Islands, the Bismarcks, the Solomons, the New Hebrides) nearer the center of the species range are generally

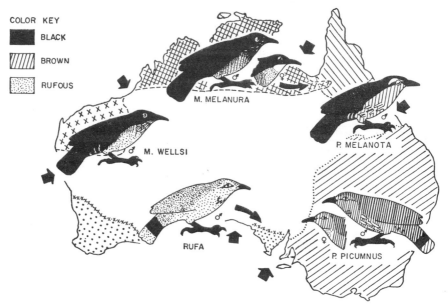

Fig. 16–3. Isolation and speciation in savannah-dwelling tree creepers (*Climacteris picumnus*) of Australia. A distinctive form is associated with each of the major savannah woodland areas. Secondary range expansion of *C. melanura* has led to an area of sympatry with *C. picumnus* in the north. The heavy black arrows indicate the distributional barriers, each a dry district from which suitable habitat is absent. The arrows on the continent indicate secondary range extensions. (From Keast 1961.)

considered only subspecifically distinct (Galbraith 1956). A large part of the taxonomic literature could be cited in this section, but I shall mention only two more studies, those of Klauber (1956) on rattlesnakes (*Crotalus*) and of Keast (1961) on peripheral isolates among Australian species of birds (Fig. 16–3). Every island or island group that is situated in front of a continent, like Britain, Ceylon, Formosa, the Ryukyus, Japan, and Tasmania has given rise to peripheral isolates. An experimental analysis of reproductive isolation sometimes proves that a form considered a species has not yet reached that stage (for example, Kawamura and Sawada 1959 for *Triturus ensigaster*) or on the contrary that forms considered only subspecies are already reproductively isolated. This was shown, for instance, by Clarke and Sheppard (1955) for the American forms of the *Papilio machaon* group and by Moore (1954) for the West Australian populations in some genera of frogs.

The majority of isolates are "borderline" cases, that is, they have some but not all attributes of new species, with isolating mechanisms more or less incompletely developed. There are far more such cases in existence than one would guess from a casual study of the systematic literature. The rigidity of zoological nomenclature forces the taxonomist to record borderline forms either as subspecies or as species. An outsider would never realize how many interesting cases of evolutionary intermediacy are concealed by the seeming definiteness of the species and subspecies designations.

(2) *Superspecies.* Allopatric populations are often so distinct that there is little doubt about their having reached species level. Rensch (1929) proposed for such groups of allopatric species the German term *Artenkreis,* for which the term superspecies (Mayr 1931) is a convenient international equivalent.

A superspecies consists of a monophyletic group of entirely or essentially allopatric species that are morphologically too different to be included in a single species. The principal feature of the superspecies is that geographically it presents essentially the picture of a polytypic species, but that the allopatric populations are so different morphologically or otherwise that reproductive isolation between them can be assumed. In some cases this has subsequently been confirmed by breeding or mate-selection experiments (for instance, Blair and Littlejohn 1960).

Superspecies are very common among birds, and many instances have been illustrated by Mayr (1940, 1942). The birds of paradise of the genera *Astrapia* (Fig. 16–4) and *Parotia,* for instance, are superspecies. Ten of the 19 species of drongos belong to two large superspecies consisting of 4 and 6 allopatric species respectively (Mayr and Vaurie 1948).

Superspecies are not an exceptional situation. They comprise a regular and sometimes rather high percentage of every fauna. There are 17 superspecies (13.6 percent) among the 135 species of Solomon Islands birds. Almost one-third of the species of Australian birds belong to superspecies (Keast 1961). Rensch (1933, 1934) has tabulated their occurrence in various animal groups and has recorded their percentages for various faunas. For instance, among the European Clausiliidae (snails) there are no less than 10 superspecies in addition to 16 polytypic species. Superspecies are also common in the mice of the genus *Peromyscus* (Blair 1950). The dis-

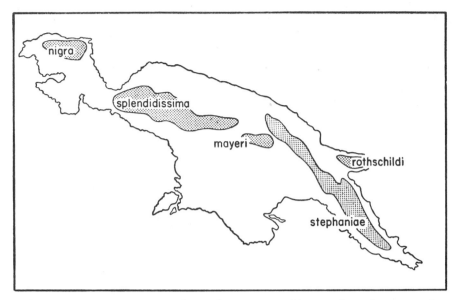

Fig. 16–4. A superspecies of paradise magpies (*Astrapia*) in the mountains of New Guinea. Each of the component species has been described as a separate genus. Hybridization has been recorded in the zone of contact between *A. mayeri* and *A. stephaniae*.

tribution pattern of many species groups of insects in the Indo-Australian island region is that of superspecies.

Superspecies seem to be even more common than in birds in groups of animals with lower dispersal facilities. Hubbs and Miller describe many groups of allopatric species in fresh-water fish, particularly in the desert areas of the western United States. Three superspecies occur, for instance, in the cyprinodont fishes of the Death Valley region (*Cyprinodon, Empetrichthys, Crenichthys*) (Miller 1948, 1961). A considerable number of sterility factors have already accumulated between species separated only since the last Pluvial. Some crosses, like *Cyprinodon salinus* ♂ × *C. macularius* ♀ (lower Colorado), fail altogether. Superspecies are common in marine animals. For example, among tropical sea urchins almost every genus contains a superspecies (Mayr 1954b).

One of the best-analyzed cases of a superspecies is that of the turbellarian species complex *Dugesia gonocephala*. The reproductive isolation of the allopatric species has been substantiated by experiment (Benazzi 1949, 1950). While *D. gonocephala* has a wide and essentially continuous range in northern Europe and Asia, many of the Mediterranean populations are isolated. Although they remain similar to each other in general appearance, they differ in the morphology of the copulatory apparatus and are sterile when crossed. Among the members of this superspecies are *D. etrusca* (Tuscany), *D. ilvana* (Elba), *D. sicula* (Sicily), and *D. benazzii* (Sardinia). Superspecies are particularly common in species with insular distribution patterns, such as occur on oceanic islands, on mountain ranges, and in caves. The problem of speciation in caves is receiving ever-increasing attention. For cave-inhabiting *Asellus*, see Bresson (1955); for distribution pattern in cave beetles, see Krekeler (1958) and Barr *et al.* (1960). The five species of flightless dung beetles of the genus *Mycotrupes* in the southeastern United States form a well-analyzed superspecies (Olson, Hubbell, and Howden 1954). They live on "islands" of sandy plains or hills separated by marshes or otherwise nonsandy habitats.

The frequency of superspecies in a group of organisms is largely a function of the physiographic features of its area of occurrence. The Indo-Australian archipelago offers an ideal situation for isolation and has resulted in the production of superspecies in almost any kind of organism that occurs there (for example, Toxopeus 1930; van der Vecht 1959). Benthonic deep-sea organisms are restricted to more or less isolated deep-sea basins and trenches and thus give rise to groups of allopatric species, and the same is true of cave animals. Superspecies among continental terrestrial animals are most common in groups in which highly specific habitat or edaphic requirements favor a fractioning of the area of distribution.

The species of a superspecies form a taxonomic or phylogenetic unit, being the descendants of a single ancestral population. For comments on the taxonomic treatment of superspecies see Mayr, Linsley, and Usinger (1953). It is advisable to define the term "superspecies" broadly and to apply it also to cases of a slight overlap in the ranges of the component species. Whenever there is a large-scale overlap of ranges, it is better to speak of "species groups." The superspecies is an interesting stage of evolution and a particularly convincing illustration of the geographical nature of speciation.

(3) *Semispecies.* The allopatric species of which a superspecies is composed have been designated by me (Mayr 1940) *semispecies.* Lorković (1953, 1958) has rightly suggested that it would be more useful to broaden the term so that it designates populations which have part way completed the process of speciation. Gene exchange is still possible among semispecies, but not as freely as among conspecific populations. I agree with

this constructive emendation. In the butterflies of the *Erebia tyndarus* group Lorković has found several populations, such as *calcarius* of the Julian Alps, that are partially reproductively isolated from their nearest relatives. Flies known among taxonomists as *Musca domestica cuthbertsoni* and *M. d. curviforceps* are in part reproductively isolated from each other, and from *M. d. domestica* (Saccà 1957; Saccà and Rivosecchi 1958). There is complete sexual isolation between *cuthbertsoni* ♂ and *curviforceps* ♀, whereas isolation is negligible in the reciprocal cross. Some of the cases of partial reproductive isolation among races of *Drosophila* could also be listed under this heading.

(4) *Secondary Contact Zones and Incomplete Speciation.* When a geographical isolate reestablishes contact with the parental species (owing to a breakdown of the isolating barrier) before the isolating mechanisms have been perfected, a hybrid zone will develop in the contact zone (Chapter 13). Hybridization may be either random or much restricted, as in the *Sphyrapicus* case described by Howell (1952). Acquisition of different habitat preferences may reduce the width of the hybrid belt or prevent hybridization altogether (Chapter 6). Even though geographic speciation did not progress in these cases to completion, such cases of secondary intergradation are nevertheless examples of the geographical nature of incipient speciation.

(5) *Partial Overlap.* The invasion of the geographic range of a parental or sister species by a newly formed species is conclusive proof of completed speciation. The general distribution pattern of the joint descendants of the parental species is often still largely allopatric, except for smaller or larger areas of marginal overlap. An almost infinite number of such cases are recorded in the taxonomic literature. For instance, Mertens (1943) produces evidence that the crocodiles of the *Crocodilus niloticus-palustris* group once formed a single superspecies in the Old World. However, *Crocodilus porosus*, presumably once restricted to Sumatra and the Malay Peninsula, has expanded its range and now overlaps widely with *mindorensis* and *novae-guineae*. In the New World, likewise, the species *Crocodilus acutus* now widely overlaps the ranges of *rhombifer* and *moreletii*, with which it was presumably formerly allopatric. Cei (1944) suggests a similar history for European species of the genus *Rana*. The closely related and evidently formerly allopatric species *arvalis, agilis,* and *temporaria* now overlap. Other recently analyzed cases occur in the fish genus *Menidia* (Gosline 1948), in the flowerpeckers of the genus *Dicaeum* (Mayr and Amadon 1947), and in kingfishers of the genus

Tanysiptera (Fig. 16–5). During the Pleistocene the species *T. hydrocharis* was isolated on an island that ran from the Aru Islands to the mouth of the Fly River and was separated from the mainland form *galatea* by a branch of the ocean. When this strait was filled by alluvial debris from the mountains of New Guinea, dry land joined the island with the mainland of New Guinea, and *galatea* was enabled to invade the range of *hydrocharis,* where the two species now live side by side without inter-

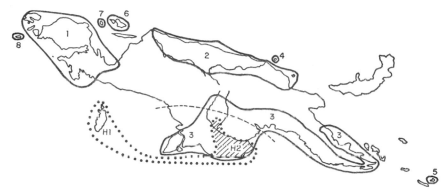

Fig. 16–5. Distribution of the kingfishers of the *Tanysiptera hydrocharis-galatea* group in the New Guinea region. The three mainland forms (1, 2, 3) are barely distinguishable. The island populations (4–8) are strikingly distinct, most originally described as separate species. The Aru Islands (H_1) and South New Guinea originally formed an island, on which the form *hydrocharis* (*H*) differentiated. When South New Guinea became attached to the main island (along the broken line) the southeast New Guinea subspecies (3) invaded the area and the fact that it does not interbreed with *hydrocharis* (*H*) proves that the latter had become a species during its isolation. (From Mayr 1954a.)

breeding or without obvious ecological competition. In lakes and oceans, likewise, the fusion of previously isolated seas and embayments has led to overlap between newly evolved species, as shown by Svärdson (1961) for the Pleistocene of Europe.

The Pied Wagtail of Europe and Asia, *Motacilla alba,* has developed two essentially allopatric species on the periphery of its range. However, the Indian *M. maderaspatensis* has invaded the range of *M. alba* in Kashmir, and *M. alba* has invaded the range of the Japanese *M. grandis* in Hokkaido and northern Honshu. There is no interbreeding in these areas of overlap. Serventy (1953) and Keast (1961) have listed several cases in Australia such as those of *Myzantha flavigula* and *melanotis,* and of *Petroica multicolor* and *goodenovii.* Toxopeus (1930) has made a particularly detailed study of species groups among the butterflies of the family

Lycaenidae. In the *Acytolepis puspa* group, in the *Surendra virarna* group, and in the *Horsfieldia narada* group there are numerous marginal overlaps. Several recently arisen species of *Triturus* in Europe show range overlap in their zone of contact (Spurway 1953), for instance Iberian *marmoratus* and Central European *cristatus* in France, Iberian *helveticus* and *vulgaris* in France, Iberian *boscai* and *alpestris* (in Spain) and possibly *c. cristatus* and *c. carnifex* in upper Austria. The tsetse flies (*Glossina*) present several such overlaps indicating recently completed speciation (for example, Machado 1959:67).

These are a few selected samples from a vast literature. Marginal overlaps in otherwise allopatric groups of species are of widespread occurrence in marine organisms. It is not claimed that each one of these overlaps proves recently completed speciation; yet this is certainly true for the majority of the cases listed. Marginal overlaps often indicate the pathway of geographical speciation.

(6) *Multiple Invasions*. Well-isolated areas, such as islands, mountain tops, or caves, are sometimes inhabited by two or more species of a widespread species group which elsewhere is represented at a given place by only a single species. Such situations have often been interpreted in the past as indicating sympatric speciation. It is now evident that a different interpretation is the correct one: the occurrence of two or more species in an isolated habitat is the result of multiple invasions. This interpretation was already given in 1872 by Weismann to explain the coexistence of two similar species of *Papilio* (*machaon* and *hospiton*) on Corsica and Sardinia. Later it was particularly Stresemann (1927–1934) who emphasized the importance of double invasions. Mayr (1942, 1951a) has given long lists and additional instances have been listed by Lack (1947a) for the Galapagos and by Ripley (1949) for India and Ceylon. There is hardly a well-isolated island from which such double invasions have not been reported. As regards birds, they are known from the Canary Islands, Samoa, Australia, Tasmania, Kauai, Ceylon, Andamans, Luzon, Celebes, Flores, Kulambangra, Comoros, Tristan da Cunha, Juan Fernandez, Norfolk Island, Antipodes, and elsewhere. There is no known case, however, in such easily accessible recent continental islands as Great Britain.

Double invasions are known not only in birds but also in lizards, snails, and butterflies, beetles, *Drosophila*, and other insects. Zimmerman (1948) shows that a good deal of the speciation in the Hawaiian Islands is due to this factor. Isolated mountains such as the Pic of Bonthain (south Celebes), Mount Kinabalu (Borneo), and the Nilgiri Hills (south India) also manifest this phenomenon. Faunal interchange between humid southeastern and southwestern Australia across an arid corridor has resulted in numerous multiple invasions (Fig. 16–6; Keast 1961; Littlejohn 1961). No thorough studies of multiple invasions in caves and isolated fresh-water lakes are so far available, but their frequency is evident.

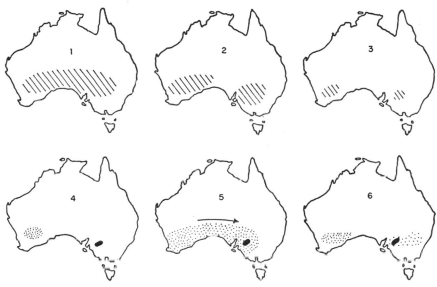

Fig. 16–6. Successive stages (1–6) in the speciation of the Australian mallee thickheads (*Pachycephala*). The present distribution is shown in map 6 with *P. rufogularis* indicated by the black area and the eastern and western areas of *P. inornata* by stippled areas. The range expansions and contractions are correlated with vegetational changes caused by shifts of rainfall belts. (From Keast 1961, which see for further details.)

The occasional occurrence in a single ocean of two sympatric species belonging to a genus that otherwise contains only allopatric species suggests that double invasions are not infrequent among marine animals (Mayr 1954b). Again, systematic investigations are not available.

The interpretation is the same in all these cases. The isolated locality was first colonized by a group of immigrants from the parental population on the mainland or an adjacent island but this isolated population diverged genetically from the parental population to such an extent that it had acquired isolating mechanisms as well as ecological compatibility when the second set of colonists reached the island. In exceptional cases there may be even a third colonization, as with the white-eyes of Norfolk

Island (*Zosterops albogularis, Z. tenuirostris,* and *Z. lateralis norfolkiensis*).

The interpretation that repeated invasions are involved, rather than sympatric speciation, has not been accepted without question. It is obviously an interpretation that can only be inferred, not "proved." Among the many lines of evidence in its favor only the following may be cited: (1) the second wave of colonists is often still so similar (if not identical) to the source population on the adjacent mainland that no question of derivation can be raised; (2) double colonizations do not occur on islands that are so close to the nearest continent that free gene exchange is possible, although ecological conditions on these islands would be just as favorable for sympatric speciation as on remote islands; (3) the number of multiple invasions is closely correlated with opportunity (compare archipelagos with isolated islands), a correlation that is not consistent with the hypothesis of sympatric speciation. Darwin's finches (Geospizidae), for instance, speciated freely on the Galapagos Islands by multiple invasions, but were unable to do so on isolated Cocos Island where the single immigrant (*Pinaroloxias inornata*) could evolve but not speciate.

Archipelago speciation. It is only a small step from double invasions to the spectacular speciation that has occurred on archipelagos like the Galapagos Islands (Lack 1947a) or the Hawaiian Islands (Zimmerman 1948; Amadon 1950). In these islands a single pair or ancestral flock has many times given rise to a vigorous new phyletic line. By colonizing one island after another, and after reaching species level recolonizing the islands from which they have come, the fauna of such an archipelago may become progressively enriched. There is little doubt that all of Darwin's finches had such a monophyletic origin (Stresemann 1936; Lack 1947a), and, likewise, the entire highly diversified family of Hawaiian honey creepers (Drepanididae), with 22 species and 45 subspecies, has descended from not more than 1 or 2 species of ancestral immigrants. The Vangashrikes (Vangidae) of the Madagascar region are a further example (Dorst 1960). Even more amazing is a similar archipelago speciation in certain groups of insects, as described by Zimmerman. The 3722 known species of Hawaiian insects have descended from between 233 and 254 original colonizations. The number of derived species is greatest among the beetles, where each successful immigrant has given rise to about 32 descendant species.

Less extreme cases are found in archipelagos that consist of only two or three islands, such as the Juan Fernandez group west of Chile, the

Tristan da Cunha group in the South Atlantic, or the islands in the Gulf of Guinea, West Africa (Amadon 1953a). Speciation in the genus *Tristanodes* is tabulated in Table 16–1.

The equivalent of archipelago speciation is sometimes found in mountain regions, as for instance in the butterflies of the genus *Calisto* in the

Table 16–1. Geographic speciation in the beetle genus *Tristanodes*, Tristan da Cunha Islands (from Brinck 1948).

Species group	Species	Island		
		Tristan	Inaccessible	Nightingale
integer	*integer*		+	
	sivertseni			+
attai	*attai*	+		
	medius		+	
	minor			+
craterophilus	*craterophilus*	+		
	echinatus		+	
	insolidus			+
reppetonis	*scirpophilus*	+		
	reppetonis		+	
	conicus		+	

mountains of Hispaniola (Munroe 1950) or in the bird genus *Psittirostra* on the island of Hawaii (Baldwin 1953). Speciation by inhabitants of fresh-water lakes often corresponds to archipelago speciation and accounts for some of the species swarms found in ancient lakes (Brooks 1950).

It would be instructive to know how rapidly such species swarms may develop. Unfortunately, it is impossible to determine this because islands are characterized by faunas of very unequal age and because speciation on islands is suspected to proceed at far more rapid rates than speciation on mainlands (Mayr 1954a).

(7) *Circular Overlaps*. The perfect demonstration of speciation is presented by the situation in which a chain of intergrading subspecies forms a loop or overlapping circle of which the terminal links have become sympatric without interbreeding, even though they are connected by a complete chain of intergrading or interbreeding populations. These cases, incidentally, are also a perfect demonstration of "speciation by distance." The development of isolating mechanisms has been possible in these cases because of a retardation of gene flow through a very long chain of popu-

lations. Although only one of many proofs for geographic speciation, circular overlaps have always been rightly considered a particular convincing one.

Circular overlaps can obviously develop only under highly exceptional constellations of geographical factors. It is therefore rather surprising how relatively common they are. Nine cases were described by Mayr (1942): *Larus argentatus, Halcyon chloris, Parus major, Lalage nigra, Pernis ptilorhynchus, Phylloscopus trochiloides, Ph. collybita, Junonia lavinia,* and *Peromyscus maniculatus.*

Additional information has become available for several of these cases. For *Halcyon cinnamomina* (Mayr 1942, Fig. 20) it was possible to show that *H. chloris matthiae* (Bismarck Archipelago) has all the prerequisites for being the ancestral population (Mayr 1950c). Dice (1949) and Blair (1950) have given additional detail on circular overlaps in *Peromyscus.* In the case of nearly all previously described circular overlaps the new information tends to show that matters are not quite as diagrammatic as originally believed. This is true, for instance, for the classical case of circular overlap, in which two sympatric species of European gulls, *Larus argentatus* and *L. fuscus,* are considered the terminal links of a chain of subspecies circling the north temperate region (Fig. 16–7). Here Stresemann and Timoféeff (1947) have shown that there is no complete continuity of populations. The gaps in the series of populations between western Europe and eastern Asia are frequent and some of them are sufficiently pronounced to induce these authors to suggest breaking up the *Larus argentatus* complex into three species. Subsequent work by Voipio (1954), Kist (1961), Goethe (1960), Macpherson (1961), and others has further clarified the situation. It is apparent that the range of *argentatus* was split during part of the Pleistocene into a number of refuges. The yellow-footed *cachinnans* group evolved in the Aralo-Caspian region and later gave rise to the Atlantic *fuscus* group. A pink-footed group (*vegae* and relatives) evolved on the Pacific coast of Asia and gave rise to the closely related typical *argentatus* in North America. The transatlantic invasion of *argentatus* into Europe is a comparatively recent event, as confirmed by the morphological investigations of Voous (1959). Where *vegae* and *cachinnans* meet (*mongolicus*) they exchange genes, likewise where *cachinnans* and *argentatus* meet in the north Baltic (*omissus*). However, where *argentatus* and *fuscus* meet along the coasts of Europe they live unmixed side by side, hybridizing only quite rarely. The behavioral and ecological differences that permit the two forms to coexist like good spe-

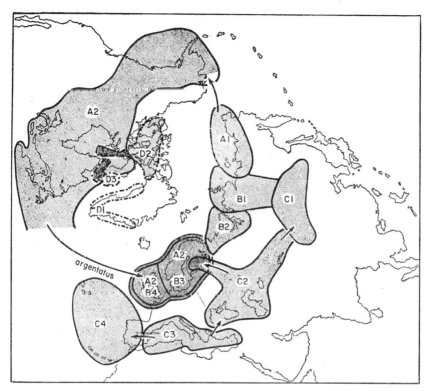

Fig. 16–7. Circular overlap in gulls of the *Larus argentatus* group. The sub-species of A, B, C evolved in Pleistocene refuges; D evolved in North America into a separate species (*L. glaucoides*). When A expanded, post-Pleistocene, probably from a north Pacific refuge (? Yukon, ? Alaska, ? Kamtchatka), it spread across all of North America and into western Europe (*argentatus*). Here it became sympatric with *fuscus* (B3, B4), the westernmost of a chain of Eurasian populations.

cies have been described by Paludan (1951) and Goethe (1955). The *Larus argentatus* group was separated during the Pleistocene into several refuges not only in Europe and Asia but also in North America. The American isolate became the species *L. glaucoides*, which is characterized by the purplish-red color of the orbital rings (not yellow as in sympatric *argentatus*). This species has three races, *glaucoides* (Greenland), *kum-lieni* (southern Baffinland), and *thayeri* (Arctic Archipelago) (Salomonsen 1950–1951; Macpherson 1961). When its range was invaded by *L. argen-tatus*, during the retreat of the ice, hybridization did not occur, but rather habitat exclusion. *L. glaucoides* is essentially a cliff nester along the sea-shore, while *L. argentatus* more often nests inland and at less precipitous locations. Where the two species nevertheless coexist in the same colonies, they do not interbreed owing to ethological isolating mechanisms (N. Smith MS).

This new interpretation of speciation in the *Larus argentatus* complex confirms the geographic origin of the various populations even though it shows that the story is somewhat more complicated than was first believed. The same is true of the often-cited circular overlap of the Greater Titmouse (*Parus major*) of the Old World. Here a ring of populations is reported to surround the central Asiatic plateau (not occupied by the species) with the end links overlapping in the Amur River basin of easternmost Asia (Fig. 16–8). The most recent investigations (Vaurie 1959) show that three groups of populations are actually involved, all three interbreeding in their zones of secondary contact. The subspecies group *major* (back green, abdomen yellow) interbreeds in Persia with (and gradually changes into) the *cinereus* group of South Asia (gray back and white abdomen). This group interbreeds and intergrades in Fokien with the East Asiatic *minor* group (green back and white abdomen), which in turn hybridizes with *major* in the Amur basin.

It is immaterial whether these instances of circular overlap present themselves in the simplest and most diagrammatic manner or whether they are somewhat more complicated, as indicated by the recent reinvestigation of *Larus argentatus*. In either case the process of geographic speciation can be followed step by step. A more dramatic demonstration of geographic speciation cannot be imagined than cases of circular overlap.

Since 1942 a casual scanning of the literature has revealed quite a few additional cases. Lack (1947a) showed how the Galapagos Finch *Camarhynchus psittacula* colonized Charles Island twice, the colonizing populations descending from the two ends of a chain of subspecies and now living on Charles Island as good species. In the solitary bee *Hoplitis producta* the Californian subspecies *gracilis* is connected through the subspecies *subgracilis* (Washington and elsewhere) with *interior* (Montana, Utah, and elsewhere), which lives around the Big Basin (Michener 1947a; Fig. 16–9). Crossing the Big Basin two *interior* populations have reinvaded the southern Californian mountains where they form the subspecies *panamintana* and *bernadina,* which live in immediate contact with *gracilis*. In fact, in the San Bernadino Mountains *gracilis* and *bernadina* have been caught on the same flowers without any signs of intergradation. A similar case was described by R. Stebbins (1949, 1957) for the sala-

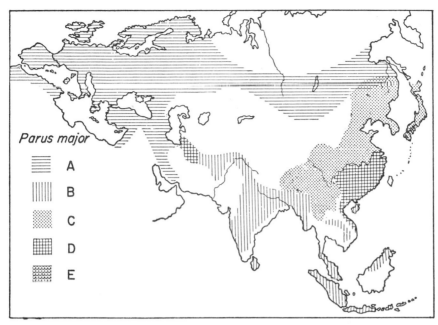

Fig. 16-8. Incomplete speciation in *Parus major*. The green-backed *major* group (*A*) of western Eurasia, the gray-backed *cinereus* group (*B*) of southern Asia, and the green-backed *minor* group (*C*) of eastern Asia are still completely allopatric. They form hybrid populations (*D*) where they meet in Iran and south China. In the Amur region (*E*) they were formerly reported to demonstrate circular overlap, but form a hybrid population (like *D*) according to more recent investigations. (After Delacour and Vaurie 1950.)

mander *Ensatina*. One chain of populations came into southern California through the coastal ranges and another one through the Sierra. The two streams of populations, which completely intergrade north of the San Joaquin Valley, meet south of it in the San Bernardino Mountains without any sign of interbreeding. The invasion of Honshu Island (Japan) by *Rana nigromaculata* from the north and by *R.* (*nigromaculata*) *brevipoda* from the south is still another case (Moriya 1960).

Additional circular overlaps have been described in the following species: *Alauda arvensis* (Vaurie 1951a), *Glossina morsitans* (Vanderplanck 1948), *Euploea tulliolus* (Corbet 1943), *Phylloscopus collybita* (second overlap) (Johansen 1947), *Charadrius hiaticula* (Bock 1959a), *Drosophila paulistorum* (Dobzhansky and Spassky 1959), *Platycercus elegans* (Cain 1955), and have been shown to be probable for three species of ducks and geese in the Perry River region of arctic North America

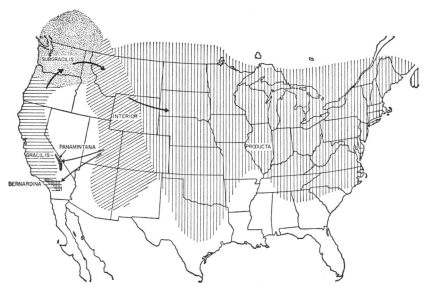

Fig. 16–9. Circular overlap in the bee *Hoplitis* (*Alcidamea*) *producta*. The subspecies *interior*, derived from *gracilis* via *subgracilis*, has twice reinvaded the range of *gracilis* across the desert barrier. The derived populations, *panamintana* and *bernardina*, now coexist with *gracilis* without interbreeding. (From Michener 1947a.)

(Gavin 1947). The seemingly greater frequency of circular overlaps in birds (15 of the 22 listed cases) is an artifact of the greater maturity of avian taxonomy.

The cases of circular overlap are considered by evolutionists as evidence for "speciation by distance." As has been shown by various authors, particularly by Wright (1943a), distance is a powerful mechanism for achieving genetic divergence. Yet gene flow and genetic homeostasis are correspondingly powerful cohesive forces (Mayr 1954a). Morphological uniformity is usually remarkably great in continuous chains of natural populations that are nowhere interrupted by extrinsic barriers. In nearly all the well-analyzed cases of circular overlap, there are major gaps in the chain of populations or at least evidence for the former existence of such gaps. The study of the geographic variation of sterility factors indicates the feasibility of speciation by distance in completely continuous series of populations, but I believe that not a single case has been proved unequivocally.

THE UNIVERSALITY OF GEOGRAPHIC SPECIATION

The widespread occurrence of geographic speciation is no longer seriously questioned by anyone. The problem now has shifted to the question whether there are also other processes of speciation among sexually reproducing animals. In Chapter 15 we examined the feasibility of the alternative modes of speciation. Here we shall try a more positive approach. If species originate geographically, it will require an abundant supply of incipient species, because only a fraction of all incipient species will ever move up to the rank of full species. Most habitats are saturated with species at any given time, and there is room only for so many new species as are needed to fill newly opened niches. Most isolates are ephemeral; they become extinct before they have had an opportunity to function as full species. Most ecological shifts in isolates are failures. When discussing the evolutionary role of peripheral isolates it is of cardinal importance to ask the right questions. The following two questions sound almost identical:

(1) Do peripheral isolates frequently (or usually) produce new species and evolutionary novelties?

(2) Are new species and evolutionary novelties usually produced by peripheral isolates?

Yet I would answer (1) negatively and (2) affirmatively. The reason is that peripheral isolates are produced 50 or 100 or 500 times as frequently as new species. Hence most peripheral isolates do *not* evolve into new species, but *when* a new species evolves, it is almost invariably from a peripheral isolate.

The crucial question concerning geographical isolates, thus, has to be framed somewhat differently: "Are there enough geographical isolates in all stages of divergence to supply the need for new species?" This question has been answered in Chapter 13 in the affirmative. What is now needed is a systematic analysis of many families in different orders and classes of animals, along the lines of the analysis prepared by me for a few avian families (Table 16–2). In the Dicaeidae, for instance, there are 3.82 isolates per species, more than there are subspecies. The data gathered in this table indicate that there are indeed more than enough isolates available to jump into the gaps caused either by the extinction of formerly existing species or by the opening up of new niches. Keast (1961) comes to the same conclusion in his analysis of geographical isolates on the

Table 16–2. Frequency of various components of species structure in four families of birds.

Family	Superspecies			Species		Subspecies		Isolates		
	Super-species	Ex-super-species	Semi-species	Poly-typic	Mono-typic	Well-defined	Clinal	Strong	Sub-specific	Weak
Dicaeidae	6	6	20	37	18	135	42	11	86	113
Dicruridae	2	—	5	11	8	36	48	10	37	44
Paradisaeidae	6	—	19	27	13	38	57	1	39	78
Ptilinorhynchidae	2	1	2	9	7	16	15	4	8	9

Australian continent. A study of the number of geographical isolates in other kinds of organisms would be highly desirable.

We can summarize the findings of this chapter by saying that innumerable aspects of the geographic variation of species, of distributional patterns, and of the ranking of lower taxonomic categories give evidence for the widespread occurrence of geographic speciation. Indeed, all the known phenomena in these areas are consistent with the theory of geographic speciation.

17 ~ The Genetics of Speciation

The evidence presented in the two preceding chapters overwhelmingly supports the conclusion that speciation in sexually reproducing higher animals is not possible without geographic isolation. Although naturalists had asserted this fact for nearly 100 years, it was until recently a strictly empirical finding. When Moritz Wagner proclaimed in 1868 that geographic isolation was a necessary prerequisite for species formation, he failed to submit any real proof. Worse than that, some of his early explanations were so obviously absurd that they endangered his thesis as a whole and were in part responsible for its cool reception. Although he gave a much sounder interpretation of his thesis in a revised edition of his work (1889), the theory of geographic speciation was still more than 50 years ahead of a rational genetic interpretation. A belief in blending inheritance of continuously varying differences was dominant in the early post-Darwinian period. Those who accepted the need for isolation did so on the basis of this belief, although their reasoning ("protection of new types against blending") does not seem nearly so absurd today, in terms of the genetics of integrated gene complexes, as it did in the early Mendelian days.

The 1890's ushered in a drastic change in genetic thinking. De Vries and Bateson championed discontinuous variation against continuous variation and thus, together with Weismann, prepared the ground for the rediscovery of Mendel's genetic laws. Mendelism, the theory of particulate inheritance, seemed to remove all need for isolation and was believed by its early adherents to have struck the theory of geographic speciation a fatal blow. The early Mendelians and, in particular, De Vries and Bateson, completely ignored this theory. Bateson (1894) was convinced that continuous and discontinuous variation "are distinct essentially . . . that they are manifestations of distinct processes." He implied that new species or even genera and higher categories might arise as intraspecific variants. His chief conclusion was "*that the discontinuity of species results from the discontinuity of variation*" (p. 568, italics his). This is strict saltationism based on a purely morphological interpretation of species. De Vries (1906) likewise made a sharp distinction between ordinary variability, which is subject to natural selection, and mutational variation: "According to the mutation theory the two [ordinary and mutational variability] are completely independent. As I hope to show, the usual variability cannot lead to a real overstepping of the species limits even with a most intense steady selection." On the other hand, each mutation "sharply and completely separates the new form, as an independent species, from the species from which it arose."

Genetics during the early decades of this century was typological, antienvironmental, and macromutational. The early Mendelians thought of speciation merely as a problem of individuals, of mutations, and of the acquisition of sterility. As late as 1922, in his famous Toronto address, Bateson said, "That particular and essential bit of the theory of evolution which is concerned with the origin and nature of *species* remains utterly mysterious . . . The production of an indubitably sterile hybrid from completely fertile parents which have arisen under critical observation from a single common origin is the event for which we wait." He completely missed the point that speciation is the origin of discontinuity not between individuals but between populations. The assertions of De Vries and Bateson, that single mutations are sufficient to produce new species and that ordinary variation has nothing to do with speciation, provoked naturalists to statements such as the often-quoted one of Osborn (1927): "Speciation is a normal and continuous process; it governs the greater part of the origin of species; it is apparently always adaptive. Mutation is an abnormal and irregular mode of origin, which while not infrequently occurring in nature is not essentially an adaptive process; it is, rather, a disturbance of the regular course of speciation." For a more detailed documentation and analysis of the history of concepts during early Mendelism, see Mayr (1959b).

A study of the statements of Fisher (1930), Dobzhansky (1937), and Muller (1940) on the genetic aspects of the multiplication of species shows how recent the developments in this area are. In the 1930's these authors were still primarily concerned with mutation and selection, that is, with evolutionary change as such. Only after these problems had been solved were the population geneticists ready to work out a genetic theory of speciation. Only now was it possible to search for the genetic reasons that necessitate geographic isolation during speciation.

THE GENETIC PROBLEM OF SPECIATION

The essence of speciation, as we now realize, is the production of two well-integrated gene complexes from a single parental one. All early attempts to explain the genetics of speciation missed this essential point, being concerned entirely with the problem of the origin of difference. To be sure, the differences between species are due to mutation and selection, but this does not explain the splitting. Much of selection is stabilizing and negates the effects of mutation. Species are kept intact by isolating mechanisms, consisting to a large extent of differences in reproductive behavior and physiology. Any mutation that would initiate an incipient difference in reproductive behavior among individuals of a population would be a particularly vulnerable target of normalizing selection.

The real problem of speciation is not how to produce difference but rather how to escape from the cohesion of the gene complex. No one will comprehend how formidable this problem is who does not understand the power of the cohesive forces that are responsible for the coadapted harmony of the gene pool. It was the object of Chapter 10 to establish this point.

It is now evident that there is only one situation in which a gene pool can be completely reconstituted genetically (with reference to a parental population) while all of its elements remain well integrated and co-adapted: spatial isolation. Most students of the speciation of sexual animals from M. Wagner on realized clearly the indispensability of this condition, but they based their conclusion strictly on empirical findings. Why isolation was needed remained a puzzle until the genetics of integrated gene complexes had replaced the old "beanbag" genetics.

Gene Flow and Genetic Cohesion

To what extent and under what circumstances are the different populations of a species held together by cohesive forces? Conversely, under what conditions can this cohesion be broken? The most important question of all which we have to ask is, what role does gene flow play in the maintenance of genetic cohesion among populations? Is it advisable to classify populations according to the amount of gene flow to which they are exposed?

Closed and Open Populations. In spite of all that has been learned during recent decades on the interaction of genes within experimental populations, there still is considerable uncertainty as to the nature of the genetic interaction in natural populations. It must be recalled that all selection experiments, whether on mice, *Drosophila*, or domestic animals, were made on closed populations with a negligible genetic input. Population size was far too small in all these cases for mutation to have played a major role unless it was artificially induced, as it was in the work of Buzzati-Traverso, of Scossiroli, and of Wallace. The response of the selected stocks, no matter how spectacular it was, must have been largely due to recombination of the initial gene complement. This source of variation is bound to dry up eventually. Natural populations (except the most rigidly isolated ones), in contradistinction, are open populations with a steady input through gene immigration. This difference in genetic input causes a number of fundamental differences between closed and open populations (Mayr 1955a, 1959b). Available genetic variation is not only infinitely greater in the open system, but it is also of a different kind. The large and continuous influx of alien genes into every local population, as well as the diversity of the environment in space and time, will never permit the gene complex to reach complete stability. The response to selection in an open system will be very different from that in a closed one. It is therefore not admissible to apply automatically the findings made on closed laboratory populations to natural populations. This qualification must be kept in mind when one wants to construct models of species structure.

It is not even certain that "closed" and "open" are the only alternatives. There may be also a major difference, with respect to genetic structure, between small and large closed populations. The unresolved problem is posed by the evidence that the same gene may have different selective values in a small closed, in a large closed, and in an open population. The relative importance of epistatic and allelic interactions differs in the three types of populations, but we still do not quite understand how. The total genetic background remains relatively constant in a closed population and the gene pool has an opportunity to utilize residual variability for the selection of a number of specialized epistatic and overdominant combinations. Overdominance for genes or gene arrangements evolved in a number of small closed experimental populations, where the founders had been taken from open populations without such overdominance (for example, Spiess 1961; Dobzhansky and Spassky 1962). It seems that in highly variable open populations there is a greater premium for a harmonious epistatic interaction than for the evolution of allelic overdominance. In such a population, where each gene has to function in each generation with an ever-new constellation of "alien" genes at various loci, there is a high premium on universality of goodness in all imaginable genetic combinations. In a closed population the epistatic interactions become far more standardized and the residual genetic variability can be used to create or improve overdominance at certain loci or chromosome sections. Closed populations usually live also in far more uniform environments than open populations, which favors their genetic homogeneity even more. These suspected differences between closed and open populations are of great importance for the problem of speciation.

Wright (1931a, 1940, 1951a, and so forth) has asserted that one particular species structure is specially suited (*a*) to carry "an enormous store of rather easily available potential variability," and (*b*) to "present the most favorable condition for transformation as a single species." This favored species structure is "a large total species population split up into a large number of partially isolated local populations . . . without marked environmental differences." This statement remains vague and meaningless as long as the term "partially isolated" is not defined. In all widespread, successful species of more mobile sexual organisms, there seems to be sufficient gene flow to maintain great similarity in the gene pools of all local populations. The capacity of the species as a whole for the storage of genetic diversity is thereby sharply reduced. Simpson (1953a:123) has come to a similar conclusion:

> The main weakness of this [Wright's] model is probably the fact that *m* (transfer of genetic materials from one deme to another) must be very low, on the order of 0.01 to 0.001, or the separation into demes will be ineffective for evolution and the species will evolve as if panmictic, i.e. slowly, adaptively, but with little potentiality for rapid change or shift in direction of adaptation.

Population geneticists, who have worked all their lives with closed populations in which all genetic input is due to mutation, tend to underestimate the magnitude of genetic input in open populations. To be sure, it is immaterial for certain aspects of evolution whether mutation or immigration is responsible for new genes in a population. Yet it would be a great mistake to lump these two sources of variation in calculations of their effect, because they are of totally different orders of magnitude. I estimate that genetic change per generation due to mutation in a local population rarely exceeds 10^{-5} per locus, while the exchange due to normal gene flow is at least as high as 10^{-3} to 10^{-2} for open populations that are normal components of species. With an effective local breeding population often as low as 2×10 and usually not higher than 3×10^2, this difference in order of magnitude becomes of vital importance. For example, assuming a population of 200 individuals, each with 100,000 loci mutating at the rate of 1 in 10,000, would give 2000 mutants for the 20 million loci, almost certainly a maximum figure. Dispersal, however, might bring 40 percent of new individuals into the population, that is, 8 million genes, of which perhaps 200,000 might be new for the population. Assuming that there will be the same amount of duplication among mutated and immigrant genes, gene replacement by immigration would thus be 100 times that by mutation in this deme.

Sewall Wright, in his discussions of the evolutionary transformation of species, allows for two alternatives, (1) "panmictic species," and (2) "species subdivided into many partially isolated local populations." Actually, there is probably a greater difference between extremes of (2) than there is between (1) and (2). In view of the steady selection in favor of genes that coadapt easily with immigrant genes, there may well be nearly as much cohesion in a partially isolated system as there is in a panmictic one. The genuinely sharp break is not between the panmictic and the partially isolated system, but between the partially and the virtually fully isolated system. The importance of complete isolation becomes evident as soon as the extensive epistatic effects of genes are properly taken into consideration. As a consequence, a population cannot change drastically as long as it is exposed to the normalizing effects of gene flow.

The absence or drastic reduction of gene flow among populations is usually indicated by morphological discontinuity, each isolate having a different phenotype (Chapter 13). Conversely, the occurrence of an appreciable amount of gene flow, even where the existing distributional barriers appear to be rather efficient, is indicated by only minor or gradual variation. The stabilizing effect of gene flow is best documented by phenotypic uniformity or at least not more than clinal variation over wide areas. Cook (1961) has calculated that it requires very high immigration rates to neutralize selection for single genes, but the inertia of the total epigenotype cannot, of course, be included in such calculations. The existence of groups of sibling species proves that not all phenotypic uniformity is due to gene flow. Efficient canalization will produce the same effect.

The steady and high genetic input caused by gene flow is the main factor responsible for genetic cohesion among the populations of a species. It not only produces the observed similarity among contiguous populations, but also is one of the principal reasons for the slow rate of evolution of common widespread species. Many authors have emphasized the slowness of evolutionary change (for example, Haldane 1949b, 1954a, 1957). Neither the steady "rain" of mutations on a population nor the incessant eroding force of selection seems to have nearly as much effect on the genotypic composition and phenotypic change of populations as one might expect. Even in relatively "rapidly evolving lines like the dinosaurs and the ancestors of the horses, measurable lengths, such as those between homologous points on homologous teeth or body lengths in general, changed by quantities of the order of 1–10 percent per million years. For a ratio of lengths, which can be regarded as a measure of shape, 2 percent is a representative figure" (Haldane 1954b). In many lines evolution is a good deal slower even than that. Such slowness of much of evolution stands in marked contrast to the outbursts of speciation that have been recorded in the fossil record of some lines, on tropical archipelagoes (Hawaii, Galapagos, West Indies), and in fresh-water lakes. Evidence is accumulating for drastically unlike rates of evolution and speciation. Evidence is, likewise, accumulating that this difference is correlated with and presumably caused by differences in the population structure of the respective species.

The adjustment of local populations to the local environment through race or ecotype formation has been stressed so much in previous chapters (Chapters 9, 11, 12), that it would now seem important to stress the basic uniformity of most continuously distributed species. I illustrated this phenomenon (Mayr 1954a) with the New Guinea Kingfisher *Tanysiptera galatea,* which displays no significant geographic variation in the vast area of that island, with its strong climatic contrasts. Yet each of the adjacent islands inhabited by this species has a markedly differentiated race even though they are in the same climatic zone as the neighboring mainland. Every taxonomist can cite literally hundreds of similar cases. There is, for instance, the butterfly *Maniola jurtina,* which has been studied so intensely by E. B. Ford and his associates:

One of the most striking features of the *M. jurtina* females is their remarkable uniformity across most of southern England [except Cornwall]. This area includes some of the greatest variations in temperature, rainfall, and geology to be found in Britain. Evidence from Cornwall and the Scillies shows that the spotting is capable of marked variation. The fact that it is so stable elsewhere

[= across most of southern England] indicates not only that natural selection is holding it at an optimum value but also that the species is in some way insensitive to environmental variation in this part of its range (Dowdeswell 1956).

The fact, taken for granted by every taxonomist, that he can identify individuals of a species (unless its range is dissected by geographical isolation) regardless of where in the range of the species they may come from is further illustration of this phenomenon. Physiologists and embryologists, likewise, have published evidence for a remarkable uniformity of physiological constants throughout the ranges of most species. The essential genetic unity of species cannot be doubted. Yet the mechanisms by which this unity is maintained are still largely unexplored. Gene flow is not nearly strong enough to make these species anywhere nearly panmictic. It is far more likely that all the populations share a limited number of highly successful epigenetic systems and homeostatic devices which place a severe restraint on genetic and phenotypic change.

Gene Flow and Species Structure

Before discussing the conditions under which a genetic turnover can nevertheless take place, we must discuss some of the effects of gene flow on the population structure and phenotypic variation of species. These concern phenomena, long familiar to naturalists, that had remained unexplained until the recent findings of the genetic cohesion of populations permitted a solution.

The Problem of the Species Border. The range of a species is delimited by a line beyond which the selective factors of the environment prevent successful reproduction. This line is called the species border. Single individuals may appear annually in considerable numbers beyond this line, yet fail to establish themselves permanently. Even if they succeed in founding new colonies, these are sooner or later eliminated in an adverse season. As a result, the species border, though fluctuating back and forth, remains a dynamically stable line. The species border is one of the most interesting phenomena of evolution and ecology, yet as a scientific problem it has been almost totally ignored. Kalela (1944) and other Finnish authors have studied range expansions (shifts of the species border) as correlates of changing climatic conditions. In the border region there is a never-ending race between reproductive capacity and mortality due to adverse conditions. Population density is far below the saturation point and the border region is a place in the area of a species where density-dependent factors are of minor, if not negligible, importance (Haldane 1956; Birch 1957).

The essential stability of the species border, on which the annual and secular fluctuations are superimposed, would seem to contradict our belief in the power of natural selection. One would expect that a few individuals would survive in a zone immediately outside the species border and form a new local population which becomes gradually better adapted under the continuous shaping influence of local selection. One would expect the species range to grow by a process of annual accretion like the rings of a tree. That this does not happen is particularly astonishing in the frequent cases where conditions beyond the borderline differ only slightly and in degree from conditions inside the species border (and where there are no drastic barriers).

The solution of this puzzle is probably that this process of local adaptation by selection is annually disrupted by the immigration of alien genes and gene combinations from the interior of the species range (Mayr 1954a). This prevents the selection of a stabilized gene complex adapted to the conditions of the border region. The border populations presumably barely maintain themselves and the new colonists beyond the species border (in mobile species such as birds and insects) come from farther inside the species range, where conditions permit a greater surplus of individuals and the increased population density in turn may stimulate emigration, but the gene complex is not adapted for the conditions of the border region.

Wallace (1959) has discussed how intimately the problem of the species border is associated with the problem of speciation. A peripheral population will not succeed in occupying a new environment beyond the present species range, owing to the relentless destruction of suitable new gene complexes, unless it can escape recombination. This involves inevitably some degree of geographic isolation. It also necessitates, where this isolation is less than perfect, some chromosomal mechanism that protects the adapted gene complexes of the border populations (see below).

Stabilized Hybrid Belts. The narrowness of many belts of allopatric hybridization has long been a puzzle (Chapter 13). Even though there seems to be random mating and no external barriers to dispersal, some of these belts have remained very narrow for thousands of years (Meise 1928a). It appears that the parental gene complexes that meet in these hybrid belts are so well balanced within themselves that recombination between them produces genotypes of inferior fitness which are eliminated by selection. The steady counterselection against introgressing genes reduces gene flow drastically and prevents a gradual broadening of the hybrid belts. Narrowness is a characteristic feature of many belts of secondary intergradation between subspecies. In spite of the absence of isolating mechanisms and the essential randomness of mating in the zone of contact, the gene complexes have reached in these cases a degree of incompatibility equal to that of two full species.

The fact that no amount of selection pressure seems to be able to complete the process of speciation in these zones of secondary intergradation is highly damaging to the theory of *semigeographic speciation*. According to this theory, "selection . . . must act gradually and progressively to minimize the diffusion of germ plasm between regions requiring different specialized aptitudes . . . until a line of distinction is produced across which there is a relatively sharp contrast in the genetic-composition of the species . . . this . . . will allow the two main bodies of the species to evolve almost in complete independence" (Fisher 1930). Other authors have gone one step further and have postulated that the species would acquire reproductive isolation along the line of genetic divergence. Such an assumption overlooks not only gene flow but also the cohesive factors in gene pools. Genetic differences between populations are built up not by the independent changes of gene frequencies at different loci, but by the harmonious reconstruction of the total genotype. The minor ecotypic adaptations on either side of a habitat border can be built up without a disturbance of the basic epigenotype of the species; indeed this is the only way in which it can be built up. If recombination between the two adaptive genotypes leads to inviable recombinants, protective cytological mechanisms will be favored (see below). There is no evidence for, or indeed likelihood of, semigeographic speciation, except in the special cases of chromosomal incompatibility mechanisms (p. 537).

Clines and Peripheral Populations. The students of species structure have long emphasized two elements, geographical isolates and clinally varying continuates. They have, furthermore, stressed that clines are regular only where populations are contiguous, and that the variation of characters is irregular wherever populations are isolated. Finally, they have emphasized the greater variability of central populations compared to peripheral ones. We have discussed in detail in Chapter 13 the various genetic interpretations that have been advanced to explain these differences. Carson (1959) suggested that structural homozygosity is favored in marginal populations for the reason that selection may favor the fitting of the whole genotype into a particular, rather restricted ecological niche. Such homozygosity favors free recombination "and thus perhaps aids in the synthesis of the genetic basis of a true adaptive novelty." There are, however, a number of considerations which indicate that Carson's interpretation of the cause of structural homozygosity in marginal populations of *Drosophila* is not necessarily the most important among several contributory causes. First of all, there are, even in *Drosophila*, numerous species (such as *virilis*) in which the central populations are structurally uniform, while there are species of grasshoppers (White 1957a), of *Drosophila* (Goldschmidt and Stumm-Zollinger 1959), and *Chilocorus* beetles (Smith 1960) in which the chromosomes of peripheral populations are not appreciably less heterogeneous structurally than those of central populations. Admittedly such observations pose difficulties of interpretation to any theory of adaptation. Stone *et al.* (1960) raise other questions. They point out that the measurement of "free" chromosomes, employed by Carson, is not a precise index of the amount of recombination because there is an increase of cross-over frequency in the "free" chromosome to compensate for the amount of chromosome tied up in inversions (Schultz and Redfield 1951; Oksala 1958). It should be mentioned parenthetically that the chromosomal variability of natural populations does not always reflect degree of adaptedness. Chromosomal variability may be sharply increased in a secondary hybrid zone (Chapter 13) between chromosomally differing populations. The contact zone between the 13-chromosome and 18-chromosome races of *Thais lapillus* (Staiger 1954) is a typical illustration. It is, as yet, known for only few natural populations to what extent such hybridization may have contributed to observed chromosomal variation.

It appears highly probable that it is increased resistance to genetic diversity rather than a selective advantage of increased recombination that is largely responsible for chromosomal uniformity in peripheral areas. This conclusion is supported by the observation (Chapter 11) that genic polymorphism is likewise greatest in central populations and is often replaced by monomorphism in peripheral populations. This cannot be correlated with the amount of recombination. It is far more likely that the monomorphism of peripheral populations, where it occurs, is a consequence of the genetic cohesion of species. These marginal populations share the homeostatic system, the epigenotype, of the species as a whole. They are under the severe handicap of having to remain coadapted with the gene pool of the species as a whole (and to assimilate a steady stream of immigrants from more central areas) and yet be adapted to local conditions. The basic gene complex of the species (with all the species-specific canalizations and feedbacks) functions optimally in the area for which it had evolved by selection, usually somewhere near the center. Here it is in balance with the environment and here it can afford much superimposed genetic variation and experimentation in niche invasion. Toward the periphery this basic genotype of the species is less and less appropriate and the leeway of genetic variation that it permits is increasingly narrowed down until much uniformity is reached. Such an impoverished and yet dependent population is not the ideal starting point for speciation. As we shall see, some degree of isolation is indispensable.

The Genetic Reconstitution of the Isolated Population

The argument in the preceding section was essentially negative. It demonstrated that contiguous populations of a species are held together by such a close tie of genetic cohesion that it is impossible to divide this essentially single gene pool into two. There is no mechanism (except instantaneous speciation through polyploidy) by which the cohesion might be loosened to the extent that it would permit the sympatric development of two independent gene complexes which would give reproductive isolation and ecological compatibility. We shall now take up the other half of the argument and show what genetic events take place in spatially isolated populations and how this leads to the formation of isolating mechanisms while leaving the genetic integration of the gene pool at all times undisturbed.

The Isolated Population. Let us study the genetic history of a newly founded population that is spatially isolated from the parental population from which it branched off. We shall at first make the two simplifying assumptions that the new population is at the beginning completely identical genetically with the parental population and that the environments of the two populations are identical. Yet, even so, as soon as they are separated, the two populations will drift apart in their genetic contents, for a number of reasons. The probability is nil that the same mutations will occur in the two populations in the same sequence. Each incorporated mutation changes the genetic background of the population and thus affects the selective value of all subsequent mutations. Furthermore, recombination will produce different genotypes in the two gene pools, and, since the same gene may have different selective values in different genotypes, this will likewise lead to a gradual shifting of gene frequencies.

A third factor leading to the divergence of the gene pools is "genetic indeterminacy" (Chapter 8). If several gene combinations have equal selective values (with respect to a given selection pressure), pure chance or some irrelevant pleiotropic effect may decide which becomes established in a given gene pool. The changes in gene frequency due to "genetic indeterminacy" will again not be the same in two independent populations. Each divergence of the two gene pools increases the difference in the genetic background of all the genes of the two populations and will thus tend to set up new selection pressures. The drifting apart is thus evidently an accelerating process.

This change is, however, opposed by various genetic mechanisms, particularly balanced heterotic systems (genetic homeostasis). It appears that the basic epigenotype of a species, the system of developmental canalizations and feedbacks, is so well integrated that it often resists change with remarkable tenacity. There are numerous cases known where populations have remained amazingly similar to the parental population in spite of long periods of complete isolation. Unless environmental conditions as well as the genetic composition and structure of the isolated population favor the speciation process, the genetic change toward species level will progress only slowly.

This consideration brings us back to the two simplifying assumptions made above. Neither, of course, is true. The selective agents to which the two separated populations are exposed are not the same, since there are no two places on the face of the earth where even the physical environment is quite identical. Where a population is completely isolated, the composition of the biota is invariably different and this shift of the biotic environment results in an additional powerful difference in selection pressure. This is true for competition, predation, and whatever other ecological interactions may exist.

These local conditions exert selection pressures leading to a steady change of gene contents and to the development of numerous new adjustments. Having to become adapted to a new environment may lead to a sufficiently drastic genetic reconstruction to affect the nature of isolating mechanisms, including potential sterility barriers with sister populations. Goldschmidt assumed that such "existential adaptations," as he called them, had no permanence, but all the available evidence contradicts this assumption. When such a "reconstructed" population returns to the area of origin, it will adapt itself by the superimposition of a new set of genetic modifications. There is abundant evidence from the field of microbial genetics that, if selection pressure favors return of a mutated strain to the original reaction norm, this is rarely due to a back mutation, but usually is achieved by a compensatory mutation at a second locus. For instance, if AB designates a streptomycin-independent strain and aB a dependent strain, return to independence would be accomplished through mutation at B, resulting in the genotype ab. The restored phenotype thus differs actually by two genetic factors from the original one. Finally, the two diverging populations are not identical even initially.

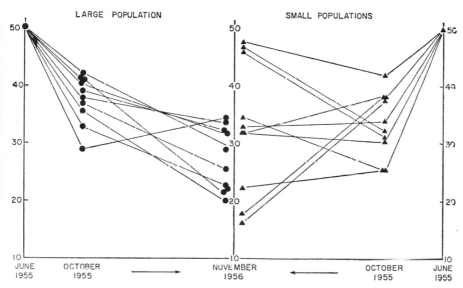

Fig. 17–1. The frequency (percent; vertical scales) of PP chromosomes in 20 replicate experimental populations of mixed geographic origin (Texas by California). The populations that have gone through a bottleneck of small population size show far greater variance after 17 months than the continuously large populations. (From Dobzhansky and Pavlovsky 1957.)

Founder Principle. The founders of a new colony of a species contain inevitably only a small fraction of the total variation of the parental species (see Chapter 8). All subsequent evolution will proceed from this original endowment. How important this restriction is, is evident from recent selection experiments in which several parallel lines were exposed to the same selection pressure. Almost invariably the end results were different in the different lines. The smaller the starting populations, the greater the degree of indeterminacy. Ten experimental populations, each descended from only twenty founders, diverged far more from each other than ten other populations founded by 4000 individuals each, all twenty populations having been derived from the same parental populations (Dobzhansky and Pavlovsky 1957; Fig. 17–1).

The founders carry such a small reservoir of genetic diversity with them that the population founded by them is highly vulnerable to the dangers of inbreeding (homozygosity). The situation is aggravated in most cases by the ecological uniformity of the insular environment and the resulting one-sidedness of selection. The frequent extinction of island populations, established by founders, can largely be attributed to this cause. Extinction under adverse conditions is also the fate of most peripheral isolates. The history of extinction indicates that a drop in population size below a certain level is fatal in many species. Not enough information is available to indicate whether the critical level consisted of 100, 50, 12, or 6 individuals in any given case. It has, however, been established beyond doubt (Lerner 1954) that a serious loss of viability ("inbreeding depression") usually occurs if a population from a normally outbreeding species is suddenly subjected to intense inbreeding (Chapter 9).

The Chances of Success of a Founder Population

Awareness of the frequency of extinction among island species and of the severe inbreeding depression observed by animal breeders might induce one to take a dim view of the prospects of a founder population. However, one must be cautious when generalizing from highly artificial selection experiments. Selection in nature, even in the smallest population, is primarily for over-all fitness. The same is true of so-called "unselected" inbred lines in the laboratory. Brother-sister matings can be continued in certain organisms for hundreds of generations without serious depression of viability and fertility, while in other organisms such a breeding plan leads to rapid extinction. On the whole, loss of genetic variance through inbreeding occurs far more slowly than one might expect (Tantawy and Reeve 1956).

There is abundant evidence in the literature for an occasional phenomenally successful founder population. Hutton (1897) pointed out long ago that the rather considerable amount of inbreeding in new colonies is not necessarily injurious. As examples, he lists a great number of animals introduced to New Zealand which were the offspring of merely a few individuals and yet were highly successful. This is true of the honeybee, the pheasant, the black swan, many other birds, the rabbit, the Tasmanian opossum, the red deer, and all accidentally introduced insects. The same is true of numerous other animal and plant immigrants on all continents (Elton 1958). Inbreeders, of course, are vastly more successful colonizers than are outbreeders.

Sometimes an exceedingly small population maintains itself successfully over a long series of generations. Some species of fish in desert springs have maintained themselves for thousands of years even though their population occasionally drops to well below 100 (Miller 1948, 1961; Hubbs and Springer 1957). A self containing "wild" herd of cattle formerly averaging about 60 head has existed in the park of Chillingham Castle in Northumberland, England, for some 700 years. For 30 years prior to 1947, the herd had been fairly constant at about 35–40 head. Only 13 individuals survived the severe winter of 1947. In 1956, the herd consisted of 5 bulls, 10 cows, 1 bull calf, and 3 heifer calves (Bilton 1957). It is only in recent years that the herd has shown definite signs of inbreeding depression. The American bison, the European ibex, and other mammals have successfully overcome an extreme reduction in numbers (Buzzati-Traverso 1950). The millions of golden hamsters (*Mesocricetus auratus*) in the laboratories around the world are supposedly all descendants of a single pregnant female. The establishment of highly successful colonies by single founders is not only feasible; it seems to be the normal method of spreading in many species of animals and plants.

These observations also have a bearing on the problem of the genetic composition of rare species. Here again one would expect a priori great homozygosity owing to inbreeding. The available information is slight, but opposed to this assumption. In the very rare grasshopper *Pedioscirtetes nevadensis*, White and Nickerson (1951) found that one or two pairs of autosomes were heterozygous in most individuals of a local population. Drift had thus been unable to override an obvious but not yet analyzed selective advantage of the heterozygous condition. In *Moraba scurra*, an Australian grasshopper reduced to isolated populations in cemeteries, many of them quite small, White (1957c) found that only a few of the usually polymorph karyotypes of this species had become homozygous. There are numerous records in the genetic literature of the tenacity with which genetic variability is maintained in small laboratory populations through scores of generations of inbreeding. The role of overdominance in preventing the loss of genetic variability from populations has been explored by Tantawy and Reeve (1956) in carefully designed experiments.

When speaking of "homozygous" founder populations, we must thus keep in mind that this is only relatively true. It is extraordinary how much genetic variability even a single diploid individual carries in its genotype. No less than 11 of 21 wild pairs of *Drosophila melanogaster*, when inbred, gave rise (among 1000 F_2 progeny) to at least one individual with the rare genetic defect "cross-veinlessness" (Milkman 1960c). Perhaps nothing demonstrates the amount of concealed genetic variability better than the work on recombination by Dobzhansky and his associates (Spassky *et al.* 1958; Spiess 1959).

The Genetic and Biotic Environment of the Founder Population

The founder population differs from the parental population not only in the drastic reduction of the diversity of its gene pool, but also by being

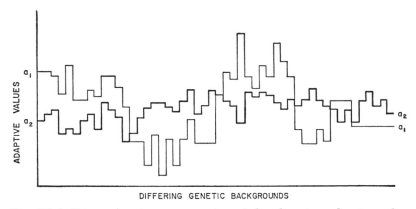

Fig. 17–2. Diagrammatic representation of the changing adaptive value of genes a_1 and a_2 on different genetic background. Gene a_1 is very good on some backgrounds and very poor on others; gene a_2 fluctuates only slightly around the mean (From Mayr 1954a.)

exposed to a totally new constellation of environmental factors, biotic as well as genetic ones. The most important of these is the sudden conversion from an open to a closed—and at that to a *small closed*—population. It is the suddenness and completeness of this shift that is decisive. The new population is at once completely emancipated from the parental population. In an open population there is a steady and rather high input of alien genes. "Little or no thought [has been given in the past] to the effect of these alien genes on the relative viability of the genes of the gene complex into which they were introduced. It appears probable that the frequent introduction of alien genes into a gene pool will lead to the selection of such 'native' genes as are tolerant to combination with these alien genes, that is, which produce viable heterozygotes with a great assortment of alien genes or gene combinations" (Mayr 1954a; Fig. 17–2). I have referred to such genes as "jack-of-all-trades" genes or "good mixers." As soon as they are cut off from the steady inflow of alien genes, such "good mixers" lose their very special selective advantage.

A similar shift in the selective value of genes is observed when a new closed experimental population is established with individuals taken from an open population. For instance, it is evident (Chapter 9) that the potential for heterosis is not maximally utilized in open populations. This is proved by interpopulation crosses (provided the populations are not too distant) in which the F_1 are nearly always heterotic. As Wallace (1955) has said rightly, "The higher average viability of interpopulation F_1 hybrids is a measure of the price paid by local populations for an integrated gene pool capable of being transmitted successfully from generation to generation." The need for maintaining harmonious epistatic interactions, in the face of gene flow, is one of several reasons for the reduced number of overdominant loci in an open population. After converting an open population into a closed one, it will always be possible to improve by natural selection the coadaptation of the limited and stable number of genes.

The observed facts are, however, controversial and subject to different, indeed to contradictory, interpretations. Even though a high number of loci in an open population is heterozygous, the number of strongly overdominant loci is apparently small. After conversion of the population into a closed one, homozygosity increases inevitably, and yet overdominance may likewise increase on a few favored loci. Selection for increased overdominance in these loci is made possible by eliminating the disturbing effect of large genetic input, generation for generation (Spiess 1961; Dobzhansky and Spassky 1962).

The effect of the increased homozygosity at the other loci of the founder population is, perhaps, more important.

As a consequence of their increased frequency in the founder population, homozygotes will be much more exposed to selection and those genes will be favored which are specially viable in homozygous condition. Thus, the "soloist" is now the favorite, rather than the "good mixer." We come thus to the important conclusion that *the mere change of the genetic environment may change the selective value of a gene very considerably* [and particularly the change from an open to a closed population]. This change, in fact, is the most drastic genetic change (except for polyploidy and hybridization) which may occur in a natural population, since it may affect all loci at once. Indeed, it may have the character of a veritable "genetic revolution." Furthermore, this "genetic revolution," released by the isolation of the founder population, may well have the character of a chain reaction. Changes in any locus will in turn affect the selective values at many other loci, until finally the system has reached a new state of equilibrium (Mayr 1954a).*

There are several other reasons for the potentiality of rapid genetic turnover in populations passing through the bottleneck of reduced population size. One is Haldane's (1957) conclusion that (as long as selection is not too intense) the number of deaths needed to secure the substitution, by natural selection, of one gene for another does not depend on the selective value of the gene but only on the natural logarithm of its initial frequency. Since this initial frequency of a new gene in a population is normally a very small fraction of 1 percent, and since large numbers of genes are substituted during evolution, Haldane postulates extremely low evolutionary rates. However, genes that elsewhere are rare may start in a founder population with an initially high frequency. This would accelerate their fixation.

There are still other factors that favor a rapid genetic turnover, for instance a possible synergistic action of deleterious genes (Chapter 9). Also, in the new environment of the founder population, there will indeed be very high selection intensities against certain genes that had been favorable in the parental population.

Haldane (1956) has called attention to still another difference between central and peripherally isolated populations: the role of density independent mortality. In the central populations of a species, the physical as well as the biotic environment is optimal, and much, if not most, of the mortality is somehow connected with the high population density. In peripheral and peripherally isolated populations, conditions are usually near the minimum for the species. In bad years the populations will be wiped out, or nearly wiped out, in good years they may build up to large numbers (in view of the scarcity of species-specific predators and pathogens). There will thus be strong population fluctuations which also favor genetic turnover. The strength of density-dependent factors in the central part of the species range damps such fluctuations (see Chapter 13).

Finally, as White (1959) has pointed out, the reduction in polymorphism and heterozygous balances that we find in peripheral (and particularly isolated) populations reduces the genetic homeostasis and

---

* The object of this paper was recorded by one critic as the introduction of "another variety of the genetic drift theory, the founder principle." I should like to correct this casual reader. The founder principle was introduced twelve years earlier (Mayr 1942:237) and the very purpose of the 1954 paper was to emphasize the importance of selection: every gene may or will have a new selective value in the drastically altered genetic environment of the founder population.

evolutionary inertia of these populations. They are far better capable of responding to new selection pressures and, consequently, to new evolutionary opportunities than populations from the "dead heart" of the species.
They are more immediately in the position to utilize new gene combinations that are generated during the genetic revolution than are populations
in which the genes are tightly knitted together by numerous balancing
mechanisms. That speciating populations are relatively monomorphic is
indicated not only by the observed reduction of polymorphism in peripheral populations (Mayr and Stresemann 1950; Carson 1955b), but also by
the observation that closely related, chromosomally polymorphic species
have only little of this polymorphism in common. For example, the two
sibling species *Drosophila pseudoobscura* and *D. persimilis,* both of
which are highly polymorphic for inversions of the third chromosome,
have only one inversion in common. The same is, by and large, true of related species in the *virilis* (Stone *et al.* 1960), *repleta* (Wasserman 1960,
1962) and *willistoni* (Carson 1954) groups of *Drosophila.*

Chromosomal Changes During Speciation

The genetic revolution in peripherally isolated populations has been
interpreted, so far, in terms of "beanbag genetics," in terms of shifts in
gene frequencies. Such an interpretation is in distinct contrast to the
belief of many early cytogeneticists that speciation is the result of a
structural repatterning of chromosomes. This hypothesis was based on
three separately valid observations: (1) that speciation is the origin of
discontinuities, (2) that related species often differ in their chromosomal
pattern, and (3) that certain chromosomal changes, such as reciprocal
translocations, are partially isolating on account of considerable sterility
among the heterozygotes. These elements were combined into the theory
that a structural chromosomal change might lead in a single step to the
origin of an individual that is reproductively isolated from the members
of the parental species. This would have provided for the instantaneous
origin of isolating mechanisms, an event that was being eagerly looked
for at that time. Further study has revealed that such a hypothesis is not
tenable. Not only does it run into the same difficulties as all other attempts
to establish new species through single individuals (Chapter 15), but it
is also contradicted by all the known facts of chromosomal variation
(White 1954, 1957, 1959).

Many species are polymorphic for the very chromosomal differences
that in other cases differentiate closely related species. In *Drosophila,*

where much of the cytological difference between related species seems to be due to paracentric inversions, polymorphism for these inversions is widespread in natural populations. In certain genera of orthopterans, species differ from each other by pericentric inversions, and these are likewise frequent as a polymorphic condition within species. Many species are individually variable for chromosome number, usually owing to the centric fusion of acrocentric chromosomes. In the extreme case of *Thais* (*Purpura*) *lapillus,* a single population may be polymorph for no less than five such centric fusions (Staiger 1954). Evidently not even this conspicuous type of chromosomal change serves as an interspecific isolating mechanism.

Conversely, well-defined and reproductively isolated species may completely agree structurally in their chromosomes and differ only in their gene contents. It can no longer be doubted that speciation can be completed by genic differentiation without structural repatterning.

Even though the old cytogenetic hypothesis of instantaneous speciation through chromosomal mutation could not be substantiated, there has been a renewed interest in recent years in the role of chromosome structure in speciation. Wallace (1959) points out, extending Mayr's (1954a) argument, that any mechanism would facilitate the local adaptation of peripheral populations that would protect their gene pools against the disruptive effect of gene immigration from more central populations. Since reproductive isolation does not exist between representatives of conspecific populations, some sort of "mechanical" protection is needed. Any chromosomal rearrangement that prevents crossing over will serve as such a protective device. It will permit the "locking up" of coadapted gene sequences and protect them against recombination. Such a "lock-up" device will be strongly selected for, because a mating between two indigenous individuals of the periphery will produce offspring of greater fitness than a mating between a peripheral and a central individual.

Any inversion, Robertsonian chromosome dissociation, or translocation, which reduces recombination between "normal" and peripheral chromosomes, will secure the preservation of local gene complexes and facilitate the continued selection of the contents of the protected chromosome sections for increased adaptation to the local environment. As Wallace points out, bearers of such protected chromosomes or chromosome sections, being members of peripheral populations, are better adapted for the marginal environment than are genotypes with the gene contents of the central population. As a consequence of their partial adaptation to the border environment, they are favored to spread beyond the present species border, as soon as a genetic mechanism arises that would perfect the adaptation. The important aspect of Wallace's view is that it stresses the adaptive significance of the chromosomal reorganization in terms of protection against disruptive recombination.

White (1959), similarly, arrives at the conclusion of a considerable importance of structural changes in the chromosomes for speciation, but on the basis of a different set of observations and premises. Among the many kinds of chromosomal rearrangements—paracentric and pericentric inversions, translocations, and centric fusions or dissociations—only one kind usually tends to give rise to balanced polymorphism in a given group of animals. This is illustrated by populations in *Drosophila* polymorphic for paracentric inversions, certain grasshoppers (Morabinae, Trimerotropi) for pericentric inversions, some scorpions and roaches for translocations, and some beetles and mantids for fusions and dissociations. Structural heterozygotes are frequent in such populations; indeed, their selective superiority is responsible for the maintenance of the polymorphism. However, if closely related species are studied, it is often found that they differ in chromosomal rearrangements, and that hybrids between them exhibit no heterozygote superiority or only a temporary or local one. It must be remembered, of course, that no genetic difference, whether genic or chromosomal, can ever become established in a population or species without passing through a stage of heterozygosity. White believes that a speciating population enters a temporary and perhaps somewhat "improbable" stage of heterozygosity while going through the "genetic revolution" described above. This always entails a replacement of a previously existing balanced polymorphism by monomorphism and results in a more or less drastic change of the genetic background.

Loss of such a polymorphism is thus likely to lead to the collapse of other genetic polymorphisms adaptively linked with it. The general level of balanced polymorphism falls sharply, and with it the genetic burden [see Chapter 9], thus probably facilitating the establishment of entirely new polymorphisms . . . Looking at the matter in this way it is *not* difficult to imagine that, at least in many groups of animals, the fixation of chromosomal rearrangements may play a rather special role in the speciation process (White 1959).

What this explanatory model leaves unsolved, the selective premium on the establishment of entirely new polymorphisms, is explained by Wallace's model. The two hypotheses together provide a convincing explanation of the reconstruction of the structural karyotype which, so often accompanies speciation. The possible role it may play in the origin of isolating mechanisms will be discussed below. It is interesting that students of plant speciation have independently developed a very similar model of rapid species formation in peripheral populations (for example, Lewis and Raven 1958; Lewis 1962).

Consequences of the Genetic Revolution

The genetic changes that occur in an isolated population have manifold consequences. With the cohesion of the parental gene pool disrupted, conditions are favorable for a departure in new directions. The direction of such departure is largely unpredictable, since chance enters the picture at many separate levels, gametic, zygotic, developmental, behavioral, and environmental, even though the results of these chance events are continuously guided by natural selection. These results concern the genotype as well as every aspect of the phenotype. The passing of the population through a bottleneck permits rather drastic shifts which would be resisted by the homeostatic system of the well-integrated continental parent population. We shall now discuss the nature of these changes.

During a genetic revolution, the population will pass from one well-integrated and stable condition through a highly unstable period to another period of balanced integration. The passing through the bottleneck and the reaching of the new balance is characterized by the occurrence of various genetic processes. Most conspicuous among these is a great loss of genetic variability (Fig. 17–3). For this loss there are a number of reasons: (1) the founders represent only a fraction of the variability of the species; (2) owing to inbreeding, more recessives will become homozygous and thus be exposed to selection; (3) owing to the reduced population size, there will be changes in the selective value of alleles and certain alleles will be eliminated (loss of "good mixers"); (4) during the reconstitution of the epigenotypes, many genes will lose the advantage of being part of a balanced system and will be selected against; (5) as long as the new population is small it may lose additional genes through errors of sampling.

Not all of this loss of genetic variability is deleterious. It reduces the genetic load quite drastically and gives the surviving population a "clean start." The capacity of small populations for getting rid of deleterious genetic factors is quite remarkable. The small natural populations of *Drosophila novamexicana* and *D. hydei* in the desert areas of the American Southwest carry very few detrimental factors. The small population

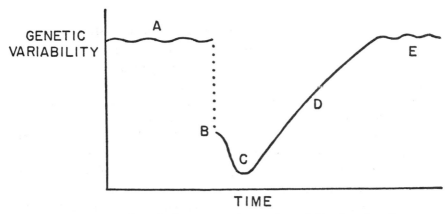

Fig. 17–3. Loss and gradual recovery of genetic variation in a founder population. The founders (B) have only a fraction of the genetic variation of the parental population (A) and further genes are lost during the ensuing genetic revolution (B to C). Variation is gradually recovered (D) if the population can find a suitable niche, until a new level (E) is reached. (From Mayr 1954a.)

of D. ananassae on Rongerik Island (Marshall Islands) recovered within about 26 generations, and the somewhat larger population on Rongelap Island within 40 generations from mutational damage due to radioactive fallout (Stone and Wilson 1958). In experimental populations, likewise, a reduction in the frequency of deleterious factors was found after a reduction in population size (Wallace 1959).

A genuine genetic revolution is characterized by a breakdown of genetic homeostasis through a loss or a reconstitution of previously existing balancing systems. The population will go through a labile state. The situation is made still more acute by the fact (to be discussed below) that the population has to cope with new selection pressures owing to the changed physical and biotic environment in the isolate.

Those populations that succeed in surviving the genetic revolution, presumably very much of a minority, will now enter a new state, which is characterized by the renewed accumulation of genetic variability and the acquisition of new and usually very different balancing systems. Recent cytogenetic studies have shown that the amount of chromosomal polymorphism with which a population can cope is usually severely limited (Wallace 1954a; Lewontin and White 1960), but there are some species, like Drosophila willistoni, in which a single individual may be heterozygous for more than a dozen inversions. An existing successful gene arrangement, functioning as a supergene, frequently interferes for a variety of reasons with the introduction of additional, new gene arrangements. The temporary state of chromosomal monomorphism during the genetic revolution favors the establishment of new chromosomal polymorphisms. In *Drosophila pseudoobscura,* where normally only the third chromosome is heterozygous for gene arrangements, polymorphism for the second chromosome occurs only in populations that are homozygous for the third chromosome (Terzaghi and Knapp 1960).

A comparison of chromosomal polymorphism of closely related species is most revealing. A total of 144 inversions were found in 46 studied species of the *Drosophila repleta* group. Even though 45 of the 144 arrangements were polymorphic within one or another species, only three arrangements were simultaneously polymorphic in two species (Wasserman 1960, 1962). In the *virilis* group, likewise, of 92 inversions only two were simultaneously polymorphic in two species (Stone *et al.* 1960; Stone 1962). It is as if speciation were normally associated with a change of the system of polymorphism. Where populations or closely related species differ more drastically in chromosomal rearrangements, particularly such as involve translocations and pericentric inversions, the need for passing through the heterozygous stage quickly is even more obvious. Such heterozygotes are often of reduced fitness owing to various difficulties during meiosis. Yet there are numerous mechanisms (Smith 1960) by which the deleterious effects of translocation heterozygosity are reduced. The translocation that led from the eastern 15-chromosome race of the grasshopper *Moraba scurra* to the western 17-chromosome race could have been successful only in a small peripherally isolated population. "In this way a colony containing a majority of dissociation homozygotes [A, B] could have arisen rapidly and could have persisted without being swamped by immigration of fused AB homozygotes from all sides" (White and Chinnick 1957). The two chromosomal races are now widely replacing each other geographically without forming heterozygous populations. In other cases, it has been shown that chromosomal differences formerly ascribed to translocation may have originated in a different manner. White (1961), for instance, gave such an interpretation (homozygosity for reduplicated chromosome ends) for the differences between the Italian newt *Triturus cristatus carnifex* and its northern European relative *T. c. cristatus.* Regardless of the cytological interpretation, such chromosomal novelties are perhaps most easily incorporated in small founder populations (Spurway and Callan 1960).

The study of chromosomal variation within species and between species has reached such a level of precision in recent years that it sometimes permits the reconstruction of the sequence of chromosomal events which accompanied speciation. The studies by Stone (1955) and Stone *et al.* (1960) of the *Drosophila virilis* group, by Wasserman (1960, 1962) of the *D. repleta* group, and by Manna and Smith (1959) of the beetles of the *Pissodes stigma* group are examples of such analyses. Under optimal conditions it is possible to determine not only the nature of the differences, but also the phylogenetic sequence in which they have occurred. I must refer to the cited papers, as well as to White (1957a,b, 1959) and Smith (1960), for a detailed treatment of the many fascinating problems of chromosomal speciation. These recent papers contain bibliographic references to the classical papers of Sturtevant and Dobzhansky, who first developed these methods.

The sharpness of the border in the intra- as well as interspecific situations proves in every case that during a very transitory genetic revolution a new balanced system was acquired. I use the term "genetic revolution" advisedly, because the genetic reconstitution was in all these cases far too drastic to be described in terms of a change of gene frequencies. Surely, a change of gene frequencies also occurred, but it was associated with a sufficiently drastic replacement of "supergenes" that it resulted in the acquisition of a new epigenotype.

Speciation and Continuous Ranges

In order to demonstrate the importance of spatial isolation, the extreme situation of a completely isolated founder population was the model on which the preceding discussion was based. This extreme, indeed, may not be too far from the truth in most cases of speciation. "Peripheral populations are not outstandingly different if they are part of a continuous series of populations. Only peripherally isolated populations show a pronounced deviation from the species 'type'" (Mayr 1954a). Stone *et al.* (1960) come to the same conclusion for *Drosophila*. *Drosophila robusta*, lacking true geographical isolates, is not speciating. The *D. repleta* group, with its numerous isolates in the American Southwest, is the most actively speciating species group so far known in the genus *Drosophila*.

However, not all isolates are established by founders. Sometimes they arise by the contraction of a previously continuous species range into isolated pockets. Mutation, recombination, and selection will henceforth be different and independent in the two areas and an increasing genetic divergence is inevitable. Yet, if the separated populations remain large throughout, they will not pass through a genetic revolution and will continue to share the same balancing systems, the same epigenotype. Normalizing selection will tend to eliminate the same deviants in both daughter gene pools, which, although now independent, will continue to act as if they were parts of a single cohesive system. No one knows how long such a "parallel cohesion" can be maintained. The case of the American and Asiatic sycamores (*Platanus*), which have failed to acquire reproductive isolation after millions of years of separation (Stebbins 1950), gives one pause.

Throughout this volume, I have stressed the tremendously cohesive effect of gene flow. Yet, when one tries to calculate the time it takes for genes to percolate from one end of the range of a widespread species to the other, one arrives at rather astronomical figures. This is particularly true for flightless species or, indeed, any organisms with poor dispersal facilities (White 1959a). Without wanting to depreciate the importance of gene flow, I advance the thesis that the cohesion of the species is also due to the fact that all those of its populations that have not undergone a genetic revolution share the same homeostatic systems and that these systems give great stability.

We have listed in Chapter 16 numerous cases of distant populations of a species that had advanced more or less far along the path of reproductive isolation. These observations would seem to contradict the implications of the above discussion, as do Wright's (1943a) calculations on isolation by distance. Wright, however, deals strictly with gene frequencies and does not deal with the cohesive properties of integrated gene complexes. And, as far as partial reproductive isolation of conspecific populations is concerned, we must admit that in most cases we know very little about the actual contiguity of the populations. What is shown on a more or less diagrammatic distribution map as a continuous range may well be the result of colonization through a series of founder populations, as in the case of *Triturus cristatus carnifex* or the western race of *Moraba scurra*. The range of *Larus argentatus*, (Fig. 16–7), a well-known case of circular overlap, certainly consists of a series of isolated colonies.

Speciation by mere distance seems far less well established now than it did 10 or 20 years ago. That it does occur, however, in rather sedentary organisms in which each local population is partially separated by barriers from other neighboring populations, seems possible. Even a rather minor and temporary barrier may occasionally permit a genetic revolution.

The Genetics of Species Differences

The recent studies on the integration of species-specific gene complexes have considerably changed our ideas on the nature of species differences. After the claims of the early Mendelians that one or a few mutations "made" species had been refuted, we had come to think of the genetic difference between species as a matter of quantity: if enough gene substitutions were piled on top of each other, there would eventually be a different species. Early authors spoke of dozens or scores of gene differences between species, but, when it was realized that even individuals within a single population (including the human species) may differ by hundreds of genes, one began to talk in bigger numbers. Haldane (1957) said recently, "Good species, even when closely related, may differ at several thousand loci," and this order of magnitude would probably be supported by most current investigators. Actually such figures tell us relatively little. Indeed, it is becoming increasingly evident that an approach which merely counts the number of gene differences is meaningless, if not misleading.

For a while it was thought that the genetic analysis of species hybrids would permit the analysis of species differences. This approach, indeed, yielded valuable information (see Hertwig 1936 for a summary of the older literature). Of more recent studies on animal hybrids, one might mention the work of Shull (1949) on coccinellid beetles, of Gordon and Rosen (1951) on xiphophorin fishes, of Steiner (1952, 1958) on weaver finches, of Spurway (1953) on newts, of Clarke and Sheppard (1955) on butterflies of the *Papilio machaon* group, and of Sadoglu (1957) on cave fishes. The results of these studies are in almost complete agreement with each other. A few species differences appear to be controlled by single genes or by a few genes with large phenotypic effects. Where single gene differences distinguish species, the very same genes may be polymorphic in related species, such as black versus yellow wing color in *Papilio*. Most species differences, however, seem to be controlled by a large number of genetic factors with small individual effects. The genetic basis of the isolating mechanisms, in particular, seems to consist largely of such genes. The results of these hybrid analyses are, thus, in substantial agreement with those derived from an analysis of population differences within species.

The blood-group genes (cellular antigens) with their highly specific cross reactions permit a genetic comparison of closely and more distantly related species. An analysis of such genes in the dove genus *Streptopelia* shows that even the most closely related species differ in scores of such genes, and the number of differences increases with the remoteness of relationship (Irwin 1953, 1955). Yet even rather distant relatives often retain a surprising degree of similarity in their antigens.

Each isolated gene pool is a different biological system, and the organization that is the result of the coadaptation of the genes may add a new dimension to the differences, which cannot be stated as the arithmetic sum of the individual gene differences. It is easy to imagine two conspecific populations that share the same species-specific isolating mechanisms and essential chromosome structure, and yet differ from each other by more individual gene substitutions than some good species. It is evident that a purely quantitative approach may well be misleading. Nor can species difference be expressed in terms of the genetic bits of information, the nucleotide pairs of the DNA. This would be quite as absurd as trying to express the difference between the Bible and Dante's *Divina Commedia* in terms of the difference in the frequency of the letters of the alphabet used in the two works. The meaningful level of integration is well above that of the basic code of information, the nucleotide pairs.

What, then, makes a species different from an intraspecific variant? I believe that Harland hit the nail on the head when he said many years ago that species are characterized by their modifiers. Today we would perhaps use a slightly different terminology for what Harland had in mind. We might say that it is the total system of developmental interactions, the totality of feedbacks and canalizations, which makes a species. Two individuals of *Drosophila melanogaster* that differ in five conspicuous mutations affecting eye color, pigmentation, wing shape, bristle structure, and haltere formation may look strikingly different from each other, yet they still share their "modifiers," their total developmental system, and are thus still *Drosophila melanogaster*. Two wild-type individuals of *D. melanogaster* and *D. simulans*, which are hardly distinct visibly, nevertheless differ from each other by hundreds, if not thousands, of genes and are the possessors of totally different developmental systems. The important point is that different species are different systems of gene interaction or different epigenetic systems.

Morphological Consequences. That the genetic reconstitution of isolated populations is often rather drastic and affects major homeostatic systems is supported by studies on morphological and physiological characters. Peripheral isolates, no matter how close to the main range of the species, almost always are noticeably different, in contrast to the essential uniformity in the contiguous range of the species. This is well shown by *Tanysiptera galathea* (Mayr 1954a), by *Dicrurus hottentottus* (Vaurie 1949), and by other species cited in Chapter 13. In the butterfly *Maniola jurtina,* the populations on the small islands of the Scilly Islands differ from each other and from the rather uniform populations of the three large islands (Dowdeswell, Ford, and McWhirter 1957). The populations of *Drosophila willistoni* on the islands of the West Indies are greatly impoverished as far as the diversity of gene arrangements is concerned, and the various islands differ considerably from each other (Dobzhansky 1957b). The rapidity with which morphological changes take place in peripheral isolates confirms our conclusion that the genetic reconstitution permits or induces shifts in the previously existing developmental homeostasis.

The degree of morphological distinctness acquired during the period of isolation is not an accurate measure of the degree either of general genetic difference or more specifically of the degree of reproductive isolation. This is equally true where, as with the snails of the genus *Cerion* (Mayr and Rosen 1956), a high degree of morphological difference is associated with lack of reproductive isolation, or for sibling species, where the reverse is true (Chapter 3).

Ecological Consequences. The marginal environment of the geographical isolate and the unfavorable properties of the "normal" niche of the species under these marginal conditions often reinforce the genetic revolution. The genetic changes, in turn, may profoundly affect the ecological preferences and adaptations of a population.

The environment in the peripheral isolate is almost always rather unlike the optimal environment of the species in the center of its range. The biotic environment, in particular, is usually somewhat unbalanced at isolated locations. The new isolate will thus be exposed to a considerably changed selection pressure. The physical environment may not be very different in the peripheral isolate from nearby peripheral areas of the species. But the response of the isolated population to this selection pressure will be quite different from that of a population which is part of a contiguous array of populations held together by gene flow and all the cohesive devices discussed in Chapter 10. A population that is part of a continuum of populations is forced to compromise between becoming adapted to local conditions and remaining in coadaptation with the gene pool of the species as a whole. The more distant a population is from the optimal center of the species range, the less suitable its genetic equipment will be to cope with the species-specific niche. The species border represents the stalemate between the efforts of local ecotypic adaptation on one hand and coadaptation with the gene pool of the species as a whole. In contrast, the isolated population can respond to its local adaptive needs without having to compromise with the solutions found by other populations.

The only answer that is possible in many cases is a shift into a new niche. Such a shift is greatly facilitated by a genetic revolution and the special properties of isolated populations. In particular, the genetic lability of such populations and the pronounced population fluctuations (in the absence of strong density-dependent factors) facilitate such shifts. In no other situation in evolution is there a greater opportunity for adaptive shifts or evolutionary novelties (Mayr 1954a).

That this is not merely a hypothesis is documented by the numerous actually observed ecological shifts in peripherally isolated populations (Chapters 11, 13, and 19). Indeed, nearly all aberrant populations of species are peripherally isolated. The ecological shifts on oceanic archipelagoes (for example, Amadon 1947; Zimmerman 1948) illustrate this phenomenon dramatically.

Requirements for Successful Speciation

Each species is an independent genetic system, which has the properties of being reproductively isolated from and ecologically compatible with other sympatric species. Speciation means the acquisition of these properties. This may take place almost instantaneously, as in the case of polyploidy, or gradually, as in the case of geographic speciation. The process by which reproductive isolation and ecological compatibility are acquired is sufficiently important to deserve detailed analysis.

The term isolation has been applied in evolutionary biology to two very different kinds of phenomena (Mayr 1959a). Even some recent authors have confounded these two phenomena and have thereby been led to erroneous theories of speciation. The two phenomena are geographic isolation and reproductive isolation (Chapter 5). *Geographic isolation* refers to the division of a single gene pool into two by strictly extrinsic factors. It is a reversible phenomenon which in itself has no effect whatsoever on the two separated gene pools. What it does is to guarantee their independent development and to permit the accumulation of genetic differences. *Reproductive isolation* refers to the protective devices of a harmoniously coadapted gene pool against destruction by genotypes from other gene pools. These protective devices are known under the term isolating mechanisms. Speciation is characterized by the acquisition of these devices.

Ecological Compatibility. An incipient species, in order to complete the process of speciation, must acquire sufficient differences in niche utilization to be able to exist sympatrically with sister species without fatal competition (see Chapter 4). Such differences arise during ecological

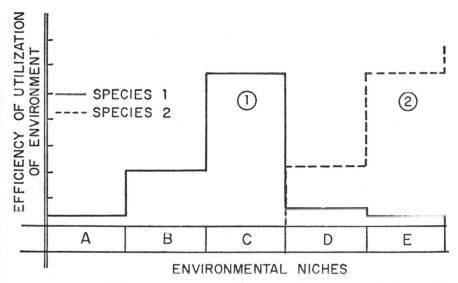

Fig. 17–4. Niche utilization by two different species. Species 1 finds optimal conditions in environmental niche *C*; it utilizes niche *B* inefficiently and niches *A*, *D*, and *E* very poorly. Species 2 cannot utilize niches *A*, *B*, and *C* at all, but finds optimal conditions in niches *E* and *F*. The absence of competition in niche *A* will invite the evolution of a species adapted for this niche. (From Mayr 1949a.)

shifts in the isolate. The longer the populations have been isolated and the more drastic the genetic revolution, the greater will be the probability of ecological differences. Ecological compatibility ("exclusion") need be only initiated during geographic isolation. Even if the ecological divergence is only slight when the species begin to overlap, selection can continue to widen the gap. Such selection will be strongly centrifugal (Fig. 17–4), since it will be directed against the individuals in the zone of ecological overlap. That such selection actually takes place is substantiated both by observation (Chapter 4) and by experiment. When *Dro-*

sophila melanogaster and *D. simulans* are put together in the same culture at 25° C, *D. melanogaster* always eliminates the competing species sooner or later. If, after 10 or 15 generations of competition, some of the experimental *simulans* flies are placed in competition with a new (unselected) batch of *D. melanogaster*, they prove to be considerably improved as competitors (Moore 1952a,b). This indicates that genes had accumulated in the experimental *D. simulans* population which enhanced their status as competitors. The work of Pittendrigh (1950) on ecological divergence of two species of *Anopheles* in an area of overlap on Trinidad is an excellent illustration of such exclusion. The differences in ecological tolerance and habitat preference of species are presumably built up polygenically.

The Origin of Isolating Mechanisms

The most indispensable step in speciation is the acquisition of isolating mechanisms. Isolating mechanisms have no selective value as such until they are reasonably efficient and can prevent the breaking up of gene complexes. They are *ad hoc* mechanisms. It is therefore somewhat difficult to comprehend how isolating mechanisms can evolve in isolated populations. This problem has been the subject of considerable discussion during recent years, and we are approaching agreement only now.

There are essentially two theories on the origin of isolating mechanisms. According to one, originally proposed by A. R. Wallace (Mayr 1959a), subsequently maintained by Weismann (1902) and Fisher (1930), and elaborated by Sturtevant (1938) and Dobzhansky (1940), natural selection is responsible for the establishment of isolating barriers. This hypothesis is based on the observation that hybrids between two species are usually of lowered fitness. It argues, therefore, that individuals with inefficient isolating mechanisms will be susceptible to hybridization in areas of contact between the parental and the incipient new species. These genotypes will be eliminated from both populations as a consequence of selection against the hybrids that they produce. Such selection will tend to spare genotypes with better-developed isolating mechanisms. This process, it is postulated, will in due time lead to an improvement and final perfection of the isolating mechanisms.

Cases in which reproductive isolation between sympatric species is greater than between allopatric species are usually cited in favor of the theory of the sympatric origin of isolating mechanisms. Dobzhansky and Koller (1938) found that the reproductive isolation of *Drosophila mi-*

randa was greater against those populations of the two related species
D. persimilis and *D. pseudoobscura* that occurred in the zone of overlap
with *miranda* than against populations that came from outside the range
of *miranda*. King (1947) found that among three very closely related species of *Drosophila* the two species that occur sympatrically in Brazil,
D. guaru and *D. guarani*, were completely isolated reproductively. Both
species, however, interbred in the laboratory rather freely with an allopatric third species, *D. subbadia*, which is known only from Mexico.

Additional support is found in the behavior and systematic literature
on other kinds of animals. Several authors have presented evidence that
two species may be more different in coloration or in call notes in areas
of sympatry than where they are allopatric (Sibley 1957; Marler 1957;
Blair 1955, 1958a; and Mecham 1961). Perhaps the best-substantiated evidence for the reinforcement of isolating mechanisms is the recent overlap
in western Russia between the two titmice *Parus caeruleus* and *Parus
cyanus*, which, when it first occurred (1870–1900), led to frequent production of hybrids ("*Parus pleskei*"). After 50 years of overlap, hybridization has greatly decreased (Vaurie 1957). Similarly, Hubbs and Delco
(1960) found more mating discrimination between species of *Gambusia*
previously sympatric than between allopatric species.

The proponents of the theory of the sympatric origin of isolating
mechanisms have not held that all isolating mechanisms are perfected
through hybridization, yet they seem to assume that this happens frequently. This viewpoint has been challenged on the basis of objections
ably stated by Moore (1957). Perhaps the most convincing argument
against the power of natural selection is supplied by the "old" hybrid
belts (Chapter 13). These have existed in many cases for thousands of
years, and the narrowness of the belts proves that the hybrids are being
selected against. Yet there is no indication that this has led to a strengthening of the isolation in any of the cases. The hypothesis that isolating
mechanisms originate or are markedly improved by natural selection has
not been able to solve the difficulties raised by introgression. If the hybrids
are not sterile, some of them will backcross with the parental species, a
process that ought to lead to a further weakening of isolation rather than
to its strengthening.

Another argument comes from a comparison of the strength of isolating mechanisms within zones of overlap and outside these zones. The
genetic factors responsible for reproductive isolation should, on this
theory, be restricted to zones of overlap between the related species, since

there would be no selective advantage to have these *ad hoc* mechanisms spread beyond the area where they are favored by selection (Moore 1957). However, there is no evidence that the isolating mechanisms are geographically thus confined. The list of the cases cited above in which the reproductive isolation of sympatric populations of two species is stronger than that between allopatric populations may well be due to biased sampling. If isolating mechanisms are a by-product of genetic divergence and if—as is well established—different populations of a species differ in the level of isolation reached, it is to be expected that in a comparison of pairs of closely related species a higher degree of reproductive isolation may be found in some cases between sympatric and in other cases between allopatric populations. This is exactly what Patterson and Stone (1952) have found. Most frequently there are no differences at all between sympatric and allopatric populations of two species with respect to the degree of reproductive isolation. This has been confirmed by Hubbs and Strawn (1957) in a considerable number of crosses between species of fishes belonging to the subfamily Etheostominae and by Minamori (1955) for the sibling species of fishes of the *Cobitis taenia* group. That selection is not necessary for the perfecting of isolating mechanisms is demonstrated by the numerous cases in which efficiently functioning isolating mechanisms have undoubtedly evolved in geographic isolation without any possibility of their improvement by subsequent selection.

Patterson and Stone (1952) cite several instances in *Drosophila* in which degree of reproductive isolation could not have been due to selection:

In tests with *D. americana* and *texana*, *virilis* from Hanchow, China, crossed readily, while *virilis* from Victoria, Texas, crossed poorly; but *virilis* from Galveston, Texas, crossed just as readily as the stock from Hanchow. *Drosophila littoralis* from Europe is much more isolated from the *americana, texana, novamexicana* complex than is *montana* and its relatives, although both cross about the same to *virilis*. . .

As a further case, we have a nice example of the accumulation of isolation by distance (Wright 1943a) in the *funebris* group. In this group *macrospina* and *ohioensis* are color phases without serious isolation. *Drosophila macrospina* is partially isolated from the eastern strains of *limpiensis*, with which it might conceivably cross occasionally; it is much more effectively isolated from the western strains of *limpiensis* (which cross freely with the eastern strains), and it is probably unable to exchange genes with *subfunebris* or even *trispina* from California, except perhaps through the intermediacy of *limpiensis*. The cline of isolation increases with distance in this case and agrees with the expected degree of differentiation correlated with restriction on gene flow with distance.

These facts are taken into account in an alternative theory of the origin of isolating mechanisms, according to which they arise as an incidental by-product of genetic divergence in isolated populations (Muller 1940). This was the thesis of Darwin (see Mayr 1959a), who could not see how natural selection could produce interspecific sterility. Poulton (1908) likewise defended the thesis of the origin of isolating mechanisms as a by-product of genetic divergence. Patterson and Stone (1952) agree "that the factors favoring sexual isolation toward other species were fixed in populations incidentally, rather than as a result of selection pressure in a zone of contact."

This hypothesis is supported by three additional sets of observations. First, there is much evidence for the geographic variation of isolating mechanisms (Chapter 16), including incipient sterility and ethological isolation. The beginnings of such isolation have been observed even in separate cultures of laboratory stocks, for instance in *Drosophila* (Bösiger 1957; Hoenigsberg and Santibanez 1960). Second, in view of the highly composite and polygenic character of the isolating mechanisms (Chapter 5), it would be unlikely for them not to be affected by the genetic reconstitution. Third, many isolating mechanisms have ecological components. The ecological shifts in incipient species are bound to have an effect on their isolating mechanisms. The thesis of the origin of reproductive isolation as a by-product of the total genetic reconstitution of the speciating population is consistent with all the known facts.

Nevertheless natural selection does play a role in the improvement of some of the isolating mechanisms, only it concerns subsidiary isolating mechanisms. The primary, basic one must be fully efficient when contact is first established.

Much of the apparent conflict between the opposing theories disappears if the large category "isolating mechanisms" is subdivided. It is quite evident that one isolating mechanism or several must be acquired in geographical isolation before contact is established. In most cases this is cross sterility. When such cross-sterile species first come into contact, it matters little whether or not they produce (sterile!) hybrids owing to the incompleteness of the other isolating mechanisms, particularly ethological ones. Such hybrids, being sterile, cannot reproduce and thus there is no danger of a breakdown of the species barrier. However, there will be strong selection in favor of the acquisition of additional isolating mechanisms to prevent such wastage of gametes. Blair (1955, 1958a), Sibley (1957), Volpe (1960), and Hubbs (1960), among others, have described cases in which such selection is operating. The selection theory is, thus, valid as far as the strengthening of secondary isolating mechanisms is concerned. That the reproductive isolation between two species can be strengthened by artificial selection (removal of all F_1 hybrids) was shown by Koopman (1950) for *Drosophila pseudoobscura* and *D. persimilis*.

Sterility is not always the first isolating mechanism to be perfected. In many families of birds, the duck family (Anatidae) for instance, sympatric species may still be quite fertile with each other and yet not hybridize in nature, because of the perfection of the ethological barriers. Occasional hybrids occur, but at such a low rate that the elimination of the introgressing genes is not too severe a burden on the parental species. Most of the hybrids are excluded anyhow from further reproduction owing to behavioral incompatibility. Such "behavioral sterility" of hybrids has also been observed in *Drosophila* (Ehrman 1960b). Which type of isolating mechanism is first perfected may depend on the particular group of organisms. In most groups it seems to be the sterility barrier; in birds it may well be the behavioral barrier.

Perdeck (1957) shows that reproductive isolation between *Chortippus brunneus* and *C. biguttulus*, two sibling species of grasshoppers, is maintained exclusively by the difference in display song. No other isolating mechanism could be detected. The fertility of the hybrids seems to be the same as that of intraspecific crosses, likewise that of the backcross hybrids (although the material is limited). The only handicap of the hybrid males is that they are discriminated against by females. The functioning of this ethological isolating mechanism must have been virtually perfect before contact was established, because the essentially fully viable hybrids would serve as a channel of gene flow between the two species, if they occurred at all frequently. A number of additional cases among amphibians and insects have been described in which the vocalization of the males seems to be the primary and by far the most important isolating barrier.

Several attempts have been made in recent years to select for sexual isolation in artificial populations. Two strains of *Drosophila melanogaster*, each homozygous for an easily visible recessive gene (for instance, *straw* and *sepia*), were brought together. All interstrain matings produced heterozygotes which were easily recognized by having the wild-type phenotype. Every hybrid was removed by the experimenter in each generation, as if it were totally inviable, thus creating a powerful selection pressure in favor of sexual isolation between the strains. After 73 generations, Wallace (1954a) found a considerable isolation for one type of female and none for the other (Table 17–1). Knight, Robertson, and Waddington (1956) had similar results. The slowness and incompleteness with which the isolation is acquired, in spite of the severity of the selection pressure, is further evidence for the highly polygenic basis of most perfected isolating mechanisms. The dramatic reconstitution of the gene pool during a genetic revolution may, however, speed up the process.

An attempt was made in these experiments to assemble isolating genes around diagnostic marker genes. Thoday and Gibson (1962) chose a different and, as it appears, far more successful approach. They selected for two very different polygenic character complexes (low and high

Table 17–1. Sexual preference of *Drosophila melanogaster* females in multiple-choice experiments after 73 generations of selection for sexual isolation (from Wallace 1954a).

Type of female	Type of fertilization		Number
	Homogamic (percent)	Heterogamic (percent)	
Straw	48.8	51.2	215
Sepia	90.3	9.7	237

bristle number) and found (after a series of generations) a considerable amount of reproductive isolation between the two extreme phenotypes. Evidently the selection simultaneously favored genes that result in homogamic mating.

The picture that emerges from our study of gradual speciation is that an isolated population acquires, during its isolation, the primary isolating mechanisms which guarantee its integrity after establishment of contact with a sister or parent species. It also acquires a minimal amount of that "adaptive property" which permits the two species to be ecologically compatible. Selection pressures, after the establishment of sympatry, will help to improve the isolating mechanisms to such an extent that no more wastage of gametes (at least not of female gametes) occurs, and ecological exclusion will be steadily improved at the same time. The total difference (SpD) between two species can be expressed in an equation (Mayr 1951a) which modifies one originally proposed by Pittendrigh (1950):

$$SpD = In + Ip + Is + An + Ap + As.$$

In this formula I stands for isolating mechanisms and A for the adaptive properties (particularly those that reduce competition); n means neutral (with respect to the interaction of the species), *p* means preadaptive (that is, existing already prior to the establishment of contact between the species), and *s* indicates those properties that were acquired through natural selection after the establishment of contact. It is obvious that two incipient species cannot become sympatric unless *Ip* and *Ap* are strong enough to prevent frequent interbreeding and strong competition. It is equally obvious that there will be a selective premium on the acquisition of *Is* and *As* until reproductive isolation and exclusion are perfected.

The Probability of Successful Speciation

Most species bud off peripheral isolates at regular intervals. Nearly all of them either reestablish contact with the parental species or else die out. "The odds are very much against a successful passing through the bottleneck of reduced variability as well as the reaching of a new level of high variability and of an unoccupied ecological niche" (Mayr 1954a). Speciation is another illustration of the opportunism of evolution. What does it matter if 98 or 99 among 100 founder populations or other isolates fall by the wayside? All is well and evolutionary progress assured, as long as one of them once in a while discovers a new niche. And if this niche should turn out to be a whole new adaptive zone, such as was found by the first amphibian, the first bird, the first weevil, the first cichlid fish in the African lakes, or the first honey creeper on Hawaii, then there will be a whole avalanche of successful speciations until that zone is completely filled by adaptive radiation. The continued presence of numerous isolates is a guarantee of the occurrence of speciation whenever the ecological situation is opportune.

CONCLUSIONS

We are now ready to summarize our findings on the genetics of speciation.

Geographic isolation is a purely extrinsic and completely reversible factor which does not by itself lead to the formation of species. Its role is simply to permit the undisturbed genetic reconstruction of populations that is the prerequisite for the building up of isolating mechanisms.

The need for coadaptation and for the harmonious integration of genes sets severe upper limits to the number of genes that can be accommodated in a gene pool, since many genetic combinations are incompatible. The rapid elimination of disharmonious combinations after hybridization is proof of this conclusion. There is a tendency in the integrated gene complex to establish an ever-greater cohesion, to achieve a steady improvement of developmental and of genetic homeostasis. Numerous feedbacks permit the individual as well as the population to compensate for the unsettling impact of the environment. Heterosis, in particular, tends to diminish the effectiveness of selection for specific effects by raising viability and by increasing independence of the environment. A well-integrated genetic system may come into perfect balance with its environment and become so well stabilized that evolutionary change will no longer occur. Such a system will be able to cope with the regular input of mutations and the normal environmental fluctuations without having to undergo any change. Its future is at best evolutionary inertia and, more likely, eventual extinction.

Speciation is potentially a process of evolutionary rejuvenation, an escape from too rigid a system of genetic homeostasis. Speciation disrupts the cohesion of the gene pool by temporarily depleting its gene contents and by inevitably forcing the population into a slightly or drastically different environment. If the genetic shake-up is sufficiently severe, it may start a chain reaction, a veritable genetic revolution. The greater the genetic change, the greater the probability that the daughter species can enter a new ecological niche and be successful in it. The genetic chain reaction may thus start an evolutionary chain reaction. The stated process is most likely to occur in its purest form in peripherally isolated populations. Speciation is a risky process. The impoverishment of the gene pool and the genetic instability that accompanies it are far more likely to lead to disaster than to success. Even though most incipient species will die out, an occasional one not only completes the process, but succeeds in entering a new niche or adaptive zone.

The importance of speciation is that it invites evolutionary experimentation. It creates new units of evolution, particularly those that are important for potential macroevolution. Speciation is a progressive, not a retrogressive, process.

18 ~ The Ecology of Speciation

Geographic speciation means the genetic reconstruction of a population during a period of geographic (spatial) isolation. The genetic factors involved in this reconstruction were discussed in the last chapter. In the present chapter I will attempt to analyze the role of the environmental factors that influence the origin and maintenance of discontinuities between populations. A clear distinction will be made between factors that induce the formation of geographic isolates and factors that control the rate of change in the isolates, keeping always in mind that speciation is not mere evolutionary change but the multiplication of species.

An exhaustive treatment of the indicated subject matter would require an entire book, for there is hardly an ecological factor that does not affect speciation directly or indirectly, actually or potentially. A general survey is all that can be presented here and we must refer to the ecological and biogeographical literature for more exhaustive treatments (see, for instance, Hesse, Allee, and Schmidt 1951; Andrewartha and Birch 1954; Kendeigh 1961).

In this study of the factors that control the establishment and fate of geographical isolates, I want to single out for special discussion: (1) factors determining the effectiveness of geographical isolation; (2) factors affecting shifts into new ecological niches; (3) factors affecting the frequency with which geographic isolates are established, and (4) factors favoring the genetic turnover within isolates. All four sets of factors have an effect on the rate of speciation.

THE EFFECTIVENESS OF GEOGRAPHICAL ISOLATION

One of the basic properties of species, and of the individuals of which species are composed, is the capacity to spread. Every species has at least one dispersal stage in its life cycle. A study of the geographical barriers that surround every species and every geographical isolate must take this ability for dispersal into consideration. To be sure, "geographical isolation" means the interruption of gene flow by external barriers. But we must realize that the physical nature of these barriers (extrinsic factors) is only one aspect of this isolation. The numerous physiological and psychological characteristics (intrinsic factors) of the individuals that encounter these barriers during their dispersal stage are of crucial importance. Indeed, to a large extent, they determine the effectiveness of these barriers.

The Role of Extrinsic Factors

An understanding of the functioning of the natural barriers that are responsible for the discontinuities between geographical isolates is an indispensable prerequisite for an understanding of speciation. The study of geographical barriers is therefore as important for the evolutionist as it is for the biogeographer and ecologist.

Kinds of Barriers. Various authors have attempted to work out a logical classification of distributional barriers (Grinnell 1914a; Gentilli 1949; Hesse, Allee, and Schmidt 1951, who have devoted an entire chapter to the topic). These studies permit the broad generalization that any area which is unsuitable for occupation by a species may serve as a distributional barrier. The action of such barriers in the process of speciation may be illustrated by some examples.

The effectiveness of the sea as a speciation mechanism was fully appreciated by Darwin, Wallace, and other founders of the science of evolution. The birds of the Galapagos Islands (Lack 1947a), of the Hawaiian Islands (Amadon 1950), and of the Papuan region (Mayr 1942) are classical examples. The six islands or island groups of the central Solomons (Fig. 18–1) have permitted subspeciation in five of the 53 species of Passerine birds occurring there (Table 18–1). Eleven of the 19 potential barriers (three species are absent from Gizo) have permitted subspeciation. Straits as narrow as 2 kilometers (1.2 miles) have resulted in the evolution of strikingly different populations in several species. Ocean barriers are even more efficient for lizards and mice. Every small island off the coast of British Columbia has a distinct population of *Peromyscus maniculatus* (McCabe and Cowan 1945). Speciation of the reptiles and mammals on the islands of the Gulf of California was studied by several authors and other cases were summarized by Mayr (1942).

Speciation can also occur on islands within large lakes. Endemic sub-

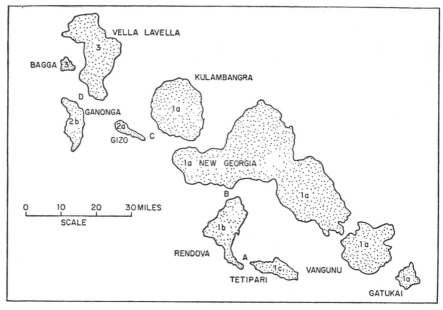

Fig. 18–1. Speciation in the white-eye *Zosterops rendovae* in the central Solomon Islands. 1, *rendovae* group; 2, *luteirostris* group; 3, *vellalavella.* Groups 1, 2, and 3 are considered full species by some authors, subspecies by others. The shortest distances between the islands are: $A = 1.7$ km; $B = 2$ km; $C = 6$ km; $D = 5$ km. (From Mayr 1942.) See Table 11–1.

species of birds and butterflies have been described from the islands of Lake Victoria in eastern Africa and several endemic subspecies of mammals have already evolved on the recently formed islands of Great Salt Lake in Utah (Marshall 1940). Lakes themselves, particularly if they are long ones like Tanganyika and Nyasa in eastern Africa, may form important barriers between populations on opposite shores.

Rivers likewise may be potent distributional barriers. This is true not only for terrestrial animals, but sometimes also for birds. The Amazon River and its tributaries are particularly effective as barriers for those birds that live in the undergrowth of the dark tropical jungle. The Ant Bird *Phlegopsis nigromaculata* occurs along the entire south bank from the Andes to the Atlantic but is absent on the north bank of the Amazon. On the south bank it is divided into four subspecies by the Madeira, Xingu, and Tocantins Rivers. Celebrated in the mammalian literature is the case of the squirrel *Callosciurus sladeni* which breaks up into a series of seven strikingly different races along the east bank of the upper Chindwin River in Burma, but is virtually without geographic variation

Table 18–1. Speciation in birds of the central Solomon Islands.[a]

Species	Vella Lavella	I	Ganonga	II	Gizo	III	New Georgia group	IV	Rendova	V	Tetipari
Ripidura cockerelli	a		a				b		b		b
Myzomela eichhorni	a		b		c		c		c		c
Monarcha barbata	a		b				c		d		d
Pachycephala pectoralis	a		b				c		d		d
Zosterops rendovae	a		b		c		d		e		f

[a] The letters a–f refer to distinct subspecies of each of the species listed in the left-hand column; for names see Mayr, *Birds of the Southwest Pacific* (Macmillan, New York, 1945). Many of these subspecies have been described as full species.
[b] Width of water barriers: I, 5 km; II, 10 km; III, 6 km; IV, 2 km; V, 1.7 km.

on the west bank (Carter 1943). Not only the Grand Canyon (Grinnell 1914b), but also the upper Colorado River (Kelson 1951), the Snake River in Idaho, and the Columbia River (Fox 1948c) are very effective geographical barriers for small mammals. On the other hand, some of the large rivers appear to be less effective. The Mississippi River is not known to be a conspicuous barrier, perhaps because it changes its course so often as to lead to a continuous fusion of the faunas of both banks. It is of historical interest to note that M. Wagner had his first clue to the importance of geographic barriers for speciation from a study of beetle populations on both banks of North African rivers.

Various other features of the landscape may likewise function as effective distributional barriers. Hooper (1944) shows that San Francisco Bay and the adjacent Sacramento River Valley form an important barrier between the mammals of northern and southern California. Of the 24 species involved, 11 (46 percent) occur both north and south of the Bay area, while 8 are found only in the north and 5 only in the south. Among 43 subspecies, only 8 occur jointly in the north and the south (18.6 percent), 18 are restricted to the north, and 17 to the south.

Mountain ranges may be important barriers, particularly if they simultaneously separate climatic zones, as the Alps separate the Mediterranean from the Central European area, and the Himalayas India from Tibet. But even in such a climatically uniform area as the equatorial island of New Guinea, the central mountain range serves in some of the very sedentary tropical species as an important barrier between strikingly different northern and southern populations.

As mountains are efficient barriers for lowland animals, so are valleys for mountain species. Each of the mountain species of birds of paradise is separated from its nearest relatives by valleys or low passes (Fig. 16–4). Snails on tropical islands may be isolated from each other either by valleys or by mountain ridges. The effect of this isolation on divergence has been discussed by Gulick (1905). The steep cliffs bordering tropical table lands (mesas) may separate very different populations on top and at the foot of the cliff (Chapman 1931).

The Pleistocene ice masses of the northern continents were among the most potent barriers in the recent history of the earth. In Europe the Scandinavian ice cap and the Alpine glaciers approached each other within 300 miles, separated merely by a cold steppe which formed an efficient barrier between the unglaciated Atlantic region and the unglaciated areas in the eastern Mediterranean and the Near East. This led to much subspeciation (Mayr 1942) but it cannot be shown with certainty that even a single glacial isolate among European birds reached full species level (Mayr 1951a). In America the glaciation seems to have been more effective. Rand (1948) lists numerous species pairs that apparently split into a western and an eastern component during one or the other phase of the Pleistocene. There were mesophytic refuges in the southeast, presumably two in the west, and one in the north in the Yukon area. The role of a Plio-Pleistocene embayment and of various physiographic features on speciation in Californian amphibians and reptiles is described by Peabody and Savage (1958). The analysis of the Pleistocene speciation has only begun (Deevey 1949). The respective roles of each of the four or five glaciations have not yet been determined and no comparison has been made between the effects of the glacial advances and those of the intervening warm dry periods of the interglacials. Nor has it been determined in what groups of animals the Pleistocene barriers have been most effective. The precision in the reconstruction of these events is greatly increased where a fossil record is available. It permitted Kurtén (1959) to show that the early Pleistocene European bear *Ursus arctos deningeri* was split into an eastern and a western population during the Mindel glaciation. The western (more peripheral!) population apparently became the Cave Bear (*Ursus spelaeus*), while the eastern population remained *U. arctos*.

A study of the barriers which now exist between geographical isolates shows that the conventional barriers (water, mountains) are no more important, at least on continents, than are vegetation zones. The borders of vegetation belts are often exceedingly sharp in the tropical and subtropical zones and form effective geographical barriers for many animals. This has been shown by Chapin (1932, 1948) for the birds of the African rain forest and by Keast (1961) for the birds of Australia. In Australia, for instance, the wet sclerophyll forests are broken up into a number of isolated areas along the periphery of the continent, separated by drier areas. Each pocket of forest has served as a local center of differentiation for incipient or fully formed species.

Even very narrow belts of unsuitable habitat can be effective geographical barriers. In the San Francisco Bay region a narrow unsuitable strip of country (1–3 miles wide) separates the salt-marsh races of the Song Sparrow (*Passerella melodia*) from the upland races of this species (Marshall 1948; Miller 1956). Forest belts that separate altitudinal grassland belts in the mountains of New Guinea likewise serve as effective barriers between populations (Rand 1936; Mayr 1942). The effectiveness of vegetation as a barrier is nowhere more evident than in areas where wooded country interdigitates with savannah or desert.

A study of the different types of barriers shows that it is not permissible to make a distinction between geographical and ecological barriers. Is a mountain a barrier because of the difficulty of overcoming it physically or because it is unsuitable for colonization or both? The Amazonian lowlands are for an Andean mountain species as much of a physical as an ecological barrier. Any terrain that is unsuitable for a species is at once a geographical as well as an ecological barrier. It is not legitimate to make distance a criterion, distinguishing between macrogeographical and microgeographical barriers, and to equate the latter with ecological barriers. It would force one to classify rivers and narrow ocean straits as ecological barriers, since they are surely microgeographical barriers as far as distance is concerned.

Where would one classify unsuitable soils for animals whose distribution is governed by preference for certain types of soil? The distribution of eight different endemic subspecies of pocket gophers (*Thomomys*) in southern Washington State on isolated "prairies" is a result of the distribution of soils and forest (Dalquest and Scheffer 1944). Similar effects of soils on the distribution of gophers have been described from several other areas, for instance, California (Ingles 1950).

Caves also are habitats that are usually very efficiently isolated. Most old caves have endemic faunas that resemble in many respects insular faunas. The literature on cave endemics is enormous and we have to refer to it for further details (for example, Barr *et al.* 1960). The factors that separate cave populations from the outside populations are often not at all clear. An endemic cave fish in Brazil, discussed by Pavan (1946), does not seem to be isolated from the outside. It is possible that some now accessible caves were at one time completely isolated underground basins and have established connection with outside streams only rather recently, after the speciation of their inhabitants had been completed.

Barriers in Fresh Water. The barriers permitting speciation in fresh-water fish have been discussed by Hubbs (1940, 1941) and by R. Miller (1950, 1961). On the whole, each stream or stream basin is a population unit separated by land from adjacent ones.

Lakes are for water animals what islands are for land animals. Every old fresh-water lake has its own endemic fauna. These faunas are either relatively young, like that of Lake Waccamaw in North Carolina (Hubbs and Raney 1946), or much older and rich in peculiar endemics, like those of Lakes Baikal, Nyasa, Tanganyika, and other ancient fresh-water lakes (Brooks 1950; Poll 1950). Each lake in turn consists of an archipelago of suitable areas, with each habitat island (such as a rocky shore) separated by a barrier (such as a sandy or muddy shore) from other suitable areas (Fryer 1959).

Barriers in the Oceans. A detailed treatment of marine barriers has been given by Ekman (1953) and Hesse, Allee, and Schmidt (1951). Evidently barriers are different for pelagic (Brinton 1959), benthonic, and deep-sea animals. Mayr (1954b) has discussed the relative efficiency of various barriers for shallow-water echinoids. Pelagic animals and such as attach themselves or their eggs to floating algae or other flotsam have usually very wide ranges.

Barriers for Parasites. Lack of contact between various potential host species reduces gene flow between parasite populations as effectively as other spatial barriers. As Clay (1949) has pointed out correctly, this spatial isolation between hosts is the exact equivalent to geographic isolation in free-living animals. The nature of the barriers, permitting speciation in parasites, has been discussed in detail in Chapter 15.

Climatic Barriers. Borders of the range of species and of geographical isolates that are determined by the climatic tolerance of populations tend to be particularly labile. They are controlled by temporary conditions of temperature and rainfall. Isotherms are of decisive importance in temperate and polar regions, and rainfall (and correlated vegetational aspects) in the subtropical and tropical regions. Climates are subject to short-term and long-term fluctuations and this causes advances and retreats of such borders. Relict populations are often left behind in favorable situations during periods of range contractions. If conditions are otherwise favorable, these relics may, like other geographical isolates, reach species level. The ever-present expansions and contractions of species ranges in response to changes of climate are sometimes ignored in the evolutionary literature. A careful study of any climatically determined species border shows that it is in a state of dynamic stability. The general amelioration of the climate in the Northern Hemisphere during the first half of this century (Hustich 1952; Shapley 1953) illustrates this quite graphically. Numerous species of birds, mammals, and other animals have greatly expanded their breeding range toward the north, while some of the northern species have retreated from the south during the same period (Kalela 1942, 1944). Many existing isolates can be explained only as the result of former range expansions during periods climatically favorable for such expansion.

There is a class of transgressions of the former species border that cannot easily be explained either in terms of the breakdown of a barrier or a climatic change of the environment. Occasionally a species suddenly starts to expand its range explosively and advances in the course of this movement far into areas which only a few years previously had seemed totally unsuitable for the species. The spread of the Mediterranean Serin (*Serinus serinus*) through most of Europe is typical of this class (Mayr 1926), although a climatic change may have facilitated the expansion. The Collared Turtle Dove (*Streptopelia decaocto*), coming from India via Turkey to the Balkans, has similarly overrun most of Europe within a few decades (Fisher 1953). In such cases one must suspect that some genotypic change occurred in a peripheral population which permitted the precipitant expansion. The genotypic change, obviously, resulted in a change in climatic or habitat tolerance that permitted the subtropical species to enter the temperate zone.

The Efficiency of Barriers. Most geographic isolation is relative, few barriers being 100-percent efficient. It would be interesting to know how large the distance between isolated populations must be in order to permit the completion of speciation. How much gene flow is permissible? No one knows. The answer depends largely on the dispersal facilities of the particular kind of organism. Even the wide Atlantic is not a complete barrier. Every year some American birds and insects appear on the coast of Ireland and England after having been blown across the North Atlantic. An Old World bird, the Cattle Egret (*Bubulcus ibis*), colonized northern South America across the Atlantic around 1930 and has since then expanded its range into the Caribbean and the southern United States. Elton (1958) records the history of many invasions across considerable barriers. The very isolated Hawaiian Islands have received their entire fauna through transoceanic colonization from Polynesia and America. The Hawaiian bird fauna is the result of 14 colonizations. The rich insect fauna of the Hawaiian Islands, consisting of some 10,000 species, is the result of no more than about 300 colonizations (Zimmerman 1948). Simpson (1940) showed that the entire mammal life of Madagascar can be interpreted as being due to no more than five colonizations across the ocean separating Madagascar from Africa.

If barriers to gene flow are as important a factor in speciation as is now generally believed, one should find the greatest amount of active speciation in areas richest in geographic barriers. This is indeed the case. Regions which in any sense of the word are insular always show active speciation, whereas continental regions show speciation only where physiographic or climatic barriers produce discontinuities among populations. The thesis that speciation should be most active and rapid where natural barriers are most frequent and most efficient is richly substantiated by all the known facts (for further examples see Mayr 1942).

The Role of Intrinsic Factors

Geographic barriers are sometimes regarded as purely mechanical devices, like a dam that holds the water in a reservoir. Such an emphasis on the mechanical aspects of barriers is one-sided. The geographical features of a given area, its mountains, rivers, ocean straits, and treeless plains, affect the population structure of different species in very different ways. This can be explained only in part by the physical means of dispersal. To a large extent these differences are the result of "intrinsic factors" (Mayr 1942:238, 1949a:288), that is, of physiological and psychological properties, which cause every species to react differently to such barriers. The dispersal ability of individuals of a species, that is, their faculty of moving longer or shorter distances from their birthplace, controls to a large extent the establishment and maintenance of geographical isolates.

Dispersal in plants is normally the task of the seeds. In animals it may occur at almost any stage of the life cycle, but there are a number of regularities (Ghilarov 1945). If the adults are sedentary, as in many marine organisms, dispersal will take place during the larval stage, preferably through free-swimming larvae. If the adults are mobile, as is the case with most insects, the larvae tend to be sedentary. If the adults are subject to passive dispersal (for instance, aphids), leading to much waste and loss, there are often found means for accelerated reproduction (parthenogenesis).

Dispersal on land, in water, and in the air (Wolfenbarger 1946; Gislén 1948) is a universal phenomenon. Indeed, the dispersal facility of most animals is so great that one is sometimes surprised that there are biogeographic barriers at all. Even oceanic islands as isolated as the Hawaiian Islands have relatively so rich a fauna and flora that some biogeographers insist that these islands must have had continental connections. Still, it is obvious that there are limits to the dispersibility of every species.

No reliable method is known for measuring dispersal ability. Individual mobility is definitely one component of it, and this is why so many work-ers in recent years have attempted to calculate the average amount of dispersal per individual per generation, for example, Dice and Howard (1951) for *Peromyscus,* A. H. Miller (1949), von Haartman (1949), and Kluijver (1951) for birds, Stickel (1950) for the tortoise *Terrapene,* Burla and Greuter (1959), and Dobzhansky (1951) for *Drosophila,* Schoof *et al.* (1952) for the housefly, and Dowdeswell *et al.* (1949) for butter-flies, to quote a few publications in a fast-growing literature. Yet these investigations give us only an incomplete, indeed a decidedly misleading, answer to our question. Such dispersal studies are based on the arbitrary assumption that the individuals of a population obey in their dispersal the same laws that control the scattering of inanimate objects (Skellam 1951). Dispersal curves, however, are rarely normal (Bateman 1950); indeed, most animal populations seem to be composed of three classes of individuals: (1) those that scatter slowly and at random like inanimate objects, (2) those that have a definite tendency to remain where they are (philopatry), and (3) those that travel far greater distances than one would expect. Classes (2) and (3), which are responsible for the kurtosis and skewness of dispersal curves, are the manifestation of intrinsic factors. Some of these intrinsic factors facilitate the overcoming of barriers, others reinforce them. The capacity of a group of animals to speciate depends to a considerable extent on the relative strength of the two sets of factors.

Factors Facilitating Dispersal and the Crossing of Barriers. Dispersal may be active, passive, or a mixture of both. Probability of passive dis-persal is increased by numerous factors, some of which are listed by Simpson (1952a:168), such as small size, low specific gravity, protective coating, a dormant stage, and so forth. Species that are optimal for all these factors may have world-wide ranges, such as certain tardigrades (Fig. 18–2), rotifers, and fresh-water crustaceans. A successful cosmopoli-tan with an essentially panmictic species population is evidently barred from geographic speciation. A high dispersal ability is a necessity for occu-pants of temporary habitats, such as most bodies of fresh water. It is like-wise characteristic of most marine organisms. More than 70 percent of the bottom-living marine invertebrates have a pelagic larval stage. The ability to overcome barriers (for instance, the deep sea between shallow waters) depends on the length of the larval stage. Marine gastropods with long larval stages, such as *Cypraea, Conus,* and *Mitra* have far wider ranges in the Pacific than pelecypods with short larval periods. Most planulae of the reef-building coral *Galaxia aspera* settle within a week, but a few

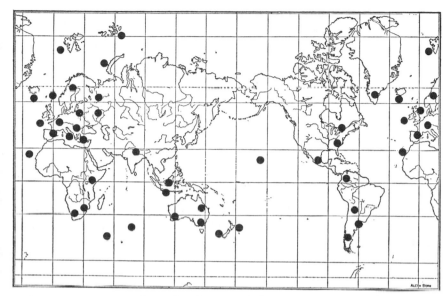

Fig. 18–2. Cosmopolitan distribution of the tardigrade *Macrobiotus hufelandii.*

float for more than 60 days (Atoda 1951). In an ocean current of 2 miles per hour (48 miles per day), such larvae could cover 3000 miles before settling. It is not surprising that species with such dispersal facilities do not speciate even in the vast area of the Pacific and that littoral species are found on such remote oceanic islands as Ascension, Saint Helena, and the Hawaiian Islands (Mayr 1954b).

Differences between species in the degree of active dispersal are even greater than differences in passive dispersal. Various ill-defined and poorly understood properties of species are responsible for these differences. For instance, the ability to cross barriers, and thus to establish or swamp geographical isolates, is often determined by psychological factors in the higher vertebrates. In 1856 the small white-eye *Zosterops lateralis* jumped the 2000-kilometer gap between Tasmania and New Zealand and proceeded to become the most common native songbird of New Zealand. In subsequent decades it successfully colonized all the outlying islands of the stormy New Zealand seas. Yet a close relative, the white-eye of the central Solomons, *Zosterops rendovae*, refuses to cross barriers only a few kilometers wide (Fig. 18–1) even though its flying equipment is essentially the same as that of the New Zealand bird. Similar differences in dispersal facility of morphologically equivalent species of drongos (Dicruridae) were described by Mayr and Vaurie (1948). The active fliers (*Himatione, Vestiaria,* and *Psittirostra*) among the Hawaiian honey creepers show no subspeciation from island to island, while the weak fliers form very pronounced isolates (Baldwin 1953). Fruit- and nectar-feeding birds which have to follow shifting food supplies show greater dispersal and less subspeciation than the more sedentary insect eaters.

The number of instances in which a species has actively crossed an important geographical barrier and established a new isolate, as the New Zealand white-eye has done, is probably quite large. However, only rarely will a naturalist be present to record the details of the invasion. Such a case is the invasion of Greenland in January 1937 by a flock of the Eurasian thrush *Turdus pilaris,* carefully documented by Salomonsen (1951). By 1949 the bird had become a common breeding bird of southwest Greenland. The histories of many other invasions (most of them due to passive dispersal) have been described by Elton (1958).

It is difficult to generalize on the dispersal facility of a species, not only because the dispersal drive changes with age, but also because populations appear to be polymorphic for dispersal restlessness. Long-distance travel of a minority of individuals has been established for mice, birds, and even flies. Such exceptionally restless individuals are, of course, far more important for the crossing of barriers than the average individuals of a population. The colonization of remote islands owes much to such enterprising individuals.

Factors Reducing Dispersal. The probability of passive dispersal is reduced by the reverse of the characters enumerated above as facilitating dispersal. Flightlessness increases the rate of speciation in wingless genera of carabid and tenebrionid beetles. Retreat into caves sets up new barriers and induces speciation. Reduction of the pelagic larval stage in marine animals makes local barriers more effective.

Again it appears that in many animals, particularly the higher vertebrates, psychological factors are the most important means of reinforcing geographical barriers. The Rendova White-eye (*Zosterops rendovae*) is a typical example of a species showing *philopatry* (an urge to stay at its native locality), even though a flight of only a few minutes' duration would carry the birds to the next island. Such a preference for staying on the home ground is widespread among animals. It has been documented for mice, lizards, turtles, snakes, fish, snails, butterflies, and, in fact, for nearly all species of animals in which the movements of marked individuals were carefully recorded.

Other intrinsic factors that reduce gene flow are the maintenance of territories and the ability of homing. The importance of homing in birds has long been known and explains the extreme localization of the ranges of many subspecies of geese (*Anser, Branta*), in spite of the enormous migrations of these species. Homing is well substantiated not only for birds and mammals, but also for reptiles, amphibians, and fishes. It occurs even among invertebrates, as, for instance, limpets (Villee and Groody 1940), snails (Edelstam and Palmer 1950), chitons (Crozier 1921), and crustaceans (Creaser and Travis 1950). It has been reported to occur in the tsetse fly *Glossina* (Jackson 1944) and may well be more widespread among insects than is now recognized. The ability of homing, combined with philopatry, severely restricts dispersal and helps to reinforce extrinsic barriers.

Degree of philopatry may differ in different populations of the same species. In Finnish populations of the Pied Flycatcher (*Ficedula hypoleuca*) only about 1 percent of the juveniles return to the area around their birthplace, while in Central European populations 20 to 35 percent return (von Haartman 1949). Finland is the most recently settled area in the range of the species and the Finnish population consists of descendants of immigrants that appear to have a low genetic component for philopatry and homing facility.

Any factor that reduces dispersal may facilitate speciation. Parental care, for instance, tends to increase philopatry and reduce dispersal sharply. The extremely late emancipation of the young in the geese is a good illustration. Mouth breeding in the cichlid fishes is another. By the time the young cichlids become independent, they have become thoroughly habituated to the very localized station of their parents. As a consequence, mouth breeding, philopatry, and great efficiency of habitat barriers are closely correlated in the cichlids (Fryer 1959).

The extreme localization of populations in potentially mobile species can be explained only by these intrinsic factors. This is true for cave bears (Kurtén 1955), butterflies (Lorković 1955), and certain snails, for instance of the genera *Achatinella, Partula,* and *Cerion.* The study of the sensory mechanisms and "psychological factors" facilitating such a regard for barriers is still at the very beginning.

Habitat Selection. Perhaps the most important of all the intrinsic factors leading to the localization of populations and to a restriction of species to their species-specific ecological niches is a phenomenon usually referred to as "habitat selection." Members of every species have the ability, as well as the urge, at the end of the dispersal phase in their life cycle, to choose as domicile an area that shows a constellation of environmental factors characteristic of the species. Every naturalist is familiar with habitat selection and takes it for granted. Occasionally it can be studied by direct observation. Often, when an individual of a species is forced out of its proper habitat, it makes every effort to return to it as quickly as possible. Habitat selection is particularly conspicuous in all animals that are cryptically colored and thus agree in their color with that of their substrate. If a pale, cryptically colored beach grasshopper, *Trimerotropis maritima*, is forced to take flight and is chased toward a grassy spot, it will not settle on the grass but circles back to the sand. Niethammer (1940), who experimented with the color preference of South African larks, made the following observations:

> It is very striking in southwest Africa that reddish larks are found only on red soil, and dark ones on dark soil, even where two completely different types of soil meet, as at Waltersdorf, for example, where dark soil, rich in humus, comes in contact with the red Kalahari sand. *Mirafra sabota hoeschi* stayed entirely on the dark soil, in spite of the fact that the area of the red sand began only a few hundred meters from its territories. On the other hand, I met *Mirafra africanoides* on the red sand up to its very edge but never on the dark soil, which was inhabited exclusively by *Mirafra sabota hoeschi*. Similar conditions prevailed at the Farm "Spatzenfeld" with the exception that the red sands here border light lime pans, where *Spizocorys starki* lives. I tried to chase little groups of *Spizocorys starki* to the red sands, but in vain—they turned before the beginning of the red soil and flew back directly, as if they knew, to their accustomed light lime soil. The red *Mirafra fasciolata deserti* which in Spatzenfeld inhabits the red sands did not go astray either in the lime pans. The reverse experiment I made in Lidfontein. I tried to chase the red *Mirafra africanoides gobabisensis* from a red dune to the light lime soil—again in vain. I do not believe that the experiment will ever have a different result because it is obvious that the birds are conscious of the color that corresponds to their own coloration.

This deliberate choosing of the proper habitat and the avoidance of unsuitable habitats serves as a most powerful reinforcement of geographical barriers (Palmgren 1938; Miller 1942). Habitat selection prevents the wastage of individuals in unsuitable habitats, as pointed out by Grinnell (1928:429), and is thus the ecological equivalent of the isolating mechanisms.

The nature and importance of the psychological factors underlying habitat selection were discussed among ornithologists first by Schnurre, Sunkel, Lorenz, Lack, and Palmgren, and more recently by Peitzmeier

(1942) and Svärdson (1949b). That marine invertebrates have an ability to settle on the appropriate substrate has long been known to marine ecologists (Verwey 1949; Thorson 1950). The question of the genetic contribution to this ability is largely unanswered. Do animals become conditioned or "imprinted" to the species-specific habitat during the juvenile or larval stages or do they get the "information" on the nature of the species-specific habitat in their genes? The work of Thorpe (1945) shows that a certain amount of larval conditioning is possible, more in some species, less in others. Yet a strong genetic component is becoming increasingly evident. Harris (1952) found that individuals of forest and prairie subspecies of the deermouse, *Peromyscus maniculatus*, bred and raised in standard laboratory cages, differ in their preference for various artificial habitats. *P. m. bairdi* prefers habitats that resemble grasslands, *P. m. gracilis* such as resemble woodland. The innate preference of parasites for their species-specific hosts and of many insects for their normal host plants is, of course, well established (Dethier 1954).

Habitat selection is a conservative factor in speciation since it reduces the probability that new isolates will be established beyond the present species border. The normal habitat of the species usually does not occur beyond the species border. If a species has the ability to change its habitat preference, it not only can expand its range but can also change genetically under the pressure of the new environment in the newly established geographical isolate. This creates conditions that are unusually favorable for rapid speciation. The great importance for speciation of such shifts in habitat preference was correctly recognized by Thorpe (1945) and other authors but has been quite often erroneously interpreted in the past as evidence for sympatric speciation. All speciation involves a greater or lesser amount of ecological transformation. It is important to realize that this is superimposed on the geographical changes, as Fryer (1960b) points out for the fishes of Lake Nyasa, and not an alternative to geographical speciation.

In sum, dispersal within the range of the species and beyond the species borders is not a purely mechanical phenomenon to be explained in terms of the physical means of dispersal. Intrinsic factors, such as philopatry, homing ability, restlessness, length of dispersal stage, parental care, habitat selection, and other psychological and physiological characteristics strongly influence the amount and the distance of dispersal. Together they control the establishment and maintenance of geographical isolates and are thus one of the determining factors in speciation.

COLONIZATION AND THE INVASION OF NEW NICHES

Dispersal facility alone is not sufficient to make the crossing of a barrier a success. It must be supplemented by an ability to find suitable habitats beyond the barrier and to colonize them. In this respect there appears to be a pronounced difference between plants and higher animals. Plants have high dispersal facilities but also make highly specific demands on their habitat and seem to find it difficult to colonize novel areas (weeds excepted). Animals, particularly vertebrates, find it more difficult to cross barriers but this is compensated for by a greater ability to cope with new conditions. The result is that the distribution of floras tends to be rather different from that of the vertebrate faunas. In the East Indies, for instance, the entire tropical belt from Malaya to New Guinea is essentially a floristic unit, more or less sharply distinguished from Australia. The higher animals, on the other hand, are clearly separated into a Malayan fauna to the west and an Australo-Papuan fauna to the east of Weber's line (Mayr 1953b). (Some animals, however, like those insects that depend on definite plants, may share the distribution pattern of their plant hosts.) The mobility of animals, which permits habitat selection, is one of the causes of their ability to cope with new areas. A newborn or young animal can search for a proper niche, whereas a plant seedling must succeed wherever it germinates. To what extent this difference in colonizing power is responsible for differences in the speciation pattern of animals and plants is still unknown.

The Invasion of New Niches

A colonizing organism that has crossed the current species border is often unable to find a habitat or ecological niche identical with that which it has left. It will not be able to establish itself unless it has the capacity for a shift in its ecological requirements or its niche. To understand such a shift, it must be realized that a species does not necessarily have a typologically fixed "demand" on its environment. There is no such thing as *the* ecology of a given species. As we have seen in Chapter 16, there is often geographical variation in habitat or niche requirements, and peripheral populations deviate especially often from the "norm" of the species. A typological approach is particularly inapplicable in all those species— and they are presumably the majority of plant and animal species—in which the species range is occupied by numerous discrete local populations, each adapted to its local environment and only loosely held together by gene flow (Mayr 1942, 1949a, 1951a). Even within a single population, individuals may differ in their ecological preferences and tolerances. It is, thus, evident that sufficient raw material is available in most species to permit some of the colonizing individuals to shift into new niches. A species must have a considerable amount of such adaptability, as well as a capacity for conditioning, in order to be able to make the jump into a new niche. Yet such a jump will not be successful unless the founders of the new population have a genetic constitution that will give success in the new environment.

The shift into a new niche must pass, in many cases, through a stage of ecological polymorphism. The new preference or tolerance is simply added to the already existing ecological characteristics of the population (Ludwig 1950). There is, however, a limit to the ecological versatility of a population and, if a new niche preference is superior to a previously existing one, it may lead to the displacement of the older one. When such a shift occurs in a geographical isolate and the isolate acquires reproductive isolation during a lasting period of geographic isolation, it will have acquired all the prerequisites of a new species.

The Environmental Constellation

The shift into a new ecological niche, the acquisition of a new ecological trait, is facilitated by some circumstances and definitely impeded by others. As was pointed out in Chapter 17, a shift into a new niche is favored near the species border. "The ecologically most aberrant populations are nearly always found along the periphery of the range of the species" (Mayr 1947). Two factors are principally responsible for this: the genetic reconstruction of these peripheral populations (Mayr 1954a) and the fact that the habitat, which in the center of the range is the preferred station, may not offer optimal living conditions near the periphery. A shift into a different niche may restore optimal living conditions under these circumstances. Geographic isolation of the population may facilitate the genetic consolidation of the ecological shift. Two factors of the biotic environment have been singled out by several recent authors as of particular importance for the chance of success of ecological shifts.

Competitors. The role of competitors in speciation has been stressed by several recent authors. It is obvious that a population will find it difficult to enter a new niche if this niche is already occupied by another species. Yet there is evidence that there are vacant or partly vacant niches even in well-balanced faunas. The spectacular success of faunal transfers illustrates this point (Elton 1958), as does the sudden breakdown of faunal barriers. After the Suez Canal was opened in 1869, fifteen Red Sea species of fish colonized the Mediterranean and some have become quite abundant. It is probable that these species filled vacant niches, because there is no apparent decrease in the frequency of any of the Mediterranean species. No Mediterranean fish succeeded in entering the Red Sea (Kosswig 1950).

There seems to be a pronounced difference between continents and islands in the availability of niches. Abundant empty niches are available on diversified islands when the first colonists arrive and this invites rapid adaptive radiation. The Hawaiian honey creepers (Drepanididae) had this opportunity and evolved finchlike, honey-eaterlike, creeperlike, and woodpeckerlike forms (Amadon 1950). The occupation of these diverse niches by the honey creepers prevented a similar adaptive radiation by the postdrepanid immigrants consisting of thrushes, flycatchers, and honey eaters. Speciation among these later arrivals was inhibited by the presence of already well-established competitors. A similar sequence of events has been described by Lack (1947a) for Darwin's finches on the Galapagos and by Zimmerman (1948) for the Hawaiian insects. We see again and again that an incipient species can complete the process of speciation only if it can find a previously unoccupied niche. This is rather difficult on continents with their rich faunas, and yet there is evidence of a continuous increase in the number of species even in old faunas (unless reversed by catastrophes). This is particularly true for the humid tropics, where seasonal stability and low demands by the physical environment permit strong niche specialization, resulting in great faunal richness and considerable radiation of closely related species.

Predators. The effect of predator pressure on the invasion of new niches is still in doubt. Worthington (1940) suggested that the difference in the number of species of endemic fish in various African lakes was due to differences in predation pressure. Mayr (1942, 1947) questioned this hypothesis and pointed out that the size of the various lakes and their ages were closely correlated with the number of species. Fryer (1959) shows that, in addition to the predators listed by Worthington, certain cichlid species also have become predators. Indeed Fryer believes "that the presence of predacious fishes in Lake Nyasa and other lakes has probably hastened and assisted speciation of nonpredatory species . . . the most abundant species will furnish the major portion of the diet of the predators, and this will thus facilitate the survival of those species whose ability to increase is less." The weakness of this argument is that it confuses absolute numbers and percentages. If among the fish of a certain place 950 individuals are of species *A* and 50 individuals of species *B*, and if the predator takes 95 *A* ("major portion of the diet of the predator") and 10 *B*, it takes only 10 percent of *A*, but 20 percent of *B*. Such a difference in rate of predation might be fatal to *B*. The truth is, presumably, somewhere between the extremes of Worthington and Fryer. Some retardation of speciation by predation is highly probable. An individual that attempts to shift into a novel niche is presumably far more vulnerable to predation than the member of a well-established population. Predators may well prevent populations from entering niches for which they are not particularly well adapted. The rather great frequency of successful shifts into radically new niches in the absence of predators that have occurred on oceanic archipelagos such as the Galapagos and the Hawaiian Islands seems to confirm this thesis. The simultaneous absence of competitors, however, on these oceanic islands makes a complete analysis impossible.

The shift into the new ecological niche and the need to become fully adapted to it sets up a considerable selection pressure. This leads to a genetic reconstruction of the population until it is ultimately genetically as well as ecologically so different from the parental population that the new species can return to its point of origin and coexist with the parental species without being seriously in competition with it. Although prevention of interbreeding with the parental species is dependent on the acquisition of isolating mechanisms, one should not minimize the importance of ecological factors in the process of geographic speciation.

ECOLOGICAL ASPECTS OF RATE OF SPECIATION

There is perhaps no other aspect of speciation of which we know as little as its rate.[*] Indeed, we shall probably never have very accurate information on this phenomenon. The splitting of one species into two is a short-time event which, as such, is not preserved in the fossil record. For information we rely entirely on inference.

The rate of speciation depends on three sets of factors: (1) the frequency of barriers, that is, of factors producing geographical isolates, (2) the rates at which geographical isolates become genetically transformed and more specifically at which they acquire isolating mechanisms, and (3) the degree of ecological diversity offering vacant ecological niches

[*] Rate of evolutionary change as such is considered outside the scope of this treatment. For excellent recent discussions see Simpson (1953a) and Kurtén (1958, 1959).

to newly arising species. Among these three sets of factors only the second has a genetic component. Yet even the rate at which a geographical isolate is genetically transformed depends only to the smallest extent on the rate of mutation in the gene pool of the isolate. Far more important are the degree of isolation (prevention of swamping by immigrants) and the "adaptation pressure," that is, the degree of selection pressure required to achieve perfect adaptation in the environment of the isolate. "Environment" includes not only the physical environment, but also the entire biotic one, such as new predators or the absence of old ones, and new competitors or the absence of old ones. It is evident from these considerations that ecological factors play a far greater role in determining rates of speciation than genetic factors. Nothing could illustrate this conclusion better than the explosive multiplication of species where the environment favors such a process, as in the Hawaiian Islands and in certain freshwater lakes.

Rates of speciation, broadly speaking, can be partitioned into rates of "transformation of isolates" and rates of "formation of discontinuities." Neither can be measured directly, although estimates are possible which may perhaps permit a determination of their order of magnitude.

Transformation of Isolates

Geological Evidence. So little direct evidence is available on the time required to transform a geographical isolate into a separate species that we must fall back on the geological evidence. This has been treated authoritatively by Zeuner (1946), Rensch (1960a), and particularly Simpson (1944, 1949, 1953a). In spite of the enormous difficulties encountered by such studies, the order of magnitude of the rates is beginning to become established. The age of geological strata is being determined increasingly precisely. Certain somewhat unrealistic assumptions cannot be avoided in such calculations, such as that the more recent forms (in a series of strata) are the direct descendants of the earlier forms and that the amount of morphological change reflects accurately the degree of speciation. This is often quite improbable, as, for instance, in the so-called rapid evolution (and speciation) of the ammonites. Presumably one must equate a much greater amount of phenotypic change with "speciation" in a group as morphologically plastic as the ammonites than in a group as morphologically static as the pelecypods.

The outstanding, yet somewhat disappointing, result of these studies is the extreme inequality of rates of transformation. The brachiopod genus

Lingula, which was very common during the Ordovician and Silurian, being known from these periods in several hundred species, extends into the present. It is often claimed that the soft parts of the modern genus are presumably very different from those of the fossil genus. Certain differences undoubtedly exist, but the impressions left by the complicated system of muscles, which permit locomotion and the closing of the shell, are virtually identical in the fossil and modern forms. It can be stated with conviction that there has been no conspicuous change in 440 million years, notwithstanding the fact that the specialist may propose generic separation. Some of the genera of Foraminifera are known from the Cambrian to the present. The horseshoe crab, *Limulus,* goes back at least to the lowest Triassic, 200 million years ago, and is known from most of the intermediate strata. Genera that extend back from the present to the Triassic are found also in other groups of marine animals, such as corals, ostracods, bivalves, gastropods, and bryozoans.

The more "primitive" the group of animals, the greater is usually the number of forms that exist for long periods. However, even among the more complicated vertebrates and higher arthropods, long-lived forms are recorded, such as are found, for instance, among the fishes: *Latimeria, Polypterus, Lepidosteus, Amia, Neoceratodus,* and the cyclostomes. There has been very little morphological change in the coelacanth and dipnoan fishes since the end of the Carboniferous, 250 million years ago (Schaeffer 1952; Westoll 1949). Some plant genera likewise have not changed for long periods, like *Gingko* for 80–100 million years. Long life is known even for species. The fairy shrimp *Triops cancriformis* is known from the upper Triassic to the present, which gives this species a life span of about 180 million years.

The opposite of such evolutionary inertia is the occurrence of veritable bursts of speciation, such as are found by paleontologists (Henbest 1952) and by the students of fresh-water lakes. The foraminiferan family Fusulinidae originated in the latest Mississippian, flourished in the Pennsylvanian, and died out in the Permian, not much longer than 50 million years after its origin. In this short period it developed 6 subfamilies, 48 genera, and more than 1000 species (Thompson 1948). Fusulinids are often more abundant in fossil faunas of this period than all other contemporary invertebrates combined.

Rates of evolution may be very different among close relatives living under identical conditions. Weismann (1902) pointed out long ago that among the Alpine butterflies which are obvious Pleistocene relicts some have reached subspecific difference since the separation, while others are still inseparable from their Arctic relatives. The same is true of the marine species which were divided into a Pacific and a Caribbean population when the last Central American portal was closed at the end of the Pliocene, about 2 million years ago. Among the crabs, for instance, 11 species have stayed identical while 13 species have split into allopatric pairs of subspecies or species.

Fossil mammals show perhaps the fastest known transformation of species (Zeuner 1946; Simpson 1953a), birds apparently changing far more slowly. Most Miocene nonpasseres belong to modern genera and many Pleistocene birds cannot even be separated specifically.

Geographical Evidence. Instead of using fossil evidence, the rate of transformation can also be determined by comparing allopatric populations that have been separated for a known period. Here one runs into the same three difficulties one had to face in the evaluation of the fossil evidence: (1) the date of separation can usually be arrived at only by inference, (2) only phenotypes can be studied, leaving the genetic basis of the differences undecided, and (3) the degree of correlation between the acquisition of morphological differences and genetic isolating mechanisms is unknown. The age of subspecies, thus, does not give us by extrapolation the age of species. Let us illustrate these difficulties on the basis of some concrete cases. The British Red Deer (*Cervus elaphus scoticus*) evolved during the 8000 years since the submergence of the English Channel. Yet when this form was introduced into New Zealand, it assumed within a generation or two the phenotype of the Carpathian Deer, thus becoming far more different from its parental stock than is the British Red Deer from the Continental race. Much subspeciation and perhaps even speciation (Gentilli 1949) in Australia is due to the isolation of populations, during an arid period, in more humid pockets ("drought refuges") along the periphery of the continent. Yet the timing of the arid period is still uncertain; it may have happened 4000 or 20,000 years ago (Keast 1961). The dating of eastern and western races and species of European birds is likewise highly controversial. The east-west pairs of races in *Sitta europaea, Aegithalos caudatus, Pyrrhula pyrrhula,* and *Corvus corone* are believed by Steinbacher (1948) to be of post-Pleistocene origin, while Stresemann (1919) accepts a Pleistocene date which seems more probable to me (Mayr 1951a). Some pairs of forest species, like *Certhia familiaris* and *C. brachydactyla,* or *Regulus regulus* and *R. ignicapillus,* often considered the product of glacial barriers, may well have arisen in a drought period either during an Interglacial or during the Pliocene.

Huxley (1942), Mayr (1942), Zeuner (1946), and Rensch (1947) give concrete dates on the probable ages of various isolates. A few examples may be cited. Cameron (1958) points out that the island of Newfoundland at the mouth of the Saint Lawrence River became habitable for mammals less than 12,000 years ago. Of the 14 species of native mammals, 10 have evolved well-defined subspecies in that period, some, like the Newfoundland beaver (*Castor canadensis caecator*), almost distinct enough to be considered a separate species. Some beaver populations on Newfoundland have become adapted to life in the virtually treeless barrens where they inhabit bog ponds and feed primarily on the root stocks (tubers) of the yellow pond lily. Several species of mammals have developed endemic subspecies on the islands of Great Salt Lake in Utah, isolated for 8000–10,000 years (Marshall 1940). Even more distinct are some of the rodents (*Clethrionomys*) on the Inner Hebrides, isolated for some 7000–9000 years (Steven 1953). Introduced European rabbits (*Oryctolagus cuniculus*) have become subspecifically distinct on several islands within about 500 years. Endemic races of the house mouse (*Mus musculus*) developed on a number of islands in historic times, in one case in as little as 100 years (Huxley 1942). The Faeroe Island form, introduced as recently as 300 years ago, has become so different that it has been considered a full species by some authors. The West African Green Monkey (*Cercopithecus aethiops*), introduced on Saint Kitts Island in the West Indies some 300 years ago, has become different in a number of characters (Ashton and Zuckerman 1950). A land-locked seal (*Phoca vitulina mellonae*) in the lakes east of Hudson Bay acquired its characters within about 4000 years (Doutt 1942). One may assume that similarly rapid rates of racial divergence occur among human isolates but no reliable data seem to exist.

Rates of subspeciation in birds (Mayr 1951a) may well be slower. However, the endemic races that live on the Anamba and Natuna Islands between Borneo and Sumatra, which owe their isolation to the post-Pleistocene drowning of the Sunda shelf, are almost certainly less than 10,000 years old (Stresemann 1939). The races that developed in Europe (Salomonsen 1931) and North America (Rand 1948) as a result of glaciation are considerably older, and this is presumably also true of some of the races of birds that owe their origin to vegetational changes in Australia (Keast 1961).

The number of estimates on the age of endemic subspecies is legion and only a few more will be presented. Four endemic subspecies of fish evolved on Isle Royale in Lake Superior after an isolation of 12,000–25,000 years (Hubbs and Lagler 1949). Far more rapid rates of speciation are made probable by the fish faunas of the African lakes (see Chapter 15) and of Lake Lanao in the Philippines (Myers 1960). Very rapid speciation since the Wisconsin glaciation is indicated by the fish fauna in the creeks and springs of the western North American desert (Hubbs and Miller 1948; Miller 1950, 1961). Numerous local races and even species seem to have originated since the last pluvial. The precise time of isolation can unfortunately not be determined. Extremely rapid evolution in isolated areas in the desert occurs in many groups of animals and even in plants. Iltis (1957) believes that the evolution of the plant genus *Oxystylis*, endemic in Death Valley, took place there in the extremely short time of 15,000 years.

New subspecies of grasshoppers, butterflies, and beetles have formed in northern Europe during a period of about 8000–15,000 years. The order of magnitude of the time interval is, thus, roughly the same as for vertebrates. Unfortunately, none of these rates of subspeciation tell us much about rates of speciation. Morphological differentiation, leading to the recognition of subspecies, is not a halfway point toward the acquisition of isolating mechanisms. Even in a species where it takes only 10,000 years to develop a well-defined island subspecies, it might well take 100,000 or perhaps 1,000,000 years for the completion of the speciation process. Our ignorance is nearly complete.

Rate of Acquisition of Isolating Mechanisms

Since speciation basically means the acquisition of isolating mechanisms, one way of determining rate of speciation is to determine the rate at which effective isolating mechanisms are acquired in geographically isolated populations. This, it turns out, varies from case to case. The frequency of zones of allopatric hybridization (Chapter 13) indicates that speciation is normally completed only slowly. Stebbins (1950:241) illustrates this by the fact that sterility barriers have not yet developed between the American and Asiatic populations in certain genera of plants such as *Platanus* and *Catalpa*, in spite of a geographic isolation that presumably has existed for many millions of years. Allopatric species of Australian frogs, believed to have been isolated since the Würm glaciation, have relatively slight call differences (Littlejohn 1959).

The other extreme is shown by species in fresh-water lakes where whole species swarms seem to have evolved in periods of apparently less than 1,000,000 years (Chapter 15). The same appears true of species in desert springs (Miller 1948, 1961). The time scale, unfortunately, is highly conjectural in these cases, yet it seems entirely possible that new species may originate under specially favorable circumstances in periods considerably shorter than 100,000 years, perhaps several orders of magnitude shorter.

Also unsolved is the problem of the degree of correlation between rate of phyletic change and rate of speciation. Phyletic change, as such, is affected by length of generation, rate of mutation, phenotypic variability, structure of the genotype (genetic system), population size, breeding system, and relation to the environment (see Chapter 14 and Simpson 1953a). To what extent rate of speciation is affected by these factors is still largely unknown.

It appears probable that speciation proceeds slowly, all other things being equal, in the absence of genetic revolutions. Where mere isolation by distance is involved or where the continuity between two parts of a large formerly continuous range is broken, selection will tend to prevent the destruction of the genetic cohesion, in spite of the centrifugal force of mutation and ecotypic adaptation. Even here the ultimate evolution of isolating mechanisms as a by-product of the steady genetic divergence is inevitable. However, it may well be a matter of millions of years.

The situation is totally different in the case of populations that have experienced a genetic revolution. Where a population, while passing through a bottleneck, has also acquired translocations, chromosomal fusions (or dissociations), pericentric inversions, and other structural chromosomal changes, the acquisition of isolating mechanisms is presumably greatly accelerated (Chapter 17). It would not surprise me if new species could arise, under these circumstances, in a period measured only in thousands or even hundreds of years. The constellation of factors that would have to be just right to permit such rapid speciation is, however, sufficiently improbable (in the statistical sense) that such a rapid rate will be exceedingly rare. This much, however, is certain, that there is no "standard" rate of speciation. Each case is different and the range between the possible extremes is enormous.

Factors Affecting the Rate of the Multiplication of Species

Formation of Discontinuities. Perhaps the broadest generalization which one can derive from all the figures on the rate of speciation is that this rate depends on an interaction of various factors and is therefore different in every species and genus. Some of these factors have already been discussed; others will be mentioned here. Among birds, for instance, large-sized species such as herons, storks, and hawks usually have fewer but more widely ranging subspecies and isolates than small songbirds. They also seem to extend further back into the geological past without changes transgressing the species level (Miller 1944b; Howard 1950). Sedentary birds have on the average twice as many subspecies as migratory species. Arboreal birds, such as vireos and warblers, have fewer subspecies per species than birds such as buntings or larks that live on the ground. The number of isolates would be a more significant figure than the number of subspecies but is unfortunately not available for many species (but see Keast 1961). A high number of subspecies indicates, however, a considerable localization of populations and reduced gene flow, both factors favorable to speciation. In a study of desert birds Hoesch (1953) showed that five species of larks and buntings that were dependent on water had only one subspecies each in southwestern Africa while six other species that were independent of water had an average of four subspecies in the same area. The species that are independent of water are far more sedentary than the dependent species. In a comparison of the rates of speciation in two groups of salamanders, Wahlert (1957) found that the Ambystomatoids, which normally spawn in standing waters and consequently generally live in comparatively flat country, have formed only one family with six genera and about 40 forms altogether. The Plethodontoids, however, which live in mountainous terrain and are easily separated by barriers, have formed several families and subfamilies with about 26 genera and more than 200 species and subspecies. The difference in population structure in the two groups of salamanders is reflected in the amount of speciation.

No other group of animals has such a large number of species as the insects. The reason must be an exceptionally high frequency of successful speciation, since the survival of species is quite evidently not nearly as long as in some other types of organisms, for instance brachiopods or pelecypods. The great potentiality of insects for specialization may hold the answer to this puzzling situation. A high percentage of species are either "host specific" (tied to a single species of plants) or at least oligophagous. Indeed, several different species may exist in different parts of the same host plant, in the flower buds, in the upper stem, in the lower stem, or in the root stalk. Three different species of mites of the genus *Acarapis* occupy three different body areas of the same host species, the honeybee. Several species can thus utilize resources that would in different circumstances be occupied by a single polyphagous or otherwise ecologically versatile species. Loss of wings and thus of mobility has resulted in a burst of speciation in flightless beetles and grasshoppers.

Mobility and the ability to select the proper niche have given animals the opportunity for high specialization without risk of the loss of colonizing zygotes in unsuitable locations. Insects have utilized this potentiality to a greater extent than any other group of animals and this is the reason for their high rate of speciation. At the other extreme is the species *Homo sapiens*, which can live in every environment from the Pole to the Equator, which flourishes on a pure meat diet (Eskimos) or on a virtually pure vegetarian diet, which can live the life of a hunter, a nomad, a farmer, an industrial worker, a minister, or a theoretical physicist. What other niche would there be available, if man were ready to speciate?

Fresh-water organisms seem to fall into two groups. Those that live in permanent bodies of water, with their comparatively high isolation from each other, tend to have a high rate of speciation. On the other hand, the species that are successful in colonizing temporary bodies of water have in general such superb dispersal abilities that their entire world population may well be nearly panmictic. Small islands are the equivalent of small bodies of water. The small coral islets of the Pacific are occupied to a considerable extent by a fauna that is exceedingly widespread and apparently steadily on the move from island to island. The small local populations are continuously polluted by immigrants from elsewhere; there is little isolation and not much speciation.

If one postulates that the frequency of speciation depends to a considerable extent on the frequency of unoccupied niches, one would expect very little speciation in the oceans, with their exceedingly stable environment and scarcity of barriers. What little we know about speciation in marine animals seems to indicate the truth of this assumption. Yet barriers are not altogether absent. Benthonic species, either of shallow waters or of the deep sea, have species populations that are often split into isolates by geographic barriers (Mayr 1954b). Although speciation is slow, it is regular and steady in these faunas. Speciation in pelagic animals is more of a problem. However, they are adapted to waters of particular temperatures, salinities, and pressures and thus associated with well-defined water masses which in turn are confined to definite ocean basins. It is probable that the physical conditions of these water masses produce distributional barriers which in turn are responsible for speciation in these pelagic forms (Brinton 1959). The relative importance of the various potential factors has not yet been determined (Harding and Tebble 1962).

Species that differ in their ecological tolerance differ in their potential for speciation.

A species which is so narrow in its requirements that it can survive anywhere only in the same narrow niche will have . . . little chance to break up into several species. The same is true at the other extreme, for a species with high dispersal faculties which is ecologically so tolerant that it will prosper in every conceivable niche, as for instance man. A species that has rather specific requirements at any one given locality, but which is sufficiently plastic to be able to shift into a new niche when colonizing new geographical regions, has the best chance for speciation (Mayr 1951a).

Miller (1956), in an analysis of speciation in western North America, has independently come to similar conclusions. Species with an intermediate amount of ecological tolerance show the greatest amount of incipient speciation.

Perhaps one can go one step further. There is evidence that species occasionally go through a phase of rapid expansion and greatly increased ecological tolerance. At the end of such a period of rapid range expansion such a species, as for instance among Australian birds the Sacred Kingfisher (*Halcyon sancta*) or the Rufous Whistler (*Pachycephala rufiventris*), are spread so widely and evenly within their habitat zone that there is little opportunity for the formation of isolates. Yet the special conditions that make a species successful at a given time are bound to change. The species may then lose some of its ecological tolerance and may have to retreat into isolated refuges. The continuous range will be converted into a fragmented range and in place of a single array of interconnected populations there will be a set of geographic isolates. There are indications that much speciation is due to such an alternation between a successful and a fragmented stage in the cycle of a species.

In periods of rapidly changing conditions several or many species will have synchronous cycles of expansion or contraction. This in part may explain some of the periods of rapid or of stagnant speciation described by paleontologists (Henbest 1952). A period of deteriorating conditions may initiate vast extinction, such as was experienced by the terrestrial tetrapods at the end of the Triassic and of the Cretaceous (Colbert 1958) and by many marine invertebrates during the Permian. Such extinctions, whatever their cause, must have left many vacant niches which favored an increase in the rate of speciation in the ensuing period. The importance of the availability of empty ecological niches cannot be better illustrated than by the veritable outburst of insect speciation that followed the rise of the angiosperms.

The question is sometimes raised whether there has really been enough time available in the geological history of the earth to produce the millions of species of animals and plants that are known to live and to have lived on the earth. A few simple calculations show, however, that this doubt has no foundation. If a species were to produce only four new species every 3 million years, half of which became extinct without further speciation, there would already be 65,000 species after 50 million years and this sum would double every 3 million years. This is by no means a particularly rapid rate of speciation since species seem to arise quite often in less than 1 million years in normally speciating groups and species may be budded off at the periphery of a parental species at many different places. The large number of geographic isolates (Chapter 16) that has been revealed by systematic analysis substantiates this point.

A number of broad generalizations seem to emerge from the analysis of the pattern of speciation in the various groups of animals. The rapidity with which an isolate is converted into a separate species depends on the size of the isolated area (number of contained demes), the ability of the population to shift into a new niche, the selection pressure to which the isolated population is exposed, and the effectiveness of the isolation. This in turn depends on the distance from potential sources of immigration and on the dispersal efficiency of the species. To this are to be added the genetic and cytological factors that have been discussed in Chapter 17. Although each of these factors is important, it is equally certain that their relative importance differs from case to case.

19 ~ Species and Transpecific Evolution

The nature and cause of transpecific evolution has been a highly controversial subject during the first half of this century. The proponents of the synthetic theory maintain that all evolution is due to the accumulation of small genetic changes, guided by natural selection, and that transpecific evolution (Rensch 1947) is nothing but an extrapolation and magnification of the events that take place within populations and species. A well-informed minority, however, which includes such outstanding authorities as the geneticist Goldschmidt (1940, 1948a, 1952a), the paleontologist Schindewolf (1950b), and the zoologists Jeannel (1950), Cuénot (1951), and Cannon (1958), maintain that neither evolution within species nor geographic speciation can explain the phenomena of "macroevolution," or, as it is better called, "transpecific evolution." These authors maintain that the origin of new "types" and of new organs cannot be explained by the known facts of genetics and systematics. As alternatives they advance two explanations which are in conflict with the synthetic theory: saltations (the sudden origin of new types) and intrinsic (orthogenetic) trends.

It is not the task of this volume, which centers around the evolutionary problems of the species, to refute these theories and to cover in detail the entire area of transpecific evolution. This has been done superbly by Simpson (1953a) with emphasis on the paleontological evidence and by Rensch (1947, 1954, 1960a) with emphasis on the general zoological evidence. Other aspects have been covered by Heberer (1957) and by contributors to the Princeton (Jepsen, Mayr, and Simpson 1949), Cold Spring Harbor (1959), and Chicago (Tax 1960a) conferences. They all agree that essentially the same genetic and selective factors are responsible for evolutionary changes on the specific and on the transpecific levels and that it is misleading to make a distinction between the causes of micro- and of macroevolution. If used at all, these terms should be considered purely descriptive. The manifestations of transpecific evolution are, of course, in many respects different from those of infraspecific evolution, even though the underlying mechanisms are the same. Only those macroevolutionary phenomena will be discussed in detail in this chapter to the elucidation of which the study of the species can contribute. For the treatment of other aspects of transpecific evolution, such as rates, trends, degeneration, the origin of phyla in the pre-Cambrian, and the origin of life, we refer to the above-listed authorities.

Simpson not only has shown very clearly that the transpecific phenomena can be interpreted in terms of the known facts and theories of genetics, but has also demonstrated that most of the deviant evolutionary theories are based on misconceptions of the actual course of evolutionary change. Some of the most glaring of these misconceptions are that evolution is normally rectilinear (orthogenetic), that new types come into existence with all the characteristics which the "type" shows after it has reached evolutionary maturity, that evolutionary rates are constant, and that ancestral forms of new lines are simple and unspecialized.

A consideration by the systematist (of recent species) of the problems connected with the origin of new higher categories supplements in many respects the approach of the paleontologist and the comparative anatomist. To state the problems of macroevolution in terms of species and populations as "units of evolution" reveals previously neglected problems and sometimes leads to an emphasis on different aspects. Thoday (1953) points out that the species is not necessarily the essential unit of evolution. For instance, where isolating barriers break down easily so that introgressive hybridization or allopolyploidy is frequent, the unit of evolution will comprise more than one species. On the other hand, in asexual organisms it is the individual that is the unit of evolution. All possible intermediates exist between these extremes. The lack of coincidence between taxonomic species and units of evolution is, on the whole far greater among plants than among the higher animals, where permanent asexuality is rare and attainment of species level is usually an irreversible step.

THE SPECIES AS A POTENTIAL EVOLUTIONARY PIONEER

The key role that the species plays in the evolutionary process is based on the following facts. Every species (1) is a different aggregate of genes which controls a unique epigenetic system, (2) occupies a unique niche, having found its own specific answer to the demands of the environment, (3) is to some extent polymorphic and polytypic, thus able to adjust to changes and variations of its total environment, and (4) is ever ready to bud off populations which experiment with new niches. Entire species or separate populations of species may, at any time, (*a*) acquire a new combination of genes, a new epigenetic system, which constitutes a novel, more successful adaptation to the environment, or (*b*) shift to a new ecological niche which is so favorable that it becomes an entirely novel adaptive zone (Simpson 1944); the genetic and the ecological shift usually go hand in hand. Every population that makes such a shift is an evolutionary pioneer and may become the founder of a new type, a new higher category.

It cannot be emphasized too strongly that the population is ultimately the key to every evolutionary problem and that any evolutionary theory that attempts to side-step the consideration of populations is doomed to failure. It is the fatal weakness of the saltationist theories that they operate with mutated individuals. In actuality it is not the individual but the local population in which selection leads to the increase or decrease of certain genes and gene combinations; it is the local population which becomes conditioned to new habits and which may acquire new ecological preferences; it is the local population in which all intraspecific competition takes place and which in turn is in competition with its sympatric competitors. In short, to repeat, it is the local population which is the key to the solution of all evolutionary problems. Every higher category must have ultimately originated as the local population of a species. The task before us, then, is to analyze the steps by which a population may become a new higher category, a new type.

THE ORIGIN OF A NEW TYPE

Different forms of life were referred to as "types" by the comparative anatomists of the last century and even earlier. Bats, whales, birds, penguins, snails, sea urchins, and all the other well-known kinds of animals and plants are such types. Remane (1952), Heberer (1957), and Simpson (1953a, 1961) have discussed the various categories of types recognized in the phylogenetic literature. Purely metaphysical constructs like the archetypes of the idealistic morphologists, which are based on Plato's concept of the eidos, have led to such confusion that systematists now use the term "type" * (if they use it at all) only in the strictly descriptive sense

---

* For the use of the term in zoological nomenclature, see Mayr, Linsley, and Usinger (1953).

of the totality of characteristics of a given taxon. There is thus essential coincidence between types, the systematist's taxa that they represent, and the structures that characterize them. It is evident that the origin of new types, the origin of new morphological and other biological characteristics, and the origin of the higher taxa are three problems that cannot be separated from each other. Indeed, they are merely three different aspects of the same problem. Whatever contribution is made to one of the three problems is also a contribution to the other two.

When trying to explain the origin of a new, different type, we are at once stymied by an inability to define unambiguously the term "different type." A bird is surely a different type from a terrestrial reptile, and a penguin a different type from an albatross. But is not the Emperor Penguin (*Aptenodytes forsteri*), with its superb adaptation to the rigorous life on the Antarctic ice, also a type quite different from the small, hole-nesting penguins of the southern temperate zone, and even from its subantarctic cousin, the King Penguin (*A. patagonicus*)? Is not every good species a separate type, considering all the numerous morphological, physiological, and other biological adaptations for the unique niche it occupies (Chapter 4)? Indeed, does not all the available evidence suggest that the differences between minor and major types are merely matters of degree? Every species is an incipient new genus, every genus an incipient new family, and so forth. And the same is true on the ecological side: there is an insensible gradation from the new niche to the new adaptive zone. And it can probably never be predicted which new niche is a cul-de-sac and which the entrance to a new adaptive zone. Simpson (1953a) says quite rightly, "The event that leads, forthwith or later, to the development of a higher category is the occupation of a new adaptive zone," and he continues, "the broader the zone the higher the category *when fully developed*," yet the new occupant when first entering the zone will hardly, if at all, differ from the parental population.

The higher categories with which the student of phylogeny deals are orders, classes, and phyla, but these are far too "fully developed" to tell us much about the first origin of a higher category. For this we have to study those lower categories that are incipient higher categories,* such as families, genera, and species. For species we have shown already (Chapter 16) how they originate from geographical isolates by acquiring not

---

* "Higher" in connection with categories is a very subjective designation. A paleontologist would hardly ever designate a category below the order a higher category, while neontologists are known to have referred to polytypic species as higher categories.

only isolating mechanisms but also sufficient niche differentiation to be able to coexist with the parental or with sister species. And some of these species are indeed different enough to qualify as potential higher categories. When we map the distribution of monotypic genera we find occasionally that they are nothing but highly differentiated geographical isolates. Mayr (1942) has cited such cases of the geographic variation of generic characters. For instance, the pigeon genus *Serresius* in the Marquesas Islands is nothing but a glorified subspecies of *Ducula pacifica;* in *Myiagra ferrocyanea* of the Solomon Islands, the width of the bill ranges in various subspecies from that of a narrow-billed warbler to that of a very broad-billed flycatcher. Vaurie (1952) reports the case of the geographic variation of the flycatcher species *Rhinomyias gularis* which completely bridges the gap between the families of flycatchers (Muscicapidae) and of thrushes (Turdidae). In fact, if the close relationship of the subspecies were not evident from the color pattern and the distribution, one might be tempted to classify some of the subspecies of this species with the flycatchers and others with the thrushes. In the family of the drongos (Dicruridae), every generic character varies geographically (Mayr and Vaurie 1948) and several races of *Dicrurus hottentottus* were originally described as separate genera. On the island of Celebes, the Sarasins (1899:229) found that several so-called genera or subgenera of snails were connected by intermediate populations and turned out to be nothing but subspecies of a single polytypic species as in *Nanina cincta* (Lea). In the sloth *Choloepus didactylus* geographic races differ in the number of vertebrae, usually considered in mammals a distinguishing feature of higher categories. In some groups of mammals there is geographic variation in the numbers of premolars or molars, while in others this number is a fixed character of whole higher categories. Geographic races of *Terrapene carolina* differ in the number of claws and phalanges (Baur 1893).

One of the most spectacular cases of the geographic variation of a basic character was described by Amadon (1947). In the Hawaiian genus *Hemignathus* (honey creepers) the bill is long and strongly curved (Fig. 19–1). In the species *H. obscurus* the lower mandible is about as long as the upper; in *H. lucidus* it is shortened and thickened. When creeping along the trunks of trees in search of insects, the bird uses this unique bill to pry or chip off bits of bark with the heavier lower mandible. The development is carried one step further in *H. (lucidus) wilsoni,* the geographical representative of *lucidus* on the island of Hawaii. Here the

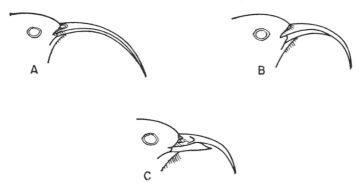

Fig. 19–1. Geographic variation of bill function in the Hawaiian honey creeper *Hemignathus lucidus*. $A = H. o. obscurus$; $B = H. lucidus hanapepe$; $C = H. (lucidus) wilsoni$. (From Amadon 1947.)

lower mandible has become straight and heavy and is used in woodpecker fashion like a chisel, opening an entirely new niche to the bird. This population was far more numerous than any *lucidus* population on the other islands and had all the earmarks of a new "type."

The Ocellated Turkey ("*Agriocharis*") of Yucatan is clearly a very divergent geographical representative of the common Turkey (*Meleagris*). In the sea-urchin genus *Brissus* two of the six allopatric forms of the Pacific, those of New Zealand and Japan, differ so much from the others in the teeth of the globiferous pedicellariae that specialists have considered placing them in a separate genus. The description of geographic races as different genera has also been reported in systematic botany. A case I have recently encountered is the cruciferan species *Smelowskia borealis* (Drury and Rollins 1952).

The objection may be raised that such genera do not count, that they are not true genera but merely artifacts of the taxonomist. To be sure, they are genera in a restricted taxonomic sense, based on taxonomic characters which, if poorly chosen, may be misleading. Yet there is no way in which it can be determined, in most cases, to what selection pressure the isolate owes its peculiar "generic" character and whether it may not be the key to the door of a new adaptive zone. Even so-called artificial characters quite often have the earmarks of an adaptive complex.

The geographic origin of genera is far more common than one would infer from the literature. The concealment of this interesting fact is largely the fault of the taxonomist. If an allopatric population is strikingly different, he may record it in the literature as a separate genus without stressing its obvious origin by geographic speciation. If it is not quite as different, he may list it as a subspecies, as the Sarasins did for *Nanina cincta,* without emphasizing that geographic variation has produced what is normally considered a generic character. It would be well worth while if taxonomists would go through their material and bring additional cases of this phenomenon to the attention of the evolutionists. In the case of most genera geographic overlap among species is completed before the generic characters develop.

Adaptiveness of Taxonomic Characters

The question of the adaptiveness, the "selective value," of taxonomic characters has been a source of bitter argument since Darwin's days. It is now possible to make a far more objective analysis than was formerly possible. A close analysis has revealed again and again that "useless," "deleterious," and "neutral" taxonomic characters give cryptic selective advantages. On the other hand, it is equally true that not all differences between species and genera nor certain diagnostic characters of higher categories are necessarily the result of an *ad hoc* selection for that particular component of the phenotype.

To use *Drosophila* as an illustration: the more than 600 known species of this genus all have three orbital bristles on either side of their heads, and the anterior of these bristles is always proclinate (bent forward), while the other two are reclinate (bent backward). Now, why should this character be retained so tenaciously in so many species? Is it really important for the flies of this genus to have one proclinate and two reclinate orbital bristles? (Dobzhansky 1956a).

There are usually numerous pathways open to achieve a certain biological end and it will depend on the particular genetic constitution of the incipient species which particular pathway is chosen. Nothing can be incorporated into the phenotype that definitely lowers fitness, but, to cite one example, the choice of prey in a parasitic wasp, the way it is handled, the building and provisioning of the brood cells, and many other aspects of the insect's life cycle and behavior have components that were "permitted" by natural selection rather than dictated by it. These are "correlated responses" of selection for other components of the phenotype. Only the adherents of the one-gene–one-character hypothesis insist that every aspect of the phenotype is the result of *ad hoc* selection.

Preadaptation

In order to be able to enter a new niche or adaptive zone successfully a species must be preadapted for it. The term "preadaptation" was coined by Cuénot during the heyday of mutationism. All evolutionary change was believed by the early Mendelians to be due to major saltations and the new "hopeful monster" (as Goldschmidt later called it) was either preadapted for a new niche or doomed to immediate extinction. This saltational concept of preadaptation, refuted by Simpson (1951a) and others, has been abandoned by the modern theory of gradual evolution.[*] A redefined concept of preadaptation is, however, useful in evolutionary biology: an organism is said to be preadapted if it is able to shift into a new habitat; a structure is preadapted if it can assume a new function without interference with the original function. Bock (1959b) has critically discussed the preadapted aspects of structures. Either type of shift may lead to the occupation of a new adaptive zone.

The Invasion of New Adaptive Zones. Habitat shifts of a minor nature, such as those of geographical isolates discussed in Chapter 16, require little special preadaptation and have little evolutionary potential. At the other extreme are shifts of fundamental significance such as those from aquatic to terrestrial, or from terrestrial to aerial life. Such shifts are possible only to the possessor of a highly unlikely combination of characteristics and this is the reason for the infrequency of such shifts. Let us look, for instance, at the shift from water to land. Although the surface of the earth is such a favorable and diverse environment that 85 to 90 percent of all species of animals live on land, they all belong to only 3 of the 35 known phyla of animals, the arthropods, the mollusks, and the higher vertebrates. The reasons for the infrequency of this ecological shift are obvious on the basis of the following considerations (Rensch 1947).

(1) *Weight.* Water is 800 times as heavy as air. Only animals with a strong skeleton or armor can become terrestrial. A jellyfish needs no support in water, but collapses completely as soon as it is brought on land. Ciliary locomotion, so common in water, is useless in air. Locomotion on land requires strong muscles.

(2) *Protection against the environment.* A land animal must be protected against the danger of drying out and against strong fluctuations of

---

[*] Cases of preadaptation in the sense of Cuénot do occur occasionally, such as the gene for melanism in moths from industrial districts and certain physiological genes (Huxley 1942).

temperature. There is a high selective premium in favor of a tough skin, armor, or scales.

(3) *The excretory system.* Permanent life on land as compared to an amphibious existence requires the excretion of metabolic end products in such a manner as to reduce water loss to a minimum.

(4) *The respiratory system.* It must be possible for the animal to take up oxygen from the air.

(5) *Sense organs.* On land, there is much greater need than in water for sense organs that are effective at long range, particularly among the rapidly moving land animals; hence the development of long-distance vision and hearing.

Although all of these preadaptations will continue to be perfected owing to greatly increased selection pressure, as soon as the amphibious mode of life is adopted, yet the very beginning of an amphibious life is impossible without minimal preadaptation. Individual species, genera, and families of many other phyla have gone on land, such as certain turbellarians (land planarians), nemerteans, rotifers, ostracods, nematodes, oligochaetes (earthworms), and land leeches. However, all of these are forced to live in a moist environment; none has succeeded in becoming truly terrestrial. In spite of all preadaptations the shift of the tetrapods from water to land was a slow and painful process. The first truly terrestrial reptiles did not appear until some 60–75 million years after the origin of the tetrapods (Romer 1957).

Each of the major groups of land animals has, sooner or later, attempted to reinvade the ocean. Among the reptiles, there are, for instance, the sea turtles, the sea snakes, several partly marine crocodilians, a marine iguanid on the Galapagos Islands, and four marine orders of fossil reptiles. Among the mammals are the seals, the whales, the manatees, and the sea otter. Among the birds are the extinct *Hesperornis*, the penguins, the auks, the diving petrels, and some semimarine groups like the cormorants. It is interesting that of the more than 800,000 species of insects only a group of chironomids has been able to reinvade the ocean, although some water bugs (*Halobates*) live on the surface of the sea.

Most invasions of new adaptive zones do not require anywhere nearly as formidable a set of preadaptations as the shift from water to land. To become a member of the rich fauna that Remane (1951) discovered in the interstitial spaces of sea-bottom sand, small size, an ability to tolerate temporary oxygen deficiencies, and certain types of locomotion seem to be the major prerequisites. This extraordinary habitat has been colonized by some coelenterates, mollusks, holothurians, a mobile solitary bryozooan, various crustaceans, and nemerteans, and is one of the major areas of radiation for turbellarians, gastrotrichs, archiannelids, copepods, nematodes, and ciliates. The colonization of fresh water by salt-water forms and vice versa involves primarily preadaptations of water and salt metabolism.

Sometimes no active invasion of a new habitat is involved but only an ability to take advantage of a climatic or vegetational shift. This was the case when some line of browsing horses began to feed on grasses (grasslands had then already existed for millions of years) and gave rise to the prosperous branch of modern horses.

The origin of a new niche is invariably a stimulus for the origin of types inhabiting this niche. The evolution of land plants preceded the evolution of terrestrial faunas. The rise of the angiosperms was followed by the explosive evolution of insects and insect eaters. And the origin of any kind of higher organisms is almost invariably followed by the origin of some new species of parasitic arthropod, helminth, or protozoan. Among the 40,000 species of animals that occur in Germany (Arndt 1941) no less than 10,000 are parasites on the remaining 30,000.

Each major shift of habitat is an evolutionary experiment. Each of the successful branches of the animal kingdom, for instance the insects, the tetrapods, or the birds, is the product of such a shift, with the decisive step quite likely taken by a single species. "However, not all such shifts are equally successful. No spectacular adaptive radiation has followed the invasion of the sand niche by a coelenterate. The shift of a carnivore (Giant Panda) to a herbivorous diet has not led to a new phylogenetic breakthrough . . . The tree kangaroos, the return of a specialized line of terrestrial marsupials to arboreal life, likewise seem to have reached an evolutionary dead end" (Mayr 1960). Not every evolutionary experiment is a success; in fact most of them are failures.

Generalized or Specialized Pioneers

It is difficult to establish valid generalizations concerning the evolutionary potentialities of various types. Gigantic forms generally have little leeway left and rapidly become extinct. Impressed by the evolutionary failure of pterodactyls, dinosaurs, and titanotheres, Cope (1896:172) proposed the "law of the unspecialized," according to which every specialization is a dead-end street and genuine evolutionary advance is to be expected only from simple, unspecialized forms. So broad a generalization is not supported by the facts. Indeed, nearly all major evolutionary break-throughs were made by specially preadapted and hence highly special-ized forms. This is true of those fishes that gave rise to the tetrapods, and of the reptiles from which the birds and mammals originated. Admittedly, many of the generalized forms have survived for long periods, but most of them have not given rise to remarkable new types. A balanced evaluation might be to say that most specialization leads into blind alleys and that many unspecialized forms have given rise to successful specialized de-scendants, but, as Amadon (1943) and Romer (1946) have correctly em-phasized, some of the most highly specialized types have conquered ex-ceedingly important new adaptive zones. Cope's "law" does not have general validity. Remane (1952) has demonstrated this in detail with the help of numerous examples. On the contrary, since novel niches often require rather specialized preadaptations, it is the specialized species rather than the generalized which often has the better opportunity of entering such a niche.

Key Characters and Mosaic Evolution

According to the saltationist, "a new type comes suddenly into exist-ence and evolves harmoniously from that point on. The original ancestor, the archetype, lacks all the specializations of its descendants, and subse-quent evolution affects all structures at about the same rate. Missing links are in every respect half way between the types which they connect."

Paleontologists and comparative anatomists have shown that not one of these postulates is correct. The rates of evolution of different organs are often drastically different. Some may rush far ahead, while others stagnate. *Archaeopteryx*, the "missing link" between reptiles and birds, is a typical pseudosuchian reptile in nearly all of its characters, but in its feathers it is like a modern bird (Table 19–1). The first artiodactyl was, except for its tarsal joint, a typical member of the Condylarthra (Schaeffer 1948). The first hominid was, except for certain pelvic muscle attach-ments, a typical anthropoid. One of the oldest known amphibians, the stegocephalian *Ichthyostega* from the upper Devonian of Greenland, had as many (or more) fish characters as amphibian characters (Heberer 1957:874; Järvik 1955). The amphibians grade so insensibly into the rep-tiles that the assignment of certain fossils becomes rather arbitrary (Fig. 19–2). The same is true of the border between mammallike reptiles and the mammals. Since the various diagnostic characters of the mammals are acquired independently and successively (Simpson 1959b), assignment

Table 19–1. Mixture of reptilian and avian
characters in *Archaeopteryx* (from Heberer 1957).

Reptilian characters	Avian characters
Teeth	Feathers
Free tail vertebrae (20)	Furcula
Ribs simple, without processus uncinati	Pelvis with backward pubes
Brain simple, with small cerebellum	Large eyes
Metacarpals free	
Metatarsals free	
Ilia and ischia separated	

of a given fossil to one of the classes depends on the character chosen. Unwilling to cope with such a disorderly situation, at least one taxonomist has proposed in desperation to "segregate" all the links between reptiles and mammals into a separate taxon, in order to preserve the purity of mammals and reptiles (and the validity of the conventional class diagnoses).

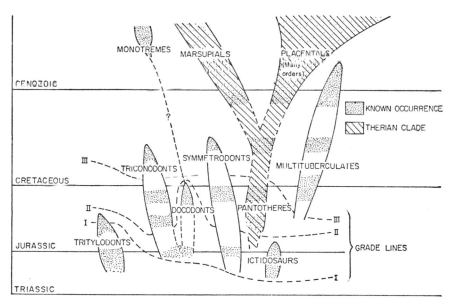

Fig. 19–2. Repeated and independent acquisition of mammalian characters (grades I, II, III) by various lines of mammallike reptiles (therapsids). The triconodonts, symmetrodonts and multituberculates became extinct. The docodonts probably gave rise to the monotremes, the pantotheres to the genuine eutherian mammals. (From Simpson 1959a; for details see Simpson 1959b.)

During the shift into a new adaptive zone, one structure or structural complex is usually under particularly strong selection pressure, like the wing in Proavis. This structure or complex then evolves very rapidly while all others lag behind. As a result there is not a steady and harmonious change of all parts of the "type," as envisioned by the school of idealistic morphology, but rather a "mosaic evolution." Every evolutionary type is a mosaic of primitive and advanced characters, of general and specialized features. A realization of this unequal evolution of the components of the type has been implicit in the work of outstanding paleontologists for nearly 100 years. Others, in their vain search for unspecialized archetypes, were forced to reconstruct quite absurd phylogenies. Remane (1952) analyzes the weaknesses of this idealistic approach, typified by Osborn's Proboscidian phylogeny where every Pliocene line is presented as a separate, direct radiation from a lower Eocene unspecialized ancestor (Fig. 19–3).

How can one recognize missing links if one fails to realize that they may be odd mixtures of advanced and primitive characters? When the South African ape-man *Australopithecus* was discovered, an outstanding anthropologist assured me that it was surely an aberrant side line and not anywhere near the direct human ancestry because it was a "disharmonious type," a composite of hominid and anthropoid characters, and consequently a failure and doomed to extinction. To be sure, *Australopithecus* was very much of a mixture of advanced and ancestral characters, yet it is now quite evident that man's ancestors must have passed through an *Australopithecus*-like stage (Chapter 20).

Which key structure is responsible for the evolutionary breakthrough is sometimes quite obvious. For birds it is the development of the feather, for flying insects the conversion of the paranota into wings (Lemche 1940), and for the artiodactyls the reconstruction of the tarsal joints (Schaeffer 1948). In other cases there are several structures involved, as lobed fins, lungs, and internal nares (Atz 1952) in the crossopterygians that gave rise to the earliest tetrapods. Sometimes it is a new behavior pattern, rather than a structure, which is the key invention that opens up the new adaptive zone. The adoption of internal fertilization in some branches of the amphibians and early reptiles was an absolute prerequisite for a truly terrestrial existence (Wahlert 1956). This, in turn, permitted the development of the hard-shelled amniote egg which, so Romer (1957) suggests, was developed by the early reptiles before they had become truly terrestrial. Perhaps adaptive shifts occur most commonly in

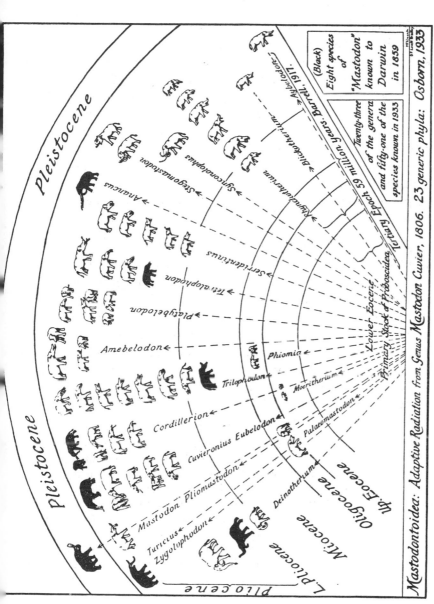

Fig. 19-3. Typological diagram of the evolution of the mastodons. None of the discovered fossils corresponded to the reconstructed archetypes and the branching of the lines had therefore to be pushed back to the lower Eocene. (From Osborn 1936.)

organisms that show—at first—no structural change at all. Many tree-trunk-climbing birds, for instance, lack the stiffened tail feathers of woodpeckers and true tree creepers. And the more primitive woodpeckers got along perfectly well with a normal perching foot; only the more advanced genera have evolved specialized foot structures (Bock and Miller 1959). If several lines evolve from such an unspecialized ancestor, the same structure may have different rates of change, even though the various lines live in essentially the same adaptive zone, such as various contemporary genera of grazing horses (Simpson 1953a). Differences in niche utilization, in biotic environment, and in the response of the various genotypes to selection account for the different responses to very similar selection pressures.

THE ORIGIN OF HIGHER CATEGORIES

The hierarchy of categories that the classifying taxonomist recognizes is an attempt to express similarity ("characters in common") and common descent. The methods employed and the pitfalls encountered in this endeavor are excellently discussed by Simpson (1953a, 1959a, 1961). The most similar species, the "most closely related" species, are combined into genera, groups of similar genera into subfamilies and families, these into orders, classes, and phyla. The origin of this hierarchy of categories has been interpreted differently by the adherent of the synthetic theory of evolution and the saltationist. The viewpoint of the latter has been stated particularly clearly by Goldschmidt (1952a):

A phylum consists of a number of classes all of which are basically recognizable as belonging to the phylum but, in addition, are different from each other. The same principle is repeated at each taxonomic level. All the genera of a family have in common the traits which characterize the family; for instance, all genera of penguins are penguins. But among themselves they differ from genus to genus. So it goes on down to the level of species. Can this mean anything but that the type of the phylum was evolved first and later separated into the types of classes, then into orders, and so on down the line? This natural, naive interpretation of the existing hierarchy of forms actually agrees with the historical facts furnished in paleontology. The phyla existing today can be followed farthest back into remote geological time. Classes are a little younger, still younger are the orders, and so on until we come to the recent species which appear only in the latest geological epochs. Thus, logic as well as historical fact tells us that the big categories exist first, and that in time they split in the form of the genealogical tree into lower and still lower categories.

With this interpretation Goldschmidt has fallen into the error of considering these categories as something natural rather than (particularly in the crucial area of branching) a man-made artifact. Our recognition of a higher category and its designation, delimitation, and placement in the hierarchy have a large arbitrary component. We would never consider *Archaeopteryx* the representative of a separate class of vertebrates if this branch of the reptiles had not given rise to the flourishing multitude of birds, outnumbering by far all the "other" living reptiles. "A higher category is higher because it *became* distinctive, varied, or both to a higher degree and not directly because of characteristics it had when it was arising" (Simpson 1953a).

The earliest representatives of a new higher category would certainly be included in the category from which they originated if it were not for their known descendants. The fossil genera of placental mammals, for instance, that are known from the early and middle Paleocene are decidedly less different from each other than all the known marsupials which always have been classed as a single order. "If we knew no placentals after the middle Paleocene we would certainly place them in a single order. *As of then* their proper comparative categorical rank was in fact that of an order. They are placed in six different orders because we recognize in them ancestors and allies of what *later* became six orders" (Simpson 1953a). It is not true that a new category arises as an order, class, or phylum. It arises as a new species and eventually becomes a new genus which we assign to a new order only because its subsequent descendants show the degree of distinctness and of discontinuity (after much extinction) which by convention is considered to signify ordinal rank. There is an early Paleocene genus, *Protogonodon*, in which on balance of resemblance in small details some species would be classified, according to Simpson, as carnivores and some as ungulates. The experienced specialist can see here the beginnings of two orders within a single genus. For the malacostracan crustaceans of the Carboniferous, Glaessner (1957) states that they were very "close to each other and to a generalised ancestral type . . . and but for our knowledge of their further differentiation into their recent descendants . . . they would have been placed in a single order of the Malacostraca."

Where a new group is the result of a broad and varied adaptation, it may arise cryptically and its distinguishing characters may not become apparent until some time after the group has branched off. The origin of the mammals is a prime example of this. In other cases, the origin of a new group is very closely correlated with, and even sometimes preceded by, an evolutionary invention, the origin of an evolutionary novelty.

The Emergence of Evolutionary Novelties

In the preceding discussion of the origin of new types, the emphasis has been on the shift into new habitats owing to a generalized preadaptation for such a shift. The more specialized the new adaptive·zone is into which an evolutionary pioneer shifts, the more likely it is that the shift will be correlated with a specific structure. This is exemplified by the rise of birds based on the "invention" of a wing consisting of individual feathers (rather than of a wing membrane, as in virtually all other flying or gliding animals). It is sometimes asked, how can such a new structure come into being, except by the saltational origin of the new type that has them? Furthermore, how can an entirely new structure originate without a complete reconstruction of the entire type? And how can it be gradually acquired when the incipient structure has no selective advantage until it has reached considerable size and complexity?

I have treated these questions in detail elsewhere (Mayr 1960) and will here only summarize my findings. We may begin by defining evolutionary novelty as any newly acquired structure or property that permits the performance of a new function, which, in turn, will open a new adaptive zone. Many evolutionary novelties, such as new habits and behavior patterns, are not primarily morphological although they may have morphological consequences. The study of the biochemistry of metabolic pathways has revealed the occurrence of evolutionary novelties that are as important as are such new structures as lungs, extremities, or brood pouches. Chemical inventions on the cellular level are the prerequisite of some of the most important adaptive shifts. Alas, our knowledge of comparative biochemistry is still far too rudimentary to tell us whether or not it was a biochemical invention that gave the mollusks, crustaceans, and other now dominant groups of marine invertebrates their ascendancy over eurypterids, trilobites, graptolites, and brachiopods, once the rulers of the seas. Biochemical inventions have in general a comparatively simple genetic basis (Wagner and Mitchell 1955; Sager and Ryan 1961) and their acquisition by natural selection is no major evolutionary problem.

The real problem concerns new structures, like the bird feather, the mammalian middle ear, the swim bladder of fishes, the wings of insects, and the sting of aculeate hymenopterans. Three modes of origin of new structures have been suggested.

(1) They may be initially the incidental phenotypic by-product of a genotype selected for other contributions to fitness. Since most if not all

genes are pleiotropic in their effects, it is possible that an evolving genotype may result in an incidental change of the phenotype that offers
selective advantages.

(2) A preexisting structure may be modified owing to an "intensification of function." Sewertzoff (1931) has made a special study of this
process. An intensification of the running function has led to the conversion of the five-toed foot of primitive ungulates into the two-toed foot
of the artiodactyls or the one-toed foot of the perissodactyls. Looking at it
somewhat differently, one could also call this an "intensification of selection pressure." Since this pressure is directed against an existing structure, "intensification" does not lead to the emergence of anything that is
basically new and yet it may result in such a drastic reorganization of
the phenotype that the first impression is that of the emergence of an entirely new organ. During this process there is never a stage in which the
incipient structure is "not yet of selective value," as is sometimes claimed
by antiselectionists.

When we compare birds or mammals with their strikingly different
reptilian ancestors, we have trouble in finding any truly new structures.
Most differences are merely shifts in proportions, fusions, losses, secondary duplications, and other modifications which do not materially
affect what the morphologist calls the "plan" of the particular type. An
intensification of function of one component as compared to others has
played a decisive role in most of these modifications.

(3) The most important cause of the origin of new structures is a
"change of function." This factor, clearly described by Darwin (1872),
was specially emphasized by Dohrn (1875) and analyzed in detail by
Plate (1924) and Sewertzoff (1931). The origin of a new structure
through change of function depends on two prerequisites: the capacity of
a structure to perform simultaneously two functions (and more specifically its preadaptation for a second function) and the duplication of
function. Many primitive fishes, for instance, had two independent organs
of respiration, gills and primitive lungs. In the terrestrial tetrapods the
simple baglike lungs were converted into the highly complex organ of
mammals and birds, and the gill arches and pouches into endocrine
glands and accessory organs of the digestive system. In the later fishes the
gills were elaborated and the lungs often converted into a swim bladder
or accessory sense organ. Cases of drastic changes of function of an organ
are found in every phyletic line of animals. To give a complete catalogue
would mean listing a good portion of all animal structures. The change

of the ovipositor of bees into a sting, of scales into teeth, of the Daphnia antennae into paddles, and of arthropod legs into mouth parts or copulatory organs are a few examples. In all these cases the structure evolved under the selection pressure of the primary function until it was large enough to take on the additional secondary function. The finding that not every function requires its own executive structure invalidates one of the favorite arguments of the antiselectionists. A new function can make use of a structure that had evolved for a different purpose.

The analysis of the origin of new structures is still in its beginning. The arthropods, in particular, are rich in structures the history of which is still quite obscure. How did the pockets for mites that are found in certain bees (*Xylocopa*) and wasps evolve? The auxiliary structures that are often needed to facilitate aberrant behavior patterns in reproduction, such as secondary genital openings in the females of various groups of insects and various receptacles for sperm, are particularly difficult to explain.

The Role of Behavior in Evolutionary Shifts

A shift into a new niche or adaptive zone is, almost without exception, initiated by a change in behavior. The other adaptations to the new niche, particularly the structural ones, are acquired secondarily (Mayr 1958, 1960). With habitat and food selection—behavioral phenomena—playing a major role in the shift into new adaptive zones, the importance of behavior in initiating new evolutionary events is self-evident. Sibling species, in spite of their morphological similarity, often show remarkable behavioral differences. Most recent shifts into new ecological niches are, at first, unaccompanied by structural modifications (Robson and Richards 1936). Where a new habit develops, structural reinforcements follow sooner or later. Miller (1950) has shown how the characteristics of the thrasher genus *Toxostoma* can be gradually traced from a mockingbirdlike arboreal ancestor. As the various species of thrashers became more and more terrestrial and more and more adapted to scratch on the ground and to use their bills for digging and removal of leaf litter, there was a selective premium for a strengthening of tarsus and toes, and for an increased curvature of the bill.

The same is true of special structures that facilitate specific behavior elements (Lorenz 1943; Tinbergen 1951). There are many birds that raise their nape feathers, but only some of these have developed elongated crests in this area. However, there are no birds known that have developed a crest without the ability to raise it. This principle is most easily demonstrated with respect to courtship patterns in birds and the accumulation of morphological characteristics that make the courtship more conspicuous and stimulating.

This is not the place to discuss how the behavior changes themselves originate, a problem still poorly understood. Explanations and interpretations are given by Tinbergen (1954), Thorpe (1956), Hinde (1959), and various authors in Roe and Simpson (1958), to mention merely a few titles. The point that is important for us is that new habits and behavior always start in a concrete local population. If the new behavior adds to fitness, it will be favored by selection and so will be all genes that contribute to its efficiency. That new habits occur all the time in natural populations is abundantly documented in the natural history literature. A particularly striking example is the recently developed habit of British titmice, mostly *Parus major* (Fisher and Hinde 1948), of opening milk bottles and drinking the cream. If the milk bottles had been a natural unoccupied niche, it is evident that a selection pressure would have been set up on one hand for the titmice to develop a more efficient milk-bottle opener, and for the milk bottles to become less easily opened, assuming the milk bottles to be organic material that could be modified with the help of selection. The phylogeny of breathing behavior has been studied by Spurway and Haldane (1953). The enormous role played by behavior in initiating transpecific evolution is increasingly being appreciated by evolutionists.

An analysis of complex interactions of instincts and structures has shown again and again that they also fit our explanation. This is true of the New Zealand cave gnat (*Araschnocampa*) with its luminous larvae (Goldschmidt 1948b; Mayr 1960), the Yucca moth (*Pronuba*), and all the marvelous adaptations of parasites and symbionts. Two aspects are characteristic for most of these complex systems. One is that there is a key structure or key behavior for which all other components of the adaptation are merely facilitating features, and the other that the selection pressure in favor of a smooth operation of the species interaction is usually far greater on one than the other partner. To find a chemical that will induce a plant to grow a protective gall is of far greater immediate selective advantage to the gall insect than it would be to the plant not to have to produce the gall. Myrmecophily, owing to the development of mechanisms for escaping predation by ants, will not be strongly selected against, as long as the myrmecophiles remain infrequent (Hölldobler 1949). Any adaptive complex, no matter how "purposive" at first sight, can be interpreted in terms of a balance of selection pressures.

Ontogeny and Evolutionary Novelties

Many preevolutionary embryologists and philosophers thought that development was an unfolding (*evolutio*) of the (preformed) type and that imperfectly formed types had existed in former geological periods. It was Haeckel who combined these earlier ideas of Bonnet, L. Agassiz, and others into a formal "biogenetic law" according to which ontogeny is a recapitulation of phylogeny. The invalidity of this law has been demonstrated so often and so conclusively that it is easy to fall into the opposite extreme and ignore the fact that many organisms that are highly dissimilar as adults go through similar larval or embryonic stages. The gill arches of bird and mammal embryos are a typical example. An "explanation" has been advanced that the genes which lead to the adult stages are the last that were incorporated into the genotype, and, being superimposed on the "older" genes, they are unable to affect the course of ontogeny. This explanation, though in the right direction, does not really explain anything. It is more likely that the longer a gene is part of the genotype, the more completely it becomes a part of the epigenetic system and the earlier its pleiotropic effects will play a role in ontogeny. Such "deep-seated" characters as the gill arches of the embryonic tetrapods are developmental by-products of a well-balanced complex of pleiotropic genes which control vital developmental processes more or less as a unit. Even if we assume that the structures which are produced by the embryonic gill arches could be produced directly, it is, so to speak, far simpler for the organism to retain these unnecessary aspects of the phenotype than to destroy the harmonious gene complex that controls development. Where ancestral components of the phenotype are lost, owing to a reconstruction of the genotype (in connection with newly acquired specializations), they may be restored when these specializations are later lost again. The secondarily flightless Ratites show a number of primitive avian characters which are usually explained by the claim that the Ratites are flightless descendants of very primitive birds. I wonder whether some of the primitive characters did not reemerge during the genetic reconstruction that accompanied the loss of flight.

The larval and juvenile stages are sometimes more drastically affected by evolutionary innovations than the adults. The pelagic larvae of marine invertebrates, for instance, are exposed to high selection pressures. Corresponding larval stages of different groups may show convergences or differences that are not indicative of degree of relationship. The primary biological significance of free-swimming larvae is that of dispersal. However, most of the larvae feed during dispersal and this opens a new niche for the species. The more favorable it is, the greater will be the pressure to prolong this stage until finally the more or less fixed adult stage may be altogether eliminated and reproduction taken over by the "larvae" (neoteny). The origin of the chordates from sessile (Berrill 1955; Romer 1958) or burrowing (Steiner 1956) prechordates is explained by recent authors in this manner. Innovations in the form of growth, straight versus coiled, symmetrical versus asymmetrical, simple versus branched, such as distinguish related forms of marine invertebrates, are often originated in early juvenile stages, as particularly emphasized by Schindewolf (1950b). Later growth stages many revert to the ancestral growth pattern, until the proper genetic background has been selected that will permit carrying the new growth form to adulthood.

When discussing the role of ontogeny in evolution some phylogenists treat the adults as if they were the descendants of the juveniles. In fact there are diagrams in books of "typogenesis" in which evolution is pictured as a spiral in which each adult stage is the direct descendant of the juvenile stage. These authors forget that the juvenile phenotype and the adult phenotype are expressions of the same genotype. It is a matter of selection pressure (including that exercised by the epigenetic system as a whole) which determines whether the juvenile or the adult phenotype is modified more rapidly and more drastically.

Rensch (1954, 1960a) has given the best modern discussion of the manifestations of evolution at the various stages of the ontogenetic cycle. He describes the manifold ways in which evolution can affect development and considers critically the various evolutionary theories based on ontogenetic phenomena.

EVOLUTIONARY POTENTIAL AND PREDISPOSITION

The reevaluation of mutation as the source of variability rather than a direct evolutionary force and of the character as a product of the whole genotype rather than of a single gene has changed the interpretation of many evolutionary phenomena. The new concepts make it easier to understand why there are such definite limits to variation in any given group of animals and why a change in one character often produces such serious "correlated effects" (Chapter 10), phenomena that had puzzled

Darwin greatly. Every group of animals is "predisposed" to vary in certain of its structures, and to be amazingly stable in others. I have pointed out earlier (Mayr and Vaurie 1948) that whenever the plumage of drongos (Dicruridae) varies it is either the frontal crest, the tail, or both that are affected. Certain groups of mammals have a predisposition to develop horns on the forehead, others on the top of the head, others not to have horns at all. In some the premolars are very stable and the molars highly variable in size and number; in other groups the reverse is true. In some groups of animals striped patterns are common, in others spots. The characters that vary in *Drosophila* are totally different from those that vary, let us say, in grasshoppers (Spurway 1949). Indeed, taxonomy would be difficult if it were not for these inherent differences between groups.

Only part of these differences can be explained by the differences in selection pressures to which the organisms are exposed; the remainder are due to the developmental and evolutionary limitation set by the organisms' genotype and its epigenetic system. Earlier authors (for example, Haecker 1925) had a far greater interest in the phenotypic potential bestowed on an evolutionary line by its epigenetic heritage than recent evolutionists. This must be kept in mind when we speak of the randomness of mutations. Mutations are random with respect to the environmental constellation. However, the epigenotype sets severe limits to the phenotypic expression of such mutations; it restricts the phenotypic potential. The understanding of this limitation facilitates the understanding of evolutionary parallelism and polyphyletic evolution. The evolution from therapsid reptiles to mammals in the Triassic and Jurassic proceeded so slowly and gradually that the four diagnostic differences between reptiles and mammals changed individually rather than as a block and the mammalian grade was acquired independently by at least four and possibly as many as seven to nine lines (Simpson 1959b). In the strict technical sense mammals thus have originated polyphyletically, yet all the lines that led to the mammalian grade came from the same group of therapsid reptiles and shared, no doubt, a gene complex and epigenetic system that was responsible for the subsequent parallel evolution in the jaw and middle-ear region. The argument whether or not the mammals have a monophyletic basis becomes, under these circumstances, largely a matter of semantics. Does the questioner refer to the phenotype or the genotype?

The felicitous term "grade" was introduced into the evolutionary literature by Huxley (1958), following Simpson (1949), to designate "a step of anagenetic advance, or unit of biological improvement." Several related lines may reach the same adaptive or structural grade independently. It is, of course, quite unreasonable to demand that major levels in phyletic evolution should always coincide with the branching of lineages. Often they do, but often they do not. Arkell and Moy-Thomas (1940) have listed many examples where parallel lines go through the same stages of evolutionary change independently and where a "horizontal" classification according to grade is far more practical than a phylogenetically more correct "vertical" classification. There are numerous cases of such parallel evolution in the animal and plant kingdoms (see, for example, Schaeffer 1956; Osche 1958; Shideler 1952).

True parallelism is due to response of a common heritage to similar demands of the environment (similar selection pressures). Where no common heritage exists evolutionary parallelism is more correctly called convergence. The animal world is full of convergences (and so is the plant world!) where similar demands by the environment have evoked similar phenotypic responses in unrelated or at least not closely related organisms. Kosswig (1948), for instance, has shown how many times internal fertilization, live-bearing, and other biological specializations have occurred independently among fishes and have led to numerous structural similarities. Many of the higher categories are unnatural groupings of unrelated animals that have become very similar owing to convergence. In the early days of Mendelism there was much search for homologous genes that would account for such similarities. Much that has been learned about gene physiology makes it evident that the search for homologous genes is quite futile except in very close relatives (Dobzhansky 1955a). If there is only one efficient solution for a certain functional demand, very different gene complexes will come up with the same solution, no matter how different the pathway by which it is achieved. The saying, "Many roads lead to Rome," is as true in evolution as in daily affairs.

And where there is a vacant adaptive zone, it serves as a powerful magnet for those preadapted for it. Various groups of aquatic arthropods have colonized land independently at least eight times. It is particularly significant that most of these invasions took place in the Devonian and Carboniferous, as soon as the development of a flourishing land flora made the land favorable as a habitat for terrestrial animals. Vacant, or at least only partially filled, niches are the reason for much evolutionary convergence. The evolution of plants placed a premium on herbivores. The availability of abundant herbivores sped the evolution of carnivores. When grass seeds became abundant after the evolution of grasses some five or six groups of birds became grass-seed eaters, that is, "finches."

Baldwin Effect

It is frequently stated that the development of a new evolutionary type or the adaptation to a new niche is facilitated by the so-called Baldwin effect, first proposed by Baldwin (1896). (Lloyd Morgan and H. F. Osborn published the same thesis in the same year.) The Baldwin effect designates the condition in which, owing to a suitable modification of the phenotype, an organism can stay in a favorable environment until selection has achieved the genetic fixation of this phenotype. Baldwin describes his "organic selection" (as he calls it) as a strict alternative to natural selection, which he disclaims as an evolutionary force owing to its purely negative effects (as he says). The principle was proposed four years before the rediscovery of particulate inheritance as an attempt at a reconciliation between neo-Lamarckism and Weismann's neo-Darwinism. Simpson (1953b), who gives an excellent critical discussion of the Baldwin effect, points out that it involves three steps (rephrased by me):

(1) The genetically determined reaction norms of individual organisms permit the development of behavioral, physiological, or structural modifications of the phenotype which are not hereditary as such, but which are advantageous for survival and permit the descendants of the organisms that have them to continue in the given environment;

(2) Mutations (and gene combinations) occur in this population that produce the favored phenotype obligatorily and rigidly rather than as a facultative modification;

(3) The genetic factors under (2) are favored by natural selection and therefore spread in the population over the course of generations until the facultative character becomes obligatory and fixed.

Simpson points out correctly that this hypothesis is, of course, no reconciliation between Lamarckism and Darwinism, as is still believed by some French and Russian evolutionists. If the Baldwin effect occurs and if there were a direct effect of the phenotypic modification on the induction of the genetic factors reinforcing the favored phenotype, then we would have Lamarckism pure and simple. If there is no such induction, then we have simply natural selection, that is, the synthetic theory of evolution. Other objections to the Baldwin effect are well summarized by Simpson, who shows how unproved various claims of support by Huxley (1942:304ff) and Gause (1947) are. Yet even Simpson leaves the door open for a role of the Baldwin effect in the evolutionary acquisition of certain characters, for instance, calluses. It seems to me, however, that a more detailed analysis shows that the conceptual assumptions underlying the hypothesis of the Baldwin effect make it desirable to discard this concept altogether. Three assumptions, in particular, seem to be fatal to the hypothesis:

(1) The argument is always stated in terms of the individual genotype. If it is phenotypically plastic it may be able to withstand the stress exerted by the environment, until relief appears in the form of a mutation or gene combination that protects the organism against the environment. This is essentially a typological argument. What is really exposed to the selection pressure is a phenotypically and genetically variable population in which no two individuals are the same. Selection often favors heterozygotes even though they produce, by segregation, deleterious (phenodeviant) homozygotes. Those genes will be selected in such a population which produce genotypes with an optimal modifiability of the phenotype.

(2) It is not nearly as strongly emphasized (as is important for the correct interpretation) that the degree of modification of the phenotype is in itself genetically controlled. All recent experiments on the production of phenocopies, as well as Waddington's revealing experiments on "genetic assimilation" (discussed in Chapter 8), have identical results. If a population (even an inbred laboratory population!) is exposed to a heat shock, an ether shock, or some other treatment that interferes with normal development, different individuals will show different phenotypic responses. In contrast to the Baldwin effect, the phenotypic response does not serve as a "reprieve" until favorable mutations can occur, but as an indicator of the genotypes to be selected. What is selected is a gene pool with a maximal penetrance of the desired character.

(3) The Baldwin effect makes the tacit assumption that phenotypic rigidity is selectively superior to phenotypic flexibility. This is certainly often not true. Furthermore, the phenotypic rigidity may well be the result of developmental flexibility (Thoday 1953), a factor with high selective value leading to an apparent stability of the phenotype that is quite different from the genetic fixity of the phenotype postulated by the Baldwin effect. It would seem to me that either an epigenetic system such as is found in many plants, in which the phenotype is highly modifiable, or one like that described by Thoday (1953) and Lerner (1954), which achieves great phenotypical stability in spite of great genetic variability,

through developmental flexibility, would be selectively superior to the system postulated in the Baldwin effect. Underwood (1954), Warburton (1956), and Stern (1959) have criticized other aspects of the Baldwin effect and some of its interpretation by Waddington (1953b, 1956b; see also 1957, 1960a).

What then is the final verdict on the evolutionary significance of the Baldwin effect? It seems to me that it has no validity, in the typological form originally proposed by Baldwin, and that it is not legitimate to transfer the term to the phenomenon of the selection of polygenic threshold shifts. It seems to me, furthermore, that it beclouds the issue to introduce a separate term, "genetic assimilation" (Waddington), for the accumulation of such polygenic threshold genes by selection. The unwary might think, as Baldwin did for his organic selection, that this is an alternative to natural selection or at least a special and rare case of selection. Actually what Waddington designates as genetic assimilation is one of the normal aspects of the process of natural selection. Since (a) virtually all characters are highly polygenic, and (b) all genotypes tend to vary phenotypically according to the existing environmental conditions, natural selection will act most strongly on the extreme phenotypes (of the character in question), whatever genetic or environmental constellation produced them. It will result in an accumulation and integration of all the genes that will produce the favored phenotype to an optimal extent in the greatest number of encountered environments. This, I believe, is the normal process of selection of a polygenic character and requires no special terminology.

THE GENETICS AND EPIGENETICS OF TRANSPECIFIC EVOLUTION

The nature of the genetic differences between populations can be analyzed experimentally only when fertile hybrids can be produced. An experimental analysis of the differences between higher categories is thus impossible. Some evolutionists, including some geneticists, have based on this fact the theory that there are two types of genetics, intraspecific genetics (the genetics of microevolution) and transpecific genetics (the genetics of macroevolution). They have claimed that the findings of classical genetics are not consistent with the observed phenomena of transpecific evolution. In this they are quite correct. The picture of random mutations (mostly deleterious at that!) as prime movers of evolution and the typological flavor of the one-gene–one-character–one-selective-value theory were certainly a poor basis on which to interpret macroevolution. The newer concepts of genetics, developed within the last decades, make such a dualism unnecessary. To be sure, it is still impossible to crossbreed higher categories, but this now seems increasingly irrelevant, since the modern interpretation of intraspecific genetics (see Chapters 9 and 17) is quite in harmony with the phenomena of transpecific evolution. The fact that the genetic differences between higher categories cannot be analyzed by the Mendelian method is not a special difficulty of macroevolution, because complex polygenic systems cannot be fully analyzed even within a single population (Mather 1943). It would seem quite legitimate, therefore, to study the genetics of macroevolution by extrapolation (see also Kosswig 1960).

It is most important to clear up first some misconceptions still held by a few, not familiar with modern genetics:

(1) Evolution is *not* primarily a genetic event. Mutation merely supplies the gene pool with genetic variation; it is selection that induces evolutionary change.

(2) A character is *not* (normally) the product of a single gene and a change of this character an indication that this gene has mutated. Virtually all characters are highly polygenic and, since most genes are pleiotropic, the change of a character indicates a minor or major reconstruction of the genotype.

(3) The concept that genes can be classified into those that are superior and will automatically be incorporated into the gene pool and those that are inferior and will inexorably be eliminated is an oversimplification that is actually quite misleading. The selective value of a gene is not absolute, but in an individual case is determined to a large extent by the external environment and the epigenetic system in which it operates. Natural selection consequently is not a problem of simple arithmetic. It must, furthermore, be assumed that the genes of a gene pool, being members of a successful team, will make on the average a greater contribution to fitness than randomly added new genes.

Transpecific evolution is not a matter of isolated genes and mutations, but of whole coadapted gene complexes. As soon as this is clearly understood, the interpretation of macroevolutionary phenomena causes much less difficulty. The genotype is not a beanbag full of unconnected genes. Since most actions of genes are interactions during development, the genotype is an epigenetic system, or, as Waddington called it, an epigenotype.

The "wholeness" of the genotype, owing to its epigenetic system, explains many other phenomena that are difficult to interpret in terms of beanbag genetics, for instance, "tendencies" in certain families and orders that are absent in others. So-called orthogenetic trends are due to the fact that the evolutionary changes of the phenotype owing to natural selection are limited by the possible amplitude of response of the epigenotype. This epigenetic system further explains why phenotypic characters sometimes appear in species hybrids that are known in other species of the family but do not occur in either parent (Steiner 1958). A mixing up of two gene complexes reveals potential canalizations that are completely concealed in the parents. It is this set of phenomena which suggests that the occurrence of so-called primitive characters in birds that have lost the power of flight (Ratites) may not be a case of retention but rather of restoration of such characters.

The evolution of the genotype, as a whole, may explain also the well-known phenomenon of evolutionary inertia. The larger the number of genes that contribute to the shaping of a phenotypic trait, a "character," the less likely it will be that such a character will respond to natural selection. For many of these genes will be pleiotropic and will simultaneously affect components of fitness. This is the explanation of the morphological similarity of sibling species and of the existence of generic, family, and ordinal characters. Other factors that contribute to stability, on the level of the population, are allelic balance (heterosis) and epistatic (internal) balance, as discussed in Chapters 9 and 10. Almost any change, but particularly a major change, of the phenotype in such a well-balanced system will be deleterious. Although minor gene substitutions may be frequent, the well-buffered system of developmental canalizations shields the phenotype against major changes. There is opportunity for speciation, but a major alteration is impossible as long as the epigenotype is intact. A gene pool rich in genetic variation and with a well-buffered epigenetic system has great fitness, but there is much evidence to indicate that its evolutionary potential is not as great as is generally believed. We find again and again in the fossil record abundant and apparently highly successful genera that remained essentially unchanged for millions of years while far less common types simultaneously underwent a process of rapid evolution. One of the reasons for the absence in the fossil record of some of the links between major groups is that they belonged to such rarer lineages.

Evolutionary stagnation has always been a puzzle. To mention a few bradytelic types (Simpson 1944), how could *Lingula* have stayed so nearly the same for 450 million years, or *Limulus* and *Triops* for more than 200 million years? Morphological uniformity and evolutionary stagnation are often ascribed to a "depletion" of genetic variability. It is far more probable that a too well-balanced epigenotype is responsible for such phenotypic and genotypic stability. Indeed, it appears to me that to store genetic variability is perhaps not nearly as great an evolutionary problem as to escape the strait jacket of a too well-balanced genotype. This is the problem that Goldschmidt sensed quite properly, even though his solution ("systemic mutations") was unrealistic. Since new characters are not produced by mutations (as the typologists thought) but by a reorganization of the genotype, it may require a "genetic revolution" (Chapter 17) to break up the perfectly buffered genotype.

The loosening of a tightly knit, coadapted gene complex can presumably be achieved in many ways. One way is a rapid change of population size accompanied by a temporary depletion of the gene pool, resulting in a drastic change of selective values of the included genes (Chapter 17). In plants, hybridization might achieve the same end. A biochemical invention might affect a sufficient number of different metabolic pathways to result in a genetic shake-up. Finally, the emergence of any new structure, for the reasons given above, may set up a selection pressure sufficiently strong and varied to break down the genetic homeostasis. Much of this is obviously speculative, but it seems to me that the interpretation advanced here fits many of the facts of transpecific evolution better than the more orthodox interpretations.

It is evident that there are numerous genetic systems, some favoring specialization, some broad, general adaptation, and some evolutionary pioneering. A lineage must be able to switch back and forth between these systems to achieve the greatest evolutionary success. A pioneering line, after having made a breakthrough, must develop broadly adapted species to have a future. A further discussion of genetic systems and kinds of species was presented in Chapter 14.

The Life History of a Higher Category

Some evolutionists have compared the evolution of a higher category with the life cycle of an individual. They contend that a higher category goes through a series of stages that can be designated by the terms birth, growth, maturity, senescence, and death. Furthermore they claim that such life cycles are sometimes paralleled in various groups to such an extent that the entire fauna simultaneously goes through such a cycle. The more detailed analysis of fossil faunas shows, however, that such claims are greatly exaggerated and that the stated regularities are largely absent. To be sure, every group of animals has originated at some time and has eventually died out unless it persists into the Recent fauna. But, as Rensch (1954, 1960a) and Simpson (1953a, 1959a, 1960a) have shown, this is about all the regularity one can find. Most evolutionary lines with an adequate fossil record have had one or several periods of flowering ("bursts"), while a few lines seem to have persisted from beginning to end, for periods of sometimes more than 100 million years, without marked ups and downs in generic and familial diversification. The great period of proliferation of lower categories often comes immediately after the origin of the new type, as with the trilobites, nautiloids, and some of the brachiopods and corals (Henbest 1952; Nicol, Desborough, and Solliday 1959). The reptiles, however, although well diversified when first recorded in the Middle Pennsylvanian, did not have their great period of dominance and diversity until Triassic and Jurassic times. Mammals have existed since the Triassic, but their flowering did not take place until more than 100 million years later, in the Paleocene and Eocene.

Even less correspondence with a life cycle is indicated by the fact that many groups have had repeated periods of eruptive evolution, such as the crinoids in the Silurian, Mississippian, and Permian, or various groups of mollusks. The period of eruptive evolution may be short, such as that of the mammals in the early Tertiary or that of the land pulmonates, while that of the marine prosobranchs extended for 150 million years. Periods of strong eruptive evolution in marine animals occur in different geological periods from those of land animals; in the oceans in the Ordovician, Triassic, and Jurassic, on land in the Carboniferous and Permian.

All these facts prove that the resemblance of the life cycle of a higher category to that of an individual is quite spurious. Indeed, the rise and fall of higher categories is merely the total effect of the superposition of a number of largely independent evolutionary phenomena. Five of these are clearly distinguishable and the existence of others is indicated by the facts. These five phenomena are: (1) the origin of the new type, (2) a great multiplication of species of this type, (3) their divergent evolution and adaptive radiation, (4) increases or decreases in genetic homeostasis, and (5) a gradual or sudden decrease and extinction. Each of these phenomena can be analyzed individually, as has been done by Simpson (1953a, 1960a), Rensch (1954, 1960a), Stebbins (1950), and the authors cited in their bibliographies. It is evident from these studies that numerous details are still a mystery. Why did the dinosaurs on land and the ammonites in the sea suddenly become extinct in the Cretaceous? Why did some classes and phyla of invertebrates have their great burst in the Ordovician, others in the Silurian, or in the Mississippian? Why were such dominant groups as the trilobites, graptolites, brachiopods, fusulinid foraminifera, and nautiloids displaced by later groups? Many of these events took place at such a remote time that we may never get an answer. Yet it has become clear that there is nothing in the past history of the earth that cannot be interpreted in terms of the processes that are known to occur in the Recent fauna. There is no need to invoke unknown vital forces, mutational avalanches, or cosmic catastrophes.

Geographic speciation, adaptation to the available niches (guided by selection), and competition are largely responsible for the observed phenomena (see Chapters 16 and 18). In other words, the interaction between organism and environment is the most important single determinant in the rise and fall of evolutionary types. For instance, the alternate flooding and drying up of the continental shelves has been, apparently, the most important factor in the rate of speciation of shallow-water marine animals. When the continental shelves are deeply flooded or completely dry, few geographical barriers exist in the oceans and there is little opportunity for a multiplication of species. In periods with extensive shallow-shelf seas there are numerous partially isolated embayments which provide abundant opportunity for speciation. When the previously isolated faunas of numerous such embayments break out and intermingle, owing to changes in sea level, the result may manifest itself in the fossil record as a very drastic change of faunas. The mass extinctions of species of marine invertebrates in periods like the Devonian, Pennsylvanian, and Permian was apparently due to the withdrawal of the epicontinental seas from the continents (Newell 1952, 1956b).

The occurrence of speciation, even of a very active multiplication of species, does not necessarily signify great evolutionary activity. We have discussed above speciation in stable types, such as *Drosophila, Culex,* or the weevils. Yet, with few exceptions, a breakthrough into a new adaptive zone is followed by two events: a colossal speed-up in the rate of evolutionary change and a period of "adaptive radiation." Simpson in particular (1944, 1953a, 1959a, 1960a) has pointed out how rapidly a new type may reach a new phylogenetic "grade," to remain essentially stable afterward. The bats (Chiroptera), for instance, evolved presumably from insectivorelike ancestors sometime in the Paleocene, but the first known fossil

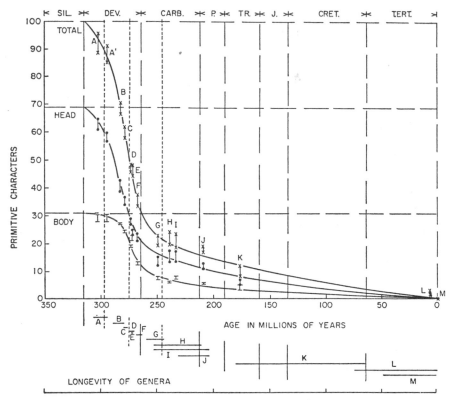

Fig. 19–4. Graphs showing rate of loss of characters of ancestral type (= acquisition of new characters) during the evolution of the Dipnoi. The lowest graph gives approximate time ranges of the main genera (A–M). There was twice as much evolutionary change in the first 30 million years as in the ensuing 250 million years. (From Westoll 1949.)

(middle Eocene) is already essentially a modern bat. About twice as much phylogenetic change occurred in the lungfishes (Dipnoi) during a period of about 30 million years in the middle and late Devonian as in the 250 million years since (Westoll 1949; Fig. 19–4). In the middle Pennsylvanian, where the reptiles are first encountered as fossils, they are already well differentiated into a number of lines. There must have been a considerable radiation right after their origin in the early Pennsylvanian. Most of the shift from the reptilian to the avian organization apparently took place in the late Jurassic–early Cretaceous, since the Tertiary birds are all modern, and most of the few known Cretaceous birds had likewise essentially completed the shift. In view of these exceedingly uneven rates of evolution, it is quite misleading to extrapolate from the rate of evolution along one part of a lineage to others. Some "extremely rapid" rates of evolution are not quite so rapid when the amount of change per million years is calculated (Simpson 1953a; Kurtén 1959). Once an ancestral type has been "loosened up," by unbalancing its structural, epigenetic, and genetic homeostasis, a rapid shift seems to be possible under the greatly increased selection pressure in the new adaptive zone. This will proceed until a new balance is achieved, and a new grade has been integrated. Simpson (1953a, 1960a) provides a fuller discussion than this limited survey.

The second aspect of every major breakthrough, and in fact even of most minor ones, is a great development of minor types in the new adaptive zone, a phenomenon usually referred to as "adaptive radiation." Every kind of animal is adapted for a certain mode of living, but there are numerous niches possible within this mode. For instance, birds are adapted for life in the air, but a warm-blooded air-inhabiting vertebrate has a choice of thousands of niches, ranging from that of a swift to that of a woodpecker, a duck, or a penguin. On the adaptive plateau "bird" there are numerous minor peaks that can be occupied. The fossil history indicates that during periods of proliferation numerous "experiments" are made which have no lasting success. Aberrant types produced during such periods disappear as rapidly as they appear. Such types may be, in part, the product of the genetic relaxation of homeostasis during the period of the general revamping of the genotype. If there are no serious competitors or predators, even rather improbable genotypes may have a chance to survive for a while. Inevitably they give way, however, to more successful types in related branches.

The origin of discontinuities (speciation) and evolutionary divergence (adaptive radiation) are the two evolutionary phenomena that (together with extinction) have the greatest effect on the size and structure of systematic categories. If a group speciates actively (develops discontinuities actively), without much adaptive radiation, it will have many species per genus. If a group radiates actively (developing strong evolutionary divergence) without much speciation, it will develop many monotypic genera and even families. In most groups there will be a mixture of both tendencies. The independence of these two processes is the main reason for the so-called "hollow curve" of the taxonomists (Mayr 1942). Another reason is, as Wright (1941) points out, that extinction and speciation tend to balance each other. The hollow curve is certainly no evidence for an origin of genera by saltation, as claimed by Willis (1940).

Extinction. With the number of living species presumably adding up to a good deal less than 1 percent of those that have ever existed, extinction is one of the most conspicuous evolutionary phenomena. Although a certain number of species become extinct at all times, some geological periods, for instance the end of the Permian, have witnessed far more extinction than others. Cosmic events, like the passing of the earth through a radioactive cloud, are sometimes invoked as explanations, but the fact that the great periods of faunal turnover on land and in the seas do not coincide deprives the cosmic theories of all probability. Climatic events and the fluctuations of sea level on the continental shelves are far more likely causes. Individual species may become extinct owing to new or newly invading diseases (Haldane 1949c) or changes in the biotic environment (particularly a loss of habitat or the arrival of a more successful competitor), yet ultimately their extinction is due to an inability of their genotype to respond to new selection pressures. Large-scale extinction often occurs when two whole faunas mingle owing to geological events (fusing of two continents or two ocean basins). The reaching of certain threshold values in the changing chemistry of ocean water may also be a factor in large-scale extinction. The actual cause of the extinction of any fossil species will presumably always remain uncertain. Indeed, we often have trouble determining the cause of the extinction of Recent species. It is certain, however, that any major epidemic of extinction is always correlated with a major environmental upheaval.

THE EVOLUTIONARY ROLE OF SPECIES

This consideration of macroevolution and of the higher categories has opened up new vistas on the evolutionary significance of species. Closely related species are variations on a theme. Even though each of them occupies a different niche, these niches are, on the whole, quite similar in related forms. Such niches are only subdivisions of a single "adaptive zone," as Simpson would call it. The adaptive zones, particularly the major ones, are occupied by distinct types, such as rodents or bats, birds or snakes, sharks or eels. Each of these adaptive types is characterized by a unique set of attributes. It requires a singular combination of genetic, physiological, and morphological properties and a unique constellation of environmental conditions to enable an animal to invade a unique major adaptive zone. Only one species in 10,000 or 100,000 will have the improbable combination of characteristics that preadapts it to undertake a major ecological shift.

The real difficulty, stressed by Simpson, is that ecological "space" is not continuous. The terrestrial insectivores and the aerial bats are separated by an adaptive discontinuity, and so are the diving petrels and the penguins. The gap between the adaptive zones is a zone of adaptive disequilibrium. To cross it is hazardous and the chance of success is slight. Every new species is an ecological experiment, an attempt to occupy a new niche. The number of successful shifts into new adaptive zones will be directly proportional to the total number of new species that come into existence.

For this reason I do not agree with Huxley (1942) when he says, "Species formation constitutes one aspect of evolution; but a large fraction of it is, in a sense, accident, a biological luxury, without bearing upon the major and continuing trends of evolutionary process." On the contrary, I feel that it is the very process of creating so many species which leads to evolutionary progress. Species, in the sense of evolution, are quite comparable to mutations. They also are a necessity for evolutionary progress, even though only one out of many mutations leads to a significant improvement of the genotype. Since each coadapted gene complex has different properties and since these properties are, so to speak, not predictable, it requires the creation of a large number of such gene complexes before one is achieved that will lead to real evolutionary advance. Seen in this light, it appears then that a prodigious multiplication of species is a prerequisite for evolutionary progress.

Each species is a biological experiment. The probability is very high that the new niche into which it shifts is an evolutionary dead-end street. There is no way to predict, as far as the incipient species is concerned, whether the new niche it enters is a dead end or the entrance into a large new adaptive zone. The tetrapods go back to a single ancestral species, as do the insects and presumably the angiosperms. If this ancestor had not experimented with a niche shift, the new adaptive empire would not have been discovered.

The evolutionary significance of species is now quite clear. Although the evolutionist may speak of broad phenomena, such as trends, adaptations, specializations, and regressions, they are really not separable from the progression of entities that display these trends, the species. The species are the real units of evolution, as the temporary incarnation of harmonious, well-integrated gene complexes. And speciation, the production of new gene complexes capable of ecological shifts, is the method by which evolution advances. Without speciation there would be no diversification of the organic world, no adaptive radiation, and very little evolutionary progress. The species, then, is the keystone of evolution.

20 ~ Man as a Biological Species

The evolution of man is so vast a subject that whole libraries of volumes have been devoted to it. It would be folly to try to summarize this information—not to mention the enormous number of conflicting interpretations—in a few paragraphs of a short chapter. I am forced to neglect large parts of this field, particularly those that have been dealt with effectively by other recent authors. My own presentation will be deliberately eclectic, with an endeavor to contribute to the solution of controversial problems. Those who want to penetrate more deeply into the subject will benefit from reading in some of the literature cited in the footnote.* The titles cited range from technical to popular, from physical anthropology to cultural anthropology, and from population problems to man's future. Until recently, so little concrete information about the evolutionary history of mankind was available that philosophers, politicians, poets, and ideologists—such men as Schopenhauer, George Bernard Shaw (*Back to Methuselah*), Bergson, Marx, Hitler, and Stalin—considered themselves qualified to publish pronouncements on the subject. Their publications have been at best incomplete and one-sided, more often misleading, and frequently highly pernicious. Treatises on the evolution of man published as recently as 20 or 30 years ago appear shockingly antiquated to the reader of today.

The rapid changes in our thinking on the evolution of man have had many sources, some of which will be discussed in this chapter. Particularly important are the radical replacement of the prevailing typological thinking by population concepts; application of the findings of modern

* Cold Spring Harbor Symposium 1950; Dobzhansky 1962; Dunn 1959; Haldane 1949a; Heberer 1954–1959; Howells 1959; Huxley 1953; Huxley and Huxley 1947; Le Gros Clark 1955, 1960; Muller 1960; Rensch 1959; Roe and Simpson 1958 (several contributors); Tax 1960b (numerous contributors); Washburn 1950; Washburn and Avis 1958; Washburn and Howell 1960; Waddington 1960b.

population genetics to man; and reinterpretation of the role of nonphysical factors, be they cultural or ecological, in shaping modern man. While until a few years ago the study of the evolution of man was the exclusive domain of anthropologists, it is increasingly being incorporated into the sphere of interest of geneticists, systematists, mammalian paleontologists, ecologists, and other biological specialists. Each specialist tends to approach the central problem from a different direction, and thus throws light on aspects previously overlooked. At first, such a many-sided approach tends to generate conflict. In the long run, however, it is the most productive method by which a balanced understanding of such a complex field can be reached.

New fossil hominid discoveries follow each other so rapidly that almost any statement made here may be obsolete within a year or two. One who is not satisfied merely to record raw data, but wants to give a causal interpretation of human evolution, takes this risk gladly. The best way to discover contradictions and gaps in our understanding is to attempt to tell a consistent story.

In no other branch of science have workers been so dominated by preconceived notions as in the science of man. Human beings seem quite incapable of speaking about themselves and their history without becoming emotional in one way or another. The student of human evolution has to be constantly aware of this. It is the specific object of this chapter to apply the more significant findings of Chapters 1–19 of this volume to the evolution of man.

Man is a species of animal, as is self-evident as soon as one applies the concept of evolution to man. He shares many characteristics with other species, and it leads to a more balanced consideration of man to approach him from the viewpoint of the biologist. But no more tragic mistake could be made than to consider man "merely an animal." Man is unique; he differs from all other animals in numerous properties such as speech, tradition, culture, and an enormously extended period of growth and parental care. This has been pointed out perceptively by Huxley, Haldane, Simpson, Dobzhansky, and other recent writers. My own discussion of some aspects of man is admittedly one-sided. I am not qualified to treat man's cultural, physiological, and sociological properties. My emphasis will be quite frankly on the biological aspects of man's evolution and on those questions that must be asked as a consequence of recent advances in our understanding of biological evolution. Let no reader forget the limitations of this approach!

THE PHYLOGENY OF MAN

Man is as much a product of evolution as is any other organism. Indeed, man is the historical creature par excellence. He has not only a biological heritage but also a cultural one, making him history-bound in two separate ways. The gradual emergence of man's being "not merely an animal," and the forces that brought about this evolution, are by no means fully understood and are a source of much controversy, in part because the reconstruction of man's phylogeny is still largely a matter of guesswork. Yet considerable progress has been made in recent years owing not only to the discovery of many additional fossil hominids in southern Asia and in Africa but also, and perhaps more importantly, to the revision of some basic concepts of evolution and phylogeny. Our thinking on missing links, the evolution of "types," irreversibility, the role of mutations and of the environment in evolution, the variability of population samples, and the meaning of species has changed so much in recent years that even the previously known fossil hominids are now regarded in a very different light than they were only 20 years ago.

Man is so strikingly similar to certain other mammals that no biologist can question the closest relationship. Even long before evolution was seriously considered, Linnaeus placed man in the order Primates together with apes and monkeys. Although not conceived in terms of evolutionary biology, this arrangement was generally accepted by post-Linnaean authors. Comparative anatomical studies fully confirmed the great similarity between man and the anthropoid apes (Pongidae).

As in all phylogenetic investigations, two aspects in the evolution of man are involved, which must be carefully distinguished in order to avoid confusion. One is the branching of the hominid from the pongid line (speciation) and the other is the reaching of the human level within the hominid line (assuming that the earlier representatives of the hominid line were still anthropoid apes). Consequently we must answer two questions:

(1) When and where did the hominid line branch off from the anthropoid line that gave rise to the Pongidae and what did the missing link look like?

(2) Through what stages did the hominid line pass, after its separation from the pongid line, before the truly human level was reached?

The Search for the Missing Link

Man's closest relatives among the living primates are undoubtedly the so-called anthropoid apes. They consist of three groups, perhaps best regarded as three genera, all belonging to the family Pongidae. These are the chimpanzee and the gorilla (genus *Pan*) in Africa, the orang (*Pongo*)

Table 20–1. Some of the differences between man and the anthropoids.

Characteristic	Man	Anthropoids
Long bones of lower extremities	Longer than those of upper extremity	Shorter than those of upper extremity
Tarsal bones	Rather long, toes rather short	Rather short, toes rather long
Trunk	Short compared to lower extremities	Long compared to lower extremities
Vertebral column	Alternately curved backward and forward	Straight or curved uniformly backward
Leg	In upright posture, straight in knee and hip joint	Curved, the knees turned outward
Joint between skull and vertebral column	Almost in center of base of skull	At back of skull
Canines	No larger than premolars	Large fangs
Crown of first lower premolar	Unspecialized (bicuspid)	Has blade-like cutting edge
Dental arch	Rounded without sharp angles	Laterally compressed with side rows of teeth almost parallel
Jaws	Short	Long and large
Face	Short, steep, under brain	Long, in front of brain, protruding
Brain	Large	Average size $\frac{1}{3}$ that of human

in the East Indies (Borneo, Sumatra), and the gibbon group (*Hylobates*) in southeastern Asia and the East Indies. Some of the differences between man and the living anthropoids are listed in Table 20–1.

The living anthropoids, in spite of all their differences, share several so-called "anthropoid characters," such as powerful canines, large incisors, a sectorial form of the first lower premolar, a simian shelf of the mandible, specialized feet, and powerful brachiating arms. These characteristics are striking and make for a pronounced gap between man and the anthropoids. Yet the theory of evolution demands that man and these anthropoids have descended from a common ancestor.

For a long time, the study of fossil man was essentially a search for a connecting form, the "missing link." At first no one knew quite what to look for. The earliest reconstructions, for instance that by Haeckel, pictured a creature which, character by character, was intermediate between man and chimpanzee. This implied that man had the chimpanzee as his direct ancestor, that the chimpanzee stopped evolving as soon as it had given rise to the human line, and that all organs evolved at the same rate! Equally unfounded is the additional assumption that the living anthropoids are primitive and that man had to go through a stage represented by these anthropoids. Numerous recent fossil discoveries have made it clear that the anthropoids have evolved as much since branching off from the common anthropoid-hominid stem as has the hominid line. In fact, in many morphological respects the anthropoid line is apparently less similar to the common ancestor than is modern man. To find the common ancestor of the Hominidae and the Pongidae, we must look for a creature that lacks the brachiating specialization of the living Pongidae and the complete bipedalism and brain development of the recent Hominidae, but one that possesses some of the characteristics by which the anthropoids (including *Homo*) differ from the cercopithecoid monkeys. Forms close to this postulated creature have been found in the lower Miocene of Africa.

Miocene Anthropoids

Fossil discoveries in eastern Africa comprise several anthropoid genera that are clearly separated from the regular cercopithecoid branch of primates. Three groups can be recognized (Le Gros Clark 1950). The genus *Limnopithecus* is related to the gibbons. Even closer to the modern gibbons, as far as teeth and jaw are concerned, is *Pliopithecus* of Europe. The discovery of some fairly complete skeletons (Zapfe 1958) has revealed that this form was not yet a specialized brachiator and that the proportions of the anterior extremities (in relation to the trunk) did not deviate materially from those of man or of the Miocene African anthropoids. This is full confirmation of the brilliant conclusion of Schlosser (1901) that "the conspicuous elongation of the upper arm of the anthropoids is merely a newly acquired specialization which certainly does not date further back than the Pliocene."

The second type of early Miocene anthropoids is represented by the genus *Proconsul,* species of which range in size from smaller than chimpanzees to the size of a gorilla. There is much in the characters of *Proconsul* to qualify this genus as being close to the ancestry of gorilla and chimpanzee even though it still lacks some of the anthropoid specializations such as brachiating arms and a simian shelf. In fact Le Gros Clark and Leakey (1951) come to the conclusion

that the characteristic features of the skulls of the African apes of today, such as the powerful supraorbital torus, the large circular and forwardly directed orbits associated with a strong development of their lateral margins, the broadening of the lower extremity of the nasal aperture, the extension downward and forward of the subnasal part of the premaxilla, and widening of the symphysial regions of the mandible in association with a relative hypertrophy of the incisor teeth, and the development of a simian shelf, were probably all secondary developments which appeared at a later stage of evolution.

Lacking all these specializations of the later anthropoids and yet clearly differing in dentition and the proportions of the facial skeleton from the cercopithecoids, the *Proconsul* group must have been fairly close to the hominid line. It is improbable that the hominid line goes back directly to *Proconsul*. However, the mere fact of the large canines would not disqualify this group from hominid ancestry since there is a period of about 20 million years between the time when *Proconsul* existed and that of *Australopithecus*. There was thus plenty of time for a reduction of the canines. However, a third group of anthropoids is known from the Miocene and Pliocene of Africa and Asia some of which seem close to the hominid line. This includes such genera as *Dryopithecus*, *Sivapithecus*, *Ramapithecus*, and *Bramapithecus*. *Ramapithecus* is by far the most likely candidate for human ancestry (Simons 1961), together with the recently discovered *Kenyapithecus*.

Finally, there is the controversial *Oreopithecus* from the lower Pliocene, found in an Italian coal bed. Whether classified as cercopithecoid or hominoid, it is too aberrant to qualify as a missing link.

Scanty though this evidence is, it permits some conclusions. There are three major theories on the branching point between pongids and hominids. According to the first, the hominid line branched off from the common stem of the living anthropoids before they split into three separate lines. This theory is based on the many similarities among the living anthropoids as well as on the many peculiarities of man that separate him from all other primates. The discovery that the brachiating adaptations of *Hylobates*, *Pongo*, and *Pan* are due to evolutionary parallelism, and the rather early occurrence of unmistakable gibbons (*Limnopithecus* of the early Miocene), make this alternative highly unlikely. The second possibility is that the hominid line branched off well after the gibbon line, but before the pongids split into the lines that eventually gave rise to *Pan* and *Pongo*. The split between hominids and pongids in this case would have taken place in the late Oligocene or early Miocene, about 25–30 million years ago. The third possibility is that the hominid line branched off from the line of the African apes (chimpanzee and gorilla) at a comparatively recent date, long after the pongid line had split into an Asiatic (*Pongo*) and an African (*Pan + Homo*) branch. No fossils are known that would clearly favor the second or third alternative. The crucial genera and species may well have been inhabitants of tropical forests which do not leave an abundant fossil record. The search for the first unmistakable fossil hominid has not yet ended. However, an analysis of hemoglobins (Zuckerkandl *et al.* 1960) and serum proteins (Goodman 1962) proves conclusively that the third alternative is correct.

Kinds of Hominids

There is a long gap in the fossil record between the early Miocene of eastern Africa, with its rich deposits of anthropoids, and the early Pleistocene, again rich in hominid fossils. The anthropoid fossils of the intervening 20–25 million years are either fragmentary (*Kenyapithecus, Ramapithecus*) or definitely pongid. It is consequently not known through what stages the hominid line went in the Miocene and Pliocene. Considering how primitive the earliest known Pleistocene hominids are, it can be concluded that hominid evolution in the Tertiary was rather slow.

Within the Pleistocene three rather well-defined stages of hominid progress can be discerned: (1) the Australopithecus stage of the late Villafranchian of Africa, but covering a considerable time span; (2) the *Homo erectus* ("*Pithecanthropus*") stage of the middle Pleistocene in Africa, Europe, and Asia; and (3) the *Homo sapiens* stage, of the late Pleistocene. Each of these three grades is characterized by a distinctive mean brain size, by other physical characteristics, and by certain stone implements (Washburn and Howell 1960).

The Australopithecines

The first discovery of this kind of hominid, the skull of an immature individual, was made in 1924 at Taung * in Bechuanaland. He was described by Dart as *Australopithecus africanus*. From 1936 on, the late Robert Broom and his associates discovered additional rich sites in Transvaal, so that of all fossil hominids this group is now the most abundantly

* This, not Taungs, is the correct spelling.

represented. These South African man-apes are characterized by an essentially hominid dentition consisting of small canines, typically bicuspid premolars, and a molar cusp pattern similar to that of Java Man. The teeth are arranged in an even arcade of elliptical form with no gap (diastema) between the canines and adjacent teeth. Pelvis and limb bones are constructed along hominid lines and indicate a posture that was upright, though not as highly perfected as that of modern man. Upright posture is also indicated by the location of the occipital condyle (the articulation of the skull with the spinal column), which faces essentially downward, as in modern man, rather than backward, as in the anthropoids. Other hominid features become apparent in a close study of the skull. However, combined with these human features are others that connect these hominids with their anthropoid ancestors. The cranial capacity was hardly larger (relative to body size) than in the modern large apes. The jaws were exceedingly large compared to the cranium, and the skull bones, particularly the jaw bones, were tremendously thick and heavy.

An impartial observer cannot escape the conclusion that these forms are clearly members of the hominid line of evolution but have not yet reached the level of man. In South Africa, Australopithecines have been found in at least five different sites, and every new find was at first described as a different species or genus. It is now evident that two distinct types existed in South Africa. The older is Australopithecus (sensu stricto), which is the only representative of South African ape-man at the three oldest sites, Taung, Makapan, and Sterkfontein. Sexual dimorphism in the skull is not conspicuous and the dentition is not specialized. Incisors and canines are comparatively large, while the molars are not excessively enlarged, and there are no large bony crests on the skull for muscle attachment. A second type, Paranthropus, found in the later deposits of Swartkrans and Kromdraai, is a larger form with more pronounced sexual dimorphism. Incisors and canines are very small, while the molars are huge, and there are pronounced bony crests on the skull, particularly in adult males.

Nothing is known about the relation of the two forms to each other but they seem to belong to two different phyletic lines, Paranthropus apparently representing a specialized, perhaps more vegetarian, side line diverging farther than Australopithecus from the human type. Leakey et al. (1961) discovered another member of the Paranthropus group in eastern Africa and named it Zinjanthropus. Since this was found in an open site (not a cave) and was accompanied by many fossils, its relative age can be given rather accurately as upper Villafranchian (contemporary with *Australopithecus* in South Africa or even a bit earlier). What this means in terms of absolute chronology is still controversial. The conventional date for upper Villafranchian is about 600,000 years ago, while Evernden and Curtis (in Leakey *et al.* 1961) have calculated an age of 1,750,000 years on the basis of the potassium-argon method. Criticism has been raised against the accuracy of this method, so that caution would seem indicated. Regardless of the absolute date, the most important aspect of this newly discovered fossil is its association with stone tools of the Oldowan (pre-Chelles-Acheul) culture, together with the earmarks of tool manufacturing: abundant waste flakes and a hammerstone. Alas, even this association is not decisive, since more recently discovered skeletal remains in the same strata are of the *Homo* type (Leakey MS). A species of *Homo* is, of course, a more likely maker of these tools than *Zinjanthropus*. Similar tools were also found at Sterkfontein in South Africa, although none were found in the lower strata richest in Australopithecines. Tools of this culture have been found in many lower Pleistocene localities in Africa, always associated with a Villafranchian fauna (Howell 1959), but nowhere outside Africa. New finds follow each other so rapidly that any inference might be invalidated by new facts before this is printed. This much seems well established, that tool-manufacturing hominids were widespread in Villafranchian Africa, possibly in several species and surely in numerous local races.

A word must be said at this point about the general size of these early man-apes. If one compares only their massive skulls and jaws and their huge teeth with those of modern man, one may conclude that these forms must have been giants. When fragments of jaws and teeth of similar forms were discovered in southeastern Asia, they were indeed described as remains of giants (Weidenreich 1945, 1946). Now that abundant material of the body skeleton has become available, it is evident that *Australopithecus* probably did not exceed 5 feet in height, being probably somewhat shorter than a modern Bushman. This evidence flatly contradicts the popular notion that man descended from a line of giants. It also negates the hypothesis that the shift of the original hominids from a more arboreal to a more terrestrial mode of life was necessitated by an excessive increase in body size.

The Australopithecines had an extraordinary mixture of apelike and human characteristics. Pelvis and lower extremities were much like those of modern man (of course *not* identical), indicating an essentially bipedal locomotion. Yet in their small brains and huge jaws they were apelike. That the human line went through a similar stage is not at all un-likely. Indeed, the Australopithecines do not have a single character or combination of characters that would clearly disqualify them from the main line of human evolution. As far as the recognition of the genus *Australopithecus* is concerned, it depends somewhat on the evaluation of its characteristics. Since it already has the essential morphological charac-ters of *Homo,* such as upright posture, reduced canines, and bicuspid premolars, I remarked previously that "not even *Australopithecus* has un-equivocal claims for [generic] separation" (Mayr 1950a). I now agree with those authors who have since pointed out not only that upright loco-motion was still imperfect but also that the tremendous evolution of the brain since *Australopithecus* permitted man to enter so completely differ-ent a niche that generic separation is definitely justified. Here, as in other cases, it is important not only to count characters but also to weight them. The brain evolution justifies the generic separation of *Australopithecus* from *Homo* no matter how similar they are in other characters. Paranthro-pus, a name here used in a vernacular sense, hardly shows the degree of difference from *Australopithecus* necessary to justify generic status.

The Pithecanthropines

The most famous fossil hominid before the discovery of *Australopithe-cus* was *Pithecanthropus.* Haeckel had coined this name for the "missing link" and thus greatly stirred the imagination of many a young man to find it. A young Dutch anatomist, E. Dubois, obtained an appointment as army doctor in the East Indies so that he could search there for the miss-ing link, and much to everyone's surprise (and probably his own) he ac-tually succeeded. In 1891 he found a skull cap at Trinil in eastern Java, and the next year, some 40 or 50 feet away in the same fluvial deposits, a thigh bone (femur) as well as other skeletal remains. Dubois's report on *Pithecanthropus erectus,* published in 1894, precipitated one of the most heated controversies in the history of anthropology (rich as it is in hot controversies!). Some authors regarded the remains as human, others as anthropoid. The latter disagreed as to whether the closest relationship was with orang, gibbon, or chimpanzee. As far as the femur was concerned, the majority opinion was that it could not belong with the skull cap since it was "of a different type." *Pithecanthropus* remained highly controversial until von Koenigswald in the 1930's systematically and with enormous en-ergy explored the fossil beds of Java, bringing to light many additional and much better specimens (Le Gros Clark 1955).

Java Man is so close to modern man in all essential anatomical features that it seems inadvisable to separate it generically. It will be referred to as *Homo erectus* in the following account. Some of the primitive characters of this form are its low forehead, the lowness of the skull as a whole, the high line of attachment of the occipital bones, the strong supraorbital ridges, and the heaviness of the skull. The most important feature is the small size of the brain, indicated by an endocranial volume of 775–1000 cm³ with an estimated average of less than 900 cm³. Yet this volume represents an enormous advance over the Australopithecines.

Java Man was of great importance in the history of the discovery of fossil hominids, being the first fossil man known outside the range of variation of polytypic *Homo sapiens* (*sensu lato*). Many characteristics that were at first puzzling and incongruous, such as the combination of an ape-like jaw and an almost modern femur, are considered quite natural now that we know *Australopithecus*. Perhaps the greatest remaining puzzle about *Homo erectus* is its age, whether only middle Pleistocene or also early Pleistocene.

Java Man and Pekin Man (originally described as *Sinanthropus pekinensis*) were long considered the only representatives of the *Homo erectus* stage of human evolution. Now it is realized that the famous *Homo heidelbergensis* (known from a single mandible found at Mauer near Heidelberg, Germany), in spite of its more modern appearance, is roughly contemporary with Java Man (end of First Interglacial or beginning of Second Glacial) and overlapping in time the last of the Australopithecines (Kromdraai). That Heidelberg Man is clearly a different type from the Asiatic Pithecanthropines is indicated by its peculiar combination of a massive jaw and relatively small, almost "modern" teeth. Yet three jaws found by Arambourg (1955) in North Africa (Ternifine Man) agree reasonably well in their main features with Pekin Man. What kind of hominids existed in the intervening areas and how are they related to "*Telanthropus*" (see below) of South Africa? These are only some of the many tantalizing questions we would like to see answered.

The *Homo erectus* stage is characterized by a body skeleton which, so far as we know, does not differ from that of modern man in any essential point. The main differences from modern man are a more massive skull and dentition and a smaller brain (though apparently brain size overlaps normal individual variants of modern man). These are good species differences, but would hardly be conceded generic rank anywhere else in vertebrate classification. Most of the fossil hominids were described by anatomists who seem to have thought that every specimen should have a combination of two distinct names, equivalent to the human Christian and family names. At best they would state in a footnote that their *Paleoanthropus, Africanthropus,* or *Meganthropus* was not to be considered equivalent to the generic names of zoologists. This is no excuse. Generic names, specific names, and subspecific names indicate distinctive systematic and evolutionary levels and it can only lead to confusion if what is merely a species or a race in time and space is designated as a separate genus. To reiterate, from the zoological viewpoint I can see no reason to consider Java Man a separate genus.

EVOLUTIONARY TRENDS IN THE HOMINID LINE

When we compare the early Miocene anthropoids with the Australopithecines, Pithecanthropines, and modern man, a definite evolutionary sequence becomes evident in spite of the vast remaining gaps in our knowledge. To achieve the rank of *Homo sapiens* man's ancestors had to pass a number of milestones which I shall try to characterize.*

Bipedal Locomotion and Tool Use

Apparently none of the anthropoids of the mid-Tertiary were specialized brachiators and man's ancestors at this period moved presumably like some of the more terrestrial (but unspecialized) ceroopithecoid monkeys of today, quite adapted to arboreal life but in a general, rather than a specialized, way. The acquisition of upright posture and bipedal locomotion was the key event in the evolution of the hominid line (alas, entirely undocumented by fossil evidence). All anthropoids are of course capable of walking bipedally but their upright posture differs from that of man in many respects. A number of shifts in construction were needed, particularly a shift in the attachment of the musculus glutaeus maximus (Washburn 1950), to achieve the particular type of foot and upright locomotion characteristic of man. Bipedal locomotion is on the whole a relatively inefficient form of locomotion for a mammal. Its advantages, particularly well discussed by Bartholomew and Birdsell (1953), are not only the opening up of the terrestrial habitat to a previously arboreal creature but more importantly the partial freeing of the forelimbs for other functions. It permitted the use of the hands for the efficient manipulation of advantageous tools such as rocks, sticks, or bones.

---

 * A draft of this section was completed in MS. when the very similar but far more authoritative account of Washburn and Howell (1960) was published, on which I lean for some of the detail.

It has been claimed that the skillful use of tools set up a strong selection pressure for increased brain size until the brain was large enough to enable its owner to manufacture his implements himself. The discovery of stone cultures among rather small-brained hominids forces us to modify our ideas. It now seems probable that the use of tools is an ancient hominid trait, an assumption supported by the readiness with which, for instance, chimpanzees adopt implements. Instead of claiming that bipedalism made the hands available for other functions, one begins to wonder whether the perfection of bipedalism was not greatly speeded up by the preoccupation of the anterior extremity with another function, that of "manipulation." The use, and perhaps even the manufacture, of simple tools apparently did not require a great increase in brain capacity. Nor did it require a major reconstruction of the anterior extremity. Arm and hand changed remarkably little from the time the hand was used largely for grasping a branch to the time it was first used for piano playing or the repair of a fine watch.

Some anthropologists are unaware of the widespread use of tools in the animal kingdom. "Not only do other primates use tools—the use of sticks and rocks by chimpanzees and baboons is generally familiar—but such unlikely animals as the sea otter and one of the Galapagos finches routinely use rocks or sticks to obtain food" (Bartholomew and Birdsell 1953). Every naturalist can cite many additional cases, such as bower birds that use paint brushes, spiders that employ throw nets, wasps that harden the soil over their nest holes by pounding it with pebbles, and so forth. In most of these cases the use of the tool is either very incidental (as with the bower birds) or is a component of one single rather rigid behavior pattern. In the case of man there is or was a direct correlation between survival and skill in the use of the tools. Bartholomew and Birdsell are right in saying that man is the only mammal that is continuously dependent on tools for survival. This dependence on the learned use of tools involves development of a previously unexploited potentiality of behavior and thus releases entirely new selection pressures.

Brain Size and Speech

The assumption that rather small-brained hominids were experienced tool users and manufacturers raises at once the question of the nature of that (tremendous) selection pressure which caused an increase of brain size during the mid-Pleistocene at an unprecedented rate (Haldane 1949b). Average cranial capacity rose from 1000 to 1400 cm³ in less than

1 million years. A rough picture of this evolution is given by a list of cranial capacities (omitting extremes):

Chimpanzee and Gorilla	325– 650 cm³
Australopithecines	450– 650
Java Man	800–1000
Pekin Man	900–1100
Neanderthal and Recent Man	1200–1600

It seems likely that the ability to make tools contributed far less to this selection pressure than did the need for an efficient system of communication, that is, speech. Foresight and capacity for leadership would be greatly enhanced by an ability for articulate communication. Many aspects of intelligence and planning would have little survival value without a medium of communication far more efficient than that of the anthropoid apes. The possible role of family structure among the early hominids for the acceleration of brain evolution will be discussed below.

The hominid line was well preadapted for the development of speech, owing to the low position of the larynx, the oval shape of the tooth row, the absence of diastemas between the teeth, the separation of the hyoid from the cartilage of the larynx, the general mobility of the tongue, and the vaulting of the palate (Kipp 1955). The transfer of the food-uptake function from the snout to the hands further facilitated the specialization of the mouth as an organ of speech. Speech does not fossilize and all we can say about the origin of language is pure conjecture. Yet it is evident that a superior ability for communication and the possession of associated brain functions to make such communication optimally effective would add enormously to fitness. The hominid evolution is an impressive example of the chain reaction of evolutionary change that results from key innovations, such as bipedalism and speech.

The increase in brain size led to a complete reconstruction of the skull, which was favored by two other developments. One is the forward shift of the support of the skull in connection with upright posture. The other is a lessening of the selection pressure in favor of strong jaws and big teeth in connection with the prepared food (tools, cooking, shift of diet) and the new means of attack and defense (weapons) made possible by the enlarged brain. All this resulted in a reduction of the jaws, the teeth, and the entire facial part of the skull, while simultaneously the cerebral part of the skull enlarged. It resulted also in reduction of the facial muscles and all bony crests and ridges to which these muscles are attached. The understanding of these trends has been long delayed by various preconceived typological and orthogenetic theories.

One of these misleading theories is the "fetalization" hypothesis of Bolk. Starting with the observation that adult man lacks the excessive specializations of adult male anthropoids (with their tremendous bony crests, huge jaws, and large canines) and that in these structures man is more similar to young anthropoids, Bolk proposes that adult man has returned —morphologically—to the fetal condition of his ancestors owing to a retardation of ontogeny. This theory that the hominid line has passed through a gorilla-orang stage and has since become "fetalized" is not supported by the known facts. The brain is ahead in ontogeny in all mammals. Indeed, since human babies have such extremely large brains, one might say exactly the opposite of what Bolk has said and state that they have become "adultified." Not only is the fetalization hypothesis in conflict with evolutionary theory, but detailed investigations by Weidenreich (1941), Kummer (1953), and Starck (1960, 1962) have shown that it is contradicted by the anatomical facts. The evolution of the human phenotype is exactly what one might expect as a compromise between various selection pressures. No mysterious *élan vital* has directed the observed trends.

Evolution of Behavior

Through the higher mammals, and most strikingly in man, there has been a trend toward replacing rigidly genetically determined behavior patterns by behavior that is subject to learning and conditioning. The "closed" program of genetic information is increasingly replaced in the course of this evolution by an "open" program, a program which is so set up that it can incorporate new information. In other words, the behavior phenotype is no longer absolutely determined genetically, but to a greater or lesser extent is the result of learning and education.

This involves not only a capacity for learning, but, as Waddington (1960b) has rightly stressed, also a readiness to accept authority: "The newborn infant has to be ready to believe what it is told." It is this system of nongenetic determination of the behavioral phenotype that permits the development of religious dogmas (based on revelation) and of ethical codes. The capacity to accept concepts, dogmas, and codes of behavior is one of the many forms of imprinting. The greater the amount of parental care and education and the more highly developed the means of communication, the more important becomes "conceptual imprinting." The acceptance of ethical systems and religions is as much testimony to this as is the success of demagogues and of the mass media.

Mosaic Evolution

The evolution of the hominids is an almost classical demonstration of mosaic evolution. Each organ and each system of organs has its own rate and pattern of evolution (Mayr 1950a). Bipedalism and hand use came first and the resulting reconstruction of the pelvis and the extremities was virtually completed at a time when brain increase and repatterning of the skull had barely started. This is why it was so completely wrong to construct the missing link by making an exact intermediate between living man and chimpanzee. This is why it was erroneous to say that "Australopithecus cannot be a hominid because in addition to some hominid features it is too apelike." This is like saying that Archaeopteryx cannot be a link between reptiles and birds because even though it has feathers it has too many reptilian features. Mosaic evolution is the characteristic form of evolution of all types that shift into a new adaptive zone (Chapter 19).

The gradualness of man's becoming man must be stressed in opposition to continuing attempts to present the origin of man as a single-step phenomenon. What stage in this continuum could one arbitrarily pick out and designate as the "real" origin of man? Would it be the branching-off point from the pongids, or the first manufacturing of tools, or the use of fire, or a speech development indicated by a brain size of 1000 cm³, or the first attainment of 1500 cm³? There is not merely one "missing link," but a whole series of grades of "missing links" in hominid history. The utter futility of interpreting the origin of man as the result of a saltation becomes obvious in the light of such considerations.

SPECIATION IN THE HOMINIDS

It was hopeless to try making sense of hominid phylogeny as long as the fossil remains of man's ancestors were considered mere anatomical "types." Named fossils (and every fossil was named!) were placed in morphological series, and any specimen that was specialized in any way was called an "aberrant side branch." The study of the geographic variation of animals and a new insight into the process of speciation have introduced into the study of fossil man new concepts which have led to a great simplification of the general picture and permit a clearer statement of the unsolved problems. The most important conceptual advance is to regard hominid fossils as samples of populations with a definite distribution in space and time (rather than as anatomical types). Furthermore, since Recent Man is a polytypic species and since most species of mammals are polytypic, it can be assumed that the species of fossil hominids likewise were polytypic.* We must furthermore assume that throughout hominid history some of the geographic isolates (Chapter 13) reached species rank (reproductive isolation) and were then able to overlap or exterminate sister species. Finally we must assume that rates of phyletic change differed in the various isolates of a polytypic species, as they do in polytypic species of living animals, so that advanced and retarded races were contemporaries. Applying all these principles to the known fossil hominids, what do we learn about speciation in the hominid phyletic line?

The basis of all scientific interpretation is the rule of parsimony, which demands in each case the simplest explanation consistent with the facts. As far as concerns fossil hominids, the simplest assumption would be that at any given time only a single polytypic species of hominid existed, and that the variety of observed types is merely a manifestation of individual and geographic variation. Let us now see into what difficulties this simplifying model runs.

Hominid Compatibility

Modern man approaches more and more the status of a panmictic species. In contradistinction, gene flow in the early hominids must have been slight. Fossil man consisted of numerous localized family groups and small bands, isolated by numerous geographical and ecological barriers. There was probably very little mixing between neighboring bands, and local differentiation must have been high, at least at the Australopithecine and Pithecanthropine levels. It is to be assumed—and the few available facts support it—that the allopatric populations of the earlier hominids were more distinct from each other than are the races of modern man. Whether or not they had reached a level of distinction justifying specific status is unknown. They evidently never reached a degree of ecological compatibility (Chapter 4) that would have permitted extensive sympatry.

This ecological incompatibility of hominids with each other is one of the reasons for our taxonomic difficulties. Two entirely different causes

---

* Coon's *Origin of Races* (1962) was published after this had been written. I refrained from making even a single change in my wording to permit a clearer perception of similarities and differences in my parallel treatment.

may be responsible for the geographical exclusion of two closely related forms: either they are subspecies or they are good species but ecologically incompatible. Both factors seem to contribute to the extreme rarity with which two different kinds of fossil hominids have been found at the same site. *Australopithecus* and *Paranthropus* exclude each other completely in the known South African sites, even though the difference in their dentition suggests that there might have been a considerable difference in diet. *Paranthropus*, however, coexisted with an Australopithecine at Oldoway and with *Telanthropus* (= *?Homo erectus*) at Swartkrans in situations that might well have been competitive. Even less is known of the apparent coexistence of *Meganthropus* (= *?Paranthropus*) and *Homo erectus* in the middle Pleistocene of Java. All these cases of sympatry of two species of hominids are from the early (or early mid-) Pleistocene. I agree with Washburn and Howell (1960), who say: "If these creatures had been fully effective hunters, as was the case among later Pleistocene peoples, the presence of two species from a single site would be most improbable." In the later Pleistocene such comparatively different types have been contemporary as Java Man and Heidelberg Man, and as Neanderthal and Recent Man, but never at the same place. Their taxonomic status cannot be deduced from their geographic replacement or morphologic difference, but only by the presence or absence of intermediate populations.

Australopithecines. The Australopithecines apparently were widespread over Africa in the later part of the Villafranchian, judging by the distribution of the pre-Chelles-Acheul implements that characterize them (Howell 1959). Yet so far we have fossils only from eastern Africa ("*Zinjanthropus*") and two series from South Africa, an earlier (*Australopithecus*) and a later (*Paranthropus*). The two South African types, even though not found in the same deposits, are sufficiently distinct to be considered different species. The later one, *Paranthropus*, with its powerful dentition and jaw musculature, appears less human than the earlier and is presumably a side branch. "*Zinjanthropus*," with its very similar dentition, appears not separable from *Paranthropus*, and *Paranthropus* could well be generically combined with *Australopithecus*. Although I consider it highly likely that man has passed through an Australopithecine stage, it is unlikely that a South African population was ancestral to the more advanced types. There have presumably been considerable morphological differences between extreme populations of the widespread, polytypic *Australopithecus* complex. Nominate *africanus,* being a peripheral form in

South Africa, might well have been one of the more extreme and aberrant races of the species. The population that gave rise to the next higher grade of hominid presumably lived elsewhere, perhaps in Ethiopia, Morocco, or east of the Mediterranean, and it might have been more similar to *Homo erectus*. Leakey's recent discovery of a *Homo*-like type in the *"Zinjanthropus"* strata of Olduvai proves conclusively that the separation of *Homo* and *Australopithecus* must have taken place at an earlier period. Nevertheless, the common ancestor must have been Australopithecine in character.

There are two further complications. Fragments of giant hominid jaws (*"Meganthropus"*) have been found in Java which Robinson (1955) considers as belonging to *Paranthropus*. Better specimens are, however, needed before the taxonomic status of *Meganthropus* can be considered established. Finally, at Swartkrans among more than 300 specimens of *Paranthropus* five fragments of a smaller form (*"Telanthropus"*) were found, which plainly belong to another species. They are not clearly different from *Homo erectus;* in particular their similarity to the Heidelberg jaw is most suggestive, nor are they very different from *Australopithecus africanus*.

Pithecanthropines. As the hominids moved up the evolutionary scale they became more mobile and their social groups larger. They became more independent of the environment and the ecological barriers consequently became less formidable. All this favored gene flow within the polytypic species and reduced the probability of successful speciation.

Pekin Man was originally described as a separate genus, *Sinanthropus*. Now there is quite general agreement that he does not differ from Java Man generically, indeed that there is little justification for separating this form even specifically. He agrees with Java Man in so many respects that the two are best combined into a single polytypic species, with Pekin Man designated as *Homo erectus pekinensis*. He is not only a geographic but also a chronologic subspecies, having lived somewhat later in the Pleistocene than did Java Man.

These Pithecanthropines of eastern Asia lived in the Middle Pleistocene, and Heidelberg Man as well as Ternifine Man (in Algeria) belong in the same period. Not enough is known of these western representatives to draw substantial conclusions. The mid-Pleistocene artifacts found in Africa, Europe, and western Asia are different from those of eastern Asia, which indicates lack of gene flow but does not necessarily prove specific distinction. The jaws of Ternifine Man (*"Atlanthropus"*) are rather similar to those of Pekin Man, while Heidelberg Man has remarkably small teeth for his massive jaw. The total morphological variation in mid-Pleistocene is far less than it was in the Villafranchian, but the biological relation of the various populations represented by fossil remains is still quite obscure. There must have been populations of Pithecanthropines in the Old World all the way from the Atlantic to the Pacific, but whether or not they formed an essentially unbroken gradient between Heidelberg Man and Pekin Man only further discoveries can establish. The relationship of the African Pithecanthropines is even more obscure.

Homo sapiens. Finds have turned up from before the Second Interglacial that are strangely similar to Recent Man, or, perhaps more precisely, that combine features of Recent Man and Neanderthal: Swanscombe and Steinheim. They could be remnants of a widespread polytypic species that gave rise to both Neanderthal and Recent Man, and this has been suggested repeatedly. No facts are known that would clearly disprove this possibility, and the "early" Neanderthals, those of the Riss-Würm Interglacial, are indeed considerably more *sapiens*-like than is the later "classical" Neanderthal of the first stage of the Würm glaciation.

Neanderthal Man has long been a bone of contention (Howell 1957). There are those who consider him the brutish stage through which man passed on his way to becoming *sapiens.* Yet all evidence indicates that Neanderthal did not gradually change into Recent Man but died out rather rapidly before or during the first Würm Interstadial. The other extreme is to deny all connection between these two kinds of *Homo* and assert that Neanderthal was an aberrant side branch of the hominid line, who lived as a good species sympatrically with *Homo sapiens.* This alternative likewise is not substantiated by the facts. Although Neanderthal differed from *sapiens* by various skull characters such as a flattened brain case, bun-shaped protuberant occiput, marked projection of jaws, virtual absence of chin, high attachment of muscles on occipital bone, immense supraorbital ridges, very large orbits, and a powerful mandible, he had a cranial capacity at least as large as that of Recent Man and a highly developed paleolithic culture. More important is the distributional picture. In not a single paleolithic site has Neanderthal been found associated with Recent Man. Neanderthal, on the whole, is a western type, with his center of distribution in Europe, although some finds in North Africa, Palestine (Tabun), Iraq (Shanidar), and Turkestan (Teshik-Tash) vastly enlarge his area of distribution. When only the European sites of classical Neanderthal were known, it was tempting to consider him an arctic ecotype, the "Eskimo" of the first stage of the Würm glaciation. However, the finds south and east of the Mediterranean disprove such an assumption. Even so, in view of the absence of established sympatry, he might have been a more northerly and westerly geographical representative of *Homo sapiens sapiens*. The time progression from Steinheim through Fontechevade, Ehringsdorf, and Sacco-Pastore to classical Neanderthal shows that the Neanderthals are not primitive. I agree with Howell (1952), who is "convinced that, at least in the structure of the classic Neanderthal facial skeleton and cranial base, selective forces have been the major contributing evolutionary factor at work."

Whether or not this climatically adapted geographic race of the western Palearctic had reached species status is one of the many Neanderthal problems. The ultimate fate of this type is equally puzzling. Wherever it is found, it is associated with artifacts of a flake culture (Mousterian). It was suddenly replaced by typical modern *Homo sapiens sapiens*, associated with a blade culture (Perigordian). In Europe there is no clear evidence of hybridization between Neanderthal and the invaders. Whether Neanderthal had become extinct before Cro-Magnon Man arrived or whether the latter exterminated Neanderthal is unknown. Remnants of Neanderthal might have been absorbed by Cro-Magnon without leaving demonstrable traces. To add to the difficulties, there are some deposits in which flake cultures seem to evolve into blade cultures or be contemporary with them (Bordes 1960). Alas, no fossil remains seem to be known from these sites.

The best evidence of mixture comes from the two caves of Mount Carmel in Palestine. Both caves were inhabited early in the Würm glaciation. The older cave (Tabun) was inhabited by Neanderthals with a slight admixture of modern characters, the younger cave (Skhul) by an essentially modern population with distinct Neanderthaloid characters. The date is too late for these populations to have belonged to the ancestral stock that gave rise to both Neanderthal and Recent Man. The differences between Tabun and Skhul are too great for the caves to have been inhabited by a single population coming from the area of geographical intergradation between Neanderthal and Recent Man, although this could be true of the Skhul population. Hybridization between invading Cro-Magnon Man and Neanderthal remnants is perhaps still the most plausible interpretation for the Skhul population, while there is no good reason for not considering that of Tabun a regular Neanderthal population, particularly in view of its similarity to the Shanidar specimens (Stewart 1960).

If the skull shape of Neanderthal has a special adaptive significance (but what is it?), it could have evolved repeatedly independently. Two so-called Neanderthaloids, Rhodesian Man in Africa and Solo Man in Java, share the huge supraorbitals, but differ strongly in other features of the skull. They do not seem to be related to European–West Asian Neanderthal. Yet the solution of the Neanderthal problem does not solve the problem of the origin of *Homo sapiens sapiens*. He must have originated in some isolate in Africa or Asia, but where?—Ethiopia? Arabia? India? All we know is that he suddenly broke out of his isolation and invaded Europe some 35,000 years ago. A single lucky discovery may end all our speculating.

THE POLYTYPIC SPECIES HOMO SAPIENS

All the different kinds of living man on the face of the earth belong to a single species. They form a single set of intercommunicating gene pools. As a matter of fact, the various races of man are less different from each other than are the subspecies of many polytypic species of animals. Yet a few misguided individuals have applied a typological species definition to man and have divided him into five or six separate species by using such artificial criteria as white, yellow, red, or black skin color. Such a division not only leaves a considerable portion of mankind unclassified as intermediates or relic primitives, but also is completely contrary to the biological species concept (Chapters 2 and 12). There are no genetic isolating mechanisms separating any of the races of mankind, and even the social barriers function inefficiently where different races come into contact.

It is often asked whether man is in the process of speciating and whether the races of man should be considered incipient species. In attempting to answer this question one must recall that the hominids occupy one of the most spectacularly distinctive adaptive zones on earth. In the animal kingdom the invasion of a new adaptive zone usually results in a burst of adaptive radiation into various subniches. This has not happened in the history of the family Hominidae. Mayr (1950a) has pointed out that this failure of man to speciate is due to two causes. "It seems to me that [one] reason is man's great ecological diversity. Man has, so to speak, specialized in despecialization. Man occupies more different ecological niches than any known animal. If the single species man occupies successfully all the niches that are open for *Homo*-like creatures, it is obvious that he cannot speciate." The second reason is that isolating mechanisms in hominids apparently develop only slowly. There have been many isolates in the polytypic species *Homo sapiens* and in the species ancestral to it, but isolation never lasted sufficiently long for isolating mechanisms to become perfected. Man's great mobility and independence of the environment have made perfect geographic isolation impossible. As a consequence all parts of the globe, including all climatic zones, are now occupied by a single species. What other species of animal includes populations adapted to the Arctic as well as to the tropics, and ranging from almost pure vegetarians to almost pure carnivores? The probability of man's breaking up into several species has become smaller and smaller with the steady improvement of communication and means of transport. The internal cohesion of the genetic system of man is being strengthened constantly.

The Races of Man

There is no agreement yet on the formal classification of the subdivisions of *Homo sapiens*. Of two highly competent treatises on the races of man, both published in 1950, one (Boyd 1950) recognizes 6 races, the other (Coon *et al.* 1950) recognizes 30. Both classifications are equally legitimate. Yet even the division into 30 races is by no means exhaustive. Race number 9 ("Negrito"), of Coon *et al.* lumps together numerous relic populations from the Congo in Africa and from southeast Asia, the Philippines, and New Guinea, populations that (if they are related at all) are less closely related than are the four European races recognized by these authors. The same is true, to a lesser extent, of nearly all the other races. All are collective groupings of more or less differentiated local populations. Even Lundman's recognition (1952) of 37 races and some 30 additional subraces contains several heterogeneous groupings. And yet this is only one of the difficulties. All these are contemporary races of Recent Man. If we go back in history we find chronological subdivisions of *Homo sapiens* such as, let us say, Cro-Magnon Man or more differentiated Neanderthal and finally the Steinheim-Swanscombe Man, the earliest *Homo* that cannot clearly be separated from the polytypic species *Homo sapiens*. Biologically, it is immaterial how many subspecies and races of man one wants to recognize. The essential point is to recognize the genetic and biological continuity of all these gene pools, localized in space and time, and to recognize the biological meaning of their adaptations and specializations.

Adaptiveness of Human Races

A large component of the geographic variation of animals is adaptive, that is, each local race is to a lesser or greater extent adapted to the climatic and other environmental conditions of the given area (Chapter 11). There is no reason why man should be immune to this type of natural selection, and yet there have been many authors who ascribe the differences between human races to "accidents of variation." That this is not the case can be demonstrated, and Coon, Garn, and Birdsell (1950) and Lundman (1952) have accumulated a great deal of evidence for trends of adaptive variation in human races. Pigmentation is almost without exception more intense in the humid tropics than in the more arid and cooler regions (Gloger's rule). Prominent body parts, and the heat-radiating body surface as a whole, are reduced in races exposed to the full brunt of cold winters (central Asia, Arctic), while the opposite trend can be observed in tribes that live in subtropical and tropical savannahs; they have comparatively small bodies with elongated extremities. The effect of selection is particularly obvious where a single racial group has invaded different climatic zones, as the American Indians of Tierra del Fuego and of the Amazon Valley have done. Another case is that of the Lapps in northern Scandinavia. By their blood groups, facial height, and other characters they can be diagnosed rather decisively as a European race. And yet, owing to living in an Arctic climate, they have by convergent evolution acquired many of the features of Mongoloid races living in similar climates (Lundman 1952). Unfortunately it is difficult to get precise figures on the relation between body volume and body surface because body surface is so difficult to measure accurately. Yet whatever data exist (Schreider 1950) show that there is a close parallel between body surface and climate (see also Newman 1953). Information on the physiological adaptations of human races is now also beginning to accumulate (Barnicot 1959).

A special problem is that of the origin of the white race, particularly in its extreme blue-eyed, blond form. This combination of characteristics has a distinctly negative selective value in tropical areas. In order to have become the dominant type in certain regions one must assume that it has had a positive selective value. The suggestion has been made that the cloudy, foggy climate of western Europe with cold rainy summers that prevailed during the last Interstadial and after the retreat of the ice might have favored the origin of this racial type. Nothing further can be said on this subject until we know more about the differential susceptibility to diseases of blond, blue-eyed individuals under the stated environmental conditions. Reduced susceptibility to, let us say, colds or arthritis might be such a factor.

The Amount of Difference Among Human Races

It is relatively easy to describe the differences among human races in terms of dimensions, proportions, pigmentation, hair shape, and other morphological characters. Yet the question is raised again and again: just how meaningful are such morphological data? First of all, they have only a partial genetic basis since, as is well known, pigmentation may depend on exposure to the sun and size on nutrition. More serious is the objection that neither size nor pigmentation is a critical human characteristic. The critical characteristics are intelligence, inventiveness, imagination, compassion, and other traits that are difficult to measure and to compare. As stated above, the extreme viewpoint has been to deny that such differences among human races exist. A more conservative view would be to assume that as a consequence of the manifold genetic differences among human populations there will be average differences for any kind of trait that has at least in part a genetic basis. All attempts to separate genetic and nongenetic characters (or contributions to the characters) have until now been quite unsuccessful. So far as I know, there is not one single mental trait for which a clear-cut racial difference has been established, in spite of the high probability that such differences exist. Twin studies are so far the only reliable proof of the partial genetic determination of mental traits. This much is certain, however, that the differences between individuals of a single population or race are usually larger than those between populations or races.

The Population Concept in Man

As in biology, the outstanding conceptual revolution that has occurred in physical anthropology is the replacement of typological thinking by population thinking. This shift has affected every concept in anthropology, although none as strongly as the race concept. The typological race concept of the racists is something thoroughly odious; the statistically defined race of the botanist and zoologist is a fact of nature. The basis for race formation is the same for all sexually reproducing organisms and consists in the fact that no two individuals are identical nor are any two local populations. No individual can therefore be "typical" of a race. Indeed, in polymorphic races different individuals may be strikingly different. To look for and speak of "pure races" is sheer nonsense. Variability is inherent in any natural population and is favored by natural selection on account of the frequent superiority of heterozygotes and the diversity of the environment (Chapters 9 and 10). What differs from race to race is the degree of variability and this depends on the size of the population, the amount of gene exchange with adjacent populations, the variability of the habitat, and other factors discussed in previous chapters. Much of the phenotypic variability of mankind is presumably due to the occasional production of homozygotes by heterozygous parents. This is particularly true of constitutional extremes, individuals that are exceptionally large, small, obese, or thin.

There are other reasons for man's phenotypic variability. Man is a restless creature and since prehistoric times he has made large-scale migrations. The numerous colonizations of America by Asiatic tribes, the conquest of the South Seas by the Polynesians, the great Bantu migration, and the massive movements of Slavic and Germanic tribes in the dying days of the Roman Empire are only a few spectacular examples. Conquerors almost invariably absorb part of the defeated tribe or nation or are absorbed by it. On this basis one might expect man to have exceptionally high individual variability, but this is not the case. Schultz (1944, 1947) has shown that some of the anthropoids far exceed man in this respect, as do many other animals. Some human populations that are clearly the product of hybridization do not seem to have significantly higher variability than have unmixed races (Trevor 1953).

Polymorphism in Man

In addition to ordinary quantitative variation, which has a polygenic inheritance, we also find genuine polymorphism in man (Chapter 7). The human blood groups (Mourant 1954; Mourant et al. 1958) are the best-known examples of characters with a simple genetic basis. There are now some nine loci recognized, of which the ABO and the Rh are the best known, which determine blood-group characters (Race and Sanger 1954). At first these genes were described as neutral genes, but it is now known that the different genotypes have different viabilities. For instance, individuals with blood group A are more likely to develop cancer of the stomach, and individuals with blood group O duodenal ulcers, than other individuals. Roberts (1957) has given a summary of this literature and Levine (1958) has reviewed the interactions of mother and fetus. Another interesting case of human polymorphism is that of a deficient type of hemoglobin which produces the so-called sickle-cell anemia in regions with subtertian malaria (Chapter 9).

The assumption that superiority of the heterozygotes is responsible for most cases of human polymorphism (Ford 1945) is based mainly on analogy with the better-analyzed cases of polymorphism in lower organisms. The average type, the average individual, seems very often best adapted and this is particularly true of the human "constitution." Extreme constitutional types are quite obviously less well adapted and often visibly more susceptible to illness. Still they are a normal component of every human population, since their production (through recombination of parental gene combinations), is the inevitable by-product of the turnover of the gene pool in every generation. In addition, much human polymorphism may be selectively favored by the diversity of the environment (Chapter 9).

Identity versus Equality

That no two individuals are alike is as true of the human population as of all other sexually reproducing organisms. Every individual is unique and differs in a large number of morphological, physiological, and psychological characteristics from all other individuals. Each individual is a different combination of characters and of the genetic factors on which these characters are based. Much evidence for this high individuality of man has been gathered by Williams (1953, 1956). The humanitarian answer to the challenge of genetic variability within and between human populations is the principle of equality. Simply stated, equality means equal status before the law and equal status in human social relations in spite of genetic difference. Equality, as Dobzhansky has stressed, is a social and ethical concept, not a biological one. Equality means equal opportunity to make the best of one's genetic endowment.

That society enjoys the greatest amount of liberty in which the greatest number of human genotypes can develop their peculiar abilities. It is generally admitted that liberty demands equality of opportunity. It is not equally realized that it also demands a variety of opportunities and a tolerance of those who fail to conform to standards which may be culturally desirable but are not essential for the functioning of society . . . If a nation were a pure line there would be little scope for liberty . . . There would be no freedom, no deviants, and no progress (Haldane 1949a).

Equality in spite of evident nonidentity is a somewhat sophisticated concept and requires a moral stature of which many individuals seem to be incapable. They rather deny human variability and equate equality with identity. Or they claim that the human species is exceptional in the organic world in that only morphological characteristics are controlled by genes and that all other traits of the mind or character are due entirely to "conditioning" or other nongenetic factors. Such authors conveniently ignore the results of twin studies and of the genetic analysis of nonmorphological traits in animals. An ideology built on such obviously wrong premises can lead only to disaster. Its championship of human equality is based on a claim of human identity. As soon as it is proved that the latter does not exist, the support for equality is likewise lost.

The denial of genetic difference among human beings with respect to intellectual and character traits is based on fallacy. This ideology is particularly pernicious when applied to education. The widely preached principle of "the same education for everybody" denies equal opportunities because differently endowed pupils would undoubtedly obtain different kinds, rates, and degrees of education if truly given "equal opportunities." Educational identicism is antidemocratic. According to the concepts of true democracy, as developed in the period of the Enlightenment, the free spirits of young men and women should not be shackled by the leveling restrictions of a false identicism. Every school class is a variable population and true equality (of opportunity) for the pupils can be achieved only by making allowance for these differences, not by suppressing them. There is still much to be learned about the relative contribution of inheritance and environment to individual human traits. This much, however, is already known: inheritance *does* play a considerable role and it can only do harm to ignore this role.

Claims of human identity are the outcome of typological thinking, of a belief that within the human type there is "no essential variation." Political theorizers have invariably applied such typological formulas when trying to resolve the difficulties posed by man's variability. The racism of the Nazis, for instance, was an outcome of such thinking. They rigidly defined each race by absolute characteristics: the X race "is lazy," the Y race is "of great intelligence," the Z race is "musical," and, worst of all, the A race is "superior." This allowed neither for the fact that many of the characteristics mentioned have only a partial (often a very small) genetic component, nor for the fact that many members of the various races do not have these characteristics at all. Another fallacy of typological racism is that it claims perfect correlation between the various characteristics ascribed to each race. Accordingly it claims an association between a particular color of the eyes or the hair and certain traits of the mind or the character. Actually, all available evidence negates the existence of such absolute correlations.

Every politician, clergyman, educator, or physician, in short, anyone dealing with human individuals, is bound to make grave mistakes if he ignores these two great truths of population zoology: (1) no two individuals are alike, and (2) both environment and genetic endowment make a contribution to nearly every trait.

BIOLOGICAL FACTORS IN THE EVOLUTION OF MAN

The phylogeny and population structure of man are now understood in their broad outlines. But much as we may know about the "how" of human evolution, the "why" is still a great puzzle. High intelligence and harmonious social integration are undoubtedly attributes of high selective value, so much so, indeed that we may ask with Etkin (1954) "why all animals are not as intelligent as Einstein and as moral as Albert Schweitzer?" What has enabled man to break the chains of his animal heritage and to evolve from a primate to his present level? The crucial stages of this development took place in the past and like all historical events can never be tested experimentally. However, as Bartholomew and Birdsell (1953), Etkin (1954), and others have pointed out, an analysis of the ecological conditions under which primitive hominids lived and of their social and population structure permits the reconstruction of a fairly realistic model of the crucial selection pressures.

The most astounding phenomenon of human evolution is the rapid increase in brain size during the Pleistocene, discussed above. Hominid evolution up to and including the *Australopithecus* stage is characterized by a shift into the terrestrial niche, the assumption of upright posture, the freeing of the hand for manipulation, and the increasing use and manufacture of tools. Man's ancestor was a highly vulnerable creature at this stage, since its locomotion was none too efficient and its niche none too hospitable. What the original hominids lacked in speed, in natural weapons, in nocturnal secretiveness, and in arboreal safety they had to counterbalance by the development of the various facilities now considered peculiarly human. These include the widespread use of tools, the use or manufacture of devices for protection against the inclemency of the weather, and particularly vocal communication (speech). Mortality from enemies, famine, exposure, and intraspecific strife must have been high. The premium on the ability to cope with these difficulties must have been correspondingly high.

There are two biological factors that may have facilitated this development. One is the population structure of primitive hominids. They faced their adverse environment not as individuals but as family groups or small bands. The unit of evolution was not the individual but the population. As soon as this is appreciated it becomes apparent that the description of natural selection so prevalent in popular writings and among post-Darwinian philosophers and sociologists is not correct. To describe natural selection in terms of brute force—"nature red in tooth and claw" and "survival of the fittest"—places a misleading emphasis on only a single aspect of natural selection. Inventiveness, foresight, leadership, and, in many cases, cooperation, far more than brute force, will be favored by natural selection in an animal with the social population structure of the primitive hominids, as indeed was already pointed out by Darwin himself. As Julian Huxley has demonstrated so convincingly, there is no conflict between natural selection and human ethics. He who makes the greatest contribution to the harmony and well-being of the group may thereby become the ancestor of the greatest number of surviving descendants. "Ethical" qualities, in a social organism, are apt to be important components of fitness.

Polygyny is more or less developed in nearly all anthropoid apes. There are good reasons for postulating that it was characteristic of the primitive hominids * (Bartholomew and Birdsell 1953). What effect on evolutionary rate would polygyny have? If the leader of a group has several wives (perhaps even all the mature females of the family group), he will contribute a far greater than average share to the genetic composition of the next generation of his group. Such tremendous reproductive advantage of a leader in a self-contained family group or tribe would favor the very characteristics that have made man what he is. We have discussed in Chapter 8 the inherent dangers of mere "reproductive success" if it does not add to the adaptedness of the population. However, the condition described for primitive hominids is completely different from the situation in birds of paradise or other organisms in which the favored male does not contribute in any manner to the survival of his offspring. Reproductive advantage in the primitive hominid society, we may speculate, was not a matter of any bizarre secondary sex characters, but one of social position within the group which depended on defi-

---

* It may still be an original condition in a few living primitive tribes. Most cases of polygyny among contemporary peoples were, however, secondarily derived from a preceding monogamy.

nite qualities, physical as well as mental. These in turn depended to a considerable extent on the genetic endowment of the individual. In this case, then, reproductive advantage results in a maximal contribution to the fitness of the entire group.

The second biological factor responsible for the acceleration of hominid evolution is parental care. Much of the mortality in animals, particularly at the immature level, is accidental and haphazard. When 95 or 99 percent of the zygotes are killed by such accidents as are caused by weather or indiscriminate predators (such as a whale straining out planktonic larvae), it leads to a considerable reduction in the efficiency of selection. The institution of parental care permits a marked decrease in such random mortality. Survival now depends increasingly on the goodness of the care bestowed upon the child by the parents. The high selection pressure in favor of brain development in such a breeding system is obvious. It is reinforced by an interesting feedback. The increase in brain size results in a slowing down of the development of the human infant and a lengthening of the period during which parental care is required (Portmann 1956). This, again, increases the selective value of parental care and once more exerts a selection pressure in favor of increased brain size in the parents.

This mutual reinforcement of selection pressures can account quite efficiently for the rapid increase of brain size in the hominids but it does not tell us why this evolutionary trend came to such a sudden halt. There has been no increase in brain size since the time of Neanderthal (an average of about 1350 cm^3 nearly 100,000 years ago!). To be sure, there may have been an improvement of the brain without an enlargement of the cranial capacity but there is no real evidence of this. Something must have happened to weaken the selection pressure drastically. We cannot escape the conclusion that man's evolution toward manness suddenly came to a halt. As evolutionists we must attempt to find the cause of this drastic reduction of the selective advantage of increased brain size after the *Homo sapiens* level had been reached. It is suggested that among several possible and potential factors two are most important. One is an increase in the size of the "unit of selection," that is, the family group, tribe, or nation which as a whole has a selective advantage in competition with similar units. The larger such a unit is, the relatively less will the genes of its leader contribute to the gene pool of the next generation and the more protected (biologically) will be the average or below-average individual of the group. Reproductive success will no longer be closely correlated with genetic superiority. There is evidence for an even greater reduction of the selective premium on genetic superiority in recent times. Add to this the dysgenic effect of urbanization and of density-dependent diseases, and it becomes apparent why the trend that created man has not continued to produce superman. The social structure of contemporary society no longer awards superiority with reproductive success. Another consequence of the increased population size is that it slows down genetic turnover (Chapter 15). The larger a population is, the smaller the initial frequency of new genes and gene combinations and the slower their ascendancy. The development of cultural tradition and the steady improvement in means of communication finally contribute to still another factor complex that has reduced selection pressure. All the members of a community benefit from the technological and other achievements of the superior individuals and this helps the below-average individual, provided he is not too far below average, to make a living and to reproduce as successfully as the above-average one. We are not interested at this point in discussing whether this development is good or bad or whether anything should be done about it. All I want to point out is the highly interesting phenomenon of the almost abrupt flattening out of an exceedingly steep evolutionary advance, a phenomenon for which the evolutionist must attempt to find a causal explanation.

BIOLOGICAL PROBLEMS OF MODERN MAN

Students of man have long been aware that no biological phenomenon can be fully understood unless its evolutionary aspects are also understood. As a consequence they have given much consideration to man's evolution, including his evolutionary future (for instance Huxley 1953; Huxley and Huxley 1947; Medawar 1960; Dobzhansky 1962; Simpson 1949, 1960c; and Tax 1960a). Evolutionary thinking is no longer considered the source of all evil in economic and sociological theory, as it was only a few decades ago (Loewenberg 1957; Tax 1960b). One no longer finds such aberrant views as those of George Bernard Shaw who wrote in 1920 (*Back to Methuselah*, p. x): "Neo-Darwinism in politics has produced a European catastrophe of a magnitude so appalling . . ." When Shaw promoted Neo-Darwinism to the role of his favorite bogy, he merely revealed that he did not understand evolutionary theory at all. There is much in various typological creeds that was responsible for the catastrophic European politics of the period 1900 to 1950, but nothing in Darwinism. European politics has been typological, dysgenic, and counterselective, in complete contrast to everything that Darwinism would demand or postulate. We have advanced far beyond the antievolutionary prejudices of the early decades of this century, but there is still confusion in the contemporary literature. Applying the findings of population genetics and population systematics to man might help to clarify some of our thinking.

The evolution of an organism has two aspects which, although they always go hand in hand in nature, should not be confused (Mayr 1956). One is evolution toward ecotypic adaptation and the other is progressive or phyletic evolution. The often posed question, "Is man going downhill biologically?" cannot be answered until it is phrased more precisely. Which evolutionary aspect is meant: whether man as an organism is becoming less well adapted to his environment, or whether man is losing some of his most characteristically human attributes? It seems to me that in the current literature these two questions are frequently confused.

Is Man Becoming Less Well-Adapted to His Environment?

Modern man is an almost supreme master of his environment. Clothing, housing, heating, travel, transport, food production, and food storage have made him independent of the environment to a degree unparalleled elsewhere in the animal kingdom. The environment as such has become a far less severe selective factor than in any wild animal, and even than in the ancestral hominids. The conquest of disease and the mitigation of the effects of aging have achieved spectacular results. And these developments will continue to make steady progress regardless of minor changes in the genetic composition of mankind. The genetic results will be the survival of genotypes, one might say a normalization of genotypes, that formerly were highly deleterious, such as those of diabetics, sufferers from Addison's disease, or poor antibody formers. It seems to me that in discussions of man's future far too much space is given to the role of genes controlling such diseases or, let us say, to the various metabolic disturbances that characterize most genetic diseases. A rise in the frequency of such genes will have no drastic effect on the future of mankind as long as adequate medical facilities are available. The increase in frequency of such genes in the human species indicates a relaxation of normalizing selection (Chapter 8) which has exceedingly little to do with man's phyletic evolution. The futility of eugenics based on endeavors to eliminate such "bad" genes has been brilliantly demonstrated by Dunn and Dobzhansky (1946). Such genes will become a real problem for mankind only if sufferers from such diseases should occur in a large number of families and thereby interfere with the proper functioning of family life. This indeed could conceivably happen as a result of greatly increased atomic-radiation damage. A breakdown of civilization owing to overpopulation or wartime destruction, resulting in a widespread unavailability of modern heating, housing, food, and medicines, might likewise convert many now "normal" genes into lethal ones. Muller (1960) considers the accumulation of deleterious genes a real threat to man.

One of the major discoveries of population genetics is that population size and population structure may have a considerable effect on fitness. Marriage in primitive man usually takes place between members of the same tribe or group and leads to moderate inbreeding, even though there are nearly always taboos against extreme inbreeding as well as provisions for intertribal marriages. Civilized man has been far more endogamous, until recently, than is usually realized. Not only was there a high frequency of marriages among known relatives (cousins and others), but there were all sorts of social, religious, and economic barriers that tended to limit the choice of mates. That this was not necessarily deleterious is shown by rather inbred families that were nevertheless healthy and rich in children and men of achievement.

The amount of outbreeding has dramatically increased in modern man, leading to a great enlargement of the effective breeding population and greatly increased genetic variability of the individual deme. This quite rightly is generally considered a favorable development, since it reduces the danger of deleterious genes becoming homozygous. It is, unfortunately, not an unmixed blessing, since such outbreeding merely postpones the day of reckoning. "Genetic death" is controlled by rate of mutation regardless of population size. It is often ignored in discussions of outbreeding that the beneficial effects may not materialize if the mixed gene complexes are too different. Hybridization between species leads almost invariably to unbalance through deleterious combinations of genes. We have presented evidence in Chapter 10, based on the work of Stone, Brncic, Vetukhiv, and Wallace, that in *Drosophila* even the hybridization of races may lead to destruction of well-integrated gene combinations. For man Medawar (1960) has cautiously stated the case as follows: "Hybridization between people of different races need not be expected to lead to an improvement, because both races will probably have adopted the well-balanced genetic constitution that matches their own environment." To what extent the findings on F_2 incompatibility can be applied to man is still uncertain. All investigations of race mixtures in man, such as those of Fischer (1913) on the Rehoboth Bastaards and of Shapiro on the Pitcairn Islanders (1929), have failed to produce any evidence of decreased fitness. On the contrary, these mixed populations seem to be exceptionally vigorous and well adapted. It must be admitted, however, that the data are not precise enough (mortality and fertility data for F_2, F_3, and so on are scarce or absent) to permit far-reaching conclusions; nor is there much evidence on nonphysical traits. Indeed, none of these hybrid populations has produced an eminent person.

Natural Selection in Man

In view of his general emancipation from the environment, it is legitimate to ask whether natural selection still operates in man. Let us remember that the tremendous ecological evolution of man from a semistarved, hunting and hunted caveman to the machine man of the atomic age has taken place without visible biological evolution (Stebbins 1952). Cro-Magnon man, who entered history about 30,000 years ago, differs physically from modern man no more than do various modern races of man from each other. The great humanization of man during recent millennia is primarily a result of his ability to transmit nongenetic components of culture,* including all sorts of scientific and technological information. This has been interpreted by some authors to mean that man has emancipated himself from natural selection. For instance, La Barre (1954) has asserted that "within his own species he has almost abolished the evolutionary significance of heredity and natural selection." This assertion is based on either or both of two unspoken assumptions: that there are no genetic differences among human individuals in any but physical features, and that whatever differences there are, are without selective significance.

---

* Cultural tradition is not altogether absent elsewhere in the animal kingdom. A well-developed family structure is found not only in primates but also in certain groups of birds. Male and female are permanently mated in many species of tropical birds, but this does not lead to a well-knit family group because the young leave the companionship of the parents soon after fledging. True family groups are known in the temperate zone among the geese (Anserinae) and the cranes (Gruidae). Here the young of the year not only migrate southward with the parents, but remain closely attached to them during the entire winter and during most of the homeward migration in spring. The family unit apparently does not break up until shortly before the beginning of the new breeding season. In both groups of birds, curiously, only a minimum of tradition seems to have evolved. The only known traditions have to do with migratory routes and winter quarters and, if the observations of the naturalists are correct, a taboo on incest. A comparison of the family structure in the lower primates with the family structure in these birds might reveal interesting differences of biological significance.

Neither of these assumptions is tenable. The considerable mortality of human zygotes before the age of reproduction (Crow 1958) and the highly differential fertility among human individuals indicate clearly how active natural selection still is today. The importance of natural selection for modern man has been reaffirmed by several recent authors (Dobzhansky and Allen 1956; Neel 1958). This is not to deny that the nature of selection pressures changes from generation to generation and that the intensity of selection is relaxed in times of rapid expansion of the human population and of great medical discoveries. Yet wherever there is an acute shortage of food or poor hygienic conditions—and this applies to more than 50 percent of mankind—even the crudest form of natural selection is still a highly effective agent. And where prereproductive mortality has been largely eliminated as a selective factor, it has been replaced by unequal rates and ages of reproduction, both of which make an enormous potential contribution to Darwinian fitness (Crow 1961). Opportunity for selection owing to mortality before the end of the reproductive age has dropped in the United States by about 90 percent between the late nineteenth century and 1950. The importance of selection by unequal reproduction has increased correspondingly. Let me illustrate this by one example. The living Ramah Navaho Indians go back to 29 founders. Fourteen of these produced 84.48 percent of the descendants, 14 others a total of only 13.56 percent of the descendants (Spuhler 1959). This certainly represents an enormous selection differential for whatever genetic differences existed between the two groups of founders. Many comparisons of rates of reproduction of different social classes or sympatric human races indicate that differential reproduction nowadays makes a far greater contribution to selection than prereproductive mortality. Together, the two components are so large that claims that natural selection is unimportant for modern man appear quite absurd.

We are now prepared to answer the question whether modern man is less well adapted than was man in former generations. I think we can say confidently that, allowing for the change in environment, there is no appreciable difference. Perhaps we are losing the ecotypic adaptations that permitted human races to flourish in special environments. We may be losing the extremes of pigmentation, body proportions, and resistance to cold or heat that make the Eskimos successful in the Arctic and the Watusi in tropical savannahs. The number of genes involved in ecotypic adaptation must be very large, considering how many generations of counterselection are required to modify such genotypes after migration to a different climatic zone. Yet, with the increasing emancipation of civilized man from the selective forces of the physical environment, the importance of these ecotypic adaptations is being steadily reduced. Their gradual loss will not make modern man less well adapted in the new environment he occupies. Nor can relaxation of the pressure of normalizing selection under the conditions of civilization be considered a serious loss of adaptation.

The danger that man will become extinct is negligible, unless he exterminates himself through atomic war or some other act of stupidity. No other organism can live successfully in so many climatic zones and in so many habitats. Man is sufficiently polymorphic that even the most devastating diseases should leave survivors. The semiisolated life of many primitive societies of man increases this probability of survival. On purely biological grounds, then, there is not much reason to worry about the genetic continuity of mankind. But this still leaves our second question unanswered.

Is Man Losing His Most Typically Human Characteristics?

Man's phyletic evolution was characterized by the acquisition of upright posture, the adjustment of the hand for all the skills for which the hand is now used, and, most importantly, the development of speech, the ability of abstract thinking, and the many other uniquely human characteristics associated with man's large brain and cultural evolution. Considering the magnificent achievements of human art, literature, science, and technology, man has good reason to be proud of past progress. Looking at these positive aspects of man's evolutionary history, one can understand the arrogant assertion sometimes made that man is good enough as he is. Yet, when one considers with what futility the human brain has tackled the major problems of the universe, one is forced to display more humility. And when one observes with what speed human individuals and whole societies can degrade themselves, one cannot avoid feeling that man could still go a long way on the road toward improvement. Nevertheless, there is no evidence of any biological improvement in at least the last 30,000 years. On the contrary, as Huxley (1953) has correctly pointed out, it is probable that

man's genetic nature has degenerated and is still doing so . . . There is also the fact that modern industrial civilization favors the differential decrease of the genes concerned with intelligence. It seems now to be established that, both in communist Russia and in most capitalist countries, people with higher intelligence have, on the average, a lower reproductive rate than the less intelligent; and that some of this difference [in intelligence] is genetically determined. The genetic differences are slight, but as I pointed out in my second chapter, such slight differences speedily multiply to produce large effects. If this process were to continue, the results would be extremely grave.

There is hardly a more controversial question in the field of human biology than the question of the genetic contribution to intelligence and its correlation with fertility. Typologists are utterly incapable of discussing this subject because they fail to realize that these are statistical phenomena with highly incomplete correlations. Everyone now recognizes the inadequacy of IQ tests, particularly the verbal-facility tests. Everyone likewise realizes the considerable contribution made by home background, motivation, and general education to performance in such tests. Yet, when all these factors are duly considered, there is no doubt that there is some correlation between performance in IQ tests and genuine intelligence. Nor can there be any question that intelligence has in part a genetic basis, and it is quite immaterial for our argument whether the heritability of intelligence is 25 percent or 75 percent. Finally, there is abundant statistical evidence that in most communities those people whose professions require high intelligence produce on the average smaller families and at a later age than do people like unskilled laborers, whose professions do not make such requirements. Even though it is still being heatedly denied by identicists, the weight of the available evidence fully supports Huxley's conclusion that those who are intellectually best endowed contribute less to the gene pool of the next generation than do the average and, indeed, most of the less than average. Here we have another illustration of the principle discussed in Chapter 8, that natural selection is unable to discriminate between reproductive success as such and reproductive success owing to the possession of characteristics benefiting the adaptation of the species as a whole.

It has been claimed that the higher intellects are nothing but a homozygous "fringe" due to the segregation of previously balanced heterozygous gene complexes and that their lower reproductivity is a result of homozygous inferiority. I have no doubt that this is true to some extent but it seems to me that there is much evidence to discredit it as a broad generalization. First of all, this claim is based on the assumption that the low fertility of those with higher intelligence is a biological characteristic. All special investigations, however, indicate that the low fertility of professionals is largely due to family planning, and not to neurosis or physiological deficiencies. A study of exceptionally high IQ's made by Terman and Oden (1959) in California revealed that they surpassed the average of the population in mental and physical health, indeed, that they were highly superior. Further refutation of the claim that most of the low fertility of the superior intellects is biological consists in the fact that in the same circle of intellectuals in which it was fashionable in the 1920's and 1930's to have two children it became in the 1950's and 1960's fashionable to have four or five children. There is no evidence whatsoever to suggest that the lowered fertility of the higher intellects is comparable, let us say, to the decreased fertility of *Drosophila* strains selected for high bristle number. Intelligence, at least in the human species, is in itself a strong viability characteristic and will therefore be different in its response to selection from such quantitative aspects of the phenotype as size or bristle number which, to a considerable extent, are pleiotropic by-products of genes forming the general genetic background of the viability genes. This consideration is another objection to the balanced-polymorphism theory of intelligence and fertility.

The assumption that all directional selection leads to a loss of fitness is a fallacy widely accepted among students of selection. It is based on experiments, discussed in Chapter 10, in which closed populations were exposed to extreme selection pressures. Such completely artificial conditions should not be used as the basis for broad generalizations. In natural populations, no matter how strong the directional selection pressure, there is simultaneously always extremely high selection in favor of general fitness. These being open populations, ever new gene combinations can be tested for the desired phenotype. The rapid increase in brain size between the Australopithecines and *Homo sapiens* proves how drastic a change of phenotype is possible without any loss of fitness. The spectacular achievement of selection in domestic animals and cultivated plants is further proof of the possibility of combining progressive selection with retention of fitness in the respective environment. There is thus no genetic reason why the increase in brain size could not be continued if there were a selective premium on such a process. The possibility exists that the plateau in brain size reached in the curve of hominid brain evolution is indicative of an upper ceiling. The head of the human baby has to pass at birth through the pelvis of the mother and too large a brain unquestionably adds to the perils of the birth act and would be selected against. Yet natural selection would have alternative pathways if the selection for large brain size were strong enough: (1) increase in the size of the female pelvis, (2) shortening of the length of pregnancy, (3) shift of a greater portion of the growth period of the brain to the postnatal stage. These and other considerations support the conclusion that increase of brain size stopped not because it ran into a road block of direct counterselection, but rather because it was no longer awarded with a reproductive premium.

Eugenic Measures

Those who are convinced that counterselection is operating in modern man and that the most desirable genes and gene combinations are not maintained at current frequencies in the total gene pool of the human species are at present vastly in the minority. They may even be in a minority among the geneticists and students of human populations. But let us assume, for the sake of the argument, that they are right. It would then be our duty to propose countermeasures. Many such measures have indeed been proposed. Unhappily, most of them are unpalatable to liberal-minded, freedom-loving modern man. Worse than that, many of them are merely negative eugenics and will not materially contribute toward the desired end. However, the situation is not altogether hopeless.

Animal breeding has long abandoned all attempts to discover superior genes individually. In fact, such desirable economic features as high egg production in chickens or high milk output in dairy cows are exceedingly difficult to analyze, not only genetically but even physiologically. All sorts of generalized factors, such as resistance to disease, superior utilization of food, and so forth, contribute largely to the goal of selection. One could readily translate this into terms of desirable goals for human biological progress. Perhaps it is not unreasonable to assume that a person with a good record of achievement in certain areas of human endeavor has on the average a more desirable gene combination than a person whose achievements are less spectacular. In our present society, the superior person is punished by the government in numerous ways, by taxes and otherwise, which make it more difficult for him to raise a large family. Why, for instance, should tax exemption for children be a fixed sum rather than a percentage of earned income? Why should tuition in school be based, in large part, on the ability of the father to pay rather than inversely on the achievement of the student? Innumerable administrative rules and laws of the government discriminate inadvertently against the most gifted members of the community. Changing these laws so as to place a premium on performance (the "opportunity" of true democracy, rather than identicism) is entirely different from distributing privileges according to the artificial, arbitrary criteria of the racists, such as blond hair and blue eyes. I firmly believe that such positive measures would do far more toward the increase of desirable genes in the human gene pool than all the negative measures proposed by eugenicists of former generations. More rapid progress would presumably be made if the sperm-bank proposal of Muller (1960) were adopted, but its psychological feasibility remains to be demonstrated.

Overpopulation

Alas, the progressive loss of valuable genes is not the only danger facing the human species. Indeed, overpopulation is a far more serious problem in the immediate future. I am not speaking of the material aspects such as the exhaustion of mineral and soil resources and the increasing difficulty of food supply for 6, 8, or 10 billion people. Human technology may find answers to all these difficulties. Yet I cannot see how all the best things in man can prosper—his spiritual life, his enjoyment of the beauty of nature, and whatever else distinguishes him from the animals—if there is "standing room only," as one writer on the subject has put it. It seems to me that long before that point has been reached man's struggle and preoccupation with social, economic, and engineering problems would become so great, and the undesirable by-products of crowded cities so deleterious, that little opportunity would be left for the cultivation of man's highest and most specifically human attributes. Nor do I see where natural selection could enter the picture to halt this trend. Man may continue to prosper physically under these circumstances, but will he still be anywhere near the ideal of man? Let us hope that the biological aspects of man's evolution are duly taken into consideration by those entrusted with the task of planning for the future of mankind.

Glossary

Acrocentric chromosomes. Chromosomes with the centromere at or near one of the ends; rod-shaped chromosomes.

Adaptive radiation. Evolutionary divergence of members of a single phyletic line into a series of rather different niches or adaptive zones.

Additive variance. That due to the average value of the different genes.

Allele. Any of the alternative expressions (states) of a gene (locus).

Allometric growth. Growth in which the growth rate of one part of an organism is different from that of another part or of the body as a whole.

Allopatric. Of populations or species, occupying mutually exclusive (but usually adjacent) geographical areas.

Allopatric hybridization. The hybridization of two allopatric populations or species in a zone of contact.

Allopolyploid. A polyploid that originated by the doubling of the chromosomes of a zygote with two unlike chromosome sets, usually owing to hybridization of two species.

Amphiploid. Allopolyploid.

Anadromous. Wandering upstream (from the sea) in order to spawn, like the salmon.

Anagenesis. Progressive ("upward") evolution.

Arrhenotoky. The haplodiploid parthenogenesis in which males arise from unfertilized, hence haploid, egg cells.

Autopolyploid. A polyploid originating by the doubling of a chromosome set.

Balanced load. That which depresses the over-all fitness of a population owing to the segregation of inferior genotypes the component genes of which are maintained in the population because they add to fitness in different combinations (for instance, as heterozygotes).

Balanced polymorphism. A polymorphism maintained by a selective superiority of the heterozygotes over either type of homozygotes.

Batesian mimicry. The mimicking (similarity) of a species distasteful or dangerous to a predator by unrelated edible species.

Biological species concept. A concept of the category species based on the reproductive isolation of the constituent populations from other species.

Biota. Fauna and flora together.

Biotype. A group of genetically identical individuals.

Blending inheritance. The complete fusion of the genetic factors of father and mother in the offspring.

Canalization. The property of developmental pathways of achieving a standard phenotype in spite of genetic or environmental disturbances.

Canalizing selection. The selection of genes that would stabilize the developmental pathways so as to make the phenotype less susceptible to the effect of environmental or genetic disturbances.

Category (taxonomic). One of a hierarchy of levels to which taxa are assigned, such as subspecies, species, genus, and so forth.

Cenospecies. All the ecospecies that are so related that they may exchange genes among themselves to a limited extent through hybridization.

Centers of diversification. Geographic areas with the greatest number of different cultivated strains.

Centric fusion. The fusion of two acrocentric (rod) chromosomes into a single metacentric (V) chromosome through translocation and loss of a centromere.

Centromere. A special region on the chromosome where it becomes attached to the spindle.

Character divergence. The name given by Darwin to the differences developing in two (or more) related species in their area of sympatry owing to the selective effects of competition.

Chiasma. An X-shaped chromosome configuration caused by the breakage, exchange, and reciprocal fusion of equivalent segments of homologous chromatids (in meiotic cell division).

Chronocline. A character gradient in the time dimension.

Circular overlap. The phenomenon in which a chain of contiguous and intergrading populations curves back until the terminal links overlap with each other and behave like good species (noninterbreeding).

Cistron. The functional gene; the totality of sites on a gene locus that jointly control a unitary function (for example, the formation of an enzyme) (as shown by noncomplementarity and a cis-trans effect of a set of recessive mutants).

Cladogenesis. Branching evolution.

Climatic rules. Rules describing regularities in geographic variation correlated with climatic gradients.

Clone. All the individuals derived by asexual reproduction from a single sexually produced individual.

Closed population. A population with no genetic input other than by mutation.

Competition. The simultaneous seeking of an essential resource of the environment that is in limited supply.

Conditioning. The process of acquisition by an animal of the capacity to respond to a given stimulus with the reaction proper to another stimulus when the two stimuli are applied concurrently for a number of times.

Controlling factor. With reference to competition, any factor the effect of which becomes more severe as the density of the population increases.

Correlated response. Change of the phenotype occurring as an incidental consequence of selection for a seemingly independent character, such as sterility resulting from selection for high bristle number.

Crossing over. The exchange, during meiosis, of corresponding segments between homologous chromosomes.

Cyclomorphosis. A cyclic, seasonal change of form in a series of genetically identical populations, as in cladocerans and rotifers.

Cytoplasmic factor. A genetic factor in the cytoplasm.

Deme. A local population, as defined in Chapter 7, pp. 130–137.

Developmental homeostasis. The capacity of the developmental pathways to produce a normal phenotype in spite of developmental or environmental disturbances.

De Vriesianism. The hypothesis that evolution in general and speciation in particular are the results of drastic mutations (saltations).

Diapause. A temporary interruption of growth in the embryos or larvae of insects, usually during hibernation or aestivation.

Dioecious. Having the male and the female reproductive organs segregated into different individuals.

Diploid. Having a double set of chromosomes, as is typical in animals of most individuals derived from a fertilized egg cell.

Disruptive selection. Selection for phenotypic extremes in a population (until a discontinuity is achieved).

Dominant. Of an allele, determining the phenotypic appearance of the heterozygote.

Dosage compensation. The effect produced by modifying genes that compensate for the difference between the dosage of major sex-linked genes present in male and female.

DNA. Deoxyribonucleic acid.

Ecogeographical rules. The formulation of regularities in geographic variation (of size, pigmentation, and so forth) correlated with environmental conditions.

Ecological race. A local race that owes its most conspicuous attributes to the selective effect of a specific environment (see Ecotype).

Ecophenotype. A nongenetic modification of the phenotype in response to an environmental condition.

Ecospecies. "A group of populations so related that they are able to exchange genes freely without loss of fertility or vigor in the offspring" (Turesson).

Ecotype. A local race that owes its most conspicuous characters to the selective effects of local environments (see Ecological race).

Ectothermal. Having the body temperature determined by the temperature of the environment; poikilothermal.

Edaphic race. A race that is affected by the properties of the substrate (soil) rather than by other environmental factors.

Eidos. Any of the fixed types (ideas) that Plato conceived to underlie the apparent variability of phenomena.

Epigamic. Serving to attract or stimulate individuals of the opposite sex during courtship.

Epigenetic. Developmental;—referring to the interaction of genetic factors during the developmental process.

Epigenotype. The total developmental system; the totality of interactions among genes resulting in the phenotype.

Epistatic interaction. An interaction of nonallelic genes.

Ethological. Behavioral, particularly with reference to species-specific behavior elements the phenotype of which is largely determined genetically.

Ethological barriers. Isolating mechanisms caused by behavioral incompatibilities of potential mates.

Evolutionary novelty. A newly acquired structure or other property that permits the performance of a new function.

Exclusion principle. The principle stating that two species cannot coexist at the same locality if they have identical ecological requirements.

Fecundity. Reproductive potential as measured by the quantity of gametes, particularly eggs, produced.

Fertility. Reproductive potential as measured by the quantity or percentage of developing eggs or of fertile matings.

Firefly. A beetle of the family Lampyridae.

Formenkreis. Kleinschmidt's term for an aggregate of geographically representative (allopatric) species and subspecies.

Founder principle. The principle that the founders of a new colony (or population) contain only a small fraction of the total genetic variation of the parental population (or species).

Gause principle. Exclusion principle.

Gene arrangements. Alternative gene sequences on chromosomes, owing to inversion, translocation, or other chromosomal changes.

Gene flow. The exchange of genetic factors between populations owing to the dispersal of gametes or zygotes.

Gene pool. The totality of the genes of a given population existing at a given time.

Genetic drift. Changes in the gene frequencies of populations not due to selection, mutation, or immigration.

Genetic homeostasis. The property of the population of equilibrating its genetic composition and of resisting sudden changes.

Genotype. The totality of genetic factors that make up the genetic constitution of an individual.

Geoffroyism. The belief in an adaptive response of the genotype to the demands of the environment; environmental induction of appropriate genetic changes;—usually (though not strictly correctly) included with Lamarckism.

Geographic barrier. Any terrain that prevents gene flow between populations.

Geographic isolate. A population that is separated by geographic barriers from the main body of the species.

Geographic isolation. The separation of a gene pool by geographic barriers; the prevention of gene exchange between a population and others by geographic barriers.

Geographic speciation. The acquisition of isolating mechanisms by a population during a period of geographic isolation.

Geographic variation. The differences between spatially segregated populations of a species; population differences in the space dimension.

Gloger's rule. "Races in warm and humid areas are more heavily pigmented than those in cool and dry areas."

Gonochorism. The possession of gonads of only one sex (either male or female) in an individual.

Gonochoristic. Of individuals, having functional gonads of only one sex; of breeding populations, composed of male and female individuals.

Grade. A group of animals similar in level of organization; a level of anagenetic advance.

Gynogenesis. Parthenogenetic development of the egg cell after the egg membrane has been penetrated by a male gamete.

Habitat selection. The capacity of a dispersing individual to select an appropriate (the species-specific) habitat.

Haploid. Having only a single set of chromosomes;—gametes are usually haploid.

Hardy-Weinberg formula. The statement in mathematical terms (p. 167) that the frequency of genes in a population remains constant in the absence of selection, of nonrandom mating, and of accidents of sampling.

Heritability. The genetic component of phenotypic variability.

Hermaphroditism. The occurrence of gonads of both sexes in a single individual.

Heterochromatic. Of chromosomal sections: staining differently from the major portions of the chromosomes.

Heterogamy. The mating of unlike individuals; the preference of an individual to mate with an individual of unlike phenotype or genotype (opposed to Homogamy).

Heterosis. Selective superiority of heterozygotes.

Heterostyly. A polymorphism of flowers in which styles and stamens are of two (or more) unequal lengths, a system that insures cross-fertilization.

Heterozygote. An individual with different genetic factors (alleles) at the homologous (corresponding) loci of the two parental chromosomes.

Higher category. A taxonomic category of higher rank than the species: genus, family, order, and so forth.

Holometabolic. Of an insect, undergoing a complete metamorphosis between larval and adult stage.

Homeotic mutant. In insects, the mutational change of one in a series of structures to the form of another structure in the series, as wing into haltere, arista into leg, wing into leg, and the like.

Homogamy. The preference of a mating individual for another with similar phenotype or genotype (opposed to Heterogamy).

Homostyly. A system of flowers in which styles and stamens are of equal length.

Homozygote. An individual with identical genetic factors (alleles) at the homologous (corresponding) loci of the two parental chromosomes.

Hybridization. The crossing of individuals belonging to two unlike natural populations.

Inbreeding depression. The reduction of viability caused by increased homozygosity resulting from inbreeding.

Industrial melanism. The increase in the frequency of melanistic (blackish) individuals (morphs) in populations of lepidopterans in sooty areas.

Input load. The load of inferior alleles in a gene pool, caused by mutation and immigration.

Instantaneous speciation. The production of a single individual that is reproductively isolated from the species to which the parents belong and is reproductively and ecologically capable of establishing a new species population.

Internal balance. The harmonious epistatic interactions of genes at different loci.

Introgression. The incorporation of genes of one species into the gene pool of another species.

Introgressive hybridization. Any hybridization leading to introgression.

Inversion. Reversal of the linear order of the genes in a segment of a chromosome.

Irreversibility. The inability of an evolving group of organisms (or a structure of an organism) to return to an ancestral condition;—the theory of irreversibility is that a given structure or adaptation which has been lost in evolution cannot be restored exactly to its prior condition.

Isoalleles. Alleles that produce such slight phenotypic differences that special techniques are required to reveal their presence.

Isolate. A population or group of populations that is separated from other populations.

Isolating mechanisms. Properties of individuals that prevent successful interbreeding with individuals that belong to different populations.

Isophenes. Lines on a map that connect points of equal expression of a clinally varying character.

Karyotype. The chromosome complement.

Lamarckism. The theory, advocated by Lamarck, that evolution is brought about by volition or by environmental induction.

Levels of integration. Levels of complexity in structures, patterns, or associations at which new properties emerge that could not have been predicted from the properties of the component parts.

Locus. The location of a given gene on a chromosome.

Ludwig theorem. The theory that new genotypes can be added to a population if they can utilize new components of the environment (occupy a new subniche) even if they are inferior in the ancestral niche.

Luxuriance. Somatic vigor of hybrids (heterozygotes) that does not add to their fitness.

Macrogenesis. The sudden origin of new types by saltation.

Mechanical isolation. Reproductive isolation owing to mechanical incompatibility of male and female genitalic structures.

Meiosis. Two consecutive special cell divisions in the developing germ cells characterized by the pairing and segregation of homologous chromosomes. The resulting gametes will have reduced, that is, haploid, chromosome sets.

Meiotic drive. A force able to alter the mechanics of meiotic cell division in such a manner that the two kinds of gametes produced by a heterozygote do not occur with equal frequency.

Mendelism. Particulate inheritance; see also De Vriesianism.

Meristic variation. Variation in characters that can be counted, like number of vertebrae, scales, fin rays, and so forth.

Metacentric. Having the centromere somewhere along the chromosome, but not at or near the tip; characteristic of chromosomes that are J- or V-shaped in metaphase.

Metamorphosis. A drastic change of form during development, as when a tadpole changes into a frog, or an insect larva into an imago.

Mimetic polymorphism. Polymorphism in which the various morphs resemble other species distasteful or dangerous to a predator;—often restricted to females.

Modifiers. Genes that affect the phenotypic expression of genes at other loci.

Monotypic species. A species containing only a single (= the nominate) subspecies.

Morph. Any of the genetic forms (individual variants) that account for polymorphism.

Morphism. Polymorphism.

Mullerian mimicry. Similarity among several species distasteful (inedible) to a predator.

Multifactorial. Controlled by several gene loci.

Mutationism. De Vriesianism.

Myrmecophily. The utilization by other insects, mostly beetles, of ant colonies as domicile and source of food.

Neo-Darwinism. Weismann's theory of evolution; sometimes, any modern evolutionary theory featuring natural selection.

Neontology. The study of Recent organisms;—antonym of *paleontology.*

Neoteny. The elimination of metamorphosis into the adult stage, breeding taking place in the larval or juvenile stage.

Niche (ecological). The constellation of environmental factors into which a species (or other taxon) fits; the outward projection of the needs of an organism, its specific way of utilizing its environment.

Nondimensional species. The species concept characterized by the noninterbreeding of two coexisting demes, uncomplicated by the dimensions of space and time.

Nondisjunction. The failure of separation of paired chromosomes at meiosis and their passage to the same spindle pole, resulting in unequal chromosome numbers in the daughter cells.

Normalizing selection. The removal by selection of all genes that produce deviations from the normal (= average) phenotype of a population.

Oligogenic. Of a character, determined by few genes.

Oligolectic. Of bees, collecting the pollen of only a few kinds of flowers.

Oligophagous. Feeding on few species of food plants.

Ontogenetic. Pertaining to the development of the individual, particularly the embryogenesis.

Open population. A population freely exposed to gene flow and subject to much input of alien genes owing to immigration.

Organization effect. An interaction among adjacent loci owing to some features of organization of the chromosomes.

Orthogenesis. Evolution of phyletic lines following a predetermined rectilinear pathway, the direction not being determined by natural selection.

Overdominance. Superiority of the heterozygote over both kinds of homozygotes.

Panmictic. Of populations, randomly interbreeding, the whole population or species forming a single deme.

Paracentric inversions. An inversion that does not include the centromere.

Parthenogenesis. The development of eggs without fertilization.

Particulate inheritance. Mendel's theory that the genetic factors received from mother and father do not blend or fuse, but retain their integrity from generation to generation.

Penetrance. The frequency with which a (dominant or homozygous) gene manifests itself in the phenotype;—most genes have 100-percent penetrance.

Pericentric inversions. An inversion that includes the centromere.

Peripheral isolate. A population isolated at or beyond the periphery of the species range.

Phage. A bacterial virus.

Phenocopy. A modification of the phenotype (owing to special environmental conditions) that resembles a change of the phenotype caused by a mutation.

Phenodeviants. Phenotypes that deviate from the population (or species) mean (or norm), owing to special gene combinations, for instance extreme homozygosity.

Phenotype. The totality of characteristics of an individual (its appearance) as a result of the interaction between genotype and environment.

Philopatry. The drive (tendency) of an individual to return to (or stay in) its home area (birthplace or another adopted locality).

Plasmagenes. Genetic factors located in the cytoplasm (outside the nucleus).

Pleiotropy. The capacity of a gene to affect several characters, that is, several aspects of the phenotype.

Pleistocene refuges. Favorable areas south of the borders of the ice, where species and populations survived periods of glaciation.

Ploidy. A term referring to the number of chromosome sets.

Polygenes. Genes that jointly with several or many other genes control a character.

Polymorphism. The simultaneous occurrence of several discontinuous genetically controlled phenotypes in a population, with the frequency even of the rarest type higher than can be maintained by recurrent mutation.

Polyphagous. Feeding on many different kinds of food, such as species of host plants.

Polyphenism. The occurrence of several phenotypes in a population, the differences between which are not the result of genetic differences.

Polyploidy. A condition in which the number of chromosome sets in the nucleus is a multiple (greater than 2) of the haploid numbers.

Polytopic subspecies. A subspecies composed of widely separated but phenotypically identical populations.

Polytypic. Of a category, containing two or more immediately subordinate categories, for instance a species with several subspecies.

Poikilothermal. Ectothermal.

Position effect. The difference in the phenotypic expression of a gene caused by a change in its spatial relation to other genes on the chromosome.

Preadaptation. The possession of the necessary properties to permit a shift into a new niche or habitat. A structure is preadapted if it can assume a new function without interference with the original function.

Preformism. The belief that the egg (or sperm or zygote) contains a preformed adult in miniature, to be "unfolded" during development.

Procaryota. Those microorganisms (viruses, bacteria, blue-green algae) that lack well-defined nuclei and meiosis.

Pseudoalleles. Genes at closed adjacent loci that react physiologically in many ways as if they were alleles and between which crossing over is rare.

Pseudogamy. Gynogenesis.

Pseudopolyploidy. A numerical relation of chromosome sets in groups of related species which leads to their erroneous interpretation as polyploids.

Random fixation. The loss of an allele (and fixation of the other allele) in a population owing to accidents of sampling.

Rassenkreis. The German equivalent of *polytypic species;*—not "a circle of races."

Recessiveness. The failure of a gene to express its presence in the phenotype of the heterozygote.

Reductionism. The erroneous belief that complex phenomena can be entirely explained by reducing them to the smallest possible component parts and by explaining these.

Salivary chromosomes. Giant chromosomes (with highly specific patterns of dark and light bands) in the salivary glands of the larvae of certain kinds of dipterans (flies, mosquitoes, midges, and the like).

Saltation. A change by a leap across a discontinuity.

Secondary intergradation. The intergradation or hybridization of two distinct populations (or groups of populations) along a zone of secondary contact.

Semigeographic speciation. The splitting apart of species along lines of secondary intergradation or along lines of strong ecological contrast.

Semispecies. The component species of superspecies (Mayr); also, populations that have acquired some, but not yet all, attributes of species rank; borderline cases between species and subspecies.

Sibling species. Morphologically similar or identical populations that are reproductively isolated.

Sickle-cell anemia. An anemia due to a hemoglobin mutation found mostly in tropical areas and lethal to homozygotes.

Speciation. The acquisition of reproductive isolation by a population or group of populations; the multiplication of species.

Species group. A group of closely related species, usually with partially overlapping ranges.

Species recognition. The exchange of appropriate (species-specific) stimuli and responses between individuals (particularly during courtship).

Specific modifier. A gene that has the specific and perhaps exclusive function of modifying the expression of a gene at another locus.

Spontaneous generation. The sudden, spontaneous origin of organisms from inert matter;—now discredited.

Stabilizing selection. The elimination by selection of all phenotypes deviating too far from the population mean, and hence also of genes producing such deviating phenotypes.

Subspecies. An aggregate of local populations of a species inhabiting a geographic subdivision of the range of the species, and differing taxonomically from other populations of the species.

Substitutional load. The cost to a population of replacing an allele by another in the course of evolutionary change.

Substrate race. A local race selected to agree in its coloration with that of the substrate, for example, a black race on a lava flow.

Supergene. A chromosome segment that is protected against crossing over and is transmitted as if a single gene.

Superoptimal stimuli. Sensory stimuli to which an animal responds more strongly than to the natural stimuli for which the response has been selected.

Superspecies. A monophyletic group of entirely or essentially allopatric species that are too distinct to be included in a single species.

Suppressor gene. A gene that suppresses the phenotypic expression of a gene at another locus.

Switch gene. A gene that causes the epigenotype to switch to a different developmental pathway.

Sympatry. The occurrence of two or more populations in the same area; more precisely, the existence of a population in breeding condition within the cruising range of individuals of another population.

Sympatric speciation. Speciation without geographic isolation; the acquisition of isolating mechanisms within a deme.

Synthetic lethals. Lethal chromosomes derived from normally viable chromosomes by recombination (as a result of crossing over).

Synthetic theory. The current evolutionary theory which is a synthesis of the best components of many previously proposed theories, with mutation and selection as the basic elements.

Systemic mutation. A mutation, postulated by R. Goldschmidt, that would fundamentally reorganize the germ plasm, and permit the origin of wholly new types of organisms.

Taxon. A group of organisms recognized as a formal unit at any level of a hierarchic classification; for instance, polar bears, bears, carnivores, mammals, and chordates constitute taxa.

Telomere. The postulated material at the tip of a chromosome, having certain properties not found in the rest of the chromosome.

Territory. An area defended by an animal against other members of its species (and occasionally members of other species).

Tetraploid. A polyploid with four haploid chromosome sets, normally the result of the doubling of the diploid chromosome number.

Thelytoky. Parthenogenesis of the type in which females give rise to female progeny without fertilization.

Transient polymorphism. Polymorphism existing during the period when an allele is being replaced by a superior one.

Translocation. The shift of a segment of a chromosome to another chromosome.

Typological thinking. A concept in which variation is disregarded and the members of a population are considered as replicas of the "type," the Platonic *eidos.*

Variety. An ambiguous term of classical (Linnaean) taxonomy for a heterogeneous group of phenomena including nongenetic variations of the phenotype, morphs, domestic breeds, and geographic races.

Zygote. A fertilized egg; the cell (individual) that results from the fertilization of an egg cell.

Bibliography

Adriaanse, A.
    1947.   *Ammophila campestris* Latr. und *Ammophila adriaansei* Wilcke. *Behaviour* 1:1–34.

Aeppli, E.
    1952.   Natürliche Polyploidie bei den Planarien *Dendrocoelum lacteum* (Müller) und *Dendrocoelum infernale* (Steinmann). *Z. ind. Abst.-u. Vererb.* 84:182–212.

Aird, I., H. H. Bentall, and J. A. Fraser Roberts
    1953.   Relationship between cancer of stomach and the ABO groups. *Brit. Med. J.* 1:799–801.

Aitken, T. H. G.
    1945.   Studies on the anopheline complex of western America. *Univ. Calif. Publ. Entomol.* 7:273–304.

Aksiray, F.
    1952.   Genetical contributions to the systematical relationship of anatolian cyprinodont fishes. *Publ. Hydrobiol. Res. Inst. Univ. Istanbul* (B) 1:33–81.

Alberti, B.
    1943.   Zur Frage der Hybridisation zwischen *Colias erate, hyale,* und *edusa* und über die Umgrenzung der drei Arten. *Mitt. Münch. Entomol. Ges.* 33:606–624.

Albonico, R.
    1948.   Die Farbvarietäten der grossen Wegschnecke, *Arion empiricorum* Fer., und deren Abhängigkeit von den Umweltbedingungen. *Rev. Suisse Zool.* 55:347–425.

Aldrich, J. W.
    1946.   Significance of racial variation in birds to wildlife management. *J. Wildl. Mgmt.* 10:86–93.

Alexander, R. D.
    1957a. The taxonomy of the field crickets of the eastern United States (Orthoptera: Gryllidae: *Acheta*). *Ann. Entomol. Soc. Amer.* 50:584–602.
    1957b. The song relationships of four species of ground crickets (Orthoptera: Gryllidae: *Nemobius*). *Ohio J. Sci.* 57:153–163.

1960. Sound communication in Orthoptera and Cicadidae. In W. E. Lanyon and W. N. Tavolga, eds., *Animal sounds and communication* (AIBS Publ. No. 7), 38–92.

1961. Aggressiveness, territoriality, and sexual behavior in field crickets (Orthoptera: Gryllidae). *Behaviour* 17:130–223.

Alexander, R. D., and R. S. Bigelow

1960. Allochronic speciation in field crickets, and a new species, *Acheta veletis*. *Evolution* 14:334–346.

Alexander, R. D., and T. E. Moore

1962. The evolutionary relationships of 17-year and 13-year cicadas, etc. *Misc. Publ. Mus. Zool., Univ. Mich.*, No. 121:1–59.

Alexander, R. D., and E. S. Thomas

1959. Systematic and behavioral studies on the crickets of the *Nemobius fasciatus* group (Orthoptera: Gryllidae: Nemobiinae). *Ann. Entomol. Soc. Amer.* 52:591–605.

Allard, H. A.

1929. Physiological differentiation in overwintering individuals of certain musical Orthoptera. *Canad. Entomol.* 61:195–198.

Allee, W. C., A. E. Emerson, O. Park, T. Park, and K. P. Schmidt

1949. *Principles of animal ecology* (Saunders, Philadelphia).

Allen, J. A.

1877. The influence of physical conditions in the genesis of species. *Radical Rev.* 1:108–140.

Allison, A. C.

1955. Aspects of polymorphism in man. *Cold Spring Harbor Symp. Quant. Biol.* 20:239–255.

1959. Metabolic polymorphisms in mammals and their bearing on problems of biochemical genetics. *Amer. Nat.* 93:5–17.

Alpatov, W. W.

1929. Biometrical studies on variation and races of the honey-bee (*Apis mellifera*). *Quart. Rev. Biol.* 4:1–58.

Altenburg, E.

1934. A theory of hermaphroditism. *Amer. Nat.* 68:88–91.

Altevogt, R.

1957. Untersuchungen zur Biologie, Ökologie und Physiologie indischer Winkerkrabben. *Z. Morphol. u. Ökol. Tiere* 46:1–110.

Amadon, D.

1943. Specialization and evolution. *Amer. Nat.* 77:133–141.

1947. Ecology and the evolution of some Hawaiian birds. *Evolution* 1:63–68.

1950. The Hawaiian honeycreepers (Aves, Drepaniidae). *Bull. Amer. Mus. Nat. Hist.* 95:151–262.

1953a. Avian systematics and evolution in the Gulf of Guinea. *Bull. Amer. Mus. Nat. Hist.* 100:397–451.

1953b. Remarks on the Asiatic hawk-eagles of the genus *Spizaëtus*. *Ibis* 95:492–500.

1959. The significance of sexual differences in size among birds. *Proc. Amer. Phil. Soc.* 103:531–536.

Anderson, E.
1937. Supra-specific variation in nature and in classification from the viewpoint of botany. *Amer. Nat.* 71:223–235.
1949. *Introgressive hybridization* (Wiley, New York).
1953. Introgressive hybridization. *Biol. Rev.* 28:280–307.

Anderson, E., and L. Hubricht
1938. Hybridization in *Tradescantia*. III. The evidence for introgressive hybridization. *Amer. J. Botany* 25:396–402.

Andrewartha, H. G.
1961. *Introduction to the study of animal populations* (Methuen, London).

Andrewartha, H. G., and L. C. Birch
1954. *The distribution and abundance of animals* (University of Chicago Press, Chicago).

Arambourg, C.
1955. A recent discovery in human paleontology. *Atlanthropus* of Ternifine (Algeria). *Amer. J. Phys. Anthropol.* (n.s.) 13:191–201.

Arkell, W. J., and J. A. Moy-Thomas
1940. Palaeontology and the taxonomic problem. In J. S. Huxley, ed., *The new systematics* (Clarendon Press, Oxford), 395–410.

Armstrong, T.
1945. Differences in the life history of the codling moth *Carpocapsa pomonella* (L.), attacking pear and apple. *Canad. Entomol.* 77:231–233.

Arndt, W.
1941. Die Anzahl der bisher in Deutschland (Altreich) nachgewiesenen rezenten Tierarten. *Zoogeographica* 4:28–92.

Ashton, E. H., and S. Zuckerman
1950. The influence of geographic isolation on the skull of the green monkey (*Cercopithecus aethiops sabaeus*). *Proc. Roy. Soc., London* (B) 137:212–238.

Atoda, K.
1951. The larva and postlarval development of the reef-building corals. IV. *Galaxea aspera* Quelch. *J. Morphol.* 89:17–35.

Atz, J. W.
1952. Narial breathing in fishes and the evolution of internal nares. *Quart. Rev. Biol.* 27:366–377.

Avery, O. T., C. M. MacLeod, and M. McCarty
1944. Studies on the chemical nature of the substance inducing transformation of pneumococcal types. Induction of transformation by a desoxyribonucleic acid fraction isolated from pneumococcus type III. *J. Exptl. Med.* 79:137–158.

Babcock, E. B.
1947a. The genus *Crepis*, Part I. The taxonomy, phylogeny, distribution, and evolution of *Crepis*. *Univ. Calif. Publ. Botany* 21:1–197.

    1947b. The genus *Crepis*, Part II. Systematic treatment. *Univ. Calif. Publ. Botany* 22:199–1030.

Babers, F. H., and J. J. Pratt, Jr.
    1951. *Development of insect resistance to insecticides. II. A critical review of the literature up to 1951* (U.S. Dept. Agr., May, E–818), 1–45.

Bacci, G.
    1949. Ricerche su *Asterina gibbosa* (Penn.). II. L'ermafroditismo in una popolazione di Plymouth. *Arch. Zool. Ital.* 34:49–73.
    1950. Osservazioni sulla sessualità dei Nereimorfi (Anellidi Policheti). *Boll. Zool.* 17:55–61.
    1955a. Variabilità sessuale di popolazioni, razze e specie ermafrodite. *Ricerca Sci., Suppl.* 25:1–6.
    1955b. La variabilità dei genotipi sessuali negli animali ermafroditi. *Pubbl. Staz. Zool. Napoli* 26:110–137.

Baer, J. G.
    1951. *Ecology of animal parasites* (University of Illinois Press, Urbana).

Baerends, G. P., and J. M. Baerends-Van Roon
    1950. An introduction to the study of the ethology of cichlid fishes. *Behaviour, Suppl.* 1:1–243.

Bagenal, T. B.
    1951. A note on the papers of Elton and Williams on the generic relations of species in small ecological communities. *J. Animal Ecol.* 20:242–245.

Bailey, D. W.
    1956. Re-examination of the diversity in *Partula taeniata*. *Evolution* 10:360–366.

Bailey, R. M.
    1942. An intergeneric hybrid rattlesnake. *Amer. Nat.* 76:376–385.

Baker, H. G.
    1948. The ecotypes of *Melandrium dioicum* (L. emend.) Coss. & Germ. *New Phytol.* 47:131–145.
    1951. Hybridization and natural gene-flow between higher plants. *Biol. Rev.* 26:302–337.

Baker, P. T.
    1958. The biological adaptation of man to hot deserts. *Amer. Nat.* 92:337–357.

Baker, R. H.
    1960. Mammals of the Guadiana Lava Field, Durango, Mexico. *Mich. State Univ., Biol. Ser.* 1:305–327.

Baldi, E., A. Buzzati-Traverso, L. L. Cavalli, and L. Pirocchi
    1945. Frammentamento di una popolazione specifica (*Mixodiaptomus laciniatus* Lill.) in un grande lago in sottopopolazioni geneticamente differenziate. *Mem. Ist. Ital. Idrobiol.* 2:169–216.

Baldwin, J. M.
    1896. A new factor in evolution. *Amer. Nat.* 30:441–451, 536–553.

Baldwin, P. H.
1953. Annual cycle, environment and evolution in the Hawaiian honey-creepers (Aves: Drepaniidae). *Univ. Calif. Publ. Zool.* 52:285–398.
Banks, R. C., and N. K. Johnson
1961. A review of North American hybrid hummingbirds. *Condor* 63:3–28.
Barber, H. S.
1951. North American fireflies of the genus *Photuris*. *Smithsonian Misc. Coll.* 117:1–58.
Barigozzi, C.
1957. Genotipi complessi e loro trasformazioni, a proposito dei tumori melanotici di *Drosophila*. *Boll. Zool.* 24:411–425.
Barnicot, N. A.
1959. Climatic factors in the evolution of human populations. *Cold Spring Harbor Symp. Quant. Biol.* 24:115–129.
Barr, T. C., Jr., L. H. Hyman, H. H. Hobbs, Jr., C. J. and M. L. Goodnight, K. Christiansen, O. Park, J. C. Chamberlin, D. R. Malcolm, N. B. Causey, and Brother G. Nicholas
1960. Speciation and raciation in cavernicoles. *Amer. Midl. Nat.* 64:1–160.
Barter, G. W., and W. J. Brown
1949. On the identity of *Agrilus anxius* Gory and some allied species (Coleoptera: Buprestidae). *Canad. Entomol.* 81:245–249.
Bartholomew, G. A., Jr., and J. B. Birdsell
1953. Ecology and the protohominids. *Amer. Anthropol.* 55:481–498.
Basrur, V. R., and K. H. Rothfels
1959. Triploidy in natural populations of the black fly *Cnephia mutata* (Malloch). *Canad. J. Zool.* 37:571–589.
Bastock, M.
1956. A gene mutation which changes a behavior pattern. *Evolution* 10:421–439.
Bastock, M., and A. Manning
1955. The courtship of *Drosophila melanogaster*. *Behaviour* 8:7–111.
Bateman, A. J.
1948. Intra-sexual selection in *Drosophila*. *Heredity* 2:349–368.
1950. Is gene dispersion normal? *Heredity* 4:353–363.
Bateman, K. G.
1959. The genetic assimilation of four venation phenocopies. *J. Genet.* 56:443–474.
Bates, H. W.
1863. *The naturalist on the river Amazons* (John Murray, London), vol. 1.
Bates, M.
1949. *The natural history of mosquitoes* (Macmillan, New York).
Bateson, W.
1894. *Materials for the study of variation* (Macmillan, New York).
1913. *Problems of genetics* (Yale University Press, New Haven).
1922. Evolutionary faith and modern doubts. *Science* 55:55–61.

Battaglia, B.
  1956. Differenziamento fisiologico e incipiente isolamento intraspecifico in
      Tisbe reticulata Bocquet. *Atti Ist. Veneto Sci.* 114:111–125.
  1958. Balanced polymorphism in *Tisbe reticulata,* a marine copepod.
      Evolution 12:358–364.

Bauer, H.
  1947. Karyologische Notizen I. Über generative Polyploidie bei Dermap-
      teren. *Z. f. Naturforsch.* 2b:63–66.

Baur, E.
  1932. Artumgrenzung und Artbildung in der Gattung Antirrhinum . . .
      Z. ind. Abst.- u. Vererb. 63:256–302.

Baur, G.
  1893. Further notes on American box-tortoises. *Amer. Nat.* 27:677–678.

Beardmore, J. A.
  1960. Developmental stability in constant and fluctuating temperatures.
      Heredity 14:411–422.
  1961. Diurnal temperature fluctuation and genetic variance in *Drosophila*
      populations. *Nature* 189:162–163.

Beardmore, J. A., Th. Dobzhansky, and O. A. Pavlovsky
  1960. An attempt to compare the fitness of polymorphic and monomorphic
      experimental populations of *Drosophila pseudoobscura. Heredity*
      14:19–33.

Beaudry, J. R.
  1960. The species concept: its evolution and present status. *Rev. Canad.*
      Biol. 19:219–240.

Becher, E.
  1917. *Die fremddienliche Zweckmässigkeit der Pflanzengallen und die*
      Hypothese eines überindividuellen Seelischen (Veit, Leipzig).

Beddall, B. G.
  1957. Historical notes on avian classification. *Systematic Zool.* 6:129–136.

Beer, R. E.
  1955a. Biological studies in the genus *Periclista* (Hymenoptera, Tenthredi-
      nidae, Blennocompinae). *J. Kansas Entomol. Soc.* 28:19–26.
  1955b. Biological studies and taxonomic notes on the genus *Strongylogaster*
      Dahlbom (Hymenoptera: Tenthredinidae). *Univ. Kansas Sci. Bull.*
      37:223–249.

Beermann, W.
  1952. Chromosomenpolymorphismus und Bastardierung zweier Chirono-
      mus-Arten. *Verh. Deut. Zool. Ges., Freiburg, 1952,* 290–295.
  1961. Ein Balbiani-Ring als Locus einer Speicheldrüsen-Mutation. *Chro-*
      mosoma 12:1–25.

Beheim, D.
  1942. Über den taxonomischen und isolierenden Wert der Forcepsvaria-
      tionen einiger Caraboidea. *Z. Morphol. u. Ökol. Tiere* 39:21–46.

Beirne, B. P.
  1952. *British pyralid and plume moths* (Frederick Warne, London).

Benazzi, M.
    1949. Ricerche genetico-sistematiche sue tricladi. *Ricerca Sci., Suppl.* 19:1–8.
    1950. Problemi di zoogeografia tirrenica studiati nelle planarie. *Atti Soc. Toscana Sci. Nat. Mem.* 57(B):21–28.
    1957. Considerazioni sulla evoluzione cromosomica negli animali. *Boll. Zool.* 24:373–409.

Bennett, D., S. Badenhausen, and L. C. Dunn
    1959. The embryological effects of four late-lethal t-alleles in the mouse, which affect the neural tube and skeleton. *J. Morphol.* 105:105–143.

Benson, C. W.
    1946. On the change of coloration in *Lybius zombae* (Shelley). *Bull. Brit. Ornithol. Club* 67:33–35.
    1948. Geographical voice-variation in African birds. *Ibis* 90:48–71.

Benson, R. B.
    1950. An introduction to the natural history of British sawflies. *Trans. Soc. Brit. Entomol.* 10:45–142.

Benson, S. B.
    1933. Concealing coloration among some desert rodents of southwestern United States. *Univ. Calif. Publ. Zool.* 40:1–70.

Benzer, S.
    1957. The elementary units of heredity. In W. D. McElroy and B. Glass, eds., *The chemical basis of heredity* (Johns Hopkins Press, Baltimore), 70–93.

Bergmann, C.
    1847. Über die Verhältnisse der Wärmeökonomie der Thiere zu ihrer Grösse. *Göttinger Studien*, pt. 1:595–708.

Berrill, N. J.
    1955. *The origin of vertebrates* (Clarendon Press, Oxford).

Betts, M. M.
    1955. The food of titmice in oak woodland. *J. Animal Ecol.* 24:282–323.

Bigelow, R. S.
    1958. Evolution in the field cricket, *Acheta assimilis* Fab. *Canad. J. Zool.* 36:139–151.

Bilton, L.
    1957. The Chillingham herd of wild cattle. *Trans. Nat. Hist. Soc. Northumberland, Durham and Newcastle upon Tyne* 12:137–160.

Birch, L. C.
    1953. Experimental background to the study of the distribution and abundance of insects. III. The relation between innate capacity for increase and survival of different species of beetles living together on the same food. *Evolution* 7:136–144.
    1954. Experiments on the relative abundance of two sibling species. *Austral. J. Zool.* 2:66–74.
    1955. Selection in *Drosophila pseudoobscura* in relation to crowding. *Evolution* 9:389–399.

1957a. The meanings of competition. *Amer. Nat.* 91:5–18.

1957b. The role of weather in determining the distribution and abundance of animals. *Cold Spring Harbor Symp. Quant. Biol.* 22:203–218.

1961. Natural selection between two species of tephritid fruit fly of the genus *Dacus*. *Evolution* 15:360–374.

Birdsell, J. B.

1950. Some implications of the genetical concept of race in terms of spatial analysis. *Cold Spring Harbor Symp. Quant. Biol.* 15:259–314.

Biswas, B.

1950. On the shrike *Lanius tephronotus* (Vigors), with remarks on the *erythronotus* and *tricolor* groups of *Lanius schach* Linné, and their hybrids. *J. Bombay Nat. Hist. Soc.* 49:444–455.

Blair, A. P.

1941. Variation, isolation mechanisms, and hybridization in certain toads. *Genetics* 26:398–417.

1942. Isolating mechanisms in a complex of four species of toads. *Biol. Symp.* 6:235–249.

1955. Distribution, variation, and hybridization in a relict toad (*Bufo microscaphus*) in southwestern Utah. *Amer. Mus. Novitates,* No. 1722:1–38.

Blair, W. F.

1943. Ecological distribution of mammals in the Tularosa Basin, New Mexico. *Contrib. Lab. Vert. Biol., Univ. Mich.,* No. 20:1–24.

1947. Estimated frequencies of the buff and gray genes (G,g) in adjacent populations of deer-mice (*Peromyscus maniculatus blandus*) living on soils of different colors. *Contrib. Lab. Vert. Biol., Univ. Mich.,* No. 36:1–16.

1950. Ecological factors in speciation of *Peromyscus*. *Evolution* 4:253–275.

1951. Interbreeding of natural populations of vertebrates. *Amer. Nat.* 85:9–30.

1955. Mating call and stage of speciation in the *Microhyla olivacea–M. carolinensis* complex. *Evolution* 9:469–480.

1958a. Mating call in the speciation of anuran amphibians. *Amer. Nat.* 92:27–51.

1958b. Response of a green treefrog (*Hyla cinerea*) to the call of the male. *Copeia,* No. 4:333–334.

1959. Genetic compatibility and species groups in U. S. toads (*Bufo*). *Texas J. Sci.* 11:427–453.

Blair, W. F., and M. J. Littlejohn

1960. Stage of speciation of two allopatric populations of chorus frogs (*Pseudacris*). *Evolution* 14:82–87.

Blanchard, B. D.

1941. The white-crowned sparrows (*Zonotrichia leucophrys*) of the Pacific seaboard: environment and annual cycle. *Univ. Calif. Publ. Zool.* 46:1–178.

Blasing, I.
  1953. Experimentelle Untersuchungen über den Umfang der ökologischen und physiologischen Toleranz von *Planaria alpina* Dana und *Planaria gonocephala* Dugès. *Zool. Jahrb. (Physiol.)* 64:112–152.
Blaxter, J. H. S.
  1958. The racial problem in herring from the viewpoint of recent physiological, evolutionary, and genetical theory. *Rapp. Cons. Explor. Mer* 143:10–19.
Bloomer, H. H.
  1939. A note on the sex of *Pseudanodonta* and *Anodonta*. *Proc. Malacol. Soc. London* 23:285–297.
Bock, W. J.
  1959a. The status of the semipalmated plover. *Auk* 76:98–100.
  1959b. Preadaptation and multiple evolutionary pathways. *Evolution* 13:194–211.
Bock, W. J., and W. D. Miller
  1959. The scansorial foot of the woodpeckers, with comments on the evolution of perching and climbing feet in birds. *Amer. Mus. Novitates*, No. 1931:1–45.
Bocquet, C.
  1951. Recherches sur *Tisbe* (= *Idyaea*) *reticulata*, n. sp. *Arch. Zool. Exptl. Gén.* 87:335–416.
  1953. Recherches sur le polymorphisme naturel des *Jaera marina* (Fabr.) (Isopodes Asellotes). *Arch. Zool. Exptl. Gén.* 90.187–450.
Bocquet, C., C. Lévi, and G. Teissier
  1951. Recherches sur le polychromatisme de *Sphaeroma serratum*. *Arch. Zool. Exptl. Gén.* 87:245–298.
Bocquet, C., and G. Teissier
  1960. Génétique des populations de *Sphaeroma serratum* (F.). I. Stabilité du polychromatisme local. *Cahiers Biol. Marine* 1:103–111.
Bodmer, W. F., and A. W. F. Edwards
  1960. Natural selection and the sex ratio. *Ann. Human Genet.* 24:239–244.
Boettger, C. R.
  1931. Die Entstehung von Populationen mit bestimmter Variantenzahl bei der Landschneckengattung *Cepaea* Held. *Z. ind. Abst.- u. Vererb.* 58:295–316.
Bogert, C. M.
  1960. The influence of sound on the behavior of amphibians and reptiles. In W. E. Lanyon and W. N. Tavolga, eds., *Animal sounds and communication* (AIBS Publ. No. 7), 137–320.
Bøggild, O., and J. Keiding
  1958. Competition in house fly larvae. *Oikos* 9:1–25.
Bohart, G. E.
  1942. Notes on some feeding and hibernation habits of California *Polistes*. *Pan-Pacific Entomol.* 18:30.

Bordes, F. H.
1960. Evolution in the paleolithic cultures. In S. Tax, ed., *The evolution of man* (University of Chicago Press, Chicago), 99–110.

Boschma, H.
1948. The species problem in *Millepora. Zool. Verh.* No. 1:1–115.

Bösiger, E.
1953. Fréquence des mutations visibles dans deux populations naturelles de *Drosophila melanogaster. Compt. Rend.* 236:1999–2002.

1957. Sur la parade nuptiale des mâles de deux souches de *Drosophila melanogaster. Compt. Rend.* 244:2107–2110.

1960. Sur le rôle de la sélection sexuelle dans l'évolution. *Experientia* 16:270–273.

Bovey, P.
1941. Contribution à l'étude génétique et biogéographique de *Zygaena ephialtes* L. (Lep. Zygaenidae). *Rev. Suisse Zool.* 48:1–90.

Bowman, T. E.
1955. A new copepod of the genus *Calanus* from the northeastern Pacific with notes on *Calanus tenuicornis* Dana. *Pacific Sci.* 9:413–422.

Boyd, W.
1950. *Genetics and the races of man* (Little, Brown, Boston).

Božić, B.
1960. Le genre *Tigriopus* Norman (Copépodes Harpacticoïdes) et ses formes européennes; recherches morphologiques et expérimentales. *Arch. Zool. Exptl. Gén.* 98:167–269.

Breese, E. L., and K. Mather
1957. The organisation of polygenic activity within a chromosome in *Drosophila.* I. Hair characters. *Heredity* 11:373–395.

1960. The organisation of polygenic activity within a chromosome in *Drosophila.* II. Viability. *Heredity* 14:375–399.

Bresson, J.
1955. Aselles de sources et de grottes d'Eurasie et d'Amérique du Nord. *Arch. Zool. Exptl. Gén.* 92:45–77.

Brett, C. H.
1947. Interrelated effects of food, temperature, and humidity on the development of the lesser migratory grasshopper. *Oklahoma Agr. Exp. Sta. Tech. Bull.,* No. T–26:1–50.

Brian, M. V.
1956. Segregation of species of the ant genus *Myrmica. J. Animal Ecol.* 25:319–337.

Brian, M. V., and A. D. Brian
1949. Observations on the taxonomy of the ants *Myrmica rubra* L. and *M. laevinodis* Nyl. *Trans. Roy. Entomol. Soc. London* 100:393–409.

Briles, W. E., C. P. Allen, and T. W. Millen
1957. The B blood group system of chickens. I. Heterozygosity in closed populations. *Genetics* 42:631–648.

Brinck, P.
   1948. Coleoptera of Tristan da Cunha. *Results Norv. Sci. Exped. Tristan da Cunha 1937–1938*, No. 17:1–122.
Brinkhurst, R. O.
   1959. Alary polymorphism in the Gerroidea (Hemiptera-Heteroptera). *J. Animal Ecol.* 28:211–230.
Brinkmann, R.
   1929. Statistisch-biostratigraphische Untersuchungen an mitteljurassischen Ammoniten über Artbegriff und Stammesentwicklung. *Abhandl. Ges. Wiss. Göttingen, Math. Nat. Kl.* (N. F.) 13:1–249.
Brinton, E.
   1959. Geographical isolation in the pelagic environment. A discussion of the distribution of euphausiid crustaceans in the Pacific. *Intern. Oceanogr. Congr., Washington, 1959:* 255–256 (preprint).
Brncic, D.
   1954. Heterosis and the integration of the genotype in geographical populations of *Drosophila pseudoobscura*. *Genetics* 39:77–88.
Broadhead, E.
   1958. The psocid fauna of larch trees in northern England—an ecological study of mixed species populations exploiting a common resource. *J. Animal Ecol.* 27:217–263.
Brooks, J. L.
   1946. Cyclomorphosis in *Daphnia*. I. An Analysis of *D. retrocurva* and *D. galeata*. *Ecol. Monographs* 16:409–447.
   1947. Turbulence as an environmental determinant of relative growth in *Daphnia*. *Proc. Nat. Acad. Sci.* 33:141–148.
   1950. Speciation in ancient lakes. *Quart. Rev. Biol.* 25:30–176.
   1957a. The systematics of North American *Daphnia*. *Mem. Connecticut Acad. Arts Sci.* 13:1–180.
   1957b. The species problem in freshwater animals. In E. Mayr, ed., *The species problem* (Amer. Assoc. Adv. Sci., Publ. No. 50), 81–123.
Brower, L. P.
   1958. Larval foodplant specificity in butterflies of the *Papilio glaucus* group. *Lepidopterists' News* 12:103–114.
   1959a. Speciation in butterflies of the *Papilio glaucus* group. I. Morphological relationships and hybridization. *Evolution* 13:40–63.
   1959b. Speciation in butterflies of the *Papilio glaucus* group. II. Ecological relationships and interspecific sexual behavior. *Evolution* 13:212–228.
Brown, A. W. A.
   1958. *Insecticide resistance in arthropods* (World Health Organisation, Monogr. Ser., No. 38), 1–240.
Brown, W. C., and J. T. Marshall, Jr.
   1953. New scincoid lizards from the Marshall Islands, with notes on their distribution. *Copeia*, No. 4:201–207.

Brown, W. J.
  1945. Food-plants and distribution of the species of *Calligrapha* in Canada, with descriptions of new species (Coleoptera, Chrysomelidae). *Canad. Entomol.* 77:117–133.
  1956. The new world species of *Chrysomela* L. (Coleoptera: Chrysomelidae). *Canad. Entomol.* 88 (Suppl. 3):1–54.
  1958. Sibling species in the Chrysomelidae. *Proc. 10th Intern. Congr. Entomol.* 1:103–110.
  1959. Taxonomic problems with closely related species. *Ann. Rev. Entomol.* 4:77–98.

Brown, W. L.
  1950. The status of two common North American carpenter ants. *Entomol. News* 61:157–161.

Brown, W. L., and E. O. Wilson
  1956. Character displacement. *Systematic Zool.* 5:49–64.

Brues, A. M.
  1954. Selection and polymorphism in the A–B–O blood groups. *Amer. J. Phys. Anthropol.* 12:559–597.

Brues, C. T.
  1924. The specificity of food-plants in the evolution of phytophagous insects. *Amer. Nat.* 58:127–144.

Buch, L. von
  1825. *Physicalische Beschreibung der Canarischen Inseln* (Kgl. Akad. Wiss., Berlin), 132–133.

Buchanan, L. L.
  1947. A correction and two new races in *Graphognathus* (white-fringed beetles) (Coleoptera, Curculionidae). *J. Wash. Acad. Sci.* 37:19–22.

Buchholz, K. F.
  1954. Zur Kenntnis der Rassen von *Lacerta pityusensis* Bosca (Reptilia, Lacertidae). *Bonn. Zool. Beitr.* 5:69–88.

Buchner, H., and F. Mulzer
  1961. Untersuchungen über die Variabilität der Rädertiere. II. Der Ablauf der Variation im Freien. *Z. Morphol. u. Ökol. Tiere* 50:330–374.

Buchner, H., F. Mulzer, and F. Rauh
  1957. Untersuchungen über die Variabilität der Rädertiere. I. Problemstellung und vorläufige Mitteilung über die Ergebnisse. *Biol. Zentr.* 76:289–315.

Bullock, T. H.
  1957. The objectives of studying physiology as function of latitude and longitude. *Ann. Biol.* 33:199–203.

Bumpus, H. C.
  1896. The variations and mutations of the introduced sparrow, *Passer domesticus*. *Biol. Lectures, Marine Biol. Lab. Wood's Hole* (1896–1897), 1–15.

Burla, H., and M. Greuter
  1959. Einige Komponenten des Ausbreitungsvorgangs bei *Drosophila*. *Vierteljahrsschr. Naturforsch. Ges. Zürich* 104:236–245.

Burtt, E.
  1951. The ability of adult grasshoppers to change colour on burnt ground. *Proc. Roy. Entomol. Soc. London* 26:45–48.
Busvine, J. R.
  1948. The 'head' and 'body' races of *Pediculus humanus* L. *Parasitology* 39:1–16.
Butler, L.
  1947. The genetics of the colour phases of the red fox in the Mackenzie River locality. *Canad. J. Research* (D) 25:190–215.
Büttiker, W.
  1948. Beitrag zur Kenntnis der Biologie und Verbreitung einiger Stechmückenarten in der Schweiz. *Mitt. Schweiz. Entomol. Ges.* 21:1–148.
Buzzati-Traverso, A.
  1947. Genetica di popolazioni in *Drosophila*. VI. Analisi delle segregazioni di mutanti presenti in popolazioni naturali. *Mem. Ist. Ital. Idrobiol.* 4:41–64.
  1950. Genetic structure of natural populations and interbreeding units in the human species. *Cold Spring Harbor Symp. Quant. Biol.* 15:13–23.
  1952. Heterosis in population genetics. In J. W. Gowen, ed., *Heterosis* (Iowa State College Press, Ames), 149–160.
  1954. On the role of mutation rate in evolution. *Proc. 9th Intern. Congr. Genet.* 1:450–462.
Byers, C. F.
  1940. A study of the dragonflies of the genus *Progomphus* (*Gomphoides*). *Proc. Florida Acad. Sci.* 4:19–86.
Bytinsky-Salz, H., and A. Günther
  1930. Untersuchungen an Lepidopterenhybriden I. *Z. ind. Abst.- u. Vererb.* 53:1–234.
Cailleux, A.
  1954. How many species? *Evolution* 8:83–84.
Cain, A. J.
  1951a. So-called non-adaptive or neutral characters in evolution. *Nature* 168:424.
  1951b. Non-adaptive or neutral characters in evolution. *Nature* 168:1049.
  1953. Geography, ecology and coexistence in relation to the biological definition of the species. *Evolution* 7:76–83.
  1954a. *Animal species and their evolution* (Hutchinson, London).
  1954b. Subdivisions of the genus *Ptilinopus* (Aves, Columbae). *Bull. Brit. Mus. (Nat. Hist.), Zool.* 2:265–284.
  1955. A revision of *Trichoglossus haematodus* and of the Australian platy-cercine parrots. *Ibis* 97:432–479.
Cain, A. J., and P. M. Sheppard
  1950. Selection in the polymorphic land snail *Cepaea nemoralis*. *Heredity* 4:275–294.
  1954. The theory of adaptive polymorphism. *Amer. Nat.* 88:321–326.

1957. Some breeding experiments with *Cepaea nemoralis* (L.). *J. Genet.* 55:195–199.

Calhoun, J. B.
1950. Population cycles and gene frequency fluctuations in foxes of the genus *Vulpes* in Canada. *Canad. J. Research* (D) 28:45–57.

Cameron, A. W.
1958. Mammals of the islands in the Gulf of St. Lawrence. *Nat. Mus. Canada, Bull.,* No. 154:1–165.

Camin, J. H., and P. R. Ehrlich
1958. Natural selection in water snakes (*Natrix sipedon* L.) on islands in Lake Erie. *Evolution* 12:504–511.

Camp, W. H., and C. L. Gilly
1943. The structure and origin of species. *Brittonia* 4:323–385.

Cannon, H. G.
1958. *The evolution of living things* (Manchester University Press, Manchester).

Cannon, W. B.
1932. *The wisdom of the body* (Norton, New York).

Cantrall, I. J.
1943. The ecology of the Orthoptera and Dermaptera of the George Reserve, Michigan. *Misc. Publ. Mus. Zool., Univ. Mich.,* No. 54:1–182.

Carpenter, C. C.
1952. Comparative ecology of the common garter snake (*Thamnophis s. sirtalis*), the ribbon snake (*Thamnophis s. sauritus*), and Butler's garter snake (*Thamnophis butleri*) in mixed populations. *Ecol. Monographs* 22:235–258.

Carpenter, G. D. H.
1949. *Pseudacraea eurytus* (L.) (Lep. Nymphalidae): a study of a polymorphic mimic in various degrees of speciation. *Trans. Roy. Entomol. Soc. London* 100:71–133.

Carson, H. L.
1953. The effects of inversions on crossing over in *Drosophila robusta*. *Genetics* 38:168–186.

1954. Interfertile sibling species in the *willistoni* group of *Drosophila*. *Evolution* 8:148–165.

1955a. The genetic characteristics of marginal populations of *Drosophila*. *Cold Spring Harbor Symp. Quant. Biol.* 20:276–287.

1955b. Variation in genetic recombination in natural populations. *J. Cell. Comp. Physiol.* 45 (Suppl. 2):221–236.

1956. Marginal homozygosity for gene arrangement in *Drosophila robusta*. *Science* 123:630–631.

1958a. Increase in fitness in experimental populations resulting from heterosis. *Proc. Nat. Acad. Sci.* 44:1136–1141.

1958b. Response to selection under different conditions of recombination in *Drosophila. Cold Spring Harbor Symp. Quant. Biol.* 23:291–306.

1958c. The population genetics of *Drosophila robusta*. *Adv. Genet.* 9:1–40.

1959. Genetic conditions which promote or retard the formation of species. *Cold Spring Harbor Symp. Quant. Biol.* 24:87–105.

Carter, T. D.

1943. The mammals of the Vernay-Hopwood Chindwin expedition, northern Burma. *Bull. Amer. Mus. Nat. Hist.* 82:99–113.

Cartwright, W. B., and W. B. Noble

1947. Studies on biological races of the Hessian fly. *J. Agric. Research* 75:147–153.

Caspari, E.

1949. Physiological action of eye color mutants in the moths *Ephestia kühniella* and *Ptychopoda seriata*. *Quart. Rev. Biol.* 24:185–199.

1950. On the selective value of the alleles *Rt* and *rt* in *Ephestia kühniella*. *Amer. Nat.* 84:367–380.

1952. Pleiotropic gene action. *Evolution* 6:1–18.

1958. Genetic basis of behavior. In A. Roe and G. G. Simpson, eds., *Behavior and evolution* (Yale University Press, New Haven), 103–127.

Caspari, E., and G. S. Watson

1959. On the evolutionary importance of cytoplasmic sterility in mosquitoes. *Evolution* 13:568–570.

Caullery, M. J. G. C.

1952. *Parasitism and symbiosis* (Macmillan, New York).

Cei, J. M.

1944. Analisi biogeografica e ricerche biologiche e sperimentali sul ciclo sessuale annuo delle Rane rosse d'Europa. *Monitore Zool. Ital.* 54(suppl.):1–117.

1949a. Factores genetico-raciales que diferencian la regulación hormonal del ciclo sexual en *Leptodactylus ocellatus* (L.) de la Argentina. *Acta Zool. Lilloana* 7:113–134.

1949b. Sobre la biología sexual de un betracio de grande altura de la región Andina (*Telmatobius schreiteri* Vellard). *Acta Zool. Lilloana* 7:467–488.

Chabaud, A. G.

1954. *L'Ornithodorus erraticus* (Lucas 1849) multiplicité des races. *Ann. Parasitol.* 29:89–130.

Chapin, J. P.

1932. The birds of the Belgian Congo, Part I. *Bull. Amer. Mus. Nat. Hist.* 65:1–391.

1948. Variation and hybridization among the paradise flycatchers of Africa. *Evolution* 2:111–126.

Chapman, F. M.

1931. The upper zonal bird-life of Mts. Roraima and Duida. *Bull. Amer. Mus. Nat. Hist.* 63:1–135.

Chetverikov, S. S.

1926. On certain aspects of the evolutionary process from the standpoint of modern genetics. *J. Exptl. Biol.* (Russian) A2:3–54. Eng. transl. (1961), *Proc. Amer. Phil. Soc.* 105:167–195.

Clark, E.
   1959.   Functional hermaphroditism and self-fertilization in a serranid fish. *Science* 129:215–216.

Clark, E., L. R. Aronson, and M. Gordon
   1954.   Mating behavior patterns in two sympatric species of xiphophorin fishes: their inheritance and significance in sexual isolation. *Bull. Amer. Mus. Nat. Hist.* 103:135–226.

Clarke, C. A., and P. M. Sheppard
   1955.   A preliminary report on the genetics of the *machaon* group of swallowtail butterflies. *Evolution* 9:182–201.
   1959.   The genetics of some mimetic forms of *Papilio dardanus,* Brown, and *Papilio glaucus,* Linn. *J. Genet.* 56:236–260.
   1960a. The evolution of dominance under disruptive selection. *Heredity* 14:73–87.
   1960b. The evolution of mimicry in the butterfly *Papilio dardanus. Heredity* 14:163–173.

Clausen, J.
   1951.   *Stages in the evolution of plant species* (Cornell University Press, Ithaca).

Clausen, J., D. D. Keck, and W. M. Hiesey
   1945.   Experimental studies on the nature of species. II. Plant evolution through amphiploidy and autoploidy, with examples from the Madiinae. *Carnegie Inst. Wash. Publ.,* No. 564:1–174.
   1948.   Experimental studies on the nature of species. III. Environmental responses of climatic races of Achillea. *Carnegie Inst. Wash. Publ.,* No. 581:1–129.

Clausen, R. T.
   1941.   On the use of the terms "subspecies" and "variety." *Rhodora* 43:157–167.

Clay, T.
   1949.   Some problems in the evolution of a group of ectoparasites. *Evolution* 3:279–299.

Cockrum, E. L.
   1952.   A checklist and bibliography of hybrid birds in North America north of Mexico. *Wilson Bull.* 64:140–159.

Colbert, E. H.
   1958.   Tetrapod extinctions at the end of the Triassic period. *Proc. Nat. Acad. Sci.* 44:973–977.

Cold Spring Harbor Symposia on Quantitative Biology (Long Island Biological Association, Cold Spring Harbor, New York):
   1950.   Volume 15. Origin and evolution of man.
   1951.   Volume 16. Genes and mutations.
   1955.   Volume 20. Population genetics: the nature and causes of genetic variability in populations.

1958. Volume 23. Exchange of genetic material: mechanisms and consequences.

1959. Volume 24. Genetics and twentieth century Darwinism.

Connell, J. H.

1959. An experimental analysis of interspecific competition in natural populations of intertidal barnacles. *Proc. XV Intern. Congr. Zool., London,* 290–293.

Cooch, F. G., and J. A. Beardmore

1959. Assortative mating and reciprocal difference in the blue–snow goose complex. *Nature* 183:1833–1844.

Cook, L. M.

1961a. The edge effect in population genetics. *Amer. Nat.* 95:295–307.

1961b. Food-plant specialization in the moth *Panaxia dominula* L. *Evolution* 15:478–485.

Coon, C. S., S. M. Garn, and J. B. Birdsell

1950. *Races* (Thomas, Springfield, Ill.).

Cooper, K. W.

1939. The nuclear cytology of the grass mite, *Pediculopsis graminum* (Reut.), with special reference to karyomerokinesis. *Chromosoma* 1:51–103.

1953. The ecology, predation and competition of *Ancistrocerus antilope* (Panzer). *Trans. Amer. Entomol. Soc.* 79:13–35.

Cooper, R. P.

1961. Field notes on the nesting of the red-tipped pardalotes. *Emu* 61:1–6.

Cope, E. D.

1896. *The primary factors of organic evolution* (Open Court, Chicago).

Corbet, A. S.

1943. Taxonomy of the moths infesting stored food products. *Nature* 152:742–743.

Cordeiro, A. R.

1952. Experiments on the effects in heterozygous condition of second chromosomes from natural populations of *Drosophila willistoni*. *Proc. Nat. Acad. Sci.* 38:471–478.

Cory, Brother Lawrence, and J. J. Manion

1955. Ecology and hybridization in the genus *Bufo* in the Michigan–Indiana region. *Evolution* 9:42–51.

Cott, H. B.

1940. *Adaptive coloration in animals* (Methuen, London).

Craighead, F. C.

1921. Hopkins host-selection principle as related to certain cerambycid beetles. *J. Agric. Research* 22:189–220.

Crampton, H. E.

1916. Studies on the variation, distribution, and evolution of the genus *Partula*. The species inhabiting Tahiti. *Carnegie Inst. Wash. Publ.,* No. 228:1–331.

1932. Studies on the variation, distribution, and evolution of the genus *Partula. Carnegie Inst. Wash. Publ.* No. 410:1–335.

Crane, J.
1944. On the color changes of fiddler crabs (genus *Uca*) in the field. *Zoologica* 39:161–168.

1948–1950. Comparative biology of salticid spiders at Rancho Grande, Venezuela. *Zoologica* 33–35. (A series of papers.)

1957. Basic patterns of display in fiddler crabs (Ocypodidae, genus *Uca*). *Zoologica* 42:69–82.

Creaser, E. P., and D. Travis
1950. Evidence of a homing instinct in the Bermuda spiny lobster. *Science* 112:169–170.

Creighton, W. S.
1950. The ants of North America. *Bull. Mus. Comp. Zool. Harvard Coll.* 104:1–585.

Crombie, A. C.
1946. Further experiments on insect competition. *Proc. Roy. Soc., London* (B) 133:76–109.

1947. Interspecific competition. *J. Animal Ecol.* 16:44–73.

Crosby, J. L.
1949. Selection of an unfavourable gene-complex. *Evolution* 3:212–230.

Crow, J. F.
1948. Alternative hypotheses of hybrid vigor. *Genetics* 33:477–487.

1952. Dominance and overdominance. In J. W. Gowen, ed., *Heterosis* (Iowa State College Press, Ames), 282–297.

1954. Analysis of a DDT resistant strain of Drosophila. *J. Econ. Entomol.* 47:393–398.

1955. General theory of population genetics: synthesis. *Cold Spring Harbor Symp. Quant. Biol.* 20:54–59.

1957. Genetics of insect resistance to chemicals. *Ann. Rev. Entomol.* 2:227–246.

1958. Some possibilities for measuring selection intensities in man. In J. N. Spuhler, ed., *Natural selection in man* (Wayne State University Press, Detroit), 1–13.

1960. Genetics of insecticide resistance: general considerations. *Misc. Publ. Entomol. Soc.* 2:69–74.

1961. Mechanisms and trends in human evolution. *Daedalus* (summer issue):416–431.

Crow, J. F., and N. E. Morton
1960. The genetic load due to mother-child incompatibility. *Amer. Nat.* 94:413–419.

Crozier, W. J.
1921. "Homing" behaviour in chiton. *Amer. Nat.* 55:276–281.

Cuénot, L.
1936. *L'Espèce* (Doin, Paris).

1951. *L'Évolution biologique* (Masson, Paris).

Cumber, R. A.
   1949. The biology of humble-bees, with special reference to the production of the worker caste. *Trans. Roy. Entomol. Soc. London* 100:1–45.
Curio, E.
   1961. Zur geographischen Variation von Verhaltensweisen. *Vogelwelt* 82:33–48.
Curran, C. H.
   1932. New North American diptera. *Amer. Mus. Novitates*, No. 526:3–8.
da Cunha, A. B.
   1949. Genetic analysis of the polymorphism of color pattern in *Drosophila polymorpha. Evolution* 3:239–251.
   1953. A further analysis of the polymorphism of *Drosophila polymorpha. Nature* 171:887.
   1955. Chromosomal polymorphism in the diptera. *Adv. Genet.* 7:93–138.
da Cunha, A. B., H. Burla, and Th. Dobzhansky
   1950. Adaptive chromosomal polymorphism in *Drosophila willistoni. Evolution* 4:212–235.
da Cunha, A. B., and Th. Dobzhansky
   1954. A further study of chromosomal polymorphism in *Drosophila willistoni* in its relation to the environment. *Evolution* 8:119–134.
da Cunha, A. B., Th. Dobzhansky, O. Pavlovsky, and B. Spassky
   1959. Genetics of natural populations. XXVIII. Supplementary data on the chromosomal polymorphism in *Drosophila willistoni* in its relation to the environment. *Evolution* 13:389–404.
da Cunha, A. B., A. M. El-Tabey Shehata, and W. de Oliveira
   1957. A study of the diets and nutritional preferences of tropical species of Drosophila. *Ecology* 38:98–106.
Dahm, A. G.
   1958. *Taxonomy and ecology of five species groups in the family Planariidae* (Nya Litografen, Malmö).
Dalquest, W. W., and V. B. Scheffer
   1944. Distribution and variation in pocket gophers, *Thomomys talpoides,* in the state of Washington. *Amer. Nat.* 78:308–333.
d'Ancona, U.
   1942. Variabilità, differenziamento di razze locali e di specie nel genere *Niphargus. Mem. Ist. Ital. Idrobiol.* 1:145–167.
d'Ancona, U., and L. V. d'Ancona
   1949. Le dafnie di Nemi in un venticinquennio di osservazioni. *Ricerca Sci., Suppl.* 19:79–91.
Darewski, I. S., and W. N. Kulikowa
   1961. Natürliche Parthenogenese in der polymorphen Gruppe der kaukasischen Felseidechse (*Lacerta saxicola* Eversmann). *Zool. Jahrb. (Syst.)* 89:119–176.
Darlington, C. D.
   1932. *Recent advances in cytology* (Blakiston, Philadelphia).
   1939. *The evolution of genetic systems* (Cambridge University Press, Cambridge, England).

694 ～ BIBLIOGRAPHY

1940. Taxonomic species and genetic systems. In J. S. Huxley, ed., *The new systematics* (Clarendon Press, Oxford), 137–160.

1953. Polyploidy in animals. *Nature* 171:191–194.

Darlington, C. D., and K. Mather

1949. *The elements of genetics* (Allen and Unwin, London).

Darwin, C.

1859. *On the origin of species by means of natural selection, or the preservation of favoured races in the struggle for life* (John Murray, London).

1872. *The origin of species* (6th ed.; John Murray, London).

Darwin, F. (ed.)

1888. *The life and letters of Charles Darwin* (John Murray, London), vol. 3.

Dawson, R. W.

1931. The problem of voltinism and dormancy in the polyphemus moth (*Telea polyphemus* Cramer). *J. Exptl. Zool.* 59:87–131.

De Beer, G. R.

1951. *Embryos and ancestors* (2nd ed.; Clarendon Press, Oxford).

1960. Darwin's notebooks on transmutation of species. *Bull. British Mus. (Nat. Hist.) Historical Ser.* 2:23–150.

1961. The origins of Darwin's ideas on evolution and natural selection. *Proc. Roy. Soc., London* (B) 155:321–338.

Deevey, E. S., Jr.

1949. Biogeography of the Pleistocene. *Bull. Geol. Soc. Amer.* 60:1315–1416.

Degerbøl, M.

1940. Mammalia. In Ad. S. Jensen, W. Lundbeck, Th. Mortensen, and R. Spärck, eds., *The zoology of the Faroes* (Høst, Copenhagen, 1935–1942), 1–132.

Dehnel, P. A.

1955. Rates of growth of gastropods as a function of latitude. *Physiol. Zool.* 28:115–144.

1956. Growth rates in latitudinally and vertically separated populations of *Mytilus californianus. Biol. Bull.* 110:43–53.

Delacour, J.

1949. The genus *Lophura* (Gallopheasants). *Ibis* 91:188–220.

Delacour, J., and C. Vaurie

1950. Les mésanges charbonnières (révision de l'espèce *Parus major*). *L'Oiseau* 20:91–121.

Delco, E. A., Jr.

1960. Sound discrimination by males of two cyprinid fishes. *Texas J. Sci.* 12:48–54.

De Lesse, H.

1960. Spéciation et variation chromosomique chez les Lépidoptères rhopalocères. *Ann. Sci. Nat. Zool.*, (Ser. 12) 2:1–223.

Demerec, M.
  1954. What makes genes mutate? *Proc. Amer. Phil. Soc.* 98:318–322.
Demerec, M., and P. E. Hartman
  1959. Complex loci in microorganisms. *Ann. Rev. Microbiol.* 13:337–405.
Dempster, E. R.
  1949. Effects of linkage on parental-combination and recombination frequencies in F_2. *Genetics* 34:272–284.
  1955. Maintenance of genetic heterogeneity. *Cold Spring Harbor Symp. Quant. Biol.* 20:25–32.
Desneux, J.
  1948. Les nidifications souterraines des *Apicotermes,* termites de l'Afrique tropicale. *Rev. Zool. Bot. Africaines* 41:1–54.
  1952. Les constructions hypogées des *Apicotermes,* termites de l'Afrique tropicale. *Ann. Mus. Roy. Congo Belge, Zool.* 17:1–98.
Dethier, V. G.
  1947. *Chemical insect attractants and repellents* (Blakiston, Philadelphia).
  1954. Evolution of feeding preferences in phytophagous insects. *Evolution* 8:33–54.
De Vries, H.
  1901. Die Mutation und die Mutationsperioden bei der Entstehung der Arten. *Verh. Ges. Deut. Naturf. Ärzte* 73:202–212.
  1906. *Species and varieties, their origin by mutation* (Lectures delivered at the University of California; 2nd ed.; Open Court, Chicago).
Dice, L. R.
  1940. Ecologic and genetic variability within species of *Peromyscus. Amer. Nat.* 74:212–221.
  1947. Effectiveness of selection by owls of deer-mice (*Peromyscus maniculatus*) which contrast in color with their background. *Contrib. Lab. Vert. Biol., Univ. Mich.,* No. 34:1–20.
  1949. Variation of *Peromyscus maniculatus* in parts of western Washington and adjacent Oregon. *Contrib. Lab. Vert. Biol., Univ. Mich.,* No. 44:1–34.
Dice, L. R., and P. M. Blossom
  1937. Studies of mammalian ecology in southwestern North America, with special attention to the colors of desert mammals. *Carnegie Inst. Wash. Publ.,* No. 485:1–129.
Dice, L. R., and W. E. Howard
  1951. Distance of dispersal by prairie deermice from birthplace to breeding sites. *Contrib. Lab. Vert. Biol., Univ. Mich.,* No. 50:1–15.
Dickerson, G. E.
  1955. Genetic slippage in response to selection for multiple objectives. *Cold Spring Harbor Symp. Quant. Biol.* 20:213–224.
Dickinson, J. C., Jr.
  1952. Geographic variation in the red-eyed towhee of the eastern United States. *Bull. Mus. Comp. Zool. Harvard Coll.* 107:274–352.

Dilger, W. C.
1956a. Hostile behavior and reproductive isolating mechanisms in the avian genera *Catharus* and *Hylocichla*. *Auk* 73:313–353.
1956b. Adaptive modifications and ecological isolating mechanisms in the thrush genera *Catharus* and *Hylocichla*. *Wilson Bull.* 68:171–199.

Diver, C.
1929. Fossil records of Mendelian mutants. *Nature* 124:183.
1939. Aspects of the study of variation in snails. *J. Conch.* 21:91–141.
1940. The problem of closely related species living in the same area. In J. S. Huxley, ed., *The new systematics* (Clarendon Press, Oxford), 303–328.

Dixon, K. L.
1954. Some ecological relations of chickadees and titmice in central California. *Condor* 56:113–124.
1955. An ecological analysis of the interbreeding of crested titmice in Texas. *Univ. Calif. Publ. Zool.* 54:125–206.
1961. Habitat distribution and niche relationships in North American species of *Parus*. In W. F. Blair, ed., *Vertebrate speciation* (University of Texas Press, Austin), 179–216.

Dobzhansky, Th.
1933. Geographical variation in lady-beetles. *Amer. Nat.* 67:97–126.
1937. *Genetics and the origin of species* (1st ed.; Columbia University Press, New York).
1940. Speciation as a stage in evolutionary divergence. *Amer. Nat.* 74:312–321.
1941. *Genetics and the origin of species* (2nd ed.; Columbia University Press, New York).
1943. Genetics of natural populations. IX. Temporal changes in the composition of populations of *Drosophila pseudoobscura*. *Genetics* 28:162–186.
1944. Chromosomal races in *Drosophila pseudoobscura* and *Drosophila persimilis*. *Carnegie Inst. Wash. Publ.*, No. 554:47–144.
1946. Genetics of natural populations. XIII. Recombination and variability in populations of *Drosophila pseudoobscura*. *Genetics* 31:269–290.
1947a. Adaptive changes induced by natural selection in wild populations of *Drosophila*. *Evolution* 1:1–16.
1947b. Effectiveness of intraspecific and interspecific matings in *Drosophila pseudoobscura* and *D. persimilis*. *Amer. Nat.* 81:66–72.
1947c. Genetics of natural populations. XIV. A response of certain gene arrangements in the third chromosome of *Drosophila pseudoobscura* to natural selection. *Genetics* 32:142–160.
1948. Genetics of natural populations. XVI. Altitudinal and seasonal changes produced by natural selection in certain populations of *Drosophila pseudoobscura* and *Drosophila persimilis*. *Genetics* 33:158–176.
1950. Mendelian populations and their evolution. *Amer. Nat.* 84:401–418.

1951. *Genetics and the origin of species* (3rd ed.; Columbia University Press, New York).

1952. Genetics of natural populations. XX. Changes induced by drought in *Drosophila pseudoobscura* and *Drosophila persimilis*. *Evolution* 6:234–243.

1954. Evolution as a creative process. *Caryologia, Suppl.*, 435–449.

1955a. *Evolution, genetics, and man* (Wiley, New York).

1955b. A review of some fundamental concepts and problems of population genetics. *Cold Spring Harbor Symp. Quant. Biol.* 20:1–15.

1956a. What is an adaptive trait? *Amer. Nat.* 90:337–347.

1956b. Genetics of natural populations. XXV. Genetic changes in populations of *Drosophila pseudoobscura* and *Drosophila persimilis* in some localities in California. *Evolution* 10:82–92.

1957a. Mendelian populations as genetic systems. *Cold Spring Harbor Symp. Quant. Biol.* 22:385–393.

1957b. Genetics of natural populations. XXVI. Chromosomal variability in island and continental populations of *Drosophila willistoni* from Central America and the West Indies. *Evolution* 11:280–293.

1958. Genetics of natural populations. XXVII. The genetic changes in populations of *Drosophila pseudoobscura* in the American Southwest. *Evolution* 12:385–401.

1959a. Variation and evolution. *Proc. Amer. Phil. Soc.* 103:252–263.

1959b. Evolution of genes and genes in evolution. *Cold Spring Harbor Symp. Quant. Biol.* 24:15–30.

1960. Evolution and environment In S. Tax, ed., *The evolution of life* (University of Chicago Press, Chicago), 403–428.

1961. On the dynamics of chromosomal polymorphism in *Drosophila*. In J. S. Kennedy, ed., *Insect polymorphism* (Symp. Roy. Entomol. Soc. London, No. 1), 30–42.

1962. *Mankind evolving* (Yale University Press, New Haven and London).

Dobzhansky, Th., and G. Allen

1956. Does natural selection continue to operate in modern mankind? *Amer. Anthropol.* 58:591–603.

Dobzhansky, Th., H. Burla, and A. B. da Cunha

1950. A comparative study of chromosomal polymorphism in sibling species of the *willistoni* group of *Drosophila*. *Amer. Nat.* 84:229–246.

Dobzhansky, Th., and C. Epling

1944. Contributions to the genetics, taxonomy, and ecology of *Drosophila pseudoobscura* and its relatives. *Carnegie Inst. Wash. Publ.*, No. 554:1–46.

Dobzhansky, Th., and A. M. Holz

1943. A re-examination of the problem of manifold effects of genes in *Drosophila melanogaster*. *Genetics* 28:295–303.

Dobzhansky, Th., and P. Ch. Koller

1938. An experimental study of sexual isolation in *Drosophila*. *Biol. Zentr.* 58:589–607.

Dobzhansky, Th., and H. Levene
  1955. Genetics of natural populations. XXIV. Developmental homeostasis in natural populations of *Drosophila pseudoobscura*. *Genetics* 40:797–808.
Dobzhansky, Th., and C. Pavan
  1950. Local and seasonal variations in relative frequencies of species of *Drosophila* in Brazil. *J. Animal Ecol.* 19:1–14.
Dobzhansky, Th., and O. Pavlovsky
  1953. Indeterminate outcome of certain experiments on *Drosophila* populations. *Evolution* 7:198–210.
  1955. An extreme case of heterosis. *Proc. Nat. Acad. Sci.* 41:289–295.
  1957. An experimental study of interaction between genetic drift and natural selection. *Evolution* 11:311–319.
  1960. How stable is balanced polymorphism? *Proc. Nat. Acad. Sci.* 46:41–47.
  1961. A further study of fitness of chromosomally polymorphic and monomorphic populations of *Drosophila pseudoobscura*. *Heredity* 16:169–179.
Dobzhansky, Th., O. Pavlovsky, B. Spassky, and N. Spassky
  1955. Genetics of natural populations. XXIII. Biological role of deleterious recessives in populations of *Drosophila pseudoobscura*. *Genetics* 40:781–796.
Dobzhansky, Th., and B. Spassky
  1947. Evolutionary changes in laboratory cultures of *Drosophila pseudoobscura*. *Evolution* 1:191–216.
  1954a. Genetics of natural populations. XXII. A comparison of the concealed variability in *Drosophila prosaltans* with that in other species. *Genetics* 39:472–487.
  1954b. Environmental modification of heterosis in *Drosophila pseudoobscura*. *Proc. Nat. Acad. Sci.* 40:407–415.
  1959. *Drosophila paulistorum*, a cluster of species in statu nascendi. *Proc. Nat. Acad. Sci.* 45:419–428.
  1960. Release of genetic variability through recombination. V. Break-up of synthetic lethals by crossing over in *Drosophila pseudoobscura*. *Zool. Jahrb. (Syst.)* 88:57–66.
  1962. Genetic drift and natural selection in experimental populations of *Drosophila pseudoobscura*. *Proc. Nat. Acad. Sci* 48:148–156.
Dobzhansky, Th., and B. Wallace
  1953. The genetics of homeostasis in Drosophila. *Proc. Nat. Acad. Sci.* 39:162–171.
Döderlein, L.
  1902. Über die Beziehungen nahe verwandter "Thierformen" zu einander. *Z. Morphol. Anthropol.* 4:394–442.
Dohrn, A.
  1875. *Der Ursprung der Wirbelthiere und das Princip des Functionswechsels* (Engelmann, Leipzig).

Dorst, J.
  1960.   Considérations sur les Passereaux de la famille des Vangidés. *Proc XII Intern. Ornithol. Congr. Helsinki, 1958,* 173–177.
Dottrens, E.
  1953.   Contribution à la connaissance du Weissfelchen de l'Untersee (Genre Coregonus, Salmonid.) *Rev. Suisse Zool.* 60:452–461.
  1959.   Systématique des Corégones de l'Europe occidentale, basée sur une étude biométrique. *Rev. Suisse Zool.* 66:1–66.
Dottrens, E., and A. Quartier
  1949.   Les Corégones du lac de Neuchâtel. *Rev. Suisse Zool.* 56:689–730.
Dougherty, E. C.
  1955.   The origin of sexuality. *Systematic Zool.* 4:145–169.
Doutt, J. K.
  1942.   A review of the genus *Phoca. Ann. Carnegie Mus. Pittsburgh* 29:61–125.
  1955.   Terminology of microgeographic races in mammals. *Systematic Zool.* 4:179–185.
Dowdeswell, W. H.
  1956.   Isolation and adaptation in populations of the Lepidoptera. *Proc. Roy. Soc., London* (B) 145:322–329.
  1961.   Experimental studies on natural selection in the butterfly, *Maniola jurtina. Heredity* 16:39–52.
Dowdeswell, W. H., R. A. Fisher, and E. B. Ford
  1949.   The quantitative study of populations in the Lepidoptera. II. *Maniola jurtina* L. *Heredity* 3:67–84.
Dowdeswell, W. H., E. B. Ford, and K. G. McWhirter
  1957.   Further studies on isolation in the butterfly *Maniola jurtina* L. *Heredity* 11:51–65.
Drees, O.
  1952.   Untersuchungen über die angeborenen Verhaltensweisen bei Springspinnen (Salticidae). *Z. Tierpsychol.* 9:169–207.
Dreyfus, A., and M. E. Breuer
  1944.   Chromosomes and sex determination in the parasitic hymenopteron *Telenomus fariai* (Lima). *Genetics* 29:75–82.
Drummond, F. H.
  1951.   The *Culex pipiens* complex in Australia. *Trans. Roy. Entomol. Soc. London* 102:369–371.
Drury, W. H., Jr., and R. C. Rollins
  1952.   The North American representatives of Smelowskia (Cruciferae). *Rhodora* 54:85–119.
Dubinin, N. P.
  1948.   Experimental investigation of the integration of hereditary systems in the processes of evolution of populations. *Zhurn. Obshch. Biol.* 9:203–244.
Dubinin, N. P., M. A. Heptner, S. IA. Bessmertnaia, S. IU. Goldat, K. A.

Panina, E. Pogossian, S. V. Saprykina, B. N. Sidorov, L. W. Ferry, and M. G. TSubina
1934. Experimental study of the ecogenotypes of *Drosophila melanogaster*. *Biol. Zhurn.* 3:166–216.

Dubinin, N. P., and G. G. Tiniakov
1945. Seasonal cycles and the concentration of inversions in populations of *Drosophila funebris. Amer. Nat.* 79:570–572.

Dufour, L.
1844. Anatomie générale des Diptères. *Ann. Sci. Nat.* 1:244–264.

Dumas, P. C.
1956. The ecological relations of sympatry in *Plethodon dunni* and *Plethodon vehiculum. Ecology* 37:484–495.

Dunbar, R. W.
1959. The salivary gland chromosomes of seven forms of black flies included in *Eusimulium aureum* Fries. *Canad. J. Zool.* 37:495–525.

Dunn, E. R.
1940. The races of *Ambystoma tigrinum. Copeia*, No. 3:154–162.
1942. Survival value of varietal characters in snakes. *Amer. Nat.* 76:104–109.

Dunn, L. C.
1944. The possible genetic basis of the ringed and striped patterns [in Western king snakes]. *Amer. Midl. Nat.* 31:91–95.
1956. Analysis of a complex gene in the house mouse. *Cold Spring Harbor Symp. Quant. Biol.* 21:187–195.
1959. *Heredity and evolution in human populations* (Harvard University Press, Cambridge, Massachusetts).

Dunn, L. C., and Th. Dobzhansky
1946. *Heredity, race and society* (Penguin Books, New York).

DuRietz, G. E.
1930. The fundamental units of biological taxonomy. *Svensk. Bot. Tidskrift* 24:333–428.

Durrant, S. D.
1952. Mammals of Utah. Taxonomy and distribution. *Univ. Kansas Publ., Mus. Nat. Hist.* 6:1–549.

Durrant, S. D., and R. M. Hansen
1954. Distribution patterns and phylogeny of some western ground squirrels. *Systematic Zool.* 3:82–85.

DuShane, G. P., and C. Hutchinson
1944. Differences in size and developmental rate between eastern and midwestern embryos of *Ambystoma maculatum. Ecology* 25:414–423.

East, E. M.
1935. Genetic reactions in Nicotiana. III. Dominance. *Genetics* 20:443–451.

Edelstam, C., and C. Palmer
1950. Homing behaviour in gastropodes. *Oikos* 2:259–270.

Edgren, R. A.
1953. Copulatory adjustment in snakes and its evolutionary implications. *Copeia*, No. 3:162–164.

Ehrendorfer, F.
1953. Ökologisch-geographische Mikro-Differenzierung einer Population von *Galium pumilum* Muı. s. str. *Österreich. Botan. Z.* 100:616–638.

Ehrman, L.
1960a. The genetics of hybrid sterility in *Drosophila paulistorum*. *Evolution* 14:212–223.
1960b. A genetic constitution frustrating the sexual drive in *Drosophila paulistorum*. *Science* 131:1381–1382.

Eichhorn, O.
1958. Morphologische und papierchromatographische Untersuchungen zur Artentrennung in der Cattung *Dreyfusia* C. B. (*Adelgidae*). *Z. angew. Entomol.* 42:278–283.

Eiseley, L.
1958. *Darwin's century: evolution and the men who discovered it* (Doubleday, New York).
1959. Charles Darwin, Edward Blyth, and the theory of natural selection. *Proc. Amer. Phil. Soc.* 103:94–158.

Eisentraut, M.
1949. Die Eidechsen der spanischen Mittelmeerinseln und ihre Rassenaufspaltung im Lichte der Evolution. *Mitt. Zool. Mus. Berlin* 26:1 228.

Ekman, S.
1953. *Zoogeography of the sea* (Sedgwick and Jackson, London).

Eller, K.
1939. Versuch einer historischen und geographischen Analyse zur Rassen- und Artbildung. *Z. ind. Abst.- u. Vererb.* 77:135–171.

Ellerman, J. R., and T. C. S. Morrison-Scott
1951. *Checklist of Palaearctic and Indian mammals 1758 to 1946* (British Museum [Nat. Hist.], London).

Elton, C. S.
1930. *Animal ecology and evolution* (Clarendon Press, Oxford).
1946. Competition and the structure of ecological communities. *J. Animal Ecol.* 15:54–68.
1958. *The ecology of invasions by animals and plants* (Methuen, London).

Emerson, A. E.
1935. Termitophile distribution and quantitative characters as indicators of physiological speciation in British Guiana termites. *Ann. Entomol. Soc. Amer.* 28:369–395.
1949. Ecology and isolation. In W. C. Allee, A. E. Emerson, O. Park, T. Park, and K. P. Schmidt, *Principles of animal ecology* (Saunders, Philadelphia), 605–630.
1956. Ethospecies, ethotypes, taxonomy, and evolution of *Apicotermes* and *Allognathotermes* (Isoptera, Termitidae). *Amer. Mus. Novitates*, No. 1771:1–31.

Emlen, J. T., Jr.
  1957. Display and mate selection in the whydahs and bishop birds. *Ostrich* 28:202–213.
Emlen, J. T., Jr., and G. B. Schaller
  1960. Distribution and status of the mountain gorilla. *Zoologica* 45:41–52.
Epling, C.
  1944. Contributions to the genetics, taxonomy, and ecology of *Drosophila pseudoobscura* and its relatives. III. The historical background. *Carnegie Inst. Wash. Publ.*, No. 554:145–183.
  1947a. Natural hybridization of *Salvia apiana* and *S. mellifera*. *Evolution* 1:69–78.
  1947b. The genetic aspects of natural populations. Actual and potential gene flow in natural populations. *Amer. Nat.* 81:104–113.
Epling, C., and Th. Dobzhansky
  1942. Microgeographic races in *Linanthus parryae*. *Genetics* 27:317–332.
Epling, C., H. Lewis, and F. M. Ball
  1960. The breeding group and seed storage: a study in population dynamics. *Evolution* 14:238–255.
Epling, C., and W. R. Lower
  1957. Changes in an inversion system during a hundred generations. *Evolution* 11:248–256.
Epling, C., D. F. Mitchell, and R. H. T. Mattoni
  1953. On the role of inversions in wild populations of *Drosophila pseudoobscura*. *Evolution* 7:342–365.
  1957. The relation of an inversion system to recombination in wild populations. *Evolution* 11:225–247.
Ergene, S.
  1952. Farbanpassung entsprechend der jeweiligen Substratfärbung bei *Acrida turrita*. *Z. vergl. Physiol.* 34:69–74.
  1955. Über die Bevorzugung eines homochromen Milieus bei Heuschrecken und Gottesanbeterin. *Zool. Jahrb. (Syst.)* 83:185–322.
  1957. Homochromie und Dressierbarkeit nach Versuchen mit *Oedipoda coerulescens*-Imagines. *Zool. Anz.* 158:38–44.
Ericson, D.
  1959. Coiling direction of *Globigerina pachyderma* as a climatic index. *Science* 130:219–220.
Ernst, F.
  1952. Biometrische Untersuchungen an schweizerischen Populationen von *Triton alp. alpestris* (Laur.). *Rev. Suisse Zool.* 59:399–476.
Esper, E. J. C.
  1781. *De varietatibus specierum in naturae productis*. Sect. 1, 2. (Erlangen).
Etchécopar, R. D., and F. Hüe
  1957. Données écologiques sur l'avifaune de la zone désertique arabosaharienne. Écologie humaine et animale. *Unesco Publ.* (1956), 141–163.

Etkin, W.
  1954. Social behavior and the evolution of man's mental faculties. *Amer. Nat.* 88:129–142.
Evans, R. G.
  1953. Studies on the biology of British limpets of the genus *Patella* on the south coast of England. *Proc. Zool. Soc. London* 123:357–375.
Faber, A.
  1929. *Chorthippus longicornis* Latr. (= *parallelus Zett.*) *und Chorthippus montanus* Charp. (bisher nach Finot als *longicornis* Latr. bezeichnet). *Zool. Anz.* 81:1–24.
  1953. *Laut- und Gebärdensprache bei Insekten. Orthoptera (Geradflügler)* (Mitt. Mus. Naturkunde, Stuttgart).
Fabergé, A. C.
  1943. The concept of polygenes. *Nature* 151:643.
Fabricius, E.
  1950. Heterogeneous stimulus summation in the release of spawning activities in fish. *Inst. Fresh-water Research, Drottningholm,* Report No. 31:57–99.
Faegri, K.
  1937. Some fundamental problems of taxonomy and phylogenetics. *Bot. Rev.* 3:400–423.
Fage, L., and P. Drach (eds.)
  1958. Biologic comparée des espèces marines. *Intern. Union Biol. Sci.* (Ser. B), No. 24:1–327.
Falconer, D. S.
  1960. *Introduction to quantitative genetics* (Oliver and Boyd, Edinburgh and London).
Falla, R. A.
  1946. An undescribed form of the black petrel. *Rec. Canterbury (N.Z.) Mus.* 5:111–113.
Faure, J. C.
  1932. The phases of locusts in South Africa. *Bull. Entomol. Res.* 23:293–405.
Fischer, A. G.
  1960. Latitudinal variations in organic diversity. *Evolution* 14:64–81.
Fischer, E.
  1913. *Die Rehobother Bastards* (Gustav Fischer, Jena.)
Fisher, J.
  1953. The collared turtle dove in Europe. *British Birds* 46:153–181.
Fisher, J., and R. A. Hinde
  1948. The opening of milk bottles by birds. *British Birds* 42:347–357.
Fisher R. A.
  1922. On the dominance ratio. *Proc. Roy. Soc., Edinburgh* 42:321–341.
  1923. Darwinian evolution by mutations. In *Eugenics, genetics and the family* (Sci. Papers 2nd Intern. Congr. Eugenics, New York, 1921; Williams and Wilkins, Baltimore) 1:115–119.

1930. *The genetical theory of natural selection* (Clarendon Press, Oxford).

1939. Selective forces in wild populations of *Paratettix texanus*. *Ann. Eugenics* 9:109–122.

1941. Average excess and average effect of a gene substitution. *Ann. Eugenics* 11:53–63.

1950. Gene frequencies in a cline determined by selection. *Biometrics* 6:353–361.

1954. Retrospect of the criticisms of the theory of natural selection. In J. S. Huxley, A. C. Hardy, and E. B. Ford, eds., *Evolution as a process* (Allen and Unwin, London), 84–98.

Fitch, H. S.
1940. A biographical study of the *ordinoides* artenkreis of garter snakes (genus *Thamnophis*). *Univ. Calif. Publ. Zool.* 44:1–150.

1947. Predation by owls in the Sierran foothills of California. *Condor* 49:137–151.

Fleming, C. A.
1957. The genus *Pecten* in New Zealand. *New Zealand Geol. Survey, Paleontol. Bull.* 26:1–69.

Forbes, G. S., and H. E. Crampton
1942. The differentiation of geographical groups in *Lymnaea palustris*. *Biol. Bull.* 82:26–46.

Ford, C. E., J. L. Hamerton, and G. B. Sharman
1957. Chromosome polymorphism in the common shrew. *Nature* 180:392–393.

Ford, E. B.
1937. Problems of heredity in the Lepidoptera. *Biol. Rev.* 12:461–503.

1940. Polymorphism and taxonomy. In J. S. Huxley, ed., *The new systematics* (Clarendon Press, Oxford), 493–513.

1945. Polymorphism. *Biol. Rev.* 20:73–88.

1946. *Butterflies* (New Naturalist Series; Collins, London).

1949. Early stages in allopatric speciation. In G. L. Jepsen, E. Mayr, and G. G. Simpson, eds., *Genetics, paleontology, and evolution* (Princeton University Press, Princeton), 309–314.

1953. The genetics of polymorphism in the Lepidoptera. *Adv. Genet.* 5:43–87.

1955. Rapid evolution and the conditions which make it possible. *Cold Spring Harbor Symp. Quant. Biol.* 20:230–238.

1957. The study of evolution by observation and experiment. *Proc. Roy. Inst. Great Britain* 36:1–9.

Ford, H. D., and E. B. Ford
1930. Fluctuation in numbers and its influence on variation. *Trans. Entomol. Soc. London* 78:345–351.

Forsman, B.
1949. Weitere Studien über die Rassen von *Jaera albifrons* Leach. *Zool. Bidrag Uppsala* 27:449–463.

Fox, W.
   1948a. The relationships of the garter snake *Thamnophis ordinoides*. *Copeia*, No. 2:113–120.
   1948b. Effect of temperature on development of scutellation in the garter snake, *Thamnophis elegans atratus*. *Copeia*, No. 4:252–262.
   1948c. Variation in the deer-mouse (*Peromyscus maniculatus*) along the lower Columbia River. *Amer. Midl. Nat.* 40:420–452.
   1951. Relationships among the garter snakes of the *Thamnophis elegans* Rassenkreis. *Univ. Calif. Publ. Zool.* 50:485–530.
Fox, W., C. Gordon, and M. H. Fox
   1961. Morphological effects of low temperatures during the embryonic development of the garter snake, *Thamnophis elegans*. *Zoologica* 46:57–71.
Frank, F.
   1956. Beiträge zur Biologie der Feldmaus, *Microtus arvalis* (Pallas). Teil II: Laboratoriumsergebnisse. *Zool. Jahrb.* (*Syst.*) 84:32–74.
Frank, P. W.
   1952. A laboratory study of intraspecies and interspecies competition in *Daphnia pulicaria* (Forbes) and *Simocephalus vetulus* O. F. Müller. *Physiol. Zool.* 25:178–204.
   1957. Coactions in laboratory populations of two species of *Daphnia*. *Ecology* 38:510–519.
Freeman, T. N., M. R. MacKay, I. M. Campbell, C. E. Cox, S. G. Smith, and G. S. Walley
   1953. Studies on the spruce and jack-pine budworms *Canad. Entomol.* 85:121–152.
Frizzi, G.
   1952. Nuovi contributi e prospettive di recerca nel gruppo *Anopheles maculipennis* in base allo studio del dimorfismo cromosomico (ordinamento ad X invertito e tipico) nel *messeae*. *Symp. Genet.* 3:231–265.
Fryer, G.
   1957. The food of some freshwater cyclopoid copepods. *J. Animal Ecol.* 26:263–286.
   1959. The ecology and evolution of a group of rock-frequenting Nyasan cichlid fishes known as the "Mbuna." *Proc. Zool. Soc. London* 132:237–281.
   1960a. Evolution of fishes in Lake Nyasa. *Evolution* 14:396–400.
   1960b. Some controversial aspects of speciation of African cichlid fishes. *Proc. Zool. Soc. London* 135:569–578.
Fulton, B. B.
   1925. Physiological variation in the snowy tree-cricket, *Oecanthus niveus* DeGeer. *Ann. Entomol. Soc. Amer.* 18:363–383.
   1931. A study of the genus *Nemobius* (Orthoptera: Gryllidae). *Ann. Entomol. Soc. Amer.* 24:205–237.
   1933. Inheritance of song in hybrids of two subspecies of *Nemobius fasciatus* (Orthoptera). *Ann. Entomol. Soc. Amer.* 26:368–376.

1937. Experimental crossing of subspecies in *Nemobius* (Orthoptera: Gryllidae). *Ann. Entomol. Soc. Amer.* 30:201–207.

1952. Speciation in the field cricket. *Evolution* 6:283–295.

Gabriel, M. L.

1944. Factors affecting the number and form of vertebrae in *Fundulus heteroclitus*. *J. Exptl. Zool.* 95:105–143.

Gabritschevsky, E.

1924. Farbenpolymorphismus and Vererbung mimetischer Varietäten der Fliege *Volucella bombylans* und anderer "hummelähnlicher" Zweiflügler. *Z. ind. Abst.- u. Vererb.* 32:321–353.

Galbraith, I. C. J.

1956. Variation, relationships and evolution in the *Pachycephala pectoralis* superspecies (Aves, Muscicapidae). *Bull. Brit. Mus. (Nat. Hist.), Zool.* 4:133–222.

Galliker, P.

1958. Morphologie und Systematik der präimaginalen Stadien der schweizerischen Solenobia-Arten (Lep. Psychidae). *Rev. Suisse Zool.* 65:95–183.

Gause, G. F.

1934. *The struggle for existence* (Williams and Wilkins, Baltimore).

1947. Problems of evolution. *Trans. Connecticut Acad. Arts Sci.* 37:17–68.

Gause, G. F., N. P. Smaragdova, and W. W. Alpatov

1942. Geographic variation in *Paramecium* and the role of stabilizing selection in the origin of geographic differences. *Amer. Nat.* 76:63–74.

Gavin, A.

1947. Birds of Perry River district, Northwest territories. *Wilson Bull.* 59:195–203.

Gentilli, J.

1949. Foundations of Australian bird geography. *Emu* 49:85–129.

Gerhardt, U.

1939. Neue biologische Untersuchungen an Limaciden. *Z. Morphol. u. Ökol. Tiere* 35:183–202.

Gering, R. L.

1953. Structure and function of the genitalia in some American agelenid spiders. *Smithsonian Misc. Coll.* 121:1–84.

Gerould, J. H.

1946. Hybridization and female albinism in *Colias philodice* and *C. eurytheme*. A New Hampshire survey in 1943 with subsequent data. *Ann. Entomol. Soc. Amer.* 39:383–396.

Gershenson, S.

1945. Evolutionary studies on the distribution and dynamics of melanism in the hamster (*Cricetus cricetus* L.) I and II. *Genetics* 30:207–251.

1946. The role of natural selection in the distribution and dynamics of melanism in the hamster. *J. Gen. Biol., Moscow* 7:97–130. (English summary.)

Ghilarov, M. S.
1945. Influence of the character of dispersal on the ontogenesis of insects. *J. Gen. Biol., Moscow* 6:36.
Gibb, J.
1954. Feeding ecology of tits, with notes on tree-creeper and goldcrest. *Ibis* 96:513–543.
Gilbert, O., T. B. Reynoldson, and J. Hobart
1952. Gause's hypothesis: an examination. *J. Animal Ecol.* 21:310–312.
Gilliard, E. T.
1959. The ecology of hybridization in New Guinea honeyeaters (Aves). *Amer. Mus. Novitates,* No. 1937:1–26.
Gilmour, J. S. L., and J. W. Gregor
1939. Demes: a suggested new terminology. *Nature* 144:333–334.
Gilmour, J. S. L. and J. Heslop-Harrison
1954. The deme terminology and the units of micro-evolutionary change. *Genetica* 27:147–161.
Ginsburg, I.
1938. Arithmetical definition of the species, subspecies and race concept, with a proposal for a modified nomenclature. *Zoologica* 23:253–286.
Gisin, H.
1947. Le groupe *Entomobrya nivalis* (Collembola). *Mitt Schweiz. Entomol. Ges.* 20:541–550.
Gislén, T.
1948. Aerial plankton and its conditions of life. *Biol. Rev.* 23:109–126.
Glaessner, M. F.
1957. Evolutionary trends in Crustacea (Malacostraca). *Evolution* 11:178–184.
Glass, B.
1956. On the evidence of random genetic drift in human populations. *Amer. J. Phys. Anthropol.* (n.s.) 14:541–556.
1957a. A summary of the symposium on the chemical basis of heredity. In W. D. McElroy and B. Glass, eds., *Symposium on the chemical basis of heredity* (Johns Hopkins Press, Baltimore), 757–834.
1957b. In pursuit of a gene. *Science* 126:683–689.
Glass, B., M. S. Sacks, E. F. Jahn, and C. Hess
1952. Genetic drift in a religious isolate: an analysis of the causes of variation in blood group and other gene frequencies in a small population. *Amer. Nat.* 86:145–160.
Glendenning, R.
1929. Further additions to the list of aphids of British Columbia. *Proc. Entomol. Soc. British Columbia* 26:54–57.
Gloger, C. L.
1833. *Das Abändern der Vögel durch Einfluss des Klimas* (Breslau).
Goethe, F.
1955. Vergleichende Beobachtungen zum Verhalten der Silbermöwe (*Larus*

argentatus) und der Heringsmöwe (*Larus fuscus*). *Acta XI Intern. Ornithol. Congr., Basel, 1954,* 577–582.

Goetze, G., and H. Schmidt

1942. Ein neues Duftorgan der männlichen Honigbiene und seine Bedeutung für die Systematik der Bienenrassen (*Apis mellifica*). *Zool. Jahrb.* (*Syst.*) 75:337–348.

Goin, C. J.

1947. Studies on the life history of *Eleutherodactylus ricordii planirostris* (Cope) in Florida. *Univ. Florida Studies, Biol. Sci. Ser.* 4:1–66.

1950. Color pattern inheritance in some frogs of the genus *Eleutherodactylus. Bull. Chicago Acad. Sci.* 9:1–15.

Goldschmidt, E.

1952. Fluctuation in chromosome number in *Artemia salina. J. Morphol.* 91:111–134.

1953a. Multiple sex-chromosome mechanisms and polyploidy in animals. *J. Genet.* 51:434–440.

1953b. Chromosome numbers and sex mechanism in euphyllopods. *Experientia* 9:65–66.

1956. Chromosomal polymorphism in a population of *Drosophila subobscura* from Israel. *J. Genet.* 54:474–496.

Goldschmidt, R. B.

1934. *Lymantria. Biblioteca Genet.* 11:1–189.

1940. *The material basis of evolution* (Yale University Press, New Haven).

1945. Mimetic polymorphism, a controversial chapter of Darwinism. *Quart. Rev. Biol.* 20:147–164, 205–230.

1947. A note on industrial melanism in relation to some recent work with *Drosophila. Amer. Nat.* 81:474–476.

1948a. Ecotype, ecospecies and macroevolution. *Experientia* 4:465–472.

1948b. Glow worms and evolution. *Rev. Sci., Paris* 86:607–612.

1952a. Evolution as viewed by one geneticist. *Amer. Sci.* 40:84–98.

1952b. Homeotic mutants and evolution. *Acta Biotheoretica* 10:87–104.

1953. Pricking a bubble. (A critical review of N. P. Dubinin's 1948 paper on experimental investigation of the integration of hereditary systems in the process of evolution of populations.) *Evolution* 12:264–269.

1955. *Theoretical genetics* (University of California Press, Berkeley and Los Angeles).

Gontcharoff, M.

1951. Biologie de la régénération et de la reproduction chez quelques *Lineidae* de France. *Ann. Sci. Nat., Zool.* (Ser. 11) 13:149–235.

Goodman, M.

1962. Evolution of the immunologic species specificity of human serum proteins. *Human Biol.* 34:104–150.

Goodnight, C. J., and M. L. Goodnight

1953. The Opilionid fauna of Chiapas, Mexico, and adjacent areas (Arachnoidea, Opiliones). *Amer. Mus. Novitates,* No. 1610:1–81.

Gordon, H., and M. Gordon
1950. Colour patterns and gene frequencies in natural populations of a platyfish. *Heredity* 4:61–73.
1957. Maintenance of polymorphism by potentially injurious genes in eight natural populations of the platyfish, *Xiphophorus maculatus*. *J. Conol.* 55:1–44.

Gordon, M.
1937. The production of spontaneous melanotic neoplasms in fishes by selective matings. *Amer. J. Cancer* 30:362–375.
1947. Speciation in fishes. Distribution in time and space of seven dominant multiple alleles in *Platypoecilus maculatus*. *Adv. Genet.* 1:95–132.

Gordon, M., and D. E. Rosen
1951. Genetics of species differences in the morphology of the male genitalia of Xiphophorin fishes. *Bull. Amer. Mus. Nat. Hist.* 95:413–464.

Gosline, W. A.
1948. Speciation in the fishes of the genus *Menidia*. *Evolution* 2:306–313.

Gösswald, K.
1941. Rassenstudien an der roten Waldameise *Formica rufa* L. auf systematischer, ökologischer, physiologischer und biologischer Grundlage. *Z. angew. Entomol.* 28:62–124.

Götz, B.
1951. Die Sexualduftstoffe an Lepidopteren. *Experientia* 7:406–418.

Götz, W.
1959. Rassenbiometrische Studien an Insekten. *Arch. Julius Klaus-Stiftung* 34:246–252.

Gowen, J. W. (ed.)
1952. *Heterosis, a record of researches directed toward explaining and utilizing the vigor of hybrids* (Iowa State College Press, Ames).

Grant, V.
1952a. Genetic and taxonomic studies in *Gilia*. III. The *Gilia tricolor* complex. *El Aliso* 2:275–288.
1952b. Isolation and hybridization between *Aquilegia formosa* and *A. pubescens*. *El Aliso* 2:341–360.
1957. The plant species in theory and practice. In E. Mayr, ed., *The species problem* (Amer. Assoc. Adv. Sci. Publ. No. 50), 39–80.
1958. The regulation of recombination in plants. *Cold Spring Harbor Symp. Quant. Biol.* 23:337–363.

Gray, A. P.
1958. *Bird hybrids; a check list with bibliography* (Commonwealth Agricultural Bureau, Bucks, England).

Green, M. M.
1959. The discrimination of wild-type isoalleles at the white locus of *Drosophila melanogaster*. *Proc. Nat. Acad. Sci.* 45:549–553.

Greenway, J. C., Jr.
  1958. *Extinct and vanishing birds of the world* (American Committee for International Wild Life Protection, Special Publ. No. 13).

Greenwood, P. H.
  1951. Evolution of the African cichlid fishes; the *Haplochromis* species-flock in Lake Victoria. *Nature* 167:19–20.

Gregor, J. W.
  1946. Ecotypic differentiation. *New Phytol.* 45:254–270.
  1947. Presidential address: some reflections on intra-specific ecological variation and its classification. *Trans. Proc. Botan. Soc. Edinburgh* 34:377–391.

Gregor, J. W., and J. M. S. Lang
  1950. Intra-colonial variation in plant size and habit in sea plantains. *New Phytol.* 49:135–141.

Gregor, J. W., and P. J. Watson
  1954. Some observations and reflexions concerning the patterns of intra-specific differentiation. *New Phytol.* 53:291–300.

Grinnell, J.
  1914a. An account of the mammals and birds of the lower Colorado Valley with especial reference to the distributional problems presented. *Univ. Calif. Publ. Zool.* 12:51–294.
  1914b. Barriers to distribution as regards birds and mammals. *Amer. Nat.* 48:248–254.
  1925. Risks incurred in the introduction of alien game birds. *Science* 61:621–623.
  1926. Geography and evolution in the pocket gopher. *Univ. Calif. Chronicle* 30:429–450.

Grosch, D. S.
  1947. The importance of antennae in mating reaction of male Habrobracon. *J. Comp. Physiol. Psychol.* 40:23–29.

Grüneberg, H.
  1952. *The genetics of the mouse* (Martinus Nijhoff, The Hague).
  1954. Variation within inbred strains of mice. *Nature* 173:674–676.

Gulick, J. T.
  1873. On diversity of evolution under one set of external conditions. *Linn. Soc. J., Zool., London* 11:496–505.
  1887. Divergent evolution through cumulative segregation. *Linn. Soc. J., Zool.* 20:189–274.
  1905. Evolution, racial and habitudinal. *Carnegie Inst. Wash. Publ.*, No. 25:1–265.

Gustafsson, A.
  1953. The cooperation of genotypes in barley. *Hereditas* 39:1–18.
  1954. Mutations, viability, and population structure. *Acta Agr. Scandinavica* 4:601–632.

Haartman, L. von
1949. Der Trauerfliegenschnäpper. I. Ortstreue und Rassenbildung. *Acta Zool. Fenn.* 56:1–104.

Hadorn, E.
1951. Chromatographische Trennung und Messung fluoreszierender Stoffe bei Augenfarb-Mutanten von *Drosophila melanogaster*. *Arch. Julius Klaus-Stiftung* 26:470–475.
1956. Patterns of biochemical and developmental pleiotropy. *Cold Spring Harbor Symp. Quant. Biol.* 21:363–373.

Haecker, V.
1925. *Pluripotenzerscheinungen. Synthetische Beiträge zur Vererbungs- und Abstammungslehre* (Gustav Fischer, Jena).

Hagedoorn, A. L., and A. C. Hagedoorn
1917. Rats and evolution. *Amer. Nat.* 51:385–418.
1921. *The relative value of the processes causing evolution* (Martinus Nijhoff, The Hague).

Hairston, N. G.
1958. Observations on the ecology of *Paramecium*, with comments on the species problem. *Evolution* 12:440–450.

Hairston, N. G., and C. H. Pope
1948. Geographic variation and speciation in Appalachian salamanders (*Plethodon jordani* group). *Evolution* 2:266–278.

Hairston, N. G., F. E. Smith, and L. B. Slobodkin
1960. Community structure, population control, and competition. *Amer. Nat.* 94:421–425.

Haldane, J. B. S.
1932. *The causes of evolution* (Harper, London and New York).
1937. The effect of variation on fitness. *Amer. Nat.* 71:337–349.
1948. The theory of a cline. *J. Genet.* 48:277–284.
1949a. Human evolution: past and future. In G. L. Jepsen, E. Mayr, and G. G. Simpson, eds., *Genetics, paleontology, and evolution* (Princeton University Press, Princeton), 405–418.
1949b. Suggestions as to quantitative measurement of rates of evolution. *Evolution* 3:51–56.
1949c. Disease and evolution. *Ricerca Sci., Suppl.* 19:68–76.
1953. Animal populations and their regulation. *New Biol.*, No. 15:9–24.
1954a. The statics of evolution. In J. Huxley, A. C. Hardy, and E. B. Ford, eds., *Evolution as a process* (Allen and Unwin, London), 109–121.
1954b. The measurement of natural selection. *Caryologia, Suppl.* 6:480–487.
1956. The relation between density regulation and natural selection. *Proc. Roy. Soc., London* (B) 145:306–308.
1957. The cost of natural selection. *J. Genet.* 55:511–524.

Hall, E. R.
1943. Intergradation versus hybridization in ground squirrels of the western United States. *Amer. Midl. Nat.* 29:375–378.

1951. American weasels. *Univ. Kansas Publ., Mus. Nat. Hist.* 4:1–466.

Hamann, O.
1892. *Entwicklungslehre und Darwinismus. Eine kritische Darstellung der modernen Entwicklungslehre und ihrer Erklärungsversuche mit besonderer Berücksichtigung der Stellung des Menschen in der Natur* (Hermann Costenoble, Jena).

Hamilton, T. H.
1959. Adaptive variation in the genus *Vireo. Wilson Bull.* 70:307–346.
1961. The adaptive significances of intraspecific trends of variation in wing length and body size among bird species. *Evolution* 15:180–195.
1962. Species relationships and adaptations for sympatry in the avian genus *Vireo. Condor* 64:40–68.

Hardin, G.
1960. The competitive exclusion principle. *Science* 131:1291–1297.

Harding, J. P.
1950. [Cytology, genetics and classification.] *Nature* 166:769–771.

Harding, J. P., and N. Tebble
1962. Speciation in the sea. *Nature* 193:24–26.

Harland, S. C.
1936. The genetical conception of the species. *Biol. Rev.* 11:83–112.

Harland, S. C., and O. M. Atteck
1933. Breeding experiments with biological races of *Trichogramma minutum* in the West Indies. *Z. ind. Abst.- u. Vererb.* 64:54–76.

Harris, V. T.
1952. An experimental study of habitat selection by prairie and forest races of the deermouse, *Peromyscus maniculatus. Contrib. Lab. Vert. Biol., Univ. Mich.,* No. 56:1–53.

Harrison, G. A.
1958. The adaptability of mice to high environmental temperatures. *J. Exptl. Biol.* 35:892–901.
1959. Environmental determination of the phenotype. In A. J. Cain, ed., *Function and taxonomic importance* (Systematics Assoc. Publ. No. 3), 81–86.

Hart, J. S.
1952 Geographic variations of some physiological and morphological characters in certain freshwater fish. *Univ. Toronto Biol. Ser.,* No. 60:1–79.

Hartman, P. E.
1956. Linked loci in the control of histidine synthesis in *Salmonella typhimurium. Carnegie Inst. Wash. Publ.,* No. 612:35–62.

Hartmann, M.
1953. Die Rassenaufspaltung der Balearischen Inseleidechsen. *Zool. Jahrb. (Syst.)* 64:1–96.

Hasebroek, K.
1934. Industrie und Grosstadt als Ursache des neuzeitlichen vererblichen

Melanismus der Schmetterlinge in England und Deutschland. *Zool. Jahrb.* (*Physiol.*) 53:411–460.

Haskell, G.
1954. Correlated responses to polygenic selection in animals and plants. *Amer. Nat.* 88:5–20.

Haskins, C. P., E. F. Hoskins, and R. E. Hewitt
1960. Pseudogamy as an evolutionary factor in the poeciliid fish *Mollienisia formosa. Evolution* 14:473–483.

Haskins, C. P., E. F. Haskins, J. J. A. McLaughlin, and R. E. Hewitt
1961. Polymorphism and population structure in *Lebistes reticulatus*, an ecological study. In W. F. Blair, ed., *Vertebrate speciation* (University of Texas Press, Austin), 320–395.

Hauenschild, A., and C. Hauenschild
1951. Untersuchungen über die stoffliche Koordination der Paarung des Polychäten *Grubea clavata. Zool. Jahrb.* (*Physiol.*) 62:429–440.

Hayne, D. W.
1950. Reliability of laboratory-bred stocks as samples of wild populations, as shown in a study of the variation of *Peromyscus polionotus* in parts of Florida and Alabama. *Contrib. Lab. Vert. Biol., Univ. Mich.,* No. 46:1–56.

Heberer, G.
1954–1959. ed., *Die Evolution der Organismen* (Gustav Fischer, Stuttgart).
1957. Theorie der additiven Typogenese. In G. Heberer, ed., *Die Evolution der Organismen* (Gustav Fischer, Stuttgart), 857–914.

Hecht, M. K.
1952. Natural selection in the lizard genus *Aristelliger. Evolution* 6:112–124.

Hecht, M. K., and D. Marien
1956. The coral snake mimic problem: a reinterpretation. *J. Morphol.* 98:335–364.

Hecht, M. K., and B. M. Matalas
1946. A review of the middle North American toads of the genus *Microhyla. Amer. Mus. Novitates,* No. 1315:1–21.

Heed, W. B., and N. B. Krishnamurthy
1959. Genetic studies on the Cardini group of Drosophila in the West Indies. *Biol. Contrib., Univ. Texas,* No. 5914:155–179.

Heim de Balsac, H.
1936. Biogéographie des mammifères et des oiseaux de l'Afrique du Nord. *Bull. Biol. France, Suppl.* 21:1–447.

Heiser, C. B.
1949. Natural hybridization with particular reference to introgression. *Bot. Rev.* 15:645–687.

Hellmich, W. C.
1951. On ecotypic and autotypic characters, a contribution to the knowl-

edge of the evolution of the genus *Liolaemus* (Iguanidae). *Evolution* 5:359–369.

Hemmingsen, A. M.
  1960. Energy metabolism as related to body size and respiratory surfaces
         and its evolution. *Repts. Steno Memorial Hospital, Nordisk Insulin-
         lab., Denmark* 9:1–110.

Henbest, L. G. (ed.)
  1952. Distribution of evolutionary explosions in geologic time. A sympo-
         sium. *J. Paleontol.* 26:298–394.

Hendrickson, J. R.
  1954. Ecology and systematics of salamanders of the genus *Batrachoseps*.
         Univ. Calif. Publ. Zool. 54:1–46.

Hennig, W.
  1950. *Grundzüge einer Theorie der phylogenetischen Systematik* (Deut-
         scher Zentralverlag, Berlin).

Herter, K.
  1934. Studien zur Verbreitung der europäischen Igel. *Arch. Naturg.* (N.F.)
         3:313–382.

Hertwig, P.
  1936. Artbastarde bei Tieren. In E. Baur and M. Hartmann, eds., *Hand-
         buch der Vererbungswiss.* 2 (B):1–140.

Heslop-Harrison, J. W.
  1927. Experiments on the egg-laying instincts of the sawfly, *Pontania salicis*
         Christ., and their bearing on the inheritance of acquired characters;
         with some remarks on a new principle in evolution. *Proc. Roy. Soc.,
         London* (B) 101:115–126.

Heslop-Harrison, J.
  1954. Botany. Genecology and orthodox taxonomy: some theoretical as-
         pects. *Sci. Progress,* No. 167:484–494.
  1958. Ecological variation and ethological isolation. *Uppsala Univ. Arsskrift*
         6:150–158.

Hesse, R., W. C. Allee, and K. P. Schmidt
  1951. *Ecological animal geography* (2nd ed.; Wiley, New York).

Heuts, M. J.
  1947. The phenotypical variability of *Gasterosteus aculeatus* (L.) popula-
         tions in Belgium. *Verh. Kon. Vlaamsche Akad. Wet., Belgie* 9:1–63.
  1949. On the mechanism and nature of adaptive evolution. *Ricerca Sci.,
         Suppl.* 19:1–12.
  1951. Les théories de l'évolution devant les données expérimentales. *Rev.
         Quest. Sci.* 122:58–89.
  1952. Theorien und Tatsachen der biologischen Evolution. *Verh. Deut.
         Zool. Ges., Freiburg, 1952,* 409–429.
  1956. Temperature adaptation in *Gasterosteus aculeatus* L. *Pubbl. Staz.
         Zool. Napoli* 28:44–61.

Heydemann, F.
  1943. Die Bedeutung der sogenannten Dualspecies (Zwillingsarten) für

unsere Kenntnis der Art- und Rassenbildung bei Lepidopteren. *Stettiner Entomol. Ztg.* 104:116–142.

1944. Zur Kenntnis der Gattung *Aplecta* Guen. und zweier "Dualspezies" in derselben. (Lep. Noct.). *Stettiner Entomol. Ztg.* 105:12–33.

Hildreth, Ph. E.

1956. The problem of synthetic lethals in *Drosophila melanogaster*. *Genetics* 41:729–742.

Hile, R.

1937. Morphometry of the cisco, *Leucichthys artedi* (Le Sueur), in the lakes of the northeastern highlands, Wisconsin. *Intern. Rev. ges. Hydrol. Hydrog.* 36:57–130.

Hill, I. R.

1954. The taxonomic status of the mid-Gulf Coast *Amphiuma*. *Tulane Studies Zool.* 1:191–215.

Hinde, R. A.

1955. A comparative study of the courtship of certain finches (Fringillidae). *Ibis* 97:706–745.

1959. Behaviour and speciation in birds and lower vertebrates. *Biol. Rev.* 34:85–128.

Hinde, R. A., and N. Tinbergen

1958. The comparative study of species-specific behavior. In A. Roe and G. G. Simpson, eds., *Behavior and evolution* (Yale University Press, New Haven), 251–268.

Hindwood, K. A., and E. Mayr

1946. A revision of the striped-crowned pardalotes. *Emu* 46:49–67.

Hiraizumi, Y., L. Sandler, and J. F. Crow

1960. Populational implications of the segregation distorter locus. *Evolution* 14:433–444.

Hirschmann, H.

1951. Über das Vorkommen zweier Mundhöhlentypen bei *Diplogaster lheritieri* Maupas und *Diplogaster biformis* n. sp. und die Entstehung dieser hermaphroditischen Art aus *Diplogaster lheritieri*. *Zool. Jahrb. (Syst.)* 80:1–188.

Hoare, C. A.

1943. Biological races in parasitic protozoans. *Biol. Rev.* 18:137–144.

1952. The taxonomic status of biological races in parasitic protozoa. *Proc. Linn. Soc. London, Session 163*, Pt. 1:44–47.

Hobbs, H. H., Jr.

1942. The crayfishes of Florida. *Univ. Florida Publ. Biol. Sci.* 3:1–179.

1945. The subspecies and intergrades of the Florida burrowing crayfish *Procambarus rogersi* (Hobbs). *J. Wash. Acad. Sci.* 35: 247–260.

1953. On the ranges of certain crayfishes of the Spiculifer group of the genus *Procambarus*, with the description of a new species (Decapoda: Astacidae). *J. Wash. Acad. Sci.* 43:412–416.

Hobbs, H. H., Jr., and L. J. Marchand
  1943. A contribution towards a knowledge of the crayfishes of the Reelfoot
          Lake area. *J. Tenn. Acad. Sci.* 18:6–35.
Hoenigsberg, H. F., and S. Koref-Santibañez
  1960. Courtship and sensory preferences in inbred lines of *Drosophila
          melanogaster. Evolution* 14:1–7.
Hoesch, W.
  1953. Über die Rassenbildung der s.w.-afrikanischen Bodenvögel unter
          Berücksichtigung von Wasserabhängigkeit, Niederschlagsmenge und
          Bodenfärbung. *J. Ornithol.* 94:274–281.
  1956. Das Problem der Farbübereinstimmung von Körperfarbe und Unter-
          grund. *Bonn. Zool. Beitr.* 7:59–83.
Hoestlandt, H.
  1958. Comparaison des fréquences raciales d'un crustacé littoral, *Sphaeroma
          serratum*, aux Canaries et sur d'autres côtes atlantiques insulaires ou
          continentales. *Anuar. Estud. Atlánticos* No. 4:17–36.
Hoffman, R. L.
  1951. Subspecies of the milliped *Apheloria trimaculata* (Wood) (Poly-
          desmida: Xystodesmidae). *Nat. Hist. Miscellanea,* No. 81:1–6.
Hoffmeister, D. F.
  1956. Mammals of the Graham (Pinaleno) Mountains, Arizona. *Amer.
          Midl. Nat.* 55:257–288.
Hogben, L.
  1946. *An introduction to mathematical genetics* (Norton, New York).
Hölldobler, K.
  1949. Über ein parasitologisches Problem. Die Gastpflege der Ameisen und
          die Symphilieinstinkte. *Z. Parasitenkunde* 14:3–26.
Holm, Å.
  1956. Notes on Arctic spiders of the genera *Erigone* Aud. and *Hilaira* Sim.
          Arkiv Zool. 9:453–467.
Hooijer, D. A.
  1949. Mammalian evolution in the Quaternary of southern and eastern Asia.
          Evolution 3:125–128.
Hooke, R.
  1675 (ca.). *Posthumous works,* p. 411 (originally written 1670–1680), quoted
          in (1950) *Proc. Roy. Soc., London* (B) 137:182.
Hooper, E. T.
  1941. Mammals of the lavafields and adjoining areas in Valencia County,
          New Mexico. *Misc. Publ. Mus. Zool., Univ. Mich.,* No. 51:1–47.
  1944. San Francisco Bay as a factor influencing speciation in rodents. *Misc.
          Publ. Mus. Zool., Univ. Mich.,* No. 59:1–89.
Hovanitz, W.
  1942. Genetic and ecologic analyses of wild populations in Lepidoptera.
          I. Pupal size and weight variation in some California populations of
          Melitaea chalcedona. Ecology 23:175–188.

1944. The distribution of gene frequencies in wild populations of *Colias*. *Genetics* 29:31–60.

1948a. Ecological segregation of interfertile species of *Colias*. *Ecology* 29:461–469.

1948b. Differences in the field activity of two female color phases of *Colias* butterflies at various times of the day. *Contrib. Lab. Vert. Biol., Univ. Mich.*, No. 41:1–37.

1949. Increased variability in populations following natural hybridization. In G. L. Jepsen, E. Mayr, and G. G. Simpson, eds., *Genetics, paleontology, and evolution* (Princeton University Press, Princeton), 339–355.

1953. Polymorphism and evolution. *Symp. Soc. Exptl. Biol.*, No. 7:238–253.

Howard, H.
1950. Fossil evidence of avian evolution. *Ibis* 92:1–21.

Howell, F. C.
1952. Pleistocene glacial ecology and evolution of "classic Neandertal" man. *Southwestern J. Anthropol.* 8:337–410.

1957. The evolutionary significance of variation and varieties of "Neanderthal" man. *Quart. Rev. Biol.* 32:330–347.

1959. The Villafranchian and human origins. *Science* 130:831–844.

Howell, T. R.
1952. Natural history and differentiation in the yellow-bellied sapsucker. *Condor* 54:237–282.

Howells, W.
1959. *Mankind in the making* (Doubleday, New York).

Hrubant, H. E.
1955. An analysis of the color phases of the eastern screech owl, *Otus asio*, by the gene frequency method. *Amer. Nat.* 89:223–230.

Hsu, T. C., and T. T. Liu
1948. Microgeographic analysis of chromosomal variation in a Chinese species of *Chironomus* (Diptera). *Evolution* 2:49–57.

Hsu, W. S.
1956. Oogenesis in the Bdelloidea rotifer, *Philodina roseola* Ehrenberg. *Cellule* 57:281–296.

Hubbell, T. H.
1954. The naming of geographically variant populations. *Systematic Zool.* 3:113–121.

1956. Some aspects of geographic variation in insects. *Ann. Rev. Entomol.* 1:71–88.

Hubbs, C. L.
1922. Variations in the number of vertebrae and other meristic characters of fishes correlated with the temperature of water during development. *Amer. Nat.* 56:360–372.

1940. Speciation of fishes. *Amer. Nat.* 74:198–211.

1941. The relation of hydrological conditions to speciation in fishes. In A *symposium on hydrobiology* (University of Wisconsin Press, Madison), 182–195.

1955. Hybridization between fish species in nature. *Systematic Zool.* 4:1–20.

1961. Isolating mechanisms in the speciation of fishes. In W. F. Blair, ed., *Vertebrate speciation* (University of Texas Press, Austin), 5–23.

Hubbs, C. L., and L. C. Hubbs

1932. Apparent parthenogenesis in nature, in a form of fish of hybrid origin. *Science* 76:628–630.

Hubbs, C. L., L. C. Hubbs, and R. E. Johnson

1943. Hybridization in nature between species of Catostomid fishes. *Contrib. Lab. Vert. Biol., Univ. Mich.*, No. 22:1–76.

Hubbs, C. L., and K. E. Lagler

1949. Fishes of Isle Royale, Lake Superior, Michigan. *Papers Mich. Acad. Sci.* 33:73–133.

Hubbs, C. L., and R. R. Miller

1943. Mass hybridization between two genera of Cyprinid fishes in the Mohave Desert, California. *Papers Mich. Acad. Sci.* 28:343–378.

1948. The Great Basin with emphasis on glacial and postglacial times. II. The zoological evidence. Correlation between fish distribution and hydrographic history in the desert basins of western United States. *Bull. Univ. Utah* 38:18–166.

Hubbs, C. L., and E. C. Raney

1946. Endemic fish fauna of Lake Waccamaw, North Carolina. *Misc. Publ. Mus. Zool., Univ. Mich.*, No. 65:1–30.

Hubbs, Cl.

1960. Duration of sperm function in the Percid fishes *Etheostoma lepidum* and *E. spectabile,* associated with sympatry of the parental populations. *Copeia*, No. 1:1–8.

Hubbs, Cl., and E. A. Delco

1960. Mate preference in males of four species of Gambusiine fishes. *Evolution* 14:145–152.

Hubbs, Cl., and V. G. Springer

1957. A revision of the *Gambusia nobilis* species group, with descriptions of three new species, and notes on their variation, ecology, and evolution. *Texas J. Sci.* 9:279–327.

Hubbs, Cl., and K. Strawn

1956. Infertility between two sympatric fishes, *Notropis lutrensis* and *Notropis venustus. Evolution* 10:341–344.

1957. Survival of F_1 hybrids between fishes of the subfamily Etheostominae. *J. Exptl. Zool.* 134:33–62.

Hubendick, B.

1951. Recent Lymnaeidae. Their variation, morphology, taxonomy, nomenclature, and distribution. *Kungl. Svenska Vetenskapsakad. Handlingar* 3:1–223.

1952. On the evolution of the so-called thalassoid molluscs of Lake Tanganyika. *Arkiv Zool. Stockholm* (Ser. 2) 3:319–323.

1960. The Ancylidae of Lake Ochrid and their bearing on intralacustrine speciation. *Proc. Zool. Soc. London* 133:497–529.

Huntington, C. E.

1952. Hybridization in the purple grackle, *Quiscalus quiscula*. *Systematic Zool.* 1:149–170.

Hustich, I. (ed.)

1952. The recent climatic fluctuation in Finland and its consequences. *Fennia* 75:1–128.

Hutchinson, G. E.

1957. Concluding remarks. *Cold Spring Harbor Symp. Quant. Biol.* 22:415–427.

1959. Homage to Santa Rosalia *or* why are there so many kinds of animals? *Amer. Nat.* 93:145–159.

Hutchinson, G. E., and R. H. MacArthur

1959. A theoretical ecological model of size distribution among species of animals. *Amer. Nat.* 93:117–125.

Hutton, F. W.

1897. The place of isolation in organic evolution. *Nat. Sci.* 11:240–246.

Huxley, J. S.

1939. Clines: an auxiliary method in taxonomy. *Bijdr. Dierk.* 27:491–520.

1940. ed., *The new systematics* (Clarendon Press, Oxford).

1942. *Evolution, the modern synthesis* (Allen and Unwin, London).

1953. *Evolution in action* (Harper, New York).

1955a. Morphism in birds. *Acta XI Intern. Ornithol. Congr., Basel, 1954*, 309–328.

1955b. Morphism and evolution. *Heredity* 9:1–52.

1958. Evolutionary processes and taxonomy with special reference to grades. *Uppsala Univ. Arsskr.* (1958), 21–39.

Huxley, T. H., and J. S. Huxley

1947. *Touchstone for ethics* (Harper, New York).

Hynes, H. B. N.

1950. The food of freshwater sticklebacks, (*Gasterosteus aculeatus* and *Pygosteus pungitius*), with a review of methods used in studies of the food of fishes. *J. Animal. Ecol.* 19:36–58.

1954. The ecology of *Gammarus duebeni* Lilljeborg and its occurrence in fresh water in western Britain. *J. Animal Ecol.* 23:38–84.

Iltis, H. H.

1957. Studies in the Capparidaceae. III. Evolution and phylogeny of the western North American Cleomoideae. *Ann. Missouri Bot. Garden* 44:77–119.

Imbrie, J.

1957. The species problem with fossil animals. In E. Mayr, ed., *The species problem* (Amer. Assoc. Adv. Sci. Publ. No. 50), 125–153.

Inger, R. F.
1961. Problems in the application of the subspecies concept in vertebrate taxonomy. In W. F. Blair, ed., *Vertebrate speciation* (University of Texas Press, Austin), 262–285.

Ingles, L. G.
1950. Pigmental variations in populations of pocket gophers. *Evolution* 4:353–357.

Ingles, L. G., and N. J. Biglione
1952. The continuity of the ranges of two subspecies of pocket gophers *Evolution* 6:204–207.

Ingram, V. M.
1956. A specific chemical difference between the globins of normal human and sickle cell anemia hemoglobin. *Nature* 178:792–794.

Irving, L.
1957. The usefulness of Scholander's views on adaptive insulation of animals. *Evolution* 11:257–260.

Irwin, M. R.
1947. Immunogenetics. *Adv. Genet.* 1:133–159.
1953. Evolutionary patterns of antigenic substances of the blood corpuscles in Columbidae. *Evolution* 7:31–50.
1955. On interrelationships of the cellular antigens of several species of *Streptopelia*. *Evolution* 9:261–279.

Iseley, F. B.
1938. Survival value of Acridian protective coloration. *Ecology* 19:370–389.
1946. Differential feeding in relation to local distribution of grasshoppers. *Ecology* 27:128–138.

Ives, P. T.
1950. The importance of mutation rate genes in evolution. *Evolution* 4:236–252.
1954. Genetic changes in American populations of *Drosophila melanogaster*. *Proc. Nat. Acad. Sci.* 40:87–92.

Jackson, C. H. N.
1944. The analysis of a tsetse fly population. *Ann. Eugenics* 12:176–205.

Jacob, F., and J. Monod
1961a. Genetic regulatory mechanisms in the synthesis of proteins. *J. Molec. Biol.* 3:318–356.
1961b. On the regulation of gene activity. *Cold Spring Harbor Symp. Quant. Biol.* 26:193–211.

Jacobi, A.
1900. Lage und Form biogeographischer Gebiete. *Z. Ges. Erdk. Berlin* 35:147–238.

Jacobs, W.
1950. Vergleichende Verhaltensstudien an Feldheuschrecken. *Z. Tierpsychol.* 7:169–216.

Jacobson, M., M. Beroza, and W. A. Jones
 1960. Isolation, identification, and synthesis of the sex attractant of gypsy
   moth. *Science* 132:1011–1012.
Jahn, H.
 1942. Zur Ökologie und Biologie der Vögel Japans. *J. Ornithol.* 90:1–302.
Janzer, W.
 1950. Versuche zur Entstehung der Höhlentiermerkmale. *Naturw.* 37:286.
Janzer, W., and W. Ludwig
 1952. Versuche zur evolutorischen Entstehung der Höhlentiermerkmale.
   Z. ind. Abst.- u. Vererb. 84:462–479.
Jarvik, E.
 1955. The oldest tetrapods and their forerunners. *Sci. Monthly* 80:141–
   154.
Jeannel, R.
 1950. *La marche de l'évolution* (Presses Universitaires de France, Paris).
Jennings, H. S.
 1938. Sex reaction types and their interrelations in *Paramecium bursaria*.
   I and II. *Proc. Nat. Acad. Sci.* 24:112–120.
Jensen, Ad. S.
 1941. On subspecies and races of the lesser sand eel (*Ammodytes lancea*
   s. lat.). *Biol. Medd., K. Danske Vidensk. Selskab* 16:1–33.
Jepsen, G. L.
 1943. Systematics and the origin of species, from the viewpoint of a zoologist, a discussion. *Amer. J. Sci.* 251:521–528.
Jepsen, C. L., E. Mayr, and G. G. Simpson (eds.)
 1949. *Genetics, paleontology, and evolution* (Princeton University Press,
   Princeton).
Johansen, H.
 1947. Notes on the geographical variation of the chiffchaff (*Phylloscopus
   collybita*). *Dansk Ornithol. Foren. Tidsskr.* 41:198–215.
 1955. Die Jennissei-Faunenscheide. *Zool. Jahrb.* (*Syst.*) 83:185–322.
Johnsen, S.
 1944. Variation in fish in North-European waters. I. Variation in size.
   Bergens Mus. Årbok, No. 4:1–129.
Johnsgard, P. A.
 1960a. Hybridization in the Anatidae and its taxonomic implications. *Condor* 62:25–33.
 1960b. A quantitative study of sexual behavior of mallards and black ducks.
   Wilson Bull. 72:133–155.
Johnson, Charles W.
 1916. The *Volucella bombylans* group in America. *Psyche* 23:159–163.
 1925. The North American varieties of *Volucella bombylans*. *Psyche*
   32:114–117.
Johnson, Clifford
 1959. Genetic incompatibility in the call races of *Hyla versicolor* Le Conte
   in Texas, *Copeia*, No. 4:327–335.

Johnson, D. H.
  1943.   Systematic review of the chipmunks (genus *Eutamias*) of California. *Univ. Calif. Publ. Zool.* 48:63–148.
Iohnson, D. S.
  1952.   A thermal race of *Daphnia atkinsoni* Baird, and its distributional significance. *J. Animal Ecol.* 21:118–119.
  1960.   Subspecific and infraspecific variation in some freshwater prawns of the Indo-Pacific. *Proc. Cent. Bicent. Congr. Biol., Singapore (1958)*, 259–267.
Johnson, M. L.
  1947.   The status of the *elegans* subspecies of *Thamnophis* with description of a new subspecies from Washington state. *Herpetologica* 3:159–165.
Johnston, R. F.
  1954.   Variation in breeding season and clutch size in song sparrows of the Pacific coast. *Condor* 56:268–273.
Jordan, D. S.
  1905.   The origin of species through isolation. *Science* 22:545–562.
  1908.   The law of geminate species. *Amer. Nat.* 42:73–80.
Jordan, K.
  1896.   On mechanical selection and other problems. *Novit. Zool.* 3:426–525.
  1897.   Reproductive divergence: a factor in evolution? *Nat. Sci.* 11:317–320.
  1898.   Reproductive divergence not a factor in the evolution of new species. *Nat. Sci.* 12:45–47.
  1903.   Bemerkungen zu Herrn Dr. Petersen's Aufsatz: Entstehung der Arten durch physiologische Isolierung. *Biol. Zentr.* 33:660–664.
  1905.   Der Gegensatz zwischen geographischer und nichtgeographischer Variation. *Z. wiss. Zool.* 83:151–210.
  1938.   Where subspecies meet. *Novit. Zool.* 41:103–111.
Kachkarov, D. N., and E. P. Korovine
  1942.   *La vie dans les déserts* (Payot, Paris).
Kalela, O.
  1942.   Die Ausbreitung der kulturbedingten Vogelfauna als Glied der spät-quartären Faunengeschichte Europas. *Ornis Fenn.* 19:1–23.
  1944.   Zur Frage der Ausbreitungstendenz der Tiere. *Ann. Zool. Soc. Vanamo* 10:1–23.
Karlson, P., and A. Butenandt
  1959.   Pheromones (ectohormones) in insects. *Ann. Rev. Entomol.* 4:39–58.
Kaston, B. J.
  1936.   The senses involved in the courtship of some of the vagabond spiders. *Entomol. Amer.* 16:97–167.
Kawamura, T., and M. Kobayashi
  1959.   Studies on hybridization in amphibians. VI. Reciprocal hybrids between *Rana temporaria temporaria* L. and *Rana temporaria ornativentris* Werner. *J. Sci. Hiroshima Univ.* (B) 18:1–15.

1960. Studies on hybridization in amphibians. VII. Hybrids between Japanese and European brown frogs. *J. Sci. Hiroshima Univ.* (B) 18:221–238.

Kawamura, T., and S. Sawada
1959. On the sexual isolation among different species and local races of Japanese newts. *J. Sci. Hiroshima Univ.* (B) 18:17–30.

Keast, A.
1959. The Australian environment. In A. Keast, R. L. Crocker, and C. S. Christian, eds., *Biogeography and ecology in Australia* (Junk, The Hague).
1961. Bird speciation on the Australian continent. *Bull. Mus. Comp. Zool. Harvard Coll.* 123:305–495.

Keleher, J. J.
1952. Growth and Triaenophorus parasitism in relation to taxonomy of Lake Winnipeg ciscoes (*Leucichthys*). *J. Fish. Research Board Canada* 8:469–478.

Kelson, K. R.
1951. Speciation in rodents of the Colorado River drainage. *Univ. Utah Biol. Ser.* 11:1–125.

Kempthorne, O.
1957. *An introduction to genetic statistics* (Wiley, New York).

Kendeigh, S. C.
1934. The role of environment in the life of birds. *Ecol. Monographs* 4:299–417.
1961. *Animal ecology* (Prentice-Hall, Englewood Cliffs, New Jersey).

Kennedy, J. S. (ed.)
1961. Insect polymorphism. *Symp. Roy. Entomol. Soc. London*, No. 1:1–115.

Kermack, K. A.
1954. A biometrical study of *Micraster coranguinum* and *M.* (*Isomicraster*) *senonensis*. *Phil. Trans. Roy. Soc. London* (B) 237:375–428.

Kerr, W. E.
1950. Evolution of caste determination in the genus *Melipona*. *Evolution* 4:7–13.

Kettle, D. S.
1950. The seasonal distribution of *Culicoides impunctatus* Goetghebuer [Diptera: Heleidae (Ceratopogonidae)] with a discussion on the possibility that it may be composed of two or more biological races. *Trans. Roy. Entomol. Soc. London* 101:125–146.

Kettle, D. S., and G. Sellick
1947. The duration of the egg stage in the races of *Anopheles maculipennis* Meigen (Diptera, Culicidae). *J. Animal Ecol.* 16:38–43.

Kettlewell, H. B. D.
1956. Further selection experiments on industrial melanism in the Lepidoptera. *Heredity* 10:287–301.

1961. The phenomenon of industrial melanism in Lepidoptera. *Ann. Rev. Entomol.* 6:245–262.

Key, K. H. L.
1950. A critique on the phase theory of locusts. *Quart. Rev. Biol.* 25:363–407.

Kiefer, F.
1952. Copepoda, Calanoida und Cyclopoida, *Explor. Parc Nat. Albert, Mission H. Damas*, Fasc. 21:1–136.

Kimura, M.
1955. Stochastic processes and distribution of gene frequencies under natural selection. *Cold Spring Harbor Symp. Quant. Biol.* 20:33–53.
1960. Optimum mutation rate and degree of dominance as determined by the principle of minimum genetic load. *J. Genet.* 57:21–34.

King, J. C.
1947. A comparative analysis of the chromosomes of the *guarani* group of *Drosophila. Evolution* 1:48–62.
1956. Evidence for the integration of the gene pool from studies of DDT resistance in *Drosophila. Cold Spring Harbor Symp. Quant. Biol.* 20:311–317.

King, J. C., and L. Sømme
1958. Chromosomal analyses of the genetic factors for resistance to DDT in two resistant lines of *Drosophila melanogaster. Genetics* 43:577–593.

Kinne, O.
1954. Die *Gammarus*-Arten der Kieler Bucht. *Zool. Jahrb.* (*Syst.*) 82:405–496.

Kinsey, A. C.
1936. The origin of higher categories in Cynips. *Indiana Univ. Publ., Sci. Ser.,* No. 4:1–334.
1937. An evolutionary analysis of insular and continental species. *Proc. Nat. Acad. Sci.* 23:5–11.

Kipp, F. A.
1942. Über Flügelbau und Wanderzug der Vögel. *Biol. Zentr.* 62:289–299.
1948. Über die Eierzahl der Vögel. *Biol. Zentr.* 67:250–267.
1955. Die Entstehung der menschlichen Lautbildungsfähigkeit als Evolutionsproblem. *Experientia* 11:89–94.

Kirikov, S. V.
1940. On the connection between the red crossbills and the coniferous trees. *Bull. Acad. Sci. USSR* (*Biol.*) *1940,* 359–376. (Russian with English summary.)

Kist, J.
1961. "Systematische" beschouwingen naar aanleiding van de waarneming van Heuglins Geelpootzilvermeeuw, *Larus cachinnans heuglini* Bree, in Nederland. *Ardea* 49:1–51.

Kitzmiller, J. B.
1953. Mosquito genetics and cytogenetics. *Rev. Bras. Malariol. Trop.* 5:285–359.

1959. Race formation and speciation in mosquitoes. *Cold Spring Harbor Symp. Quant. Biol.* 24:161–165.

Kitzmiller, J. B., and W. L. French

1961. Chromosomes of *Anopheles quadrimaculatus. Amer. Zool.* 1:366.

Klauber, L. M.

1939. A further study of pattern dimorphism in the California king snake. *Bull. Zool. Soc. San Diego,* No. 15:1–23.

1941. The long-nosed snakes of the genus *Rhinocheilus. Trans. San Diego Nat. Hist. Soc.* 9:289–332.

1944. The California king snake: a further discussion. *Amer. Midl. Nat.* 31:85–87.

1946. The glossy snake, *Arizona,* with descriptions of new subspecies. *Trans. San Diego Nat. Hist. Soc.* 10:311–398.

1947. Classification and ranges of the gopher-snakes of the genus *Pituophis* in the western United States. *Bull. Zool. Soc. San Diego,* No. 22:1–83.

1956. *Rattlesnakes: their habits, life histories, and influence on mankind* (University of California Press, Berkeley and Los Angeles).

Kleiber, M.

1947. Body size and metabolic rate. *Physiol. Rev.* 27:511–541.

Kleinschmidt, O.

1900. Arten oder Formenkreise? *J. Ornithol.* 48:134–139.

Klopfer, P. M., and R. H. MacArthur

1960. Niche size and faunal diversity. *Amer. Nat.* 94:293–300.

Klots, A. B., and H. K. Clench

1952. A new species of *Strymon* Huebner from Georgia (Lepidoptera, Lycaenidae). *Amer. Mus. Novitates,* No. 1600:1–19.

Kluijver, H. N.

1951. The population ecology of the great tit, *Parus m. major* L. *Ardea* 39:1–135.

Knight, G. R., A. Robertson, and C. H. Waddington

1956. Selection for sexual isolation within a species. *Evolution* 10:14–22.

Kobelt, W.

1881. Exkursionen in Süditalien. *Jahrb. Deut. malak. Ges.* 8:50–67.

Kohn, A. J.

1958. Problems of speciation in marine invertebrates. In A. A. Buzzati-Traverso, ed., *Perspectives in marine biology* (University of California Press, Berkeley and Los Angeles), 571–588.

1959. The ecology of *Conus* in Hawaii. *Ecol. Monographs* 29:47–90.

Kolman, W. A.

1960. The mechanism of natural selection for the sex ratio. *Amer. Nat.* 94:373–377.

Komai, T., M. Chino, and Y. Hosino

1950. Contributions to the evolutionary genetics of the lady-beetle, *Harmonia.* I. Geographic and temporal variation in the relative frequencies of the elytral pattern types and in the frequency of elytral ridge. *Genetics* 35:589–601.

1951. Contributions to the evolutionary genetics of the lady-beetle *Harmonia*. II. Microgeographic variations. *Genetics* 36:382–390.

Kontkanen, P.
1953. On the sibling species in the leafhopper fauna of Finland (Homoptera, Auchenorrhyncha). *Arch. Soc. Zool. Vanamo* 7:100–106.

Koopman, K. F.
1950. Natural selection for reproductive isolation between *Drosophila pseudoobscura* and *Drosophila persimilis*. *Evolution* 4:135–148.

Koref-Santibañez, S., and E. del Solar O.
1961. Courtship and sexual isolation in *Drosophila pavani* Brncic and *Drosophila gaucha* Jaeger and Salzano. *Evolution* 15:401–406.

Korringa, P.
1958. Water temperature and breeding throughout the geographical range of *Ostrea edulis*. In L. Fage and P. Drach, eds., *Biologie comparée des espèces marines* (I. U. B. S., Ser. B, No. 24), 1–17.

Kosswig, C.
1944. Zur Evolution der Höhlentiermerkmale. *Rev. Fac. Sci. Istanbul* (B) 9:285–287.

1947. Selective mating as a factor for speciation in Cichlid fish of East African lakes. *Nature* 159:604.

1948. Genetische Beiträge zur Präadaptationstheorie. *Rev. Fac. Sci. Istanbul* (B) 13:176–209.

1950. Erythräische Fische im Mittelmeer und an der Grenze der Agäis. *Syllegomena Biol., Festschr. Kleinschmidt, Wittenberg, 1950*, 203–212.

1953. Über die Verwandtschaftsbeziehungen anatolischer Zahnkarpfen. *Hidrobiologi, Istanbul* (B) 1:186–198.

1960. Genetische Analyse stammesgeschichtlicher Einheiten. *Zool. Anz., Suppl.* 23:42–73.

Kosswig, C., and A. Sengün
1945. Über arttrennende Mechanismen. *Rev. Fac. Sci. Istanbul* (B) 10:164–214.

Kramer, G.
1941. Über das "Concolor"-Merkmal (Fehlen der Zeichnung) bei Eidechsen und seine Vererbung. *Biol. Zentr.* 61:1–15.

1946. Veränderungen von Nachkommenziffer und Nachkommengrösse sowie der Altersverteilung von Inseleidechsen. *Z. Naturforsch.* 1:700–710.

1949. Über Inselmelanismus bei Eidechsen. *Z. ind. Abst.- u. Vererb.* 83:157–164.

1951. Body proportions of mainland and island lizards. *Evolution* 5:193–206.

1960. Funktionsgerechte Allometrien. *Proc. XII Intern. Ornithol. Congr., Helsinki, 1958* 1:426–436.

Kramer, G., and R. Mertens
1938. Rassenbildung bei west-istrianischen Inseleidechsen in Abhängigkeit

von Isolierungsalter und Arealgrösse. *Arch. Naturg.* (N. F.) 7:189–234.

Krekeler, C. H.
1958. Speciation in cave beetles of the genus *Pseudanophthalmus* (Coleoptera, Carabidae). *Amer. Midl. Nat.* 59:167–189.

Krimbas, C. B.
1960. Synthetic sterility in *Drosophila willistoni*. *Proc. Nat. Acad. Sci.* 46:832–833.

Kühn, A.
1955. *Vorlesungen über Entwicklungsphysiologie* (Springer, Berlin).

Kullenberg, B.
1947. Der Kopulationsapparat der Insekten aus phylogenetischem Gesichtspunkt. *Zool. Bidrag Uppsala* 25:79–90.

1956. Field experiments with chemical sexual attractants on aculeate Hymenoptera males 1. *Zool. Bidrag Uppsala* 31:253–354.

1961. Studies in *Ophrys* pollination. *Zool. Bidrag Uppsala* 34:1–340.

Kummer, B.
1953. Untersuchungen über die Entwicklung der Schädelform des Menschen und einiger Anthropoiden. *Abhandl. Exacten Biol.*, Fasc. 3:1–44.

Kunze, L.
1959. Die funktionsanatomischen Grundlagen der Kopulation der Zwergzikaden, untersucht an *Euscelis plebejus* (Fall.) und einigen Typhlocybinen. *Deut. Entomol. Z.* (N. F.) 6:322–387.

Kupka, E.
1948. Chromosomale Verschiedenheiten bei schweizerischen Coregonen (Felchen). *Rev. Suisse Zool.* 55:285–293.

1950. Die Mitosen- und Chromosomenverhältnisse bei der grossen Schwebrenke, *Coregonus wartmanni* (Bloch), des Attersees. *Österreich. Zool. Z.* 2:605–623.

Kurtén, B.
1955. Sex dimorphism and size trends in the cave bear, *Ursus spelaeus* Rosenmüller and Heinroth. *Acta Zool. Fenn.* 90:1–48.

1958. A differentiation index, and a new measure of evolutionary rates. *Evolution* 12:146–157.

1959. Rates of evolution in fossil mammals. *Cold Spring Harbor Symp. Quant. Biol.* 24:205–215.

Kusnezov, N. N.
1956. A comparative study of ants in desert regions of Central Asia and of South America. *Amer. Nat.* 90:349–360.

La Barre, W.
1954. *The human animal* (University of Chicago Press, Chicago).

Lack, D.
1942. Ecological features of the bird faunas of British small islands. *J. Animal Ecol.* 11:9–36.

1944. Ecological aspects of species-formation in passerine birds. *Ibis* 86:260–286.

1945. The ecology of closely related species with special reference to cormorant (*Ph. carbo*) and shag (*P. aristotelis*). *J. Animal Ecol.* 14:12–16.

1946. Competition for food by birds of prey. *J. Animal Ecol.* 15:123–129.

1947a. *Darwin's finches* (Cambridge University Press, Cambridge, England).

1947b. The significance of clutch-size. *Ibis* 89:302–352.

1949. The significance of ecological isolation. In G. L. Jepsen, E. Mayr, and G. G. Simpson, eds., *Genetics, paleontology, and evolution* (Princeton University Press, Princeton), 299–308.

1954a. *The natural regulation of animal numbers* (Clarendon Press, Oxford).

1954b. The evolution of reproductive rates. In J. Huxley, A. C. Hardy, and E. B. Ford, eds., *Evolution as a process* (Allen and Unwin, London), 143–156.

Lack, D., and H. N. Southern

1949. Birds on Tenerife. *Ibis* 91:607–626.

Lagler, K. F., and R. M. Bailey

1947. The genetic fixity of differential characters in subspecies of the percid fish, *Boleosoma nigrum. Copeia,* No. 1:50–59.

Lal, K. B.

1934. Insect parasites of Psyllidae. *Parasitol.* 26:325–334.

Lamarck, M. de

1815. *Histoire naturelle des animaux sans vertèbres* (Verdière, Paris).

Lamotte, M.

1951. Recherches sur la structure génétique des populations naturelles de *Cepaea nemoralis* (L.) *Bull. Biol. France, Suppl.* 35:1–239.

1952. Le rôle des fluctuations fortuites dans la diversité des populations naturelles de *Cepaea nemoralis* (L.). *Heredity* 6:333–343.

1959. Polymorphism of natural populations of *Cepaea nemoralis. Cold Spring Harbor Symp. Quant. Biol.* 24:65–86.

Lancefield, D. E.

1929. A genetic study of crosses of two races or physiological species of *Drosophila obscura. Z. ind. Abst.- u. Vererb.* 52:287–317.

Lanyon, W. E.

1957. The comparative biology of the meadowlarks (*Sturnella*) in Wisconsin. *Publ. Nuttall Ornithol. Club,* No. 1:1–67.

1960. The Middle American populations of the crested flycatcher *Myiarchus tyrannulus. Condor* 62:341–350.

Lanyon, W. E., and W. N. Tavolga (eds.)

1960. *Animal sounds and communication* (AIBS Publ. No. 7).

Lapage, G.

1951. *Parasitic animals* (Cambridge University Press, Cambridge, England).

Lasker, G. W.
  1952. Mixture and genetic drift in ongoing human evolution. *Amer. Anthropol.* 54:433–436.
Lattin, G. de
  1951. Über die Bestimmung und Vererbung des Geschlechts einiger Oniscoideen (Crust., Isop.). I. Untersuchungen über die geschlechtsbeeinflussende Wirkung von Farbfaktoren bei Porcellio und Tracheoniscus. *Z. ind. Abst.- u. Vererb.* 84:1–37.
Laurent, R. F.
  1952. Sur les notions d'espèce et de relation spécifique, de sous-espèce et de relation subspécifique. *Ann. Soc. Roy. Zool. Belgique* 83:201–210.
Laven, H.
  1953. Reziprok unterschiedliche Kreuzbarkeit von Stechmücken (Culicidae) und ihre Deutung als plasmatische Vererbung. *Z. ind. Abst.- u. Vererb.* 85:118–136.
  1959. Speciation by cytoplasmic isolation in the *Culex pipiens*–complex. *Cold Spring Harbor Symp. Quant. Biol.* 24:166–173.
Leakey, L. S. B.
  1959. A new fossil skull from Olduvai. *Nature* 184:491–493.
Leakey, L. S. B., J. F. Evernden, and G. H. Curtis
  1961. Age of bed I, Olduvai Gorge, Tanganyika. *Nature* 191:478–479.
Le Gare, M., and W. Hovanitz
  1951. Genetic and ecologic analyses of wild populations in Lepidoptera. II. Color pattern variation in *Melitaea chalcedona*. *Wasmann J. Biol.* 9:257–310.
Legg, K., and F. A. Pitelka
  1956. Ecologic overlap of Allen and Anna hummingbirds nesting at Santa Cruz, California. *Condor* 58:393–405.
Le Gros Clark, W. E.
  1950. New palaeontological evidence bearing on the evolution of the Hominoidea. *Quart. J. Geolog. Soc. London* 105:225–264.
  1955. *The fossil evidence for human evolution* (University of Chicago Press, Chicago).
  1960. *The antecedents of man; an introduction to the evolution of the primates* (Quadrangle Books, Chicago).
Le Gros Clark, W. E., and L. S. B. Leakey
  1951. *Fossil mammals of Africa, No. 1. The Miocene Hominoidea of East Africa* (Brit. Mus. [Nat. Hist.], London).
Lemche, H.
  1940. The origin of winged insects. *Vidensk. Medd. fra Dansk naturh. Foren.* 104:127–168.
  1948. Northern and arctic tectibranch gastropods. I. The larval shells. *K. danske vidensk. Selsk. (Biol. Skr.)* 5:1–28.
Leopold, A. S.
  1944. The nature of heritable wildness in turkeys. *Condor* 46:133–197.

Lerner, I. M.
1950.   *Population genetics and animal improvement* (Cambridge University Press, Cambridge, England).
1954.   *Genetic homeostasis* (Oliver and Boyd, Edinburgh).
1958.   *The genetic basis of selection* (Wiley, New York).
1959.   The concept of natural selection: a centennial view. *Proc. Amer. Phil. Soc.* 103:173–182.

Lerner, I. M., and E. R. Dempster
1962.   Indeterminism in interspecific competition. *Proc. Nat. Acad. Sci.* 48:821–826.

Levene, H.
1953.   Genetic equilibrium when more than one ecological niche is available. *Amer. Nat.* 87:331–333.

Levene, H., and Th. Dobzhansky
1958.   New evidence of heterosis in naturally occurring inversion heterozygotes in *Drosophila pseudoobscura. Heredity* 12:37–49.
1959.   Possible genetic difference between the head louse and the body louse (*Pediculus humanus* L.). *Amer. Nat.* 93:347–353.

Lévi, C.
1956.   Étude des *Halisarca* de Roscoff. Embryologie et systématique des Démosponges. *Arch. Zool. Exptl. Gén.* 93:1–181.

Levi, H.
1959.   The spider genus *Latrodectus* (Araneae, Theridiidae). *Trans. Amer. Microscop. Soc.* 78:7–43.

Levine, P.
1958.   The influence of the ABO system on Rh hemolytic disease. *Human Biol.* 30:14–28.

Levitan, M.
1952.   Experiments on chromosomal variability in *Drosophila robusta. Genetics* 36:285–305.
1955.   Studies of linkage in populations. I. Associations of second chromosome inversions in *Drosophila robusta. Evolution* 9:62–74.
1958.   Non-random associations of inversions. *Cold Spring Harbor Symp. Quant. Biol.* 23:251–268.

Levitan, M., H. L. Carson, and H. D. Stalker
1954.   Triads of overlapping inversions in *Drosophila robusta. Amer. Nat.* 88:113–114.

Levitan, M., and F. M. Salzano
1959.   Studies of linkage in populations. III. An association of linked inversions in *Drosophila guaramunu. Heredity* 13:243–248.

Lewis, E. B.
1950.   The phenomenon of position effect. *Adv. Genet.* 3:73–115.
1951.   Pseudoallelism and gene evolution. *Cold Spring Harbor Symp. Quant. Biol.* 16:159–174.

Lewis, H.
1956.   Specific and infraspecific categories in plants. In E. Mayr, ed., *Bio-*

logical systematics (Proc. 16th Ann. Biol. Colloquium, Oregon State College, Corvallis), 13–20.

1962. Catastrophic selection as a factor in speciation. *Evolution* 16:257–271.

Lewis, H., and P. H. Raven
1958. Rapid evolution in *Clarkia*. *Evolution* 12:319–336.

Lewis, T. H.
1949. Dark coloration in the reptiles of the Tularosa Malpais, New Mexico. *Copeia,* No. 3:181–184.

Lewontin, R. C.
1953. The effect of compensation on populations subject to natural selection. *Amer. Nat.* 87:375–381.

1957. The adaptations of populations to varying environments. *Cold Spring Harbor Symp. Quant. Biol.* 22:395–408.

1958. Studies on heterozygosity and homeostasis. II. Loss of heterosis in a constant environment. *Evolution* 12:494–503.

1959. On the anomalous response of *Drosophila pseudoobscura* to light. *Amer. Nat.* 93:321–328.

Lewontin, R. C., and L. C. Dunn
1960. The evolutionary dynamics of a polymorphism in the house mouse. *Genetics* 45:705–722.

Lewontin, R. C., and K. Kojima
1960. The evolutionary dynamics of complex polymorphisms. *Evolution* 14:458–472.

Lewontin, R. C., and M. J. D. White
1960. Interaction between inversion polymorphisms of two chromosome pairs in the grasshopper, *Moraba scurra*. *Evolution* 14:116–129.

L'Héritier, Ph., and G. Teissier
1935. Recherches sur la concurrence vitale. Etude de populations mixtes de *Drosophila melanogaster* et de *Drosophila funebris*. *Compt. Rend. Soc. Biol., Paris* 118:1396.

1937. Elimination des formes mutantes dans les populations de Drosophiles. Cas des Drosophiles "bar." *Compt. Rend. Soc. Biol., Paris* 124:880.

Li, C. C.
1955a. *Population genetics* (University of Chicago Press, Chicago).

1955b. The stability of an equilibrium and the average fitness of a population. *Amer. Nat.* 89:281–296.

Lieftinck, M. A.
1949. The dragonflies (*Odonata*) of New Guinea and neighbouring islands. *Nova Guinea* (n.s.) 5:1–271.

Lindeborg, R. G.
1952. Water requirements of certain rodents from xeric and mesic habitats. *Contrib. Lab. Vert. Biol., Univ. Mich.,* No. 58:1–32.

Linnaeus, C.
1751. *Philosophia Botanica* (Godofr. Kiesewetter, Stockholmiae).

1758. *Systema Naturae. Regnum Animale* (10th ed. tomus I; L. Salvii, Holminae).

Linsdale, J. M.
  1938. Environmental responses of vertebrates in the Great Basin. *Amer. Midl. Nat.* 19:1–206.

Linsley, E. G., and J. W. MacSwain
  1958. The significance of floral constancy among bees of the genus *Diadasia* (Hymenoptera, Anthophoridae). *Evolution* 12:219–223.

Littlejohn, M. J.
  1959. Call differentiation in a complex of seven species of *Crinia*. (Anura, Leptodactylidae). *Evolution* 13:452–468.
  1961. Age and origin of some southwestern Australian species of *Crinia* (Anura: Leptodactylidae). In W. F. Blair, ed., *Vertebrate speciation* (University of Texas Press, Austin), 514–536.

Littlejohn, M. J., and T. C. Michaud
  1959. Mating call discrimination by females of Strecker's chorus frog (*Pseudacris streckeri*). *Texas J. Sci.* 11:86–92.

Livingston, F. B.
  1958. The distribution of the sickle cell gene in Liberia. *Amer. J. Human Genet.* 10:33–41.

Lloyd, B. E.
  1912. *The growth of groups in the animal kingdom* (Longmans, Green, London).

Loewenberg, B. J.
  1957. *Darwinism. Reaction or reform?* (Rinehart, New York).

Loosanoff, V. L., and C. A. Nomejko
  1951. Existence of physiologically different races of oysters, *Crassostrea virginica. Biol. Bull.* 101:151–156.

Lord, R. D., Jr.
  1960. Litter size and latitude in North American mammals. *Amer. Midl. Nat.* 64:488–499.

Lorenz, K.
  1941. Vergleichende Bewegungsstudien an Anatinen. *J. Ornithol.* 89:19–29.
  1943. Die angeborenen Formen möglicher Erfahrung. *Z. Tierpsychol.* 5:235–409.
  1950. The comparative method in studying innate behaviour patterns. *Symp. Soc. Exptl. Biol.*, No. 4:221–268.

Lorković, Z.
  1942. Studien über den Speciesbegriff. II. Artberechtigung von *Everes argiades* Pall., *E. alcetas* Hffg. und *E. decolorata* Stgr. *Mitt. Münch. Entomol. Ges.* 32:599–624.
  1943. Modifikationen und Rassen von *Everes argiades* Pall. und ihre Beziehungen zu den klimatischen Faktoren ihrer Verbreitungsgebiete. *Mitt. Münch. Entomol. Ges.* 33:431–478.
  1949. Chromosomenzahlen-Vervielfachung bei Schmetterlingen und ein neuer Fall fünffacher Zahl. *Rev. Suisse Zool.* 56:243–249.

1953. Spezifische, semispezifische und rassische Differenzierung bei *Erebia tyndarus* Esp. I und II. *Rad l'Acad. Yougoslave* 294:269–309, 315–358.

1955. Die Populationsanalyse zweier neuen stenochoren Erebia-Rassen aus Kroatien. *Biol. Glasnik* 8:53–76.

1958. Die Merkmale der unvollständigen Speziationsstule und die Frage der Einführung der Semispezies in die Systematik. *Uppsala Univ. Arsskr. 1958* 6:159–168.

Lowther, J. K.
1961. Polymorphism in the white-throated sparrow, *Zonotrichia albicollis* (Gmelin). *Canad. J. Zool.* 39:281–292.

Lucas, C. E.
1947. The ecological effects of external metabolites. *Biol. Rev.* 22:270–295.

Ludwig, W.
1940. Selektion und Stammesentwicklung. *Naturw.* 28:689–705.

1950. Zur Theorie der Konkurrenz. Die Annidation (Einnischung) als fünfter Evolutionsfaktor. *Neue Ergeb. Probleme Zool., Klatt-Festschrift 1950,* 516–537.

1954. Die Selektionstheorie. In G. Heberer, ed., *Die Evolution der Organismen* (Gustav Fischer, Stuttgart), 662–712.

Lukin, E. I.
1940. *Darwinism and geographic regularities in variation of organisms* (Acad. Sci. U.S.S.R., Moscow-Leningrad).

Lundman, B.
1947. Maps of the racial geography of some Partulae of the Society Islands based upon the material published by H. E. Crampton. *Zool. Bidrag Uppsala* 25:517–533.

1952. *Umriss der Rassenkunde des Menschen in geschichtlicher Zeit* (Ejnar Munksgaard, Copenhagen).

MacArthur, J. W.
1949. Selection for small and large body size in the house mouse. *Genetics* 34:194–209.

MacArthur, R. H.
1957. On the relative abundance of bird species. *Proc. Nat. Acad. Sci.* 43:293–295.

1958. Population ecology of some warblers of north-eastern coniferous forests. *Ecology* 39:599–619.

Machado, A. de Barros
1959. Nouvelles contributions à l'étude systématique et biogéographique des Glossines (Diptera). *Publ. cult. Co. Diam. Angola, Lisbon,* No. 46:13–90.

Macpherson, A. H.
1961. Observations on Canadian arctic *Larus* gulls, and on the taxonomy of *L. thayeri* Brooks. *Arctic Inst. North Amer., Tech. Paper,* No. 7:1–40.

Main, A. R.
   1957.   Studies in Australian amphibia. I. The genus *Crinia* Tschudi in south-western Australia and some species from south-eastern Australia. *Austral. J. Zool.* 5:30–55.
Main, A. R., A. K. Lee, and M. J. Littlejohn
   1958.   Evolution in three genera of Australian frogs. *Evolution* 12:224–233.
Manna, G. K., and S. C. Smith
   1959.   Chromosomal polymorphism and inter-relationships among bark weevils of the genus *Pissodes* Germar. *Nucleus* 2:179–208.
Manning, A.
   1959a.   Comparison of mating behavior in *Drosophila melanogaster* and *Drosophila simulans*. *Behaviour* 15:123–146.
   1959b.   The sexual isolation between *Drosophila melanogaster* and *Drosophila simulans*. *Animal Behaviour* 7:60–65.
Maramorosch, K.
   1958.   Studies of aster yellows virus transmission by the leafhopper species *Macrosteles fascifrons* Stål and *M. laevis* Ribaut. *Proc. 10th Intern. Congr. Entomol.* 3:221–228.
Marien, D.
   1950.   Notes on some Asiatic Meropidae (birds). *J. Bombay Nat. Hist. Soc.* 49:151–164.
   1951.   Notes on the bird family Prunellidae in southern Eurasia. *Amer. Mus. Novitates*, No. 1482:1–28.
   1958.   Selection for developmental rate in *Drosophila pseudoobscura*. *Genetics* 43:3–15.
Marks, E. N.
   1954.   A review of the *Aedes scutellaris* subgroup with a study of variation in *Aedes pseudoscutellaris* (Theobald) (Diptera: Culicidae). *Bull. Brit. Mus. (Nat. Hist.), Entomol.* 3:347–414.
Marler, P.
   1957.   Specific distinctiveness in the communication signals of birds. *Behaviour* 11:13–39.
   1960.   Bird songs and mate selection. In W. E. Lanyon and W. N. Tavolga, eds., *Animal sounds and communication* (AIBS Publ. No. 7), 348–367.
Marshall, J. T., Jr.
   1948.   Ecologic races of song sparrows in the San Francisco Bay region. Part II. Geographic variation. *Condor* 50:233–256.
   1960.   Interrelations of Abert and Brown towhees. *Condor* 62:49–64.
Marshall, W. H.
   1940.   A survey of mammals of the islands in Great Salt Lake, Utah. *J. Mammal.* 21:144–159.
Maslin, T. P.
   1962.   All-female species of the lizard genus *Cnemidophorus*, Teiidae. *Science* 135:212–213.

Mather, K.
1941. Variation and selection of polygenic characters. *J. Genet.* 41:159–193.
1943. Polygenic inheritance and natural selection. *Biol. Rev.* 18:32–64.
1949. *Biometrical genetics* (Methuen, London).
1950. The genetical architecture of heterostyly in *Primula sinensis. Evolution* 4:340–352.
1953. The genetical structure of populations. *Symp. Soc. Exptl. Biol.,* No. 7:66–95.
1954. The genetical units of continuous variation. *Caryologia, suppl. vol.,* 106–123.
1955. Polymorphism as an outcome of disruptive selection. *Evolution* 9:52–61.
1956. Polygenic mutation and variation in populations. *Proc. Roy. Soc., London* (B) 145:293–297.
Mather, K., and B. J. Harrison
1949. The manifold effect of selection. *Heredity* 3:1–52, 131–162.
Matsunaga, E., and Y. Hiraizumi
1962. Changes in titer of ecdysone in *Bombyx mori* during metamorphosis. *Science* 135:432–434.
Matthew, W. D.
1915. Climate and evolution. *Ann. N. Y. Acad. Sci.* 24:171–318.
Matthey, R.
1949. *Les chromosomes des vertébrés* (Rouge, Lausanne).
1953. A propos de la polyploidie animale: réponse à un article de C. D. Darlington. *Rev. Suisse Zool.* 60:466–471.
1954. Chromosomes et systématique des Canidés. *Mammalia* 18:225–230.
1958. Les chromosomes des mammifères euthériens. Liste critique et essai sur l'évolution chromosomique. *Arch. Julius Klaus-Stiftung* 33:253–297.
1961. Études de cytogénétique et de taxonomie chez les Muridae (Rodentia) *Reithrodontomys megalotis dychei* Allen, *Hypogeomys antimena* Grand., *Neofiber alleni* True. *Mammalia* 25:145–161.
Mattingly, P. F., L. E. Rozeboom, K. L. Knight, H. Laven, F. H. Drummond, S. R. Christophers, and P. G. Shute
1951. The *Culex pipiens* complex. *Trans. Roy. Entomol. Soc. London* 102:331–382.
Mattingly, P. F., P. G. Shute, H. Laven, and K. L. Knight
1953. The *Culex pipiens* complex. *Trans. 9th Intern. Congr. Entomol., Amsterdam* 2:283–300.
Mayer, A. G.
1910. Medusae of the world. III. The Scyphomedusae. *Carnegie Inst. Wash. Publ.,* No. 109:499–735.
Mayfield, H.
1960. *The Kirtland's Warbler* (Bull. 40, Cranbrook Institute of Science, Bloomfield Hills, Michigan).

Mayr, E.
1926. Die Ausbreitung des Girlitz. *J. Ornithol.* 74:571–671.
1931. Birds collected during the Whitney South Sea expedition. XII. Notes on *Halcyon chloris* and some of its subspecies. *Amer. Mus. Novitates,* No. 469:1–10.
1932a. Notes on thickheads (*Pachycephala*) from the Solomon Islands. *Amer. Mus. Novitates,* No. 522:1–22.
1932b. Notes on thickheads (*Pachycephala*) from Polynesia. *Amer. Mus. Novitates,* No. 531:1–23.
1940. Speciation phenomena in birds. *Amer. Nat.* 74:249–278.
1941. Die geographische Variation der Färbungstypen von *Microscelis leucocephalus. J. Ornithol.* 89:377–392.
1942. *Systematics and the origin of species* (Columbia University Press, New York).
1944. The birds of Timor and Sumba. *Bull. Amer. Mus. Nat. Hist.* 83:123–194.
1945. Symposium on age of the distribution pattern of gene arrangements in *Drosophila pseudoobscura.* Introduction and some evidence in favor of a recent date. *Lloydia* 8:69–83.
1946a. History of the North American bird fauna. *Wilson Bull.* 58:3–41.
1946b. The naturalist in Leidy's time and today. *Proc. Acad. Nat. Sci. Philadelphia* 98:271–276.
1947. Ecological factors in speciation. *Evolution* 1:263–288.
1948. The bearing of the new systematics on genetical problems. The nature of species. *Adv. Genet.* 2:205–237.
1949a. Speciation and systematics. In G. L. Jepsen, E. Mayr, and G. G. Simpson, eds., *Genetics, paleontology, and evolution* (Princeton University Press, Princeton), 281–298.
1949b. Speciation and selection. *Proc. Amer. Phil. Soc.* 93:514–519.
1950a. Taxonomic categories in fossil hominids. *Cold Spring Harbor Symp. Quant. Biol.* 15:109–118.
1950b. The role of the antennae in the mating behavior of female *Drosophila. Evolution* 4:149–154.
1950c. Artbildung und Variation in der *Halcyon-chloris*-Gruppe. *Ornithol. biol. Wiss.* (Festschrift zum 60. Geburtstag von Erwin Stresemann), 55–60.
1950d. Notes on the genus *Neositta. Emu* 49:282–291.
1951a. Speciation in birds. *Proc. Xth Intern. Ornithol. Congr., Uppsala, 1950,* 91–131.
1951b. Notes on some pigeons and parrots from western Australia. *Emu* 51:137–145.
1953a. Concepts of classification and nomenclature in higher organisms and microorganisms. *Ann. N. Y. Acad. Sci.* 56:391–397.
1953b. Fragments of a Papuan ornithogeography. *Proc. 7th Pacific Sci. Congr., 1949* 4:11–19.
1954a. Change of genetic environment and evolution. In J. Huxley,

A. C. Hardy, and E. B. Ford, eds., *Evolution as a process* (Allen and Unwin, London), 157–180.

1954b. Geographic speciation in tropical echinoids. *Evolution* 8:1–18.

1954c. Notes on nomenclature and classification. *Systematic Zool.* 3:86–89.

1955a. Integration of genotypes: synthesis. *Cold Spring Harbor Symp. Quant. Biol.* 20:327–333.

1955b. Karl Jordan's contribution to current concepts in systematics and evolution. *Trans. Roy. Entomol. Soc. London* 107:45–66.

1956. Geographical character gradients and climatic adaptation. *Evolution* 10:105–108.

1957a. Species concepts and definitions. In E. Mayr, ed., *The species problem* (Amer. Assoc. Adv. Sci. Publ. No. 50), 1–22.

1957b. Difficulties and importance of the biological species. In E. Mayr, ed., *The species problem* (Amer. Assoc. Adv. Sci. Publ. No. 50), 371–388.

1957c. Die denkmöglichen Formen der Artentstehung. *Rev. Suisse Zool.* 64:219–235.

1958. Behavior and systematics. In A. Roe and G. G. Simpson, eds., *Behavior and evolution* (Yale University Press, New Haven), 341–362.

1959a. Isolation as an evolutionary factor. *Proc. Amer. Phil. Soc.* 103:221–230.

1959b. Where are we? *Cold Spring Harbor Symp. Quant. Biol.* 24:1–14.

1959c. Darwin and the evolutionary theory in biology. In *Evolution and anthropology: a centennial appraisal* (Anthropol. Soc. Washington), 3–12.

1959d. Agassiz, Darwin, and evolution. *Harvard Library Bull.* 13:165–194.

1960. The emergence of evolutionary novelties. In S. Tax, ed., *The evolution of life* (University of Chicago Press, Chicago), 349–380.

Mayr, E., and D. Amadon

1947. A review of the Dicaeidae. *Amer. Mus. Novitates,* No. 1360:1–32.

1951. A classification of recent birds. *Amer. Mus. Novitates,* No. 1496:1–42.

Mayr, E., and E. T. Gilliard

1952a. The ribbon-tailed bird of paradise (*Astrapia mayeri*) and its allies. *Amer. Mus. Novitates,* No. 1551:1–13.

1952b. Altitudinal hybridization in New Guinea honeyeaters. *Condor* 54:325–337.

Mayr, E., E. G. Linsley, and R. L. Usinger

1953. *Methods and principles of systematic zoology* (McGraw-Hill, New York).

Mayr, E., and M. Moynihan

1946. Evolution in the *Rhipidura rufifrons* group. *Amer. Mus. Novitates,* No. 1321:1–21.

Mayr, E., and S. D. Ripley

1941. Birds collected during the Whitney South Sea expedition. XLIV. Notes on the genus *Lalage* Boie. *Amer. Mus. Novitates,* No. 1116:1–18.

Mayr, E., and C. B. Rosen
  1956. Geographic variation and hybridization in populations of Bahama snails (*Cerion*). *Amer. Mus. Novitates*, No. 1806:1–48.
Mayr, E., and E. Stresemann
  1950. Polymorphism in the chat genus *Oenanthe* (Aves). *Evolution* 4:291–300.
Mayr, E., and C. Vaurie
  1948. Evolution in the family Dicruridae (birds). *Evolution* 2:238–265.
McAtee, W. L.
  1937. Survival of the ordinary. *Quart. Rev. Biol.* 12:47–64.
McCabe, T., and B. D. Blanchard
  1950. *Three species of* Peromyscus (Rood Associates, Santa Barbara).
McCabe, T., and I. M. Cowan
  1945. *Peromyscus maniculatus macrorhinus* and the problem of insularity. *Trans. Roy. Canad. Inst.* 25:117–215.
McCamey, F.
  1950. A puzzling hybrid warbler from Michigan. *Jack-Pine Warbler* 28:67–72.
McCarley, W. H.
  1954. Natural hybridization in the *Peromyscus leucopus* species group of mice. *Evolution* 8:314–323.
McCarthy, M. D.
  1945. Chromosome studies on eight species of *Sciara* (Diptera) with special reference to chromosome changes of evolutionary significance. I and II. *Amer. Nat.* 79:104–121, 228–245.
McClintock, B.
  1950. The origin and behavior of mutable loci in maize. *Proc. Nat. Acad. Sci.* 36:344–355.
McDunnough, J.
  1946. A study of the Caryaefoliella group of the family Coleophoridae (Lepidoptera). *Canad. Entomol.* 78:1–14.
McKinney, F.
  1961. An analysis of the displays of the European eider *Somateria mollissima mollissima* (Linnaeus) and the Pacific eider *Somateria mollissima v. nigra* Bonaparte. *Behaviour, Suppl.* 7:1–124.
Meacham, W. R.
  1962. Factors affecting secondary intergradation between two allopatric populations in the *Bufo woodhousei* complex. *Amer. Midl. Nat.* 67:282–304.
Mead, G. W.
  1960. Hermaphroditism in archibenthic and pelagic fishes of the order Iniomi. *Deep-Sea Research* 6:234–235.
Mecham, J. S.
  1961. Isolating mechanisms in anuran amphibians. In W. F. Blair, ed., *Vertebrate speciation* (University of Texas Press, Austin), 24–61.

Medawar, P. B.
1957.   The uniqueness of the individual (Methuen, London).
1960.   The future of man: the Reith lectures (Methuen, London).
Meglitsch, P. A.
1954.   On the nature of the species. Systematic Zool. 3:49–65.
Meinertzhagen, R.
1954.   Desert coloration. Birds of Arabia (Oliver and Boyd, Edinburgh),
        8–31.
Meise, W.
1928a. Die Verbreitung der Aaskrähe (Formenkreis Corvus corone L.).
        J. Ornithol. 76:1–203.
1928b. Rassenkreuzungen an den Arealgrenzen. Verh. Deut. Zool. Ges.,
        München, 1928, 96–105.
1936a. Über Artenstehung durch Kreuzung in der Vogelwelt. Biol. Zentr.
        56:590–604.
1936b. Zur Systematik und Verbreitungsgeschichte der Haus- und Weiden-
        sperlinge, Passer domesticus (L.) und hispaniolensis (T.). J. Orni-
        thol. 84:631–672.
1938.   Fortschritte der ornithologischen Systematik seit 1920. Proc. VIII
        Intern. Ornithol. Congr., Oxford, 1934, 49–189.
Melander, Y., and E. Montén
1950.   Probable parthenogenesis in Coregonus. Hereditas 36:105–106.
Melchinger, H.
1955.   Unterschiede im Gehäusebau und Verhalten der beiden Heide-
        schnecken Helicella ericetorum und Helicella obvia. Zool. Jahrb.
        (Syst.) 83:185–322.
Mell, R.
1941.   Die Beziehungen zwischen Entwicklungszyklus und Klima bei Go-
        nepteryx (Lep. Rhop.). Biol. Zentr. 61:603–607.
Mertens, R.
1928.   Über den Rassen- und Artenwandel auf Grund des Migrations-
        prinzipes, dargestellt an einigen Amphibien und Reptilien. Sencken-
        bergiana 10:81–91.
1934.   Die Insel-reptilien, ihre Ausbreitung, Variation, und Artbildung.
        Zoologica 32:1–209.
1943.   Die rezenten Krokodile des Natur-Museums Senckenberg. Sencken-
        bergiana 26:252–312.
1947.   Studien zur Eidonomie und Taxonomie der Ringelnatter (Natrix
        natrix). Abhandl. senckenberg. naturforsch. Ges. 476:1–38.
1950.   Über Reptilienbastarde. Senckenbergiana 31:127–144.
1952.   Schwarzblaue Insel-Eidechsen und die neueren Ansichten über ihr
        Farbkleid. Natur u. Volk 82:386–394.
Mettler, L. E.
1957.   Studies on experimental populations of Drosophila arizonensis and
        Drosophila mojavensis. Univ. Texas Publ., No. 5721:157–181.

Meylan, A.
1961. Contribution à l'étude du polymorphisme chromosomique chez *Sorex araneus* L. (*Mamm. Insectivora*). *Rev. Suisse Zool.* 67:258–261.

Michener, C. D.
1947a. A revision of the American species of *Hoplitis* (Hymenoptera, Megachilidae). *Bull. Amer. Mus. Nat. Hist.* 89:257–318.
1947b. A character analysis of a solitary bee, *Hoplitis albifrons* (Hymenoptera, Megachilidae). *Evolution* 1:172–185.

Milani, R.
1956. Ricerche genetiche sulla resistenza degli insetti alla azione delle sostanze tossiche. *Riv. Parassitol.* 17:223–246.

Milkman, R. D.
1960a. The genetic basis of natural variation. I. Crossveins in *Drosophila melanogaster*. *Genetics* 45:35–48.
1960b. The genetic basis of natural variation. II. Analysis of a polygenic system in *Drosophila melanogaster*. *Genetics* 45:377–391.
1960c. Potential genetic variability of wild pairs of *Drosophila melanogaster*. *Science* 131:225–226.
1961. The genetic basis of natural variation. III. Developmental lability and evolutionary potential. *Genetics* 46:25–38.

Miller, A. H.
1941. Speciation in the avian genus *Junco*. *Univ. Calif. Publ. Zool.* 44:173–434.
1942. Habitat selection among higher vertebrates and its relation to intraspecific variation. *Amer. Nat.* 76:25–35.
1944a. Specific differences in the call notes of chipmunks. *J. Mammal.* 25:87–89.
1944b. An avifauna from the Lower Miocene of South Dakota. *Univ. Calif. Publ., Bull. Dept. Geol. Sci.* 27:85–100.
1949. Some concepts of hybridization and intergradation in wild populations of birds. *Auk* 66:338–342.
1950. Some ecologic and morphologic considerations in the evolution of higher taxonomic categories. *Ornithol. biol. Wiss.* (Festschrift zum 60. Geburtstag von Erwin Stresemann), 84–88.
1951. An analysis of the distribution of the birds of California. *Univ. Calif. Publ. Zool.* 50:531–644.
1955. Concepts and problems of avian systematics in relation to evolutionary processes. In A. Wolfson, ed., *Recent studies in avian biology* (University of Illinois Press, Urbana), 1–22.
1956. Ecologic factors that accelerate formation of races and species of terrestrial vertebrates. *Evolution* 10:262–277.
1960. Adaptation of breeding schedule to latitude. *Proc. XII Intern. Ornithol. Congr., Helsinki, 1958*, 513–522.

Miller, R. R.
1948. The cyprinodont fishes of the Death Valley system of eastern Cali-

fornia and southwestern Nevada. *Misc. Publ. Mus. Zool., Univ. Mich.*, No. 68:1–155.

1950. Speciation in fishes of the genera *Cyprinodon* and *Empctrichthys*, inhabiting the Death Valley region. *Evolution* 4:155–163.

1955. A systematic review of the middle American fishes of the genus *Profundulus. Misc. Publ. Mus. Zool., Univ. Mich.*, No. 92:1–64.

1961. Speciation rates in some fresh-water fishes of western North America. In W. F. Blair, ed., *Vertebrate speciation* (University of Texas Press, Austin), 537–560.

Miller, R. R., and R. J. Schultz
1959. All-female strains of the teleost fishes of the genus *Poeciliopsis. Science* 130:1656–1657.

Millicent, E., and J. M. Thoday
1960. Gene flow and divergence under disruptive selection. *Science* 131:1311–1312.

Milne, A.
1961. Definition of competition among animals. *Symp. Soc. Exptl. Biol.*, No. 15:40–71.

Milstead, W. M.
1957. Some aspects of competition in natural populations of whiptail lizards (genus *Cnemidophorus*). *Texas J. Sci.* 9:410–447.

Minamori, S.
1952. Physiological isolation in Cobitidae. I. Two races of the striated spinous loach with special reference to the difference in their embryonic respiration. *J. Sci. Hiroshima Univ.* (B) 13:1–14.

1955. Physiological isolation in Cobitidae. III. Hybrid sterility and hybrid breakdown in contact regions of two races of the striated spinous loach. *Jap. J. Genet.* 30:243–251.

1956. Physiological isolation in Cobitidae. IV. Speciation of two sympatric races of Lake Biwa of the striated spinous loach. *Jap. J. Zool.* 12:89–104.

Montalenti, G.
1958. Perspectives of research on sex problems in marine animals. In A. A. Buzzati-Traverso, ed., *Perspectives in marine biology* (University of California Press, Berkeley and Los Angeles), 589–602.

1960. Alcune considerazioni sull'evoluzione della determinazione del sesso. *Accad. Naz. Lincei, Rome*, Quaderno N. 47:153–181.

Moore, J. A.
1944. Geographic variation in *Rana pipiens* Schreber of eastern North America. *Bull. Amer. Mus. Nat. Hist.* 82:345–370.

1946. Incipient intraspecific isolating mechanisms in *Rana pipiens. Genetics* 31:304–326.

1949. Geographic variation of adaptive characters in *Rana pipiens* Schreber. *Evolution* 3:1–24.

1950. Further studies on *Rana pipiens* racial hybrids. *Amer. Nat.* 84:247–254.

1952a. Competition between *Drosophila melanogaster* and *Drosophila simulans*. I. Population cage experiments. *Evolution* 6:407–420.

1952b. Competition between *Drosophila melanogaster* and *Drosophila simulans*. II. The improvement of competitive ability through selection. *Proc. Nat. Acad. Sci.* 38:813–817.

1954. Geographic and genetic isolation in Australian amphibia. *Amer. Nat* 88:65–74.

1955. Abnormal combinations of nuclear and cytoplasmic systems in frogs and toads. *Adv. Genet.* 7:139–182.

1957. An embryologist's view of the species concept. In E. Mayr, ed., *The species problem* (Amer. Assoc. Adv. Sci. Publ. No. 50), 325–338.

Moreau, R. E.

1944a. Clutch-size: a comparative study, with special reference to African birds. *Ibis* 86:286–347.

1944b. Clutch-size in introduced birds. *Auk* 61:583–587.

1947. Relations between number in brood, feeding-rate, and nestling period in nine species of birds in Tanganyika territory. *J. Animal Ecol.* 16:205–209.

1948. Ecological isolation in a rich tropical avifauna. *J. Animal Ecol.* 17:113–126.

Morgenthaler, O.

1934. Krankheitserregende und harmlose Arten der Bienenmilbe *Acarapis*, zugleich ein Beitrag zum Species-Problem. *Rev. Suisse Zool.* 41:429–446.

Moriya, K.

1951. On isolating mechanisms between the two subspecies of the pond frog, *Rana nigromaculata*. *J. Sci. Hiroshima Univ.* (B) 12:47–56.

1960. Studies on the five races of the Japanese pond frog, *Rana nigromaculata* Hallowell. III. Sterility in interracial hybrids. *J. Sci. Hiroshima Univ.* (B) 18:125–156.

Morpurgo, G., and B. Nicoletti

1956. Sull'importanza del matrimonio selettivo per la evoluzione di popolazioni artificiali de *Drosophila melanogaster*. *Ricerca Sci.*, *Suppl.* 26:1–11.

Moser, H.

1958. The dynamics of bacterial populations maintained in the chemostat. *Carnegie Inst. Wash. Publ.*, No. 614:1–136.

Mourant, A. E.

1954. *The distribution of the human blood groups* (Blackwell, Oxford).

Mourant, A. E., A. C. Kopeć, and K. Domaniewska-Sobczak

1958. *The ABO blood groups: comprehensive tables and maps of world distribution* (Blackwell, Oxford).

Mozley, A.

1935. The variation of two species of *Lymnaea*. *Genetics* 20:452–465.

Muller, H. J.
1918. Genetic variability: twin hybrids and constant hybrids in a case of balanced lethal factors. *Genetics* 3:422–499.
1925. Why polyploidy is rarer in animals than in plants. *Amer. Nat.* 59:346–353.
1929. The method of evolution. *Sci. Monthly* 29:481–505.
1940. Bearings of the Drosophila work on systematics. In J. S. Huxley, ed., *The new systematics* (Clarendon Press, Oxford), 185–268.
1942. Isolating mechanisms, evolution and temperature. *Biol. Symp.* 6:71–125.
1950a. Our load of mutations. *Amer. J. Human Genet.* 2:111–176.
1950b. *Evidence of the precision of genetic adaptation* (The Harvey Lectures, Ser. 43.; Thomas, Springfield, Ill.), 165–229.
1960. The guidance of human evolution. In S. Tax, ed., *The evolution of man* (University of Chicago Press, Chicago), 423–462.
Muller, H. J., E. Carlson, and A. Schalet
1961. Mutation by alteration of the already existing gene. *Genetics* 46:213–226.
Muller, H. J., and G. Pontecorvo
1941. Recessive genes causing interspecific sterility and other disharmonies between *Drosophila melanogaster* and *Drosophila simulans*. *Genetics* 27:157.
Muller, H. J., S. Wright, G. L. Jepsen, G. L. Stebbins, Jr., and E. Mayr
1949. Symposium on natural selection and adaptation. *Proc. Amer. Phil. Soc.* 93:459–519.
Munroe, E. G.
1950. The systematics of *Calisto* (Lepidoptera, Satyrinae), with remarks on the evolutionary and zoogeographic significance of the genus. *J. N. Y. Entomol. Soc.* 58:211–240.
1951. Subspeciation in the Microlepidoptera. *Lepidopterists' News* 5:29–31.
Münzing, J.
1959. Biologie, Variabilität und Genetik von *Gasterosteus aculeatus* L. (Pisces). Untersuchungen im Elbegebiet. *Intern. Rev. Ges. Hydrobiol.* 44:318–382.
Murphy, R. C.
1938. The need of insular exploration as illustrated by birds. *Science* 88:533–539.
Murphy, R. C., and J. M. Pennoyer
1952. Larger petrels of the genus *Pterodroma*. *Amer. Mus. Novitates,* No. 1580:1–43.
Myers, G. S.
1950. The systematic status of *Hyla septentrionalis*, the large tree frog of the Florida Keys, the Bahamas and Cuba. *Copeia,* No. 3:203–214.
1960. The endemic fish fauna of Lake Lanao, and the evolution of higher taxonomic categories. *Evolution* 14:323–333.

Neave, F.
  1944. Racial characteristics and migratory habits in *Salmo gairdneri*. *J. Fish. Research Board Canada* 6:245–251.
Neel, J. V.
  1958. The study of natural selection in primitive and civilized human populations. *Human Biol.* 30:43–72.
Newell, N. D.
  1947. Infraspecific categories in invertebrate paleontology. *Evolution* 1:163–171.
  1952. Periodicity in invertebrate evolution. *J. Paleontol.* 26:371–385.
  1956a. Fossil populations. In P. C. Sylvester-Bradley, ed., *The species concept in palaeontology* (Syst. Assoc. Publ. No. 2), 63–82.
  1956b. Catastrophism and the fossil record. *Evolution* 10:97–101.
Newman, M. T.
  1953. The application of ecological rules to the racial anthropology of the aboriginal new world. *Amer. Anthropol.* 55:311–327.
Nice, M. M.
  1937. Studies in the life history of the song sparrow. I. *Trans. Linn. Soc. New York* 4:1–247.
Nicol, D., G. A. Desborough, and J. R. Solliday
  1959. Paleontology: paleontologic record of the primary differentiation in some major invertebrate groups. *J. Wash. Acad. Sci.* 49:351–366.
Nichols, D.
  1959. Changes in the chalk heart-urchin *Micraster* interpreted in relation to living forms. *Phil. Trans. Roy. Soc., London* (B) 242:347–437.
Nicholson, A. J.
  1954. An outline of the dynamics of animal populations. *Austral. J. Zool.* 2:9–65.
  1957. The self-adjustment of populations to change. *Cold Spring Harbor Symp. Quant. Biol.* 22:153–173.
Niethammer, G.
  1940. Die Schutzanpassung der Lerchen. In W. Hoesch and G. Niethammer, *Die Vogelwelt Deutsch-Südwestafrikas. J. Ornithol.*, Sonderheft 88:75–83.
  1959. Die Rolle der Auslese durch Feinde bei Wüstenvögeln. *Bonn. Zool. Beitr.* 10:179–197.
Nilsson, N.-A.
  1955. Studies on the feeding habits of trout and char in north Swedish lakes. *Inst. Freshwater Research, Drottningholm*, No. 36:163–225.
  1960. Seasonal fluctuations in the food segregation of trout, char and whitefish in 14 North-Swedish lakes. *Inst. Freshwater Research, Drottningholm*, No. 41:185–205.
Nilsson-Ehle, H.
  1909. *Kreuzungsuntersuchungen an Hafer und Weizen* (Acta Universita, Lund).

Nørrevang, A.
1959. Double invasions and character displacement. *Vidensk. Medd. fra Dansk naturh. Foren.* 121:171–180.

Nümann, W.
1948. Artbildungsvorgänge bei Forellen (*Salmo lacustris* und *S. Carpio*). *Biol. Zentr.* 66:77–81.

Oksala, T.
1954. Genetics of the dark phases of the red fox in experiment and in nature. *Papers Game Research, Helsinki* 11:1–16.
1958. Chromosome pairing, crossing over, and segregation in meiosis in *Drosophila melanogaster* females. *Cold Spring Harbor Symp. Quant. Biol.* 23:197–210.

Oliver, J. A., and C. E. Shaw
1953. The amphibians and reptiles of the Hawaiian Islands. *Zoologica* 38:71–73.

Olson, A. D., T. H. Hubbell, and H. F. Howden
1954. The burrowing beetles of the genus *Mycotrupes* (Coleoptera: Scarabacidae: Geotrupinae). *Misc. Publ. Mus. Zool., Univ. Mich.,* No. 84:1–59.

Omodeo, P.
1951. Gametogenesi e sistematica intraspecifica come problemi connessi con la poliploidia nei *Lumbricidae*. *Atti Soc. Toscana Sci. Nat.* (B) 58:1–12.
1952. Cariologia dei *Lumbricidae*. *Caryologia* 4:173–275.

Osborn, H. F.
1927. The origin of species. V. Speciation and mutation. *Amer. Nat.* 61:5–42.
1936. *Proboscidea* (American Museum Press, New York), vol. 1.

Osche, G.
1952. Systematik und Phylogenie der Gattung *Rhabditis* (Nematoda). *Zool. Jahrb.* (*Syst.*) 81:190–280.
1954. Über die gegenwärtig ablaufende Entstehung von Zwillings- und Komplementärarten bei Rhabditiden (Nematodes). *Zool. Jahrb.* (*Syst.*) 82:617–654.
1958. Die Bursa- und Schwanzstrukturen und ihre Aberrationen bei den Strongylina (Nematoda). *Z. Morphol. u. Ökol. Tiere* 46:571–635.

Oshima, C.
1958. Studies on DDT-resistance in *Drosophila melanogaster*. *J. Heredity* 49:22–31.

Palmén, E.
1949. The *Diplopoda* of eastern Fennoscandia. *Ann. Zool. Soc. Vanamo* 13:1–54.

Palmgren, P.
1938. Zur Kausalanalyse der ökologischen und geographischen Verbreitung der Vögel Nordeuropas. *Arch. Naturg.* (N. F.) 7:235–269.

Paludan, K.
1940. Contributions to the ornithology of Iran. *Danish Sci. Invest. Iran*, Pt. 2:11–54.
1951. Contributions to the breeding biology of *Larus argentatus* and *Larus fuscus. Vidensk. Medd. fra naturh. Foren.* 114:1–128.

Papi, F.
1954. Aspetti del differenziamento razziale e specifico nei Turbellari Rabdoceli. *Boll. Zool.* 21:357–377.

Paraense, W. L.
1959. One-sided reproductive isolation between geographically remote populations of a planorbid snail. *Amer. Nat.* 93:93–101.

Park, O.
1949. Application of the converse Bergmann principle to the carabid beetle, *Dicaelus purpuratus. Physiol. Zool.* 22:359–372.

Park, T.
1948. Interspecies competition in populations of *Tribolium confusum* Duval and *Tribolium castaneum* Herbst. *Ecol. Monographs* 18:265–308.
1954. Experimental studies of interspecies competition. II. Temperature, humidity, and competition in two species of *Tribolium. Physiol. Zool.* 27:177–238.

Parkes, K. C.
1951. The genetics of the golden-winged × blue-winged warbler complex. *Wilson Bull.* 63:5–15.
1961. Intergeneric hybrids in the family Pipridae. *Condor* 63:345–350.

Parson, P. A., and W. F. Bodmer
1961. The evolution of overdominance: natural selection and heterozygote advantage. *Nature* 190:7–12.

Pateff, P.
1947. On the systematic position of the starlings inhabiting Bulgaria and the neighbouring countries. *Ibis* 89:494–507.

Patterson, J. T.
1942. Isolating mechanisms in the genus *Drosophila. Biol. Symp.* 6:271–287.
1947. Studies in the genetics of *Drosophila*. V. Isolating mechanisms. *Univ. Texas Publ.*, No. 4720:1–184.
1953. Studies in the genetics of *Drosophila*. VII. Revision of the *montana* complex of the *virilis* species group. *Univ. Texas Publ.*, No. 5204:20–34.

Patterson, J. T., and Th. Dobzhansky
1945. Incipient reproductive isolation between two subspecies of *Drosophila pallidipennis. Genetics* 30:429–438.

Patterson, J. T., and W. S. Stone
1952. *Evolution in the genus* Drosophila (Macmillan, New York).

Pavan, C.
1946. Observations and experiments on the cave fish *Pimelodella kronei* and its relatives. *Amer. Nat.* 80:343–361.

Peabody, F. E., and J. M. Savage
1958. Evolution of a coast range corridor in California and its effect on the origin and dispersal of living amphibians and reptiles. In C. L. Hubbs, ed., *Zoogeography* (Amer. Assoc. Adv. Sci. Publ. No. 51), 159–186.

Peitzmeier, J.
1942. Die Bedeutung der oekologischen Beharrungstendenz für faunistische Untersuchungen. *J. Ornithol.* 90:311–322.

Pejler, B.
1956. Introgression in planktonic Rotatoria with some points of view on its causes and conceivable results. *Evolution* 10:246–261.
1957a. Taxonomical and ecological studies on planktonic Rotatoria from northern Swedish Lapland. *K. svenska VetenskAkad. Handl.* 6;1–68.
1957b. On variation and evolution in planktonic Rotatoria. *Zool. Bidrag Uppsala* 32:1–66.

Penn, G. H.
1957. Variation and subspecies of the crawfish *Orconectes palmeri* (Faxon) (Decapoda, Astacidae). *Tulane Studies Zool.* 5:231–262.

Perdeck, A. C.
1957. The isolating value of specific song patterns in two sibling species of grasshoppers (*Chorthippus brunneus* Thunb. and *C. biguttulus* L.). *Behaviour* 12:1–75.

Petersen, B.
1947a. Die geographische Variation einiger Fennoskandischer Lepidopteren. *Zool. Bidrag Uppsala* 26:329–531.
1947b. On the difference in species between *Boloria pales* Schiff. and *Boloria arsilache* Esp. *Zool. Bidrag Uppsala* 25:335–343.
1952. Studies on geographic variation of allometry in some European Lepidoptera. *Zool. Bidrag Uppsala* 29:1–38.
1955. Geographische Variation von *Pieris* (*napi*) *bryoniae* durch Bastardierung mit *Pieris napi*. *Zool. Bidrag Uppsala* 30:355–397.

Petersen, B., and O. Tenow
1954. Studien am Rapsweissling und Bergweissling (*Pieris napi* L. und *Pieris bryoniae* O.). *Zool. Bidrag Uppsala* 30:169–198.

Petersen, W.
1903. Entstehung der Arten durch physiologische Isolierung. *Biol. Zentr.* 23:468–477.
1905. Über beginnende Art-Divergenz. *Arch. Rass.- u. Ges. Biol.* 2:641–662.

Petit, C.
1958. Le déterminisme génétique et psycho-physiologique de la compétition sexuelle chez *Drosophila melanogaster*. *Bull. Biol.* 92:248–329.

Petrunkevitch, A.
    1952. Macro-evolution and the fossil record of Arachnida. *Amer. Sci.* 40:99–122.

Peus, F.
    1950. Der Formenkreis des *Ctenophthalmus agyrtes* Heller (Insecta, Aphaniptera). *Syllegomena Biol., Festschr. Kleinschmidt, Wittenberg 1950*, 286–318.

Philip, J. R.
    1955. Note on the mathematical theory of population dynamics and a recent fallacy. *Austral. J. Zool.* 3:287–294.

Phillips, J. C.
    1915. Experimental studies of hybridization among ducks and pheasants. *J. Exptl. Zool.* 18:69–143.

Pickford, G. E., and B. H. McConnaughey
    1949. The *Octopus bimaculatus* problem: a study in sibling species. *Bull. Bingham Oceanographic Coll., Peabody Mus. Nat. Hist., Yale Univ.* 12:1–66.

Pictet, A.
    1938. Les races physiologiques de *Nemeophila* (*Parasemia*) *plantaginis* L. au Parc National Suisse et dans les massifs limitrophes. *Bull. Soc. Entomol. Suisse* 17:373–391.

Pierce, G. W.
    1948. *The songs of insects* (Harvard University Press, Cambridge, Massachusetts).

Pitelka, F. A.
    1951a. Speciation and ecologic distribution in American jays of the genus *Aphelocoma*. *Univ. Calif. Publ. Zool.* 50:195–464.
    1951b. Ecologic overlap and interspecific strife in breeding populations of Anna and Allen hummingbirds. *Ecology* 32:641–661
    1951c. Principles of animal ecology. *Evolution* 5:81–84.

Pittendrigh, C. S.
    1950. The ecoclimatic divergence of *Anopheles bellator* and *A. homunculus*. *Evolution* 4:43–63.
    1958. Adaptation, natural selection, and behavior. In A. Roe and G. G. Simpson, eds., *Behavior and evolution* (Yale University Press, New Haven), 390–416.

Plate, L.
    1913. *Selektionsprinzip und Probleme der Artbildung* (Engelmann, Leipzig and Berlin).
    1924. *Allgemeine Zoologie und Abstammungslehre* (Gustav Fischer, Jena).

Plessen, V. Baron von
    1926. Verbreitung und Lebensweise von *Leucopsar rothschildi* Stres. *Ornithol. Monatsber.* 34:71–73.

Poll, M.
    1950. Histoire du peuplement et origine des espèces de la faune ichthyo

logique du Lac Tanganyika. *Ann. Soc. Roy. Zool. Belgique* 81:111–140.

Pontecorvo, G.
1958. *Trends in genetic analysis* (Columbia University Press, New York).

Pontin, A. J.
1961. Population stabilization and competition between the ants *Lasius flavus* (F.) and *L. niger* (L.) *J. Animal Ecol.* 30:47–54.

Popham, E. J.
1942. The variation in the colour of certain species of *Arctocorisa* (*Hemiptera*) and its significance. *Proc. Zool. Soc. London* (A) 111:135–172.

1947. Ecological studies of the mating habits of certain species of Corixidae and their significance. *Proc. Zool. Soc. London* 116:692–706.

Portmann, A.
1956. *Zoologie und das neue Bild vom Menschen* (Rowohlts deutsche Enzyklopädie, Hamburg).

Portugal-Araújo, V., and W. E. Kerr
1959. A case of sibling species among social bees. *Rev. Brasil. Biol.* 19:223–228.

Poulson, D. F., and B. Sakaguchi
1961. Nature of "sex-ratio" agent in Drosophila. *Science* 133:1489–1490.

Poulton, E. B.
1903. What is a species? *Proc. Entomol. Soc. London,* lxxvi–cxvi.

1908. *Essays on evolution, 1889–1907* (Clarendon Press, Oxford).

Prevosti, A.
1954. Variación geográfica de varios caracteres cuantitativos en poblaciones catalanas de *Drosophila subobscura. Genét. Ibérica* 6:33–68.

1955a. Variación geográfica de caracteres cuantitativos en poblaciones Británicas de *Drosophila subobscura. Genét. Ibérica* 7:3–11.

1955b. Geographical variability in quantitative traits in populations of *Drosophila subobscura. Cold Spring Harbor Symp. Quant. Biol.* 20:294–299.

Prosser, C. L.
1955. Physiological variation in animals. *Biol. Rev.* 30:229–262.

1957. The species problem from the viewpoint of a physiologist. In E. Mayr, ed., *The species problem* (Amer. Assoc. Adv. Sci. Publ. No. 50), 339–369.

Pryer, H.
1886. *Rhopalocera nihonica: a description of the butterflies of Japan* (Yokohama).

Pyburn, W. F.
1961. The inheritance and distribution of vertebral stripe color in the cricket frog. In W. F. Blair, ed., *Vertebrate speciation* (University of Texas Press, Austin), 235–261.

Race, R. R., and R. Sanger
1954. *Blood groups in man* (2nd ed.; Blackwell, Oxford).

Radovanović, M.
  1956. Rassenbildung bei den Eidechsen auf Adriatischen Inseln. Österr.
        Akad. Wiss. 110:1–82.
  1959. Zum Problem der Speziation bei Inseleidechsen. Zool. Jahrb. (Syst.)
        86:395–436.

Ramme, W.
  1930. Revisionen und Neubeschreibungen in der Gattung Pholidoptera
        Wesm. (Orth., Tettigon.). Mitt. Zool. Mus. Berlin 16:789–821.
  1951. Zur Systematik, Faunistik und Biologie der Orthopteren von Südost-
        Europa und Vorderasien. Mitt. Zool. Mus. Berlin 27:1–431.

Rand, A. L.
  1936. Altitudinal variation in New Guinea birds. Amer. Mus. Novitates, No.
        890:1–14.
  1938. Results of the Archbold expeditions. No. 22. On the breeding habits
        of some birds of paradise in the wild. Amer. Mus. Novitates, No.
        993:1–8.
  1948. Glaciation, an isolating factor in speciation. Evolution 2:314–321.
  1952. Secondary sexual characters and ecological competition. Fieldiana:
        Zool. (Chicago) 34:65–70.
  1958. Notes on African bulbuls. Family Pycnonotidae: class Aves. Fieldi-
        ana: Zool. (Chicago) 35:143–220.

Rao, K. P.
  1953. Rate of water propulsion in Mytilus californianus as a function of
        latitude. Biol. Bull. 104:171–181.

Rau, P.
  1942. The nesting habits of Polistes wasps as a factor in taxonomy. Ann.
        Entomol. Soc. Amer. 35:335–338.
  1946. The nests and the adults of colonies of Polistes wasps. Ann. Entomol.
        Soc. Amer. 39:11–27.

Rawson, G. W., and J. B. Ziegler
  1950. A new species of Mitoura Scudder from the pine barrens of New
        Jersey (Lepidoptera, Lycaenidae). J. N. Y. Entomol. Soc. 58:69–82.

Ray, C.
  1960. The application of Bergmann's and Allen's rules to the poikilotherms.
        J. Morphol. 106:85–108.

Ray, J.
  1686. Historia plantarum (London), vol. 1.

Reed, S. C., and E. W. Reed
  1948. Morphological differences and problems of speciation in Drosophila.
        Evolution 2:40–48.
  1950. Natural selection in laboratory populations of Drosophila. II. Compe-
        tition between a white-eye gene and its wild type allele. Evolution
        4:34–42.

Reed, S. C., C. M. Williams, and L. E. Chadwick
  1942. Frequency of wing-beat as a character for separating species, races
        and geographic varieties of Drosophila. Genetics 27:349–361.

Reeve, E. C. R., and F. W. Robertson
 1953. Studies in quantitative inheritance. II. Analysis of a strain of *Drosophila melanogaster* selected for long wings. *J. Genet.* 51:276–316.
Reid, J. A.
 1953. The *Anopheles hyrcanus* group in south-east Asia (Diptera: Culicidae). *Bull. Entomol. Res. London* 44:5–76.
 1960. Mosquitoes, insecticides and evolution. *Proc. Centenary (Darwin-Wallace) and Bicentenary (Linnaeus) Congr., Univ Malaya, Singapore 1958*, 217–219.
Reinig, W. F.
 1939. Die Evolutionsmechanismen, erläutert an den Hummeln. *Zool. Anz., Suppl.* 12:170–206.
Reiser, O. L.
 1958. The concept of evolution in philosophy. In R. Buchsbaum, ed., *A book that shook the world* (University of Pittsburgh Press, Pittsburgh), 38–47.
Remane, A.
 1951. Die Besiedelung des Sandbodens im Meere und die Bedeutung der Lebensformtypen für die Ökologie. *Verh. Deut. Zool. Ges., Wilhelmshaven, 1951*, 327–359.
 1952. *Die Grundlagen des natürlichen Systems, der vergleichenden Anatomie und der Phylogenetik* (Geest and Portig, Leipzig).
Remington, C. L.
 1954. The genetics of *Colias* (Lepidoptera). *Adv. Genet.* 6:403–450.
 1958. Genetics of populations of Lepidoptera. *Proc. 10th Intern. Congr. Entomol. 1956*, 2:787–805.
Remington, C. L., M. M. Cary, A. B. Klots, B. P. Beirne, E. Munroe, and L. P. Grey
 1951. Geographic subspeciation in the Lepidoptera: a symposium. *Lepidopterists' News* 5:17–35.
Rendel, J. M.
 1959. Canalization of the scute phenotype of *Drosophila*. *Evolution* 13:425–439.
Rensch, B.
 1929. *Das Prinzip geographischer Rassenkreise und das Problem der Artbildung* (Borntraeger, Berlin).
 1932. Über die Abhängigkeit der Grösse, des relativen Gewichtes und der Oberflächenstruktur der Landschneckenschalen von den Umweltfaktoren. *Z. Morphol. u. Ökol. Tiere* 25:757–807.
 1933. Zoologische Systematik und Artbildungsproblem. *Verh. Deut. Zool. Ges., Köln, 1933*, 19–83.
 1934. *Kurze Anweisung für zoologische-systematische Studien* (Akademische Verlagsgesellschaft, Leipzig).
 1936. Studien über klimatische Parallelität der Merkmalsausprägung bei Vögeln und Säugern. *Arch. Naturg.* (N. F.) 5:317–363.
 1938. Bestehen die Regeln klimatischer Parallelität bei der Merkmalsaus-

prägung von homöothermen Tieren zu Recht? *Arch. Naturg.* (N. F.) 7:364–389.

1939. Klimatische Auslese von Grössenvarianten. *Arch. Naturg.* (N. F.) 8:89–129.

1940. Die ganzheitliche Auswirkung der Grössenauslese am Vogelskelett. *J. Ornithol.* 88:373–388.

1943. Studien über Korrelation und klimatische Parallelität der Rassenmerkmale von Carabus-Formen. *Zool. Jahrb.* (*Syst.*) 76:103–170.

1947. *Neuere Probleme der Abstammungslehre* (1st ed.; Enke, Stuttgart).

1948. Organproportionen und Körpergrösse bei Vögeln und Säugetieren. *Zool. Jahrb.* (*Physiol.*) 61:337–450.

1954. *Neuere Probleme der Abstammungslehre* (2nd ed.; Enke, Stuttgart).

1959. *Homo sapiens. Vom Tiere zum Halbgott* (Vandenhoeck and Ruprecht, Göttingen).

1960a. *Evolution above the species level* (Columbia University Press, New York).

1960b. [Discussion comment.] *Zool. Anz., Suppl.* 23:63.

Reynoldson, T. B.

1948. British species of *Polycelis* (Platyhelminthes). *Nature* 162:620.

1956. The population dynamics of host specificity in *Urceolaria mitra* (Peritricha) epizoic on freshwater triclads. *J. Animal Ecol.* 25:127–143.

Richards, O. W.

1927. Sexual selection and allied problems in the insects. *Biol. Rev.* 2:298–364.

Ricker, W. E.

1938. "Residual" and Kokanee salmon in Cultus Lake. *J. Fish. Research Bd. Canada* 4:192–218.

1940. On the origin of Kokanee, a freshwater type of sockeye salmon. *Trans. Roy. Soc. Canada, Biol. Sci.* (3) 34:121–135.

Riech, E.

1937. Systematische, anatomische, ökologische und tiergeographische Untersuchungen über die Süsswassermollusken Papuasiens und Melanesiens. *Arch. Naturg.* (N. F.) 6:37–153.

Riley, H. P.

1952. Ecological barriers. *Amer. Nat.* 86:23–32.

Ripley, S. D.

1942. A revision of the kingfishers, *Ceyx erithacus* and *rufidorsus*. *Zoologica* 27:55–59.

1949. Avian relicts and double invasions in peninsular India and Ceylon. *Evolution* 3:150–159.

1959. Character displacement in Indian nuthatches (*Sitta*). *Postilla, Yale Peabody Mus.*, No. 42:1–11.

1961. Aggressive neglect as a factor in interspecific competition in birds. *Auk* 78:366–371.

Ripley, S. D., and H. Birkhead
1942. Birds collected during the Whitney South Sea expedition 51. On the fruit pigeons of the *Ptilinopus purpuratus* group. *Amer. Mus. Novitates,* No. 1192:1–14.

Rizki, M. T. M.
1951. Morphological differences between two sibling species, *Drosophila pseudoobscura* and *Drosophila persimilis. Proc. Nat. Acad. Sci.* 37:156–159.

Roberts, J. A. Fraser
1957. Blood groups and susceptibility to disease: a review. *Brit. J. Prev. Social Med.* 11:107–125.

Robertson, A.
1955. Selection in animals: synthesis. *Cold Spring Harbor Symp. Quant. Biol.* 20:225–229.

Robertson, F. W., and E. C. Reeve
1952a. Heterozygosity, environmental variation and heterosis. *Nature* 170:296.
1952b. Studies in quantitative inheritance. I. The effects of selection of wing and thorax length in *Drosophila melanogaster. J. Genet.* 50:414–448.

Robertson, W. R. B.
1916. Chromosome studies. I. Taxonomic relationships shown in the chromosomes of Tettigidae and Acrididae. V-shaped chromosomes and their significance in Acrididae, Locustidae and Gryllidae: chromosomes and variation. *J. Morphol.* 27:179–331.

Robinson, J. T.
1955. Further remarks on the relationship between "Meganthropus" and Australopithecines. *Amer. J. Phys. Anthropol.* (n. s.) 13:429–445.

Robson, G. C., and O. W. Richards
1936. *The variations of animals in nature* (Longmans, Green, London).

Rodgers, T. L., and H. S. Fitch
1947. Variation in the skinks (Reptilia: Lacertilia) of the *skiltonianus* group. *Univ. Calif. Publ. Zool.* 48:169–220.

Roe, A., and G. G. Simpson (eds.)
1958. *Behavior and evolution* (Yale University Press, New Haven).

Rogers, W. P.
1962. *The nature of parasitism; the relationship of some metazoan parasites to their hosts* (Academic Press, New York).

Romanes, G. J.
1897. *Darwin, and after Darwin.* (Open Court, Chicago), vol. 3.

Romer, A. S.
1946. The early evolution of fishes. *Quart. Rev. Biol.* 21:33–69.
1957. Origin of the amniote egg. *Sci. Monthly* 85:57–63.
1958. Phylogeny and behavior with special reference to vertebrate evolution. In A. Roe and G. G. Simpson, eds., *Behavior and evolution* (Yale University Press, New Haven), 48–75.

Rosen, D. E.
    1960. Middle-American poeciliid fishes of the genus *Xiphophorus. Bull. Florida State Mus. Biol. Sci.* 5:57–242.
Rosen, D. E., and M. Gordon
    1953. Functional anatomy and evolution of male genitalia in poeciliid fishes. *Zoologica* 38:1–47.
Rosin, S., J. K. Moor-Jankowski, and M. Schneeberger
    1958. Die Fertilität im Bluterstamm von Tenna (Hämophilie B). *Acta Genet.* 8:1–24.
Ross, H. H.
    1957. Principles of natural coexistence indicated by leafhopper populations. *Evolution* 11:113–129.
    1958. Evidence suggesting a hybrid origin for certain leafhopper species. *Evolution* 12:337–446.
Roth, L. M.
    1948. A study of mosquito behavior. An experimental laboratory study of the sexual behavior of *Aedes aegypti* (Linnaeus). *Amer. Midl. Nat.* 40:265–352.
Roth, L. M., and E. R. Willis
    1952. A study of cockroach behavior. *Amer. Midl. Nat.* 47:66–129.
    1954. The reproduction of cockroaches. *Smithsonian Misc. Coll.* 122:1–49.
Rothfels, K. H.
    1956. Black flies: siblings, sex, and species groupings. *J. Heredity* 47:113–122.
Rothschild, W., E. Hartert, and K. Jordan
    1894. Note of the editors. *Novit. Zool.* 1:1.
Rozeboom, L. E.
    1952. The significance of Anopheles species complexes in problems of disease transmission and control. *J. Econ. Entomol.* 45:222–226.
Rudd, R. L.
    1955. Population variation and hybridization in some Californian shrews. *Systematic Zool.* 4:21–34.
Ruibal, R.
    1955. A study of altitudinal races in *Rana pipiens. Evolution* 9:322–338.
    1957. An altitudinal and latitudinal cline in *Rana pipiens. Copeia,* No. 3:212–221.
    1961. Thermal relations of five species of tropical lizards. *Evolution* 15:98–111.
Ruiter, L. de
    1955. Countershading in caterpillars, an analysis of its adaptive significance. *Arch. Néerl. Zool.* 11:285–341.
    1958. Natural selection in *Cepaea nemoralis. Arch. Néerl. Zool.* 12:571–573.
Rümmler, H.
    1938. Die Systematik und Verbreitung der Muriden Neuguineas. *Mitt. Zool. Mus. Berlin* 23:1–298.

Runnström, S.
1927. Über die Thermopathie der Fortpflanzung und Entwicklung mariner Tiere in Beziehung zu ihrer geographischen Verbreitung. *Bergens Mus. Årbok*, No. 2:1–67.
1929. Weitere Studien über die Temperaturanpassung der Fortpflanzung und Entwicklung mariner Tiere. *Bergens Mus. Årbok*, No. 10:1–46.
1936. Die Anpassung der Fortpflanzung und Entwicklung mariner Tiere an die Temperaturverhältnisse verschiedener Verbreitungsgebiete. *Bergens Mus. Årbok*, No. 3:1–36.
Russell, W. L.
1951. X-ray-induced mutations in mice. *Cold Spring Harbor Symp. Quant. Biol.* 16:327–336.
Ryan, F. J.
1953. Natural selection in bacterial populations. *Atti VI Congr. Intern. Microbiol. Roma* 1:649–657.
Saccà, G.
1952. Variabilità fenotipica in *Musca domestica* L. *Boll. Zool.* 19:293–296.
1953. Contributo alla conoscenza tassonomica del "gruppo" *domestica* (Diptera, Muscidae). *Rend. Ist. Sup. Sanità* 16:442–464.
1956. Speciation in the house-fly. I. Recent views on the taxonomic problem (Diptera, Muscidae gen. *Musca*). *Selected Sci. Papers Ist Sup. Sanità* 1:141–154.
1957. Ricerche sulla speciazione nelle mosche domestiche. IV. Esperienze sull' isolamento sessuale fra le sub-specie di *Musca domestica* L. *Rend. Ist. Sup. Sanità* 20:702–712.
1958. Ricerche sulla speciazione nelle mosche domestiche. VI. Ibridismo naturale e ibridismo sperimentale fra le subspecie di *Musca domestica* L. *Rend. Ist. Sup. Sanità* 21:1170–1184.
Saccà, G., and L. Rivosecchi
1958. Ricerche sulla speciazione nelle mosche domestiche. V. L'areale di distribuzione delle subspecie di *Musca domestica* L. (Diptera, Muscidae). *Rend Ist. Sup. Sanità* 21:1149–1169.
Sachs, H.
1950. Die Nematodenfauna der Rinderexkremente. *Zool. Jahrb. (Syst.)* 79:209–272.
Sachs, L.
1952. Polyploid evolution and mammalian chromosomes. *Heredity* 6:357–364.
Sadoglu, P.
1957. Mendelian inheritance in the hybrids between the Mexican blind cave fishes and their overground ancestor. *Verh. Deut. Zool. Ges., Graz, 1957*, 432–439.
Sager, R., and F. J. Ryan
1961. *Cell heredity.* (Wiley, New York).

Sailer, R. I.
  1954. Interspecific hybridization among insects with a report on cross-breeding experiments with stink bugs. *J. Econ. Entomol.* 47:377–383.
Salmon, J. T.
  1955. Parthenogenesis in New Zealand stick insects. *Trans. Roy. Soc. New Zealand* 82:1189–1192.
Salomonsen, F.
  1931. Diluviale Isolation und Artbildung. *Proc. VII Intern. Ornithol. Congr., Amsterdam, 1930*, 413–438.
  1949. The European hybrid-population of the great gray shrike (*Lanius excubitor* L.). *Vidensk. Medd. fra Dansk naturh. Foren.* 111:149–161.
  1950–1951. *The birds of Greenland* (Ejnar Munksgaard, Copenhagen).
  1951. The immigration and breeding of the fieldfare (*Turdus pilaris* L.) in Greenland. *Proc. Xth Intern. Ornithol. Congr. Uppsala, 1950*, 515–526.
  1955. The evolutionary significance of bird-migration. *Dan. Biol. Medd.* 22:1–62.
Sammalisto, L.
  1956. Secondary intergradation of the blue-headed and grey-headed wagtails (*Motacilla f. flava* L. and *Motacilla f. thunbergi* Billb.) in South Finland. *Ornis Fenn.* 33:1–19.
Sandler, L., and E. Novitski
  1957. Meiotic drive as an evolutionary force. *Amer. Nat.* 91:105–110.
Sandler, L., Y. Hiraizumi, and I. Sandler
  1959. Meiotic drive in natural populations of *Drosophila melanogaster*. I. The cytogenetic basis of segregation-distortion. *Genetics* 44:233–250.
Sarasin, F., and P. Sarasin
  1899. *Die Landmollusken von Celebes* (Kreidel, Wiesbaden).
Sauter, W.
  1956. Morphologie und Systematik der schweizerischen *Solenobia*-Arten (Lep. Psychidae). *Rev. Suisse Zool.* 63:451–544.
Savage, J. M.
  1958. The concept of ecologic niche, with reference to the theory of natural coexistence. *Evolution* 12:111–121.
Savile, D. B. O.
  1960. Limitations of the competitive exclusion principle. *Science* 132:1761.
Schaeffer, B.
  1948. The origin of a mammalian ordinal character. *Evolution* 2:164–175.
  1952. Rates of evolution in the coelacanth and dipnoan fishes. *Evolution* 6:101–111.
  1956. Evolution in the subholostean fishes. *Evolution* 10:201–212.
Schaller, F., and H. Schwalb
  1961. Attrappenversuche mit Larven und Imagines einheimischer Leuchtkäfer (Lampyrinae). *Zool. Anz., Suppl.* 24:154–166.

Schilder, F. A.
  1952. *Einführung in die Biotaxonomie* (*Formenkreislehre*) (Gustav Fischer, Jena).
Schindewolf, O. H.
  1930. *Paläontologie, Entwicklungslehre und Genetik* (Borntraeger, Berlin).
  1950a. *Grundfragen der Paläontologie* (Schweizerbart, Stuttgart).
  1950b. *Der Zeitfaktor in Geologie und Paläontologie* (Schweizerbart, Stuttgart).
Schlegel, H.
  1844. *Kritische Übersicht der europäischen Vögel* (Arnz u. Comp., Leiden).
Schlieper, C.
  1957. Comparative study of *Asterias rubens* and *Mytilus edulis* from the North Sea and the western Baltic Sea. *L'Ann. Biol.* 33:117–127.
Schlosser, M.
  1901. Die menschenähnlichen Zähne aus dem Bohnerz der schwäbischen Alb. *Zool. Anz.* 24:261–271.
Schmalhausen, I. I.
  1949. *Factors of evolution; the theory of stabilizing selection* (Blakiston, Philadelphia).
Schmidt, J.
  1918. Racial studies in fishes. I. Statistical investigations with *Zoarces viviparus* L. *J. Genet.* 7:105–118.
Schmidt, K. P.
  1950. The concept of geographic range. *Texas J. Sci.* 2:326–334.
Schmidt, R. S.
  1955a. The evolution of nest-building behavior in *Apicotermes* (Isoptera). *Evolution* 9:157–181.
  1955b. Termite (*Apicotermes*) nests—important ethological material. *Behaviour* 8:344–356.
  1958. The nest of *Apicotermes trägårdhi* (Isoptera)—new evidence on the evolution of nest-building. *Behaviour* 12:76–94.
Schnakenbeck, W.
  1931. Zum Rassenproblem bei den Fischen. *Z. Morphol. u. Ökol. Tiere* 21:409–556.
Schneirla, T. C.
  1957. A comparison of species and genera in the ant subfamily Dorylinae with respect to functional pattern. *Insectes Sociaux* 4:259–298.
Schnetter, M.
  1951. Veränderungen der genetischen Konstitution in natürlichen Populationen der polymorphen Bänderschnecken. *Zool. Anz., Suppl.* 15:192–206.
Schnitter, H.
  1922. Die Najaden der Schweiz. *Z. Hydrol.* 2 (suppl.):1–200.
Scholander, P. F.
  1955. Evolution of climatic adaptation in homeotherms. *Evolution* 9:15–26.

Schoof, H. F., R. E. Siverly, and J. A. Jensen
  1952. House fly dispersion studies in Metropolitan areas. *J. Econ. Entomol.* 45:675–683.
Schrader, F., and S. Hughes-Schrader
  1956. Polyploidy and fragmentation in the chromosomal evolution of various species of *Thyanta* (Hemiptera). *Chromosoma* 7:469–496.
Schreider, E.
  1950. Geographical distribution of the body-weight/body-surface ratio. *Nature* 165:286.
  1957. Ecological rules and body-heat regulation in man. *Nature* 179:915–916.
Schultz, A. H.
  1944. Age changes and variability in gibbons. *Amer. J. Phys. Anthropol.* (n. s.) 2:1–129.
  1947. Variability in man and other primates. *Amer. J. Phys. Anthropol.* 5:1–14.
Schultz, J., and H. Redfield
  1951. Interchromosomal effects on crossing over in Drosophila. *Cold Spring Harbor Symp. Quant. Biol.* 16:175–197.
Schultz, R. J.
  1961. Reproductive mechanism of unisexual and bisexual strains of the viviparous fish *Poeciliopsis. Evolution* 15:302–325.
Schuster, O.
  1950. Die klimaparallele Ausbildung der Körperproportionen bei Poikilothermen. *Abhandl. Senckenb. Naturf. Ges.,* No. 482:1–89.
Schwarz, M.
  1956. Über die Variationsbreite der Camargue-Schafstelzen (*Motacilla flava*) und die Schafstelzen-Einwanderung in die Schweiz. *Ornithol. Beob.* 53:61–72.
Scossiroli, R. E.
  1954. Effectiveness of artificial selection under irradiation of plateaued populations of *Drosophila melanogaster. I. U. B. S. Symp. Genet. Population Structure, Pavia 1953,* 42–66.
Sedlmair, H.
  1956. Verhaltens-, Resistenz- und Gehäuseunterschiede bei den polymorphen Bänderschnecken *Cepaea hortensis* (Müll.) und *Cepaea nemoralis* (L.). *Biol. Zentr.* 75:281–313.
Seebohm, H.
  1888. *The geographical distribution of the family Charadriidae* (Henry Sotheran, London).
Seiler, J.
  1946. Die Verbreitungsgebiete der verschiedenen Rassen von *Solenobia triquetrella* (Psychidae) in der Schweiz. *Rev. Suisse Zool.* 53:529–533.
  1961. Untersuchungen über die Entstehung der Parthenogenese bei *Sole-*

nobia triquetrella F. R. (Lepidoptera, Psychidae). III. Z. *Vererb.-
Lehre* 92:261–316.

Seiler, J., and O. Puchta
  1956. Die Fortpflanzungsbiologie der Solenobien (Lepid. Psychidae), Ver-
        halten bei Artkreuzungen und F,–Resultate. *Roux' Arch. Entwick-
        lungs.* 149:115–246.

Seitz, A.
  1951. Vergleichende Verhaltensstudien an Buntbarschen. Z. *Tierpsychol.*
        6:202–255.

Selander, R. K., and D. R. Giller
  1959a. Sympatry of the jays *Cissilopha beecheii* and *C. san-blasiana* in
        Nayarit. *Condor* 61:52.
  1959b. Interspecific relations of woodpeckers in Texas. *Wilson Bull.* 71:107–
        124.
  1961. Analysis of sympatry of great-tailed and boat-tailed grackles. *Condor*
        63:29–86.

Semenov-Tian Shansky, A.
  1910. *Die taxonomischen Grenzen der Art und ihrer Unterabteilungen*
        (Friedländer, Berlin).

Semper, K. G.
  1881. *Animal life as affected by the natural conditions of existence* (Apple-
        ton, New York).

Sengün, A.
  1944. Experimente zur sexuell-mechanischen Isolation. *Rev. Fac. Sci. Istan-
        bul* (B) 9:239–253.

Serventy, D. L.
  1951. Inter-specific competition on small islands. *Western Austral. Nat.*
        3:59–60.
  1953. Some speciation problems in Australian birds. *Emu* 53:131–145.

Setzer, H. W.
  1949. Subspeciation in the kangaroo rat, *Dipodomys ordii*. *Univ. Kansas
        Publ., Mus. Nat. Hist.* 1:473–573.

Sewertzoff, A. N.
  1931. *Morphologische Gesetzmässigkeiten der Evolution* (Gustav Fischer,
        Jena).

Shapiro, H. L.
  1929. Descendants of the mutineers of the bounty. *Mem. Bishop Mus.,
        Honolulu* 2:1–106.

Shapley, H. (ed.)
  1953. *Climatic change: evidence, causes, and effects* (Harvard University
        Press, Cambridge, Massachusetts).

Sharpe, R. B.
  1909. *A handlist of the genera and species of birds* (British Museum, Lon-
        don), vol. 5.

Shaw, R. F.

1959. Equilibrium for the sex ratio factor in *Drosophila pseudoobscura*. *Amer. Nat.* 93:385–386.

1961. The effect of polygamy and infanticide on the sex-ratio. *Amer. J. Phys. Anthropol.* 19:79–83.

Sheppard, P. M.

1951a. Fluctuations in the selective value of certain phenotypes in the polymorphic land snail *Cepaea nemoralis* (L.). *Heredity* 5:125–134.

1951b. A quantitative study of two populations of the moth, *Panaxia dominula*. *Heredity* 5:349–378.

1952a. Natural selection in two colonies of the polymorphic land snail *Cepaea nemoralis*. *Heredity* 6:233–238.

1952b. A note on non-random mating in the moth *Panaxia dominula* (L.). *Heredity* 6:239–241.

1953a. Polymorphism and population studies. *Symp. Soc. Exptl. Biol.*, No. 7:274–289.

1953b. Polymorphism, linkage and the blood groups. *Amer. Nat.* 87:283–294.

1958. *Natural selection and heredity* (Hutchinson, London).

1961. Some contributions to population genetics resulting from the study of the Lepidoptera. *Adv. Genet.* 10:165–216.

Shideler, W. H.

1952. Paleontology and evolution. *Ohio J. Sci.* 52:177–186.

Shorten, M.

1954. *Squirrels* (Collins, London).

Shull, A. F.

1946. The form of the chitinous male genitalia in crosses of the species *Hippodamia quinquesignata* and *H. convergens. Genetics* 31:291–303.

1949. Extent of genetic differences between species of *Hippodamia* (Coccinellidae). *Proc. 8th Intern. Congr. Genet.* (*Hereditas, Suppl. vol., 1949*), 417–428.

Shultz, F. T., and W. E. Briles

1953. The adaptive value of blood group genes in chickens. *Genetics* 38:34–50.

Sibley, C. G.

1950. Species formation in the red-eyed towhees of Mexico. *Univ. Calif. Publ. Zool.* 50:109–194.

1954a. Hybridization in the red-eyed towhees of Mexico. *Evolution* 8:252–290.

1954b. The contribution of avian taxonomy. *Systematic Zool.* 3:105–110.

1957. The evolutionary and taxonomic significance of sexual dimorphism and hybridization in birds. *Condor* 59:166–191.

1959. Hybridization in birds: taxonomic and evolutionary implications. *Bull. British Ornithol. Club* 79:154–158.

1961. Hybridization and isolating mechanisms. In W. F. Blair, ed., *Vertebrate speciation* (University of Texas Press, Austin), 69–88.

Sibley, C. G., and L. L. Short, Jr.
1959. Hybridization in the buntings (*Passerina*) of the Great Plains. *Auk* 76:443–463.

Sibley, C. G., and D. A. West
1958. Hybridization in the red-eyed towhees of Mexico: the eastern plateau populations. *Condor* 60:85–104.

Silas, E. G.
1956. Speciation among the freshwater fishes of Ceylon. *Bull. No. VII, Nat. Inst. Sci. India, New Delhi,* 248–259.

Simmons, K. E. L.
1951. Interspecific territorialism. *Ibis* 93:407–413.

Simons, E. L.
1961. The phyletic position of *Ramapithecus. Postilla, Yale Peabody Mus.,* No. 57:1–9.

Simpson, G. G.
1940. Mammals and land bridges. *J. Wash. Acad. Sci.* 30:137–163.
1941. The role of the individual in evolution. *J. Wash. Acad. Sci.* 31:1–20.
1943. Criteria for genera, species and subspecies in zoology and paleozoology. *Ann. New York Acad. Sci.* 44:170–177.
1944. *Tempo and mode in evolution* (Columbia University Press, New York).
1947. The problem of plan and purpose in nature. *Sci. Monthly* 64:481–495.
1949. *The meaning of evolution* (Yale University Press, New Haven).
1950. History of the fauna of Latin America. *Amer. Sci.* 38:361–389.
1951a. Some principles of historical biology bearing on human origins. *Cold Spring Harbor Symp. Quant. Biol.* 15:55–66.
1951b. The species concept. *Evolution* 5:285–298.
1952a. Probabilities of dispersal in geologic time. *Bull. Amer. Mus. Nat. Hist.* 99:163–176.
1952b. How many species? *Evolution* 6:342.
1953a. *The major features of evolution* (Columbia University Press, New York).
1953b. The Baldwin effect. *Evolution* 7:110–117.
1959a. The nature and origin of supraspecific taxa. *Cold Spring Harbor Symp. Quant. Biol.* 24:255–271.
1959b. Mesozoic mammals and the polyphyletic origin of mammals. *Evolution* 13:405–414.
1960a. The history of life. In S. Tax, ed., *The evolution of life* (University of Chicago Press, Chicago), 117–180.
1960b. The world into which Darwin led us. *Science* 131:966–974.
1960c. Man's evolutionary future. *Zool. Jahrb. (Syst.)* 88:125–134.
1961. *Principles of animal taxonomy* (Columbia University Press, New York).

Sims, R. W.
1959. The *Ceyx erithacus* and *rufidorsus* species problem. *J. Linn. Soc. London, Zool.* 44:212–221.

Sinnott, E. D., L. C. Dunn, and Th. Dobzhansky
  1958.   *Principles of genetics* (McGraw Hill, New York).
Skellam, J. G.
  1951.   Random dispersal in theoretical populations. *Biometrika* 38:196–218.
Skutch, A. F.
  1949.   Do tropical birds rear as many young as they can nourish? *Ibis* 91:430–455.
  1951.   Congeneric species of birds nesting together in Central America. *Condor* 53:3–15.
Smith, S. G.
  1949.   Evolutionary changes in the sex chromosomes of Coleoptera. I. Wood borers of the genus *Agrilus. Evolution* 3:344–357.
  1953.   Reproductive isolation and the integrity of two sympatric species of *Choristoneura* (Lepidoptera: Tortricidae). *Canad. Entomol.* 85:141–151.
  1954.   A partial breakdown of temporal and ecological isolation between *Choristoneura* species (Lepidoptera: Tortricidae). *Evolution* 8:206–224.
  1960.   Cytogenetics of insects. *Ann. Rev. Entomol.* 5:69–84.
Snow, D. A.
  1954a. Trends in geographical variation in palaearctic members of the genus *Parus. Evolution* 8:19–28.
  1954b. The habitats of Eurasian tits (*Parus* spp). *Ibis* 96:565–585.
Sokal, R. R.
  1959.   A morphometric analysis of strains of *Drosophila melanogaster* differing in DDT-resistance. *J. Kansas Entomol. Soc.* 32:155–172.
Sokal, R. R., and P. E. Hunter
  1954.   Reciprocal selection for correlated quantitative characters in *Drosophila. Science* 119:649–651.
Sokal, R. R., and T. Hiroyoshi
  1959.   The supposed correlation between the ratio X_5 and DDT-resistance in house flies. *J. Econ. Entomol.* 52:1077–1080.
Sokoloff, A.
  1955.   Competition between sibling species of the *pseudoobscura* subgroup of *Drosophila. Ecol. Monographs* 25:387–409.
  1957.   [Discussion on competition in *Drosophila.*] *Cold Spring Harbor Symp. Quant. Biol.* 22:268–270.
Sonneborn, T. M.
  1938.   Mating types in *Paramecium aurelia:* diverse conditions for mating in different stocks, occurrence, number, and interrelations of the types. *Proc. Amer. Phil. Soc.* 79:411–434.
  1957.   Breeding systems, reproductive methods, and species problems in protozoa. In E. Mayr, ed., *The species problem* (Amer. Assoc. Adv. Sci. Publ. No. 50), 155–324.

Southern, H. N.
   1939.   The status and problem of the bridled guillemot. *Proc. Zool. Soc. London* (A) 109:31.
   1954.   Mimicry in cuckoos' eggs. In J. Huxley, A. C. Hardy, and E. B. Ford, eds., *Evolution as a process* (Allen and Unwin, London), 218–232.
Spassky, B.
   1957.   Morphological differences between sibling species of *Drosophila*. *Univ. Texas Publ.*, No. 5721:48–61.
Spassky, B., N. Spassky, H. Levene, and Th. Dobzhansky
   1958.   Release of genetic variability through recombination. I. *Drosophila pseudoobscura. Genetics* 43:844–867.
Spencer, W. P.
   1944.   Iso-alleles at the bobbed locus in *Drosophila hydei* populations. *Genetics* 29:520–536.
   1947a. Mutations in wild populations in *Drosophila. Adv. Genet.* 1:359–402.
   1947b. Genetic drift in a population of *Drosophila immigrans. Evolution* 1:103–110.
Spiess, E. B.
   1959.   Release of genetic variability through recombination. II. *Drosophila persimilis. Genetics* 44:43–58.
   1961.   Chromosomal fitness changes in experimental populations of *Drosophila persimilis* from Timberline in the Sierra Nevada. *Evolution* 15:340–351.
Spiess, E. B., and A. C. Allen
   1961.   Release of genetic variability through recombination. VII. Second and third chromosomes of *Drosophila melanogaster. Genetics* 46:1531–1553.
Spiess, F. B., and B. Langer
   1961.   Chromosomal adaptive polymorphism in *Drosophila persimilis*. III. Mating propensity of homokaryotypes. *Evolution* 15:535–544.
Spieth, H. T.
   1941.   Taxonomic studies on the Ephemeroptera. II. The genus *Hexagenia. Amer. Midl. Nat.* 26:233–280.
   1947.   Taxonomic studies on the Ephemeroptera. IV. The genus *Stenonema. Ann. Entomol. Soc. Amer.* 40:87–122.
   1949.   Sexual behavior and isolation in *Drosophila*. II. The interspecific mating behavior of species of the *willistoni* group. *Evolution* 3:67–81.
   1951.   Mating behavior and sexual isolation in the *Drosophila virilis* species group. *Behaviour* 3:105–145.
   1952.   Mating behavior within the genus *Drosophila* (Diptera). *Bull. Amer. Mus. Nat. Hist.* 99:399–474.
   1958.   Behavior and isolating mechanisms. In A. Roe and G. G. Simpson, eds., *Behavior and evolution* (Yale University Press, New Haven), 363–389.

Spieth, H. T., and T. C. Hsu
    1950. The influence of light on the mating behavior of seven species of the *Drosophila melanogaster* species group. *Evolution* 4:316–325.

Spofford, J. B.
    1956. The relation between expressivity and selection against eyeless in *Drosophila melanogaster*. *Genetics* 41:938–959.

Spring, A. F.
    1838. *Über die naturhistorischen Begriffe von Gattung, Art und Abart und über die Ursachen der Abartungen in den organischen Reichen* (Friedrich Fleischer, Leipzig).

Spuhler, J. N.
    1959. Physical anthropology and demography. In P. M. Hauser and O. D. Duncan, eds., *The study of population* (University of Chicago Press, Chicago), 728–758.

Spurway, H.
    1949. Remarks on Vavilov's law of homologous variation. *Ricerca Sci., Suppl.*, 18–24.

    1953. Genetics of specific and subspecific differences in European newts. *Symp. Soc. Exptl. Biol.*, No. 7:200–237.

    1955. The sub-human capacities for species recognition and their correlation with reproductive isolation. *Acta XI Congr. Intern. Ornithol., Basel, 1954*, 340–349.

    1957. Hermaphroditism with self-fertilization, and the monthly extrusion of unfertilized eggs, in the viviparous fish *Lebistes reticulatus*. *Nature* 180:1248–1251.

Spurway, H., and H. G. Callan
    1960. The vigour and male sterility of hybrids between the species *Triturus vulgaris* and *T. helveticus*. *J. Genetics* 57:84–118.

Spurway, H., and J. B. S. Haldane
    1953. The comparative ethology of vertebrate breathing. I. Breathing in newts, with a general survey. *Behaviour* 6:8–34.

Srb, A. M., and R. D. Owen
    1952. *General genetics* (Freeman, San Francisco).

Stadler, L. J.
    1954. The gene. *Science* 120:811–819.

Staiger, H.
    1954. Der Chromosomendimorphismus beim Prosobranchier *Purpura lapillus* in Beziehung zur Ökologie der Art. *Chromosoma* 6:419–478.

Staiger, H., and Ch. Bocquet
    1954. Cytological demonstration of female heterogamety in isopods. *Experientia* 10:64–66.

Stalker, H. D.
    1960. Chromosomal polymorphism in *Drosophila paramelanica* Patterson. *Genetics* 45:95–114.

    1961. The genetic systems modifying meiotic drive in *Drosophila paramelanica*. *Genetics* 46:177–202.

Stalker, H. D., and H. L. Carson
  1947. Morphological variation in natural populations of *Drosophila robusta* Sturtevant. *Evolution* 1:237–248.
  1948. An altitudinal transect of *Drosophila robusta* Sturtevant. *Evolution* 2:295–305.
Standfuss, M.
  1896. *Handbuch der paläarktischen Gross-Schmetterlinge für Forscher und Sammler* (Gustav Fischer, Jena).
Starck, D.
  1960. Das Cranium eines Schimpansenfetus (*Pan troglodytes* [Blumenbach 1799]) von 71 mm SchStlg., nebst Bemerkungen über die Körperform von Schimpansenfeten. *Morphol. Jahrb.* 100:559–647.
  1962. *Der heutige Stand des Fetalisationsproblems* (Paul Parey, Hamburg and Berlin).
Stauber, L. A.
  1950. The problem of physiological species with special reference to oysters and oyster drills. *Ecology* 31:109–118.
Stebbins, G. L., Jr.
  1942. The genetic approach to problems of rare and endemic species. *Madroño* 6:241–258.
  1950. *Variation and evolution in plants* (Columbia University Press, New York).
  1952. Organic evolution and social evolution. *Idea and Experiment* 11:3–7.
  1959. The role of hybridization in evolution. *Proc. Amer. Phil. Soc* 103:231–251.
  1960. The comparative evolution of genetic systems. In S. Tax, ed., *The evolution of life* (University of Chicago Press, Chicago), 197–226.
Stebbins, R. C.
  1949. Speciation in salamanders of the Plethodontid genus *Ensatina. Univ. Calif. Publ. Zool.* 48:377–526.
  1957. Intraspecific sympatry in the lungless salamander *Ensatina eschscholtzi. Evolution* 11:265–270.
Stebbins, R. C., and H. B. Robinson
  1946. Further analysis of a population of the lizard *Sceloporus graciosus gracilis. Univ. Calif. Publ. Zool.* 48:149–168.
Steere, J. B.
  1894. On the distribution of genera and species of non-migratory landbirds in the Philippines. *Ibis* 6:411–420.
Stegmann, B.
  1935. Unterschiede im oekologischen Verhalten als taxonomisches Kriterium. *Ornithol. Monatsber.* 43:17–21.
Stehr, G.
  1959. Hemolymph polymorphism in a moth and the nature of sex-controlled inheritance. *Evolution* 13:537–560.

Stein, G. H. W.
1950. Grössenvariabilität und Rassenbildung bei *Talpa europaea* L. *Zool. Jahrb.* (*Syst.*) 79:321–349.
1951. Populationsanalytische Untersuchungen am europäischen Maulwurf. II. Über zeitliche Grössenschwankungen. *Zool. Jahrb.* (*Syst.*) 79:567–590.
1958. Über den Selektionswert der Simplex-Zahnform bei der Feldmaus, *Microtus arvalis* (Pallas). *Zool. Jahrb.* (*Syst.*) 86:27–34.
1959. Ökotypen beim Maulwurf, *Talpa europaea* L. (Mammalia). *Mitt. Zool. Mus. Berlin* 35:5–43.
1960. Schädelallometrien und Systematik bei altweltlichen Maulwürfen (Talpinae). *Mitt. Zool. Mus. Berlin* 36:1–48.
Stein, R. C.
1958. The behavioral, ecological and morphological characters of two populations of the alder flycatcher, *Empidonax traillii* (Audubon). *New York State Mus. Bull.* 37:1–63.
Steinbacher, G.
1948. Der Einfluss der Eiszeit auf die europäische Vogelwelt. *Biol. Zentr.* 67:444–456.
Steiner, H.
1952. Vererbungsstudien an Vogelbastarden. III. Die Kreuzung *Amauresthes fringilloides* (Lafr.) × *Spermestes nigriceps* Cass. innerhalb des Formenkreises der Spermestinae; eine art- oder gattungsmässige Kreuzung? *Arch. Julius Klaus-Stiftung* 27:119–137.
1956. Gedanken zur Initialgestaltung der Chordaten. *Rev. Suisse Zool.* 63:330–341.
1958. Artspezifische Merkmalsphänokopien bei australischen Prachtfinken, Spermestidae, insbesondere beim Zebrafinken, *Taeniopygia castanotis* Gould. *Arch. Julius Klaus-Stiftung* 38:62–70.
Steinmann, P.
1952. Polytypie und intraspezifische Evolution bei Süsswassertieren. *Schweiz. Z. Hydrol.* 14:313–332.
Stephens, S. G.
1950. The genetics of 'Corky.' II. Further studies on its genetic basis in relation to the general problem of interspecific isolating mechanisms *J. Genet.* 50:9–20.
Stephens, S. G., M. M. Green, E. B. Lewis, J. R. Laughnan, and C. Stormont
1955. Pseudoallelism and the theory of the gene. *Amer. Nat.* 89:65–122.
Stern, C.
1936. Interspecific sterility. *Amer. Nat.* 70:123–142.
1958. Selection for subthreshold differences and the origin of pseudo-exogenous adaptations. *Amer. Nat.* 92:313–316.
1959. Variation and hereditary transmission. *Proc. Amer. Phil. Soc.* 103:183–189.
1960. *Principles of human genetics* (2nd ed.; Freeman, San Francisco and London).

Stern, C., and E. W. Schaeffer
1943.   On wild-type iso-alleles in *Drosophila melanogaster. Proc. Nat. Acad. Sci.* 29:361–367.

Steven, D. M.
1953.   Recent evolution in the genus *Clethrionomys. Symp. Soc. Exptl. Biol.,* No. 7:310–319.

Stickel, L. F.
1950.   Populations and home range relationships of the box turtle, *Terrapene c. carolina* (Linnaeus). *Ecol. Monographs* 20:351–378.

Stone, F. L.
1947.   Notes on two darters of the genus *Boleosoma. Copeia,* No. 1:92–96.

Stone, W. S.
1942.   Heterosis in *Drosophila hydei. Univ. Texas Publ.,* No. 4228:16–22.
1955.   Genetic and chromosomal variability in *Drosophila. Cold Spring Harbor Symp. Quant. Biol.* 20:256–270.
1962.   The dominance of natural selection and the reality of superspecies (species groups) in the evolution of *Drosophila. Univ. Texas Publ.,* No. 6205:507–537.

Stone, W. S., W. C. Guest, and F. D. Wilson
1960.   The evolutionary implications of the cytological polymorphism and phylogeny of the *virilis* group of *Drosophila. Proc. Nat. Acad. Sci.* 46:350–361.

Stone, W. S., and F. D. Wilson
1958.   Genetic studies of irradiated natural populations of *Drosophila.* II. 1957 tests. *Proc. Nat. Acad. Sci.* 44:505–575.

Storey, H. H.
1932.   The inheritance by an insect vector of the ability to transmit a plant virus. *Proc. Roy. Soc., London* (B) 112:46–60.

Storr, G. M.
1958.   Are marsupials "second-class" mammals? *Western Austral. Nat.* 6:179–183.

Strenzke, K.
1959.   Revision der Gattung *Chironomus* Meig. *Arch. Hydrobiol.* 56:1–42.
1960.   Die systematische und ökologische Differenzierung der Gattung *Chironomus. Ann. Entomol. Fenn.* 26:111–138.

Stresemann, E.
1919.   Über die europäischen Baumläufer. *Verh. Ornithol. Ges. Bayern* 14:39–74.
1926.   Übersicht über die "Mutationsstudien" und ihre wichtigsten Ergebnisse. *J. Ornithol.* 74:377–385.
1927–1934.   *Aves* (Walter de Gruyter, Berlin and Leipzig).
1936.   Zur Frage der Artbildung in der Gattung *Geospiza. Orgaan Club Ned. Vogelk.* 9:13–21.
1939.   Die Vögel von Celebes. *J. Ornithol.* 88:1–135, 389–487.
1943.   Oekologische Sippen-, Rassen- und Artunterschiede bei Vögeln. *J. Ornithol.* 91:305–328.

768 — BIBLIOGRAPHY

1948. Nachtigall und Sprosser: ihre Verbreitung und Ökologie. *Ornithol. Ber.* 1:193–222.

1951. *Die Entwicklung der Ornithologie* (Peters, Berlin).

Stresemann, E., and N. W. Timoféeff-Ressovsky
1947. Artentstehung in geographischen Formenkreisen. I. Der Formenkreis *Larus argentatus-cachinnans-fuscus. Biol. Zentr.* 66:57–76.

Stubbe, H., and K. Pirschle
1940. Über einen monogen bedingten Fall von Heterosis bei *Antirrhinum majus. Ber. Deut. Bot. Ges.* 58:546–558.

Stumm-Zollinger, E., and E. Goldschmidt
1959. Geographical differentiation of inversion systems in *Drosophila subobscura. Evolution* 13:89–98.

Sturtevant, A. H.
1929. The genetics of *Drosophila simulans. Carnegie Inst. Wash. Publ.*, No. 399:1–62.

1938. Essays on evolution. III. On the origin of interspecific sterility. *Quart. Rev. Biol.* 13:333–335.

1942. The classification of the genus *Drosophila*, with descriptions of nine new species. *Univ. Texas Publ.*, No. 4213:5–51.

Suchetet, A.
1897. *Des hybrides à l'état sauvage* (Baillière et Fils, Paris), vol. 1, Classe des oiseaux.

Sumner, F. B.
1909. Some effects of external conditions upon the white mouse. *J. Exptl. Zool.* 7:97–155.

1932. Genetic, distributional and evolutionary studies of the subspecies of deer-mice (*Peromyscus*). *Biblio. Genet.* 9:1–106.

1934. Does "protective coloration" protect? Results of some experiments with fishes and birds. *Proc. Nat. Acad. Sci.* 20:559–564.

1935. Studies of protective color change. III. Experiments with fishes both as predators and prey. *Proc. Nat. Acad. Sci.* 21:345–353.

Suomalainen, E.
1941. Vererbungsstudien an der Schmetterlingsart *Leucodonta bicoloria. Hereditas* 27:313–318.

1950. Parthenogenesis in animals. *Adv. Genet.* 3:193–253.

1958. On polyploidy in animals. *Proc. Finnish Acad. Sci. Letters 1958:* 1–15.

1961. On morphological differences and evolution of different polyploid parthenogenetic weevil populations. *Hereditas* 47:309–341.

Sutton, G. M.
1938. Oddly plumaged orioles from western Oklahoma. *Auk* 55:1–6.

Svärdson, G.
1945. Chromosome studies on Salmonidae. *Medd. St. undersökn.- o. försöksanst. f. sötvattensfisket* Nr. 23:1–151.

1949a. The Coregonid problem. I. Some general aspects of the problem. *Inst. Freshwater Research, Drottningholm,* No. 29:89–101.

1949b. Competition and habitat selection in birds. *Oikos* 1:157–174.

1950. The Coregonid problem. II. Morphology of two Coregonid species in different environments. *Inst. Freshwater Research, Drottningholm*, No. 31:151–162.

1952. The Coregonid problem. IV. The significance of scales and gill-rakers. *Inst. Freshwater Research, Drottningholm*, No. 33:201 232.

1957. The Coregonid problem. VI. The Palearctic species and their intergrades. *Inst. Freshwater Research, Drottningholm*, No. 38:267–356.

1961. Young sibling fish species in northwestern Europe. In. W. F. Blair, ed., *Vertebrate speciation* (University of Texas Press, Austin), 498–513.

Swan, E. F.
1953. The Strongylocentrotidae (Echinoidea) of the northeast Pacific. *Evolution* 7:269–273.

Sweadner, W. R.
1937. Hybridisation and the phylogeny of the genus *Platysamia*. *Ann. Carnegie Mus.* 25:163–242.

Sylvester-Bradley, P. D.
1951. The subspecies in palaeontology. *Geol. Mag.* 88:88 102.

1956. ed., *The species concept in palaeontology* (Syst. Assoc. Publ. No. 2), 1–145.

Syme, P. D., and D. M. Davies
1958. Three new Ontario black flies of the genus *Prosimulium* (Diptera: Simuliidae). I. Descriptions, morphological comparisons with related species, and distribution. *Canad. Entomol.* 90:607–710.

Tadini, G. V.
1958. Il probabile significato biologico della monogenia. *Rend. Accad. Naz. Lincei* 24:562–566.

Talbot, M.
1934. Distribution of ant species in the Chicago region, with reference to ecological factors and physiological toleration. *Ecology* 15:416–439.

Tan, C. C.
1946. Genetics of sexual isolation between *Drosophila pseudoobscura* and *D. persimilis*. *Genetics* 31:558–573.

Taning, Å. V.
1952. Experimental study of meristic characters in fishes. *Biol. Rev.* 27:169–193.

Tantaway, A. O., and E. C. R. Reeve
1956. Studies on quantitative inheritance. IX. The effects of inbreeding at different rates in *Drosophila melanogaster*. *Z. ind. Abst.- u. Vererb.* 87:648–667.

Tax, S. (ed.)
1960a. *Evolution after Darwin* (University of Chicago Press, Chicago), vol. 1, The evolution of life.

1960b. *Evolution after Darwin* (University of Chicago Press, Chicago) vol. 2, The evolution of man.

Taylor, W. P.
1934. Significance of extreme or intermittent conditions in distribution of species and management of natural resources, with a restatement of Liebig's law of the minimum. *Ecology* 15:374–379.

Teissier, G.
1954. Conditions d'équilibre d'un couple d'allèles et supériorité des hétérozygotes. *Compt. Rend.* 238:621–623.
1958. Distinction biométrique des *Drosophila melanogaster* françaises et japonaises. *Ann. Génét.* 1:2–10.

Terman, L. M., and M. H. Oden
1959. *The gifted group at mid-life* (Stanford University Press, Stanford).

Terzaghi, E., and D. Knapp
1960. Pattern of chromosome variability in *Drosophila pseudoobscura*. *Evolution* 14:347–350.

Test, A. R.
1945. Ecology of California *Acmaea*. *Ecology* 26:395–405.
1946. Speciation in limpets of the genus *Acmaea*. *Contrib. Lab. Vert. Biol., Univ. Mich.*, No. 31:1–24.

Tetley, J.
1947. Increased variability accompanying an increase in population in a colony of *Argynnis selene* (Lep. Nymphalidae). *Entomologist, London* 80:177–179.

Thoday, J. M.
1951. Evolutionary trends and classification. *Proc. Leeds Phil. Soc.* 6:1–4.
1953. Components of fitness. *Symp. Soc. Exptl. Biol.*, No. 7:96–113.
1955. Balance, heterozygosity and developmental stability. *Cold Spring Harbor Symp. Quant. Biol.* 20:318–326.

Thoday, J. M., and T. B. Boam
1959. Effects of disruptive selection. II. Polymorphism and divergence without isolation. *Heredity* 13:205–218.

Thoday, J. M., and J. B. Gibson
1962. Isolation by disruptive selection. *Nature* 193:1164–1166.

Thomas, H. T.
1950. Field notes on the mating habits of *Sarcophaga* Meigen (Diptera). *Proc. Roy. Entomol. Soc. London* (A) 25:93–98.

Thompson, E. Y., J. Bell, and K. Pearson
1911. A third cooperative study of *Vespa vulgaris*. Comparison of queens of a single nest with queens of the general autumn population. *Biometrika* 8:1–12.

Thompson, M. L.
1948. Studies of American fusulinids. *Paleontol. Contrib. Univ. Kansas*, No. 4 (art. 1):1–184.

Thornton, W. A.
1955. Interspecific hybridization in *Bufo woodhousei* and *Bufo valliceps*. *Evolution* 9:455–468.

Thorpe, W. H.
1928. Biological races in *Hyponomeuta padella* L. *J. Linn. Soc., Zool.* 36:621–634.
1930. Biological races in insects and allied groups. *Biol. Rev.* 5:177–212.
1939. Further studies on pre-imaginal olfactory conditioning in insects. *Proc. Roy. Soc., London* (B) 127:424–433.
1940. Ecology and the future of systematics. In J. Huxley, ed., *The new systematics* (Clarendon Press, Oxford), 341–364.
1945. The evolutionary significance of habitat selection. *J. Animal Ecol.* 14:67–70.
1956. *Learning and instinct in animals* (Methuen, London).

Thorson, G.
1936. The larval development, growth and metabolism of Arctic marine bottom invertebrates, etc. *Medd. Grønland* 100:1–155.
1950. Reproductive and larval ecology of marine bottom invertebrates. *Biol. Rev.* 25:1–45.

Thorsteinson, A. J.
1960. Host selection in phytophagous insects. *Ann. Rev. Entomol.* 5:193–218.

Timoféeff-Ressovsky, N. W.
1935. Über geographische Temperaturrassen bei *Drosophila funebris* F. *Arch. Naturg.* (N. F.) 4:245–357.
1940a. Mutations and geographical variation. In. J. Huxley, ed., *The new systematics* (Clarendon Press, Oxford), 73–136.
1940b. Zur Analyse des Polymorphismus bei *Adalia bipunctata*. *Biol. Zentr.* 60:130–137.
1943. [Erbliche und oekologische Isolation.] *J. Ornithol.* 91:326–327.

Tinbergen, L.
1939. Zur Fortpflanzungsethologie von *Sepia officinalis* L. *Arch. Néerl. Zool.* 3:323–364.

Tinbergen, N.
1948. Social releasers and the experimental method required for their study. *Wilson Bull.* 60:6–61.
1951. *The study of instinct* (Clarendon Press, Oxford).
1954. The origin and evolution of courtship and threat display. In A. C. Hardy, J. S. Huxley, and E. B. Ford, eds., *Evolution as a process* (Allen and Unwin, London), 1–71.
1960. Behaviour, systematics, and natural selection. In S. Tax, ed., *The evolution of life* (University of Chicago Press, Chicago), 595–613.

Tinbergen, N., and A. C. Perdeck
1950. On the stimulus situation releasing the begging response in the newly hatched herring gull chick (*Larus argentatus argentatus* Pont.). *Behaviour* 3:1–39.

Tomilin, A. G.
1946. Thermoregulation and the geographical races of cetaceans. *Compt. Rend.* (*Doklady*) *Acad. Sci. U.R.S.S.* 54:465–468.

Tonolli, V.
1949. Isolation and stability in populations of high altitude diaptomids. *Ricerca Sci., Suppl.* 19:123–127.

Tortonese, E.
1948. Variazioni fenotipiche e biologia della popolazione di *Astropecten aranciacus* (Echinodermi) del Golfo di Napoli, con riferimenti a specie congeneri. *Boll. Ist. Mus. Zool. Univ. Torino* 1:87–123.
1950. Differenziazione geografica ed ecologica negli Asteroidi. *Boll. Zool., Suppl.* 17:339–354.

Townsend, J. I., Jr.
1952. Genetics of marginal populations of *Drosophila willistoni*. *Evolution* 6:428–442.
1954. Cryptic subspeciation in Drosophila belonging to the subgenus Sophophora. *Amer. Nat.* 88:339–351.

Toxopeus, L. J.
1930. *De soort als functie van plaats en tijd* (H. J. Paris, Amsterdam).

Traylor, M. A.
1950. Altitudinal variation in Bolivian birds. *Condor* 52:123–126.

Tretzel, E.
1955. Intragenerische Isolation und interspezifische Konkurrenz bei Spinnen. *Z. Morphol. u. Ökol. Tiere* 44:43–162.

Trevor, J. C.
1953. *Race crossing in man: the analysis of metrical characters* (Cambridge University Press, London; Eugenics Lab. Mem., No. 36).

Trewavas, E.
1947. Speciation in cichlid fishes of east African lakes. *Nature* 160:96–97.

Turesson, G.
1922. The genotypic response of the plant species to the habitat. *Hereditas* 3:211–350.
1936. Rassenökologie und Pflanzengeographie. *Bot. Notiser* häfte 3–4:420–437.

Turrill, W. B., O. Richards, C. D. Darlington, T. M. Harris, J. S. Huxley, E. B. Ford, K. Mather, J. R. Norman, W. L. C. Lawrence, A. J. Wilmott, and J. Ramsbottom
1942. Differences in the systematics of plants and animals and their dependence on differences in structure, function and behaviour in the two groups. *Proc. Linn. Soc. London* 153:272–287.

Twitty, V. C.
1959. Migration and speciation in newts. *Science* 130:1735–1743.
1961. Experiments on homing behavior and speciation in *Taricha*. In W. F. Blair, ed., *Vertebrate speciation* (University of Texas Press, Austin), 415–459.

Udvardy, M. F. D.
1959. Notes on the ecological concepts of habitat, biotope and niche. *Ecology* 40:725–728.

Underwood, G.
1954. Categories of adaptation. *Evolution* 8:365–377.
Ursin, E.
1952. Occurrence of voles, mice, and rats (Muridae) in Denmark. *Vidensk. Medd. Dansk naturh. Foren.* 114:217–244.
1956. Distribution and growth of the queen, *Chlamys opercularis* (Lamellibranchiata), in Danish and Faroese waters. *Med. Danmarks Fiskeri- og Havundersøgelser.* (n. s.) 1:1–32.
Utida, S.
1957. Population fluctuation, an experimental and theoretical approach. *Cold Spring Harbor Symp. Quant. Biol.* 22:139–151.
Vachon, M.
1958. Scorpionidea (Chelicerata) de l'Afghanistan. *Vidensk. Medd. Dansk naturh. Foren.* 120:121–187.
Vallée, L.
1959. Recherches sur *Triturus blasii* de l'Isle, hybride naturel de *Triturus cristatus* Laur. × *Triturus marmoratus* Latr. *Mem. Soc. Zool. France* 31:1–95.
Vandel, A.
1940. La parthénogenèse géographique. IV. Polyploidie et distribution géographique. *Bull. Biol. France* 74:94–100.
1947. Recherches sur la génétique et la sexualité des isopodes terrestres. X. Etude des garnitures chromosomiques de quelques espèces d'isopodes marins, dulçaquicoles et terrestres. *Bull. Biol. France* 81:154–176.
1951. Le genre *Porcellio* (Crustacés; Isopodes: Oniscoidea) évolution et systématique. *Mem. Mus. Hist. Nat., Paris* (n. s., A) 3:81–192.
1953. L'évolution considérée comme phénomène de développement. Les variations de *Phymatoniscus tuberculatus* Racovitza (Crustacé; Isopode terrestre). *Bull. Biol. France* 87:414–430.
Vanderplank, F. L.
1948. Experiments in cross-breeding tsetse-flies (Glossina species). *Ann. Trop. Med. Parasit., Liverpool* 42:131–152.
Vasseur, E.
1952. Geographic variation in the Norwegian sea-urchins, *Strongylocentrotus droebachiensis* and *S. pallidus. Evolution* 6:87–100.
Vaughan, E.
1947. Time of appearance of pink salmon runs in southeastern Alaska. *Copeia*, No. 1:40–50.
Vaurie, C.
1949. A revision of the bird family Dicruridae. *Bull. Amer. Mus. Nat. Hist.* 93:199–342.
1950. Notes on some Asiatic nuthatches and creepers. *Amer. Mus. Novitates*, No. 1472:1–39.
1951a. A study of Asiatic larks. *Bull. Amer. Mus. Nat. Hist.* 97:431–526.

1951b. Adaptive differences between two sympatric species of nuthatches (*Sitta*). *Proc. Xth Intern. Ornithol. Congr., Uppsala, 1950,* 163–166.
1952. A review of the bird genus *Rhinomyias* (Muscicapini). *Amer. Mus. Novitates,* No. 1570:1–36.
1955. Pseudo-subspecies. *Acta XI Congr. Intern. Ornithol., Basel, 1954,* 369–380.
1957. Systematic notes on Palearctic birds. No. 26. Paridae: the *Parus caeruleus* complex. *Amer. Mus. Novitates,* No. 1833:1–15.
1959. *The birds of the Palearctic fauna* (Witherby, London).

Vavilov, N. I.
1926. Studies on the origin of cultivated plants. *Bull. Appl. Bot. Plant. Breed., Leningrad* 16:1–248.
1951. The origin, variation, immunity and breeding of cultivated plants. *Chron. Bot.* 13:1–364.

Vecht, J. van der
1953. The carpenter bees (*Xylocopa* Latr.) of Celebes, with notes on some other Indonesian *Xylocopa* species. *Idea* 9:57–69.
1959. On *Eumenes arcuatus* (Fabricius) and some allied Indo-Australian wasps (Hymenoptera, Vespidae). *Zool. Verh., Leyden,* No. 41:1–71.

Vernon, H. N.
1897. Reproductive divergence: an additional factor in evolution. *Nat. Sci.* 11:181–189.

Verwey, J.
1949. Habitat selection in marine animals. *Folia Biotheoretica,* No. IV:1–22.

Vetukhiv, M.
1954. Integration of the genotype in local populations of three species of *Drosophila. Evolution* 8:241–251.
1956. Fecundity of hybrids between geographic populations of *Drosophila pseudoobscura. Evolution* 10:139–146.

Villee, C. A., and T. C. Groody
1940. The behavior of limpets with reference to their homing instinct. *Amer. Midl. Nat.* 24:190–204.

Villwock, W.
1958. Weitere genetische Untersuchungen zur Frage der Verwandtschafts-beziehungen anatolischer Zahnkarpfen. *Mitt. Hamburg Zool. Mus. Inst.* 56:81–152.

Vogt, O.
1947. Ethnos, ein neuer Begriff der Populations-Taxonomie. *Naturw.* 34:45–52.

Voipio, P.
1950. Evolution at the population level with special reference to game animals and practical game management. *Papers Game Research, Helsinki* 5:1–176.
1952a. Subspecific boundaries and genodynamics of populations in mammals and birds. *Ann. Zool. Soc. Vanamo* 15:1–32.

1952b. Ökogenetische Differenzierung und Populationsdynamik. *Ornis Fenn.* 29:1–26.

1953. The *hepaticus* variety and the juvenile plumage types of the cuckoo. *Ornis Fenn.* 30:97–117.

1954. Über die gelbfüssigen Silbermöwen Nordwesteuropas. *Acta Soc. Fauna Flora Fenn.* 71:1–56.

1957. Über die Polymorphie von *Sciurus vulgaris* L. in Finnland. *Ann. Zool. Soc. Vanamo* 18:1–24.

Volpe, E. P.
1952. Physiological evidence for natural hybridization of *Bufo americanus* and *Bufo fowleri*. *Evolution* 6:393–406.

1953. Embryonic temperature adaptations and relationships in toads. *Physiol. Zool.* 26:344–354.

1955. A taxo-genetic analysis of the status of *Rana kandiyohi* Weed. *Systematic Zool.* 4:75–82.

1956. Experimental F_1 hybrids between *Bufo valliceps* and *Bufo fowleri*. *Tulane Studies Zool.* 4:61–75.

1957. Embryonic temperature adaptations in highland *Rana pipiens*. *Amer. Nat.* 91:303–309.

1959. Experimental and natural hybridization between *Bufo terrestris* and *Bufo fowleri*. *Amer. Midl. Nat.* 61:295–312.

1960. Evolutionary consequences of hybrid sterility and vigor in toads. *Evolution* 14:181–193.

1961. Polymorphism in anuran populations. In W. F. Blair, ed., *Vertebrate speciation* (University of Texas Press, Austin), 221–234.

Voous, K. H.
1950. On the evolutionary and distributional history of *Malacopteron*. *Sarawak Mus. J.* 5:300–320.

1951. Distributional and evolutionary history of the kingfisher genus *Ceyx* in Malaysia. *Ardea* 39:182–196.

1955. Origin of the avifauna of Aruba, Curaçao, and Bonaire. *Acta XI Congr. Intern. Ornithol., Basel, 1954,* 410–414.

1959. Geographical variation of the herring-gull, *Larus argentatus*, in Europe and North America. *Ardea* 47:176–187.

Waagen, W.
1869. Die Formenreihe des Ammonites subradiatus. *Beneckes Geognost.-paläontol. Beitr.* 2:179–256.

Waddington, C. H.
1942. Canalization of development and the inheritance of acquired characters. *Nature* 150:563–565.

1948. The genetic control of development. *Symp. Soc. Exptl. Biol.*, No. 2:145–154.

1953a. Genetic assimilation of an acquired character. *Evolution* 7:118–126.

1953b. The "Baldwin effect," "genetic assimilation" and "homeostasis." *Evolution* 7:386–387.

1956a. *Principles of embryology* (Macmillan, New York).

1956b. Genetic assimilation of the *bithorax* phenotype. *Evolution* 10:1–13.
1957. *The strategy of the genes* (Allen and Unwin, London).
1960a. Evolutionary adaptation. In S. Tax, ed., *The evolution of life* (University of Chicago Press, Chicago), 381–402.
1960b. *The ethical animal* (Allen and Unwin, London).
1960c. Experiments on canalizing selection. *Genet. Res. Cambridge* 1:140–150.

Waddington, C. H., B. Woolf, and M. M. Perry
1954. Environment selection by *Drosophila* mutants. *Evolution* 8:89–96.

Wagler, E.
1951. Die Felchen (Coregonen) des Laacher Sees. *Zool. Anz.* 147:180–187.

Wagner, M.
1868. *Die Darwin'sche Theorie und das Migrationsgesetz der Organismen* (Duncker and Humblot, Leipzig).
1889. *Die Entstehung der Arten durch räumliche Sonderung* (Benno Schwalbe, Basel).

Wagner, R. P.
1944. Nutritional differences in the *mulleri* group. *Univ. Texas Publ.*, No. 4920:39–41.

Wagner, R. P., and H. K. Mitchell
1955. *Genetics and metabolism* (Wiley, New York).

Wahlert, G. v.
1956. Tatsachen und Begriffe zur Stammesgeschichte der Schwanzlurche. *Zool. Anz., Suppl.* 19:274–280.
1957. Weitere Untersuchungen zur Phylogenie der Schwanzlurche. *Zool. Anz., Suppl.* 20:347–352.

Wahrman, J., and A. Zahavi
1955. Cytological contributions to the phylogeny and classification of the rodent genus *Gerbillus. Nature* 175:600–602.

Wallace, A. R.
1889. *Darwinism. An exposition of the theory of natural selection* (Macmillan, London).

Wallace, B.
1948. Studies on "sex-ratio" in *Drosophila pseudoobscura*. I. Selection and "sex-ratio." *Evolution* 2:189–217.
1953. On coadaptation in Drosophila. *Amer. Nat.* 87:343–358.
1954a. Coadaptation and the gene arrangements of *Drosophila pseudoobscura*. In A. A. Buzzati-Traverso, ed., *Symposium on genetics of population structure, Pavia, 1953* (I. U. B. S. Ser. B, No. 15), 67–94.
1954b. Genetic divergence of isolated populations of *Drosophila melanogaster. Atti IX Congr. Intern. Genet.* [*Caryologia, vol. suppl. 1954*], 761–764.
1955. Inter-population hybrids in *Drosophila melanogaster. Evolution* 9:302–316.
1956. Studies on irradiated populations of *Drosophila melanogaster. J. Genet.* 54:280–293.

1958. The role of heterozygosity in Drosophila populations. *Proc. Xth Intern. Congr. Genet.* 1:408–419.

1959. The influence of genetic systems on geographical distribution. *Cold Spring Harbor Symp. Quant. Biol.* 24:193–204.

Wallace, B., and J. C. King

1951. Genetic changes in populations under irradiation. *Amer. Nat.* 85:209–222.

Wallace, B., J. C. King, C. V. Madden, B. Kaufmann, and E. C. McGunnigle

1953. An analysis of variability arising through recombination. *Genetics* 38:272–307.

Wallace, B., and M. Vetukhiv

1955. Adaptive organization of the gene pools of Drosophila populations. *Cold Spring Harbor Symp. Quant. Biol.* 20:303–310.

Wallgren, H.

1954. Energy metabolism of two species of the genus *Emberiza* as correlated with distribution and migration. *Acta. Zool. Fenn.* 84:1–110.

Walters, V.

1955. Fishes of western Arctic America and eastern Arctic Siberia. *Bull. Amer. Mus. Nat. Hist.* 106:261–368.

Warburton, F. E.

1956. Genetic assimilation: adaptation versus adaptability. *Evolution* 10:337–339.

Washburn, S. L.

1950. The analysis of primate evolution with particular reference to the origin of man. *Cold Spring Harbor Symp. Quant. Biol.* 15:07–78.

Washburn, S. L., and V. Avis

1958. Evolution of human behavior. In A. Roe and G. G. Simpson, eds., *Behavior and evolution* (Yale University Press, New Haven), 421–436.

Washburn, S. L., and F. C. Howell

1960. Human evolution and culture. In S. Tax, ed., *The evolution of man* (University of Chicago Press, Chicago), 33–56.

Wasserman, A. O.

1957. Factors affecting interbreeding in sympatric species of spadefoots (genus *Scaphiopus*). *Evolution* 2:320–338.

Wasserman, M.

1960. Cytological and phylogenetic relationships in the *repleta* group of the genus *Drosophila*. *Proc. Nat. Acad. Sci.* 46:842–859.

1962. Cytological studies of the *repleta* group of the genus *Drosophila*: III–VI. *Univ. Texas Publ.*, No. 6205:63–134.

Watson, G. S.

1960. The cytoplasmic "sex-ratio" condition in *Drosophila*. *Evolution* 14:256–265.

Watson, J. D., and F. H. C. Crick

1953. The structure of DNA. *Cold Spring Harbor Symp. Quant. Biol.* 18:123–131.

Webb, G. R.
   1947.   The mating-anatomy technique as applied to polygyrid landsnails. *Amer. Nat.* 81:134–147.

Weidenreich, F.
   1941.   The brain and its role in the phylogenetic transformation of the human skull. *Trans. Amer. Phil. Soc.* (n. s.) 31:321–442.
   1945.   Giant early man from Java and South China. *Anthropol. Papers Amer. Mus. Nat. Hist.* 40:5–134.
   1946.   *Apes, giants and man* (University of Chicago Press, Chicago).

Weismann, A.
   1872.   *Ueber den Einfluss der Isolirung auf die Artbildung* (Wilhelm En'gelmann, Leipzig).
   1902.   *Vorträge über Descendenztheorie* (Gustav Fischer, Jena).

Welch, D'A. A.
   1938.   Distribution and variation of *Achatinella mustelina* Mighels in the Waianae Mountains, Oahu. *Bishop Mus. Bull., Honolulu* 152:1–164.
   1942.   Distribution and variation of the Hawaiian tree snail *Achatinella apexfulva* Dixon in the Koolau range, Oahu. *Smithsonian Misc. Coll.* 103:1–236.
   1958.   Distribution and variation of the Hawaiian tree snail *Achatinella bulimoides* Swainson on the windward slope of the Koolau range, Oahu. *Proc. Acad. Nat. Sci. Philadelphia* 110:123–212.

Went, F. W.
   1953.   Gene action in relation to growth and development. I. Phenotypic variability. *Proc. Nat. Acad. Sci.* 39:839–848.

Wenzel, F.
   1955.   Über eine Artentstehung innerhalb der Gattung *Spathidium* (Holotricha, Ciliata). *Arch. Protistenkunde* 100:515–540.

Westoll, T. S.
   1949.   On the evolution of the Dipnoi. In G. L. Jepsen, E. Mayr, and G. G. Simpson, eds., *Genetics, paleontology, and evolution* (Princeton University Press, Princeton), 121–184.

White, F.
   1962.   Geographic variation and speciation in Africa with particular reference to *Diospyros*. In D. Nichols, ed., *Taxonomy and geography* (Syst. Assoc. Publ. No. 4), 71–103.

White, M. J. D.
   1946.   The evidence against polyploidy in sexually reproducing animals. *Amer. Nat.* 80:610–618.
   1954.   *Animal cytology and evolution* (2nd ed.; Cambridge University Press, Cambridge, England).
   1957a. Some general problems of chromosomal evolution and speciation in animals. *Survey Biol. Progress* 3:109–147.
   1957b. Cytogenetics of the grasshopper *Moraba scurra*. I. Meiosis of interracial and interpopulation hybrids. *Austral. J. Zool.* 5:285–304.

1958. Restrictions on recombination in grasshopper populations and species. *Cold Spring Harbor Symp. Quant. Biol.* 23:307–317.

1959. Speciation in animals. *Austral. J. Sci.* 22:32–39.

1961. The role of chromosomal translocations in urodele evolution and speciation in the light of work on grasshoppers. *Amer. Nat.* 95:315–321.

White, M. J. D., and L. J. Chinnick

1957. Cytogenetics of the grasshopper *Moraba scurra.* III. Distribution of the 15- and 17-chromosome races. *Austral. J. Zool.* 5:338–347.

White, M. J. D., and K. H. L. Key

1957. A cytotaxonomic study of the pusilla group of species in the genus *Austroicetes* Uv. (Orthoptera: Acrididae). *Austral. J. Zool.* 55:56–87.

White, M. J. D., and N. H. Nickerson

1951. Structural heterozygosity in a very rare species of grasshopper. *Amer. Nat.* 85:239–246.

Wielinga, D. T.

1958. *Deductive investigations on the nature and origin of relations between biological characters (a biodynamic study on herring).* (Groningen University, Groningen), 1–72.

Wieser, W.

1958. Problems of species formation in the benthic microfauna of the deep sea. In A. A. Buzzati-Traverso, ed., *Perspectives in marine biology* (University of California Press, Berkeley and Los Angeles), 513–518.

Wigan, L. G.

1944. Balance and potence in natural populations. *J. Genet.* 16:150–160.

Wilkes, A.

1947. The effects of selective breeding on the laboratory propagation of insect parasites. *Proc. Roy. Soc., London* (B) 134:227–244.

Williams, C. M., and S. C. Reed

1944. Physiological effects of genes: the flight of *Drosophila* considered in relation to gene mutations. *Amer. Nat.* 78:214–223.

Williams, R. J.

1953. *Free and unequal* (University of Texas Press, Austin).

1956. *Biochemical individuality* (Wiley, New York).

Williamson, K.

1958. Bergmann's rule and obligatory overseas migration. *British Birds* 51:209–232.

Williamson, M. H.

1957. An elementary theory of interspecific competition. *Nature* 180:422–425.

1958. Selection, controlling factors and polymorphism. *Amer. Nat.* 92:329–335.

Willis, J. C.

1940. *The course of evolution by differentiation* . . . (Cambridge University Press, Cambridge, England).

1949. The birth and spread of plants. *Boissiera* 8:1–561.

Wilson, E. O.

1955. A monographic revision of the ant genus *Lasius*. *Bull. Mus. Comp. Zool. Harvard Coll.* 113:1–199.

Wilson, E. O., and W. L. Brown

1953. The subspecies concept and its taxonomic application. *Systematic Zool.* 2:97–111.

Witschi, E.

1930. The geographical distribution of the sex races of the European grass frog (*Rana temporaria*). *J. Exptl. Zool.* 56:149–165.

Wolfenbarger, D. O.

1946. Dispersion of small organisms. Distance dispersion rates of bacteria, spores, seeds, pollen, and insects; incidence rates of diseases and injuries. *Amer. Midl. Nat.* 35:1–152.

Wollaston, T. V.

1856. *On the variation of species* (John Van Voorst, London).

Wollman, E. L., F. Jacob, and W. Hayes

1956. Conjugation and recombination in *Escherichia coli* K–12. *Cold Spring Harbor Symp. Quant. Biol.* 21:141–162.

Womble, W. H.

1951. Differential systematics. *Science* 114:315–322.

Wood, S. L.

1954. A revision of North American Cryphalini (Scolytidae, Coleoptera). *Univ. Kansas Sci. Bull., Lawrence* 36:959–1089.

Worthington, E. B.

1940. Geographical differentiation in fresh waters with special reference to fish. In J. Huxley, ed., *The new systematics* (Clarendon Press, Oxford), 287–302.

Wright, S.

1931a. Evolution in Mendelian populations. *Genetics* 16:97–159.

1931b. Statistical theory of evolution. *Amer. Statistical J.*, March suppl., 201–208.

1940. The statistical consequences of Mendelian heredity in relation to speciation. In J. Huxley, ed., *The new systematics* (Clarendon Press, Oxord), 161–183.

1941. The "age and area" concept extended. *Ecology* 22:345–347.

1943a. Isolation by distance. *Genetics* 28:114–138.

1943b. An analysis of local variability of flower color in *Linanthus Parryae*. *Genetics* 28:139–156.

1949a. Population structure in evolution. *Proc. Amer. Phil. Soc.* 93:471–478.

1949b. Adaptation and selection. In G. L. Jepsen, E. Mayr, and G. G. Simpson, eds., *Genetics, paleontology and evolution* (Princeton University Press, Princeton), 365–389.

1951a. The genetical structure of populations. *Ann. Eugenics* 15:323–354.

1951b. Fisher and Ford on "the Sewall Wright effect." *Amer. Sci.* 39:452–479.

1955. Classification of the factors of evolution. *Cold Spring Harbor Symp. Quant. Biol.* 20:16–24D.
1956. Modes of selection. *Amer. Nat.* 90:5–24.
1960. Physiological genetics, ecology of populations, and natural selection. In S. Tax, ed., *The evolution of life* (University of Chicago Press, Chicago), 429–475.

Wüst, E.
1930. Die Bedeutung der geographischen Rassen für die Geschichte der diluvialen Säugetierfaunen. *Palaeontol. Z.* 12:6–13.

Yamashina, Y.
1948. Notes on the Marianas mallard. *Pacific Sci., Honolulu* 11:121–124.

Young, F. N.
1958. Some facts and theories about the broods and periodicity of the periodical cicadas. *Proc. Indiana Acad. Sci.* 68:164–170.

Zapfe, H.
1958. The skeleton of *Pliopithecus* (*Epipliopithecus*) *vindobonensis* Zapfe and Hürzeler. *Amer. J. Phys. Anthropol.* (n. s.) 16:441–445.

Zeuner, F. E.
1946. *Dating the past, an introduction to geochronology* (Methuen, London).

Zimmerman, E. C.
1948. *Insects of Hawaii* (University of Hawaii Press, Honolulu), vol. 1.
1960. Possible evidence of rapid evolution in Hawaiian moths. *Evolution* 14:137–138.

Zimmermann, K.
1950. Die Randformen der mitteleuropäischen Wühlmäuse, *Syllegomena Biol., Festschr. Kleinschmidt, Wittenberg 1950*, 454–471.
1952. Die simplex-Zahnform der Feldmaus, *Microtus arvalis* Pallas. *Verh. Deut. Zool. Ges., Freiburg, 1952*, 492–498.
1961. "Proteus," a new colour gene in bank voles *Clethrionomys* Tilesius (Mammalia: Rodentia). *Bull. Research Council Israel, Sect. B Zool.* 10B:7–11.

Zimmermann, W.
1938. *Vererbung "erworbener Eigenschaften" und Auslese* (Gustav Fischer, Jena).

Zippelius, H.–M.
1949. Die Paarungsbiologie einiger Orthopteren-Arten. *Z. Tierpsychol.* 6:372–390.

Zirkle, C.
1941. Natural selection before the "origin of species." *Proc. Amer. Phil. Soc.* 84:71–123.
1959. Species before Darwin. *Proc. Amer. Phil. Soc.* 103:636–644.

Zuckerkandl, E., R. T. Jones, and L. Pauling
1960. A comparison of animal hemoglobins by tryptic peptide pattern analysis. *Proc. Nat. Acad. Sci.* 46:1349–1360.

Index

Acarapis, 582; *woodi*, 47
accident, 203
Accipiter, 152
Acentropus, 248
Achatinella, 204, 301, 310, 397
Acheta, 45; see also *Gryllus*
Acmaea, 72, 452
Acris, 52
Acrocephalus, 494
Adalia, 192, 239
adaptation, 60; by mutation, 181
adaptive peak, 288
adaptive polymorphism, 251
adaptive radiation, 574, 617, 619
adaptive significance, lack of, 310
adaptive zone, 590, 620
Adriaanse, A., 51–52
Aedes, 41
Agassiz, Louis, 4
age variation, 139
aggression, 87
Agrilus, 55
Aird, I., 161
Alexander, R. D., 44, 45, 99, 476
Allee, W. C., 60
Allen's rule, 323
Allison, A. C., 227
allometry, 324
allopatry, 345
allopolyploidy, 112, 440
altitudinal variation, 320–326
altruistic traits, 199
Amadon, D., 83, 248, 507, 574, 590, 596
Ambystoma, 248, 316
Ammodytes, 38–39
ammonites, 576
Ammophila, 51–52
Amphiuma, 115
Anas, 90, 102, 430; *platyrhynchos*, 32
Anatini, 405
Anatolichthys, 124, 430
Anderson, E., 111, 117, 130
Andrewartha, H. G., 67, 70, 80
animals and plants, 129; differences, 418
Anodonta, 32
Anolis, 64

Anopheles, 35, 41, 54, 57, 73, 78, 492, fertility of, 37; *quadrimaculatus*, 233, 257
Anser, 242, 417, 468
anthropoids, 625–626
ants, 46, 50, 64, 80, 83, 87
Aphelocoma, 73
aphids, 56, 434
Apicotermes, 308
Apis, 494
Araschnocampa, 605
Archaeopteryx, 596–597
archetypes, 599
archipelago speciation, 506, 546
arctic adaptation, 62
Arion, 144
Aristelliger, 283
Arkell, W. J., 609
army ants, 65
Arndt, W., 595
arrhenotoky, 409
Artemia, 404
Artenkreis, 499
artiodactyls, 596, 598
Ascaris, 446
Asellus, 197, 247, 421, 501
asexual organisms, 49, 56; speciation, 433
asexuality, 27, 181, 410–411; see also reproduction
Astrapia, 372, 374, 499
Atoda, K., 567
Atz, J. W., 598
Australopithecines, 628, 639
Australopithecus, 598, 629
Austroicetes, 55, 125, 313
autogamy, 65, 407
automixis, 410
autopolyploidy, 56, 440
Avery, O. T., 171

Bacci, G., 248, 316, 407
Bagenal, T. B., 78
Bailey, R. M., 115
balance, internal, 275, 284; relational, 275
balanced lethals, 233
balanced load, 254

Baldi, E., 144
Baldwin effect, 190, 610–612
Barber, H. S., 51, 98
Barbus binotatus, 465
Barigozzi, C., 213
barnacles, 67
Barnicot, N. A., 323
Barr, T. C., Jr., 562
barriers, climatic, 563; distributional, 385; ecological, 562; ethological, 95; in fresh water, 562; kinds of, 557–565; in the oceans, 563; seasonal, 94
Bartholomew, G. A., Jr., 633
Basrur, V. R., 444
Bastock, M., 102, 108
Bateman, A. J., 177, 566
Bates, H. W., 484
Bates, M., 41
Bateson, W., 330, 432, 516
Batrachoseps, 123
bats, 437, 617
Battaglia, B., 229, 493
Baur, G., 590
bdelloid rotifers, 29, 411, 433
Beardmore, J. A., 229, 244, 251, 290
Beaudry, J. R., 14
Becher, E., 196
Beermann, W., 276
bees, 248, 307, 452, 462, 494
beetles, 44
behavior, 415; evolution, 636; genetics, 108; geographical variation, 494; role of evolutionary, 604
Beheim, D., 104
Benazzi, M., 443, 500
Benson, R. B., 410
Benzer, S., 172
Bergmann's rule, 319–320
Bigelow, R. S., 45, 476
bill length, 324
biochemical inventions, 602
biogenetic law, 606
biological races, 54, 56, 453–460
biological species, value of, 29
biometric studies, 50
biometricians, 148
biosystematy, 401
bipedal locomotion, 633
Birch, L. C., 44, 50, 67, 70, 76, 80, 87, 132, 241
birds, 106, 126, 153, 618; hybrid swarms, 119
birds of paradise, 126
Birdsell, J. B., 205, 633
Biston, 192
bithorax, 278
Blair, W. F., 26, 93, 94, 123, 177, 328, 361, 499
Blanchard, B. D., 65
Blasing, T., 72
blood-group genes, 153, 161, 543
blood groups, 205, 210, 237, 255, 647
Blyth, E., 183

Bock, W. J., 593, 600
Bocquet, C., 47
Bogert, C. M., 495
Boleosoma, 375
Bombina, 94
Bombus, 247, 249, 367
Bombyx, 104
Bonellia viridis, 32
borderline cases, 496–504
Bos taurus, 133
Boschma, H., 48
Bösiger, E., 102, 108, 164, 551
Bovey, P., 331
Boyd, W., 644
brain size, 634; rapid increase, 652
breathing, 605
breeding seasons, 52, 318
breeding system, 404, 412
Briles, W. E., 223
Brinton, E., 563, 583
Brissus, 591
bristles, number, 285; patterns, 280; scutellar, 281
broods, number, 313; unisexual, 417
Brooks, J. L., 47, 124, 141, 465, 507
Brower, L. P., 107
Brown, W. C., 38, 73
Brown, W. J., 38, 44, 461
Brown, W. L., 67, 83, 348
Brues, A. M., 210
Buchanan, L. L., 56, 455
budworm, 40, 107
Bufo, 26, 52, 93, 94, 103, 105, 106; hybridization, 115, 123
Bumpus, H. C., 283
butterflies, 162, 248, 313, 326, 362, 503, 569
Buzzati-Traverso, A., 176, 225
Byers, C. F., 93

Cailleux, A., 436
Cain, A. J., 23, 28, 206, 250, 306
Calandra, 76; *oryzae*, 44, 50
Calanus, 47
Callan, H. G., 540
Calligrapha, 44
Callosciurus, 558
Callosobruchus, 79
calluses, 189, 611
Cambarus, 72, 397
Cameron, A. W., 579
Camp, W. H., 402
canalization, 219–221, 281
Canis, 133
Cantrall, I. J., 44
Carabus, 104, 326
Carpenter, G. D. H., 249
Carson, H. L., 294, 392, 525, 535
Caspari, E., 108, 159, 200, 232
caste specialization, 248
Castle, W. E., 266
Castor, 579
categories, 600

Catharus, 17–18
Catostomus, 116
cattle, 531
Cattle Egret, 564
cave beetles, 501
caves, 562
Coi, J. M., 38, 61, 502
Celerio, 90
cenospecies, 20, 405
centers of diversification, 386
centric fusion, 444
Cepaea, 103, 161, 240, 241, 246, 247, 301, 356; neolithic, 244; polymorphism, 208
Cerion, 26, 375, 398, 545
Cervus, 578
Ceuthophilus, 388
Ceyx, 123
chance, 203
change of function, 603
changes in gene frequencies, 241
Chapin, J. P., 26, 119, 561
Chapman, F. M., 560
character, 613–614
character displacement, 83
character divergence, 82
character gradients. *See* clines
characters, conservative, 281; genetic basis, 158; geographically variable, 304; physiological, 60; taxonomic, 592
Charadrius hiaticula, size variation, 322
chemoreception, 100
Chetverikov, S. S., 265, 279
chiasma, 180, 235
Chilocorus, 541
chipmunks, 98
Chironomus, 44, 55, 90, 301
Chlamys, 315
Choristoneura, 40, 94, 107
chromosomal change, phylogeny, 541
chromosomal mutations, 170
chromosome fusion, 154, 444
chromosome number, 234, 404
chromosome types, 376
chromosomes, 55, 308; lethal, 272; polytene, 446; structural changes, 537
chronoclines, 302, 364
Chrysomelidae, 44
Cicada, 474
Cicadulina, 454
cichlids, 465, 569
Cicindela, 396, 418
Cincinnati Warbler, 127
Ciona, 408
circular overlaps, 507
Cissilopha, 73, 82
cistron, 172
Citellus, 74
Clarke, C. A., 249, 499
classification, horizontal, 609
Clausen, R. T., 346
Clay, T., 463, 563
Clethrionomys, 496, 579

Climacteris picumnus, 498
climate, 563
climatic factors, 311
climatic races, 313
climatic rules, 320
clines, 361–366, 380–381
clone, 433, 434, 454
Clupea, 300
clutch size, 194, 325
Cnemidophorus, 70
coadaptation, 272, 295
Cobitis taenia, 38
coccinellids, 104
Codling Moth, 458, 459
Colaptes, 372, 377
Colbert, E. H., 584
Coleophoridae, 40
Colias, 103, 117, 124, 160
collective species, 433
Collocalia, 50
colonization, 491
coloration, variation, 305
Columba, 134, 241
comparative studies, 10
comparative systematics, 394
compatibility, ecological, 547
compensation load, 254
competition, 66, 71, 74, 548; in *Drosophila,* 76; evidence for, 72; evolutionary consequences, 81; experimental, 76; of faunas, 75
competitors, 573
conditioning, 108, 458, 470
constitutional types, 648
continuous ranges, 542
continuous variation, 149
Conus, 71
convergence, 388, 609
Coon, C. S., 644
Cooper, K. W., 77
Cope, E. D., 595
copepods, 79, 144, 444
coral, 566
Cordeiro, A. R., 273
Coregonus, 39, 142, 147, 411, 448
correlated effects, 287, 290, 607
Corvus, 370, 379
cosmopolitan distribution, 567
cost of evolution, 259
Coturnix, 75
courtship, 127; quantitative aspects, 102
cowbirds, 109
Crampton, H. E., 310
Cricetus, 223, 242
Crick, F. H. C., 171
crickets, 44, 52; see also *Gryllus*
Crinia, 38, 50, 52, 461
Criniger, 123
Crocodilus, 502
Cro-Magnon, 642
Crombie, A. C., 67, 76
crossing over, 179, 235
crossveinless flies, 190, 532

Crotalus, 115, 498
Crow, J. F., 225, 254, 255
crustaceans, 601
Ctenophthalmus, 378
cuckoos, 109, 250
Cuénot, L., 33
Culex pipiens, 41, 472
Culicoides, 475
cyclomorphosis, 141
Cyclops, 55
Cynips, 396, 435
cytology, 308
cytoplasmic factors, 42

da Cunha, A. B., 71, 223, 225
Dacus, 76, 132
Dahm, A. G., 72
d'Ancona, U., 147
Daphnia, 47, 76, 124, 141, 147
Darlington, C. D., 234, 235, 402, 448
Darwin, Charles, 2, 6, 11, 12, 14, 66, 68, 75, 83, 165, 183, 201, 290, 309, 414, 424, 429, 449, 483, 489, 551, 603, 608, 651
Darwin's finches, 506, 574
DDT resistance, 159, 191, 193, 290
deficiencies, 172
De Lesse, H., 446
deleterious structures, 190
deme, 137, 177, 301, 358, 421
Dempster, E. R., 112, 216, 228
Dendrocoelum, 442
Dendroica, 70, 76, 117, 304
density-dependent factors, 67, 250, 252, 393, 414, 534
density-dependent mortality, 260
Dermaptera, 447
desert adaptations, 323
desert animals, 186
desert birds, 582
desert coloration, 329
Desneux, J., 50, 308
Dethier, V. G., 571
development, 220, 276
De Vries, H., 169, 298, 332, 432, 517
Dice, L. R., 508
Dickerson, G. E., 217
Dicruridae, 301, 362, 608
Dicrurus, 496, 590; *hottentottus*, 305; *leucophaeus*, 388; *ludwigii*, 23
Dilger, W. C., 17
Dipnoi, 618
Dipodomys, 64, 323
Diptera, 41
disease, 75, 194
disease resistance, 252
dispersal, 177, 398, 416, 477, 565; active, 567; factors facilitating, 566; factors reducing, 568; passive, 566; rate of, 198, 566
dispersal stage, length, 566
disruptive selection, 156
Diver, C., 71, 209, 310

diversity of environment, 238
Dixon, K. L., 70, 87, 379
DNA, 171
Dobzhansky, Th., 19, 71, 86, 89, 112, 131, 147, 156, 161, 174, 175, 215, 239, 272, 273, 291, 529, 548, 609
Döderlein, L., 340
dog, 133
Dohrn, A., 603
domestic animals, 133
dominance, 219, 225, 231
dominant groups, 602
Dorst, J., 506
dosage compensation, 271
Dottrens, E., 39
double colonizations in parasites, 464
double invasions, 504
Dougherty, E. C., 27, 181, 412
Dowdeswell, W. H., 228
dragonflies, 326
Drepanididae, 574
Drosophila, adaptations, 62; *americana*, 108, 430; *bocainensis*, 90; chromosomal polymorphism, 390, 535; competition, 71, 76, 78; cyclical fluctuations, 240; heterozygote superiority, 222, 223, 231, 274; hybridization, 112, 131; isolating mechanisms, 95, 96, 100, 101, 107, 550; lethals, 218, 274; *melanogaster*, courtship, 102; *miranda*, 548; *mulleri*, 78; *pallidipennis*, 493; polymorphism, 222, 245; *pseudoobscura*, 34, 50, 244, 540; *repleta*, 540; *robusta*, 390, 392; *simulans*, 280, 548; speciation, 489; *subobscura*, 392; *tropicalis*, 233; variation, 164, 192, 302, 326; *virilis*, 157; *willistoni*, 86, 331, 390
Dubinin, N. P., 192
Dubois, E., 631
ducks, 90, 97
Dufour, L., 103
Dugesia, 367, 500
Dunkers, 205, 210
Dunn, E. R., 282, 316
Dunn, L. C., 153
duplication, 235
Du Rietz, G. E., 89

East, E. M., 173
echinoids, 563
Eciton, 64–65
ecogeographical rules, 318
ecological differences of populations, 375
ecological factors, speciation, 451
ecological load, 261
ecological mosaicism, 244
ecological races, 355, 455
ecological rules in insects, 326; in reptiles and amphibians, 325
ecological shifts, 382, 537
ecological specialization, 414
ecospecies, 351, 401
ecotype, 351

ecotypic variation, 415
egg, amniote, 598
Ehrendorfer, F., 352
Ehrman, L., 552
eidos, 5, 16, 139
Eiseley, L., 1, 183
Eisentraut, M., 367
Elton, C. S., 71, 74, 208, 530, 564
Emberiza, 61
Emberizidae, 394, 418
Emerson, A. E., 50, 54, 94, 308
Emoia, 38, 73
Empidonax, 52
Ensatina, 511
environment, 7, 310, 319; changes in, 192,
    285; diversity, 77, 244
Ephestia, 160, 232
epigenetic system, 588, 611–613; *see also*
    epigenotype
epigenetics, 277
epigenotype, 6, 528, 542, 544, 608
epistatic interactions, 270, 293, 519
Epling, C., 91, 133, 208, 223, 241
equality, 648
Erebia, 502
Ergene, S., 144, 247
Erinaceus, 118
Erythroneura, 77
espèces jumelles, 33
Esper, E. J. C., 335
Etheostoma, 50, 125
ethics, 651
ethological isolating mechanisms, weak,
    103
ethological isolation, 95, 493
Etkin, W., 650
eugenics, 661
Eutamias, 52, 93, 98
evolution, eruptive, 616; term, 4; theories,
    2; transpecific, 586–621
evolutionary novelties, 546, 602–621
evolutionary potential, 607
evolutionary rates, uneven, 618
evolutionary theories, 1
exclusion, 68–72, 493, 547; absence of,
    77, 80; on islands, 73; on mountains, 73
extinction, 74, 530, 584, 617, 620

Faber, A., 52, 99
Fabergé, A. C., 267
Faegri, K., 352
Falconer, D. S., 266, 287
fertility, 15, 405
fertilization, internal, 125, 598
fetalization, 636
Ficedula, 569
fiddler crabs, 97, 494
fireflies, 51, 98
fish, 38, 94, 97, 128, 143, 145, 259, 407,
    410, 413, 465, 500, 550, 574, 579
Fisher, J., 564, 605
Fisher, R. A., 68, 191, 204, 219, 231, 235,
    258, 481, 525

fishes, hybridization, 124
Fitch, H. S., 71, 343
fitness, 271, 279, 289; loss of, 260
Fleming, C. A., 351
flightlessness, 568
fluctuating environment, 229
food, 71, 78
Ford, C. E., 154
Ford, E. B., 161, 221, 239, 242, 648
Formenkreis, 339
Formica, 46, 62, 64
founder populations, 261, 532, 542
founder principle, 209, 211, 529, 534
Fox, W., 146
foxes, 90
Frank, P. W., 76, 200
fresh-water animals, adaptive variation,
    315
fresh-water lakes, 464
fresh-water organisms, 583
Fringilla, 74, 82
Frizzi, G., 37
frogs, 38, 61, 63, 90, 98, 499
Fryer, G., 79, 465, 466, 569, 571, 574
Fulton, B. B., 44–45
function, 603
fusion of two species, 429
Fusulinidae, 577

Gabritschevsky, E., 249
Galapagos, 83, 506, 574–575
Galbraith, I. C. J., 361, 366
gall insects, 196
game animals, 243, 318
gammarids, 465
Gammarus, 47, 82
gaps in geological record, 436
Gasterosteus, 79, 376
Gause principle, 68
gene, and character, 264; nature, 170;
    physiology, 158, 275; rare, 252; selec-
    tive value, 613
gene action, 264, 275
gene arrangements, 154, 207, 233, 391
gene complexes, coadapted, 294
gene flow, 111, 176, 178, 252, 361, 365,
    387, 393, 496, 507, 512, 520, 521, 542;
    effect of, 178; and genetic cohesion,
    518; and species structure, 523–527
gene mutation, 170–171
gene pool, 137, 264, 278, 289, 296, 527;
    cohesion, 518; storage capacity, 293
gene substitution, 261, 281, 521, 534
genera, monotypic, 619
generations, 141; alternation, 434; and
    recombination, 180; sequence, 416
generic characters, geographic variation,
    590–591
genes, interaction, 263; neutral, 159, 206
genetic assimilation, 190, 612
genetic background, 263
genetic changes between generations, 314

genetic drift, 204–214; definition, 204
genetic homeostasis, 288, 534
genetic inertia, 288
genetic load, 253–261; definition, 256
genetic mortality, 105
genetic revolution, 533, 538, 541, 615
genetic storage capacity, 294
genetic system, 402
genetic variability, 165; factors influencing, 166; loss, 538
genetics, 10; of higher categories, 613; of speciation, 516–555
genotype, 277–279; stable, 304; unity of, 263–296
Gentilli, J., 578
Geoffroyism, 2, 10
geographic isolate, 366, 496; ecology, 545
geographic isolation, effectiveness, 556
geographic speciation, 481–515
geographic variation, 297–333; absence of, 302; adaptive, 311–333; amount, 301; conclusions, 333; generic characters, 590; in behavioral characters, 308; in fresh-water organisms, 300; in genitalia, 495; of balanced polymorphism, 329; of coloration, 317; of marine animals, 300, 316; of neoteny, 316; of song, 494; physiological characters, 312
Geospiza, 83–84
Geospizidae, 506
Gerbillus, 154; pyramidum, 404
Gerhardt, U., 52
Gershenson, S., 223, 242
Geschwisterarten, 33
Ghilarov, M. S., 565
Gila, 116
gill arches, 606
Giller, D. R., 73, 87
Gilliard, E. T., 82, 121, 374
Gilmour, J. S. L., 137, 357
Ginsburg, I., 15
Glaessner, M. F., 601
Glass, B., 205, 210, 270
Gloger, C. L., 309, 338, 482
Gloger's rule, 324
Glossina, 98, 504
Goethe, F., 508–509
Goldschmidt, E., 392, 447
Goldschmidt, R. B., 249, 269, 298, 312, 332, 600, 605
gonochorism, 248, 316, 407, 441
Gordon, M., 112, 489
gorilla, 383
Gosline, W. A., 502
grade, 608, 617
Grant, V., 26, 28, 234
Graphognathus, 455
grasshoppers, 52, 313, 495, 526, 531, 552, 570
Greenwood, P. H., 465
Gregor, J. W., 352
Grinnell, J., 68, 75, 310
Grubea, 100

Grüneberg, H., 159
Gryllotalpa, 447
Gryllus, 45, 52, 99, 476
Gulick, J. T., 204, 301, 310, 485, 560
gulls, 97, 509; see also Larus
Gustafsson, A., 174, 253
gynogenesis, 443

Haartman, L. von, 569
habitat, disturbance, 128; exclusion, 82; geographic variation, 492; isolation, 92; preference, 470; selection, 246, 419, 456, 569–571; utilization, 414; variation, 142
Habrobracon, 100
Hadorn, E., 218
Haeckel, Ernst, 606
Haecker, V., 608
Hairston, N. G., 28, 49, 80
Halcyon chloris, 508; sancta, 584
Haldane, J. B. S., 75, 86, 164, 194, 198, 216, 252, 253, 259, 393, 522, 534, 543, 620, 634, 648
Halisarca, 49, 52
Hall, E. R., 114, 317
Hamilton, T. H., 70, 319, 323
hamster, 447
Hardy-Weinberg law, 167–168
Harland, S. C., 271, 544
Harmonia, 243, 247
Harrison, G. A., 145, 321
Hasebroek, K., 242
Hawaiian Islands, 564, 575
Hayne, D. W., 364
Heberer, G., 436
Hecht, M. K., 196, 283
Hedylepta, 260
Heidelberg Man, 623
Helicella, 64
Hemignathus, 590
hemophilia, 224
Henbest, L. G., 577, 584, 616
herbivores, 80
heritability, 217
hermaphroditism, 248, 406–408
Heslop-Harrison, J., 357
Hesperoleucus, 124
Hessian Fly, 459
heterochromatin, 173
heterogamy, 469
heterosis, 222, 274, 291; causes, 225; double advantage, 228; in interpopulation crosses, 230–231; loss, 229; origin, 230
heterostyly, 201
heterotic systems, interaction, 294
heterozygosity, 279; selected, 221
heterozygotes, 218; biochemical versatility, 229; bridge effect, 478; multiple, 258; phenotypic variance, 229; superiority, 221, 223, 224, 232
heterozygote superiority, 223, 224, 232; see also overdominance
Heuts, M. J., 1–2, 294, 312

Hexagenia, 474
higher categories, 588, 615; origin, 600
Hildreth, Ph. E., 274
Hinde, R. A., 102, 605
Hiraizumi, Y., 237
Hobbs, H. H., Jr., 72, 379
Hoesch, W., 64, 186, 328, 582
Hogben, L., 460
Hölldobler, K., 605
hollow curve, 619
homeostasis, developmental, 220, 230; genetic, 288, 534
homeostatic mechanisms, 61
homeostatic systems, 542
homing, 569
hominids, 624–662; population structure, 651
Homo erectus, 632, 640; *sapiens,* 641–662; as polytypic species, 643
homogamy, 102, 252, 468
homologous genes, 609
homologous morphs, 157
homozygosity, 283, 526, 530
honeybee, 326
Hooijer, D. A., 320
Hooke, R., 168
Hooper, E. T., 178, 560
hopeful monster, 438
Hoplitis, 326; *producta,* 510
host preference, 52, 571
host races, 41, 250, 349, 458
host-selection principle, 459
host specificity, 458
Hovanitz, W., 160, 326
Howell, F. C., 630, 642
Howell, T. R., 374
Hsu, T. C., 96, 301
Hubbell, T. H., 372, 388
Hubbs, C. L., 115, 124, 125, 128, 410, 500, 549, 550, 562, 579
Hubendick, B., 32, 48
human races, adaptiveness, 645
Hutchinson, G. E., 78, 85, 88
Hutterites, 205
Hutton, F. W., 530
Huxley, J. S., 2, 151, 163, 361, 362, 426, 579, 608, 610, 621, 651
hybrid belts, 369, 371; narrow, 378; secondary, 293; stabilized, 524
hybrid inferiority, 106
hybrid populations, artificial, 293; stabilized, 121, 132, 378
hybrid sterility, 106, 111
hybrid swarms, 118, 124
hybrid vigor, 111, 224
hybridization, 26, 110–135, 296, 526; allopatric, 369, 371; in amphibians, 115; causes, 125; definition, 110; in *Drosophila,* 131; evolutionary role, 130; in fishes, 115; genetics, 111; insular, 373; in invertebrates, 124; kinds, 113; in man, 656; occasional, 114; population aspects, 113

hybrids, 280, 614; artificial, 112; *Drosophila,* 112; viability, 274
Hyla, 52, 115
Hylobates, 625
Hylocichla, 18
Hymenoptera, 46
Hynes, H. B. N., 79, 82
hypsodonty, 238

identicism, 649
Imbrie, J., 23
immigration, 521, 530
immigration load, 254
imprinting, 636
inbreeding, 283, 404, 421; close, 408; in man, 655
inbreeding depression, 224, 530
incipient species, 151
incompatibility load, 255
indeterminacy, 214
individual, 137
individual variation, 230
inertia, evolutionary, 614
Ingram, V. M., 171
inheritance, blending, 165; particulate, 167
input, genetic, 520
input load, 254
input variability, 218
insects, 582–583
insemination reaction, 105
instantaneous speciation, 432
insular variation, 385
integration of genotype, 264
intelligence, 650
intensification of function, 603
interactions, allelic, 269; epistatic, 270
interference, aggressive, 87
intergradation, 368–384; primary, 369, 380; secondary, 369, 380
interspecific crosses, 105
introgression, 111, 123, 130, 132, 178, 429; in animals, 116
introgressive hybridization. *See* introgression
invasion of land, 609
invasions, 74, 564; multiple, 504; of new adaptive zones, 594
inversions, 155, 294
Irish Elk, 191
irradiation, 272, 539
Irwin, M. R., 153, 544
isoalleles, 220, 224, 233, 258, 271, 284
isolate, 27, 205, 210; geographical, 366–368, 491
isolated population, 527
isolates, ages, 579; fate, 368; number, 582; peripheral, 368, 496, 512; transformation, 576
isolating mechanisms, 89–109; breakdown, 110–135; classification, 91; in *Drosophila,* 96; ethological, 103; genetics, 107; geographic variation, 493; im-

isolating mechanisms (*cont.*)
  provement by selection, 551; loss, 109; origin, 548; role, 109
isolation, 393, 483, 485, 546; between hosts, 563; geographic, 91, 484; reproductive, 90–91; seasonal, 94
isophenes, 362
isopods, 340, 444
Ives, P. T., 173, 244

Jacob, F., 276
Jacobs, W., 99
Jaera, 47, 55; *albifrons*, 90
Järvik, E., 596
Jennings, H. S., 49
Jepsen, G. L., 435
Johansen, H., 371
Johnson, Charles W., 250
Johnson, D. H., 93
Jordan, D. S., 169
Jordan, K., 103, 338, 378, 472, 486
Jordan's law, 487
Junco, 375, 378

Kalela, O., 523, 563
karyotype, reconstruction, 537
Keast, A., 361, 367, 372, 498, 499, 503, 513, 579, 582
Keleher, J. J., 40, 54
Kendeigh, S. C., 61
Kettlewell, H. B. D., 192, 242
key characters, 596; structure, 605
Kimura, M., 228, 259
king snakes, 152
Kinne, O., 47, 356
Kinsey, A. C., 367, 435
Kirtland's Warbler, 303
Kitzmiller, J. B., 41, 233
Kleinschmidt, O., 338
Kobelt, W., 337
Koelreuter, J. G., 15
Komai, T., 243
Koopman, K. F., 552
Kosmoceras, 467
Kosswig, C., 128, 247, 574, 609, 613
Kramer, G., 38, 194, 208, 311, 324, 326
Kühn, A., 146
Kummer, B., 636
Kurtén, B., 561, 569

Lacerta, 38, 208, 326, 367, 384, 411
Lack, D., 69, 70, 71, 74, 79, 81, 83, 84, 194, 195, 325, 504, 506, 574
Lagopus, 141
Lake Baikal, 464
Lake Lanao, 465
Lake Ochrid, 466
lakes, 563
Lalage, 306
Lamarck, J. B., 297
Lamarckism, 2, 165, 310, 610
Lamotte, M., 161, 209, 301
Lampropeltis, 152
Lapps, 645

larks, 329, 570
Larus, 94, 196, 375, 381; *argentatus*, 508; *fuscus*, 508; *glaucoides*, 509
larvae, 607
Lasius, 46, 80, 83
latitude effect, 323
Latrodectus, 47
lava races, 178, 328
Laven, H., 42, 473
law of the unspecialized, 595
leafhoppers, 46
Leakey, L. S. B., 627, 629, 640
Lemche, H., 317, 598
Leopold, A. S., 291
Lepidoptera, 40, 100, 302, 446
Leptodactylus, 38
Lerner, I. M., 57, 77, 146, 183, 224, 267, 270, 282, 285, 288, 530, 611
lethal chromosomes, 272
lethal genes, 223
lethal homozygotes, 233
Leucichthys, 40, 142
Leucopsar, 302, 413
Lévi, C., 49
Levi, H., 47
Levine, P., 647
Levitan, M., 277, 294
Lewis, E. B., 277, 278
Lewontin, R. C., 35, 62, 220, 229, 235, 237, 253, 258, 278, 288, 294
L'Héritier, Ph., 68, 77
Liguus, 151
Limulus, 304, 577
Linanthus, 208
Lineus, 48
Lingula, 577, 614
linkage, 235, 236, 278
Linnaeus, C., 4, 13, 334, 346
Linsley, E. G., 22, 462
Liolaemus, 325
litter size, 325
Littlejohn, M. J., 499
Littorina, 245
lizards, 194, 208, 311, 367, 557; see also *Lacerta*
Lloyd, B. E., 204
load, genetic, 253–261
local population, 136
loci, number, 174
Lophophaps, 178, 366
Lorenz, K., 604
Lorković, Z., 26, 50, 118, 446, 501, 569
louse, 464; see also *Pediculus*
Loxia, 324
Loxops, 83
Ludwig, W., 185, 573
Ludwig effect, 245, 250, 258, 391, 454
Lumbricidae, 442
Lundman, B., 644–645
lungs, 603
Luscinia, 38
Lygosoma, 74
Lymantria, 298, 312, 495

Lymnaea, 32, 48, 385
lyrebird, 98
Lysandra, 446

MacArthur, R. H., 70, 79, 88
MacDowell, E. C., 266
macroevolution, 586
macrogenesis, 435
Madagascar, 564
Magicicada, 52, 99, 474
Main, A. R., 38
Malacopteron, 73
Malacostraca, 601
malaria, 227
malaria-mosquito, 35
mallard, 90
mallophaga, 463
mammal evolution, 608
mammals, 55, 447, 616; fossil, 601; origin, 596; polyploidy, 447
man, causes of evolution, 650; ecotypic adaptations, 657–658; fertility, 659; independence of environment, 654; loss of adaptedness, 654; natural selection, 656; phylogeny, 624; polymorphism, 647; a polytypic species, 638; races, 644; selection pressure, 650; as a species, 622–662; variability, 647; versatility, 644
Maniola, 227
Manning, A., 96, 100
Manucodia, 126
marginal overlaps, 504
Marien, D., 70, 82, 196, 292
marine animals, 315
marine vertebrates, 594
Marshall, J. T., Jr., 80, 375, 456
Marshall, W. H., 558, 579
Mather, K., 146, 149, 235, 236, 266, 285, 613
mating, assortative, 252; selective, 374
mating bond, 126
mating types, 408
Matthew, W. D., 75, 386
Matthey, R., 55, 133, 444
Mayfield, H., 303
McAtee, W. L., 182, 280
McCabe, T., 65, 557
Mecham, J. S., 92
mechanical isolation, 103–104
Medawar, P. B., 164, 655
Meganthropus, 639–640
Megapodius, 373
Meglitsch, P. A., 28
meiosis, 179
meiotic drive, 236
Meise, W., 119, 370, 371
Melania, 32
melanism, industrial, 242
Meleagris, 290, 591
Melidectes, 121
Melipona, 248
Mell, R., 313

Melospiza. See *Passerella*
Mendelians, 10, 266, 516
Menidia, 502
meristic characters, 148
Merops, 81
Mertens, R., 94, 115, 208, 282, 367, 502
Mesocricetus, 531
Metrioptera, 99
Mettler, L. E., 132, 293
Michener, C. D., 326, 510
Micraster, 24, 468
Microhyla, 118
microspecies, 433
Microtus, 80, 200, 323
migration, 417
Milkman, R. D., 190
Millepora, 48
Miller, A. H., 52, 98, 375, 380, 584, 604
Miller, R. R., 500, 580
Milstead, W. M., 70
mimicry of snakes, 196
Minamori, S., 38
missing links, 436, 596, 614, 625–637
mites, 47
mobility, 416
modifiers, 269; accumulation, 174
mollusks, 47
molts, 318
monophagy and speciation, 462
monotypic, 339
Montalenti, G., 421
Moore, J. A., 24, 38, 50, 61, 63, 94, 314, 461, 499, 540
Moraba, 222, 258, 294, 301; *scurra*, 531, 540
Moreau, R. E., 70, 73
Moriya, K., 38, 511
morphism, 151
morphological changes, 290
morphs, 453; homologous, 157
mosaic evolution, 596, 637
mosquitoes, 98; see also *Anopheles, Culex*
Motacilla alba, 503
moths, 40, 90
Mount Carmel, 642
mountains as barriers, 560
Mourant, A. E., 153, 647
Muller, H. J., 218, 228, 232, 254, 293, 440, 517, 551, 662
multifactoral inheritance, 267
multiple factors, 266
multiple invasions, 504
multiplication of species, 424–480
Munroe, E. G., 417, 461, 507
Murella, 337
Murphy, R. C., 475
Mus, 237, 379, 579
Musca, 159, 502
Mustela, 141
mutability, 172
mutation, 8, 168–176, 181, 432; effect, 174; as evolutionary force, 175; rate, 173, 198; systemic, 438

mutation theory, 169, 332, 517
mutational load, 254
mutations, beneficial, 174; induced, 175; randomness, 176; small, 169
Mycotrupes, 501
Myers, G. S., 259, 580
Myiagra, 590
Myiarchus, 52
myrmecophily, 605
Myrmica, 71
Myzomela, 73

natio, 136
Natrix, 282
natural selection, 182–201; history of concept, 183; in man, 652, 656; nature, 183; objections, 186; *see also* selection
Neanderthal, 641–643
nematodes, 48, 56, 71, 434
Nemeophila, 313
Nemobius, 44, 52, 131; *fasciatus*, 45
neo-Darwinism, 3
Neositta, 372–373
neoteny, 607
neutral genes, 161, 207
new structures, 604
new types, 437
niche, 69, 78; differences, 76; diversity, 245; empty, 87; invasion, 572–575; new, 621; selection, 471; shift into new, 546; utilization, 248, 547
Nicholson, A. J., 67
Niethammer, G., 329, 570
Nilsson-Ehle, H., 266
nongenetic variation in fish, 142
Notropis, 90, 98
novelty, evolutionary, 602–621
nucleic acid, 171
nuthatches, 83

Octopus, 54
Oecanthus niveus, 44
Oenothera, 169
Omodeo, P., 442
Oncorhynchus, 39
ontogeny, 606
Oreopithecus, 627
organization effect, 277
orthogenesis, 2
Orthoptera, 44, 99
Osborn, H. F., 517, 599
Osche, G., 56, 434
outbreeding, 404, 415, 418, 420
overdominance, 221, 225, 231, 256, 284, 519, 633; origin of, 232
overlap, 73, 81, 86; partial, 502; zones, 549
overpopulation, 662
oysters, 315

Pachycephala, 318, 366, 378, 396, 397, 498
paleontology, 11, 23

Pallas, P. S., 335
Paludan, K., 94, 509
Paludestrina, 441
Pan, 625
Panaxia, 244, 469, 494
Papilio, 74, 107, 118, 249, 345, 504; *machaon*, 499
Paradisaea, 364
Paradisaeidae, 494
parallelism, 609
Paramecium, 50, 65, 76, 407, 418; *aurelia*, 49
Paranthropus, 629
parasites, 54, 72, 349, 462, 464, 563, 595
Paratettix, 222, 258
Pardalotus, 430
parental care, 199, 416, 652
Park, T., 67, 77
Parkes, K. C., 117, 127
parthenogenesis, 409–411, 435, 442; advantage, 422
particulate inheritance, 167
Parulidae, 394, 417
Parus, 70, 76, 79, 87, 109, 374, 379, 492, 605; *major*, 510; *pleskei*, 549
Passer, 119–120
Passerculus, 355
Passerella, 335; *melodia*, 335, 336, 375, 396, 456, 561
Passerherbulus, 97
Passerina, 118
Patella, 47
Patterson, J. D., 50, 57, 91, 131, 550
Pavan, C., 71, 562
Pavlovsky, O., 251, 529
Payne, F., 266
Pea Aphid, 454
Pecten, 351
Pediculus, 455, 463–464
Pejler, B., 73, 124, 356
penetrance, 219, 270
Perdeck, A. C., 552
peripheral isolates, 496, 544
peripheral populations, 80, 261, 386–393, 526
Perognathus, 64
Peromyscus, 64–65, 118, 177, 195, 298, 328, 364, 499, 508, 557
Petersen, B., 50, 107, 302, 313, 326, 362–363, 456
Petersen, W., 89
Petit, C., 108
Petroica, 317
Phaeornis, 83
Phalacrocorax, 69
phasmids, 411
phenocopy, 189
phenodeviants, 282–284
phenotype, 6, 184, 190, 219, 268, 279–296, 304, 311, 331, 607, 611; a compromise, 194; constancy, 280; deviant, 239; discontinuity, 155
phenotypic flexibility, 147

phenotypic plasticity, 316
phenotypic response, 189
Philomachus, 230
philopatry, 416, 566–569
Phlegopsis, 558
Phoca, 579
phototaxis, 62
Photuris, 51, 98
Phylloscopus, 52
physiological characters, cryptic, 209
Pieris, 107, 118, 313, 363; *bryoniae*, 456
pigeons, 134
pigmentation, 324
Pinaroloxias, 506
pintail, 90
pioneers, 595
Pipilo, 80, 121–122
Pitelka, F. A., 74, 87
Pithecanthropus, 631, 640
Pittendrigh, C. S., 35, 62, 78, 548
Planaria, 72, 316
plants, 93, 129, 147, 234, 352, 401, 408, 418, 440, 572, 577, 580, 591, 615; and animals, 419
plasmagenes, 172
Plasmodium, 227
Platanus, 542, 580
Plate, L., 185, 267, 603
Plato, 5, 16
Platysamia, 103
pleiotropy, 159, 264; and fitness, 159
Pleistocene, 579
Pleistocene ice as barrier, 560
Pleistocene size increase, 320
Plethodon, 79
Pliopithecus, 626
Ploceinae, 70
Polistes, 50–51
Polyarthra, 73
Polycelis, 54
polychaete, 100
polygenic character, 612
polygeny, 265–268
polygyny, 651
Polygyrinae, 103
polymorph ratios, shifts, 242
polymorphism, 150–158, 415; adaptive nature, 250; balanced, 221, 222; chromosomal, 222, 535; cytological, 154; ecological, 247; genetics, 153; geographic variation, 329; limits, 539; mimetic, 248; neutral, 162, 207; and niche utilization, 245; occurrence, 152; recognition, 152; replacement by different, 537; stability, 209, 244; transient, 241
polyphemus moth, 313
polyphenism, definition, 150
polyploidy, 409, 439–449, 455; in fishes, 448
polytopic races, 356, 388
polytopic subspecies, 348–349
polytypic species, 334–359; definition,

339; history, 337; among marine animals, 340; in plants, 341; occurrence, 339
Polyxenus, 410
Pongo, 625
Pontecorvo, G., 27, 170
Popham, E. J., 144, 247, 469
population, asexual, 28; closed, 177, 518, 532; concept of, in man, 640; fluctuations, 200; genetic reconstruction, 527; open, 284, 289, 518; size, 212, 239; small, 531; structure, 360; as unit of selection, 197; variation, 136–261
population thinking, 5
populations, mixing, 477; peripheral, 386–393, 573; peripherally isolated, 393, 535
Portmann, A., 652
position effects, 277
Potomac Warbler, 117, 127
Poulton, E. B., 13, 551
preadaptation, 593–595
predator selection, 329
predators, 71, 80, 574
preformism, 4
Prevosti, A., 302, 326
primitive characters, 606
Primula, 201
procaryota, 412
Procellaria, 475
Proconsul, 626
Progomphus, 93
Prosimulium, 55
Prosser, C. L., 59, 60, 315
protozoans, 40
Prunellidae, 70
Pryer, H., 33
Pseudacraea, 249
pseudogamy, 442
psocids, 79
Pterodroma, 475

Quiscalus, 87, 379

rabbits, 579
race, 350
race concept, 646
races, ecological, 355; geographic, 455; of man, 644; physiological, 455; seasonal, 474
racial crosses, 495
racism, 649
radiation. See adaptive radiation
radiation damage, 272, 539
Radovanović, M., 73
Ramapithecus, 627
Ramme, W., 33, 144
Rana, 61, 63, 90, 94, 316, 495, 502; *brevipoda*, 38; *nigromaculata*, 511, *pipiens*, 24, 314
Rand, A. L., 127, 248, 371, 561
random fixation, 203, 210, 237
random fluctuations, 205
range expansions, 563

Rassenkreis, 339
rate, of evolution, 259, 522, 575–576; of speciation, 575–585
Ratites, 606
Ray, C., 316, 327
Ray, J., 14
reaction norm, 146, 189
recapitulation, 606
recessiveness, 217
recombination, 171, 172, 178, 392, 526; definition, 234
reductionism, 6
Reed, S. C., 34, 200
refuges, 508; drought, 578; glacial, 371
Regulus, 109
Reid, J. A., 41, 57, 73, 259
reinforcement, 549
Remane, A., 594, 596
Rendel, J. M., 281
Rensch, B., 242, 319, 339, 417, 465, 499, 586, 607, 616
reproduction, asexual, 406, 410; differential, 184; vegetative, 435
reproductive isolation, selection for, 552
reproductive success, 199, 651, 659
reptiles, 115
resistance to insecticides, 259
resistant strains, 193
reticulate evolution, 135, 420
Rhabditis, 48, 406, 434
Rhagoletis, 460
Rhinocheilus, 375
Rhinomyias, 590
Rhipidura, 397; *rufifrons,* 492
Rhizopertha, 76
Richards, O. W., 60
Ripley, S. D., 73, 83, 306, 504
rivers as barriers, 558
roaches, 100
Robertson, A., 289
Robertson, F. W., 229
Robertson's rule, 444
rodents, 396
Romer, A. S., 594, 596
Rosen, D. E., 131
Rothfels, K. H., 55, 444
rotifers, 142, 356
Ruibal, R., 64, 314
Ruiter, L. de, 209
Runnström, S., 315

salamanders, 582
Salmo, 39, 145, 467
salmon, 193
Salomonsen, F., 321, 509, 568
saltation, 435
saltationism, 517
Sarasin, F. and P., 590
Sarcophaga, 101
saw flies, 78
Scaphiopus, 93
Schaeffer, B., 596
Schindewolf, O. H., 607

Schlegel, H., 338
Schlosser, M., 626
Schmalhausen, I. I., 282
Schmidt, J., 298
Schneirla, T. C., 64–65
Schnetter, M., 247
Scholander, P. F., 186, 319
Schopenhauer, Arthur, 13, 437
Schrader, F., 446
Schreider, E., 323, 645
Sciara, 55
Sciurus, 74, 330
Scossiroli, R. E., 176
scutellar bristles, 281
seasonal adjustments, 318
seasonal barriers, 94
seasonal changes, 239
seasonal cycles, 192
seasonal races, 474
seasonal variation, 141
sea urchins, 500
secondary contact zones, 502
Sedlmair, H., 161, 247
Seebohm, H., 338, 425, 486
segregation, unequal, 236
Seiler, J., 40, 443
Selander, R. K., 73, 87
selection, 282, 285; artificial, 286–287, 291; canalizing, 283; creative, 201; cryptic, 212; delayed response, 195, 287; disruptive, 472; force, 191; interaction with chance, 214; normalizing, 182, 282, 283; and phenotype, 189; of populations, 197; rate, 238; response to, 289; stabilizing, 282–284; undesirable, 193; unit, 199
selection coefficients, 260
selection experiments, parallel, 292
selection pressures, changes, 238–239; different responses, 600; opposing, 196, 227, 238, 321
selective advantage, 242
selective equivalence, 212
selective values, 222
self-fertilization, 406, 408, 411
Semenov-Tianshansky, A., 136
semispecies, 118, 455, 501
Semper, K. G., 309
Serinus, 564
Serventy, D. L., 503
Sewall Wright effect, 203
Sewertzoff, A. N., 603
sex, 405
sex attractants, 100
sex chromosomes, 440
sex determination, 440
sex races, 316
sex ratio, 197, 236
sexual dimorphism, 32, 108, 199, 317; loss, 317
sexual drive, 102
sexual isolation, incipient, 493
sexuality, 405, 412; significance, 179

sexual selection, 201
Shaw, R. F., 653
Sheppard, P. M., 219, 232, 235, 240, 242, 244, 246, 249, 278
shift, ecological, 620; to new niche, 588; unsuccessful, 595; from water to land, 593
Sibley, C. G., 121, 127, 372
sibling species, 25, 31, 33–58, 220, 460–461; definition, 34; in *Drosophila*, 35; in insects, 40; in invertebrates, 47; nature, 57; recognition, 50; significance, 57; in vertebrates, 38
sickle-cell anemia, 227
Silver Fox, 208, 242
Simpson, G. G., 1, 2, 14, 20, 22, 23, 28, 30, 75, 202, 238, 436, 520, 564, 586, 588, 589, 593, 596, 597, 600, 608, 610, 616
Simuliidae, 55, 444
Sinanthropus, 632, 640
Siphateles, 116
Sitta, 83
size variation, in invertebrates, 327; on islands, 321
skull, evolution, 635
Skutch, A. F., 70, 325
slugs, 52
snails, 337, 398, 590
Snow, D. A., 323–324
snow geese, 468
soil as barrier, 562
Solenobia, 40, 56, 443
song, geographic variation, 494
Song Sparrow, 335, 336, 375, 390, 450, 561
Sonneborn, T. M., 49, 64, 407, 418
Sorex, 154, 373, 375
sound, 98
Spassky, B., 34, 174
specialization, 596
speciation, 424–480; allochronic, 476; on archipelagos, 506; bursts, 577; in caves, 501; by changing host specificity, 473; chromosomal, 535; by chromosomal changes, 439; and continuous ranges, 541; by cytoplasmic sterility, 472; by disruptive selection, 471; by distance, 507, 512, 542; ecological, 456; ecology of, 556–585; in fish, 467, 491, 571; genetics, 516–555; geographic, 481; history of theory, 482–488; in hominids, 637; by hybridization, 430; incomplete, 24, 502; instantaneous, 432–439; on islands, 557; levels, 489–491; model, 425; modes, 426; by mutation, 432; in ocean, 617; in parasites, 462; pattern, 585; in pelagic animals, 583; in Pleistocene lakes, 466; potential for, 584; rapid, 259, 580; rate, 575–585; by seasonal isolation, 474; semigeographic, 525; successful, 546; sympatric, 449–480; unit, 491

species, 11; adaptive significance, 60; biological properties, 59–88; as category, 21; coexistence, 80; collective, 433; cyclic expansion, 584; ecological types, 414; evolutionary characteristics, 587–588; evolutionary role, 422, 620–621; fossil, 24; frequency, 412; historical, 13; incipient, 489, 585; kinds, 400–423; Linnaean, 13; monotypic, 417; multidimensional, 19; multiplication, 424–480; panmictic, 304; parthenogenetic, 56; polytypic, 334–359, 394–395; population structure, 360–399; rare, 412, 531; role, 422; subdivisions, 346; successive, 24; total number, 436; uniformity, 523
species borders, 61, 523
species characters, geographic variation, 492; morphological, 31
species concepts, 12–30; application, 21; biological, 19–20; difficulties, 21; genetic, 20; history, 15; morphological, 16; nondimensional, 17; typological, 16
species criteria, 14, 402
species definitions, 14, 19, 20
species differences, genetics, 543
species groups, 501
species hybrids, genetic analysis, 543
species pairs, 81
species recognition, 95
species structure, 384, 520, 523–527
species swarms, 464
speech, 634
Sphaeroma, 210, 240
Sphingidae, 103
Sphyrapicus, 374
spiders, 47, 97, 100
Spiess, E. B., 274, 519, 533
Spieth, H. T., 35, 95, 96, 100, 101, 376, 474
Spizaetus, 73
sponges, 48
spontaneous generation, 4, 13
Spring, A. F., 168
Spurway, H., 21, 504, 540, 605, 608
stagnation, evolutionary, 289, 614
Staiger, H., 47, 376, 444, 526
Standfuss, M., 494
Starck, D., 636
Stebbins, G. L., Jr., 27, 91, 93, 94, 111, 128, 129, 181, 194, 234, 401, 412, 413, 418, 441, 510, 580, 656
Steere, J. B., 93
Stein, G. H. W., 85, 187, 323
Steinberg, A., 205
Steiner, H., 614
Steinmann, P., 356
Stenonema, 376
Stephens, S. G., 170
sterility, 90, 286, 290, 552; behavioral, 552; between geographic races, 493
Stern, C., 190, 284, 478
stickleback, 376

stimuli, auditory, 98; chemical, 99; olfactory, 99
Stone, W. S., 50, 57, 91, 230, 526, 535, 539, 541, 550
strand segregation, 446
Streptopelia, 564
Stresemann, E., 38, 482, 506, 578
Strongylocentrotus, 116
structures, new, 602
Sturnella, 52, 74
Sturnus, 74, 87, 195
Sturtevant, A. H., 280
subniches, 245
subspeciation, 579
subspecies, 347–350; area of, 396; definition, 348; frequency, 394; number, 396; in paleontology, 350; temporal, 350
subspecies borders, 368
substitutional load, 259
substrate, 361
substrate adaptation, 144, 327, 570
substrate-adapted polymorphism, 245, 331
substrate race, 177, 381
substrate selection, 247
Suez Canal, 574
Sumner, F. B., 298
Suomalainen, E., 153, 409, 443
supergene, 155, 222, 235, 256–257, 284, 294, 539
supernumeraries, 447
superoptimal stimuli, 196
superspecies, 499
suppressor genes, 270
Svärdson, G., 39, 142, 448, 466, 503, 571
Swainson's Warbler, 355
switch gene, 220, 249, 268
Sylvester-Bradley, P. D., 14, 23, 29
symbionts, 54
sympatric speciation, 449–480; assumptions, 468; definition, 451; difficulties, 477; in fossils, 467; genetic difficulties, 478; models, 471; by preadaptation, 471
sympatric species, 23
synthetic lethals, 273
synthetic theory, 7
systematics, comparative, 384, 402, 417, 427

Talpa, 187, 323, 356
Tanysiptera, 503, 522
tardigrade, 567
Taricha, 90
taxonomic characters, adaptiveness, 592
taxonomist, 11
Teissier, G., 68, 77, 222, 253, 302
Telanthropus, 639–640
Telmatobius, 61
temperature preference, 160–161
temperature races, 315
temperature tolerance, 61–62

temporal subspecies, 351
tendencies, 614
termites, 54, 308
Ternifine Man, 632, 640
Terpsiphone, 26, 119
Terrapene, 590
territory, 87
Test, A. R., 72, 452
Tetrao, 127
Thais lapillus, 376, 445, 526
thalassemia, 227
Thamnophis, 38, 70, 146, 343–344
thelytoky, 409
theories of evolution, 2
Thermocyclops, 397
Thoday, J. M., 30, 147, 414, 472, 553, 587, 611
Thomomys, 376, 396, 562
Thorpe, W. H., 56, 453, 571, 605
Thorson, G., 62, 100, 317, 571
threshold effect, 189
time, available, 585
Timoféeff-Ressovsky, N. W., 239, 508
Tinaea, 40
Tinbergen, N., 96, 101, 196, 604, 605
Tisbe, 493
toads, 98
tool use, 630, 633–634
towhee, 121
Toxostoma, 604
transformation of species, 428–429
translocation, 235, 278, 540
transpecific evolution, 586–621
trends, 241; in hominids, 633; orthogenetic, 614
Tribolium, 77
Trichogramma, 56
Trichoniscus, 56, 443
Trigona, 51
Triops, 577
Tristanodes, 507
Triturus, 115, 493, 504, 540; *ensigaster*, 498
Troglodytes, 61
tropical faunas, 88
tropics, 574
trout, 356
turbellarians, 356, 442–443, 500
Turdus, 381; *pilaris*, 568; *poliocephalus*, 492
Turesson, G., 351, 357, 401
turkey, 290
Tympanuchus, 117
type, definition, 588; origin of new, 436, 588–592
typological thinking, 5
typologist, 184, 187

Udvardy, M. F. D., 78
uniparental reproduction, 27
unisexual broods, 417
unit of evolution, 587
unrepeatability, 291

Ursus arctos, 561
Utida, 79

Vandel, A., 340, 444
Vanderplank, F. L., 98, 495
Vangidae, 506
variability, ecological, 477; genetic, 212, 414, 532, 539; of hybrid populations, 131–132, 377; limits, 292
variation, chronological, 302; clinal, 361; concealed, 164; continuous, 266; geographic, 297–333; individual, 32, 138; intraspecific, 32; kinds, 138; microgeographic, 301; nonadaptive, 309; nongenetic, 139–144, 299; in parthenogenetic clones, 410; in peripheral populations, 389; seasonal, 141; sources and maintenance, 164–181; storage, 215
variety, 334, 346
Vaurie, C., 81, 83, 388, 549, 590
Vavilov, N. I., 386
Vecht, J. van der, 307
Vermivora, 117
Vespa, 283
Vetukhiv, M., 231
Villwock, W., 124, 430
Vireo, 70
visual stimuli, 96
vocalization, 52
Voipio, P., 243, 250, 330, 369, 375, 381
Volpe, E. P., 105, 123
Volucella, 162, 249
Voous, K. H., 388
Vulpes, 242

Waagen, W., 168
Waddington, C. H., 6, 10, 57, 148, 185, 189, 220, 247, 280, 283, 612, 636
Wagler, E., 39, 147
Wagner, M., 484
Wagner, R. P., 78, 277
Wahlert, G. v., 582, 598
Wallace, A. R., 309, 548
Wallace, B., 175, 198, 231, 237, 272, 290, 294, 524, 536, 553

Wallgren, H., 61
water requirements, 64
weevils, 56
Weidenreich, F., 630, 636
Weismann, A., 4, 129, 165, 405, 485, 504, 548, 577
Westoll, T. S., 577
whales, 320
White, M. J. D., 56, 125, 154, 222, 234, 258, 393, 402, 404, 409, 531, 534, 537, 540, 542
Williamson, K., 324
Williamson, M. H., 67, 250
Willis, J. C., 619
Wilson, E. O., 67, 83, 348
Wilsonia, 492
wing rule, 324
Wollaston, T. V., 482
woodpeckers, 87
Worthington, E. B., 574
Wright, Sewall, 136, 203, 207, 208, 213, 235, 280, 427, 520, 521, 619

xiphophorin, 97
Xiphophorus, 105, 112, 210, 489, 490; *maculatus,* polymorphism, 156
Xylocopa, 307
Xylotrechus, 458

Young, F. N., 476

Zapfe, H., 626
Zimmerman, E. C., 260, 564, 574
Zimmermann, W., 185
Zinjanthropus, 629
Zirkle, C., 4, 13, 183
Zoarces, 298
zones of intergradation, 368–384; classification, 380; steep, 381; width, 378
Zonotrichia, 318
Zosterops, 506, 567–568; *rendovae,* 305, 558
Zuckerkandl, E., 628
Zygaena, 390; *ephialtes,* 331
zygotic mortality, 105